THE ART OF
COMPUTER PROGRAMMING

DONALD E. KNUTH *Stanford University*

 ADDISON-WESLEY PUBLISHING COMPANY

Volume 2 / **Seminumerical Algorithms**

THE ART OF
COMPUTER PROGRAMMING

Reading, Massachusetts • Menlo Park, California
London • Amsterdam • Don Mills, Ontario • Sydney

This book is in the
ADDISON-WESLEY SERIES IN
COMPUTER SCIENCE AND INFORMATION PROCESSING

Consulting Editors
RICHARD S. VARGA and MICHAEL A. HARRISON

ISBN 0-201-03802-1
KLMNO-MA-798

PREFACE

The algorithms discussed in this book deal directly with *numbers;* yet I believe they are properly called *seminumerical,* because they lie on the borderline between numeric and symbolic calculation. Each algorithm not only computes the desired answers to a problem, it also is intended to blend well with the internal operations of a digital computer. In many cases a person will not be able to appreciate the beauty of such an algorithm unless he also has some knowledge of a computer's machine language; the efficiency of the corresponding machine program is a vital factor which cannot be divorced from the algorithm itself. The problem is to find the best ways to make computers deal with numbers, and this involves tactical as well as numerical considerations. Therefore the subject matter of this book is unmistakably a part of computer science, as well as of numerical mathematics.

Some people working in "higher levels" of numerical analysis will regard the topics treated here as the domain of system programmers. Other people working in "higher levels" of system programming will regard the topics treated here as the domain of numerical analysts. But I hope that there are some people left who will be content to look carefully at these basic methods; although they are perhaps on a low level, they underlie all of the more grandiose applications of computers to numerical problems, so it is important to know them well. We are concerned here with the interface between numerical mathematics and computer programming, and it is the mating of both types of skills which makes the subject so interesting.

There is a noticeably higher percentage of mathematical material in this book than in other volumes of this series, because of the nature of the subjects treated. In most cases the necessary mathematical topics are developed here starting almost from scratch (or from results proved in Volume 1), but in some easily-recognizable sections a knowledge of calculus has been assumed.

This volume comprises Chapters 3 and 4 of the complete series. Chapter 3 is concerned with "random numbers": it is not only a study of various methods for generating random sequences, it also investigates statistical tests for randomness, as well as the transformation of uniform random numbers into other types of random quantities; the latter subject illustrates how random numbers are used in practice. I have also included a section about the nature of randomness itself. Chapter 4 is my attempt to tell the fascinating story of what mankind has been able to learn about the processes of arithmetic, after centuries of progress. It discusses various systems for representing numbers, and how to convert between them; and it treats arithmetic on floating-point numbers, high-precision integers, rational fractions, and polynomials, including the questions of factoring and finding greatest common divisors.

Each of Chapters 3 and 4 can be used as the basis of a one-semester college course at the junior to graduate level. Although courses on "Random Numbers" and on "Arithmetic" are not presently a part of many college curricula, I believe the reader will find that the subject matter of these chapters lends itself nicely to a unified treatment of material which has real educational value. My own experience has been that these courses are a good means of introducing elementary probability theory and number theory to college students; nearly all of the topics usually treated in such introductory courses arise naturally in connection with computer applications, and the presence of these applications can be an important motivation which helps the student to learn and to appreciate the theory. Furthermore, each chapter gives a few hints of more advanced topics which will whet the appetite of many students for further mathematical study.

In order to make this book reasonably independent of Volume 1, two sections about the MIX computer and its assembly language have been reproduced in Appendix A. Readers who are unfamiliar with MIX are encouraged to scan through this material briefly before starting the main part of the text. A few nonstandard notations, which were introduced in Volume 1, are defined in Appendix C.

In addition to the acknowledgments made in the preface to Volume 1, I would like to express deep appreciation to Elwyn R. Berlekamp, John Brillhart, George E. Collins, Stephen A. Cook, D. H. Lehmer, M. Donald MacLaren, Mervin E. Muller, Kenneth B. Stolarsky, and H. Zassenhaus, who have generously devoted considerable time to reading portions of the preliminary manuscript, and who have suggested many valuable improvements.

Princeton, New Jersey   D.E.K.
October 1968

(*Author's note:* In this second printing, I have incorporated several hundred improvements to the text and exercises while retaining the original page numbering.)

Notes on the Exercises

The exercises in this set of books have been designed for self-study as well as classroom study. It is difficult, if not impossible, for anyone to learn a subject purely by reading about it, without applying the information to specific problems and thereby forcing himself to think about what has been read. Furthermore, we all learn best the things that we have discovered for ourselves. Therefore the exercises form a major part of this work; a definite attempt has been made to keep them as informative as possible and to select problems that are enjoyable to solve.

In many books, easy exercises are found mixed randomly among extremely difficult ones. This is sometimes unfortunate because the reader should have some idea about how much time it ought to take him to do a problem before he tackles it (otherwise he may just skip over all the problems). A classic example of this situation is the book *Dynamic Programming* by Richard Bellman; this is an important, pioneering book in which a group of problems is collected together at the end of some chapters under the heading "Exercises and Research Problems," with extremely trivial questions appearing in the midst of deep, unsolved problems. It is rumored that someone once asked Dr. Bellman how to tell the exercises apart from the research problems, and he replied, "If you can solve it, it is an exercise; otherwise it's a research problem."

Good arguments can be made for including both research problems and very easy exercises in a book of this kind; therefore, to save the reader from the possible dilemma of determining which are which, *rating numbers* have been provided to indicate the level of difficulty. These numbers have the following general significance:

Rating   Interpretation

00   An extremely easy exercise which can be answered immediately if the material of the text has been understood, and which can almost always be worked "in your head."

10   A simple problem, which makes a person think over the material just read, but which is by no means difficult. It should be possible to do this in one minute at most; pencil and paper may be useful in obtaining the solution.

20   An average problem which tests basic understanding of the text material but which may take about fifteen to twenty minutes to answer completely.

30   A problem of moderate difficulty and/or complexity which may involve over two hours' work to solve satisfactorily.

40   Quite a difficult or lengthy problem which is perhaps suitable for a term project in classroom situations. It is expected that a student will be able to solve the problem in a reasonable amount of time, but the solution is not trivial.

50   A research problem which (to the author's knowledge at the time of writing) has not yet been solved satisfactorily. If the reader has found an answer to this problem, he is urged to write it up for publication; furthermore, the author of this book would appreciate hearing about the solution as soon as possible (provided it is correct)!

By interpolation in this "logarithmic" scale, the significance of other rating numbers becomes clear. For example, a rating of *17* would indicate an exercise that is a bit simpler than average. Problems with a rating of *50* which are subsequently solved by some reader may appear with a *45* rating in later editions of the book.

The author has earnestly tried to assign accurate rating numbers, but it is difficult for the person who makes up a problem to know just how formidable it will be for someone else; and everyone has more aptitude for certain types of problems than for others. It is hoped that the rating numbers represent a good guess as to the level of difficulty, but they should be taken as general guidelines, not as absolute indicators.

This book has been written for readers with varying degrees of mathematical training and sophistication; and, as a result, some of the exercises are intended only for the use of more mathematically inclined readers. Therefore the rating is preceded by an M if the exercise involves mathematical concepts or motivation to a greater extent than necessary for someone who is primarily interested in only the programming algorithms themselves. An exercise is marked with the letters "HM" if its solution necessarily involves a knowledge of calculus or other higher mathematics not developed in this book. An "HM" designation does *not* necessarily imply difficulty.

Some exercises are preceded by an arrowhead, "▶"; this designates problems which are especially instructive and which are especially recommended. Of course, no reader/student is expected to work *all* of the exercises, and so those which are perhaps the most valuable have been singled out. This is not meant to detract from the other exercises! Each reader should at least make an attempt to solve all of the problems whose rating is *10* or less; and the arrows may help in deciding which of the problems with a higher rating should be given priority.

Solutions to most of the exercises appear in the answer section. Please use them wisely; do not turn to the answer until you have made a genuine effort to solve the problem by yourself, or unless you do not have time to work this particular problem. *After* getting your own solution or giving the problem a decent try, you may find the answer instructive and helpful. The solution given will often be quite short, and it will sketch the details under the assumption that you have earnestly tried to solve it by your own means first. Sometimes the solution gives less information than was asked; often it gives more. It is quite possible that you may have a better answer than the one published here, or you may have found an error in the published solution; in such a case, the author will be pleased to know the details as soon as possible. Later editions of this book will give the improved solutions together with the solver's name where appropriate.

Summary of codes:

▶	Recommended	00	Immediate
M	Mathematically oriented	10	Simple (one minute)
HM	Requiring "higher math"	20	Medium (quarter hour)
		30	Moderately hard
		40	Term project
		50	Research problem

EXERCISES

▶ **1.** [00] What does the rating "*M20*" mean?

2. [10] Of what value can the exercises in a textbook be to the reader?

3. [*M50*] Prove that when n is an integer, $n > 2$, the equation $x^n + y^n = z^n$ has no solution in positive integers x, y, z.

CONTENTS

CHAPTER THREE

RANDOM NUMBERS

3.1. INTRODUCTION

Numbers which are "chosen at random" are useful in a very wide variety of applications. For example:

a) *Simulation.* When a computer is used to simulate natural phenomena, random numbers are required to make things realistic. Simulation covers many fields, from the study of nuclear physics (where particles are subject to random collisions) to system engineering (where people come into, say, a bank at random intervals).

b) *Sampling.* Often it is impractical to examine all possible cases, but a random sample will provide insight into some questions.

c) *Numerical analysis.* Ingenious techniques for solving complicated numerical problems have been devised using random numbers. Several books have been written on this subject.

d) *Computer programming.* Random values make a good source of data for testing the effectiveness of computer algorithms. This is the primary application of interest to us in this book; it accounts for the fact that random numbers are already being considered here in Chapter 3, before most of the other computer algorithms have appeared.

1

e) *Decision making.* There are reports that many executives make their decisions by flipping a coin or by throwing darts, etc. It is also rumored that some college professors prepare their grades on such a basis. Sometimes it is important to make a completely "unbiased" decision; this ability is occasionally useful in computer algorithms, for example in situations where a fixed decision made each time would cause the algorithm to run more slowly. Randomness is also an essential part of optimal strategies in the theory of games.

f) *Recreation.* Rolling dice, shuffling decks of cards, spinning roulette wheels, etc., are fascinating pastimes for just about everybody. This traditional use of random numbers has suggested the name "Monte Carlo method," which is a general term used to describe any algorithm that employs random numbers.

It is customary to devote several paragraphs at this point to a philosophical discussion of what "random" means. In a sense, there is no such thing as a random number; for example, is 2 a random number? Rather, we speak of a *sequence* of *independent* random numbers with a specified *distribution*, and this means loosely that each number was obtained merely by chance, having nothing to do with other numbers of the sequence, and that each number has a specified probability of falling in any given range.

A *uniform* distribution is one in which each possible number is equally probable. A distribution is generally understood to be uniform unless some other distribution is specifically mentioned.

Each of the 10 digits 0 through 9 will occur about $\frac{1}{10}$ of the time in a (uniform) sequence of random digits. Each pair of two successive digits should occur about $\frac{1}{100}$ of the time, etc. Yet if we take a truly random sequence of a million digits, it will not always have exactly 100,000 zeros, 100,000 ones, etc. In fact, chances of this are quite slim; a *sequence* of such sequences will have this character on the average.

Any specified sequence of a million digits is equally as probable as the sequence consisting of a million *zeros*. Furthermore, if we are choosing a million digits at random and if the first 999,999 of them happen to come out to be zero, the chance that the final digit is zero is still exactly $\frac{1}{10}$, in a truly random situation. These statements seem paradoxical to many people, but there is really no contradiction here.

There are several ways to formulate a good abstract definition of a random sequence, and we will return to this interesting subject in Section 3.5; but for the moment, let us content ourselves with an intuitive understanding of the concept.

At first, people who needed random numbers in their scientific work would draw balls out of a "well-stirred urn" or would roll dice or deal out cards. In 1927, a table of over 40,000 random digits, "taken at random from census reports," was published by L. H. C. Tippett. Since then, a number of special machines for mechanically generating random numbers have been built; the first such machine was used by M. G. Kendall and B. Babington-Smith to produce a table of 100,000 random digits in 1939, and in 1955 the RAND Corporation published a well-known table of a million random digits obtained with the help of another special machine. A famous random-number machine, called ERNIE, picks the winning numbers in the British Premium Savings Bonds lottery. [See the articles by Kendall and Babington-Smith in the *Journal of the Royal Statistical Society*, Series A, **101** (1938), 147–166, and Series B, **6** (1939), 51–61; see also the review of the RAND table in *Math. Comp.* **10** (1956), 39–43.]

Shortly after computers were introduced, people began to search for efficient ways to obtain random numbers in computer programs. A table can be used, but this method is of limited utility because of the memory space and input time requirement, because the table may be too short, and because it is a bit of a nuisance to prepare and maintain the table. Machines such as ERNIE might be attached to the computer, but this would be unsatisfactory since it would be impossible to reproduce calculations exactly a second time when checking out a program.

The inadequacy of these methods led to an interest in the production of random numbers using the arithmetic operations of a computer. John von Neumann first suggested this approach in about 1946, using the "middle-square" method. His idea was to take the square of the previous random number and to extract the middle digits; for example, if we are generating 10-digit numbers and the previous value was 5772156649, we square it to get

$$33317792380594909201,$$

and the next number is therefore 7923805949.

There is a fairly obvious objection to this technique: how can a sequence generated in such a way be random, since each number is completely determined by its predecessor? The answer is that this sequence *isn't* random, but it *appears* to be. In typical applications the actual relationship between one number and its successor has no physical significance; hence the nonrandom character is not really undesirable. Intuitively, the middle square seems to be a fairly good scrambling of the previous number.

Sequences generated in a deterministic way such as this are usually called *pseudo-random* or *quasi-random* sequences in the highbrow technical literature, but in this book we shall simply call them random sequences, with the understanding that they only *appear* to be random. Being "apparently random" is perhaps all that can be said about any random sequence anyway. Precise mathematical formulations of the concept of randomness are discussed at length in Section 3.5. Random numbers generated deterministically on computers have worked quite well in nearly every application (although they certainly cannot replace ERNIE for the lotteries).

Von Neumann's original "middle-square method" has proved to be a comparatively poor source of random numbers, however. The danger is that the sequence tends to get into a rut, a short cycle of repeating elements. For example, if zero ever appears as a number of the sequence, it will continually perpetuate itself. Several people experimented with the middle-square method in the early 1950's. Working with four-digit numbers instead of 10-digit ones, G. E. Forsythe tried 16 different starting values and found that 12 of them led to sequences ending with the cycle 6100, 2100, 4100, 8100, 6100, . . . , while two of them degenerated to zero. N. Metropolis conducted extensive tests on the middle-square method, mostly in the binary number system. Working with 20-bit numbers, he showed there are 13 different cycles into which the sequence might degenerate, the longest of which has a period of 142. It is fairly easy to restart the middle-square method on a new value when zero has been detected, but long cycles are harder to avoid. However, a clever method for detecting any cycle in a sequence such as this has been suggested by R. W. Floyd (see exercise 7); Floyd's method uses very little memory space, it takes only about three times as long to generate each number, and the procedure stops generating numbers before any number has been repeated.

A theoretical disadvantage of the middle-square method is given in exercises 9 and 10. On the other hand, working with 38-bit numbers, N. Metropolis obtained a sequence of about 750,000 numbers before degeneracy occurred, and the resulting $750,000 \times 38$ bits satisfactorily passed statistical tests for randomness. This shows that the middle-square method *can* give usable results, but it is rather dangerous to put much faith in it until after elaborate computations have been performed.

Many random-number generators in use today are not very good. There is a tendency for people to avoid learning anything about random-number generators; quite often we find that some old method which is comparatively unsatisfactory has blindly been passed down from one programmer to another, and today's users have no understanding of its limitations. We shall see in this chapter that it is not difficult to learn the most important facts about random-number generators and their proper use.

It is not easy to invent a fool-proof random-number generator. This fact was convincingly impressed upon the author several years ago, when he attempted to create a fantastically good random-number generator using the following peculiar method:

Algorithm K (*"Super-random" number generator*). Given a 10-digit decimal number X, this algorithm may be used to change X to the number which should come next in a supposedly random sequence. Although the algorithm might be expected to yield a quite random sequence, reasons given below show that it is, in fact, not very good at all. (The reader need not study this algorithm in great detail except to observe its complexity.)

K1. [Choose number of iterations.] Set $Y \leftarrow \lfloor X/10^9 \rfloor$, i.e., the most significant digit of X. (We will execute steps K2 through K13 $Y + 1$ times; that is, we will compute random numbers a *random* number of times.)

K2. [Choose random step.] Set $Z \leftarrow \lfloor X/10^8 \rfloor \bmod 10$, i.e., the second most significant digit of X. Go to step K$(3 + Z)$. (That is, we jump to a *random* step in the program!)

K3. [Ensure $\geq 5 \times 10^9$.] If $X < 5000000000$, set $X \leftarrow X + 5000000000$.

K4. [Middle square.] Replace X by $\lfloor X^2/10^5 \rfloor \bmod 10^{10}$, i.e., by the middle square of X.

K5. [Multiply.] Replace X by $(1001001001\,X) \bmod 10^{10}$.

K6. [Pseudo-complement.] If $X < 100000000$, then set $X \leftarrow X + 9814055677$; otherwise set $X \leftarrow 10^{10} - X$.

K7. [Interchange halves.] Interchange the low-order five digits of X with the high-order five digits, i.e., $X \leftarrow 10^5 \lfloor X \bmod 10^5 \rfloor + \lfloor X/10^5 \rfloor$, the middle 10 digits of $(10^{10} + 1)X$.

K8. [Multiply.] Same as step K5.

K9. [Decrease digits.] Decrease each nonzero digit of the decimal representation of X by one.

K10. [99999 modify.] If $X < 10^5$, set $X \leftarrow X^2 + 99999$; otherwise set $X \leftarrow X - 99999$.

K11. [Normalize.] (At this point X cannot be zero.) If $X < 10^9$, set $X \leftarrow 10X$ and repeat this step.

K12. [Modified middle square.] Replace X by $\lfloor X(X - 1)/10^5 \rfloor \bmod 10^{10}$, i.e., by the middle 10 digits of $X(X - 1)$.

K13. [Repeat?] If $Y > 0$, decrease Y by 1 and return to step K2. If $Y = 0$, the algorithm terminates with X as the desired "random" value. ∎

(The machine-language program corresponding to the above algorithm was intended to be so complicated that a man reading a listing of it without explanatory comments would not know what the program was doing.)

With all the precautions taken in Algorithm K, doesn't it seem plausible that it would produce at least an infinite supply of unbelievably random numbers? No! In fact, when this algorithm was first put onto a computer, it almost immediately converged to the 10-digit value 6065038420, which—by extraordinary coincidence—is transformed into itself by the algorithm (see Table 1). With another starting number, the sequence began to repeat after 7401 values, in a cyclic period of length 3178.

The moral to this story is that *random numbers should not be generated with a method chosen at random*. Some theory should be used.

In this chapter, we shall consider random-number generators which are superior to the middle-square method and to Algorithm K, in the sense that it is possible to guarantee theoretically that the sequence has certain desirable random properties, and that no degeneracy will occur. We shall explore the reasons for this random behavior in some detail, and we shall also consider

Table 1

A COLOSSAL COINCIDENCE: THE NUMBER 6065038420
IS TRANSFORMED INTO ITSELF BY ALGORITHM K.

Step	X (after)		Step	X (after)	
K1	6065038420		K9	1107855700	
K3	6065038420		K10	1107755701	
K4	6910360760		K11	1107755701	
K5	8031120760		K12	1226919902	$Y = 3$
K6	1968879240		K5	0048821902	
K7	7924019688		K6	9862877579	
K8	9631707688		K7	7757998628	
K9	8520606577		K8	2384626628	
K10	8520506578		K9	1273515517	
K11	8520506578		K10	1273415518	
K12	0323372207	$Y = 6$	K11	1273415518	
K6	9676627793		K12	5870802097	$Y = 2$
K7	2779396766		K11	5870802097	
K8	4942162766		K12	3172562687	$Y = 1$
K9	3831051655		K4	1540029446	
K10	3830951656		K5	7015475446	
K11	3830951656		K6	2984524554	
K12	1905867781	$Y = 5$	K7	2455429845	
K12	3319967479	$Y = 4$	K8	2730274845	
K6	6680032521		K9	1620163734	
K7	3252166800		K10	1620063735	
K8	2218966800		K11	1620063735	
			K12	6065038420	$Y = 0$

techniques for manipulating random numbers. For example, one of our investigations will be the shuffling of a simulated deck of cards within a computer program.

EXERCISES

▶ **1.** [*20*] Suppose that you wish to obtain a decimal digit at random, not using a computer. Which of the following methods would be suitable?

a) Open a telephone directory to a random place (i.e., stick your finger in it somewhere) and use the units digit of the first number found on the selected page.

b) Same as (a), but use the units digit of the *page* number.

c) Roll a die which is in the shape of a regular icosahedron, whose twenty faces have been labeled with the digits 0, 0, 1, 1, ..., 9, 9. Use the digit which appears on top, when the die comes to rest. (A felt table with a hard surface is recommended for rolling dice.)

d) Expose a geiger counter to a source of radioactivity for one minute (shielding yourself) and use the units digit of the resulting count. (Assume that the geiger counter displays the number of counts in decimal notation, and that the count is initially zero.)

e) Glance at your wristwatch, and if the position of the second-hand is between $6n$ and $6(n + 1)$, choose the digit n.

f) Ask a friend to think of a random digit, and use the digit he names.

g) Ask an enemy to think of a random digit, and use the digit he names.

h) Assume 10 horses are entered in a race and you know nothing whatever about their qualifications. Assign to these horses the digits 0 to 9, in arbitrary fashion, and after the race use the winner's digit.

2. [*M22*] In a random sequence of a million decimal digits, what is the probability that there are exactly 100,000 of each possible digit?

3. [*10*] What number follows 1010101010 in the middle-square method?

4. [*10*] Why can't the value of X be zero when step K11 of Algorithm K is performed? What would be wrong with the algorithm if X could be zero?

5. [*15*] Explain why, in any case, Algorithm K should not be expected to provide "infinitely many" random numbers, in the sense that (even if the coincidence given in Table 1 had not occurred) one should have known in advance that any sequence generated by Algorithm K will eventually be periodic.

▶ **6.** [*M20*] Suppose that we want to generate a sequence of integers, $X_0, X_1, X_2, \ldots$ in the range $0 \leq X_n < m$. Let $f(x)$ be any function such that if $0 \leq x < m$, then $0 \leq f(x) < m$. Consider a sequence formed by the rule $X_{n+1} = f(X_n)$. (Examples are the middle-square method and Algorithm K.)

a) Show that the sequence is ultimately periodic, in the sense that there exist numbers λ and μ for which the values $X_0, X_1, \ldots, X_\mu, \ldots, X_{\mu+\lambda-1}$ are distinct, but $X_{n+\lambda} = X_n$ when $n \geq \mu$. Find the maximum and minimum possible values of μ and λ.

b) Show that there exists an $n > 0$ such that $X_n = X_{2n}$; the smallest such value of n lies in the range $\mu \leq n \leq \mu + \lambda$; and the value of X_n is unique in the sense that if $X_n = X_{2n}$ and $X_r = X_{2r}$, then $X_r = X_n$ (hence $r - n$ is a multiple of λ).

▶ **7.** [*20*] Apply the result of the preceding exercise to design a useful algorithm to be used in conjunction with a random-number generator of the type $X_{n+1} = f(X_n)$. Your algorithm should (a) provide the ability to stop generating numbers of the sequence before any repetition of previous values has occurred, and (b) generate at least λ elements of the sequence before stopping, although the value of λ is not known in advance, and (c) require only a small amount of memory space (e.g., you should not simply store all of the computed sequence values!).

8. [*28*] Make a complete examination of the middle-square method in the case of two-digit decimal numbers. (a) We might start the process out with any of the 100 possible values $00, 01, \ldots, 99$. How many of these values lead ultimately to the repeating cycle $00, 00, \ldots$? [*Example:* Starting with 43, we obtain the sequence 43, 84, 05, 02, 00, 00, 00, . . .] (b) How many possible final cycles are there? How long is the longest cycle? (c) What starting value or values will give the largest number of distinct elements before the sequence repeats?

9. [*M14*] Prove that the middle-square method using $2n$-digit numbers to the base b has the following disadvantage: If ever a number X, whose most significant n digits are zero, appears, then the succeeding numbers will get smaller and smaller until zero occurs repeatedly.

10. [*M16*] Under the assumptions of the preceding exercise, what can you say about the sequence of numbers following X if the *least* significant n digits are zero? What if the least significant $n + 1$ digits are zero?

▶ **11.** [*M26*] Consider sequences of random-number generators having the form described in exercise 6. If we choose $f(x)$ and X_0 at random, i.e., if we assume that each of the m^m possible functions $f(x)$ is equally probable and if we assume that each of the m possible values of X_0 is equally probable, what is the probability that the sequence will eventually degenerate into a cycle of length $\lambda = 1$? (*Note:* The assumptions of this problem give a natural way to think of a "random" random-number generator of this type. A method such as Algorithm K may be expected to behave somewhat like the average generator considered here; the answer to this problem gives a measure of how "colossal" the coincidence of Table 1 really is.)

▶ **12.** [*M31*] Under the assumptions of the preceding exercise, what is the average length of the final cycle? What is the average length of the sequence before it begins to cycle? (In the notation of exercise 6, we wish to examine the average values of λ and of $\mu + \lambda$.)

13. [*M42*] If $f(x)$ is chosen at random in the sense of exercise 11, what is the average length of the *longest* cycle obtainable by varying the starting value X_0? (*Note:* This answer is known for the special case that $f(x)$ is a permutation; see exercise 1.3.3–23.)

14. [*M38*] If $f(x)$ is chosen at random in the sense of exercise 11, what is the average number of distinct final cycles obtainable by varying the starting value? [Cf. exercise 8(b).]

15. [*M15*] If $f(x)$ is chosen at random in the sense of exercise 11, what is the probability that none of the final cycles has length 1, regardless of the choice of X_0?

16. [*15*] A sequence generated as in exercise 6 must begin to repeat after at most m values have been generated. Suppose we generalize the method so that X_{n+1} depends on X_{n-1} as well as on X_n; formally, let $f(x, y)$ be a function such that if $0 \le x, y < m,$· then $0 \le f(x, y) < m$. The sequence is constructed by selecting X_0 and X_1 arbitrarily, and then letting

$$X_{n+1} = f(X_n, X_{n-1}) \qquad \text{for} \qquad n > 0.$$

What is the maximum period conceivably attainable in this case?

17. [*10*] Generalize the situation in the previous exercise so that X_{n+1} depends on the preceding k values of the sequence.

18. [*M22*] Invent a method analogous to that of exercise 7 for finding cycles in the general form of random-number generator discussed in exercise 17.

19. [*M50*] Solve the problems of exercises 11 through 15 for the more general case that X_{n+1} depends on the preceding k values of the sequence; each of the m^{m^k} functions $f(x_1, \ldots, x_k)$ is to be considered equally probable. (*Note:* The number of functions which give the *maximum* period is given in exercise 2.3.4.2–23.)

3.2. GENERATING UNIFORM RANDOM NUMBERS

In this section we shall consider methods for generating a sequence of random fractions, i.e., random *real numbers U_n, uniformly distributed between zero and one*. Since a computer can represent a real number with only finite accuracy, we shall actually be generating integers X_n between zero and some number m; the fraction

$$U_n = X_n/m \qquad (1)$$

will then lie between zero and one. Usually m is the word size of the computer, so X_n may be regarded (conservatively) as the integer contents of a computer word with the radix point assumed at the extreme right, and U_n may be regarded (liberally) as the contents of the same word with the radix point assumed at the extreme left.

3.2.1. The Linear Congruential Method

By far the most successful random number generators known today are special cases of the following scheme, introduced by D. H. Lehmer in 1948. [See *Proc. 2nd Symposium on Large-Scale Digital Computing Machinery* (Cambridge: Harvard University Press, 1951), 141–146.] We choose four "magic numbers":

$$
\begin{array}{lll}
X_0, & \text{the starting value;} & X_0 \geq 0. \\
a, & \text{the multiplier;} & a \geq 0. \\
c, & \text{the increment;} & c \geq 0. \\
m, & \text{the modulus;} & m > X_0, \quad m > a, \quad m > c.
\end{array} \qquad (1)
$$

The desired sequence of random numbers $\langle X_n \rangle$ is then obtained by setting

$$X_{n+1} = (aX_n + c) \bmod m, \qquad n \geq 0. \qquad (2)$$

This is called a *linear congruential sequence*.

For example, the sequence obtained when $X_0 = a = c = 7$, $m = 10$, is

$$7, \ 6, \ 9, \ 0, \ 7, \ 6, \ 9, \ 0, \ \ldots \qquad (3)$$

As this example shows, the sequence is not always "random" for all choices of X_0, a, c, and m; the principles of choosing these values appropriately will be investigated carefully in later parts of this chapter.

Example (3) illustrates the fact that the congruential sequences always "get into a loop"; i.e., there is ultimately a cycle of numbers which is repeated endlessly. This property is common to all sequences having the general form $X_{n+1} = f(X_n)$; see exercise 3.1–6. The repeating cycle is called the *period;* sequence (3) has a period of length 4. A useful sequence will of course have a relatively long period.

The special case $c = 0$ deserves explicit mention, since the number generation process is a little faster when $c = 0$ than it is when $c \neq 0$. We shall see later that the restriction $c = 0$ cuts down the length of the period of the sequence, but it is still possible to obtain a reasonably long period. Lehmer's original generation method had $c = 0$, although he mentioned $c \neq 0$ as a possibility; the idea of taking $c \neq 0$ to obtain longer periods is due to Thomson [*Comp. J.* **1** (1958), 83, 86] and, independently, to Rotenberg [*JACM* **7** (1960), 75–77]. The terms *multiplicative congruential method* and *mixed congruential method* are used by many authors to denote linear congruential methods with $c = 0$ and $c \neq 0$, respectively.

The letters a, c, m, and X_0 will be used throughout this chapter in the sense described above. Furthermore, we will find it useful to define

$$b = a - 1, \tag{4}$$

in order to simplify many of our formulas.

We can immediately reject the case $a = 1$, for this would mean that $X_n = (X_0 + nc) \bmod m$, and the sequence would certainly not behave as a random sequence. The case $a = 0$ is even worse. Hence for practical purposes we may assume that

$$a \geq 2, \quad b \geq 1. \tag{5}$$

Now we can prove a generalization of Eq. (2),

$$X_{n+k} = (a^k X_n + (a^k - 1)c/b) \bmod m, \quad k \geq 0, \quad n \geq 0, \tag{6}$$

which expresses the $(n + k)$th term directly in terms of the nth term. (The special case $n = 0$ in this equation is worthy of note.) The subsequence consisting of every kth term of our sequence is another linear congruential sequence, having the multiplier $a^k \bmod m$ and the increment $((a^k - 1)c/b) \bmod m$.

EXERCISES

1. [*10*] Example (3) shows a situation in which $X_4 = X_0$, so the sequence begins again from the beginning. Give an example of a linear congruential sequence with $m = 10$ for which X_0 never appears again in the sequence.

▶ **2.** [*M20*] Show that if a and m are relatively prime, the number X_0 will always appear in the period.

3. [*M10*] If a and m are not relatively prime, explain why the sequence will be somewhat handicapped and probably not very random; hence we will generally want a to be relatively prime to m.

4. [*11*] Prove Eq. (6).

5. [*M20*] Equation (6) holds for $k \geq 0$. If possible, give a formula which expresses X_{n+k} in terms of X_n for *negative* values of k.

3.2.1.1. Choice of Modulus. Let us first consider the proper choice of the number m. We want the value of m to be rather large, since the period can never have a length more than m. (Thus, even if a person wants to generate only random zeros and ones, he should *not* take $m = 2$, for the sequence at best would then have the form

$$\ldots, 0, 1, 0, 1, 0, 1, \ldots !$$

Methods for modifying uniform random numbers to get random zeros and ones are discussed in Section 3.4.)

Another factor which influences our choice of m is speed of generation: We want to pick a value so that the computation of $(aX_n + c) \bmod m$ is quite fast.

Consider MIX as an example. We can compute $y \bmod m$ by putting y in registers A and X and dividing by m; assuming that y and m are positive, we see that $y \bmod m$ will then appear in register X. But division is a comparatively slow operation, and it can be avoided if we take m to be a value which is especially convenient, such as the *word size* of the computer.

Let w be the computer's word size. The result of an addition operation is given modulo w, and multiplication mod w is also quite simple, since the desired result is the lower half of the product. Thus, the following program computes $(aX + c) \bmod w$ efficiently:

$$
\begin{array}{ll}
\text{LDA} & \text{A} \\
\text{MUL} & \text{X} \\
\text{SLAX} & 5 \\
\text{ADD} & \text{C}
\end{array}
\tag{1}
$$

The result appears in register A. Overflow might be on at the conclusion of the above sequence of instructions, and if this is undesirable, the code should be followed by, e.g., "JOV *+1" to turn it off.

A clever technique which is less commonly known can be used to perform computations modulo $(w + 1)$. For reasons to be explained later, we will generally want $c = 0$ when $m = w + 1$, so we merely need to compute $(aX) \bmod (w + 1)$. The following program does this:

$$
\begin{array}{lll}
01 & \text{LDAN} & \text{X} \\
02 & \text{MUL} & \text{A} \\
03 & \text{STX} & \text{TEMP} \\
04 & \text{SUB} & \text{TEMP} \\
05 & \text{JANN} & \text{*+3} \\
06 & \text{INCA} & 2 \\
07 & \text{ADD} & \text{=W-1=}
\end{array}
\tag{2}
$$

Register A now contains the value $(aX) \bmod (w + 1)$. Of course, this value might lie anywhere between 0 and w, so the reader may legitimately wonder how we can represent so many values in the A-register! (The register obviously cannot hold a number larger than $w - 1$.) The answer is that overflow will be on after the above program if and only if the result equals w (assuming that overflow was initially off).

To prove that code (2) actually does determine $(aX) \bmod (w + 1)$, note that in line 04 we are subtracting the lower half of the product from the upper half. No overflow can occur at this step; and if $aX = qw + r$, with $0 \le r < w$, we will have the quantity $r - q$ in register A after line 04. Now

$$aX = q(w + 1) + (r - q),$$

and since $q < w$, we have $-w < r - q < w$; hence $(aX) \bmod (w + 1)$ equals either $r - q$ or $r - q + (w + 1)$, depending on whether $r - q \ge 0$ or $r - q < 0$.

If we wish to use the congruential method with $m = w + 1$, it is customary to simply reject the value w if it appears, by adding the lines

```
00      JOV     *+1
. . .
08      JNOV    *+3                              (3)
09      LDAN    A
10      JMP     *-4
```

to code (2).

A similar technique can be used to get the product of two numbers modulo $(w - 1)$; see exercise 8.

In later sections we shall require a knowledge of the prime factors of m in order to choose the multiplier a correctly. Table 1 lists the complete factorization of $w \pm 1$ into primes for nearly every known word size. For an extension of this table, see *Factorization of $y^n \pm 1$*, by A. J. C. Cunningham and H. J. Woodall, Francis Hodgson, London, 1925; or use the methods of Section 4.5.4.

The reader may well ask why we bother to consider using $m = w \pm 1$, when the choice $m = w$ is so manifestly convenient. The reason is that *when $m = w$, the right-hand digits of X_n are much less random than the left-hand digits.* If d is a divisor of m, and if

$$Y_n = X_n \bmod d,$$ (4)

we can easily show that

$$Y_{n+1} = (aY_n + c) \bmod d.$$ (5)

(For, $X_{n+1} = aX_n + c - qm$ for some integer q, and taking both sides mod d causes the quantity qm to drop out when d is a factor of m.) To illustrate the significance of Eq. (5), let us suppose, for example, that we have a binary computer. If $m = w = 2^e$, the low-order four bits of X_n are the numbers $Y_n = X_n \bmod 2^4$. The gist of Eq. (5) is that the low-order four bits of X_n form a congruential sequence which has a period of length 16 or less.

Table 1

PRIME FACTORIZATIONS OF $w \pm 1$

$2^e - 1$	e	$2^e + 1$
$7 \cdot 31 \cdot 151$	15	$3^2 \cdot 11 \cdot 331$
$3 \cdot 5 \cdot 17 \cdot 257$	16	65537
131071	17	$3 \cdot 43691$
$3^2 \cdot 7 \cdot 19 \cdot 73$	18	$5 \cdot 13 \cdot 37 \cdot 109$
524287	19	$3 \cdot 174763$
$3 \cdot 5^2 \cdot 11 \cdot 31 \cdot 41$	20	$17 \cdot 61681$
$7^2 \cdot 127 \cdot 337$	21	$3^2 \cdot 43 \cdot 5419$
$3 \cdot 23 \cdot 89 \cdot 683$	22	$5 \cdot 397 \cdot 2113$
$47 \cdot 178481$	23	$3 \cdot 2796203$
$3^2 \cdot 5 \cdot 7 \cdot 13 \cdot 17 \cdot 241$	24	$97 \cdot 257 \cdot 673$
$31 \cdot 601 \cdot 1801$	25	$3 \cdot 11 \cdot 251 \cdot 4051$
$3 \cdot 2731 \cdot 8191$	26	$5 \cdot 53 \cdot 157 \cdot 1613$
$7 \cdot 73 \cdot 262657$	27	$3^4 \cdot 19 \cdot 87211$
$3 \cdot 5 \cdot 29 \cdot 43 \cdot 113 \cdot 127$	28	$17 \cdot 15790321$
$233 \cdot 1103 \cdot 2089$	29	$3 \cdot 59 \cdot 3033169$
$3^2 \cdot 7 \cdot 11 \cdot 31 \cdot 151 \cdot 331$	30	$5^2 \cdot 13 \cdot 41 \cdot 61 \cdot 1321$
2147483647	31	$3 \cdot 715827883$
$3 \cdot 5 \cdot 17 \cdot 257 \cdot 65537$	32	$641 \cdot 6700417$
$7 \cdot 23 \cdot 89 \cdot 599479$	33	$3^2 \cdot 67 \cdot 683 \cdot 20857$
$3 \cdot 43691 \cdot 131071$	34	$5 \cdot 137 \cdot 953 \cdot 26317$
$31 \cdot 71 \cdot 127 \cdot 122921$	35	$3 \cdot 11 \cdot 43 \cdot 281 \cdot 86171$
$3^2 \cdot 5 \cdot 7 \cdot 13 \cdot 19 \cdot 37 \cdot 73 \cdot 109$	36	$17 \cdot 241 \cdot 433 \cdot 38737$
$223 \cdot 616318177$	37	$3 \cdot 1777 \cdot 25781083$
$3 \cdot 174763 \cdot 524287$	38	$5 \cdot 229 \cdot 457 \cdot 525313$
$7 \cdot 79 \cdot 8191 \cdot 121369$	39	$3^2 \cdot 2731 \cdot 22366891$
$3 \cdot 5^2 \cdot 11 \cdot 17 \cdot 31 \cdot 41 \cdot 61681$	40	$257 \cdot 4278255361$
$13367 \cdot 164511353$	41	$3 \cdot 83 \cdot 8831418697$
$3^2 \cdot 7^2 \cdot 43 \cdot 127 \cdot 337 \cdot 5419$	42	$5 \cdot 13 \cdot 29 \cdot 113 \cdot 1429 \cdot 14449$
$431 \cdot 9719 \cdot 2099863$	43	$3 \cdot 2932031007403$
$3 \cdot 5 \cdot 23 \cdot 89 \cdot 397 \cdot 683 \cdot 2113$	44	$17 \cdot 353 \cdot 2931542417$
$7 \cdot 31 \cdot 73 \cdot 151 \cdot 631 \cdot 23311$	45	$3^3 \cdot 11 \cdot 19 \cdot 331 \cdot 18837001$
$3 \cdot 47 \cdot 178481 \cdot 2796203$	46	$5 \cdot 277 \cdot 1013 \cdot 1657 \cdot 30269$
$2351 \cdot 4513 \cdot 13264529$	47	$3 \cdot 283 \cdot 165768537521$
$3^2 \cdot 5 \cdot 7 \cdot 13 \cdot 17 \cdot 97 \cdot 241 \cdot 257 \cdot 673$	48	$193 \cdot 65537 \cdot 22253377$
$179951 \cdot 3203431780337$	59	$3 \cdot 2833 \cdot 37171 \cdot 1824726041$
$3^2 \cdot 5^2 \cdot 7 \cdot 11 \cdot 13 \cdot 31 \cdot 41 \cdot 61 \cdot 151 \cdot 331 \cdot 1321$	60	$17 \cdot 241 \cdot 61681 \cdot 4562284561$
$7^2 \cdot 73 \cdot 127 \cdot 337 \cdot 92737 \cdot 649657$	63	$3^3 \cdot 19 \cdot 43 \cdot 5419 \cdot 77158673929$
$3 \cdot 5 \cdot 17 \cdot 257 \cdot 641 \cdot 65537 \cdot 6700417$	64	$274177 \cdot 67280421310721$

$10^e - 1$	e	$10^e + 1$
$3^3 \cdot 7 \cdot 11 \cdot 13 \cdot 37$	6	$101 \cdot 9901$
$3^2 \cdot 239 \cdot 4649$	7	$11 \cdot 909091$
$3^2 \cdot 11 \cdot 73 \cdot 101 \cdot 137$	8	$17 \cdot 5882353$
$3^4 \cdot 37 \cdot 333667$	9	$7 \cdot 11 \cdot 13 \cdot 19 \cdot 52579$
$3^2 \cdot 11 \cdot 41 \cdot 271 \cdot 9091$	10	$101 \cdot 3541 \cdot 27961$
$3^2 \cdot 21649 \cdot 513239$	11	$11^2 \cdot 23 \cdot 4093 \cdot 8779$
$3^3 \cdot 7 \cdot 11 \cdot 13 \cdot 37 \cdot 101 \cdot 9901$	12	$73 \cdot 137 \cdot 99990001$
$3^2 \cdot 11 \cdot 17 \cdot 73 \cdot 101 \cdot 137 \cdot 5882353$	16	$353 \cdot 449 \cdot 641 \cdot 1409 \cdot 69857$

Similarly, the low-order five bits are periodic with a period of at most 32; and the least significant bit of X_n is either constant or it strictly alternates between zero and one.

This situation does not occur when $m = w \pm 1$; in this case, the low-order bits of X_n will behave just as randomly as the high-order bits do. If, for example, $w = 2^{35}$ and $m = 2^{35} - 1$, the numbers of the sequence will not be very random if we consider only their remainders mod 31, 71, 127, or 122921 (cf. Table 1); but the low-order bit (which represents the numbers of the sequence taken mod 2) would be satisfactorily random.

Another alternative is to let m be the largest prime number less than w. This prime may be found by using the techniques of Section 4.5.4, and a table of suitably large primes appears in that section.

In most applications, the low-order bits are insignificant, and the choice $m = w$ is quite satisfactory—provided that the programmer using the random numbers does so wisely.

EXERCISES

1. [12] In exercise 3.2.1–3 we concluded that the best congruential generators will have a relatively prime to m. Show that in this case it is possible to compute $(aX + c) \bmod w$ in just *three* MIX instructions, rather than the four in (1), with the result appearing in register X.

2. [16] Write a MIX subroutine having the following characteristics:

Calling sequence: JMP RANDM

Entry conditions: Location XRAND contains an integer number X.

Exit conditions: $X \leftarrow \text{rA} \leftarrow (aX + c) \bmod w$, $\text{rX} \leftarrow 0$, overflow off.

(Thus a call on this subroutine will produce the next random number of a linear congruential sequence.)

3. [20] How can the constant $(w - 1)$ be specified in general in the MIX assembly language, regardless of the value of the byte size?

4. [10] Explain the significance of the lines of code in (3).

5. [20] Given that m is less than the word size, and if x, y are nonnegative integers less than m, show that the difference $(x - y) \bmod m$ may be computed in just four MIX instructions, without requiring any division. What is the best code for the sum $(x + y) \bmod m$?

▶ **6.** [20] The previous exercise suggests that subtraction mod m is easier to perform than addition mod m. Discuss sequences generated by the rule

$$X_{n+1} = (aX_n - c) \bmod m.$$

Are these sequences essentially different from linear congruential sequences as defined in the text? Are they more suited to efficient computer calculation?

7. [*M24*] What patterns can you spot in Table 1?

▶ **8.** [*20*] Write a MIX program analogous to (2) which computes $(aX) \bmod (w - 1)$. The values 0 and $w - 1$ are to be treated as equivalent in the input and output of your program.

9. [*23*] Write a MIX program analogous to that of exercise 8 which computes $(aX) \bmod (w - 2)$.

3.2.1.2. Choice of multiplier. In this section we shall show how to choose the multiplier a so as to give the *period of maximum length*. A long period is essential for any sequence which is to be used as a source of random numbers; indeed, we would hope that the period contains considerably more numbers than will ever be used in a single application. Therefore we shall concern ourselves in this section with the question of period length. The reader should keep in mind, however, that a long period is only one desirable criterion for the randomness of our sequence. It is quite possible to have a very long period in a completely nonrandom sequence. For example, when $a = c = 1$, the sequence is simply $X_{n+1} = (X_n + 1) \bmod m$, and this obviously has a period of length m, yet it is anything but random. Other considerations affecting the choice of a multiplier will be given later in this chapter.

Since only m different values are possible, the period cannot be longer than m. Can we achieve the maximum length, m? The example above shows that it is always possible, although the choice $a = c = 1$ does not yield a desirable sequence. Let us investigate *all* possible choices of a and c which give a period of length m. (*Note:* When the period has length m, every number from 0 through $(m - 1)$ occurs exactly once in the period; therefore the choice of X_0 does not affect the period length in this case.)

Theorem A. *The linear congruential sequence has a period of length m if and only if*

    i) *c is relatively prime to m;*
    ii) *$b = a - 1$ is a multiple of p, for every prime p dividing m;*
    iii) *b is a multiple of 4, if m is a multiple of 4.*

The ideas used in the proof of this theorem go back at least a hundred years. The first proof of the theorem in this particular form was given by M. Greenberger in the special case $m = 2^e$ (see *JACM* **8** (1961), 383–389), and the sufficiency of conditions (i), (ii), and (iii) in the general case was shown by Hull and Dobell (see *SIAM Review* **4** (1962), 230–254). To prove the theorem we will first consider some auxiliary number-theoretic results which are of interest in themselves.

Lemma P. *Let p be a prime number, and let e be a positive integer, where $p^e > 2$. If*

$$x \equiv 1 \;(\text{modulo } p^e), \quad x \not\equiv 1 \;(\text{modulo } p^{e+1}) \qquad (1)$$

then

$$x^p \equiv 1 \;(\text{modulo } p^{e+1}), \quad x^p \not\equiv 1 \;(\text{modulo } p^{e+2}). \qquad (2)$$

Proof. We have $x = 1 + qp^e$ for some integer q which is not a multiple of p. By the binomial formula

$$x^p = 1 + \binom{p}{1} qp^e + \cdots + \binom{p}{p-1} q^{p-1} p^{(p-1)e} + q^p p^{pe}$$

$$= 1 + qp^{e+1}\left(1 + \frac{1}{p}\binom{p}{2}qp^e + \frac{1}{p}\binom{p}{3}q^2p^{2e} + \cdots + \frac{1}{p}\binom{p}{p}q^{p-1}p^{(p-1)e}\right).$$

The quantity in parentheses is an integer, and, in fact, every term in the parentheses is a multiple of p except the first term. For if $1 < k < p$, $\binom{p}{k}$ is divisible by p (cf. exercise 1.2.6–10), hence

$$\frac{1}{p}\binom{p}{k}q^{k-1}p^{(k-1)e}$$

is divisible by $p^{(k-1)e}$; and the last term is $q^{p-1}p^{(p-1)e-1}$, which is divisible by p, since $(p - 1)e > 1$ when $p^e > 2$. So $x^p = 1 + q'p^{e+1}$, where q' is an integer not divisible by p, and this completes the proof. (*Note:* A generalization of this result appears in exercise 3.2.2–11(a).) ∎

Lemma Q. *Let the decomposition of m into prime factors be*

$$m = p_1^{e_1} \cdots p_t^{e_t}. \tag{3}$$

The length, λ, of the period of the linear congruential sequence determined by (X_0, a, c, m) is the least common multiple of the lengths λ_j of the periods of the linear congruential sequences $(X_0 \bmod p_j^{e_j},\, a \bmod p_j^{e_j},\, c \bmod p_j^{e_j},\, p_j^{e_j})$, $1 \le j \le t$.

Proof. By induction on t, it suffices to prove that if r and s are relatively prime, the length λ of the linear congruential sequence determined by (X_0, a, c, rs) is the least common multiple of the lengths λ_1, λ_2 of the periods of the sequences $(X_0 \bmod r, a \bmod r, c \bmod r, r)$ and $(X_0 \bmod s, a \bmod s, c \bmod s, s)$. We observed in the previous section, Eq. (5), that if the elements of these three sequences are denoted by X_n, Y_n, and Z_n, respectively, we will have

$$Y_n = X_n \bmod r \qquad \text{and} \qquad Z_n = X_n \bmod s \qquad \text{for all } n \ge 0.$$

Therefore, by Law D of Section 1.2.4, we find that

$$X_n = X_k \qquad \text{if and only if} \qquad Y_n = Y_k \quad \text{and} \quad Z_n = Z_k. \tag{4}$$

Let λ' be the least common multiple of λ_1 and λ_2; we wish to prove that $\lambda' = \lambda$. Since $X_n = X_{n+\lambda}$ for all suitably large n, we have $Y_n = Y_{n+\lambda}$ (hence λ is a multiple of λ_1) and $Z_n = Z_{n+\lambda}$ (hence λ is a multiple of λ_2), so λ must be $\ge \lambda'$. Furthermore, we know that $Y_n = Y_{n+\lambda'}$ and $Z_n = Z_{n+\lambda'}$ for all suitably large n; therefore, by (4), $X_n = X_{n+\lambda'}$. This proves $\lambda \le \lambda'$, so $\lambda = \lambda'$. ∎

Now we are ready to prove Theorem A. Because of Lemma Q, it suffices to prove the theorem when m is a power of a prime number. For

$$p_1^{e_1} \ldots p_t^{e_t} = \lambda = \text{lcm} (\lambda_1, \ldots, \lambda_t) \leq \lambda_1 \ldots \lambda_t \leq p_1^{e_1} \ldots p_t^{e_t}$$

can be true if and only if $\lambda_j = p_j^{e_j}$ for $1 \leq j \leq t$.

Therefore, assume that $m = p^e$, where p is prime and e is a positive integer. The theorem is obviously true when $a = 1$, so we may take $a > 1$. The period can be of length m if and only if each possible integer $0 \leq x < m$ occurs in the period, since no value occurs in the period more than once. Therefore the period is of length m if and only if the period of the sequence with $X_0 = 0$ is of length m, and we are justified in supposing that $X_0 = 0$. By formula 3.2.1–(6), we have

$$X_n = \left(\frac{a^n - 1}{a - 1}\right) c \bmod m. \tag{5}$$

If c is not relatively prime to m, X_n can never be equal to 1, so condition (i) is necessary. The period has length m if and only if the smallest positive value of n for which $X_n = X_0 = 0$ is $n = m$. By (5) and condition (i), our theorem now reduces to proving the following fact:

Lemma R. *Assume that $1 < a < p^e$, where p is prime. If λ is the smallest positive integer for which $(a^\lambda - 1)/(a - 1) \equiv 0 \pmod{p^e}$, then*

$$\lambda = p^e \qquad \textit{if and only if} \qquad \begin{cases} a \equiv 1 \pmod{p} & \textit{when} \quad p > 2, \\ a \equiv 1 \pmod{4} & \textit{when} \quad p = 2. \end{cases}$$

Proof. Assume that $\lambda = p^e$. If $a \not\equiv 1 \pmod{p}$, then $(a^n - 1)/(a - 1) \equiv 0 \pmod{p^e}$ if and only if $a^n - 1 \equiv 0 \pmod{p^e}$. The condition $a^{p^e} - 1 \equiv 0 \pmod{p^e}$ then implies that $a^{p^e} \equiv 1 \pmod{p}$; but by Theorem 1.2.4F, $a^{p^e} \equiv a \pmod{p}$; hence $a \not\equiv 1 \pmod{p}$ leads to a contradiction. And if $p = 2$ and $a \equiv 3 \pmod{4}$, we have $(a^{2^{e-1}} - 1)/(a - 1) \equiv 0 \pmod{2^e}$ by exercise 8. These arguments show that it is necessary to have $a = 1 + qp^f$, where $p^f > 2$ and q is not a multiple of p, whenever $\lambda = p^e$.

It remains to be shown that this condition is *sufficient* to make $\lambda = p^e$. By repeated application of Lemma P, we find that

$$a^{p^g} \equiv 1 \pmod{p^{f+g}}, \qquad a^{p^g} \not\equiv 1 \pmod{p^{f+g+1}},$$

and therefore

$$(a^{p^g} - 1)/(a - 1) \equiv 0 \pmod{p^g},$$
$$(a^{p^g} - 1)/(a - 1) \not\equiv 0 \pmod{p^{g+1}}. \tag{6}$$

In particular, $(a^{p^e} - 1)/(a - 1) \equiv 0 \pmod{p^e}$. Now the congruential sequence $(0, a, 1, p^e)$ has $X_n = (a^n - 1)/(a - 1) \bmod p^e$; therefore it has a period of length λ, that is, $X_n = 0$ if and only if n is a multiple of λ. Hence p^e

is a multiple of λ. This can happen only if $\lambda = p^g$ for some g, and relation (6) implies that $\lambda = p^e$, completing the proof. ∎

The proof of Theorem A is now complete. ∎

We will conclude this section by considering the special case of pure multiplicative generators, when $c = 0$. Although the random number generation process is slightly faster in this case, Theorem A shows us that the maximum period length cannot be achieved. In fact, this is quite obvious, since the sequence now satisfies the relation

$$X_{n+1} = aX_n \bmod m, \tag{7}$$

and the value $X_n = 0$ should never appear lest the sequence degenerate to zero. In general, if d is any divisor of m and if X_n is a multiple of d, we will have $X_{n+1}, X_{n+2}, \ldots$, all multiples of d. So when $c = 0$, we will want X_n to be relatively prime to m for all n, and this limits the length of the period.

It may be possible to achieve an acceptably long period even if we stipulate that $c = 0$. Let us now try to find conditions on the multiplier so that the period is as long as possible in this special case.

Because of Lemma Q, the period of the sequence depends entirely on the periods of the sequences when $m = p^e$, so let us consider that situation. We have $X_n = a^n X_0 \bmod p^e$, and it is clear that the period will be of length 1 if a is a multiple of p, so we take a to be relatively prime to p^e. Then the period is the smallest integer λ such that $X_0 = a^\lambda X_0 \bmod p^e$. If the greatest common divisor of X_0 and p^e is p^f, this condition is equivalent to

$$a^\lambda \equiv 1 \pmod{p^{e-f}}. \tag{8}$$

By Euler's theorem (exercise 1.2.4–28), $a^{\varphi(p^{e-f})} \equiv 1 \pmod{p^{e-f}}$; hence λ is a divisor of

$$\varphi(p^{e-f}) = p^{e-f-1}(p - 1).$$

When a is relatively prime to m, the smallest integer λ for which $a^\lambda \equiv 1 \pmod{m}$ is conventionally called *the order of* a *modulo* m. Any value of a which gives the *maximum* possible order modulo m is called a *primitive element modulo* m.

Let $\lambda(m)$ denote the order of a primitive element, i.e., the maximum possible order, modulo m. The remarks above show that $\lambda(p^e)$ is a divisor of $p^{e-1}(p - 1)$; with a little care (see exercises 11 through 16 below) we can give the precise value of $\lambda(m)$ in all cases as follows:

$$\lambda(2) = 1, \quad \lambda(4) = 2, \quad \lambda(2^e) = 2^{e-2} \quad \text{if} \quad e \geq 3.$$
$$\lambda(p^e) = p^{e-1}(p - 1), \quad \text{if} \quad p > 2. \tag{9}$$
$$\lambda(p_1^{e_1} \ldots p_t^{e_t}) = \operatorname{lcm}\left(\lambda(p_1^{e_1}), \ldots, \lambda(p_t^{e_t})\right).$$

Our remarks may be summarized in the following theorem:

Theorem B. [R. D. Carmichael, *Bull. Amer. Math. Soc.* **16** (1910), 232–238.] *The maximum period possible when* $c = 0$ *is* $\lambda(m)$, *where* $\lambda(m)$ *is defined in* (9). *This period is achieved if*

　　i) X_0 *is relatively prime to* m;
　　ii) a *is a primitive element modulo* m. ∎

Note that we can obtain a period of length $m - 1$ if m is prime; this is just one less than the maximum length.

　　The question now is, how can we find primitive elements modulo m? The exercises at the close of this section give us the following facts:

Theorem C. a *is a primitive element modulo* p^e *if and only if*

　　i) $p^e = 2$, a *is odd; or* $p^e = 4$, $a \bmod 4 = 3$;
　　　　or $p^e = 8$, $a \bmod 8 = 3, 5, 7$; *or* $p = 2$, $e \geq 4$, $a \bmod 8 = 3$ *or* 5;

or

　　ii) p *is odd,* $e = 1$, $a \not\equiv 0$ (modulo p), *and* $a^{(p-1)/q} \not\equiv 1$ (modulo p) *for any prime divisor* q *of* $p - 1$;

or

　　iii) p *is odd,* $e > 1$, a *satisfies* (ii), *and* $a^{p-1} \not\equiv 1$ *(modulo* p^2). ∎

Conditions (ii) and (iii) of this theorem are readily tested on a computer for large values of p, by using the efficient methods for evaluating powers discussed in Section 4.6.3. Finally, if we are given values a_j which are primitive modulo $p_j^{e_j}$, it is possible to find a single value a such that $a \equiv a_j$ (modulo $p_j^{e_j}$), $1 \leq j \leq t$, using the "Chinese remainder algorithm" discussed in Section 4.3.2; hence a will be a primitive element modulo $p_1^{e_1} \ldots p_t^{e_t}$. This gives us a reasonably efficient way to construct multipliers satisfying the condition of Theorem B, for any desired value of m. The calculations are, however, somewhat lengthy in the general case.

　　In the common case $m = 2^e$, with $e \geq 4$, the conditions above simplify to the single requirement that $a \equiv 3$ or 5 (modulo 8). In this case, one-fourth of all possible multipliers give the maximum period.

　　The second most common case is when $m = 10^e$. Using Lemmas P and Q, it is not difficult to obtain necessary and sufficient conditions for the achievement of the maximum period in the case of a decimal computer:

Theorem D. *If* $m = 10^e$, $e \geq 5$, $c = 0$, *and* X_0 *is not a multiple of 2 or 5, the period of the linear congruential sequence is* $5 \times 10^{e-2}$ *if and only if* $a \bmod 200$ *equals one of the following 32 values:*

$$3, 11, 13, 19, 21, 27, 29, 37, 53, 59, 61, 67, 69, 77, 83, 91, 109, 117,$$
$$123, 131, 133, 139, 141, 147, 163, 171, 173, 179, 181, 187, 189, 197. \quad ∎ \qquad (10)$$

EXERCISES

1. [*10*] What is the length of the period of the linear congruential sequence with $X_0 = 5772156648$, $a = 3141592621$, $c = 2718281829$, and $m = 10000000000$?

2. [*10*] Are the following two conditions sufficient to guarantee the maximum length period, when $m = 2^e$ is a power of 2? "(i) c is odd; (ii) $a \bmod 4 = 1$."

3. [*13*] Suppose that $m = 10^e$, where $e \geq 3$, and suppose that c is not a multiple of 2 and not a multiple of 5. Show that the linear congruential sequence will have the maximum length period if and only if $a \bmod 20 = 1$.

4. [*20*] When a and c satisfy the conditions of Theorem A, and when $m = 2^e$, $X_0 = 0$, what is the value of $X_{2^{e-1}}$?

▶ **5.** [*20*] Find all multipliers a which satisfy the conditions of Theorem A when $m = 2^{35} + 1$. (The prime factors of m may be found in Table 3.2.1.1–1.)

6. [*20*] Find all multipliers a which satisfy the conditions of Theorem A when $m = 10^6 - 1$. (See Table 3.2.1.1–1.)

▶ **7.** [*M24*] The period of a congruential sequence need not start with X_0, but we can always find indices $\mu \geq 0$, $\lambda > 0$, such that $X_{n+\lambda} = X_n$ whenever $n \geq \mu$, and for which μ and λ are the smallest possible values with this property. (Cf. exercises 3.1–6 and 3.2.1–1.) If μ_j, λ_j are the indices corresponding to the sequence $(X_0 \bmod p_j^{e_j}, a \bmod p_j^{e_j}, c \bmod p_j^{e_j}, p_j^{e_j})$, and if μ, λ correspond to the sequence $(X_0, a, c, p_1^{e_1} \ldots p_t^{e_t})$, Lemma Q states that λ is the least common multiple of $\lambda_1, \ldots, \lambda_t$. What is the value of μ in terms of the values of $\mu_1, \ldots, \mu_t$? What is the maximum possible value of μ obtainable by varying X_0, a, and c, when $m = p_1^{e_1} \ldots p_t^{e_t}$ is fixed?

8. [*M20*] Show that if $a \bmod 4 = 3$, $(a^{2^{e-1}} - 1)/(a - 1) \equiv 0$ (modulo 2^e), when $e > 1$. (Use Lemma P.)

▶ **9.** [*M22*] (W. E. Thomson.) When $c = 0$ and $m = 2^e \geq 8$, Theorems B and C say that the period has length 2^{e-2} if and only if the multiplier a satisfies $a \bmod 8 = 3$ or $a \bmod 8 = 5$. Show that every such sequence is essentially a linear congruential sequence with $m = 2^{e-2}$, having *full* period, in the following sense:

a) If $X_{n+1} = (4c + 1)X_n \bmod 2^e$, and $X_n = 4Y_n + 1$, then

$$Y_{n+1} = ((4c + 1)Y_n + c) \bmod 2^{e-2}.$$

b) If $X_{n+1} = (4c - 1)X_n \bmod 2^e$, and $X_n = ((-1)^n(4Y_n + 1))$ mod 2^e, then

$$Y_{n+1} = ((1 - 4c)Y_n - c) \bmod 2^{e-2}.$$

(*Note:* In these formulas, c is an odd integer. The literature contains several statements to the effect that sequences with $c = 0$ satisfying Theorem B are somehow more random than sequences satisfying Theorem A, in spite of the fact that the period is only one-fourth as long in the case of Theorem B. This exercise refutes such statements.)

10. [*M21*] For what values of m is $\lambda(m) = \varphi(m)$?

▶ **11.** [*M28*] Let x be an odd integer greater than 1. (a) Show that there exists a unique integer $f > 1$ such that $x \equiv 2^f \pm 1$ (modulo 2^{f+1}). (b) Given that $1 < x < 2^e - 1$

and that f is the corresponding integer from part (a), show that the order of x modulo 2^e is 2^{e-f}. (c) In particular, this proves Theorem C(i).

12. [*M26*] Let p be an odd prime. If $e > 1$, prove that a is a primitive element modulo p^e if and only if a is a primitive element modulo p, and $a^{p-1} \not\equiv 1$ (modulo p^2). (For the purposes of this exercise, assume that $\lambda(p^e) = p^{e-1}(p - 1)$. This fact is proved in exercises 14 and 16 below.)

13. [*M22*] Let p be prime. Given that a is not a primitive element modulo p, show that either a is a multiple of p or $a^{(p-1)/q} \equiv 1$ (modulo p) for some prime number q which divides $p - 1$.

14. [*M18*] If $e > 1$ and p is an odd prime, and if a is a primitive element modulo p, prove that either a or $a + p$ is a primitive element modulo p^e. *Hint:* See exercise 12.

15. [*M29*] (a) Let a_1, a_2 be relatively prime to m, and let their orders modulo m be λ_1, λ_2, respectively. If λ is the least common multiple of λ_1 and λ_2, prove that $a_1^{\kappa_1} a_2^{\kappa_2}$ has order λ modulo m, for suitable integers κ_1, κ_2. [*Hint:* Consider first the case that λ_1 is relatively prime to λ_2.] (b) Let $\lambda(m)$ be the maximum order of any element modulo m. Prove that $\lambda(m)$ is a multiple of the order of each element modulo m, that is, prove that $a^{\lambda(m)} \equiv 1$ (modulo m) whenever a is relatively prime to m.

▶ **16.** [*M24*] Let p be a prime number. (a) Let $f(x) = x^n + c_1 x^{n-1} + \cdots + c_n$, where the c's are integers. Given that a is an integer for which $f(a) \equiv 0$ (modulo p), show that there exists a polynomial $q(x) = x^{n-1} + q_1 x^{n-2} + \cdots + q_{n-1}$ with integer coefficients such that $f(x) \equiv (x - a)q(x)$ (modulo p) for all integers x. (b) Let $f(x)$ be a polynomial as in (a). Show that $f(x)$ has at most n distinct "roots" modulo p; that is, there are at most n integers a, with $0 \le a < p$, such that $f(a) \equiv 0$ (modulo p). (c) Because of exercise 15(b), the polynomial $f(x) = x^{\lambda(p)} - 1$ has $p - 1$ distinct roots; hence there is an integer a with order $p - 1$.

17. [*M26*] Not all of the values listed in Theorem D would be found by the text's construction; for example, 11 is not primitive modulo 5^e. How can this be possible, when 11 is (according to Theorem D) a primitive element modulo 10^e? Which of the values listed in Theorem D *are* primitive elements modulo both 2^e and 5^e?

18. [*M25*] Prove Theorem D. (Cf. the previous exercise.)

19. [*40*] Make a table of some suitable multipliers, a, for each of the values of m listed in Table 3.2.1.1–1, assuming that $c = 0$.

▶ **20.** [*M24*] What is the length of the period of a linear congruential sequence for which (i) $X_0 = 0$; (ii) a is primitive modulo $p_j^{e_j}$, $1 \le j \le t$, for all prime powers in the factorization of $m = p_1^{e_1} \ldots p_t^{e_t}$ into primes; (iii) c is relatively prime to m?

3.2.1.3. Potency. In the preceding section, we showed that the maximum period can be obtained when $b = a - 1$ is a multiple of each prime dividing m (and b must also be a multiple of 4 if m is a multiple of 4). If z is the radix of the machine being used—so that $z = 2$ for a binary computer, and $z = 10$ for a decimal computer—and if m is the word size z^e, the multiplier

$$a = z^k + 1, \qquad 2 \le k < e \tag{1}$$

satisfies these conditions. Theorem 3.2.1.2A also says that we may take $c = 1$.

The recurrence relation now has the form

$$X_{n+1} = ((z^k + 1)X_n + 1) \bmod z^e, \tag{2}$$

and this equation suggests that we can avoid the multiplication; merely shifting and adding will suffice.

For example, suppose that $a = B^2 + 1$, where B is the byte size of MIX. The code

```
        LDA   X
        SLA   2
        ADD   X
        INCA  1
```
(3)

can be used in place of the instructions given in Section 3.2.1.1, and the execution time decreases from $16u$ to $7u$.

For this reason, multipliers having form (1) have been widely discussed in the literature, and indeed they have been recommended by many authors. However, some five years of experimentation with this method show that *multipliers having the simple form in* (1) *should be avoided*. A number of factors support this contention. In the first place, the execution time is not really halved as example (3) would indicate. If we add the times for "JMP, STJ, STA, JNOV" to the above, the comparative execution times become $22u$ for the multiplication method and $13u$ for the shift-and-add method. To this must be added the computation time of the main program, which uses the random numbers; the net percent savings of computation time is almost negligible. And on many modern computers, the execution time for a multiplication is actually *faster* than the time to shift and add!

The most important reason to refrain from using the multiplier $z^k + 1$ is the fact that the numbers usually fail to be sufficiently random. One of the reasons for this is related to the concept of "potency," which we shall now discuss.

The *potency* of a linear congruential sequence with maximum period is defined to be the least integer s such that

$$b^s \equiv 0 \pmod{m}. \tag{4}$$

(Such an integer s will always exist when the multiplier satisfies the conditions of Theorem 3.2.1.2A, since b is a multiple of every prime dividing m.)

We may analyze the randomness of the sequence by taking $X_0 = 0$, since 0 occurs somewhere in the period. With this assumption, we have $X_n = ((a^n - 1)c/b) \bmod m$, and if we expand $a^n - 1 = (b+1)^n - 1$ by the binomial theorem, we find that

$$X_n = c\left(n + \binom{n}{2}b + \cdots + \binom{n}{s}b^{s-1}\right) \bmod m. \tag{5}$$

All terms in b^s, b^{s+1}, etc., may be ignored, since they are multiples of m.

Equation (5) can be instructive, so we shall consider some special cases. If $a = 1$, the potency is 1; and $X_n \equiv cn$ (modulo m), as we have already observed, so the sequence is surely not random. If the potency is 2, we have $X_n \equiv cn + cb\binom{n}{2}$, and again the sequence is not very random; indeed,

$$X_{n+1} - X_n \equiv c + cbn,$$

so the differences between one generated number and the next are very simply related from one value of n to the next. If

$$d = cb \bmod m,$$

the point (X_n, X_{n+1}, X_{n+2}) always lies on one of the four planes

$$x - 2y + z = d + m,$$
$$x - 2y + z = d,$$
$$x - 2y + z = d - m,$$
$$\text{or} \qquad x - 2y + z = d - 2m,$$

in three-dimensional space.

If the potency is 3, the sequence begins to look somewhat more random, but there is a high degree of dependency between X_n, X_{n+1}, and X_{n+2}; $X_{n+1} - X_n$ is a sequence of potency 2, and tests show sequences with potency 3 are still not sufficiently good. Reasonable results have been reported when the potency is 4 or more, but these have been disputed by many people. A potency of at least 5 would seem to be required for sufficiently random values.

Suppose, for example, that $m = 2^{35}$ and $a = 2^k + 1$. Then $b = 2^k$, so we find that when $k \geq 18$, the value $b^2 = 2^{2k}$ is a multiple of m: the potency is 2. If $k = 17, 16, \ldots, 12$, the potency is 3, and a potency of 4 is achieved for $k = 11, 10, 9$. The only acceptable multipliers, from the standpoint of potency, therefore have $k \leq 8$. This means $a \leq 257$, and we shall see later that *small* multipliers are also to be avoided. We have thereby eliminated all multipliers of the form $2^k + 1$ when $m = 2^{35}$.

With larger word sizes, multipliers of the form $2^k + 1$ may be acceptable. A generator with $m = 2^{47}$, $a = 2^9 + 1$ (having a potency of 6) has been tested and reported successful in *CACM* **4** (1961), 350–352. In spite of this example, it pays to be extremely wary of multipliers such as (1), since nearly all the known examples of unreliable generators have been of this type. In fact, even this example fails the statistical test of Section 3.3.4.

When m is equal to $w \pm 1$, where w is the word size, m is generally not divisible by high powers of primes, and so a high potency is impossible (see exercise 6). So in this case, the maximum-period method should *not* be used, and the pure-multiplication method with $c = 0$ should be applied instead.

There still remains a lot of freedom in the choice of multiplier. In general, we want to keep the potency high and the multiplier reasonably large, and we also want to avoid simple patterns in the multiplier digits. Suppose that $m = 2^{35}$ and that a multiplication operation is somewhat faster if the number of "one" bits in the multiplier is small; a multiplier such as $2^{23} + 2^{14} + 2^2 + 1$ can be (tentatively) recommended. The term 2^{23} makes the multiplier suitably large; the term 2^2 gives it high potency; the term 1 is necessary to get the maximum period; and the term 2^{14} has been thrown in to keep the multiplier from being too simple for randomness (cf. exercise 8). A term like 2^{34} would not be as good here as 2^{23}, since X_n times 2^{34} makes use of only the least significant bit of X_n (which is not very random). If the speed of multiplication is not critical, a more "random" multiplier (for example, $a = 3141592621$) is likely to be much more satisfactory.

Actually the concept of potency is only one criterion for the choice of multiplier, and much more can be said. Section 3.3.4 below discusses the "spectral test" for multipliers of linear congruential sequences; this is an important criterion which includes potency and the size of multiplier as special cases. In Section 3.3.4 we shall see, for example, that $2^{23} + 2^{13} + 2^2 + 1$ is a noticeably poorer choice than $2^{23} + 2^{14} + 2^2 + 1$. Any multiplier that is to be used extensively should be subjected to the spectral test.

EXERCISES

1. [*M10*] Show that, no matter what the byte size B of MIX happens to be, the code (3) yields a random-number generator of maximum period.

2. [*10*] What is the potency of the generator represented by the MIX code (3)?

3. [*11*] When $m = 2^{35}$, what is the potency of the linear congruential sequence with $a = 3141592621$? What is the potency if the multiplier is $a = 2^{23} + 2^{13} + 2^2 + 1$?

4. [*20*] Show that if $m = 2^e \geq 8$, the maximum potency is achieved when $a \bmod 8 = 5$.

5. [*M20*] Given that $m = p_1^{e_1} \ldots p_t^{e_t}$ and $a = 1 + k p_1^{f_1} \ldots p_t^{f_t}$, where a satisfies the conditions of Theorem 3.2.1.2A and k is relatively prime to m, show that the potency is max $(\lceil e_1/f_1 \rceil, \ldots, \lceil e_t/f_t \rceil)$.

▶ **6.** [*20*] Which of the values of $m = w \pm 1$ in Table 3.2.1.1–1 can give a potency as much as 4? (Use the result of exercise 5.)

7. [*M20*] When a satisfies the conditions of Theorem 3.2.1.2A, it is relatively prime to m; hence there is a number a' such that $aa' \equiv 1$ (modulo m). Show that a' can be expressed simply in terms of b.

▶ **8.** [*M26*] A random-number generator with $X_{n+1} = (2^{17} + 3) X_n \bmod 2^{35}$ and $X_0 = 1$ was subjected to the following test: Let $Y_n = \lfloor 10 X_n / 2^{35} \rfloor$; then Y_n should be a random digit between 0 and 9, and the triples $(Y_{3n}, Y_{3n+1}, Y_{3n+2})$ should take on each of the 1000 possible values from $(0, 0, 0)$ to $(9, 9, 9)$ with equal probability. But with 30000 values of n tested, some triples hardly ever occurred, and others occurred much more often than they should have. Can you account for this failure?

3.2.2. Other Methods

Of course, linear congruential sequences are not the only sources of random numbers that have been proposed for computer use. In this section we shall review the most significant other methods; some of these are quite important, while others are interesting chiefly because they are not as good as a person might expect.

One of the common fallacies encountered in connection with random number generation is the idea that we can take a good generator and modify it a little, in order to get an "even-more-random" sequence. This is quite often not true; for example, we know that

$$X_{n+1} = (aX_n + c) \bmod m \tag{1}$$

leads to reasonably good random numbers; wouldn't the sequence produced by

$$X_{n+1} = ((aX_n) \bmod (m + 1) + c) \bmod m \tag{2}$$

be even *more* random? The answer is, the new sequence is probably a great deal *less* random. For the whole theory breaks down, and in the absence of any theory about the behavior of the sequence (2), we come into the area of generators of the type $X_{n+1} = f(X_n)$ with the function f chosen at random; exercises 3.1–11 through 3.1–15 show that these sequences do not behave nearly as well as the sequences with the more well-behaved function (1).

Let us consider another approach, in an attempt to get "more random" numbers. The linear congruential method can be generalized to, say, a quadratic congruential method:

$$X_{n+1} = (dX_n^2 + aX_n + c) \bmod m. \tag{3}$$

Exercise 8 generalizes Theorem 3.2.1.2A to obtain necessary and sufficient conditions on a, c, and d such that the sequence defined by (3) has a period of the maximum length m; the restrictions are not much more severe than in the linear method.

An interesting quadratic method has been proposed by R. R. Coveyou when m is a power of two; let

$$X_0 \bmod 4 = 2, \qquad X_{n+1} = X_n(X_n + 1) \bmod 2^e, \qquad n \geq 0. \tag{4}$$

This sequence can be computed with about the same efficiency as (1), without any worries of overflow. It has an interesting connection with von Neumann's original middle-square method: If we let Y_n be $2^e X_n$, so that Y_n is a double-precision number obtained by placing e zeros to the right of the binary representation of X_n, then Y_{n+1} consists of precisely the middle $2e$ digits of $Y_n^2 + 2^e Y_n$! So Coveyou's method is almost identical to a double-precision middle-square method, yet it is guaranteed to have a long period; further evidence of its randomness is also provable (see exercise 3.3.4–25).

Other generalizations of Eq. (1) also suggest themselves; for example, we might try to extend the period length of the sequence. The period of a linear congruential sequence is extremely long; when m is approximately the word size of the computer, we usually get periods on the order of 10^9 or more, so that typical calculations will use only a very small portion of the sequence. On the other hand, the period length influences the degree of randomness achievable in a sequence; see the remarks following Eq. 3.3.4–13. Therefore it is occasionally desirable to seek a longer period, and several methods are available for this purpose. One technique is to make X_{n+1} depend on X_n *and* on X_{n-1}, instead of just on X_n; then the period length can be as high as m^2, since the sequence will not begin to repeat until we have $(X_{n+\lambda}, X_{n+\lambda+1}) = (X_n, X_{n+1})$.

The simplest sequence in which X_{n+1} depends on more than one of the preceding values is the Fibonacci sequence,

$$X_{n+1} = (X_n + X_{n-1}) \bmod m. \tag{5}$$

This generator was considered in the early 1950's, and it usually gives a period length greater than m; but tests have shown that the numbers produced by the Fibonacci recurrence (5) are definitely *not* satisfactorily random, and so at the present time the main interest in (5) as a source of random numbers is that it makes a nice "bad example." We may also consider generators of the form

$$X_{n+1} = (X_n + X_{n-k}) \bmod m, \tag{6}$$

when k is a comparatively large value. These were introduced by Green, Smith, and Klem in *JACM* **6** (1959), 527–537. If $X_0, X_1, \ldots, X_k$ are chosen appropriately, this sequence has the promise of providing good random numbers. At first Eq. (6) may not seem to be extremely well suited to computer implementation, but in fact there is a very efficient procedure:

Algorithm A (*Additive number generator*). Initially locations Z, $Y[0]$, $Y[1]$, $\ldots$, $Y[k]$ are set to the values X_k, X_k, X_{k-1}, $\ldots$, X_0, respectively, and j is set equal to k. Successive performances of this algorithm will produce as output the sequence X_{k+1}, X_{k+2}, $\ldots$.

A1. [$j < 0$?] If $j < 0$, set $j \leftarrow k$.

A2. [Add, exchange.] Set $Z \leftarrow Y[j] \leftarrow (Z + Y[j]) \bmod m$.

A3. [Decrease j.] Decrease j by 1, output Z. ∎

This algorithm in MIX is simply

```
        J6NN    *+2         A1. j < 0?
        ENT6    K           Set j ← k.
        LDA     Z           A2. Add, exchange.
        ADD     Y,6         Z + Y[j] (overflow possible)       (7)
        STA     Y,6            → Y[j]
        STA     Z              → Z.
        DEC6    1           A3. Decrease j.
```

(provided that index register 6 is not needed in the remainder of the program). This generator is usually faster than the previous methods, since it does not require any multiplication.

At the present time, not very much is known about such an additive number generator; before its use can be recommended, it will be necessary to develop the theoretical results necessary to prove certain desirable randomness properties, and to carry out extensive tests for particular values of k and $X_0, \ldots, X_k$. The length of the period is discussed in exercise 11; it is generally not much longer than m. The article by Green, Smith, and Klem reports that, when $k \leq 15$, the sequence fails to pass the "gap test" described in Section 3.3.2, although when $k = 16$ the test was satisfactory.

A similar but far superior way to improve upon the randomness of linear congruential sequences can be used when m is *prime*. Here m can be chosen, for example, to be the largest prime number which fits in a single computer word; such a prime can be discovered in a reasonable amount of time by using the techniques of Section 4.5.4. When $m = p$ is prime, the theory of finite fields tells us that it is possible to find multipliers $a_1, \ldots, a_k$ such that the sequence defined by

$$X_n = (a_1 X_{n-1} + \cdots + a_k X_{n-k}) \bmod p \qquad (8)$$

has period length $p^k - 1$; here $X_0, \ldots, X_{k-1}$ may be chosen arbitrarily but not all zero. (The special case $k = 1$ corresponds to a multiplicative congruential sequence with prime modulus, with which we are already familiar.) The constants $a_1, \ldots, a_k$ in (8) have the desired property if and only if the polynomial

$$f(x) = x^k - a_1 x^{k-1} - \cdots - a_k \qquad (9)$$

is a "primitive polynomial modulo p," that is, it has a root which is a primitive element of the field with p^k elements (see exercise 4.6.2–16).

Of course, the mere fact that suitable constants $a_1, \ldots, a_k$ *exist* giving a period of length $p^k - 1$ is not enough for practical purposes; we must be able to *find* them, and we can't simply try all p^k possibilities, since p is on the order of the computer's word size. Fortunately there are exactly $\varphi(p^k - 1)/k$ suitable choices of $(a_1, \ldots, a_k)$, so there is a fairly good chance of hitting one after making a few random tries. But we also need a way to tell quickly whether or not (9) is a primitive polynomial modulo p; it is certainly unthinkable to generate up to $p^k - 1$ elements of the sequence and wait for a repetition! Methods of testing for primitivity modulo p are discussed by Alanen and Knuth in *Sankhyā*, Ser. A, **26** (1964), 305–328; the following criteria can be used: Let $r = (p^k - 1)/(p - 1)$.

    i) $(-1)^{k+1} a_k$ must be a primitive root modulo p. (Cf. Section 3.2.1.2.)

    ii) The polynomial x^r must be congruent to $(-1)^{k+1} a_k$, modulo $f(x)$ and p.

    iii) The degree of $x^{r/q} \bmod f(x)$, using polynomial arithmetic modulo p, must be positive, for each prime divisor q of r.

(Efficient means of computing $x^n \bmod f(x)$, using polynomial arithmetic modulo a prime p, are discussed in Section 4.6.2.)

In order to carry out this test, we need to know the prime factorization of $r = (p^k - 1)/(p - 1)$, and this is the limiting factor in the calculation; r can be factored in a reasonable amount of time when $k = 2, 3$, and perhaps 4, but higher values of k are difficult to handle when p is large. Even $k = 2$ essentially doubles the number of "significant random digits" over what is achievable with $k = 1$, so larger values of k will rarely be necessary.

An adaptation of the spectral test (Section 3.3.4) can be used to rate the sequence of numbers generated by (8); see exercise 3.3.4–26. The considerations of that section show that we should *not* make the obvious choice of $a_1 = +1$ or -1 when it is possible to do so; it is better to pick large, essentially "random," values of $a_1, \ldots, a_k$ which satisfy the conditions, and to verify the choice by applying the spectral test. A significant amount of computation is involved in finding $a_1, \ldots, a_k$, but all known evidence indicates that the result will be a very satisfactory source of random numbers.

The special case $p = 2$ is of independent interest. Sometimes a random-number generator is desired which merely produces a random sequence of *bits* —zeros and ones—instead of fractions between zero and one. There is a simple way to generate a highly random bit sequence on a binary computer, manipulating k-bit words: We start with an arbitrary nonzero binary word $Y = (Y_1 Y_2 \cdots Y_k)_2$. To get the next random bit of the sequence, do the following operations, shown in MIX:

```
LDA   Y      (Assume that overflow is now "off")
ADD   Y      Shift left one bit.
JNOV  *+2    Jump if high bit was originally zero.        (10)
XOR   C      Otherwise adjust number with "exclusive or."
STA   Y
```

The fourth instruction here is the "exclusive or" operation found on nearly all binary computers (cf. exercise 2.5–28); it changes each bit position in which C has a "1" bit. The value in location C is the binary constant $(a_1 \ldots a_k)_2$, where $x^k - a_1 x^{k-1} - \cdots - a_k$ is a primitive polynomial modulo 2 as above. After the code (10) has been executed, the next bit of the generated sequence may be taken as the least significant bit of Y (or, alternatively, we could always use the most significant bit of Y, if it is more convenient to do so). For example, consider Fig. 1, which illustrates the sequence generated for $k = 4$, $c = (0011)_2$. (This is, of course, an unusually small value for k.) The right-hand column shows the sequence of bits of the sequence, which repeats in a period of length $2^k - 1 = 15$: 1101011110001001 $\cdots$. This sequence is quite random, considering that it was generated with only four bits of memory; to see this, consider the adjacent sets of four bits occurring in the period, namely 1101, 1010, 0101, 1011, 0111, 1111, 1110, 1100, 1000, 0001, 0010, 0100, 1001, 0011, 0110. In general, every possible adjacent set of k bits occurs exactly once in the period, except the set of all zeros, since the period length is $2^k - 1$; thus, adjacent sets of k bits are essentially independent. We shall see in Section 3.5 that this is a very strong randomness criterion when k is, say, 30 or more. Theoretical results illustrating the randomness of this sequence are given in an article by R. C. Tausworthe, *Math. Comp.* **19** (1965), 201–209.

```
1011
0101
1010
0111
1110
1111
1101
1001
0001
0010
0100
1000
0011
0110
1100
1011
```

Fig. 1. Successive contents of the computer word Y in the binary method, assuming that $k = 4$ and $c = \text{CONTENTS}(\text{C}) = (0011)_2$.

Primitive polynomials of degree ≤ 100 modulo 2 have been tabulated by E. J. Watson, *Math. Comp.* **16** (1962), 368–369. When $k = 35$, we may take

$$c = (00000000000000000000000000000000101)_2,$$

and when $k = 30$ we may take

$$c = (000000000000000000000001010011)_2;$$

but the considerations of exercises 18 and 3.3.4–26 imply that it would be preferable to find essentially "random" constants c which define primitive polynomials modulo 2.

Caution: Several people have been trapped into believing that this random-bit-generation technique can be used to generate random whole-word fractions $(.Y_0 Y_1 \ldots Y_{k-1})_2$, $(.Y_k Y_{k+1} \ldots Y_{2k-1})_2$, $\ldots$; but it is actually a poor source of random fractions, even though the bits are individually quite random! (See exercise 18.)

We have now seen that sequences with $0 \leq X_n < m$ and period $m^k - 1$ can be found, when X_n is a suitable function of $X_{n-1}, \ldots, X_{n-k}$ and when m is prime. The highest conceivable period obtainable for *any* sequence defined by a relation of the form

$$X_n = f(X_{n-1}, \ldots, X_{n-k}), \qquad 0 \leq X_n < m, \tag{11}$$

is easily seen to be m^k. M. H. Martin [*Bull. Amer. Math. Soc.* **40** (1934), 859–864] was the first person to show that functions achieving this maximum period are possible for all m and k; his method is easy to state, but unfortunately not suitable for programming (see exercise 17). From a computational standpoint, the simplest known functions f which yield the maximum period m^k appear in exercise 21; the corresponding programs are, in general, not as efficient for random-number generation as other methods we have described, but they do give demonstrable randomness (when the period as a whole is considered).

Another important class of techniques deals with the *combination* of random-number generators, to get "more random" sequences. There will always be skeptics who feel that the linear congruential methods, additive methods, etc. are all too simple to give sufficiently random sequences; and as yet it is impossible to *prove* that their skepticism is unjustified (although we believe it is), so it is pretty useless to argue the point. There are reasonably efficient methods for combining two sequences into a third so that the third one is so haphazard that only the most hardened skeptic will fail to accept it.

Suppose we have two sequences $X_0, X_1, \ldots$, and $Y_0, Y_1, \ldots$, of random numbers between 0 and $m - 1$, hopefully determined by two unrelated methods. One suggestion has been to add them together, mod m, obtaining a sequence with $Z_n = (X_n + Y_n) \bmod m$; in this case, it is desirable to have the period lengths of $\langle X_n \rangle$ and $\langle Y_n \rangle$ relatively prime to each other (see exercise 13.)

A much better method has been suggested by MacLaren and Marsaglia, and it is eminently suited to computer programming:

Algorithm M (*A quite random sequence*). Given methods for generating two sequences $\langle X_n \rangle$ and $\langle Y_n \rangle$, this algorithm will successively output the terms of a "considerably more random" sequence. We use an auxiliary table $V[0], V[1], \ldots,$ $V[k-1]$, where k is some number chosen for convenience, usually in the neighborhood of 100. Initially, the V-table is filled with the first k values of the X-sequence.

M1. [Generate X, Y.] Set X, Y equal to the next members of the sequences $\langle X_n \rangle$, $\langle Y_n \rangle$, respectively.

M2. [Extract j.] Set $j \leftarrow \lfloor kY/m \rfloor$, where m is the modulus used in the sequence $\langle Y_n \rangle$; that is, j is a random value, $0 \le j < k$, determined by Y.

M3. [Exchange.] Output $V[j]$ and then set $V[j] \leftarrow X$. ∎

This method can be highly recommended. It will give an incredibly long period, if the periods of $\langle X_n \rangle$ and $\langle Y_n \rangle$ are relatively prime; and even if the period is of no consequence, there is very little relation between nearby terms of the sequence. The reason this method is so much better than, say, the middle-square method or the method in Eq. (2), is that we know that the sequences X_n and Y_n are already quite random to start with, and there is no chance of degeneracy. The reader is urged to work exercise 3 to see how the method works in a particular case.

On MIX we may implement Algorithm M by taking k equal to the byte size. Steps M2 and M3 are readily programmed as follows:

$$
\begin{array}{lll}
\text{LD6} & \text{Y(1:1)} & j \leftarrow \text{high-order byte of } Y. \\
\text{LDA} & \text{V,6} & \text{rA} \leftarrow \text{next element of new sequence.} \\
\text{LDX} & \text{X} & \\
\text{STX} & \text{V,6} & V[j] \leftarrow X.
\end{array}
\qquad (12)
$$

As an example, assume that Algorithm M is applied to the following two sequences, with $k = 64$:

$$
\begin{aligned}
X_0 &= 5772156649, & X_{n+1} &= (3141592653X_n + 2718281829) \bmod 2^{35}; \\
Y_0 &= 1781072418, & Y_{n+1} &= (2718281829Y_n + 3141592653) \bmod 2^{35}.
\end{aligned}
$$

We contend that the sequence obtained by applying Algorithm M will satisfy virtually *anyone's* requirements for randomness in a computer-generated sequence. Furthermore, the time required to generate this sequence is only slightly more than twice as long as it takes to generate the sequence $\langle X_n \rangle$ alone.

F. Gebhardt has shown [*Math. Comp.* **21** (1967), 708–709] that satisfactory random sequences are produced by Algorithm M even when it is applied to a sequence as nonrandom as the Fibonacci sequence, with $X_n = F_{2n} \bmod m$ and $Y_n = F_{2n+1} \bmod m$. Another way to combine two sequences, based on circular shifting and "exclusive or" on a binary computer, has been suggested by W. J. Westlake, *JACM* **14** (1967), 337–340.

EXERCISES

▶ **1.** [*12*] In practice, we form random numbers using the relation $X_{n+1} = (aX_n + c) \bmod m$, where the X's are *integers*, then afterwards treat them as the *fractions* $U_n = X_n/m$. The recurrence relation for U_n is actually

$$
U_{n+1} = (aU_n + c/m) \bmod 1.
$$

Discuss the generation of random sequences using this relation *directly*, by making use of floating-point arithmetic on the computer.

▶ **2.** [*M20*] A good source of random numbers will have $X_{n-1} < X_{n+1} < X_n$ about one-sixth of the time, since each of the six possible relative orders of X_{n-1}, X_n, and X_{n+1} should be equally probable. However, show that the above ordering *never* occurs if the Fibonacci sequence (5) is used.

3. [*21*] What sequence comes from Algorithm M if $X_0 = 0$, $X_{n+1} = (5X_n + 3) \bmod 8$, $Y_0 = 0$, $Y_{n+1} = (5Y_n + 1) \bmod 8$, and $k = 4$? (Note that the potency is two, so $\langle X_n \rangle$ and $\langle Y_n \rangle$ aren't extremely random to start with.)

4. [*00*] Why is the high-order byte used in the first line of program (12), instead of some other byte?

▶ **5.** [*20*] Discuss using $X_n = Y_n$ in Algorithm M, in order to improve the speed of generation.

6. [*10*] In the binary method (10), the text states that the low-order bit of X is random, if the code is performed repeatedly. Why isn't the entire *word* X random?

7. [*20*] Show that the full sequence of length 2^e (i.e., each of the 2^e possible sets of e adjacent bits occurs just once in the period) may be obtained if program (10) is changed to the following:

```
LDA   X
JANZ  *+2
LDA   C
ADD   X
JNOV  *+3
JAZ   *+2
XOR   C
STA   X
```

8. [*M39*] Prove that the quadratic congruential sequence (3) has period length m if and only if the following conditions are satisfied:

    i) *c is relatively prime to m;*
    ii) *d and a $-$ 1 are both multiples of p, for all odd primes p dividing m;*
    iii) *d is even, and $d \equiv a - 1$ (modulo 4), if m is a multiple of 4;*
        $d \equiv a - 1$ (modulo 2), if m is a multiple of 2;
    iv) *either $d \equiv 0$, or $a \equiv 1$ and $cd \equiv 6$ (modulo 9), if m is a multiple of 9.*

[*Hint:* The sequence defined by $X_0 = 0$, $X_{n+1} = dX_n^2 + aX_n + c$ has period length m, modulo m, only if its period length is d modulo any divisor d of m.]

▶ **9.** [*M24*] (R. R. Coveyou.) Use the result of exercise 8 to prove that the modified middle-square method (4) has a period of length 2^{e-2}.

10. [*M29*] Show that if X_0 and X_1 are not both even and if $m = 2^e$, the period of the Fibonacci sequence (5) is $3 \cdot 2^{e-1}$.

11. [*M36*] The purpose of this exercise is to analyze certain properties of integer sequences satisfying the recurrence relation

$$X_n = a_1 X_{n-1} + \cdots + a_k X_{n-k}, \qquad n \geq k;$$

if we can calculate the period length of this sequence modulo $m = p^e$, when p is prime, the period length with respect to an arbitrary modulus m is the least common multiple of the period lengths for the prime power factors of m.

    a) If $f(z), a(z), b(z)$ are polynomials with integer coefficients, let us write $a(z) \equiv b(z)$ (modulo $f(z)$ and m) if $a(z) = b(z) + f(z)u(z) + mv(z)$ for some polynomials $u(z), v(z)$ with integer coefficients. Prove that when $f(0) = 1$ and $p^e > 2$, "If $z^\lambda \equiv 1$ (modulo $f(z)$ and p^e), $z^\lambda \not\equiv 1$ (modulo $f(z)$ and p^{e+1}), then $z^{p\lambda} \equiv 1$ (modulo $f(z)$ and p^{e+1}), $z^{p\lambda} \not\equiv 1$ (modulo $f(z)$ and p^{e+2})."
    b) Let $f(z) = 1 - a_1 z - \cdots - a_k z^k$, and let

$$G(z) = 1/f(z) = A_0 + A_1 z + A_2 z^2 + \cdots.$$

Let $\lambda(m)$ denote the period length of $\langle A_n \bmod m \rangle$. Prove that $\lambda(m)$ is the smallest positive integer λ such that $z^\lambda \equiv 1$ (modulo $f(z)$ and m).

c) Given that p is prime, $p^e > 2$, and $\lambda(p^e) \neq \lambda(p^{e+1})$, prove that $\lambda(p^{e+r}) = p^r\lambda(p^e)$ for all $r \geq 0$. (Thus, to find the period length of the sequence $\langle A_n \bmod 2^e \rangle$, we can compute $\lambda(4), \lambda(8), \lambda(16), \ldots$ by hand until we find the smallest $r \geq 2$ such that $\lambda(2^{r+1}) \neq \lambda(4)$; then the period length is determined mod 2^e for all e.)

d) Show that any sequence of integers satisfying the recurrence stated at the beginning of this exercise has the generating function $g(z)/f(z)$, for some polynomial $g(z)$ with integer coefficients.

e) Given that the polynomials $f(z)$ and $g(z)$ in part (d) are relatively prime modulo p (cf. Section 4.6.1), prove that the sequence $\langle X_n \bmod p^e \rangle$ has exactly the same period length as the special sequence $\langle A_n \bmod p^e \rangle$ in (b). (No longer period could be obtained by any choice of $X_0, \ldots, X_{k-1}$, since the general sequence is a linear combination of "shifts" of the special sequence.) [*Hint:* By exercise 4.6.2–22 (Hensel's lemma), there exist polynomials such that $a(z)f(z) + b(z)g(z) \equiv 1$ (modulo p^e).]

▶ **12.** [*M28*] Find integers X_0, X_1, a, b, and c such that the sequence

$$X_{n+1} = (aX_n + bX_{n-1} + c) \bmod 2^e, \qquad n \geq 1,$$

has the longest period length of all sequences of this type.
[*Hint:* $X_{n+2} = ((a+1)X_{n+1} + (b-a)X_n - bX_{n-1}) \bmod 2^e$; see exercise 11(c).]

13. [*M20*] Let $\langle X_n \rangle$ and $\langle Y_n \rangle$ be sequences of integers mod m with periods of lengths λ_1 and λ_2, and form the sequence $Z_n = (X_n + Y_n) \bmod m$. Show that if λ_1 and λ_2 are relatively prime, the sequence $\langle Z_n \rangle$ has a period of length $\lambda_1\lambda_2$.

14. [*M24*] Let X_n, Y_n, Z_n, λ_1, λ_2 be as in the previous exercise. Suppose that $\lambda_1 = 2^{e_2}3^{e_3}5^{e_5}\cdots$ is the prime factorization of λ_1, and similarly suppose that $\lambda_2 = 2^{f_2}3^{f_3}5^{f_5}\cdots$. Let $g_p = \big(\max(e_p, f_p)$ if $e_p \neq f_p$, 0 if $e_p = f_p\big)$, and let $\lambda_0 = 2^{g_2}3^{g_3}5^{g_5}\cdots$. Show that the period λ' of the sequence Z_n is a multiple of λ_0, but it is a divisor of $\lambda = $ least common multiple of λ_1, λ_2. In particular, $\lambda' = \lambda$ if $(e_p \neq f_p$ or $e_p = f_p = 0)$ for each prime p.

15. [*M46*] What can be proved about the period length of the sequence produced by Algorithm M?

▶ **16.** [*M28*] Let the constant c in method (10) be $(a_1a_2\ldots a_k)_2$ in binary notation. Show that the sequence of bits $Y_0, Y_1, \ldots$ satisfies the relation

$$Y_n = (a_1Y_{n-1} + a_2Y_{n-2} + \cdots + a_kY_{n-k}) \bmod 2.$$

[This may be regarded as another way to define the sequence, although the connection between this relation and the efficient code (10) is not apparent at first glance!]

17. [*M33*] (M. H. Martin, 1934.) Let m, $k \geq 1$ be integers. Let $X_1 = X_2 = \cdots = X_k = 0$. For $n > 0$, set X_{n+k} equal to the highest nonnegative value $y < m$ such that the k-tuple $(X_{n+1}, \ldots, X_{n+k-1}, y)$ has not already occurred in the sequence; in other words, $(X_{n+1}, \ldots, X_{n+k-1}, y)$ must not equal $(X_{r+1}, \ldots, X_{r+k})$ for $0 \leq r < n$. In this way, each possible k-tuple will occur at most once in the sequence. Eventually the process will terminate, when we reach a value of n such that $(X_{n+1}, \ldots, X_{n+k-1}, y)$ has already occurred in the sequence for all y, $0 \leq y < m$. For example, if $m = k = 3$, the sequence is

$$000222122021121020012001110100$$

and the process terminates at this point. (a) Prove that when the sequence terminates, we have $X_{n+1} = \cdots = X_{n+k-1} = 0$. (b) Prove that *every* k-tuple $(a_1, a_2, \ldots, a_k)$ of elements with $0 \leq a_j < m$ occurs in the sequence; hence the sequence terminates when $n = m^k$. [*Hint:* Prove that $(a_1, \ldots, a_s, 0, \ldots, 0)$ appears, when $a_s \neq 0$, by induction on s.]   Note that if we now define $f(X_n, \ldots, X_{n+k-1}) = X_{n+k}$ for $1 \leq n \leq m^k$, setting $X_{m^k+k} = 0$, we obtain a function of maximum possible period.

18. [*M22*] Let Y_n be the sequence of bits generated by method (10), with $k = 35$ and $c = (0000000000000000000000000000000101)_2$. Let U_n be the binary fraction $(.Y_{nk}Y_{nk+1} \cdots Y_{nk+k-1})$; show that this sequence $\langle U_n \rangle$ fails the serial test on pairs (Section 3.3.2D) when $d = 8$.

19. [*M41*] For each prime p specified in the first column of Table 1 in Section 4.5.4, find suitable constants a_1, a_2 as suggested in the text, such that the period length of (8), when $k = 2$, is $p^2 - 1$.

20. [*M40*] Calculate constants c suitable for use with method (10), having approximately the same number of zeros as ones, for $1 \leq k \leq 64$.

21. [*M35*] (D. Rees.) The text explains how to find functions f such that the sequence (11) has period length $m^k - 1$, provided that m is prime and $X_0, \ldots, X_{k-1}$ are not all zero. Show that such functions can be modified to obtain sequences of type (11) with period length m^k, for *all* m. [*Hints:* Consider Lemma 3.2.1.2Q, the trick of exercise 7, and sequences such as $\langle pX_{2n} + X_{2n+1} \rangle$.]

▶ **22.** [*M24*] The text restricts discussion of the extended linear sequences (8) to the case that m is prime. Prove that reasonably long periods can also be obtained when m is "square-free," i.e., the product of distinct primes. (Examination of Table 3.2.1.1–1 shows that $m = w \pm 1$ often satisfies this hypothesis; many of the results of the text can therefore be carried over to that case, which is somewhat more convenient for calculation.)

3.3. STATISTICAL TESTS

Our main purpose is to obtain sequences which behave as if they are random. So far we have seen how to make the period of a sequence so long that for practical purposes it never will repeat; this is an important criterion, but it by no means guarantees that the sequence will be useful in applications. How then are we to decide whether a sequence is sufficiently random?

If we were to give some man a pencil and paper and ask him to write down 100 random decimal digits, chances are very slim that he will give a satisfactory result. People tend to avoid things which seem nonrandom, such as pairs of equal adjacent digits (although about one out of every 10 digits should equal its predecessor). And if we would show someone a table of truly random digits, he would quite probably tell us they are not random at all; his eye would spot certain apparent regularities.

According to Dr. I. J. Matrix (as quoted by Martin Gardner in *Scientific American*, January, 1965), "Mathematicians consider the decimal expansion of π a random series, but to a modern numerologist it is rich with remarkable patterns." Dr. Matrix has pointed out, for example, that the first repeated two-digit number in π's expansion is 26, and its second appearance comes in the middle of a curious repetition pattern:

$$3.14159265358979323846264338327950 \tag{1}$$

After listing a dozen or so further properties of these digits, he observed that π, when correctly interpreted, conveys the entire history of the human race!

We all notice patterns in our telephone numbers, license numbers, etc., as aids to memory. The point of these remarks is that we cannot be trusted to judge by ourselves whether a sequence of numbers is random or not. Some unbiased mechanical tests must be applied.

The theory of statistics provides us with some quantitative measures for randomness. There is literally no end to the number of tests that can be conceived; we will discuss those tests which have proved to be most useful, most instructive, and most readily adapted to computer calculation.

If a sequence behaves randomly with respect to tests $T_1, T_2, \ldots, T_n$, we cannot be *sure* in general that it will not be a miserable failure when it is subjected to a further test T_{n+1}; yet each test gives us more and more confidence in the randomness of the sequence. In practice, we apply about half a dozen different kinds of statistical tests to a sequence, and if it passes these satisfactorily we consider it to be random (it is then presumed innocent until proved guilty).

Every sequence which is to be extensively used should be tested carefully, so the following sections explain how to carry out these tests in the right way. Two kinds of tests are distinguished: *empirical tests*, for which the computer manipulates groups of numbers of the sequence and evaluates certain statistics; and *theoretical tests*, for which we establish characteristics of the sequence by using number-theoretical methods based on the recurrence rule used to form the sequence.

For further references, the reader should consult the book *How to Lie With Statistics* by Darrell Huff (Norton, 1954).

3.3.1. General Test Procedures for Studying Random Data

A. "Chi-square" tests. The chi-square test (χ^2 test) is perhaps the best known of all statistical tests, and it is a basic method which is used in connection with many other tests. Before considering the method in general, let us consider a particular example of the chi-square test, as it might be applied to dice throwing. Using two "true" dice (each of which, independently, is assumed to register 1, 2, 3, 4, 5, or 6 with equal probability), the following table gives the probability of obtaining a given total, s, on a single throw:

$$
\begin{array}{cccccccccccc}
\text{value of } s = & 2 & 3 & 4 & 5 & 6 & 7 & 8 & 9 & 10 & 11 & 12 \\
\text{probability, } p_s = & \tfrac{1}{36} & \tfrac{1}{18} & \tfrac{1}{12} & \tfrac{1}{9} & \tfrac{5}{36} & \tfrac{1}{6} & \tfrac{5}{36} & \tfrac{1}{9} & \tfrac{1}{12} & \tfrac{1}{18} & \tfrac{1}{36}
\end{array} \tag{1}
$$

(For example, a value of 4 can be thrown in three ways: $1 + 3$, $2 + 2$, $3 + 1$; this constitutes $\frac{3}{36} = \frac{1}{12} = p_4$ of the 36 possible outcomes.)

If we throw the dice n times, we should obtain the value s approximately np_s times on the average. For example, in 144 throws we should get the value 4 about 12 times. The following table shows what results were *actually* obtained in a particular sequence of 144 throws of the dice:

value of $s =$	2	3	4	5	6	7	8	9	10	11	12	
observed number, $Y_s =$	2	4	10	12	22	29	21	15	14	9	6	(2)
expected number, $np_s =$	4	8	12	16	20	24	20	16	12	8	4	

Note that the observed number is different from the expected number in all cases; in fact, random throws of the dice will hardly ever come out with *exactly* the right frequencies. There are 36^{144} possible sequences of 144 throws, all of which are equally likely. One of these sequences consists of all 2's ("snake eyes"), and anyone throwing 144 snake eyes in a row would be convinced the dice were loaded. Yet the sequence of all 2's is just as probable as any other particular sequence if we specify the outcome of each throw of each die. In view of this, how can we test whether or not a given pair of dice is loaded? The answer is that we can't make a definite yes-no statement, but we can give a *probabilistic* answer. We can say how probable or improbable certain events are.

A natural way to proceed in the above example is to consider the squares of the differences between the observed numbers Y_s and the expected numbers np_s [cf. (2)]. We can add these together and compute

$$V = (Y_2 - np_2)^2 + (Y_3 - np_3)^2 + \cdots + (Y_{12} - np_{12})^2. \qquad (3)$$

A bad set of dice should result in a relatively high value of V; and for any given value of V we can ask, "What is the probability that V is this high?" If this probability is very small, say $\frac{1}{100}$, we would know that only one time in 100 the dice would give results so far away from the expected numbers, and we would have definite grounds for suspicion. (Remember, however, that even *good* dice would give such a high value of V one time in a hundred, so a cautious person would repeat the experiment again to see if the high value of V is repeated.)

The statistic V in (3) gives equal weight to $(Y_7 - np_7)^2$ and $(Y_2 - np_2)^2$, although $(Y_7 - np_7)^2$ is likely to be a good deal higher than $(Y_2 - np_2)^2$, since 7's occur about six times as often as 2's. It turns out that the "right" statistic, at least one which has proved to be most important, will give $(Y_7 - np_7)^2$ only $\frac{1}{6}$ as much importance as $(Y_2 - np_2)^2$, and we should change (3) to the following formula:

$$V = \frac{(Y_2 - np_2)^2}{np_2} + \frac{(Y_3 - np_3)^2}{np_3} + \cdots + \frac{(Y_{12} - np_{12})^2}{np_{12}}. \qquad (4)$$

This is called the "chi-square" statistic of the observed quantities $Y_2, \ldots, Y_{12}$

in this dice-throwing experiment. For the data in (2), we find that

$$V = \frac{(2-4)^2}{4} + \frac{(4-8)^2}{8} + \cdots + \frac{(9-8)^2}{8} + \frac{(6-4)^2}{4} = 7\frac{7}{48}. \tag{5}$$

The important question now is, of course, "does $7\frac{7}{48}$ constitute an improbably high value for V to assume?" Before answering that question, let us consider the general application of the chi-square method.

In general, suppose that every observation can fall into one of k categories. We take n *independent observations;* this means that the outcome of one observation must have absolutely no effect on the outcome of any of the others. Let p_s be the probability that each observation falls into category s, and let Y_s be the number of observations which actually *do* fall into category s. We form the statistic

$$V = \sum_{1 \le s \le k} \frac{(Y_s - np_s)^2}{np_s}. \tag{6}$$

In our above example, there are eleven possible outcomes of each throw of the dice, so $k = 11$. [Eq. (6) is a slight change of notation from Eq. (4), since we are numbering the possibilities from 1 to k instead of from 2 to 12.]

By expanding $(Y_s - np_s)^2 = Y_s^2 - 2np_s Y_s + n^2 p_s^2$ in (6) and using the facts that

$$\begin{aligned} Y_1 + Y_2 + \cdots + Y_k &= n, \\ p_1 + p_2 + \cdots + p_k &= 1, \end{aligned} \tag{7}$$

we arrive at the formula

$$V = \frac{1}{n} \sum_{1 \le s \le k} \left(\frac{Y_s^2}{p_s} \right) - n, \tag{8}$$

which often makes the computation of V somewhat easier.

Now we turn to the important question, what constitutes a reasonable value of V? This is found by referring to a table such as Table 1, which gives values of "the chi-square distribution with ν degrees of freedom" for various values of ν. The line of the table with $\nu = k - 1$ is to be used; *the number of "degrees of freedom" is $k - 1$, one less than the number of categories.* (Intuitively, this means that $Y_1, Y_2, \ldots, Y_k$ are not completely independent, since by Eq. (7) Y_1 can be computed if $Y_2, \ldots, Y_k$ are known; hence, $k - 1$ degrees of freedom are present. This argument is not rigorous, but the theory below justifies it.)

If the table entry in row ν under column p is x, it means, "The quantity V in Eq. (8) will be greater than x with probability p." For example, the 5 percent entry in row 10 is 18.31; this says we will have $V > 18.31$ only 5 percent of the time.

Let us assume that the above dice-throwing experiment is simulated on a computer using some sequence of supposedly random numbers, with the following results:

$$s = 2 \quad 3 \quad 4 \quad 5 \quad 6 \quad 7 \quad 8 \quad 9 \quad 10 \quad 11 \quad 12$$

Experiment 1, $Y_s = 4 \quad 10 \quad 10 \quad 13 \quad 20 \quad 18 \quad 18 \quad 11 \quad 13 \quad 14 \quad 13$ (9)

Experiment 2, $Y_s = 3 \quad 7 \quad 11 \quad 15 \quad 19 \quad 24 \quad 21 \quad 17 \quad 13 \quad 9 \quad 5$

We can compute the chi-square statistic in the first case, getting $V_1 = 29\frac{59}{120}$, and in the second case we get the value $V_2 = 1\frac{17}{120}$. Referring to the table entries for 10 degrees of freedom, we see that V_1 is *much too high; V* will be greater than 23.2 only one percent of the time! (By using more extensive tables, we find in fact that V will be as high as this only 0.1 percent of the time.) Therefore experiment 1 represents a significant departure from random behavior.

On the other hand, V_2 is quite low, since the Y_s in experiment 2 are quite close to the expected values np_s [cf. (2)]. The chi-square table tells us that V should be bigger than 2.56, 99 percent of the time. V_2 is *much too low;* the observed values are so close to the expected values, we cannot consider the result to be random! (Indeed, reference to other tables shows that such a low value of V occurs only 0.03 percent of the time when there are 10 degrees of freedom.) Finally, the value $V = 7\frac{7}{48}$ computed in (5) can be checked with our chi-square table; it falls between the entries for 75 percent and 50 percent, so we cannot consider it to be significantly high or significantly low; the observations in (2) are satisfactorily random with respect to this test.

It is somewhat remarkable that the same table entries are used no matter what the value of n is, and no matter what the probabilities p_s are. Only the number $\nu = k - 1$ affects the results. In actual fact, the table entries are not exactly correct: *they are approximations which are valid only for large enough values of n.* How large should n be? A good rule of thumb is to take n large enough so that each of the expected values np_s is five or more; preferably, however, take n much larger than this, to get a more powerful test. Note that in our above examples we took $n = 144$, so np_2 is only 4, violating the stated "rule of thumb." This was done only because the author tired of throwing the dice; it makes the entries in Table 1 less accurate for our application. Experiments run on a computer, with $n = 1000$, or 10000, or even 100000, would be much better than this.

The proper choice of n is somewhat obscure. If the dice are actually biased, the fact would be detected as n gets larger and larger. (Cf. exercise 12.) But large values of n will tend to smooth out *locally* nonrandom behavior, i.e., blocks of numbers with a strong bias followed by blocks of numbers with the opposite bias. This type of behavior would not happen when actual dice are rolled, since the same dice are used throughout the test, but a sequence of numbers generated on a computer might very well display such locally nonrandom behavior.

Table 1

SELECTED VALUES OF THE CHI-SQUARE DISTRIBUTION

(For further values, see *Handbook of Mathematical Functions*, ed. by M. Abramowitz and I. A. Stegun, U.S. Government Printing Office, 1964, Table 26.8.)

	$p = 99\%$	$p = 95\%$	$p = 75\%$	$p = 50\%$	$p = 25\%$	$p = 5\%$	$p = 1\%$
$\nu = 1$	0.00016	0.00393	0.1015	0.4549	1.323	3.841	6.635
$\nu = 2$	0.02010	0.1026	0.5753	1.386	2.773	5.991	9.210
$\nu = 3$	0.1148	0.3518	1.213	2.366	4.108	7.815	11.34
$\nu = 4$	0.2971	0.7107	1.923	3.357	5.385	9.488	13.28
$\nu = 5$	0.5543	1.1455	2.675	4.351	6.626	11.07	15.09
$\nu = 6$	0.8720	1.635	3.455	5.348	7.841	12.59	16.81
$\nu = 7$	1.239	2.167	4.255	6.346	9.037	14.07	18.48
$\nu = 8$	1.646	2.733	5.071	7.344	10.22	15.51	20.09
$\nu = 9$	2.088	3.325	5.899	8.343	11.39	16.92	21.67
$\nu = 10$	2.558	3.940	6.737	9.342	12.55	18.31	23.21
$\nu = 11$	3.053	4.575	7.584	10.34	13.70	19.68	24.73
$\nu = 12$	3.571	5.226	8.438	11.34	14.84	21.03	26.22
$\nu = 15$	5.229	7.261	11.04	14.34	18.25	25.00	30.58
$\nu = 20$	8.260	10.85	15.45	19.34	23.83	31.41	37.57
$\nu = 30$	14.95	18.49	24.48	29.34	34.80	43.77	50.89
$\nu = 50$	29.71	34.76	42.94	49.33	56.33	67.50	76.15
$\nu > 30$	approximately $\nu + 2\sqrt{\nu}x_p + \frac{4}{3}x_p^2 - \frac{2}{3}$						
$x_p =$	-2.33	-1.64	$-.675$	0.00	0.675	1.64	2.33

Perhaps a chi-square test should be made for a number of different values of n. At any rate, n should always be rather large.

We can summarize the chi-square test as follows. A fairly large number, n, of independent observations is made. (It is important to avoid using the chi-square method unless the observations are independent. See, for example, exercise 10 which considers the case when half of the observations depend on the other half.) We count the number of observations falling into each of k categories, and compute the quantity V given in Eqs. (6) and (8). Then V is compared with the numbers in Table 1, with $\nu = k - 1$. If V is less than the 99-percent entry or greater than the one-percent entry, we reject the numbers as not sufficiently random. If V lies between 99 and 95 percent or between

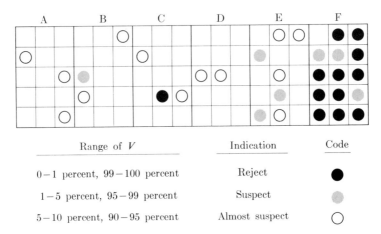

Range of V	Indication	Code
0−1 percent, 99−100 percent	Reject	●
1−5 percent, 95−99 percent	Suspect	◉
5−10 percent, 90−95 percent	Almost suspect	○

Fig. 2. Indications of "significant" deviations in 90 chi-square tests (cf. also Fig. 5).

5 and 1 percent, the numbers are "suspect"; if (by interpolation in the table) V lies between 95 and 90 percent, or 10 and 5 percent, the numbers might be "almost suspect." The chi-square test is often done at least three times on different sets of data, and if at least two of the three results are suspect the numbers are regarded as not sufficiently random.

For example, see Fig. 2, which shows schematically the results of applying five different types of chi-square tests on each of six sequences of random numbers. Each test was applied to three different blocks of numbers of the sequence. Generator A is the MacLaren-Marsaglia method (Algorithm 3.2.2M), generator E is the Fibonacci method, and the other generators are linear congruential sequences with the following parameters:

Generator B: $X_0 = 0$, $a = 3141592653$, $c = 2718281829$, $m = 2^{35}$.
Generator C: $X_0 = 0$, $a = 2^7 + 1$, $c = 1$, $m = 2^{35}$.
Generator D: $X_0 = 47594118$, $a = 23$, $c = 0$, $m = 10^8 + 1$.
Generator F: $X_0 = 314159265$, $a = 2^{18} + 1$, $c = 1$, $m = 2^{35}$.

From Fig. 2 we conclude that (so far as these tests are concerned) generators A, B, D are satisfactory, generator C is on the borderline and it should probably be rejected, while generators E and F are definitely unsatisfactory. Generator F has, of course, low potency; generators C and D have been discussed in the literature, but their multipliers are too small. (Generator D is the original multiplicative generator proposed by Lehmer in 1948; generator C is the original linear congruential generator with $c \neq 0$ proposed by Rotenberg in 1960.)

Instead of using the "suspect," "almost suspect," etc., criteria for judging the results of chi-square tests, there is a less *ad hoc* procedure available, which will be discussed later in this section.

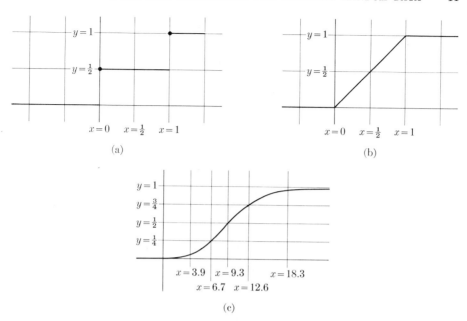

Fig. 3. Examples of distribution functions.

B. The Kolmogorov-Smirnov test. As we have seen, the chi-square test applies to the situation when observations can fall into a finite number of categories, k. It is not unusual, however, to consider random quantities which may assume infinitely many values. For example, a random real number between 0 and 1 may take on infinitely many values; even though only a finite number of these values can be represented in the computer, we want our random values to behave essentially as though they are random real numbers.

A general notation for specifying probability distributions, whether they are finite or infinite, is commonly used in the study of probability and statistics. Suppose we want to specify the distribution of the values of a random quantity, X; we do this in terms of the *distribution function $F(x)$*, where

$$F(x) = \text{probability that } (X \le x).$$

Three examples are shown in Fig. 3. First we see the distribution function for a *random bit*, i.e., for the case when X takes on only the two values 0 and 1, each with probability $\frac{1}{2}$. Part (b) of the figure shows the distribution function for a *uniformly distributed random real number* between zero and one, so the probability that $X \le x$ is simply equal to x when $0 \le x \le 1$; for example, the probability that $X \le \frac{2}{3}$ is, naturally, $\frac{2}{3}$. And part (c) shows the limiting distribution of the value V in the chi-square test (shown here with 10 degrees of freedom); this is a distribution which we have already seen represented in another way in Table 1. Note that $F(x)$ always increases from 0 to 1 as x increases from $-\infty$ to $+\infty$.

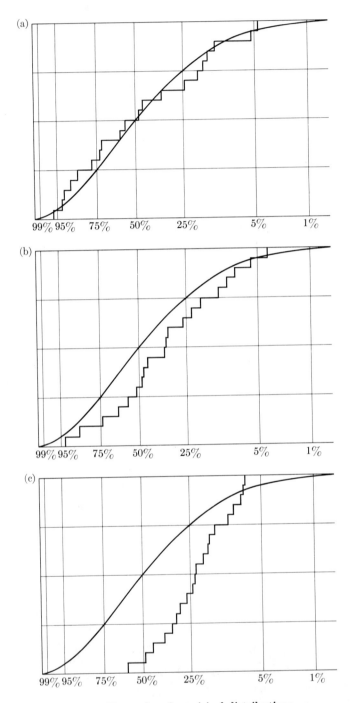

Fig. 4. Examples of empirical distributions.

If we make n independent observations of the random quantity X, thereby obtaining the values $X_1, X_2, \ldots, X_n$, we can form the *empirical distribution function* $F_n(x)$, where

$$F_n(x) = \frac{\text{number of } X_1, X_2, \ldots, X_n \text{ which are } \le x}{n}. \qquad (10)$$

Figure 4 illustrates three empirical distribution functions (shown as zigzag lines, although, strictly speaking, the vertical lines are not part of the graph of $F_n(x)$), superimposed on a graph of the assumed actual distribution function $F(x)$. As n gets large, $F_n(x)$ should be a better and better approximation to $F(x)$.

The Kolmogorov-Smirnov test (KS test) may be used when $F(x)$ has no jumps. It is based on the *difference between $F(x)$ and $F_n(x)$*. A bad random-number generator will give empirical distribution functions which do not approximate $F(x)$ sufficiently well. Figure 4(b) shows an example in which the X_i are consistently too high, so the empirical distribution function is too low. Part (c) of the figure shows an even worse example; it is plain that such great deviations between $F_n(x)$ and $F(x)$ are extremely improbable, and the KS test is used to tell us how improbable they are.

To make the test, we form the following statistics:

$$
\begin{aligned}
K_n^+ &= \sqrt{n} \max_{-\infty < x < +\infty} \left(F_n(x) - F(x) \right); \\
K_n^- &= \sqrt{n} \max_{-\infty < x < +\infty} \left(F(x) - F_n(x) \right).
\end{aligned}
\qquad (11)
$$

Here K_n^+ measures the greatest amount of deviation when F_n is greater than F, and K_n^- measures the maximum deviation when F_n is less than F. The statistics for the examples of Fig. 4 are:

	Part (a)	Part (b)	Part (c)	
K_{20}^+	0.492	0.134	0.313	(12)
K_{20}^-	0.536	1.027	2.101	

(*Note:* The factor $\sqrt{n}$ which appears in Eqs. (11) may seem puzzling at first. Exercise 6 shows that, for fixed x, the standard deviation of $F_n(x)$ from $F(x)$ is proportional to $1/\sqrt{n}$; hence the factor $\sqrt{n}$ magnifies the statistics K_n^+, K_n^- in such a way that this standard deviation is independent of n.)

As in the chi-square test, we may now look up the values K_n^+, K_n^- in a "percentile" table to determine if they are significantly high or low. Table 2 may be used for this purpose, for both K_n^+ and K_n^-. For example, the probability that K_{30}^- is greater than 0.8036 is 25 percent. Unlike the chi-square test, the table entries are *not* merely approximations which hold for large values of n; the table gives exact values for all n, and the KS test may be reliably used for any value of n.

Table 2

SELECTED VALUES OF THE DISTRIBUTION OF K_n^+ AND K_n^-

The approximations for $n > 30$ are only conjectures which have not been proved, except when $n = \infty$.

	$p = 99\%$	$p = 95\%$	$p = 75\%$	$p = 50\%$	$p = 25\%$	$p = 5\%$	$p = 1\%$
$n = 1$	0.01000	0.05000	0.2500	0.5000	0.7500	0.9500	0.9900
$n = 2$	0.01400	0.06749	0.2929	0.5176	0.7071	1.0980	1.2728
$n = 3$	0.01699	0.07919	0.3112	0.5147	0.7539	1.1017	1.3589
$n = 4$	0.01943	0.08789	0.3202	0.5110	0.7642	1.1304	1.3777
$n = 5$	0.02152	0.09471	0.3249	0.5245	0.7674	1.1392	1.4024
$n = 6$	0.02336	0.1002	0.3272	0.5319	0.7703	1.1463	1.4144
$n = 7$	0.02501	0.1048	0.3280	0.5364	0.7755	1.1537	1.4246
$n = 8$	0.02650	0.1086	0.3280	0.5392	0.7797	1.1586	1.4327
$n = 9$	0.02786	0.1119	0.3274	0.5411	0.7825	1.1624	1.4388
$n = 10$	0.02912	0.1147	0.3297	0.5426	0.7845	1.1658	1.4440
$n = 11$	0.03028	0.1172	0.3330	0.5439	0.7863	1.1688	1.4484
$n = 12$	0.03137	0.1193	0.3357	0.5453	0.7880	1.1714	1.4521
$n = 15$	0.03424	0.1244	0.3412	0.5500	0.7926	1.1773	1.4606
$n = 20$	0.03807	0.1298	0.3461	0.5547	0.7975	1.1839	1.4698
$n = 30$	0.04354	0.1351	0.3509	0.5605	0.8036	1.1916	1.4801
$n > 30$	0.07089 $-\dfrac{0.15}{\sqrt{n}}$	0.1601 $-\dfrac{0.14}{\sqrt{n}}$	0.3793 $-\dfrac{0.15}{\sqrt{n}}$	0.5887 $-\dfrac{0.15}{\sqrt{n}}$	0.8326 $-\dfrac{0.16}{\sqrt{n}}$	1.2239 $-\dfrac{0.17}{\sqrt{n}}$	1.5174 $-\dfrac{0.20}{\sqrt{n}}$

As they stand, formulas (11) are not readily adapted to computer calculation, since we are asking for a maximum over infinitely many values of x! The fact that $F(x)$ is increasing and the fact that $F_n(x)$ increases only in finite steps, however, leads to a quite simple procedure for evaluating the statistics K_n^+ and K_n^-:

Step 1. Obtain the observations $X_1, X_2, \ldots, X_n$.

Step 2. Rearrange the observations so they are sorted into ascending order, i.e., so that $X_1 \leq X_2 \leq \cdots \leq X_n$. (Efficient sorting algorithms are the subject of Chapter 5.)

Step 3. The desired statistics are now given by the formulas

$$K_n^+ = \sqrt{n} \max_{1 \leq j \leq n} \left(\frac{j}{n} - F(X_j) \right);$$

$$K_n^- = \sqrt{n} \max_{1 \leq j \leq n} \left(F(X_j) - \frac{j-1}{n} \right). \tag{13}$$

An appropriate choice of the number of observations, n, is slightly easier to make for this test than it is for the χ^2 test, although some of the considerations are similar. If the random variables X_j actually belong to the probability distribution $G(x)$, while they were assumed to belong to the distribution given by a given function $F(x)$, it will take a comparatively large value of n to prove that $G(x) \neq F(x)$; for we need n large enough that the empirical distributions $G_n(x)$ and $F_n(x)$ are expected to be observably different. On the other hand, large values of n will tend to average out locally nonrandom behavior, and such behavior is an undesirable characteristic that is of significant importance in most computer applications of random numbers; this makes a case for *smaller* values of n. The fact that all n observations are to be remembered, and sorted into ascending order, also leads us to prefer comparatively small n. A good compromise would be to take n equal to, say, 1000, and to make a fairly large number of calculations of K_{1000}^+ on different parts of a random sequence, thereby obtaining values

$$K_{1000}^+(1), \qquad K_{1000}^+(2), \qquad \ldots, \qquad K_{1000}^+(r). \tag{14}$$

We can also apply the KS test *again* to *these* results: Let $F(x)$ now be the distribution function for K_{1000}^+, and determine the empirical distribution $F_r(x)$ obtained from the observed values in (14). Fortunately, the function $F(x)$ in this case is very simple; for a large value of n like $n = 1000$, the distribution of K_n^+ is closely approximated by

$$F_\infty(x) = 1 - e^{-2x^2}, \qquad x \geq 0. \tag{15}$$

The same remarks apply to K_n^-, since K_n^+ and K_n^- have the same expected behavior. *This method of using several tests for moderately sized n, then combining the observations later in another KS test, will tend to detect both local and global nonrandom behavior.*

An experiment of this type (although on a much smaller scale) was made by the author as this chapter was being written. The "maximum of 5" test described in the next section was applied to a set of 1000 uniform random numbers, yielding 200 observations $X_1, X_2, \ldots, X_{200}$ which are supposed to belong to the distribution $F(x) = x^5$ ($0 \leq x \leq 1$). The observations were divided into 20 groups of 10 each, and the statistic K_{10}^+ was computed for each group. The 20 values of K_{10}^+, thus obtained, led to the empirical distributions shown in Fig. 4. The smooth curve shown in each of the diagrams in Fig. 4 is the actual distribution the statistic K_{10}^+ should have. Figure 4(a) shows the empirical distribution of K_{10}^+ obtained from the sequence

$$Y_{n+1} = (3141592653 Y_n + 2718281829) \bmod 2^{35}, \qquad U_n = Y_n/2^{35},$$

and it is satisfactorily random. Part (b) of the figure came from the Fibonacci method; this sequence has *globally* nonrandom behavior, i.e., it can be shown that the observations X_n in the "maximum of 5" test do not have the correct distribution $F(x) = x^5$. Part (c) came from the notorious and impotent linear congruential sequence $Y_{n+1} = ((2^{18} + 1)Y_n + 1) \bmod 2^{35}$, $U_n = Y_n/2^{35}$.

The KS test applied to the data in Fig. 4 gives the results shown in (12). Referring to Table 2 for $n = 20$, we see that the values of K_{20}^+ and K_{20}^- for Fig. 4(b) are almost suspect (they lie at about the 95 percent and 12 percent levels) but not quite bad enough to be rejected outright. The value of K_{20}^- for Part (c) is, of course, completely out of line, so the "maximum of 5" test shows a definite failure of that random-number generator.

We would expect the KS test in this experiment to have more difficulty locating global nonrandomness than local nonrandomness, since the basic observations in Fig. 4 were made on samples of only 10 observations each. If we were to take 20 groups of 1000 observations each, Part (b) would show a much more significant deviation. To illustrate this point, a *single* KS test was applied to all 200 of the observations which led to Fig. 4, and the following results were obtained:

	Part (a)	Part (b)	Part (c)	
K_{200}^+	0.477	1.537	2.819	(16)
K_{200}^-	0.817	0.194	0.058	

The global nonrandomness of the Fibonacci generator has definitely been detected here. (The generator used in Part (c) is locally nonrandom even for n as high as 1000000, so $n = 200$ does not reveal whether it is globally random or not.)

We may summarize the Kolmogorov-Smirnov test as follows. We are given n *independent observations* $X_1, \ldots, X_n$ taken from some distribution specified by a *continuous* function $F(x)$. [That is, $F(x)$ must be like the functions shown in Fig. 2(b) and (c), having no jumps like those in Fig. 2(a).] The procedure explained just before Eqs. (13) is carried out on these observations, so we obtain the statistics K_n^+ and K_n^-. These statistics should be distributed according to Table 2.

Some comparisons between the KS test and the χ^2 test should be made at this point. In the first place, we should observe that the KS test may be used *in conjunction with* the χ^2 test, to give a better procedure than the *ad hoc* method we mentioned when summarizing the χ^2 test. (That is, there is a better way to proceed than to make three tests and to consider how many of the results were "suspect"). Suppose we have made, say, 10 independent χ^2 tests on different parts of a random sequence, so that values $V_1, V_2, \ldots, V_{10}$ have been obtained. It is not a good policy to simply count how many of the V's are suspiciously large or small (although this procedure will work in extreme cases, and *very* large or *very* small values may mean the sequence has too much local non-randomness); a better general method would be to plot the empirical distribution function of these 10 values, and to compare it to the correct distribution (which may be obtained from Table 1). This gives a truer picture of the results of the χ^2 tests, and the statistics K_{10}^+ and K_{10}^- can be determined as an indication of the success or failure. With only 10 values or even as many as 100 this can all be done easily by hand, using graphical methods; with a larger number of V's, a computer subroutine for calculating the chi-square distribution would be necessary. Notice that *all 20 of the observations in Fig. 4(c) fall between the 5 and 95 percent levels,* so we would not have regarded *any* of them as suspicious, individually; yet collectively the empirical distribution shows these observations are not at all right.

An important difference between the KS test and the chi-square test is that the KS test applies to distributions $F(x)$ having no jumps, while the chi-square test applies to distributions having nothing but jumps (since all observations are divided into k categories). The two tests are thus intended for different sorts of applications. Yet it is possible to apply the χ^2 test even when $F(x)$ is continuous, if we divide the domain of $F(x)$ into k parts and ignore all variations within each part. For example, if we want to test whether or not $U_1, U_2, \ldots, U_n$ can be considered to come from the uniform distribution between zero and one, we want to test if they have the distribution $F(x) = x$ for $0 \leq x \leq 1$. This is a natural application for the KS test. But we might also divide up the interval from 0 to 1 into $k = 100$ equal parts, count how many U's fall into each part, and apply the chi-square test with 99 degrees of freedom. There are not many theoretical results available at the present time to compare the effectiveness of the KS test versus the chi-square test. The author has found some examples in which the KS test points out nonrandomness more clearly than the χ^2 test, and others in which the χ^2 test gave a more significant result. If, for example, the 100 categories mentioned above are numbered $0, 1, \ldots, 99$, and if the deviations from the expected values are positive in compartments 0 to 49 but negative in compartments 50 to 99, then the empirical distribution function will be much farther from $F(x)$ than the χ^2 value would indicate; but if the positive deviations occur in compartments $0, 2, \ldots, 98$ and the negative ones occur in $1, 3, \ldots, 99$, the empirical distribution function will tend to hug $F(x)$ much more closely. The kinds of deviations measured are therefore somewhat different. A χ^2 test was applied to the 200 observations which led to Fig. 4, with $k = 10$, and the respective values of V were 9.4, 17.7, and 39.3; so in this particular case the values are quite comparable to the KS values given in (16). Since the χ^2 test is intrinsically less accurate, and since it requires comparatively large values of n, the KS test has several advantages when a continuous distribution is to be tested.

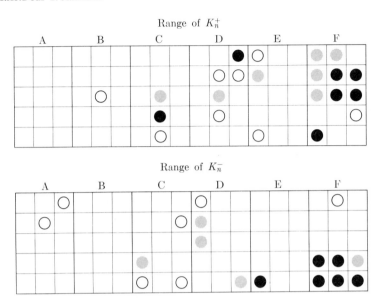

Fig. 5. The KS tests applied to the same data as Fig. 2.

A further example will also be of interest. The data which led to Fig. 2 were chi-square statistics based on $n = 200$ observations of the "maximum of t" criterion for $1 \leq t \leq 5$, with the range divided into 10 equally probable parts. KS statistics K_{200}^{+} and K_{200}^{-} can be computed from exactly the same sets of 200 observations, and the results can be tabulated in just the same way as we did in Fig. 2 (showing which KS values are beyond the 99-percent level, etc.) The results in this case are shown in Fig. 5. Note that generator D (Lehmer's method) shows up very badly in Fig. 5, while chi-square tests *on the very same data* revealed no difficulty in Fig. 2; contrariwise, generator E (the Fibonacci method) does not look so bad in Fig. 5. The good generators, A and B, passed all tests satisfactorily. The reasons for the discrepancy between Fig. 2 and Fig. 5 are primarily that (a) the number of observations, 200, is really not large enough for a good test, and (b) the "reject," "suspect," "almost suspect" ranking criterion is itself suspect.

(It is not fair to blame Lehmer for using a "bad" random-number generator in 1948, since his actual use of this generator was quite valid. The ENIAC computer was a highly parallel machine, programmed by means of a plugboard; Lehmer set it up so that one of its accumulators was repeatedly multiplying its own contents by 23, mod $10^8 + 1$, yielding a new value every few microseconds. Since this multiplier 23 is too small, we know that each value obtained by this process was too strongly related to the preceding value to be considered sufficiently random; but the durations of time between actual *uses* of the values in the special accumulator by the accompanying program were comparatively long and subject to some fluctuation. So the effective multiplier was 23^k for large, varying values of k!)

C. History, bibliography, and theory. The chi-square test was introduced by Karl Pearson in 1900 (*Philosophical Magazine*, Series 5, **50**, 157–175). Pearson's important paper is regarded as one of the foundations of modern statistics, since before that time people would simply plot experimental results graphically and assert that they were correct. In his paper, Pearson gave several interesting examples of the previous misuse of statistics; and he also proved that certain runs at roulette (which he had experienced during two weeks at Monte Carlo in 1892) were so far from the expected frequencies that odds against the assumption of an honest wheel were some 10^{29} to one! A general discussion of the chi-square test and an extensive bibliography appear in the survey article by William G. Cochran, *Annals of Mathematical Statistics* **23** (1952), 315–345.

We will now give a brief derivation of the theory behind the chi-square test. The exact probability that $Y_1 = y_1, \ldots, Y_k = y_k$ is easily seen to be

$$\frac{n!}{y_1! \cdots y_k!} p_1^{y_1} \cdots p_k^{y_k}. \tag{17}$$

If we assume that Y_s has the value y_s with the Poisson probability

$$\frac{e^{-np_s}(np_s)^{y_s}}{y_s!},$$

and that the Y's are independent, then $(Y_1, \ldots, Y_k)$ will equal $(y_1, \ldots, y_k)$ with probability

$$\prod_{1 \le s \le k} \frac{e^{-np_s}(np_s)^{y_s}}{y_s!},$$

and $Y_1 + \cdots + Y_k$ will equal n with probability

$$\sum_{\substack{y_1 + \cdots + y_k = n \\ y_1, \ldots, y_k \ge 0}} \prod_{1 \le s \le k} \frac{e^{-np_s}(np_s)^{y_s}}{y_s!} = \frac{e^{-n}n^n}{n!}.$$

If we assume they are independent *except* for the condition $Y_1 + \cdots + Y_k = n$, the probability that $(Y_1, \ldots, Y_k) = (y_1, \ldots, y_s)$ is the quotient

$$\left(\prod_{1 \le s \le k} \frac{e^{-np_s}(np_s)^{y_s}}{y_s!} \right) \Big/ \left(\frac{e^{-n}n^n}{n!} \right),$$

which equals (17). *We may therefore regard the Y's as independently Poisson distributed, except for the fact that they have a fixed sum.* Let

$$Z_s = \frac{Y_s - np_s}{\sqrt{np_s}}, \qquad V = Z_1^2 + \cdots + Z_k^2. \tag{18}$$

The condition $Y_1 + \cdots + Y_k = n$ is equivalent to requiring that

$$\sqrt{p_1}Z_1 + \cdots + \sqrt{p_k}Z_k = 0. \tag{19}$$

Let us consider the $(k-1)$-dimensional space S of all vectors $(Z_1, \ldots, Z_k)$ for which (19) holds. For large values of n, each Z_s has approximately the normal distribution (cf. exercise 1.2.10–16), and so points within a differential volume $dZ_2 \ldots dZ_k$ of S will occur with probability *approximately* proportional to $\exp(-(Z_1^2 + \cdots + Z_k^2)/2)$. (It is at this point in the derivation that the chi-square method becomes only an approximation for large n.) The probability that $V \le v$ is now

$$\frac{\int_{(Z_1,\ldots,Z_k) \text{ in } S \text{ and } Z_1^2 + \cdots + Z_k^2 \le v} \exp(-(Z_1^2 + \cdots + Z_k^2)/2)\, dz_2 \ldots dz_k}{\int_{(Z_1,\ldots,Z_k) \text{ in } S} \exp(-(Z_1^2 + \cdots + Z_k^2)/2)\, dz_2 \ldots dz_k} \qquad (20)$$

Since the plane (19) passes through the origin of k-dimensional space, the numerator in (20) is an integration over the interior of a $(k-1)$-dimensional sphere centered at the origin. An appropriate transformation to generalized polar coordinates with radius χ and angles $\omega_1, \ldots, \omega_{k-2}$ transforms (20) into

$$\frac{\int_{\chi^2 \le v} e^{-\chi^2/2} \chi^{k-2} f(\omega_1, \ldots, \omega_{k-2})\, d\chi\, d\omega_1 \ldots d\omega_{k-2}}{\int e^{-\chi^2/2} \chi^{k-2} f(\omega_1, \ldots, \omega_{k-2})\, d\chi\, d\omega_1 \ldots d\omega_{k-2}}$$

for some function f (see exercise 15), and integration over the angles $\omega_1, \ldots, \omega_{k-2}$ gives a constant factor which cancels from numerator and denominator. We finally obtain the formula

$$\frac{\int_0^{\sqrt{v}} e^{-\chi^2/2} \chi^{k-2}\, d\chi}{\int_0^\infty e^{-\chi^2/2} \chi^{k-2}\, d\chi} \qquad (21)$$

for the approximate probability that $V \le v$.

The above derivation uses the symbol χ to stand for the radial length, just as Pearson did in his original paper; this is how the χ^2 test got its name. Substituting $t = \chi^2/2$, the integrals can be expressed in terms of the incomplete gamma function, which we discussed in Section 1.2.11.3:

$$\lim_{n \to \infty} \text{probability } (V \le v) = \gamma\left(\frac{k-1}{2}, \frac{v}{2}\right) \Big/ \Gamma\left(\frac{k-1}{2}\right). \qquad (22)$$

This is the definition of the chi-square distribution with $k-1$ degrees of freedom.

We now turn to the KS test. In 1933, A. N. Kolmogorov proposed a test based on the statistic

$$K_n = \sqrt{n} \max_{-\infty < x < +\infty} |F_n(x) - F(x)| = \max(K_n^+, K_n^-). \qquad (23)$$

N. V. Smirnov gave several modifications of this test in 1939, including the individual examination of K_n^+ and K_n^- as we have suggested above. There is a large family of similar tests, but the K_n^+ and K_n^- tests seem to be most convenient for computer application. A comprehensive review of the literature concerning KS tests and their generalizations, including an extensive bibliography, has been given by D. A. Darling, *Annals of Mathematical Statistics* **28** (1957), 823–838. (The text of Prof. Darling's paper is intended primarily for experts in the field.)

To study the distribution of K_n^+ and K_n^-, we begin with the following basic fact: *If X is a random variable with the continuous distribution $F(x)$, then $F(X)$ is a uniformly distributed real number between 0 and 1.* To prove this, we need only verify that if $0 \le y \le 1$ we have $F(X) \le y$ with probability y. Since F is continuous, $F(x_0) = y$ for some x_0; thus the probability that $F(X) \le y$ is the probability that $X \le x_0$. By definition, the latter probability is $F(x_0)$, that is, it is y.

Let $Y_j = nF(X_j)$, for $1 \le j \le n$. Then the variables Y_j are uniformly distributed between 0 and n; they are independent, except for the constraint $Y_1 \le Y_2 \le \cdots \le Y_n$ [which is equivalent to the constraint $X_1 \le X_2 \le \cdots \le X_n$, a relation we may assume when using Eq. (13)], and Eq. (13) now may be transformed into

$$K_n^+ = \frac{1}{\sqrt{n}} \max (1 - Y_1, 2 - Y_2, \ldots, n - Y_n).$$

If $0 \le t \le n$, we therefore find the probability that $K_n^+ \le t/\sqrt{n}$ is the probability that $Y_j \ge j - t$ for $1 \le j \le n$. This is not hard to express as an n-dimensional integral

$$\frac{\int_{\alpha_n}^n dY_n \int_{\alpha_{n-1}}^{Y_n} dY_{n-1} \cdots \int_{\alpha_1}^{Y_2} dY_1}{\int_0^n dY_n \int_0^{Y_n} dY_{n-1} \cdots \int_0^{Y_2} dY_1}, \qquad \text{where} \qquad \alpha_j = \max (j - t, 0). \qquad (24)$$

The denominator here is immediately evaluated: it is found to be $n^n/n!$, which makes sense since the cube of all vectors $(Y_1, Y_2, \ldots, Y_n)$ with $0 \le Y_j < n$ has volume n^n, and it can be divided into $n!$ equal parts corresponding to each possible ordering of the Y's. The integral in the numerator is a little more difficult, but it yields to the attack suggested in exercise 17, and we get the general formula

$$\text{probability}\left(K_n^+ \le \frac{t}{\sqrt{n}}\right) = \frac{t}{n^n} \sum_{0 \le k \le t} \binom{n}{k}(k - t)^k(t + n - k)^{n-k-1}. \qquad (25)$$

The distribution of K_n^- is exactly the same. Equation (25) was first obtained by Z. W. Birnbaum and Fred H. Tingey; it may be used to extend Table 2.

In his original paper, Smirnov proved that

$$\lim_{n \to \infty} \text{probability } (K_n^+ \le s) = 1 - e^{-2s^2}, \qquad \text{if} \qquad s \ge 0. \qquad (26)$$

This together with (25) implies that

$$\lim_{n \to \infty} \frac{s}{\sqrt{n}} \sum_{\sqrt{n}s < k \le n} \binom{n}{k}\left(\frac{k}{n} - \frac{s}{\sqrt{n}}\right)^k \left(\frac{s}{\sqrt{n}} + 1 - \frac{k}{n}\right)^{n-k-1} = e^{-2s^2}, \qquad s \ge 0,$$

$$(27)$$

but this limiting relationship appears to be quite difficult to prove. The most elementary proof has been given by P. Whittle, *Annals of Mathematical Statistics* **32** (1961), 499–505; using a clever technique of slightly increasing or decreasing the α's in Eq. (24) in an appropriate manner, he obtains modified statistics for which the probability can be found explicitly, and he shows that these upper and lower bounds get arbitrarily close to e^{-2s^2}.

EXERCISES

1. [*00*] What line of the chi-square table should be used to check if the value $V = 7\frac{7}{48}$ of Eq. (5) is improbably high?

2. [*20*] If two dice are "loaded" so that, on one die, the value 1 will turn up exactly twice as often as any of the other values, and the other die is similarly biased towards 6, compute the probability, p_s, that a total of s will appear on the two dice, for $2 \leq s \leq 12$.

▶ **3.** [*23*] Some dice which were loaded as described in the previous exercise were rolled 144 times, and the following values were observed:

s	2	3	4	5	6	7	8	9	10	11	12
Y_s	2	6	10	16	18	32	20	13	16	9	2

Apply the chi-square test to *these* values, using the probabilities in (1), pretending it is not known that the dice are in fact faulty. Does the chi-square test detect the bad dice? If not, explain why not.

▶ **4.** [*23*] The author actually obtained the data in experiment 1 of (9) by simulating dice in which one was normal, the other was loaded so that it always turned up 1 or 6. (The latter two possibilities were equally probable.) Compute the probabilities which replace (1) in this case, and by using a chi-square test decide if the results of that experiment are consistent with the dice being loaded in this way.

5. [*22*] Let $F(x)$ be the uniform distribution, Fig. 3(b). Find K_{20}^+ and K_{20}^- for the following 20 observations:

> 0.414, 0.732, 0.236, 0.162, 0.259, 0.442, 0.189, 0.693, 0.098, 0.302,
>
> 0.442, 0.434, 0.141, 0.017, 0.318, 0.869, 0.772, 0.678, 0.354, 0.718,

and state whether these observations are significantly different from normal behavior with respect to either of these two tests.

6. [*M20*] Consider $F_n(x)$, as given in Eq. (10), for fixed x. What is the probability that $F_n(x) = s/n$, given an integer s? What is the mean value of $F_n(x)$? What is the standard deviation?

7. [*M15*] Show that K_n^+ and K_n^- can never be negative. What is the largest possible value K_n^+ can be?

8. [*00*] The text describes an experiment in which 20 values of the statistic K_{10}^+ were obtained in the study of a random sequence. These values were plotted, to obtain Fig. 4, and a KS statistic was computed from the resulting graph. Why were the table entries for $n = 20$ used to study the resulting statistic, instead of the table entries for $n = 10$?

▶ **9.** [*20*] The experiment described in the text consisted of plotting 20 values of K_{10}^+, computed from the "maximum of 5" test applied to different parts of a random sequence. We could have computed also the corresponding 20 values of K_{10}^-; since K_{10}^- has the same expected distribution as K_{10}^+, we could lump together the 40 values thus obtained (that is, 20 of the K_{10}^+'s and 20 of the K_{10}^-'s), and a KS test could be applied so that we would get new values K_{40}^+, K_{40}^-. Discuss the merits of this idea.

▶ **10.** [*20*] Suppose a chi-square test is done by making n observations, and the value V is obtained. Now we repeat the test on these same n observations over again (getting, of course, the same results), and we put together the data from both tests, regarding it as a single chi-square test with $2n$ observations. (This procedure violates the text's stipulation that all of the observations must be independent of one another.) How is the second value of V related to the first one?

11. [*20*] Solve exercise 10 substituting the KS test for the chi-square test.

12. [*M26*] Suppose a chi-square test is made to a set of n observations, assuming that p_s is the probability that each observation falls into category s; but suppose that in actual fact the observations have probability $q_s \neq p_s$ of falling into category s. (Cf. exercise 3.) We would, of course, like the chi-square test to detect the fact that the p_s assumption was incorrect. Show that this *will* happen, if n is large enough. Prove also the analogous result for the KS test.

13. [*M24*] Prove that Eqs. (13) are equivalent to Eqs. (11).

▶ **14.** [*HM26*] Let Z_s be given by Eq. (18). Show directly by using Stirling's approximation that the multinomial probability

$$ n! p_1^{Y_1} \dots p_k^{Y_k} / Y_1! \dots Y_k! = e^{-V/2} / \sqrt{(2n\pi)^{k-1} p_1 \dots p_k} + O(n^{-k/2}), $$

if $Z_1, Z_2, \dots, Z_k$ are bounded as $n \to \infty$. (This idea leads to a proof of the chi-square test which is much closer to "first principles," and requires less handwaving, than the derivation in the text.)

15. [*HM24*] Polar coordinates in two dimensions are conventionally taken to be $x = r \cos \theta$ and $y = r \sin \theta$. For the purposes of integration, we have $dx\, dy = r\, dr\, d\theta$. More generally, in n-dimensional space we can let

$$ x_k = r \sin \theta_1 \cdots \sin \theta_{k-1} \cos \theta_k, \qquad 1 \le k < n, \qquad x_n = r \sin \theta_1 \cdots \sin \theta_{n-1}. $$

Show that in this case

$$ dx_1\, dx_2 \cdots dx_n = |r^{n-1} \sin^{n-2}\theta_1 \cdots \sin \theta_{n-2}\, dr\, d\theta_1 \cdots d\theta_{n-1}|. $$

▶ **16.** [*HM35*] Generalize Theorem 1.2.11.3A to find the value of

$$ \gamma(x+1,\, x+z\sqrt{2x}+y)/\Gamma(x+1), $$

for large x and fixed y, z. Disregard terms of the answer which are $O(1/x)$. Use this result to find the approximate solution, t, to the equation

$$ \gamma\left(\frac{\nu}{2},\, \frac{t}{2}\right) \Big/ \Gamma\left(\frac{\nu}{2}\right) = p, $$

for large ν and fixed p, thereby accounting for the asymptotic formulas indicated in Table 1. [*Hint:* See exercise 1.2.11.3–8.]

17. [*HM26*] Let t be a fixed real number. For $0 \le k \le n$, let

$$P_{nk}(x) = \int_{n-t}^{x} dx_n \int_{n-1-t}^{x_n} dx_{n-1} \cdots \int_{k+1-t}^{x_{k+2}} dx_{k+1} \int_{0}^{x_{k+1}} dx_k \cdots \int_{0}^{x_2} dx_1;$$

by convention, let $P_{00}(x) = 1$. Prove the following relations:

a) $P_{nk}(x) = \int_{n}^{x+t} dx_n \int_{n-1}^{x_n} dx_{n-1} \cdots \int_{k+1}^{x_{k+2}} dx_{k+1} \int_{t}^{x_{k+1}} dx_k \cdots \int_{x}^{x_2} dx_1.$

b) $P_{n0}(x) = (x+t)^n/n! - (x+t)^{n-1}/(n-1)!.$

c) $P_{nk}(x) - P_{n(k-1)}(x) = \dfrac{(k-t)^k}{k!} P_{(n-k)0}(x-k),$ if $1 \le k \le n.$

d) Obtain a general formula for $P_{nk}(x)$, and apply it to the evaluation of Eq. (24).

18. [*M20*] Give a "simple" reason why K_n^- has the same probability distribution as K_n^+.

19. [*HM48*] Develop tests, analogous to the Kolmogorov–Smirnov test, for use with multivariate distributions $F(x_1, \ldots, x_r)$ = probability that $(X_1 \le x_1, \ldots, X_r \le x_r)$. (This could be used, for example, in connection with the "serial test" in the next section.)

20. [*HM50*] Obtain further information about the asymptotic value of the quantity on the left-hand side of (27) for large values of n. [*Notes:* Empirical evidence suggests the left-hand side is less than e^{-2s^2}, for all n, but this is not known for certain. Whittle's paper shows that the left-hand side lies roughly between $\exp(-2s^2 - (2s + s^3)/\sqrt{n})$ and $\exp(-2s^2 + 3.83s^3/\sqrt{n})$. The formulas given in Table 2 for large n are only empirical estimates based on rather limited data, and a theoretical derivation of the true behavior is desirable.]

21. [*M40*] Although the text states that the KS test should be applied only when $F(x)$ is a continuous distribution function, it is, of course, possible to try to compute K_n^+ and K_n^- even when distribution has jumps. Analyze the probable behavior of K_n^+ and K_n^- for various discontinuous distributions $F(x)$. Compare the effectiveness of the resulting statistical test with the chi-square test on several samples of random numbers.

22. [*HM42*] Discuss using the "maximum likelihood ratio" as an alternative to the chi-square method for testing random-number generators; experiment with the use of both types of statistics.

3.3.2. Empirical Tests

In this section we will describe nine kinds of specific tests that have been applied to sequences in order to investigate their randomness. The discussion of each test has two parts: (a) a "plug-in" description of how to perform the test; and (b) a study of the theoretical basis for the test. (Readers lacking mathematical training may wish to skip over the theoretical discussions. Conversely, mathematically-inclined readers may find the associated theory quite interesting, even if they never intend to test random-number generators, since some instructive combinatorial questions are involved here.)

Each test is applied to a sequence

$$\langle U_n \rangle = U_0, U_1, U_2, \ldots \tag{1}$$

of real numbers, which purports to be uniformly distributed between zero and one. Some of the tests are designed primarily for integer-valued sequences, instead of the real-valued sequence (1). In this case, the auxiliary sequence

$$\langle Y_n \rangle = Y_0, Y_1, Y_2, \ldots, \tag{2}$$

which is defined by the rule

$$Y_n = \lfloor dU_n \rfloor, \tag{3}$$

is used in the test. This is a sequence of integers which will be uniformly distributed between 0 and $d - 1$ if the sequence (1) is uniform between 0 and 1. The number d is chosen for convenience; for example, we might have $d = 64 = 2^6$ on a binary computer, so that Y_n represents the six most significant bits of the binary representation of U_n. The value of d should be large enough so that the test is meaningful, but not so large that the test becomes impracticably difficult to make on the computer.

The quantities U_n, Y_n, and d will have the above significance throughout this section. The value of d will probably be different in different tests.

A. Equidistribution test (Frequency test). The first requirement that sequence (1) must meet is that its numbers are, in fact, uniformly distributed between zero and one. There are two ways to make this test: (a) Use the Kolmogorov-Smirnov test, with $F(x) = x$ for $0 \leq x \leq 1$. (b) Let d be a convenient number, e.g., 100 on a decimal computer, 64 or 128 on a binary computer, and use the sequence (2) instead of (1). For each integer r, $0 \leq r < d$, count the number of times $Y_j = r$ for $0 \leq j < n$, and then apply the chi-square test using $k = d$ and probability $p_s = 1/d$ for each category.

The theory behind this test has been covered in the previous section.

B. Serial test. More generally, we want pairs of successive numbers to be uniformly distributed in an independent manner. To make this test, count the number of times the pair $(Y_{2j}, Y_{2j+1}) = (q, r)$ occurs, for $0 \leq j < n$; these counts are to be made for each pair of integers (q, r) with $0 \leq q, r < d$, and we apply the chi-square test to these $k = d^2$ categories with probability $1/d^2$ in each category. As with the equidistribution test, d may be chosen as any convenient number, but it will be somewhat smaller than the values suggested above since a valid chi-square test should have n large compared to k (say $n > 5d^2$ at least).

Clearly we can generalize this test to triples, quadruples, etc., instead of pairs (see exercise 2); however, the value of d must then be severely reduced in order to avoid having too many categories. When quadruples and larger numbers of adjacent elements are considered, we therefore make use of less exact tests such as the poker test or the maximum test described below.

Note that $2n$ numbers of the sequence (2) are used in this test in order to make n observations. It would be a mistake to perform the serial test on the pairs $(Y_0, Y_1), (Y_1, Y_2), \ldots, (Y_{n-1}, Y_n)$; can the reader see why? We might perform another serial test on the pairs (Y_{2j+1}, Y_{2j+2}), and expect the sequence to pass both tests. Alternatively, I. J. Good has proved that if d is prime, and if the pairs $(Y_0, Y_1), (Y_1, Y_2), \ldots, (Y_{n-1}, Y_n)$ are used, and if we use the usual chi-square method to compute both the statistics V_2 for the serial test *and* V_1 for the frequency test on $Y_0, \ldots, Y_{n-1}$ with the same value of d, then, for large n, $V_2 - 2V_1$ should have the chi-square distribution with $(d-1)^2$ degrees of freedom, although V_2 should *not* be expected to have the chi-square distribution with $d^2 - 1$ degrees of freedom. (See *Annals of Mathematical Statistics* **28** (1957), 262–264.)

C. Gap test. This test is used to examine the length of "gaps" between occurrences of U_j in a certain range. If α and β are two real numbers with $0 \leq \alpha < \beta \leq 1$, we want to consider the lengths of consecutive subsequences $U_j, U_{j+1}, \ldots, U_{j+r}$ in which U_{j+r} lies between α and β but the other U's do not. (This subsequence of $r + 1$ numbers represents a gap of length r.)

Algorithm G. (*Data for gap test*). The following algorithm, applied to the sequence (1), counts the number of gaps of lengths $0, 1, \ldots, t - 1$ and the number of gaps of length $\geq t$, until n gaps have been tabulated.

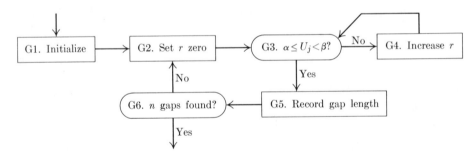

Fig. 6. Gathering data for the gap test. (Algorithms for the "coupon-collector's test" and the "run test" are similar.)

G1. [Initialize.] Set $j \leftarrow -1$, $s \leftarrow 0$, and set COUNT$[r] \leftarrow 0$ for $0 \leq r \leq t$.

G2. [Set r zero.] Set $r \leftarrow 0$.

G3. [$\alpha \leq U_j < \beta$?] Increase j by 1. If $U_j \geq \alpha$ and $U_j < \beta$, go to step G5.

G4. [Increase r.] Increase r by one, and return to step G3.

G5. [Record gap length.] (A gap of length r has now been found.) If $r \geq t$, increase COUNT$[t]$ by one, otherwise increase COUNT$[r]$ by one.

G6. [n gaps found?] Increase s by one. If $s < n$, return to step G2. ∎

After this algorithm has been performed, the chi-square test is applied to the $k = t + 1$ values of COUNT[0], COUNT[1], ..., COUNT[t], using the following probabilities:

$$p_0 = p, \qquad p_1 = p(1 - p), \qquad p_2 = p(1 - p)^2, \qquad \ldots,$$

$$p_{t-1} = p(1 - p)^{t-1}, \qquad p_t = (1 - p)^t. \qquad (4)$$

Here $p = \beta - \alpha$, the probability that $\alpha \leq U_j < \beta$. The values of n and t are to be chosen, as usual, so that each of the values of COUNT[r] is expected to be 5 or more.

The gap test is often applied with $\alpha = 0$ or $\beta = 1$ in order to facilitate the test in step G3. The special cases $(\alpha, \beta) = (0, \frac{1}{2})$ or $(\frac{1}{2}, 1)$ give rise to tests which are sometimes called "runs below the mean" and "runs above the mean," respectively.

The probabilities in Eq. (4) are easily deduced, so this derivation is left to the reader. Note that the gap test as described above observes the lengths of n gaps; it does not observe the gap lengths among n numbers. Other gap tests have been used in which a fixed number of U's is examined (see exercise 5.)

D. Poker test (Partition test). The "classical" poker test considers n groups of five successive integers, $(Y_{5j}, Y_{5j+1}, \ldots, Y_{5j+4})$, $0 \leq j < n$. We observe which of the following seven patterns each quintuple matches:

All different:	*abcde*	Full house:	*aaabb*
One pair:	*aabcd*	Four of a kind:	*aaaab*
Two pairs:	*aabbc*	Five of a kind:	*aaaaa*
Three of a kind:	*aaabc*		

A chi-square test is based on the number of quintuples in each category.

It is reasonable to ask for a somewhat simpler version of this test, to facilitate the programming involved. A good compromise would be to simply count the number of *distinct* values in the set of five. We would then have five categories:

5 different = all different;
4 different = one pair;
3 different = two pairs, or three of a kind;
2 different = full house, or four of a kind;
1 different = five of a kind.

This breakdown is easier to determine systematically, and the test is nearly as good.

In general we can consider n groups of k successive numbers, and we can count the number of k-tuples with r different values. A chi-square test is then made, using the probability

$$p_r = \frac{d(d - 1) \cdots (d - r + 1)}{d^k} \begin{Bmatrix} k \\ r \end{Bmatrix} \qquad (5)$$

that there are r different. (The Stirling numbers $\{{}^k_r\}$ are defined in Section 1.2.6, and they can be readily computed using the formulas given there.) Since the probability p_r is very small when $r = 1$ or 2, we generally lump a few categories of low probability together before the chi-square test is applied.

To derive the above formula for p_r, we must count how many of the d^k k-tuples of numbers between 0 and $d - 1$ have exactly r different elements, and divide the total by d^k. Since $d(d - 1) \ldots (d - r + 1)$ is the number of ordered choices of r things from a set of d objects, we need only show that $\{{}^k_r\}$ is the number of ways to partition a set of k elements into exactly r parts. Therefore exercise 1.2.6–64 completes the derivation of Eq. (5).

E. Coupon collector's test. This test is related to the poker test somewhat as the gap test is related to the frequency test. The sequence $Y_0, Y_1, \ldots$ is used, and we observe the lengths of segments $Y_{j+1}, Y_{j+2}, \ldots, Y_{j+r}$ required to get a "complete set" of integers from 0 to $d - 1$. Algorithm C describes this precisely:

Algorithm C (*Data for coupon collector's test*). Given a sequence of integers $Y_0, Y_1, \ldots$, with $0 \le Y_j < d$, this algorithm counts the lengths of n consecutive "coupon collector" segments. At the conclusion of the algorithm, COUNT[r] contains the number of segments with length r, for $0 \le r < t$, and COUNT[t] contains the number with length $\ge t$.

C1. [Initialize.] Set $j \leftarrow -1$, $s \leftarrow 0$, and set COUNT[r] $\leftarrow 0$ for $0 \le r \le t$.

C2. [Set q, r zero.] Set $q \leftarrow r \leftarrow 0$, and set OCCURS[$k$] $\leftarrow 0$ for $0 \le k < d$.

C3. [Next observation.] Increase r and j by 1. If OCCURS[Y_j] $\ne 0$, repeat this step.

C4. [Complete set?] Set OCCURS[Y_j] $\leftarrow 1$, $q \leftarrow q + 1$. (The subsequence observed so far contains q distinct values; if $q = d$, we therefore have a complete set.) If $q < d$, return to step C3.

C5. [Record the length.] If $r \ge t$, increase COUNT[t] by one, otherwise increase COUNT[r] by one.

C6. [n found?] Increase s by one. If $s < n$, return to step C2. ∎

For an example of this algorithm, see exercise 7. We may think of a boy collecting d types of coupons, which are randomly distributed in his breakfast cereal boxes; he must keep eating more cereal until he has one coupon of each type.

A chi-square test is to be applied to COUNT[d], COUNT[$d + 1$], $\ldots$, COUNT[t], with $k = t - d + 1$, after Algorithm C has counted n lengths. The corresponding probabilities are

$$p_r = \frac{d!}{d^r} \left\{ {r - 1 \atop d - 1} \right\}, \quad d \le r < t, \quad p_t = 1 - \frac{d!}{d^{t-1}} \left\{ {t - 1 \atop d} \right\}. \qquad (6)$$

To derive these probabilities, we simply note that if q_r denotes the probability that a subsequence of length r is *incomplete*,

$$q_r = 1 - \frac{d!}{d^r}\left\{\begin{matrix} r \\ d \end{matrix}\right\}$$

by Eq. (5); for this means we have an r-tuple of elements which do not have all d different values. Then (6) follows from the relations $p_r = q_{r-1} - q_r$ for $d \leq r < t$, $p_t = q_{t-1}$.

For formulas which arise in connection with *generalizations* of the coupon collector's test, see exercises 9 and 10 and also the paper by Hermann von Schelling, *AMM* **61** (1954), 306–311.

F. Permutation test. Divide the input sequence into n groups of t elements each, that is, $(U_{jt}, U_{jt+1}, \ldots, U_{jt+t-1})$, $0 \leq j < n$. The elements in each group can have $t!$ possible relative orderings; the number of times each ordering appears is counted, and a chi-square test is applied with $k = t!$ and with probability $1/t!$ for each ordering.

For example, if $t = 3$ we would have six categories, according to whether $U_{3j} < U_{3j+1} < U_{3j+2}$ or $U_{3j} < U_{3j+2} < U_{3j+1}$ or ... or $U_{3j+2} < U_{3j+1} < U_{3j}$. We assume in this test that equality between U's does not occur; such an assumption is justified, for the probability that two U's are equal is zero.

A convenient way to perform this test on a computer is to use the following algorithm, which is of interest in itself:

Algorithm P (*Analyze a permutation*). Given a set of distinct elements $(U_1, \ldots, U_t)$, we compute an integer $f(U_1, \ldots, U_t)$ such that

$$0 \leq f(U_1, \ldots, U_t) < t!,$$

and $f(U_1, \ldots, U_t) = f(V_1, \ldots, V_t)$ if and only if $(U_1, \ldots, U_t)$ and $(V_1, \ldots, V_t)$ have the same relative ordering. The method uses an auxiliary table $C[1], \ldots, C[t]$.

P1. [Initialize.] Set $r \leftarrow t$.

P2. [Find maximum.] Find the maximum of $\{U_1, \ldots, U_r\}$, and suppose that U_s is the maximum. Set $C[r] \leftarrow s - 1$.

P3. [Exchange.] Exchange $U_r \leftrightarrow U_s$.

P4. [Decrease r.] Decrease r by 1. If $r > 0$, return to step P2.

P5. [Compute f.] The desired function value is now given by the formula

$$f = C[t] + tC[t-1] + t(t-1)C[t-2] + \cdots + t!C[1] \tag{7}$$
$$= (\ldots ((C[1] \times 2 + C[2]) \times 3 + C[3]) + \cdots + C[t-1]) \times t + C[t]. \quad \blacksquare$$

By the construction in this algorithm, we observe that

$$0 \leq C[r] < r \tag{8}$$

(see step P2), so in particular $C[1]$ is always zero. Since $C[r]$ may take on r different values, we see there are $t!$ different settings of the C table. From the formula for f and (8), it is easy to see that $0 \leq f < t!$. When the algorithm is complete, $(U_1, \ldots, U_t)$ will have been sorted into ascending order.

To prove that the result f of Algorithm P uniquely characterizes the initial order of $(U_1, \ldots, U_t)$, we will show that the algorithm can be applied in reverse. We start with any value f and any t elements $(U_1, \ldots, U_t)$ such that $0 \leq f < t!$ and $U_1 < U_2 < \cdots < U_t$. Set $C[t] \leftarrow f \bmod t$, $C[t-1] \leftarrow \lfloor f/t \rfloor \bmod (t-1)$, $C[t-2] \leftarrow \lfloor \lfloor f/t \rfloor /(t-1) \rfloor \bmod (t-2)$, etc., thereby accomplishing step P5 in reverse. Now exchange $U_1 \leftrightarrow U_{C[1]+1}$, $U_2 \leftrightarrow U_{C[2]+1}$, ..., $U_t \leftrightarrow U_{C[t]+1}$, in that order. It is easy to see that we have thereby undone the effect of steps P1–P4. The fact that the value of f allows us to determine the original permutation proves that Algorithm P performs as advertised.

G. Run test. A sequence may be tested for "runs up" and "runs down." This means we examine the length of *monotone* subsequences of the original sequence, i.e., segments which are increasing or decreasing.

As an example of the precise definition of a run, consider the sequence of ten numbers "1298536704"; putting a vertical line at the left and right and between X_j and X_{j+1} whenever $X_j > X_{j+1}$, we obtain

$$|1 \quad 2 \quad 9|8|5|3 \quad 6 \quad 7|0 \quad 4| \qquad (9)$$

which displays the "runs up": there is a run of length 3, followed by two runs of length 1, followed by another run of length 3, followed by a run of length 2. The algorithm of exercise 12 shows how to tabulate the length of "runs up."

Unlike the gap test and the coupon collector's test (which are in many respects similar to this test), *we should not apply a chi-square test to the above data*, since adjacent runs are *not* independent. A long run will tend to be followed by a short run, and conversely. This lack of independence is enough to invalidate a straightforward chi-square test. Instead, the following statistic may be computed, when the run lengths have been determined as in exercise 12:

$$V = \frac{1}{n} \sum_{1 \leq i, j \leq 6} (\text{COUNT}[i] - nb_i)(\text{COUNT}[j] - nb_j)a_{ij}, \qquad (10)$$

where the coefficients a_{ij} and b_i are

$$
\begin{pmatrix}
a_{11} & a_{12} & a_{13} & a_{14} & a_{15} & a_{16} \\
a_{21} & a_{22} & a_{23} & a_{24} & a_{25} & a_{26} \\
a_{31} & a_{32} & a_{33} & a_{34} & a_{35} & a_{36} \\
a_{41} & a_{42} & a_{43} & a_{44} & a_{45} & a_{46} \\
a_{51} & a_{52} & a_{53} & a_{54} & a_{55} & a_{56} \\
a_{61} & a_{62} & a_{63} & a_{64} & a_{65} & a_{66}
\end{pmatrix}
=
\begin{pmatrix}
4529.4 & 9044.9 & 13568 & 18091 & 22615 & 27892 \\
9044.9 & 18097 & 27139 & 36187 & 45234 & 55789 \\
13568 & 27139 & 40721 & 54281 & 67852 & 83685 \\
18091 & 36187 & 54281 & 72414 & 90470 & 111580 \\
22615 & 45234 & 67852 & 90470 & 113262 & 139476 \\
27892 & 55789 & 83685 & 111580 & 139476 & 172860
\end{pmatrix}
$$

$$(11)$$

$$
(b_1 \quad b_2 \quad b_3 \quad b_4 \quad b_5 \quad b_6) = (\tfrac{1}{6} \quad \tfrac{5}{24} \quad \tfrac{11}{120} \quad \tfrac{19}{720} \quad \tfrac{29}{5040} \quad \tfrac{1}{840})
$$

(These values are approximate only; the exact values of the coefficients may be obtained by using formulas derived below.) *The statistic V in* (10) *should have the chi-square distribution with six* (not five) *degrees of freedom, when n is large.* The value of n should be, say, 4000 or more. The same test can be applied to "runs down."

A simpler and more practical run test appears in exercise 14, but it will be instructive from a mathematical standpoint to see how this fairly complicated test can be treated. Given any permutation on n elements, let $Z_{pi} = 1$ if position i is the beginning of an ascending run of length p or more, and let $Z_{pi} = 0$ otherwise. For example, consider the permutation (9) with $n = 10$; we have

$$Z_{11} = Z_{21} = Z_{31} = Z_{14} = Z_{15} = Z_{16} = Z_{26} = Z_{36} = Z_{19} = Z_{29} = 1,$$

and all other Z's are zero. With this notation,

$$R'_p = Z_{p1} + Z_{p2} + \cdots + Z_{pn} \tag{12}$$

is the number of runs of length $\geq p$, and

$$R_p = R'_p - R'_{p+1} \tag{13}$$

is the number of runs of length p exactly. Our goal is to compute the mean value of R_p, and also the *covariance*

$$\text{covar}(R_p, R_q) = \text{mean}((R_p - \text{mean}(R_p))(R_q - \text{mean}(R_q))),$$

which measures the interdependence of R_p and R_q. These mean values are to be computed as the average over the set of all $n!$ permutations.

Equations (12) and (13) show that the answers can be expressed in terms of the mean values of Z_{pi} and of $Z_{pi}Z_{qj}$, so as the first step of the derivation we obtain the following results (assuming that $i < j$):

$$\frac{1}{n!}\sum Z_{pi} = \begin{cases} (p + \delta_{i1})/(p+1)!, & \text{if } i \leq n - p + 1; \\ 0, & \text{otherwise.} \end{cases}$$

$$\frac{1}{n!}\sum Z_{pi}Z_{qj} = \begin{cases} (p + \delta_{i1})q/(p+1)!(q+1)!, \\ \qquad \text{if } i + p < j \leq n - q + 1; \\ (p + \delta_{i1})/(p+1)!q! - (p + q + \delta_{i1})/(p+q+1)!, \\ \qquad \text{if } i + p = j \leq n - q + 1; \\ 0, \quad \text{otherwise.} \end{cases} \tag{14}$$

The $\sum$-signs stand for a summation over all possible permutations. To illustrate the calculations involved here, we will work the most difficult case, when $i + p = j \leq n - q + 1$, and when $i > 1$. Note that $Z_{pi}Z_{qj}$ is either zero or one, so the summation consists of counting all permutations $U_1, U_2, \ldots, U_n$ for which $Z_{pi} = Z_{qj} = 1$, that is,

$$U_{i-1} > U_i < \cdots < U_{i+p-1} > U_{i+p} < \cdots < U_{i+p+q-1}. \tag{15}$$

The number of such permutations may be enumerated as follows: there are $\binom{n}{p+q+1}$ ways to choose the elements for the positions indicated in (15); there are

$$(p + q + 1)\binom{p + q}{p} - \binom{p + q + 1}{p + 1} - \binom{p + q + 1}{1} + 1 \tag{16}$$

ways to arrange them in the order (15), as shown in exercise 13; and there are $(n - p - q - 1)!$ ways to arrange the remaining elements. Thus there are $\binom{n}{p+q+1}(n - p - q - 1)!$ times (16) ways in all, and dividing by $n!$ we get the desired formula.

From relations (14), a rather lengthy calculation leads to

$$\begin{aligned}
\text{mean } (R_p') &= \text{mean } (Z_{p1} + \cdots + Z_{pn}) \\
&= (n + 1)p/(p + 1)! - (p - 1)/p!, \qquad 1 \le p \le n; \tag{17}
\end{aligned}$$

$$\begin{aligned}
\text{covar } (R_p', R_q') &= \text{mean } (R_p' R_q') - \text{mean } (R_p') \, \text{mean } (R_q') \\
&= \sum_{1 \le i, j \le n} \frac{1}{n!} \sum Z_{pi} Z_{pj} - \text{mean } (R_p') \, \text{mean } (R_q') \\
&= \begin{cases} \text{mean } (R_t') + f(p, q, n), & \text{if } p + q \le n; \\ \text{mean } (R_t') - \text{mean } (R_p') \, \text{mean } (R_q'), & \text{if } p + q > n, \end{cases} \tag{18}
\end{aligned}$$

where $t = \max (p, q)$, $s = p + q$, and

$$\begin{aligned}
f(p, q, n) = (n + 1)&\left(\frac{s(1 - pq) + pq}{(p + 1)!(q + 1)!} - \frac{2s}{(s + 1)!} \right) + 2\left(\frac{s - 1}{s!} \right) \\
&+ \frac{(s^2 - s - 2)pq - s^2 - p^2 q^2 + 1}{(p + 1)!(q + 1)!}. \tag{19}
\end{aligned}$$

This expression for the covariance is unfortunately quite complicated, but it is necessary for a successful run test as described above. From these formulas it is easy to compute

$$\begin{aligned}
\text{mean } (R_p) &= \text{mean } (R_p') - \text{mean } (R_{p+1}'), \\
\text{covar } (R_p, R_q') &= \text{covar } (R_p', R_q') - \text{covar } (R_{p+1}', R_q'), \tag{20} \\
\text{covar } (R_p, R_q) &= \text{covar } (R_p, R_q') - \text{covar } (R_p, R_{q+1}').
\end{aligned}$$

In *Annals of Mathematical Statistics* **15** (1944), 163–165, J. Wolfowitz proved that the quantities $R_1, R_2, \ldots, R_{t-1}, R_t'$ become normally distributed as $n \to \infty$, subject to the mean and covariance expressed above; this implies that the following test for runs is valid: Given a sequence of n random numbers, compute the number of runs R_p of length p for $1 \le p < t$, and also the number of runs, R_t', of length $\ge t$. Let

$$\begin{aligned}
Q_1 = R_1 - \text{mean } (R_1), \ldots, Q_{t-1} = R_{t-1} - \text{mean } (R_{t-1}), \\
Q_t = R_t' - \text{mean } (R_t'). \tag{21}
\end{aligned}$$

Form the matrix C of the covariance of the R's, for example, $C_{13} =$ covar (R_1, R_3), while $C_{1t} =$ covar (R_1, R'_t). When $t = 6$, we have

$$C = nC_1 + C_2 =$$

$$n \begin{pmatrix}
\dfrac{23}{180} & \dfrac{-7}{360} & \dfrac{-5}{336} & \dfrac{-433}{60480} & \dfrac{-13}{5670} & \dfrac{-121}{181440} \\[2mm]
\dfrac{-7}{360} & \dfrac{2843}{20160} & \dfrac{-989}{20160} & \dfrac{-7159}{362880} & \dfrac{-10019}{1814400} & \dfrac{-1303}{907200} \\[2mm]
\dfrac{-5}{336} & \dfrac{-989}{20160} & \dfrac{54563}{907200} & \dfrac{-21311}{1814400} & \dfrac{-62369}{19958400} & \dfrac{-7783}{9979200} \\[2mm]
\dfrac{-433}{60480} & \dfrac{-7159}{362880} & \dfrac{-21311}{1814400} & \dfrac{886657}{39916800} & \dfrac{-257699}{239500800} & \dfrac{-62611}{239500800} \\[2mm]
\dfrac{-13}{5670} & \dfrac{-10019}{1814400} & \dfrac{-62369}{19958400} & \dfrac{-257699}{239500800} & \dfrac{29874811}{5448643200} & \dfrac{-1407179}{21794572800} \\[2mm]
\dfrac{-121}{181440} & \dfrac{-1303}{907200} & \dfrac{-7783}{9979200} & \dfrac{-62611}{239500800} & \dfrac{-1407179}{21794572800} & \dfrac{2134697}{1816214400}
\end{pmatrix}$$

$$\tag{22}$$

$$+ \begin{pmatrix}
\dfrac{83}{180} & \dfrac{-29}{180} & \dfrac{-11}{210} & \dfrac{-41}{12096} & \dfrac{91}{25920} & \dfrac{41}{18144} \\[2mm]
\dfrac{-29}{180} & \dfrac{-305}{4032} & \dfrac{319}{20160} & \dfrac{2557}{72576} & \dfrac{10177}{604800} & \dfrac{413}{64800} \\[2mm]
\dfrac{-11}{210} & \dfrac{319}{20160} & \dfrac{-58747}{907200} & \dfrac{19703}{604800} & \dfrac{239471}{19958400} & \dfrac{39517}{9979200} \\[2mm]
\dfrac{-41}{12096} & \dfrac{2557}{72576} & \dfrac{19703}{604800} & \dfrac{-220837}{4435200} & \dfrac{1196401}{239500800} & \dfrac{360989}{239500800} \\[2mm]
\dfrac{91}{25920} & \dfrac{10177}{604800} & \dfrac{239471}{19958400} & \dfrac{1196401}{239500800} & \dfrac{-139126639}{7264857600} & \dfrac{4577641}{10897286400} \\[2mm]
\dfrac{41}{18144} & \dfrac{413}{64800} & \dfrac{39517}{9979200} & \dfrac{360989}{239500800} & \dfrac{4577641}{10897286400} & \dfrac{-122953057}{21794572800}
\end{pmatrix}$$

if $n \geq 14$. Now form $A = (a_{ij})$, the inverse of the matrix C, and compute $\sum_{1 \leq i,j \leq t} Q_i Q_j a_{ij}$. The result for large n has the chi-square distribution with t degrees of freedom.

The matrix (11) given earlier is the inverse of C_1 to five significant figures. When n is large, A will be approximately $(1/n)C_1^{-1}$. An attempt was made to give the inverse of C_1 in terms of rational numbers, but the values involved were much too large even when $t = 4$. A test with $n = 1000$ showed that each of the entries in (11) was about 1 percent lower than the exact value obtained by inverting (22). A standard method for inverting matrices is given in Section 2.2.6, exercise 18.

H. Maximum of t. For $0 \le j < n$, let $V_j = \max (U_{tj}, U_{tj+1}, \ldots, U_{tj+t-1})$. Now apply the Kolmogorov–Smirnov test to the sequence $V_0, V_1, \ldots, V_{n-1}$, with the distribution function $F(x) = x^t$, $(0 \le x \le 1)$. Alternatively, apply the equidistribution test to the sequence $V_0^t, V_1^t, \ldots, V_{n-1}^t$.

To verify this test, we must show that the distribution function for the V_j is $F(x) = x^t$. The probability that $\max (U_1, U_2, \ldots, U_t) \le x$ is the probability that $U_1 \le x$ *and* $U_2 \le x$ *and* $\ldots$ *and* $U_t \le x$, and this is the product of the individual probabilities, namely $x \cdot x \cdots x = x^t$.

I. Serial correlation. Compute the following statistic:

$$C = \frac{n(U_0 U_1 + U_1 U_2 + \cdots + U_{n-2} U_{n-1} + U_{n-1} U_0) - (U_0 + U_1 + \cdots + U_{n-1})^2}{n(U_0^2 + U_1^2 + \cdots + U_{n-1}^2) - (U_0 + U_1 + \cdots + U_{n-1})^2}. \quad (23)$$

This is the "serial correlation coefficient" which is a measure of the amount U_{j+1} depends on U_j. [When n is very large, a method which greatly reduces the time required to calculate C is given by L. P. Schmid in *CACM* **8** (1965), 115.]

Correlation coefficients appear frequently in statistics; if we have n quantities $U_0, U_1, \ldots, U_{n-1}$ and n others $V_0, V_1, \ldots, V_{n-1}$, the correlation coefficient between them is defined to be

$$C = \frac{n\sum(U_j V_j) - (\sum U_j)(\sum V_j)}{\sqrt{(n\sum U_j^2 - (\sum U_j)^2)(n\sum V_j^2 - (\sum V_j)^2)}}. \quad (24)$$

All summations in this formula are to be taken over the range $0 \le j < n$. Equation (23) is the special case $V_j = U_{(j+1) \bmod n}$. (*Note:* The denominator of (24) is zero when $U_0 = U_1 = \cdots = U_{n-1}$ or $V_0 = V_1 = \cdots = V_{n-1}$; we exclude this case from discussion.)

A correlation coefficient always lies between -1 and $+1$. When it is zero or very small, it indicates that the quantities U_j and V_j are (relatively speaking) independent of each other, but when the correlation coefficient is ± 1 it indicates total linear dependence; in fact $V_j = m \pm aU_j$ for all j in such a case, for some constants a and m. (See exercise 17.)

Therefore it is desirable to have C in Eq. (23) close to zero. In actual fact, since $U_0 U_1$ is not completely independent of $U_1 U_2$, the serial correlation coefficient is not expected to be *exactly* zero. (See exercise 18.) A "good" value of C will be between $\mu_n - 2\sigma_n$ and $\mu_n + 2\sigma_n$, where

$$\mu_n = \frac{-1}{(n-1)}, \qquad \sigma_n = \frac{1}{n-1}\sqrt{\frac{n(n-3)}{n+1}}, \qquad n > 2. \quad (25)$$

We expect C to be between these limits about 95 percent of the time.

Equations (25) are only conjectured at this time, since the exact distribution of C is not known when the U's are uniformly distributed. For the theory when the U's have the *normal* distribution, see the paper by Wilfrid J. Dixon, *Annals of Mathematical Statistics* **15** (1944), 119–144. Empirical evidence shows that we may use the formulas for the mean and standard deviation which have been derived from the assumption of the normal distribution, without much error; these are the values which have been listed in (25). It is known that $\lim_{n\to\infty} \sqrt{n}\, \sigma_n = 1$; cf. the article by Anderson and Walker, *Annals of Mathematical Statistics* **35** (1964), 1296–1303, where more general results, about serial correlations of *dependent* sequences, are derived.

J. Tests on subsequences. It frequently happens that the external program using our random sequence will call for random numbers in batches. For example, if the program has three random variables X, Y, and Z, it may consistently invoke the generation of three random numbers at a time. In such applications it is important to have the subsequence of every *third* term of the original sequence be random. If the program requires q numbers at a time, the sequences

$$U_0, U_q, U_{2q}, \ldots; \qquad U_1, U_{q+1}, U_{2q+1}, \ldots; \qquad \ldots; \qquad U_{q-1}, U_{2q-1}, \ldots; \tag{26}$$

can each be put through the tests described above for the original sequence $U_0, U_1, U_2, \ldots$.

Experience with the linear congruential sequences has shown that these derived sequences rarely if ever behave less randomly than the original sequence, unless q has a large factor in common with the modulus. On a binary computer with m equal to the word size, for example, a test of the sequences for $q = 8$ will tend to give the poorest randomness for all $q < 16$; and on a decimal computer, $q = 10$ is the most likely to be unsatisfactory. (This can be explained somewhat on the grounds of potency, since these values of q will generally lower the potency.)

K. Historical remarks and further discussion. Statistical tests arose naturally in the course of scientists' work to "prove" or "disprove" hypotheses about various observed data. The best known papers dealing with the testing of artificially generated numbers for randomness are the two articles by M. G. Kendall and B. Babington-Smith, in *Journal of the Royal Statistical Society* **101** (1938), 147–166, and the supplement to that journal, **6** (1939), 51–61. These papers were concerned with the testing of random digits between 0 and 9, rather than random real numbers; for this purpose, the authors introduced the frequency test, serial test, gap test, and poker test, although they misapplied the serial test. Kendall and Babington-Smith also used a variant of the coupon collector's test, but the method in the form described above was introduced by R. E. Greenwood in 1955.

The run test has a rather interesting history. Originally, the tests were made on runs up and down at once: a run up would be followed by a run down, and so on. Note that the run test and the permutation test do not depend on the uniform distribution of the U's, they only depend on the fact that the probability $U_i = U_j$ is zero when $i \neq j$; therefore these tests can be applied to many types of random sequences. The run test in primitive form was originated by J. Bienaymé [*Comptes Rendus* **81** (Paris: Acad. Sciences, 1875), 417–423]. Some sixty years later, W. O. Kermack and A. G. McKendrick published two extensive papers on the subject [*Proc. Royal Society Edinburgh* **57** (1937), 228–240, 332–376]; as an example they stated that Edinburgh rainfall between the years 1785 and 1930 was entirely random in character with respect to the run test (although they only examined the mean and standard deviation of run lengths). Several other people began using the test, but it was not until 1944 that the use of the chi-square method in connection with this test was shown to be incorrect. The paper by H. Levene and J. Wolfowitz, in *Annals of Mathematical Statistics* **15** (1944), 58–69, introduced the correct run test (for runs up and down, alternately) and discussed the fallacies in earlier misuses of that test. Separate tests for runs up and runs down, as proposed in the text above, are more suited to computer application, so we have not given the formulas for the alternate-up-and-down case. See the survey paper by D. E. Barton and C. L. Mallows, *Annals of Math. Statistics* **36** (1965), 236–260.

Of all the tests we have described, the frequency test and the serial correlation test seem to be the weakest, in the sense that nearly all random-number generators pass these tests. Theoretical grounds for the weakness of these tests are discussed briefly in Section 3.5 (cf. exercise 3.5–26). The run test is a rather strong test: the results of exercises 3.3.3–23, 24 suggest that linear congruential generators tend to have runs somewhat longer than normal if the multiplier is not large enough, so the run test is definitely to be recommended.

The reader probably wonders, *"Why are there so many tests?"* It has been said that more computer time has been spent testing random numbers than using them in applications! This is untrue, although it is possible to go overboard in testing.

The need for making several tests has been amply documented. It has been recorded, for example, that some numbers generated by a variant of the middle-square method have passed the frequency test, gap test, and poker test, yet flunked the serial test. Linear congruential sequences with small multipliers have been known to pass many tests, yet fail on the run test because there are too few runs of length one. The maximum-of-t test has also been used to ferret out some bad generators which otherwise performed respectably.

Perhaps the main reason for doing extensive testing on random-number generators is that people misusing Mr. X's random-number generator will hardly ever admit that their programs are at fault: they will blame the random-number generator, until Mr. X can *prove* to them that his numbers are sufficiently random. On the other hand, if the source of random numbers is only for Mr. X's personal use, he might decide not to bother to test them, since the random-number generators recommended in this chapter have a high probability of being satisfactory.

EXERCISES

1. [10] Why should the serial test described in part B be applied to (Y_0, Y_1), $(Y_2, Y_3), \ldots, (Y_{2n-2}, Y_{2n-1})$ instead of to $(Y_0, Y_1), (Y_1, Y_2), \ldots, (Y_{n-1}, Y_n)$?

2. [10] State an appropriate way to generalize the serial test to triples, quadruples, etc., instead of pairs.

▶ **3.** [M20] How many U's need to be examined in the gap test (Algorithm G) before n gaps have been found, on the average, assuming the sequence is random? What is the standard deviation of this quantity?

4. [12] Prove that the probabilities in (4) are correct for the gap test.

5. [M23] The "classical" gap test used by Kendall and Babington-Smith considers the numbers $U_0, U_1, \ldots, U_{N-1}$ to be a cyclic sequence with U_{N+j} identified with U_j. Here N is a fixed number of U's which is to be subjected to the test. If n of the numbers $U_0, \ldots, U_{N-1}$ fall into the range $\alpha \le U_j < \beta$, there are n gaps in the cyclic sequence. Let Z_r be the number of gaps of length r, for $0 \le r < t$, and let Z_t be the number of gaps of length $\ge t$; show that the quantity $V = \sum_{0 \le r \le t} (Z_r - np_r)^2/np_r$ should have the chi-square distribution with t degrees of freedom, in the limit as N goes to infinity, where p_r is given in Eq. (4).

6. [40] (H. Geiringer.) A frequency count of the first 2000 decimal digits in the representation of $e = 2.71828\ldots$ gave a χ^2 value of 1.06. This indicates that the actual frequencies for the digits $0, 1, \ldots, 9$ are much too close to their expected values of 200 to be considered randomly distributed. (In fact, $\chi^2 \ge 1.15$ with probability 99.9 percent.) The same test applied to the first 10000 digits of e gives the reasonable value $\chi^2 = 8.61$; but the fact that the first 2000 digits are so evenly distributed is still surprising. Does the same phenomenon occur in the representation of e to other bases? (See *AMM* **72** (1965), 483–500.)

7. [08] Apply the coupon collector's test procedure (Algorithm C) with $d = 3$ and $n = 7$, to the following sequence: 1101221022120202001212201010201121. What lengths do the seven subsequences have?

▶ **8.** [M22] How many U's need to be examined, on the average, in the coupon collector's test (Algorithm C) before n complete sets have been found, assuming that the sequence is random? What is the standard deviation? [*Hint:* See Eq. 1.2.9-28.]

9. [M21] Generalize the coupon collector's test so that the search stops as soon as w distinct values have been found, where w is a fixed positive integer less than or equal to d. What probabilities should be used in place of (6)?

10. [M23] Solve exercise 8 for the more general coupon collector's test described in exercise 9.

11. [00] The "runs up" in a particular permutation are displayed in (9); what are the "runs down" in that permutation?

12. [22] Let $U_0, U_1, \ldots, U_{n-1}$ be n distinct numbers. Write an algorithm which determines the lengths of all ascending runs in the sequence, as follows: When the algorithm terminates, COUNT[r] should be the number of runs of length r, for $1 \le r \le 5$, and COUNT[6] should be the number of runs of length 6 or more.

13. [M23] Show that (16) is the number of permutations of $p + q + 1$ distinct elements having the pattern (15).

14. [*M15*] If we "throw away" the element which immediately follows a run, so that when X_j is greater than X_{j+1} we start the next run with X_{j+2}, the run lengths are independent, and a simple chi-square test may be used (instead of the horribly complicated method derived in the text). What are the appropriate run-length probabilities for this simple run test?

15. [*M20*] In the "maximum of t" test, why are $V_0^t, V_1^t, \ldots, V_{n-1}^t$ supposed to be uniformly distributed between zero and one?

▶ **16.** [*15*] (a) Mr. J. H. Quick (a student) wanted to perform the "maximum-of-t" test for various values of t. If $Z_{jt} = \max(U_j, U_{j+1}, \ldots, U_{j+t-1})$, he found a clever way to go from the sequence $Z_{0(t-1)}, Z_{1(t-1)}, \ldots,$ to the sequence $Z_{0t}, Z_{1t}, \ldots,$ using very little calculation. Can the reader discover this method also?

(b) He decided to modify the "maximum-of-t" method so that $V_j = \max(U_j, \ldots, U_{j+t-1})$; in other words, he took $V_j = Z_{jt}$ instead of $V_j = Z_{(tj)t}$ as the text says. He reasoned that *all* of the Z's should have the same distribution, so the test is even stronger if each Z_{jt}, $0 \le j < n$, is used instead of just every tth one. But when he tried a chi-square equidistribution test on the values of V_j^t, he got extremely high values of the statistic V, which got even higher as t increased. Why did this happen?

17. [*M25*] (a) Given any numbers $X_1, \ldots, X_n, Y_1, \ldots, Y_n$, let

$$\bar{x} = \frac{1}{n}\sum_{1 \le k \le n} X_k, \qquad \bar{y} = \frac{1}{n}\sum_{1 \le k \le n} Y_k.$$

Let $X_k' = X_k - \bar{x}$, $Y_k' = Y_k - \bar{y}$. Show that the correlation coefficient C given in Eq. (24) is equal to

$$\sum_{1 \le k \le n} X_k' Y_k' \Big/ \sqrt{\sum_{1 \le k \le n} X_k'^2} \sqrt{\sum_{1 \le k \le n} Y_k'^2}.$$

(b) Let $C = N/D$, where N and D denote the numerator and denominator of the expression in part (a). Show that $N^2 \le D^2$; hence $-1 \le C \le +1$, and obtain a formula for the difference $D^2 - N^2$. [*Hint:* See exercise 1.2.3–30.]

(c) If $C = \pm 1$, show that $aX_k + bY_k = m$, $1 \le k \le n$, for some constants a, b, m, not all zero.

18. [*M20*] (a) Show that if $n = 2$, the serial correlation coefficient (23) is *always* equal to -1 (unless the denominator is zero). (b) Similarly, show that when $n = 3$, the serial correlation coefficient always equals $-\frac{1}{2}$. (c) Show that the denominator in (23) is zero if and only if $U_0 = U_1 = \cdots = U_{n-1}$.

19. [*M40*] What are the mean and standard deviation of the serial correlation coefficient (23) when $n = 4$ and the U's are independent and uniformly distributed between zero and one?

20. [*M50*] Find the distribution of the serial correlation coefficient (23), for general n, assuming the U_j are independent random variables uniformly distributed between zero and one.

21. [*19*] What value of f is computed by Algorithm P if it is presented with the permutation (1, 2, 9, 8, 5, 3, 6, 7, 0, 4)?

22. [*18*] For what permutation of the integers {0, 1, 2, 3, 4, 5, 6, 7, 8, 9} will Algorithm P produce the value $f = 1024$?

*3.3.3. Theoretical Tests

Although it is always possible to test any random-number generator using the methods in the previous section, it is far better to have "*a priori* tests," i.e., theoretical results which tell us in advance how well those tests will come out. Theoretical results such as this give us much more understanding about the generation methods than empirical, "trial-and-error" results do. In this section we shall study the linear congruential sequences in more detail; if we know what the results will be before we actually generate the numbers, we have a better chance of choosing a, m, and c properly.

The development of this kind of theory is quite difficult, although some progress has been made. The results obtained so far are generally for *statistical tests made over the entire period*. Not all statistical tests make sense when they are applied over a full period; for example, the equidistribution test will give results which are too perfect if it is applied over the whole period. However, the serial test, gap test, permutation test, maximum test, etc. can be fruitfully studied over a full period.

Let us begin with a proof of a simple *a priori* law, for the least complicated case of the permutation test. The gist of our first theorem is that we have $X_{n+1} < X_n$ about half the time, provided that the sequence has high potency.

Theorem P. *Let X_0, a, c, and m generate a linear congruential sequence with maximum period; let $b = a - 1$ and let d be the greatest common divisor of m and b. The probability that $X_{n+1} < X_n$ is equal to $\frac{1}{2} + r$, where*

$$r = (2(c \bmod d) - d)/2m;$$

hence $|r| < d/2m$.

The proof of this theorem involves some techniques which are of interest in themselves. First we define

$$s(x) = (ax + c) \bmod m. \tag{1}$$

Thus, $X_{n+1} = s(X_n)$, and the theorem reduces to counting the number of integers x such that $0 \le x < m$ and $s(x) < x$ (since each such integer occurs somewhere in the period). We want to show that this number is

$$\tfrac{1}{2}(m + 2(c \bmod d) - d). \tag{2}$$

We observe that the theorem is true when $a = 1$, since it is easy to see that in this case $s(x) < x$ only when $x + c \ge m$, and there are c such cases.

Lemma A. *Let α and β be real numbers and let n be an integer. Then*

$$\alpha < n \le \beta \quad \text{if and only if} \quad \lfloor \alpha \rfloor < n \le \lfloor \beta \rfloor; \qquad (3)$$

$$\alpha \le n < \beta \quad \text{if and only if} \quad \lceil \alpha \rceil \le n < \lceil \beta \rceil. \qquad (4)$$

These formulas follow immediately from the definitions of the floor and ceiling operations, as shown in exercise 1.2.4–3. ∎

For any integer x, $0 \le x < m$, let $k(x) = \lfloor (ax + c)/m \rfloor$; we have $s(x) = ax + c - k(x)m$, and $k(x) \le (ax + c)/m$; hence it follows readily that $s(x) < x$ is equivalent to

$$(k(x)m - c)/a \le x < (k(x)m - c)/b. \qquad (5)$$

Furthermore, these inequalities uniquely define $k(x)$, since they imply

$$\frac{ax + c}{m} \le k(x) < \frac{bx + c}{m} < \frac{ax + c}{m} + 1.$$

By Lemma A, we now find that $s(x) < x$ and $0 \le x < m$ if and only if one of the following mutually exclusive possibilities holds:

i) $\lceil (km - c)/a \rceil \le x < \lceil (km - c)/b \rceil$ for some k, $0 < k < a$;

ii) $\lceil m - c/a \rceil \le x < m$. $\qquad (6)$

The number of such integers x is

$$m + \sum_{0<k\le b} \lceil (km - c)/b \rceil - \sum_{0<k\le a} \lceil (km - c)/a \rceil. \qquad (7)$$

The theorem will be proved if we can show that (7) equals (2). The sums in (7) may be evaluated as explained in exercise 1.2.4–37, but we will use another method which indicates how other problems of a similar type can be attacked.

It is convenient to introduce the following functions:

$$\delta(z) = \lfloor z \rfloor + 1 - \lceil z \rceil = \begin{cases} 1, & \text{if } z \text{ is an integer}; \\ 0, & \text{if } z \text{ is not an integer}; \end{cases} \qquad (8)$$

$$((z)) = z - \lfloor z \rfloor - \tfrac{1}{2} + \tfrac{1}{2}\delta(z) = z - \lceil z \rceil + \tfrac{1}{2} - \tfrac{1}{2}\delta(z). \qquad (9)$$

The latter function is a "sawtooth" function familiar in the study of Fourier series; its graph is shown in Fig. 7. The reason for choosing to work with $((z))$ rather than $\lfloor z \rfloor$ or $\lceil z \rceil$ is that $((z))$ possesses several very useful properties:

$$((-z)) = -((z)); \qquad (10)$$

$$((z + n)) = ((z)), \quad \text{if } n \text{ is an integer}; \qquad (11)$$

$$((z)) + \left(\left(z + \frac{1}{n}\right)\right) + \cdots + \left(\left(z + \frac{n-1}{n}\right)\right) = ((nz)),$$

$$\text{if } n \text{ is a positive integer.} \qquad (12)$$

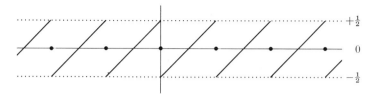

Fig. 7. The sawtooth function $((z))$.

(See exercise 2.) By changing notation, (7) is transformed into

$$\frac{1}{2}\left(m - 1 + \sum_{0 < k \leq a} \delta\left(\frac{km - c}{a}\right) - \sum_{0 < k \leq b} \delta\left(\frac{km - c}{b}\right) \right)$$
$$+ \sum_{0 < k \leq a} \left(\left(\frac{km - c}{a}\right)\right) - \sum_{0 < k \leq b} \left(\left(\frac{km - c}{b}\right)\right). \quad (13)$$

This formula looks, if anything, more formidable than the one we began with, but actually each of the sums appearing in it is easy to evaluate. Since m is relatively prime to a, we know that $(km - c) \bmod a$ takes on each of the values $0, 1, \ldots, a - 1$ in some order as $0 < k \leq a$, so

$$\sum_{0 < k \leq a} \delta\left(\frac{km - c}{a}\right) = 1, \qquad \sum_{0 < k \leq a} \left(\left(\frac{km - c}{a}\right)\right) = 0.$$

(The second formula holds because of (10) and (11).) Also, m is not relatively prime to b, but c is; hence $(km - c)/b$ is never an integer; so the second sum in (13) vanishes. Finally, we observe that

$$\left(\left(\frac{km - c}{b}\right)\right) = \left(\left(\frac{km - d\lfloor c/d \rfloor - c \bmod d}{b}\right)\right)$$
$$= \left(\left(\frac{km/d - \lfloor c/d \rfloor}{b/d}\right)\right) - \frac{c \bmod d}{b} + \frac{1}{2}\delta\left(\frac{km/d - \lfloor c/d \rfloor}{b/d}\right),$$

since $0 \leq (c \bmod d)/b < 1/(b/d)$. The sum

$$\sum_{0 < k \leq b} \left(\left(\frac{km - c}{b}\right)\right)$$

therefore becomes $\frac{1}{2}d - c \bmod d$, using the arguments above, since m/d is relatively prime to b/d. This establishes Theorem P. ∎

The proof of Theorem P gives some idea of the difficulty of making *a priori* tests, but it also illustrates the fact that such calculations are feasible. A further consequence of the theorem is that practically any choice of a and c will give a reasonable probability that $X_{n+1} < X_n$, at least over the entire period, except those which have large d. A large value of d corresponds to low potency, and we already know that generators of low potency are undesirable.

The next theorem gives us a more important criterion for choosing a and c; we will consider the *serial correlation test* applied over the entire period. The quantity C defined in Section 3.3.2, Eq. (23), is

$$C = \left(m \sum_{0 \le x < m} xs(x) - \left(\sum_{0 \le x < m} x \right)^2 \right) \Big/ \left(m \sum_{0 \le x < m} x^2 - \left(\sum_{0 \le x < m} x \right)^2 \right). \qquad (14)$$

Let x' be the element such that $s(x') = 0$. We have

$$s(x) = m \left(\left(\frac{ax + c}{m} \right) \right) + \frac{m}{2}, \qquad \text{if} \quad x \ne x'. \qquad (15)$$

The formulas we are about to derive can be expressed most easily in terms of the function

$$\sigma(h, k, c) = 12 \sum_{0 \le j < k} \left(\left(\frac{j}{k} \right) \right) \left(\left(\frac{hj + c}{k} \right) \right), \qquad (16)$$

which is an important function that arises in several mathematical problems; it is called a *generalized Dedekind sum*.

Since

$$\sum_{0 \le x < m} x = \frac{m(m-1)}{2} \qquad \text{and} \qquad \sum_{0 \le x < m} x^2 = \frac{m(m-1)(2m-1)}{6},$$

it is a straightforward matter to transform Eq. (14) into

$$C = \frac{m\sigma(a, m, c) - 3 + 6(m - x' - c)}{m^2 - 1}. \qquad (17)$$

Since m is usually very large, we may discard terms of order $1/m$, and we have the approximation

$$C \approx \sigma(a, m, c)/m, \qquad (18)$$

with an error of less than $6/m$ in absolute value.

The serial correlation test now reduces to determining the value of the Dedekind sum $\sigma(a, m, c)$. Evaluating $\sigma(a, m, c)$ directly from its definition (16), is hardly any easier than evaluating the correlation coefficient itself directly, but fortunately there are simple methods available for computing Dedekind sums quite rapidly.

Lemma B (*"Reciprocity law" for Dedekind sums*). *If $0 \le c < h$, $0 \le c < k$, and if h is relatively prime to k, then*

$$\sigma(h, k, c) + \sigma(k, h, c) = \frac{h}{k} + \frac{k}{h} + \frac{1}{hk} + \frac{6c^2}{hk} - 3. \qquad (19)$$

Proof. We leave it to the reader to prove that, under these hypotheses,

$$\sigma(h, k, c) + \sigma(k, h, c) = \sigma(h, k, 0) + \sigma(k, h, 0) + 6c^2/hk. \qquad (20)$$

(See exercise 6.) The lemma now must be proved only in the case $c = 0$.

The proof we will give, based on complex roots of unity, is essentially due to L. Carlitz; there is actually a simpler proof which uses only elementary manipulations of sums (see exercise 7), but the following method shows more of the mathematical tools which are available for problems of this kind and it is therefore much more instructive.

Let $f(x)$ and $g(x)$ be polynomials defined as follows:

$$\begin{aligned} f(x) &= 1 + x + \cdots + x^{k-1} = (x^k - 1)/(x - 1) \\ g(x) &= x + 2x^2 + \cdots + (k - 1)x^{k-1} = xf'(x) \\ &= kx^k/(x - 1) - x(x^k - 1)/(x - 1)^2. \end{aligned} \qquad (21)$$

If ω is the complex kth root of unity $e^{2\pi i/k}$, we have by Eq. 1.2.9-13

$$\frac{1}{k} \sum_{0 < j \le k} \omega^{-jr} g(\omega^j x) = rx^r, \qquad \text{if} \qquad 0 \le r < k. \qquad (22)$$

Set $x = 1$; then $g(\omega^j x) = k/(\omega^j - 1)$ if $j \ne k$, $k(k - 1)/2$ if $j = k$, and therefore

$$r \bmod k = \sum_{0 < j < k} \frac{\omega^{-jr}}{\omega^j - 1} + \tfrac{1}{2}(k - 1), \qquad \text{if} \qquad r \text{ is an integer.} \qquad (23)$$

(For (22) shows that the right-hand side equals r when $0 \le r < k$, and it is unchanged when multiples of k are added to r.) Hence

$$\left(\!\!\left(\frac{r}{k}\right)\!\!\right) = \frac{1}{k} \sum_{0 < j < k} \frac{\omega^{-jr}}{\omega^j - 1} - \frac{1}{2k} + \frac{1}{2}\delta\left(\frac{r}{k}\right). \qquad (24)$$

This important formula, which holds whenever r is an integer, allows us to reduce many calculations involving $((r/k))$ to sums involving kth roots of unity, and it brings a whole new range of techniques into the picture. In particular, we get the following formula:

$$\sigma(h, k, 0) + \frac{3(k - 1)}{k^2} = \frac{12}{k^2} \sum_{0 < r < k} \sum_{0 < i < k} \sum_{0 < j < k} \frac{\omega^{-ir}}{\omega^i - 1} \frac{\omega^{-jhr}}{\omega^j - 1}. \qquad (25)$$

The right-hand side of this formula may be simplified by carrying out the sum on r; we have $\sum_{0 \le r < k} \omega^{rs} = f(\omega^s) = 0$ if $s \bmod k \ne 0$. Eq. (25) now reduces to

$$\sigma(h, k, 0) + \frac{3(k - 1)}{k} = \frac{12}{k} \sum_{0 < j < k} \frac{1}{(\omega^{-jh} - 1)(\omega^j - 1)}. \qquad (26)$$

A similar formula is obtained for $\sigma(k, h, 0)$, with $\zeta = e^{2\pi i/h}$ replacing ω.

It is not obvious what we can do with the sum in (26), but there is an elegant way to proceed, based on the fact that each term of the sum is a function of ω^j, $0 < j < k$; hence the sum is essentially taken over the kth roots of unity other than 1. Whenever $x_1, x_2, \ldots, x_n$ are distinct complex numbers, we have the identity

$$\sum_{1 \le j \le n} \frac{1}{(x_j - x_1) \ldots (x_j - x_{j-1})(x - x_j)(x_j - x_{j+1}) \ldots (x_j - x_n)}$$
$$= \frac{1}{(x - x_1) \ldots (x - x_n)}, \quad (27)$$

which follows from the usual method of expanding the right-hand side into partial fractions. Furthermore, if $q(x) = (x - y_1)(x - y_2) \ldots (x - y_m)$, we have

$$q'(y_j) = (y_j - y_1) \ldots (y_j - y_{j-1})(y_j - y_{j+1}) \ldots (y_j - y_m); \quad (28)$$

this identity may often be used to simplify expressions like those in the left-hand side of (27). When h and k are relatively prime, the numbers $\omega, \omega^2, \ldots, \omega^{k-1}$, $\zeta, \zeta^2, \ldots, \zeta^{h-1}$ are all distinct; we can therefore consider (27) in the special case of the polynomial $(x - \omega) \ldots (x - \omega^{k-1})(x - \zeta) \ldots (x - \zeta^{h-1}) = (x^k - 1)(x^h - 1)/(x - 1)^2$, obtaining the following identity in x:

$$\frac{1}{h} \sum_{0 < j < h} \frac{\zeta^j(\zeta^j - 1)^2}{(\zeta^{jk} - 1)(x - \zeta^j)} + \frac{1}{k} \sum_{0 < j < k} \frac{\omega^j(\omega^j - 1)^2}{(\omega^{jh} - 1)(x - \omega^j)} = \frac{(x - 1)^2}{(x^h - 1)(x^k - 1)}.$$
$$(29)$$

This identity has many interesting consequences, and it leads to numerous reciprocity formulas for sums of the type given in Eq. (26). For example, if we differentiate (29) twice with respect to x and set $x = 1$, we find that

$$\frac{2}{h} \sum_{0 < j < h} \frac{\zeta^j(\zeta^j - 1)^2}{(\zeta^{jk} - 1)(1 - \zeta^j)^3} + \frac{2}{k} \sum_{0 < j < k} \frac{\omega^j(\omega^j - 1)^2}{(\omega^{jk} - 1)(1 - \omega^j)^3}$$
$$= \frac{1}{6}\left(\frac{h}{k} + \frac{k}{h} + \frac{1}{hk}\right) + \frac{1}{2} - \frac{1}{2h} - \frac{1}{2k}.$$

Replace j by $h - j$ and by $k - j$ in these sums to get [cf. (26)]

$$\frac{1}{6}\left(\sigma\,(k, h, 0) + \frac{3(h - 1)}{h}\right) + \frac{1}{6}\left(\sigma(h, k, 0) + \frac{3(k - 1)}{k}\right)$$
$$= \frac{1}{6}\left(\frac{h}{k} + \frac{k}{h} + \frac{1}{hk}\right) + \frac{1}{2} - \frac{1}{2h} - \frac{1}{2k},$$

which is equivalent to the desired result. ∎

Lemma C. *If h is relatively prime to k, and if $0 < c < k$, then*

$$\sigma(h, k, c) = \sigma(h, k, c \bmod h) + \frac{6r}{k}\,(c + c \bmod h - k) + d, \qquad (30)$$

where $r = \lfloor c/h \rfloor$, $d = 0$ *if* $c \bmod h \neq 0$, *and* $d = 3$ *if* $c \bmod h = 0$.

Proof. For any value of c, we have

$$\sigma(h, k, c + h) = 12 \sum_{0 \le j < k} \left(\!\left(\frac{j}{k}\right)\!\right)\left(\!\left(\frac{hj + c + h}{k}\right)\!\right)$$

$$= 12 \sum_{0 \le j < k} \left(\!\left(\frac{j - 1}{k}\right)\!\right)\left(\!\left(\frac{hj + c}{k}\right)\!\right)$$

$$= \sigma(h, k, c) + 6\left(\!\left(\frac{h + c}{k}\right)\!\right) + 6\left(\!\left(\frac{c}{k}\right)\!\right), \qquad (31)$$

since (9) implies that

$$\left(\!\left(\frac{j - 1}{k}\right)\!\right) = \left(\!\left(\frac{j}{k}\right)\!\right) - \frac{1}{k} + \frac{1}{2}\delta\left(\frac{j + 1}{k}\right) + \frac{1}{2}\delta\left(\frac{j}{k}\right). \qquad (32)$$

Now let $c_0 = c \bmod h$, so that $0 < c = c_0 + rh < k$. We have

$$\sigma(h, k, c) = \sigma(h, k, c_0) + 12 \sum_{0 \le j < r} \left(\!\left(\frac{c_0 + hj}{k}\right)\!\right) + 6\left(\!\left(\frac{c}{k}\right)\!\right) - 6\left(\!\left(\frac{c_0}{k}\right)\!\right)$$

$$= \sigma(h, k, c_0) + 12 \sum_{0 \le j < r} \left(\frac{c_0 + hj}{k} - \frac{1}{2}\right) + 6\left(\frac{c}{k} - \frac{1}{2}\right)$$

$$- 6\left(\frac{c_0}{k} - \frac{1}{2}\right) + d.$$

A simple summation completes the proof. ∎

     The results of Lemmas B and C give us an efficient procedure for the evaluation of $\sigma(h, k, c)$ when h is relatively prime to k:

Step 1. Add or subtract multiples of k from h and c, if necessary, to reduce them to the range $-k/2 < h \le k/2$, $-k/2 < c \le k/2$. (Alternatively, the range $0 < h \le k$, $0 \le c < k$ will suffice.) This operation is justified because it is clear from the definition that

$$\sigma(h \pm nk, k, c) = \sigma(h, k, c) = \sigma(h, k, c \pm nk). \qquad (33)$$

Step 2. If h or c is negative, change them to positive using the identities

$$\begin{aligned} \sigma(-h, k, c) &= -\sigma(h, k, c), \\ \sigma(h, k, -c) &= \sigma(h, k, c). \end{aligned} \qquad (34)$$

Step 3. If $k = 1$ or $k = 2$, $\sigma(h, k, c) = 0$ (because $((j/1)) = ((j/2)) = 0$).

Step 4. If $c > h$, reduce c to $c \bmod h$ by applying formula (30) of Lemma C.

Step 5. Now the conditions in Lemma B apply, so we have

$$\sigma(h, k, c) = -3 + \frac{h}{k} + \frac{k}{h} + \frac{1 + 6c^2}{hk} - \sigma(k, h, c). \tag{35}$$

To evaluate $\sigma(k, h, c)$, return to step 1.

The above process does not require many iterations of these steps; in fact, the successive values of h and k follow essentially the sequence of values that would be obtained if Euclid's algorithm were being applied to determine the greatest common divisor of h and k (see Section 4.5.2). We shall give several examples of the procedure.

Example 1. *Find the serial correlation when* $m = 2^{35}$, $a = 2^{34} + 1$, $c = 1$.

Solution. We have

$$C = (2^{35}\sigma(2^{34} + 1, 2^{35}, 1) - 3 + 6(2^{35} - (2^{34} - 1) - 1))/(2^{70} - 1) \tag{36}$$

by Eq. (17). By steps 1 and 2,

$$\sigma(2^{34} + 1, 2^{35}, 1) = -\sigma(2^{34} - 1, 2^{35}, 1).$$

By step 5,

$$\sigma(2^{34} - 1, 2^{35}, 1) = -3 + (2^{34} - 1)/2^{35} + 2^{35}/(2^{34} - 1)$$
$$+ 7/2^{35}(2^{34} - 1) - \sigma(2^{35}, 2^{34} - 1, 1).$$

By step 1,

$$\sigma(2^{35}, 2^{34} - 1, 1) = \sigma(2, 2^{34} - 1, 1).$$

Step 5 now gives

$$\sigma(2, 2^{34} - 1, 1) = -3 + 2/(2^{34} - 1) + (2^{34} - 1)/2$$
$$+ 7/2(2^{34} - 1) - \sigma(2^{34} - 1, 2, 1);$$

and

$$\sigma(2^{34} - 1, 2, 1) = 0.$$

Putting everything together, we find that

$$C = \tfrac{1}{4} + \epsilon, \qquad |\epsilon| < 2^{-67}. \tag{37}$$

Such a correlation is much, much too high for randomness. Of course, this generator has very low potency, and we have already rejected it as nonrandom.

Example 2. *Find the approximate serial correlation when* $m = 10^{10}$, $a = 10001$, $c = 2113248653$.

Solution. We have $C \approx \sigma(a, m, c)/m$, and the computation proceeds as follows:

$$\sigma(10001, 10^{10}, 2113248653) = \sigma(10001, 10^{10}, 7350) \\ - 6(211303)(7886743997)/10^{10};$$

$$\sigma(10001, 10^{10}, 7350) \approx -3 + 10^{10}/10001 - \sigma(10^{10}, 10001, 7350);$$

$$\sigma(10^{10}, 10001, 7350) = \sigma(100, 10001, 7350) \\ = \sigma(100, 10001, 50) - 6(73)(2601)/10001;$$

$$\sigma(100, 10001, 50) \approx -3 + 10001/100 \\ + 100/10001 - \sigma(10001, 100, 50); \qquad (38)$$

$$\sigma(10001, 100, 50) = (1, 100, 50) = -50.02.$$

$$C \approx (-3 + 999900.01 - 97.02 - 50.02 \\ + 113.91 - 99895.60)/10^{10} \\ = -0.000000003172.$$

This is a very respectable value of C indeed. But the generator has a potency of only 3, *so it is not really a very good source of random numbers in spite of the fact that it has low serial correlation.* It is necessary to have a low serial correlation, but not sufficient!

Example 3. *Estimate the serial correlation for general a, m, and c.* We can go through the first phase of the calculations above in the general case; let $c_0 = c \bmod a$.

$$\sigma(a, m, c) = \sigma(a, m, c_0) + \frac{6(c - c_0)}{am}(c + c_0 - m)$$

$$= -3 + \frac{a}{m} + \frac{m}{a} + \frac{1}{am} + \frac{6c^2}{am} - \frac{6(c - c_0)}{a} - \sigma(m, a, c_0). \qquad (39)$$

Now $|\sigma(m, a, c_0)| < a$, by exercise 12, and therefore

$$C \approx \frac{\sigma(a, m, c)}{m} \approx \frac{1}{a}\left(1 - 6\frac{c}{m} + 6\left(\frac{c}{m}\right)^2\right). \qquad (40)$$

The discarded terms in this approximation are

$$\frac{3}{m}\left(1 - 2\frac{c_0}{a} + \frac{a - x' - c}{m}\right) + \frac{1}{m}\sigma(m, a, c_0) + O\left(\frac{1}{m^2}\right),$$

so we may say the error in (40) is essentially bounded by a/m. The main contribution to the error in this approximation is $\sigma(m, a, c_0)/m$. Our results may be summarized as follows:

Theorem S. *The serial correlation for a linear congruential sequence with maximum period is given by the approximate relation* (40), *with an error of less than* $(a + 6)/m$. *The exact value may be efficiently computed from Eq.* (17) *and the use of Lemmas B and C.* ∎

Relation (40) has several noteworthy consequences. First it shows that small values of a are to be avoided, lest the serial correlation be high. On the other hand, a high value of a does not necessarily guarantee a low correlation, as shown in Example 1 above; the error in (40) may be as high as a/m, so that approximation is useless when a/m is large. If $a \approx \sqrt{m}$, the serial correlation will always be bounded by $2/\sqrt{m}$.

Relation (40) also gives us some advice about the proper choice of c. We have so far obtained no criterion for choosing c except that it be relatively prime to m. If we choose c so that, in addition to being relatively prime to m, we have

$$\frac{c}{m} \approx \frac{1}{2} - \frac{1}{6}\sqrt{3} \approx 0.21132\ 48654\ 05187\ 1$$
$$\approx (0.15414\ 54272\ 33746\ 34354\ 55716)_8, \tag{41}$$

then we obtain a very low value for the serial correlation, since the roots of the equation $1 - 6x + 6x^2 = 0$ are $\frac{1}{2} \pm \frac{1}{6}\sqrt{3}$. In the absence of any other criterion for choosing c, we might as well use this one.

The above formulas give us the correlation coefficient between X_n and X_{n+1}. It is also desirable to have low correlation between X_n and X_{n+2}, and, in general, we usually want the correlation between X_n and X_{n+t} to be low for, say, $1 \le t \le 10$. We have shown in Eq. 3.2.1-6 that

$$X_{n+t} = (a_t X_n + c_t) \bmod m, \tag{42}$$

where

$$a_t = a^t \bmod m, \qquad c_t = (a^t - 1)c/(a - 1) \bmod m. \tag{43}$$

We can compute the correlation between X_n and X_{n+t} by using a_t and c_t in the formulas above, in place of a and c. Of course, c_t will no longer satisfy condition (41), but we can't have everything.

The approximate formula which appears in Eq. (40) was first obtained by R. R. Coveyou, *JACM* **7** (1960), 72–74, by averaging over all *real numbers* x between 0 and m instead of only considering the integer values. (See exercise 21.) Methods to compute the exact formula were later given by M. Greenberger, *Math. Comp.* **15** (1961), 383–389; and extended by B. Jansson, *BIT* **4** (1964), 6–27. Their formulas do not explicitly use Dedekind sums. Jansson gave some tables of the serial correlation, but unfortunately he considered multipliers which have a form too simple to be recommended. He stated, for example, that the correlation between X_n and X_{n+t} is less than 0.000003 for the generator with $m = 2^{35}$, $a = 2^{24} + 5$, $c = 1$, for all $t \le 2500$. The spectral test (Section 3.3.4) tells us to refrain from using this particular generator, but we may consider Jansson's result as good evidence that randomly chosen linear congruential generators with high potency will tend to have low serial correlations. (*Note.* Jansson also has derived formulas for the serial correlation in sequences with

$c = 0$ and a multiplier giving the maximum period, when $m = 2^e$. These results are essentially the same as the formulas we have discussed, under the correspondence given in exercise 3.2.1.2–9.)

Dedekind sums $\sigma(h, k, c)$ and the reciprocity law, in the special case $c = 0$, were first considered by R. Dedekind as part of his work on elliptic functions in 1892. The function has been considered by many other authors, and a bibliography may be obtained by consulting the papers by Ulrich Dieter [*Journal für die reine und angewandte Mathematik* **201** (1959), 37–70] and by H. Rademacher and A. Whiteman [*American Journal of Mathematics* **63** (1941), 377–407].

The exercises below develop some other *a priori* tests. The principal conclusion to be drawn in each case is that *the multiplier in a linear congruential sequence should be reasonably large.* See also exercise 3.3.4–7 for an extension of Theorem P.

EXERCISES—First Set

1. [*M07*] Explain why (7) is the number of solutions to the relations $s(x) < x$, $0 \le x < m$.

2. [*M24*] Prove the "replicative law," Eq. (12).

3. [*HM22*] What is the Fourier series expansion (in terms of sines and cosines) of the function $f(x) = ((x))$?

▶ **4.** [*M18*] If $m = 10^{10}$, what is the highest possible value of d (in the notation of Theorem P), given that the potency of the generator is 10?

5. [*M21*] Carry out the derivation of Eq. (17).

6. [*M27*] Let $hh' + kk' = 1$. (a) Show, without using Lemmas B or C, that

$$\sigma(h, k, c) = \sigma(h, k, 0) + 12 \sum_{0 \le j < c}' \left(\left(\frac{h'j}{k}\right)\right) + 6\left(\left(\frac{h'c}{k}\right)\right)$$

for all $c \ge 0$. (b) Show that if $0 \le c < h$ and $0 \le c < k$,

$$\left(\left(\frac{h'c}{k}\right)\right) + \left(\left(\frac{k'c}{h}\right)\right) = \frac{c}{hk}.$$

(c) Under the assumptions of Lemma B, prove Eq. (20).

▶ **7.** [*M24*] Give a proof of the reciprocity law (19), when $c = 0$, by using the ideas expressed in exercise 1.2.4–45.

▶ **8.** [*M34*] Let

$$\rho(p, q, r) = 12 \sum_{0 \le j < r} \left(\left(\frac{jp}{r}\right)\right)\left(\left(\frac{jq}{r}\right)\right).$$

By generalizing the method of proof used in Lemma B, prove the following beautiful identity due to Rademacher: "If each of $p, \dot{q}, r$ is relatively prime to the other two,

$$\rho(p, q, r) + \rho(q, r, p) + \rho(r, p, q) = \frac{p}{qr} + \frac{q}{rp} + \frac{r}{pq} - 3."$$

(The reciprocity law for Dedekind sums, with $c = 0$, is the special case $r = 1$.)

9. [*M40*] Is there a simple proof of Rademacher's identity (exercise 8) along the lines of the proof in exercise 7 of a special case?

10. [*M20*] Prove the identities (34) for changing the sign of the parameters of $\sigma(h, k, c)$.

11. [*M30*] The formulas given in the text show us how to evaluate $\sigma(h, k, c)$ when h and k are relatively prime. For the general case, let $hh' \equiv d$ (modulo k), where d is the greatest common divisor of h and k. Show that

$$\sigma(h, k, c) = \sigma(h/d, k/d, \lfloor c/d \rfloor) + \delta,$$

where $\delta = 0$ if $c \bmod d = 0$,

$$\delta = 6\left(\left(\frac{\lfloor c/d \rfloor h' d}{k}\right)\right)$$

if $c \bmod d \neq 0$.

12. [*M24*] Show that if h is relatively prime to k, $|\sigma(h, k, c)| < (k - 1)(k - 2)/k$.

13. [*M22*] Equation (37) gives an *approximate* answer to the value of the serial correlation for $m = 2^{35}$, $a = 2^{34} + 1$, $c = 1$. What is the *exact* answer?

▶ **14.** [*M24*] The linear congruential generator which has $m = 2^{35}$, $a = 2^{18} + 1$, $c = 1$, was given the serial correlation test on three batches of 1000 consecutive numbers, and the result was a very high correlation, between 0.2 and 0.3, in each case. What is the serial correlation of this generator, taken over all 2^{35} numbers of the period?

15. [*M22*] Find the exact value of the serial correlation coefficient for the special case $a = 1$.

16. [*M24*] We know the linear congruential sequence is a poor source of random numbers when $a = 1$; but show that according to Theorem P we can choose c so that the probability $(X_{n+1} < X_n)$ is very nearly equal to $\frac{1}{2}$, even if $a = 1$. Similarly, we can choose c so that the serial correlation coefficient is very low. Can a single value of c be found for which both of these conditions hold simultaneously?

17. [*M21*] Explain how a value of a can be chosen so that both c and c_2 will satisfy the approximate relation (41), where c_2 is the increment for the sequence of alternate elements, given in (43).

▶ **18.** [*M35*] Let

$$S(h, k, c, z) = \sum_{0 \leq j < z} \left(\left(\frac{hj + c}{k}\right)\right).$$

Develop "reciprocity formulas" for the efficient evaluation of this function, when h is relatively prime to k, just as Lemmas B and C provide for the efficient evaluation of $\sigma(h, k, c)$. [*Hint:* Study exercise 6.]

▶ **19.** [*M35*] The *serial test* described in the preceding section can be analyzed over the whole period: Let α, β, γ, δ be integers, with $0 \leq \alpha < \beta \leq m$, $0 \leq \gamma < \delta \leq m$; derive an equation for the probability that $\alpha \leq X_n < \beta$ and $\gamma \leq X_{n+1} < \delta$.

20. [*M24*] Generalize Eq. (26) so that it gives an expression for $\sigma(h, k, c)$.

EXERCISES—Second Set

In many cases, exact computations with integers are too difficult to carry out, but we can attempt to study the probabilities which arise when we take the average over all real values of x instead of restricting the calculation to integer values. Although these results are only approximate, they shed some light on the subject.

It is convenient to deal with numbers U_n between zero and one; for linear congruential sequences, $U_n = X_n/m$, and we have $U_{n+1} = \{aU_n + \theta\}$, where $\theta = c/m$ and $\{x\}$ denotes x mod 1. For example, the formula for serial correlation now becomes

$$ C = \left(\int_0^1 x\{ax + \theta\}\, dx - \left(\int_0^1 x\, dx \right)^2 \right) \bigg/ \left(\int_0^1 x^2\, dx - \left(\int_0^1 x\, dx \right)^2 \right). $$

▶ **21.** [*HM23*] (R. R. Coveyou.) What is the value of C in the formula just given?

▶ **22.** [*M22*] Let a be an integer, and let $0 \leq \theta < 1$. If x is a real number between 0 and 1, and if $s(x) = \{ax + \theta\}$, what is the probability that $s(x) < x$? (This is the "real number" analog of Theorem P.)

23. [*M28*] The previous exercise gives the probability that $U_{n+1} < U_n$. What is the probability that $U_{n+2} < U_{n+1} < U_n$, assuming that U_n is a random real number between zero and one?

24. [*M29*] Under the assumptions of the preceding problem, except with $\theta = 0$, show that the probability $U_n > U_{n+1} > \cdots > U_{n+t-1}$ is exactly equal to

$$ \frac{1}{t!}\left(1 + \frac{1}{a} \right) \cdots \left(1 + \frac{t-2}{a} \right). $$

What is the average length of a descending run starting at U_n, assuming that U_n is selected at random between zero and one?

▶ **25.** [*M25*] Let α, β, γ, δ be real numbers with $0 \leq \alpha < \beta \leq 1$, $0 \leq \gamma < \delta \leq 1$. Under the assumptions of exercise 22, what is the probability that $\alpha \leq x < \beta$ and $\gamma \leq s(x) < \delta$? (This is the "real number" analog of exercise 19.)

26. [*M21*] Consider a "Fibonacci" generator, where $U_{n+1} = \{U_n + U_{n-1}\}$. Assuming that U_1 and U_2 are independently chosen at random between 0 and 1, find the probability that $U_1 < U_2 < U_3$, $U_1 < U_3 < U_2$, $U_2 < U_1 < U_3$, etc. [*Hint:* Divide the "unit square," i.e., the points of the plane $\{(x, y)|0 \leq x, y < 1\}$, into six parts, depending on the relative order of x, y, and $\{x + y\}$, and determine the area of each part.]

27. [*M32*] In the Fibonacci generator of the preceding exercise, let U_0 and U_1 be chosen independently in the unit square except that $U_0 > U_1$. Determine the probability that U_1 is the beginning of an upward run of length k, so that $U_0 > U_1 < \cdots < U_k > U_{k+1}$. Compare this with the corresponding probabilities for a random sequence.

▶ **28.** [*M35*] A linear congruential generator with potency 2 satisfies the condition $X_{n-1} - 2X_n + X_{n+1} \equiv (a-1)c$ (modulo m). [Cf. Eq. 3.2.1.3-5.] Consider a generator which abstracts this situation: let $U_{n+1} = \{\alpha + 2U_n - U_{n-1}\}$. As in exercise 26, divide the unit square into parts which show for each pair (U_{n-1}, U_n) what the relative order of U_{n-1}, U_n, and U_{n+1} is. Are there any values of α for which all six possible orders are achieved with probability $\frac{1}{6}$, assuming that U_{n-1} and U_n are chosen at random in the unit square?

3.3.4. The Spectral Test

An important test for the randomness of computer-generated sequences was formulated in 1965 by R. R. Coveyou and R. D. MacPherson; this test is especially significant because not only do all good random-number generators pass it, but also all linear congruential sequences now known to be bad actually *fail* it! Thus it is by far the most powerful test known, and it deserves particular attention.

The spectral test embodies aspects of both the "empirical" and "theoretical" tests studied in previous sections: it is like the theoretical tests because it considers quantities averaged over the full period, and it is like the empirical tests because it requires a computer program to determine the results.

A proper understanding of the theory behind this test involves a fair amount of mathematics, so a reader who is not mathematically inclined is advised to skip to part D at the end of this section where the spectral test is presented as a "plug-in" method.

A. Theory Behind the Test. Mathematical motivation for this test is based on the "finite Fourier transform" of a function defined on a finite set. The one-dimensional case of finite Fourier transforms was essentially used in the previous section in the proof of Lemma 3.3.3B; let us now consider the Fourier transform technique in general.

Given that $F(t_1, t_2, \ldots, t_n)$ is any complex-valued function defined for all combinations of integers t_k, where $0 \le t_k < m$ for $1 \le k \le n$, define the *Fourier transform* of F by the following rule:

$$f(s_1, \ldots, s_n) = \sum_{0 \le t_1, \ldots, t_n < m} \exp\left(\frac{-2\pi i}{m}(s_1 t_1 + \cdots + s_n t_n)\right) F(t_1, \ldots, t_n). \quad (1)$$

This function f is defined for all combinations of integers s_k; it is periodic, in the sense that $f(s_1, \ldots, s_n) = f(s_1 \bmod m, \ldots, s_n \bmod m)$. To connect this definition up with the formulas in the preceding section, note that

$$\exp\left(\frac{-2\pi i}{m}(s_1 t_1 + s_2 t_2 + \cdots + s_n t_n)\right) = \omega^{-(s_1 t_1 + s_2 t_2 + \cdots + s_n t_n)},$$

if ω is the mth root of unity $e^{2\pi i/m}$. The name "transform" is justified here since the original function $F(t_1, \ldots, t_n)$ can be reconstructed from its transform

$f(s_1, \ldots, s_n)$ as follows (see exercise 1):

$$F(t_1, \ldots, t_n) = \frac{1}{m^n} \sum_{0 \le s_1, \ldots, s_n < m} \exp\left(\frac{2\pi i}{m}(s_1 t_1 + \cdots + s_n t_n)\right) f(s_1, \ldots, s_n). \tag{2}$$

This formula can be expanded into sines and cosines so that it resembles an infinite Fourier series. The value $(1/m^n)f(s_1, \ldots, s_n)$ essentially represents the amplitude of an n-dimensional complex plane wave with frequencies $s_1/m, \ldots, s_n/m$, if $F(t_1, \ldots, t_n)$ is written as a superposition of such waves.

As a consequence of relations (1) and (2), it is possible in theory to determine any property of F from its transform f and conversely; and it is often convenient to transform a function, work with the transform, and then "untransform" the result, to deduce nonobvious properties of the original function.

Our purpose in this section is to apply this concept to random-number generation. Suppose that $X_0, X_1, X_2, \ldots$ is an infinite sequence of integers with $0 \le X_k < m$, and let n be a fixed (usually rather small) positive integer. Define

$$F(t_1, \ldots, t_n) = \lim_{N \to \infty} \frac{1}{N} \sum_{0 \le k < N} \delta_{X_k t_1} \delta_{X_{k+1} t_2} \cdots \delta_{X_{k+n-1} t_n}, \tag{3}$$

that is, $F(t_1, \ldots, t_n)$ is the limiting density of the number of appearances of the n-tuple $(t_1, \ldots, t_n)$ as n consecutive elements of the sequence $X_0, X_1, X_2, \ldots$. Since all sequences $X_0, X_1, X_2, \ldots$ of interest to us in this discussion are periodic, we may assume the limit in (3) exists, and, in fact, we may set N equal to the period length. In a truly random sequence for the uniform distribution, each possible n-tuple should appear equally often, so $F(t_1, \ldots, t_n)$ should be $1/m^n$ for all $t_1, \ldots, t_n$.

The Fourier transform of (3) has the simple form

$f(s_1, \ldots, s_n)$

$$= \lim_{N \to \infty} \frac{1}{N} \sum_{0 \le k < N} \exp\left(\frac{-2\pi i}{m}(s_1 X_k + s_2 X_{k+1} + \cdots + s_n X_{k+n-1})\right). \tag{4}$$

In a truly random sequence, this should be the transform of the constant function $1/m^n$; so *in a random sequence we should have*

$$f(s_1, \ldots, s_n) = \begin{cases} 1, & \text{if } s_1 \equiv \cdots \equiv s_n \equiv 0 \text{ (modulo } m\text{);} \\ 0, & \text{otherwise.} \end{cases} \tag{5}$$

The theoretical tests studied in the preceding section consist of averages taken over the full period. Any such test, which finds the average of some function depending only on n consecutive values of the sequence, can, in principle, be completely determined from the values of $F(t_1, \ldots, t_n)$. For example, the probability that $X_k < X_{k+1}$ is the quantity $F(t_1, t_2)$ summed over the values $0 \le t_1 < t_2 < m$; likewise, the serial correlation coefficient can be determined from the case $n = 2$ (see exercise 5).

Similarly, *any* theoretical test can in principle be determined from the transformed function $f(s_1, \ldots, s_n)$ in (4), since this function carries the same information as $F(t_1, \ldots, t_n)$ does. Therefore there is good reason to believe that the deviation of $f(s_1, \ldots, s_n)$ from the values (5) corresponding to a truly random sequence will be a useful test for randomness.

For linear congruential sequences, $f(s_1, \ldots, s_n)$ has a very simple form, while $F(t_1, \ldots, t_n)$ is not so convenient. We shall consider the case of a linear congruential sequence, defined by a, m, c, and X_0, having the *maximum period length* in accordance with Theorem 3.2.1.2A. For this sequence,

$$f(s_1, \ldots, s_n) = \frac{1}{m} \sum_{0 \le k < m} \exp\left(\frac{-2\pi i}{m} (s_1 X_k + s_2 X_{k+1} + \cdots + s_n X_{k+n-1}) \right)$$

$$= \frac{1}{m} \sum_{0 \le k < m} \exp\left(\frac{-2\pi i}{m} \left(s(a) X_k + \frac{s(a) - s(1)}{a - 1} c \right) \right), \tag{6}$$

where

$$s(a) = s_1 + s_2 a + s_3 a^2 + \cdots + s_n a^{n-1}, \tag{7}$$

since

$$X_{k+r} \equiv a^r X_k + \frac{a^r - 1}{a - 1} c \pmod{m}$$

by Eq. 3.2.1–(6). We are assuming that the sequence has maximum period, so that all values X_k occur; therefore (6) reduces to

$$\frac{1}{m} \sum_{0 \le k < m} \exp\left(\frac{-2\pi i}{m} \left(s(a)k + \frac{s(a) - s(1)}{a - 1} c \right) \right).$$

This is the sum of a geometric series, so we get the basic formula

$$f(s_1, \ldots, s_n) = \exp\left(\frac{-2\pi i c}{m} \left(\frac{s(a) - s(1)}{a - 1} \right) \right) \delta\left(\frac{s(a)}{m} \right), \tag{8}$$

where $\delta(x)$ is 1 if x is an integer, 0 otherwise.

Recall that formula (2) allows us to interpret $f(s_1, \ldots, s_n)/m^n$ physically as the amplitude of the n-dimensional complex plane wave

$$\omega(t_1, \ldots, t_n) = \exp\left(2\pi i \left(\frac{s_1}{m} t_1 + \cdots + \frac{s_n}{m} t_n \right) \right). \tag{9}$$

By convention this wave may be assigned a "wave number" ν corresponding to its "frequency", where

$$\nu = \sqrt{s_1^2 + \cdots + s_n^2}, \quad \text{when} \quad |s_k| \le \frac{m}{2} \quad \text{for} \quad 1 \le k < n. \tag{10}$$

According to Eq. (5), no waves except the constant wave (frequency zero) should appear if the sequence $X_0, X_1, X_2, \ldots$ is truly random. So a sequence which leads to wave components with nonzero frequency will be nonrandom. A very small value of $f(s_1, \ldots, s_n)$ has little effect on the randomness. For example, if we take a truly random sequence and alter only every Nth term (where N is large), the sequence remains quite random and we get values of $f(s_1, \ldots, s_n)$ whose magnitude is only of order $1/N$; such values are negligible. But note that in Eq. (8), the magnitude of $f(s_1, \ldots, s_n)$ is either zero or one, and when $|f(s_1, \ldots, s_n)| = 1$ this can have devastating effects on the randomness. In this regard it is interesting to note that *low-frequency components are more damaging to randomness than high-frequency components are.* Consider, for example, what happens if we replace X_k by $2X_k$ and m by $2m$; Eq. (4) shows that $f(s_1, \ldots, s_n)$ stays unchanged, but now components with $|s_k| \leq m$ instead of $|s_k| \leq m/2$ are significant. A study of this situation will show the reader that if we take a truly random sequence of integers $X_0, X_1, X_2, \ldots$ for $m = 2^e$, and if we truncate this sequence by setting the least significant bits of its elements to zero (in binary notation), we obtain nonzero frequency components with wave numbers 2^{e-1}, 2^{e-2}, 2^{e-3}, etc., when one, two, three, $\ldots$ bits are set zero. This fact and a further example worked in exercise 10 lead to the following intuitive interpretation: *If ν_n is the smallest nonzero value of the wave number* (10) *for which* $f(s_1, \ldots, s_n) \neq 0$ *in a linear congruential sequence with maximum period, then the sequence* $X_0/m, X_1/m, X_2/m, \ldots$ *represents a sequence of random numbers uniformly distributed between 0 and 1, having* "accuracy" *or* "truncation error" $1/\nu_n$, *with respect to the independence of n consecutive values of the sequence averaged over the entire period.* Exercise 27 gives further confirmation of this principle, together with a geometrical interpretation which reveals the significance of ν_n more clearly.

Equation (8) above gives us the "spectrum" of the linear congruential sequence, i.e., it shows us which waves appear in the Fourier transform of $F(t_1, \ldots, t_n)$. We see that $f(s_1, \ldots, s_n) = 0$ except when

$$s_1 + s_2 a + s_3 a^2 + \cdots + s_n a^{n-1} \equiv 0 \pmod{m}, \qquad (11)$$

and in this case $|f(s_1, \ldots, s_n)| = 1$. Therefore *for linear congruential sequences of maximum period, the smallest nonzero wave number in the spectrum is given by*

$$\nu_n = \min \sqrt{s_1^2 + s_2^2 + \cdots + s_n^2}, \qquad (12)$$

where the minimum is taken over all n-tuples of integers $(s_1, s_2, \ldots, s_n) \neq (0, 0, \ldots, 0)$ *satisfying* (11). Note that this condition is independent of the increment, c, of the linear congruential sequence.

One consequence of this discussion is that we can place upper limits on the possible randomness of any linear congruential sequence. For it is possible to show, using rather deep number-theoretical methods, that

$$\nu_n \leq \gamma_n m^{1/n}, \qquad (13)$$

where γ_n takes the respective values

$$1,\ (4/3)^{1/4},\ 2^{1/6},\ 2^{1/4},\ 2^{3/10},\ (64/3)^{1/12},\ 2^{3/7},\ 2^{1/2}$$

for $n = 1, 2, 3, 4, 5, 6, 7, 8$. [See exercise 9 and J. W. S. Cassels, *An Introduction to the Geometry of Numbers*, Springer, Berlin (1959), p. 332.] Actually a bound that behaves like $m^{1/n}$ is exactly what we would expect for any periodic sequence $U_0, U_1, U_2, \ldots$ of numbers between 0 and 1 whose period length is m: for the statement above that a sequence has accuracy $1/\nu$, with respect to the independence of n consecutive values, means roughly (when ν is an integer) that each of the ν^n possible values of $(\lfloor \nu U_{k+1} \rfloor, \lfloor \nu U_{k+2} \rfloor, \ldots, \lfloor \nu U_{k+n} \rfloor)$ should occur about equally often in the period; hence we get the approximate inequality $\nu^n \leq m$ on intuitive grounds.

B. Examples of the Test. An example will make the above discussion clearer. Suppose that we take the linear congruential sequence with

$$X_0 = 0, \qquad a = 3141592621, \qquad c = 1, \qquad m = 10^{10}. \qquad (14)$$

The minimum nonzero value of $s_1^2 + s_2^2$ for which

$$s_1 + 3141592621 s_2 \equiv 0 \ (\text{modulo } 10^{10})$$

occurs for $s_1 = -67654,\ s_2 = 226$, so we find

$$\nu_2 = \sqrt{67654^2 + 226^2} \approx 67654.4.$$

This means if we want the sequence

$$U_0, U_1, U_2, \ldots = X_0/m, X_1/m, X_2/m, \ldots$$

to represent random real numbers between 0 and 1 with adjacent pairs (U_k, U_{k+1}) essentially independent, we have an accuracy of about $1/67654$ when the whole period is considered; i.e., the most significant 16 bits in binary notation may be considered random in this sense. Similarly, the minimum nonzero value of $s_1^2 + s_2^2 + s_3^2$ for which

$$s_1 + 3141592621 s_2 + 3141592621^2 s_3 \equiv 0 \ (\text{modulo } 10^{10})$$

occurs for $s_1 = 227,\ s_2 = 983,\ s_3 = 130$; hence

$$\nu_3 = \sqrt{1034718} \approx 1017.2.$$

Thus when the independence of consecutive triples (U_k, U_{k+1}, U_{k+2}) is considered, we have only about 10-bit accuracy. Perhaps 3141592621 is not a very good multiplier; we will see later that it is passable but not the best (for example, the similar multiplier 3141592821 gives $\nu_3 \approx 1912$, but it has a smaller value of ν_2). By (13) we will never get $\nu_3 > 2425$ in any event, when $m = 10^{10}$.

The minimum nonzero value of $s_1^2 + s_2^2 + s_3^2 + s_4^2$ for which

$$s_1 + 3141592621 s_2 + 3141592621^2 s_3 + 3141592621^3 s_4 \equiv 0 \ (\text{modulo } 10^{10})$$

occurs for $s_1 = 52$, $s_2 = -203$, $s_3 = -54$, $s_4 = 125$, so

$$\nu_4 = \sqrt{62454} \approx 249.9.$$

We are now reduced to eight-bit accuracy (which is really quite good for most applications) with respect to independence of successive *quadruples*.

The values of ν_n for $n = 5, 6, \ldots$ are of less importance than those for $n = 1, 2, 3, 4$, since complete independence of quintuples is perhaps asking for too much randomness. For example, the serial test in Section 3.3.2 is rarely applied even to quadruples. (When the average is taken over the full period as we are doing here, it is a good policy to be rather cautious, so quadruples should not be neglected in the spectral test; but the quintuple distribution would not seem to be so important unless m is about 2^{40} or larger.) For this generator it turns out that $s_1 = -8$, $s_2 = -14$, $s_3 = 6$, $s_4 = -18$, $s_5 = 34$ gives

$$\nu_5 = \sqrt{1776} \approx 42.2;$$

we also find $\nu_6 = \sqrt{542} \approx 23.3$.

Since nobody knows what the best achievable values of ν_n are, it is difficult to decide exactly how to judge if certain values of ν_n are satisfactory for the randomness of the sequence. It seems appropriate to give, as a measure of the randomness, the volume of the ellipse in n-space defined by the relation $(x_1 m - x_2 a - x_3 a^2 - \cdots - x_n a^{n-1})^2 + x_2^2 + \cdots + x_n^2 \le \nu_n^2$, since the volume of this ellipse serves as an indication of the probability that *integer* points $(x_1, x_2, \ldots, x_n)$—corresponding to a solution of (11)—are in the ellipse. We therefore propose to calculate the quantities

$$C_n = \frac{\pi^{n/2} \nu_n^n}{(n/2)! \, m}, \tag{15}$$

in order to rate the effectiveness of the multiplier a in a linear congruential sequence of maximum period. (In this formula,

$$\left(\frac{n}{2}\right)! = \left(\frac{n}{2}\right)\left(\frac{n}{2} - 1\right) \cdots \left(\frac{1}{2}\right) \sqrt{\pi}, \qquad \text{for } n \text{ odd.}) \tag{16}$$

Thus

$$C_1 = 2\nu_1/m, \qquad C_2 = \pi \nu_2^2/m, \qquad C_3 = \tfrac{4}{3}\pi \nu_3^3/m, \qquad C_4 = \tfrac{1}{2}\pi^2 \nu_4^4/m, \quad \text{etc.}$$

Large values of C_n correspond to randomness, small values correspond to nonrandomness. Table 1 shows what sort of values occur in typical sequences. (C_1 is always equal to 2.) Lines 1 through 4 of Table 1 show generators that were the subject of Figs. 2 and 5 in Section 3.3.1. The generators in lines 1 and 2

Table 1

SAMPLE RESULTS OF THE SPECTRAL TEST

Line	a	m	C_2	C_3	C_4
1	23	$10^8 + 1$	0.000017	0.00051	0.014
2	$2^7 + 1$	2^{35}	0.000002	0.00026	0.040
3	$2^{18} + 1$	2^{35}	3.14	0.000000002	0.000000003
4	3141592653	2^{35}	0.27	0.13	0.11
5	3141592221	10^{10}	1.35	0.06	4.69
6	3141592421	10^{10}	2.68	0.34	0.54
7	3141592621	10^{10}	1.44	0.44	1.92
8	3141592821	10^{10}	0.16	2.93	0.17
9	3141592221	2^{35}	1.24	1.70	1.12
10	3141592621	2^{35}	3.02	0.17	1.26
11	2718281821	2^{35}	2.59	1.16	1.75
12	$2^{23} + 2^{12} + 5$	2^{35}	0.015	2.78	0.066
13	$2^{23} + 2^{13} + 5$	2^{35}	0.015	1.48	0.066
14	$2^{23} + 2^{14} + 5$	2^{35}	1.12	1.66	0.066
15	$2^{22} + 2^{13} + 5$	2^{35}	0.75	0.30	0.066
16	$2^{24} + 2^{13} + 5$	2^{35}	0.0008	2.92	0.066
17	5^{13}	2^{35}	3.03	0.61	1.85
18	5^{15}	2^{35}	2.02	4.02	4.03
Upper bound from Eq. (13):			3.63	5.90	9.86

suffer from too small a multiplier, and the terrible generator in line 3 has a good value of C_2 but very poor values of C_3 and C_4; indeed, $\nu_3 = 6$ and $\nu_4 = 2$ in line 3. Line 4 is a generator with a "random" multiplier, which has passed numerous empirical tests for randomness but which does not have especially high values of C_2, C_3, C_4.

Line 7 is the generator analyzed above, and the surrounding lines show similar multipliers. Note that the multiplier 3141592221 has an abnormally low value of C_3 in line 5, but the same multiplier with $m = 2^{35}$ in line 9 makes quite a good score.

Lines 12–16 show various multipliers with only four 1's in their binary representation; only the generator in line 14 is passable, and even here the value of C_4 is suspiciously low. By an odd coincidence, all five of the multipliers in lines 12–16 have the same value of C_4, and in fact the same values $s_1 = -125$, $s_2 = 75$, $s_3 = 15$, $s_4 = 1$ occur in each case! Another curiosity is the fact that line 16 has high C_3 but low C_2; in fact, $\nu_2 = \nu_3$ in line 16, since the minimum for $n = 3$ occurs when $s_1 = -2043$, $s_2 = 2047$, $s_3 = 0$.

Lines 17 and 18 show multipliers that have been used extensively since they were suggested by O. Taussky in the early 1950's; by coincidence the historically important multiplier 5^{15} actually has the best rating of all the entries in Table 1.

On the basis of the entries in Table 1 and further computational experience with many of the generators listed there, we can say the multiplier a *passes the spectral test* if C_2, C_3, and C_4 are all ≥ 0.1, and it passes the test with flying colors if all three are ≥ 1. The values of C_5, C_6, etc. might also be calculated before a multiplier is okayed for general use. The values of ν_2, ν_3, and ν_4 should also be considered, to see if the modulus m is sufficiently large for the desired accuracy of the random numbers; merely "passing the spectral test" in the above sense does not guarantee sufficient randomness for high-resolution Monte Carlo studies, when m is too small.

C. Deriving a Computation Method. The above examples illustrate the spectral test, but there is, of course, a big gap remaining in our discussion: How on earth can we determine the value of ν_n in a reasonable amount of time, using a computer? How, for example, is it possible to state that $s_1 = 227$, $s_2 = 983$, $s_3 = 130$ gives the smallest possible value of $s_1^2 + s_2^2 + s_3^2$ for which $s_1 + 3141592621 s_2 + 3141592621^2 s_3 \equiv 0$ (modulo 10^{10})? A brute-force search is obviously out of the question.

Therefore we shall now attempt to derive a reasonable computational procedure for solving this problem. It is convenient to transform Eqs. (11) and (12) into the following obviously equivalent problem: What is the minimum value of

$$(x_1 m - a x_2 - a^2 x_3 - \cdots - a^{n-1} x_n)^2 + x_2^2 + x_3^2 + \cdots + x_n^2 \quad (17)$$

for integers $x_1, x_2, \ldots, x_n$, not all zero?

It will be interesting and probably more useful if we develop a computational method for solving an even more general problem: *Find the minimum value of the quantity*

$$(a_{11} x_1 + a_{12} x_2 + \cdots + a_{1n} x_n)^2 + \cdots + (a_{n1} x_1 + a_{n2} x_2 + \cdots + a_{nn} x_n)^2 \quad (18)$$

for integers $x_1, \ldots, x_n$, *not all zero,* given any nonsingular matrix of coefficients $A = (a_{ij})$. The expression (18) is called a "positive definite quadratic form."

In the following discussion the letters $x, y, \ldots$ refer to column vectors

$$\begin{pmatrix} x_1 \\ x_2 \\ \vdots \\ x_n \end{pmatrix}, \begin{pmatrix} y_1 \\ y_2 \\ \vdots \\ y_n \end{pmatrix}, \ldots$$

The "dot product" $x \cdot y = x_1 y_1 + \cdots + x_n y_n$ may be written in matrix notation as $x^T y$, where T denotes transposition of rows and columns. For convenience we make the following definitions:

$$Q = A^T A, \quad B = A^{-1}, \quad R = Q^{-1} = B B^T. \quad (19)$$

Let A_j denote the jth *column* of A, and let B_i denote the ith *row* of B. Then

we have

$$B_i \cdot A_j = \delta_{ij}, \qquad Q_{ij} = A_i \cdot A_j, \qquad R_{ij} = B_i \cdot B_j. \qquad (20)$$

Our task is to minimize (18), i.e., to minimize $(Ax) \cdot (Ax) = x^T A^T A x = x^T Q x$, for integer vectors $x \neq 0$.

The first step is to reduce this to a finite problem, i.e., to show that not all of the infinitely many vectors x must be tested to find the minimum! Let e_k be the vector which is zero except for 1 in the kth component. Then

$$x_k = e_k^T x = e_k^T B A x = (B^T e_k) \cdot (Ax) = B_k \cdot (Ax).$$

and, by Schwarz's inequality,

$$(B_k \cdot (Ax))^2 \leq (B_k \cdot B_k)((Ax) \cdot (Ax)) = R_{kk}(x^T Q x).$$

Therefore if x is a nonzero vector which minimizes $x^T Q x$, we find

$$x_k^2 \leq R_{kk}(x^T Q x) \leq R_{kk}(e_j^T Q e_j) = R_{kk} Q_{jj}, \qquad 1 \leq j, k \leq n. \qquad (21)$$

This means only finitely many vectors x must be considered in the search for the minimum. Clearly, the argument we have just used proves more generally the following result:

Lemma A. *If*

$$x = \begin{pmatrix} x_1 \\ \vdots \\ x_n \end{pmatrix}$$

is a nonzero integer vector for which $x^T Q x$ is minimized, and if q is the value of $y^T Q y$ for some nonzero integer vector y, then

$$x_k^2 \leq R_{kk} q. \quad \blacksquare \qquad (22)$$

It is clear that the right-hand side of (22) might still be very large, much too large to make an exhaustive search feasible, so we need at least one more idea. We now apply the technique of changing variables, which is a simple but very important method for solving so many mathematical problems. Let us consider a substitution of variables of the form

$$y = Ux, \qquad (23)$$

where U is a matrix of integers with $\det U = \pm 1$. This means that whenever x is an integer column vector, so is y; and, conversely, if y is given, we can determine x by the relation $x = U^{-1}y$. (The matrix U^{-1} will have integer entries, since it equals adj $(U)/\det (U)$.) Therefore, as x runs through all integer vectors, so does $y = Ux$, and conversely; furthermore, $y = 0$ if and only if $x = 0$. Hence we may transform the problem of minimizing $(Ax) \cdot (Ax)$ for integer $x \neq 0$ into the equivalent problem of minimizing $(AU^{-1}y) \cdot (AU^{-1}y)$ for integer $y \neq 0$:

Lemma B. *If U is any integer matrix with $\det U = \pm 1$, let*

$$A' = AU^{-1}, \qquad B' = UB, \qquad Q' = (U^{-1})^T Q U^{-1}, \qquad R' = URU^T. \quad (24)$$

The minimization problem defined by the matrices A', B', Q', R' has the same solution as the minimization problem defined by A, B, Q, R. ∎

Now we can see how it may be possible to compute the minimum value efficiently: we want to transform the original problem by a suitable matrix U, as in Lemma B, and repeat this process until we obtain a problem for which the inequality in Lemma A is good enough to make an exhaustive search feasible.

For this purpose we want to look for simple integer matrices of determinant 1, and the following type of matrix suggests itself, for arbitrary integers $c_1, \ldots, c_n$:

$$U = \begin{pmatrix} 1 & & & & & & & \\ & \ddots & & & & & & \\ & & 1 & & & & & \\ c_1 & \cdots & c_{k-1} & 1 & c_{k+1} & \cdots & & c_n \\ & & & & 1 & & & \\ & & & & & \ddots & & \\ & & & & & & & 1 \end{pmatrix},$$

$$U^{-1} = \begin{pmatrix} 1 & & & & & & & \\ & \ddots & & & & & & \\ & & 1 & & & & & \\ -c_1 & \cdots & -c_{k-1} & 1 & -c_{k+1} & \cdots & & -c_n \\ & & & & 1 & & & \\ & & & & & \ddots & & \\ & & & & & & & 1 \end{pmatrix}.$$

$$(25)$$

(Here k is some fixed number; all entries not shown are zero. The relation $y = Ux$ in this case simply means $y_j = x_j$ for $j \neq k$ and $y_k = x_k + \sum_{j \neq k} c_j x_j$; this certainly is the most natural sort of substitution to try.) The effect of this matrix on the matrices A and B is readily calculated [cf. (24)]:

$$\begin{aligned} A'_j &= A_j - c_j A_k, & \text{for} \quad j \neq k, & \qquad A'_k = A_k; \\ B'_j &= B_j, & \text{for} \quad j \neq k, & \qquad B'_k = B_k + \sum_{j \neq k} c_j B_j. \end{aligned} \quad (26)$$

No matter what integers k and c_j are given, the matrix U in (25) is a potential candidate transformation. According to the inequality (21), it will be most helpful to choose the integers $c_1, \ldots, c_{k-1}, c_{k+1}, \ldots, c_n$ so that the *diagonal entries* of both Q' and R' get as small as possible. It is therefore natural to ask the following two questions:

a) *What is the best choice of real numbers c_j for $j \neq k$, to make the diagonal elements of $Q' = (U^{-1})^T Q U^{-1}$ as small as possible?*

b) *What is the best choice of real numbers c_j for $j \neq k$, to make the diagonal elements of $R' = URU^T$ as small as possible?*

In the case of problem (a), the diagonal entries of Q', which are equal to $A'_j \cdot A'_j$ by Eq. (20), will be changed by the transformation U whenever $j \neq k$. It is easy to see that the minimum value of

$$(A_j - c_j A_k) \cdot (A_j - c_j A_k) = Q_{jj} - 2c_j Q_{jk} + c_j^2 Q_{kk}$$
$$= Q_{kk}(c_j - Q_{jk}/Q_{kk})^2 + Q_{jj} - Q_{jk}^2/Q_{kk}$$

occurs when

$$c_j = Q_{jk}/Q_{kk}. \tag{27}$$

Geometrically (see Fig. 8) we are asking what multiple of vector A_k should be subtracted from A_j so that the resulting vector A'_j has minimum length, and the answer is to choose c_j so that A_j is perpendicular to A_k (i.e., so that $A'_j \cdot A'_k = Q'_{jk} = 0$). Equation (27) solves problem (a).

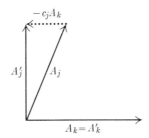

Fig. 8. Geometric derivation of (27).

In the case of problem (b), the diagonal elements of R' are exactly the same as those of R except for $R'_{kk} = B'_k \cdot B'_k$. So now we want to choose c_j so that $B_k + \sum_{j \neq k} c_j B_j$ has minimum length; geometrically this means we want to add some vector, in the $(n - 1)$-dimensional hyperplane generated by $\{B_j \mid j \neq k\}$, to the vector B_k. This problem like the one shown in Fig. 8 is solved by choosing things so that B'_k is perpendicular to the hyperplane, so that B'_k is in fact perpendicular to B'_j for all $j \neq k$. Thus we want to solve the equations $B'_k \cdot B'_j = 0$, namely

$$R_{kj} + \sum_{i \neq k} c_i R_{ij} = 0, \qquad 1 \leq j \leq n, \; j \neq k. \tag{28}$$

For a rigorous proof that a solution to problem (b) must be a solution to these equations, see exercise 12.

We have now solved problems (a) and (b), and we find ourselves in somewhat of a quandary; should we choose the c's according to (27) so the diagonal elements of Q' are minimized, or should we choose the c's according to (28) so the diagonal elements of R' are minimized? Either of these alternatives makes an improvement in the right-hand side of (21), so it is not immediately clear which choice should get priority. Fortunately, there is a very simple answer to this dilemma: Conditions (27) and (28) are exactly the same! The condition that $R' = Q'^{-1}$ means that the off-diagonal elements in row and column k of Q' are zero if and only if the off-diagonal elements in row and column k of R' are zero. Therefore problems (a) and (b) have the same solution. This happy state of affairs means we can reduce the diagonal elements of both Q and R simultaneously. (It may be worth while to point out that we have just rediscovered what is called the "Schmidt orthogonalization process.")

Of course, problems (a) and (b) ask for real values of the c_j, and we are allowed to use only integer values in the matrix U. We cannot in general make A'_j exactly perpendicular to A'_k; but if we take c_j to be the *nearest integer* to Q_{jk}/Q_{kk}, we are doing reasonably well. This choice of c_j is the best integer solution to problem (a), and it is close to (but *not* always equal to) the best integer solution to problem (b).

By performing transformations of the form (25) for various values of k, with c_j the nearest integer to Q_{jk}/Q_{kk}, it appears reasonable to assume we can ultimately reduce the problem to one in which the upper bounds given in Eq. (21) make an exhaustive search feasible. The algorithm given below therefore suggests itself. Several hundred computer runs of the method were performed by the author as this chapter was being written, and the algorithm, in fact, always converged much more rapidly than expected. Problems with $n = 6$ and with huge entries in both Q and R took only about 21 iterations of the transformation (25), until less than 500 cases remained to be examined in an exhaustive search. Thus the computation takes only a matter of seconds on a computer.

D. How to perform the spectral test. Here now is a computational procedure which follows from the considerations above.

Algorithm S (*The spectral test*). The spectral test is used to rate the choice of multiplier, a, in a linear congruential sequence of maximum period. (For extensions of the test to other linear congruential sequences, see exercises 20 and 21.) The test depends on a and the modulus m, and it tends to measure the statistical independence of adjacent n-tuples of numbers; the test is generally applied for $n = 2, 3, 4$ and perhaps a few higher values of n.

The following procedure therefore assumes a, m, and n are given as input, and the quantity $q = \nu_n^2$ [see (12)] is output. Auxiliary $n \times n$ matrices Q and R, and auxiliary n-place vectors X and c, are used. Exact integer operations are implied throughout the computation, and this generally means multiple-precision calculations are necessary; see the comments following the algorithm.

S1. [Initialize.]   Set $X[1] \leftarrow 1$, and set $X[k+1] \leftarrow (aX[k]) \bmod m$ for $1 \le k < n$. If any $X[k]$ is greater than $m/2$, set $X[k] \leftarrow X[k] - m$. Now form the matrices

$$Q = \begin{pmatrix} m^2 & -mX_2 & -mX_3 & \ldots & -mX_n \\ -mX_2 & 1+X_2^2 & X_2X_3 & \ldots & X_2X_n \\ -mX_3 & X_2X_3 & 1+X_3^2 & \ldots & X_3X_n \\ \vdots & & & & \vdots \\ -mX_n & X_2X_n & X_3X_n & \ldots & 1+X_n^2 \end{pmatrix},$$

(29)

$$R = \begin{pmatrix} \sum X_j^2 & mX_2 & mX_3 & \ldots & mX_n \\ mX_2 & m^2 & 0 & \ldots & 0 \\ mX_3 & 0 & m^2 & \ldots & 0 \\ \vdots & & & & \vdots \\ mX_n & 0 & 0 & \ldots & m^2 \end{pmatrix}.$$

Thus, set $Q[1,1] \leftarrow m^2$, $R[1,1] \leftarrow \sum_{1 \le j \le n} X[j]^2$; for $1 < j \le n$, set $Q[1,j] \leftarrow Q[j,1] \leftarrow -mX[j]$, $R[1,j] \leftarrow R[j,1] \leftarrow mX[j]$, $Q[j,j] \leftarrow 1 + X[j]^2$, $R[j,i] \leftarrow m^2$; and for $1 < j < k \le n$, set $Q[j,k] \leftarrow Q[k,j] \leftarrow X[j]\,X[k]$, $R[j,k] \leftarrow R[k,j] \leftarrow 0$. (These are the matrices Q and R which appear in (19), except that $R = m^2Q^{-1}$ instead of Q^{-1}, so that all computations below are done with integers.)

Now set $k \leftarrow n$ and $q \leftarrow m^2$.

S2. [Find minimum Q_{jj}.] For $1 \le j \le n$, if $Q[j,j] < q$, then set $q \leftarrow Q[j,j]$.

S3. [Ready for exhaustive search?] For $1 \le j \le n$, set $c[j] \leftarrow \lfloor \sqrt{qR[j,j]}/m \rfloor$. If now $\prod_{1 \le j \le n} (2c[j] + 1) \le 1000$, go to step S6. (By Lemma A, a minimum solution $X[1], \ldots, X[n]$ will have $-c[j] \le X[j] \le c[j]$ for all j, so the meaning of this step is to go to S6 if less than 1000 cases—in fact, as shown below, less than 500 cases—need to be considered to find the absolute minimum. It is of course unnecessary to evaluate $\prod(2c_j + 1)$ exactly, and overflow may occur if such an attempt is made; just see if it is > 1000 or not.)

S4. [Transform.] For $1 \le j \le n$, $j \ne k$, set $c[j]$ so that it is an integer that is as close as possible to $Q[j,k]/Q[k,k]$, for example, $\lfloor Q[j,k]/Q[k,k] + \frac{1}{2} \rfloor$. [Cf. Eq. (27).] Now for $1 \le j \le n$, if $j \ne k$ and $c[j] \ne 0$, do the operations TRANS$(Q, j, k, -c[j])$ and TRANS$(R, k, j, c[j])$.

The operation TRANS(P, i, j, t) is defined as follows when P is a matrix: For $1 \le r \le n$ set $P[i,r] \leftarrow P[i,r] + tP[j,r]$; then for $1 \le r \le n$ set $P[r,i] \leftarrow P[r,i] + tP[r,j]$. [Thus, t times row j is added to row i, then t times column j is added to column i. The effect of step S4 is to perform the transformation (24) using a matrix U as in (25).]

S5. [Modify k.] Decrease k by 1; then if $k = 0$, set $k \leftarrow n$. Go back to step S2.

S6. [Prepare for search.] (Now the absolute minimum will be determined using an exhaustive search.) Set $k \leftarrow n$ and set $X[j] \leftarrow 0$ for $1 \le j \le n$.

S7. [Advance $X[k]$.] Set $X[k] \leftarrow X[k] + 1$. If $X[k] > c[k]$, go to S9.

S8. [Advance k.] Set $k \leftarrow k + 1$. If $k \leq n$, set $X[k] \leftarrow -c[k]$ and repeat step S8. If $k > n$, set $q' \leftarrow \sum_{1 \leq i \leq n} \sum_{1 \leq j \leq n} X[i]X[j]Q[i,j]$, and if $q' < q$ set $q \leftarrow q'$.

S9. [Decrease k.] Set $k \leftarrow k - 1$. If $k \geq 1$, return to S7, otherwise the algorithm terminates. (Note: Steps S5–S9 are a simple case of the "backtrack" method, which is discussed at length in Chapter 7.) ∎

At the conclusion of this algorithm, the quantity q is the important value: Let $\nu_n = \sqrt{q}$ and determine C_n by Eq. (15). For a discussion of the significance of ν_n and C_n, and conditions under which the multiplier a has "passed" the spectral test, see Subsection B above.

Since m is generally the word size of the computer on which Algorithm S is being performed, this algorithm usually requires multiple-precision integer arithmetic. Experience shows that triple precision (i.e., three computer words per integer) is adequate when m is the word size. For example, the author prepared Table 1 using 76-bit arithmetic; this was adequate for $m = 10^{10}$ and $n \geq 6$, but overflow occurred in several cases when $m = 2^{35}$. It seems 90 or more bits of accuracy will be sufficient to handle almost all situations when $m = 2^{35}$. No theory about the size of intermediate results has yet been developed.

In practice, Algorithm S is surprisingly efficient; the observed number of iterations of steps S2–S5 was approximately 6, 10, 14, 17, 21 for $m = 2^{35}$ and $n = 2, 3, 4, 5, 6$. However, there is at present no theory that accounts for the behavior in general, and in fact *the method will get into an infinite loop* in some cases. So the name "algorithm" isn't strictly justified here. There is a simple way to modify the procedure so that infinite loops can be rigorously prevented, as shown in exercise 16: We can add another variable d, and set $d \leftarrow 0$ in step S1 and in the TRANS operation; also say "$d \leftarrow d + 1$; if $d > n$, go to S6" at the end of step S3. However, this may possibly lead to a very long search in steps S6–S9, as shown in exercise 18. The techniques discussed in exercises 22 and 23 below would appear to offer significant advantages in such a situation. The method loops when $n = 2$, $a = 1025$, $m = 2^{46}$, but such failures are rare.

The author has noticed a case for $n = 6$ in which the "expected length of search," $\prod(2c_j + 1)$, in step S3, took the successive values

$$1 \times 10^{43},\ 6 \times 10^{42},\ 2 \times 10^{42},\ 9 \times 10^{41},\ 2 \times 10^{41},\ 6 \times 10^{33},\ 4 \times 10^{33},$$
$$1 \times 10^{29},\ 1 \times 10^{20},\ 6 \times 10^{19},\ 4 \times 10^{18},\ 9 \times 10^{12},\ 4 \times 10^{10},\ 3 \times 10^{8},$$
$$1 \times 10^{8},\ 8 \times 10^{7},\ 1 \times 10^{7},\ 7 \times 10^{6},\ 1.7 \times 10^{7},\ 1.8 \times 10^{7},\ 7 \times 10^{5},$$
$$1 \times 10^{5},\ 5 \times 10^{4},\ 3825,\ 3825,\ 675.$$

Thus, this quantity decreases from 10^{43} to a value less than 1000, but it does so somewhat erratically; its value actually *increased* twice in succession. It is

Table 2

EXAMPLE OF ALGORITHM S

Line	Matrix Q		
1.	1 00000 00000 00000 00000 −31415 92621 00000 00000 36783 50359 00000 00000	−31415 92621 00000 00000 9869 60419 63216 49642 −11555 87834 52871 00939	36783 50359 00000 00000 −11555 87834 52871 00939 13530 26136 35554 28882
⋮	⋮		
7.	1160 62418 −110 45623 324 06810	−110 45623 189 42062 −70 72864	324 06810 −70 72864 99 86024
8.	114 95774 126 21707 24 48738	126 21707 147 82358 29 13160	24 48738 29 13160 99 86024

probable that further iterations would reduce the value 675 even further, so perhaps the constant 1000 in step S3 should be lowered to, say, 100. (*Note:* The actual number of sets $X[1], \ldots, X[n]$ tested in the exhaustive search, steps S6–S9, is only $\lfloor \frac{1}{2}\prod(2c_j + 1)\rfloor$, not $\prod(2c_j + 1)$, since the algorithm only considers vectors whose first nonzero entry is positive.)

Let us consider briefly an example of Algorithm S in action, when $a = 3141592621$, $m = 10^{10}$, $n = 3$. At the top of Table 2 we have the initial settings of Q and R as prepared by step S1. For these matrices a brute-force search for the answer by means of Lemma A alone would require testing over 10^{29} cases, so it is out of the question. After six iterations of steps S2–S5, the entries in Q and R have gotten much smaller (see line 7 of Table 2), and Lemma A tells us that $|x_1| \leq 3$, $|x_2| \leq 3$, $|x_3| \leq 14$ in this new problem. A further reduction using Lemma B takes us to line 8: in matrix Q of line 7, add column 3 to column 2, then row 3 to row 2, then subtract three times column 3 from column 1, and subtract three times row 3 from row 1. In matrix R, subtract column 2 from column 3, then subtract row 2 from row 3, then add three times column 1 to column 3, then add three times row 1 to row 3. This reduces Q and R so that Lemma A tells us we now only need to try $|x_1| \leq 3$, $|x_2| \leq 3$, $|x_3| \leq 1$ in a search for the absolute minimum. So the search method in steps S6–S9 goes into action, and it locates the combination $x_1 = 1$, $x_2 = -1$, $x_3 = 0$, which gives the minimum value, $x^T Q x = 1034718$. These calculations could be done by hand on a desk calculator in a few hours, although the problem that has been solved sounds quite formidable at first.

The spectral test made its first appearance in the article "Fourier Analysis of Uniform Random Number Generators" by R. R. Coveyou and R. D. MacPherson, *JACM* **14** (1967), 100–119. This article describes an algorithm essentially like Algorithm S to make the test, except with a somewhat different transformation rule in step S4.

Line	Matrix R		
1.	$\begin{pmatrix} 23399\ 86555\ 98770\ 78523 \\ 31415\ 92621\ 00000\ 00000 \\ -36783\ 50359\ 00000\ 00000 \end{pmatrix}$	$\begin{matrix} 31415\ 92621\ 00000\ 00000 \\ 1\ 00000\ 00000\ 00000\ 00000 \\ 0 \end{matrix}$	$\begin{matrix} -36783\ 50359\ 00000\ 00000 \\ 0 \\ 1\ 00000\ 00000\ 00000\ 00000 \end{matrix}$
$\vdots$	$\vdots$		
7.	$\begin{pmatrix} 13913\ 04805\ 78992 \\ -11890\ 71034\ 30888 \\ -53572\ 76149\ 67948 \end{pmatrix}$	$\begin{matrix} -11890\ 71034\ 30888 \\ 10880\ 07572\ 69932 \\ 46294\ 02921\ 32522 \end{matrix}$	$\begin{matrix} -53572\ 76149\ 67948 \\ 46294\ 02921\ 32522 \\ 2\ 07645\ 57301\ 67787 \end{matrix}$
8.	$\begin{pmatrix} 13913\ 04805\ 78992 \\ -11890\ 71034\ 30888 \\ 57\ 09301\ 99916 \end{pmatrix}$	$\begin{matrix} -11890\ 71034\ 30888 \\ 10880\ 07572\ 69932 \\ -258\ 17754\ 30074 \end{matrix}$	$\begin{matrix} 57\ 09301\ 99916 \\ -258\ 17754\ 30074 \\ 1062\ 71591\ 61243 \end{matrix}$

EXERCISES

1. [M20] Prove Eq. (2) from Eq. (1).

2. [M20] Prove Eq. (1) from Eq. (2), assuming that $0 \leq s_1, \ldots, s_n < m$.

3. [M22] (a) Let $F(t)$ be defined for integers t, $0 \leq t < m$, and extend its definition to all integers by the formula $F(t) = F(t \bmod m)$. Let $f(s)$ be the Fourier transform of $F(t)$, as in Eq. (1) with $n = 1$. Find the Fourier transform of $F(t + 1)$ in terms of $f(s)$.

(b) Find the Fourier transform of $\sum_{0 \leq k < m} F(k) G(t - k)$ in terms of the transforms of F and G.

▶ 4. [M22] Let $X_0, X_1, X_2, \ldots$ be a periodic sequence of integers, with $0 \leq X_k < m$, and let $f(s_1, s_2)$ be the function in (4). Express the probability that $X_{k+1} < X_k$ in this sequence, in terms of f. (Simplify your answer as much as possible so that the coefficient of each particular value of f is shown explicitly.)

5. [M23] Let $X_0, X_1, X_2, \ldots$ be a periodic sequence of integers, with $0 \leq X_k < m$, and with period length m. Express the quantity $\sum_{0 \leq k < m} X_k X_{k+1}$, which is an important part of the formula for the serial correlation coefficient [Eq. 3.3.2–(23)], in terms of the function F of Eq. (3) and also in terms of the function f of Eq. (4).

▶ 6. [M28] Prove Theorem 3.3.3P by using Eq. (8) and the result of exercise 4.

7. [M40] Derive formulas that allow reasonably efficient calculation of the exact probability that $X_{k+2} < X_{k+1} < X_k$ in a linear congruential sequence with maximum period. [Hint: Using the methods of exercise 6 and the result of exercise 3.3.3–20, this quantity can apparently be expressed in terms of generalized Dedekind sums. Note that without the use of Fourier transforms, this problem is considerably more difficult.]

8. [M22] Find a reasonably simple formula for $\sum_{0 \leq k < m} X_k X_{k+1}$ in a linear congruential sequence of maximum period by using the result of exercise 5 in connection with Eq. (8).

9. [*HM30*] (C. Hermite.) Given an $n \times n$ matrix A, prove that there is a nonzero integer vector x for which $Ax \cdot Ax \leq (\frac{4}{3})^{(n-1)/2}(\det A)^{2/n}$. [*Hints:* First show that for all $\epsilon > 0$ there is a matrix of integers U with determinant 1 for which $AUx \cdot AUx$ comes within ϵ of the greatest lower bound of its values when $(x_1, x_2, \ldots, x_n) = (1, 0, \ldots, 0)$. Then prove the general result by induction on n, writing $Ax \cdot Ax$ in the form $\alpha(x_1 + \beta_2 x_2 + \cdots + \beta_n x_n)^2 + g(x_2, \ldots, x_n)$, where g corresponds to an $(n-1) \times (n-1)$ matrix A'.]

10. [*HM30*] (Coveyou and MacPherson.) Let $X_0, X_1, X_2, \ldots$ be a sequence of integers with $0 \leq X_k < m$, having Fourier coefficients $f(s_1, \ldots, s_n)$ as given in Eq. (4). Let $U_k = X_k/m$, and let $V_0, V_1, V_2, \ldots$ be an independent, truly random sequence of real numbers uniformly distributed between 0 and 1. Let λ be a number less than 1 and let $W_k = (U_k + \lambda V_k) \bmod 1$. (Now W_k represents the sequence U_k with the values "smeared" randomly through a distance λ.) Define the Fourier coefficients of any sequence of real numbers between 0 and 1 as

$$\lim_{N \to \infty} \frac{1}{N} \sum_{0 \leq k < N} \exp \left(2\pi_i(s_1 U_k + \cdots + s_n U_{k+n-1})\right).$$

Find the Fourier coefficients of the sequence $\langle W_k \rangle$ in terms of those of the sequence $\langle X_k \rangle$.

11. [*M10*] Equation (8) shows that the value of c in a linear congruential sequence of maximum period does not affect the Fourier coefficients except to change the "argument" of the complex number $f(s_1, \ldots, s_n)$; in other words, the absolute value of $f(s_1, \ldots, s_n)$ is independent of c. But is it possible to choose c so that the nonrandom effect of one wave $f(s_1, \ldots, s_n)$ might be cancelled out by an "opposite" effect of another wave $f(s_1', \ldots, s_n')$?

12. [*HM23*] Prove without using geometrical arguments that any solution to "problem (b)" as stated in Subsection C of the text must also be a solution to the set of equations (28).

13. [*HM30*] The text has sidestepped a rather important issue: the tacit assumption was made that if A is any nonsingular matrix of real numbers, the quantity in (18) *does* have a minimum value which is *achieved* for some integer vector x. (a) Prove that the greatest lower bound of the quantity (18), taken over all nonzero integer vectors x, is attained for some x, when A is nonsingular. (b) Show that if A is singular, it may be impossible to find any nonzero integer vector for which the greatest lower bound of (18) is attained.

▶ **14.** [*24*] Perform Algorithm S by hand, for $m = 100$, $a = 41$, $n = 3$. Change the constant "1000" in step S3 to "3".

15. [*M18*] What would happen if the operation "$k \leftarrow n$" at the end of step S1 were changed to "$k \leftarrow 1$"?

16. [*M25*] It is conceivable (although this has not been observed yet) that Algorithm S may get into a loop, endlessly repeating steps S2–S5. Show that this can happen if and only if n consecutive executions of step S4 produce no transformations (i.e., no TRANS operations).

▶ **17.** [*M28*] Modify Algorithm S so that, in addition to computing the quantity q, it determines a set of integers $s_1, \ldots, s_n$ satisfying (11) for which $s_1^2 + \cdots + s_n^2 = q$.

[*Hint:* Algorithm S retains only the values of Q and R of (19), not A or B. If the value of A and/or B is maintained throughout the algorithm, it seems likely that the values of $s_1, \ldots, s_n$ will be available with little extra difficulty.]

18. [*M25*] Find a 3×3 matrix A for which, if $Q = A^T A$ and $R = Q^{-1}$, steps S2–S5 of Algorithm S will never exit to S6 (so the computation will never terminate). [*Hint:* Consider "combinatorial matrices," i.e., matrices whose elements have the form $a + b\delta_{ij}$; cf. exercise 1.2.3–39.]

19. [*M20*] Show that the Fibonacci sequence mod m is a bad source of random numbers, by showing that the corresponding function $f(s_1, s_2, s_3)$ as defined in (4) has large low-frequency components.

▶ **20.** [*M24*] Compute the Fourier coefficients $f(s_1, \ldots, s_n)$ for the linear congruential sequence defined by $X_0 = 1$, $c = 0$, $m = 2^e \geq 8$, and $a \bmod 8 = 5$. Discuss how to extend the spectral test (which is defined in the text only for linear congruential sequences with the *maximum* period, while the sequence defined here has a period of length $m/4$) to this type of random number generator.

21. [*M25*] Same as exercise 20, but with $a \bmod 8 = 3$.

▶ **22.** [*M30*] Design an algorithm which is like Algorithm S except it is based on a transformation U which has all of its nonzero off-diagonal elements in *column* k, instead of in *row* k as in (25). Compare this method with Algorithm S. Show that the quantity $\prod(2c_j + 1)$ in step S3 never increases from one iteration to the next in this algorithm.

23. [*M50*] Although the example in exercise 18 causes the basic reduction cycle of Algorithm S to loop indefinitely, it presents no special problem to the "dual" algorithm of exercise 22. For convenience, let us call the latter method Algorithm S'. Since Algorithm S' is essentially Algorithm S with the roles of matrices Q and R interchanged, there are also matrices which cause Algorithm S' to loop indefinitely. This suggests a combination of the two methods; for example, we can use Algorithm S until it gets stuck, then switch to Algorithm S' until *it* gets stuck, then back to Algorithm S, etc.

Using a combined algorithm such as this, if we ignore the branch in step S3, the computation procedure will ultimately reach one of two situations: (a) A cycle is established in which each of Algorithms S and S' alternately transforms Q and R in some way such that the computation never terminates; or (b) we reach matrices Q and R which are affected neither by Algorithm S nor Algorithm S'.

These observations lead naturally to the following three questions, which ought to be answered if we are to have a completely satisfactory solution to the computational problem posed in this section: Can case (a) occur? When case (b) occurs, how large can the search constant $\prod(2c_j + 1)$ in step S3 be? Is there a better general procedure for determining min $\{x^T Q x \mid \text{integer } x \neq 0\}$, for positive definite Q, than the combination of Algorithms S and S' just described?

Note: The second question above can be reduced to the following: *Let Q be a symmetric positive definite $n \times n$ matrix of real numbers whose diagonal entries are 1 and for which $|q_{ij}| \leq \frac{1}{2}$ for $i \neq j$. Let $R = Q^{-1}$, and suppose $|r_{ij}| \leq \frac{1}{2}r_{jj}$ for $i \neq j$. Under these circumstances, how large can r_{11} be?* No such matrices have yet been observed for which $r_{11} \geq 2$. If all the off-diagonal elements of Q are taken to be $-1/n$, we find that $r_{11} = 2n/(n+1)$, and this example shows that the search constant in step S3 can never be assumed to go below 3^n for a general positive definite matrix even if Algorithms S and S′ are combined; this example gives max r_{11}, when $n = 2$, but not when $n = 3$.

24. [*M20*] Compare the Fourier transform $f(s_1, \ldots, s_n)$, Eq. (1), to the n-variable *generating function* for $F(t_1, \ldots, t_n)$ defined in the usual way as

$$g(z_1, \ldots, z_n) = \sum_{0 \le t_1, \ldots, t_n < m} F(t_1, \ldots, t_n) z_1^{t_1} \ldots z_n^{t_n}.$$

25. [*HM28*] (R. R. Coveyou.)   Analyze the two-dimensional Fourier transform $f(s_1, s_2)$ of the modified middle-square method, sequence 3.2.2–4. [*Hint:* Consider the double sum for $|f(s_1, s_2)|^2 = f(s_1, s_2) f(-s_1, -s_2)$.]

▶ **26.** [*M26*] What is the value of the three-dimensional Fourier transform $f(s_1, s_2, s_3)$ of the sequence 3.2.2–8, when $k = 2$, having period length $p^2 - 1$?

▶ **27.** [*HM24*] (G. Marsaglia.) Let X_0, a, m, c generate a linear congruential sequence, and let U_0, U_1, U_2, $\ldots = X_0/m$, X_1/m, X_2/m, $\ldots$ Assume that s_1, s_2, $\ldots$, s_n are integers satisfying (11), and prove the following facts: (a) Each of the n-tuples $(U_k, U_{k+1}, \ldots, U_{k+n-1})$, treated as a point in n-dimensional space, lies on one of the parallel hyperplanes defined by the equation

$$s_1 x_1 + s_2 x_2 + \cdots + s_n x_n = N + (s(a) - s(1))c/(a - 1)m,$$

where N is an integer and $s(a)$ is defined in Eq. (7). (b) The distance between neighboring planes of this form is $1/\sqrt{s_1^2 + \cdots + s_n^2}$. (c) The number of such hyperplanes which intersect the n-cube $0 \le x_1, \ldots, x_n < 1$ is $|s_1| + \cdots + |s_n| - \delta$, where $\delta = 1$ if $s_i s_j < 0$ for some i and j, otherwise $\delta = 0$.

    Note: In the words of W. Givens, consecutive random vectors generated by linear congruential methods "stay mainly in the planes." This exercise gives a convincing demonstration of the unsuitability of linear congruential sequences for high-resolution Monte Carlo applications; for example, the three-dimensional vectors (U_k, U_{k+1}, U_{k+2}) produced by generator (14) all lie on parallel planes which are $1/\nu_3 \approx 0.001$ units apart.

3.4. OTHER TYPES OF RANDOM QUANTITIES

We have now seen how to make a computer generate a sequence of numbers U_0, U_1, U_2, $\ldots$ which behaves as though each number is independently selected at random between zero and one with the uniform distribution. Applications of random numbers often call for other kinds of distributions; for example, if we want to make a random choice between k alternatives, we want a random *integer* between 1 and k. If some simulation process calls for a random waiting time between occurrences of independent events, a random number with the "exponential distribution" is desired. Sometimes we don't even want random *numbers*—we want a random permutation (i.e., a random arrangement of n objects) or a random combination (i.e., a random choice of k elements from a collection of n).

In principle, any of these other random quantities may be obtained from the uniform deviates $U_0, U_1, U_2, \ldots$. There are a number of important "random tricks" which may be used to perform these manipulations efficiently on a computer, and a study of these techniques also gives some insight into the proper use of random numbers in any Monte Carlo application.

It is conceivable that someday someone will invent a random-number generator which produces some of these other random quantities *directly*, instead of getting them indirectly via the uniform distribution. Except for the "random bit" generator described in Section 3.2.2, no direct methods have so far proved to be practical.

The discussion in the following section assumes that $U_0, U_1, U_2, \ldots$ is a random sequence of uniformly distributed real numbers between zero and one. The letter U without a subscript denotes the current element of the sequence; a new U is generated whenever we need some random quantity. These numbers are normally represented in a computer word with the decimal point assumed at the left. Actually, of course, a computer word has only finite accuracy, and if this accuracy is not sufficient one can always combine several U's into a single value having higher precision.

3.4.1. Numerical Distributions

This section summarizes the best known techniques for producing numbers from various important distributions. Many of the methods were originally suggested by John von Neumann in the early 1950's, and these were gradually improved upon by other people, notably George Marsaglia.

A. Random choices from a finite set. The simplest and most common type of distribution required in practice is a random *integer*. An integer between 0 and 7 can be extracted from three bits of U on a binary computer; in such a case, these bits should be extracted from the *most significant* (left-hand) part of the computer word, since in many random-number generators the least significant bits are not sufficiently random. (See the discussion of this in Section 3.2.1.1.)

In general, to get a random number X between 0 and $k - 1$, we can *multiply* by k, and let $X = \lfloor kU \rfloor$. On MIX, we would write

$$
\begin{array}{ll}
\text{LDA} & \text{U} \\
\text{MUL} & \text{K}
\end{array}
\tag{1}
$$

and after these two instructions have been executed the desired integer will appear in register A. If a random integer between 1 and k is desired, we add one to this result. [The instruction "INCA 1" would follow (1).]

This method gives each integer with equal probability. (There is a slight error due to the finiteness of the computer word size; see exercise 2. The error is quite negligible if k is small, for example, if $k/m < 1/10000$.) In a more general situation we might want to give different weights to different integers. Suppose that the value $X = x_1$ is to be obtained with probability p_1, $X = x_2$ with probability $p_2, \ldots$, and $X = x_k$ with probability p_k. We can generate a uniform number U, and let

$$
X = \begin{cases}
x_1, & \text{if} & 0 \leq U < p_1; \\
x_2, & \text{if} & p_1 \leq U < p_1 + p_2; \\
\cdots \\
x_k, & \text{if} & p_1 + p_2 + \cdots + p_{k-1} \leq U < 1.
\end{cases} \tag{2}
$$

(Note that $p_1 + p_2 + \cdots + p_k = 1$.)

There is a "best possible" way to do the comparisons of U against various values of $p_1 + p_2 + \cdots + p_s$, as implied in (2); this situation is discussed in Section 2.3.4.5. A "table-lookup" instruction, present in some computers, might also be used. Special cases can be handled by more efficient methods; for example, to obtain one of the eleven values $2, 3, \ldots, 12$ with the respective probabilities $\frac{1}{36}, \frac{2}{36}, \ldots, \frac{6}{36}, \ldots, \frac{2}{36}, \frac{1}{36}$, we could compute two independent random integers between 1 and 6 and add them together.

However, none of the above techniques is really the fastest way to select $x_1, \ldots, x_k$ with the correct probabilities. A method which is considerably more efficient in most cases, at a slight increase in storage space, is explained in exercises 20 and 21.

B. General methods for continuous distributions. The most general real-valued distribution may be expressed in terms of the "distribution function" $F(x)$; we want the random quantity X to be less than or equal to x with the probability $F(x)$:

$$
F(x) = \text{probability } (X \leq x). \tag{3}
$$

This function always increases monotonically from zero to one:

$$
F(x_1) \leq F(x_2), \quad \text{if} \quad x_1 \leq x_2; \quad F(-\infty) = 0, \quad F(+\infty) = 1. \tag{4}
$$

Examples of distribution functions are given in Section 3.3.1, Fig. 3. If $F(x)$ is continuous and strictly increasing (so that $F(x_1) < F(x_2)$ when $x_1 < x_2$), it takes on all values between zero and one, and there is an *inverse function* $F^{-1}(y)$ such that, if $0 < y < 1$,

$$
y = F(x) \quad \text{if and only if} \quad x = F^{-1}(y). \tag{5}
$$

A general way to compute a random quantity X with the continuous, strictly increasing distribution $F(x)$ is to set

$$
X = F^{-1}(U). \tag{6}
$$

For the probability that $X \leq x$ is the probability that $F^{-1}(U) \leq x$, i.e., the probability that $U \leq F(x)$, and this is $F(x)$.

The problem now reduces to one of numerical analysis, for determining good methods of evaluating $F^{-1}(U)$ to the desired accuracy. Numerical analysis lies outside the scope of this book; yet there are a number of important shortcuts available to speed up this general method, and we will consider them here.

In the first place, if X_1 is a random variable having the distribution $F_1(x)$ and if X_2 is an independent random variable with the distribution $F_2(x)$, then

$$
\begin{aligned}
&\max (X_1, X_2) \quad \text{has the distribution} \quad F_1(x)F_2(x), \\
&\min (X_1, X_2) \quad \text{has the distribution} \quad F_1(x) + F_2(x) - F_1(x)F_2(x).
\end{aligned}
\tag{7}
$$

(See exercise 4.) For example, the uniform deviate U has the distribution $F(x) = x$, for $0 \le x < 1$; if $U_1, U_2, \ldots, U_t$ are independent uniform deviates, then $\max (U_1, U_2, \ldots, U_t)$ has the distribution function $F(x) = x^t, 0 \le x < 1$. This is the basis of the "maximum of t" test given in Section 3.3.2. Note that the inverse function in this case is $F^{-1}(y) = \sqrt[t]{y}$. In the special case $t = 2$, we see therefore that the two formulas

$$
X = \sqrt{U} \quad \text{and} \quad X = \max (U_1, U_2)
\tag{8}
$$

will give equivalent distributions to the random variable X, although this is not obvious at first glance. We need not take the square root of a uniform deviate.

The number of tricks like this is endless; any algorithm which employs random numbers as input will give a random quantity with *some* distribution as output. The problem is to find general methods for constructing the algorithm, given the distribution function of the output.

Using Eq. (7), we can get the *product* of two distribution functions. There is also a method for the *mixture* of two distributions: suppose that

$$
F(x) = pF_1(x) + (1 - p)F_2(x), \quad 0 < p < 1.
\tag{9}
$$

We can obtain a random variable X with distribution $F(x)$ by first taking a uniform deviate U; if $U < p$ set X to a number with distribution $F_1(x)$, and if $U \ge p$, set X to a number with distribution $F_2(x)$.

This procedure can be extremely useful when p is nearly equal to one and when $F_1(x)$ is an easy distribution to accommodate. Then even though it may be more difficult to generate random variables from the distribution $F_2(x)$ than it is for the desired overall distribution $F(x)$, the more difficult computation must only be done infrequently, with probability $(1 - p)$. We will see below how this idea can be exploited profitably in several important cases.

C. The normal distribution. Perhaps the most important nonuniform, continuous distribution is the *normal distribution with mean zero and standard deviation one:*

$$
F(x) = \frac{1}{\sqrt{2\pi}} \int_{-\infty}^{x} e^{-t^2/2} \, dt.
\tag{10}
$$

Several techniques are available for generating normal deviates:

(1) The Polar Method.

Algorithm P *(Polar method for normal deviates).* This algorithm calculates two independent normally distributed variables, X_1 and X_2, given two independent uniformly distributed variables, U_1 and U_2.

P1. [Get uniform variables.]   Generate two independent random variables, U_1, U_2, uniformly distributed between zero and one. Set $V_1 \leftarrow 2U_1 - 1$, $V_2 \leftarrow 2U_2 - 1$. (Now V_1 and V_2 are uniformly distributed between -1 and $+1$. On most computers it will be preferable to have V_1 and V_2 represented in floating-point form at this point.)

P2. [Compute S.] Set $S \leftarrow V_1^2 + V_2^2$.

P3. [Is $S \geq 1$?] If $S \geq 1$, return to step P1. (Steps P1 through P3 are executed 1.27 times on the average, with a standard deviation of 0.587; see exercise 6.)

P4. [Compute X_1, X_2.] Set X_1, X_2 according to the following two equations:

$$X_1 = V_1 \sqrt{\frac{-2 \ln S}{S}}, \qquad X_2 = V_2 \sqrt{\frac{-2 \ln S}{S}}. \qquad (11)$$

These are the normally distributed variables desired. ∎

To prove the validity of this method, we use elementary analytic geometry and calculus: If $S < 1$ in step P3, the point in the plane with Cartesian coordinates (V_1, V_2) is a *random point uniformly distributed inside the unit circle.* Transforming to polar coordinates $V_1 = R \cos \Theta$, $V_2 = R \sin \Theta$, we find $S = R^2$, $X_1 = \sqrt{-2 \ln S} \cos \Theta$, $X_2 = \sqrt{-2 \ln S} \sin \Theta$.   Using also the polar coordinates $X_1 = R' \cos \Theta'$, $X_2 = R' \sin \Theta'$, we find that $\Theta' = \Theta$ and $R' = \sqrt{-2 \ln S}$. It is clear that R' and Θ' are independent, since R and Θ are independent inside the unit circle.   Also, Θ' is uniformly distributed between 0 and 2π; and the probability that $R' \leq r$ is the probability $-2 \ln S \leq r^2$, i.e., the probability $S \geq e^{-r^2/2}$.   This equals $1 - e^{-r^2/2}$, since $S = R^2$ is uniformly distributed between zero and one.   The probability that R' lies between r and $r + dr$ is therefore the derivative of $1 - e^{-r^2/2}$, namely, $re^{-r^2/2} dr$.   Similarly, the probability that Θ' lies between θ and $\theta + d\theta$ is $(1/2\pi) d\theta$. Therefore the probability that $X_1 \leq x_1$ and that $X_2 \leq x_2$ is

$$\int_{\{(r,\theta)\mid r \cos \theta \leq x_1, r \sin \theta \leq x_2\}} \frac{1}{2\pi} e^{-r^2/2} r \, dr \, d\theta$$

$$= \frac{1}{2\pi} \int_{\{(x,y)\mid x \leq x_1, y \leq x_2\}} e^{-(x^2 + y^2)/2} \, dx \, dy$$

$$= \left(\sqrt{\frac{1}{2\pi}} \int_{-\infty}^{x_1} e^{-x^2/2} \, dx \right) \left(\sqrt{\frac{1}{2\pi}} \int_{-\infty}^{x_2} e^{-y^2/2} \, dy \right).$$

This proves X_1 and X_2 are independent and normally distributed, as desired. Algorithm P is due to G. E. P. Box, M. E. Muller, and G. Marsaglia.

(2) Marsaglia's rectangle-wedge-tail method. In this method we use the distribution

$$F(x) = \sqrt{\frac{2}{\pi}} \int_0^x e^{-t^2/2}\, dt \qquad x \geq 0, \tag{12}$$

so that $F(x)$ gives the distribution of the *absolute value* of a normal deviate. After X has been computed according to this distribution, we will attach a random sign to its value, and this will make it a true normal deviate.

Marsaglia's general approach to efficient random number manipulation will be illustrated from the distribution in (12). As we study this particular example, we will learn several important general methods at the same time.

Basically, we want to write

$$F(x) = p_1 F_1(x) + p_2 F_2(x) + \cdots + p_n F_n(x) \tag{13}$$

for certain distributions $F_1, F_2, \ldots, F_n$, and then we will generate X according to the distribution F_j with probability p_j. Some of the distributions $F_j(x)$ may be rather difficult to handle, but we can usually arrange things so that the probability p_j is very small in this case. Most of the distributions $F_j(x)$ will be quite easy to accommodate, since they are trivial modifications of the uniform distribution. The resulting method yields an extremely efficient program, since its *average* running time is very small.

It is easier to understand the method if we work with the *derivatives* of the distributions instead of the distributions themselves. Let

$$f(x) = F'(x), \qquad f_j(x) = F_j'(x);$$

these are called the *density* functions of the probability distributions. Equation (13) becomes

$$f(x) = p_1 f_1(x) + p_2 f_2(x) + \cdots + p_n f_n(x). \tag{14}$$

Each $f_j(x)$ is ≥ 0, and the total area under the graph of $f_j(x)$ is 1; so there is a convenient graphical way to display the relation (14): The area under $f(x)$ is divided into n parts, with the part corresponding to $f_j(x)$ having area p_j. See Fig. 9, which illustrates the situation in the case of interest to us here, with $f(x) = F'(x) = \sqrt{(2/\pi)}e^{-x^2/2}$; the area under this curve has been divided into $n = 37$ parts. There are 12 rather large rectangles, which represent $p_1 f_1(x), \ldots, p_{12} f_{12}(x)$; there are 12 skinny little rectangles perched on top of these, which represent $p_{13} f_{13}(x), \ldots, p_{24} f_{24}(x)$; there are 12 wedge-shaped pieces, which represent $p_{25} f_{25}(x), \ldots, p_{36} f_{36}(x)$; and the remaining part $p_{37} f_{37}(x)$ is the entire graph of $f(x)$ for $x \geq 3$.

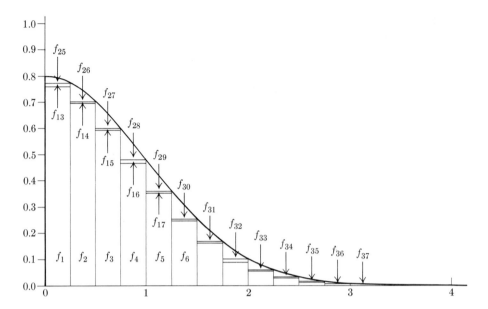

Fig. 9. The frequency function divided into 37 parts. The area of each part represents the average number of times a random number with that frequency is to be computed.

The rectangular parts $f_1(x), \ldots, f_{12}(x)$ represent *uniform distributions*. For example, $f_3(x)$ represents a random variable uniformly distributed between $\frac{1}{2}$ and $\frac{3}{4}$. We want to make it possible to determine the probabilities $p_1, p_2, \ldots, p_{12}$ efficiently as the normal deviate is being generated, so (assuming we have a binary computer—the procedure for a decimal computer is similar) let us take each of these to be multiples of, say, $\frac{1}{256}$. This means we want the area under $p_j f_j(x)$ to be a multiple of $\frac{1}{256}$, so the altitude of $f_j(x)$ is taken to be

$$\lfloor 64 f(j/4) \rfloor / 64, \qquad 1 \leq j \leq 12.$$

The following table gives the resulting probabilities:

p_1	p_2	p_3	p_4	p_5	p_6	p_7	p_8	p_9	p_{10}	p_{11}	p_{12}	$(p_{13} + \cdots + p_{37})$	
$\frac{49}{256}$	$\frac{45}{256}$	$\frac{38}{256}$	$\frac{30}{256}$	$\frac{23}{256}$	$\frac{16}{256}$	$\frac{11}{256}$	$\frac{6}{256}$	$\frac{4}{256}$	$\frac{2}{256}$	$\frac{1}{256}$	$\frac{0}{256}$	$\frac{31}{256}$.	(15)

It follows that the uniform distributions $F_1, \ldots, F_{12}$ may be used $p_1 + \cdots + p_{12} = 88$ percent of the time; 88 percent of the area under $f(x)$ is taken up by the large rectangles.

An important technique has been given by Marsaglia for efficient selection of the 13 alternatives represented in (15). Given a random binary integer between 0 and 255, whose binary representation is $b_1 b_2 b_3 b_4 b_5 b_6 b_7 b_8$, we proceed as

Table 1

TABLES FOR EFFICIENT SELECTION OF
12 ALTERNATIVES, IN METHOD (16)

$A[0] = 0$	$B[40] = \frac{1}{4}$	$C[208] = 0$
$A[1] = 0$	$B[41] = \frac{1}{4}$	$C[209] = \frac{1}{4}$
$A[2] = 0$	$B[42] = \frac{1}{4}$	$C[210] = \frac{1}{2}$
$A[3] = \frac{1}{4}$	$B[43] = \frac{1}{2}$	$C[211] = \frac{1}{2}$
$A[4] = \frac{1}{4}$	$B[44] = \frac{3}{4}$	$C[212] = \frac{3}{4}$
$A[5] = \frac{1}{2}$	$B[45] = \frac{3}{4}$	$C[213] = \frac{3}{4}$
$A[6] = \frac{1}{2}$	$B[46] = \frac{3}{4}$	$C[214] = 1$
$A[7] = \frac{3}{4}$	$B[47] = 1$	$C[215] = 1$
$A[8] = 1$	$B[48] = \frac{3}{2}$	$C[216] = 1$
$A[9] = \frac{5}{4}$	$B[49] = \frac{3}{2}$	$C[217] = \frac{3}{2}$
	$B[50] = \frac{7}{4}$	$C[218] = \frac{3}{2}$
	$B[51] = 2$	$C[219] = \frac{3}{2}$
		$C[220] = \frac{7}{4}$
		$C[221] = \frac{7}{4}$
		$C[222] = \frac{9}{4}$
		$C[223] = \frac{9}{4}$
		$C[224] = \frac{5}{2}$

follows:

if $0 \le b_1b_2b_3b_4 < 10,$ let $X = A[b_1b_2b_3b_4] + \frac{1}{4}U;$

if $40 \le b_1b_2b_3b_4b_5b_6 < 52,$ let $X = B[b_1b_2b_3b_4b_5b_6] + \frac{1}{4}U;$ (16)

if $208 \le b_1b_2b_3b_4b_5b_6b_7b_8 < 225,$ let $X = C[b_1b_2b_3b_4b_5b_6b_7b_8] + \frac{1}{4}U;$

if $225 \le b_1b_2b_3b_4b_5b_6b_7b_8,$ generate X using one of the distributions $F_{13}, \ldots, F_{37}$, as described below.

Here A, B, and C are auxiliary tables which appear in Table 1, and U is a uniform deviate between zero and one.

To see why this procedure works, consider for example the case $j = 4$. $F_4(x)$ is the uniform distribution between $\frac{3}{4}$ and 1, and $X = \frac{3}{4} + \frac{1}{4}U$ has this distribution. The probability that this distribution is used in the above method is $\frac{1}{16}$ times the number of appearances of "$\frac{3}{4}$" in the A table, plus $\frac{1}{64}$ times the number of its appearances in B, plus $\frac{1}{256}$ times the number in C; this equals $\frac{1}{16} + \frac{3}{64} + \frac{2}{256} = \frac{30}{256} = p_4$, so $F_4(x)$ is selected with the right probability. Of course, it would be even faster to use the following method:

"if $0 \le b_1b_2b_3b_4b_5b_6b_7b_8 < 225,$ let $X = T[b_1b_2b_3b_4b_5b_6b_7b_8] + \frac{1}{4}U$"

in place of the first three tests in (16). Here T would be a table with 225 entries, 49 of which are "0", 45 of which are "$\frac{1}{4}$", 38 are "$\frac{1}{2}$", etc. By comparison, the previous method uses only 39 table entries; and it is only a little slower, since the test "$b_1b_2b_3b_4b_5b_6 < 52$?" needs to be made only three-eighths of the time,

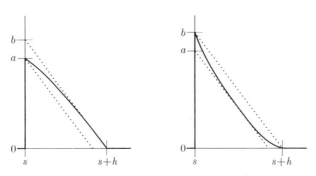

Fig. 10. Frequency functions for which Algorithm L may be used to generate random numbers.

and the test "$b_1 b_2 b_3 b_4 b_5 b_6 b_7 b_8 < 225$?" needs to be made only three-sixteenths of the time. The savings of storage space in a method like (16) is even more significant in the more general situation when there are more than 256 possibilities.

The reader should pause at this point to study the above method carefully, until he understands why method (16) selects the distributions F_j with the probability p_j, for $1 \leq j \leq 12$.

The distributions $F_{13}(x), \ldots, F_{24}(x)$ are also uniform distributions, and they may be handled similarly. They have quite small probability, however (some number between 0 and $\frac{1}{256}$), so it is most efficient to test for these probabilities in a part of our normal-deviate-generating algorithm which is not used very frequently (see step M4 in Algorithm M below). These distributions represent the correction to the large rectangles because of the truncation error in the probabilities (15). We have

$$p_j + p_{j+12} = \tfrac{1}{4}f(j/4), \qquad 1 \leq j \leq 12. \tag{17}$$

Now we turn to the question of computing random variables from the wedge-shaped distributions $F_{25}(x), \ldots, F_{36}(x)$. A typical case is shown in Fig. 10; when $x < 1$, the curved part is concave downward, and when $x > 1$ it is concave upward, but in each case the curved part is reasonably close to a straight line, and it can be enclosed in two parallel lines as shown.

Algorithm L (*Nearly linear densities*). This algorithm may be used to generate a random variable X for any distribution density $f(x)$ which satisfies the following conditions (cf. Fig. 10):

$$f(x) = 0, \qquad \text{for} \quad x < s \quad \text{and for} \quad x > s + h;$$
$$a - b(x - s)/h \leq f(x) \leq b - b(x - s)/h, \qquad \text{for} \quad s \leq x \leq s + h. \tag{18}$$

L1. [Get $U \leq V$.] Generate two independent random variables U, V, uniformly distributed between zero and one. If $U > V$, exchange $U \leftrightarrow V$.

L2. [Easy case?] If $V \leq a/b$, go to L4.

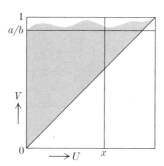

Fig. 11. Region of "acceptance" in Algorithm L.

L3. [Try again?] If $V > U + (1/b)f(s + hU)$, go back to step L1. (If a/b is close to 1, this step of the algorithm will not be necessary very often.)

L4. [Compute X.] Set $X \leftarrow s + hU$. ∎

To prove that this algorithm is valid, we observe that when step L4 is reached, the point (U, V) is a random point in the area shown in Fig. 11, namely, $0 \leq U \leq V \leq U + (1/b)f(s + hU)$. Conditions (18) ensure that

$$\frac{a}{b} \leq U + \frac{1}{b}f(s + hU) \leq 1.$$

Now the probability that $X \leq s + hx$, for $0 \leq x \leq 1$, is the ratio of area to the left of the vertical line $U = x$ in Fig. 11 to the total area, namely,

$$\int_0^x \frac{1}{b}f(s + hu)\, du \bigg/ \int_0^1 \frac{1}{b}f(s + hu)\, du = \int_s^{s+hx} f(v)\, dv;$$

therefore X has the correct distribution.

To use this algorithm, we need to determine a_j, b_j, s_j, h for the densities $f_{j+24}(x)$ of Fig. 9. It is not hard to see that when $1 \leq j \leq 12$,

$$f_{j+24}(x) = \frac{1}{p_{j+24}} \sqrt{\frac{2}{\pi}} (e^{-x^2/2} - e^{-(j/4)^2/2}), \qquad s_j \leq x \leq s_j + h;$$

$$h = \tfrac{1}{4}; \qquad s_j = (j - 1)/4; \tag{19}$$

$$p_{j+24} = \sqrt{\frac{2}{\pi}} \int_{s_j}^{s_j+h} (e^{-t^2/2} - e^{-(j/4)^2/2})\, dt.$$

Also

$$
\begin{aligned}
a_j &= f_{j+24}(s_j) & \text{when} && 1 \leq j \leq 4, \\
b_j &= f_{j+24}(s_j) & \text{when} && 5 \leq j \leq 12; \\
b_j &= -hf'_{j+24}(s_j + h) & \text{when} && 1 \leq j \leq 4, \\
a_j &= f_{j+24}(x_j) + (x_j - s_j)b_j/h & \text{when} && 5 \leq j \leq 12,
\end{aligned}
\tag{20}
$$

where x_j is the root of the equation $f'_{j+24}(x_j) = -b_j/h$.

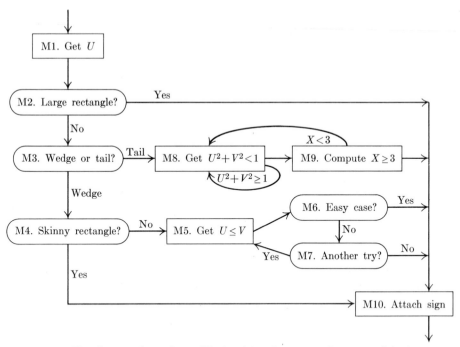

Fig. 12. The "rectangle-wedge-tail" algorithm for generating normal deviates.

The final distribution, $F_{37}(x)$, is to be computed only about one time in 400; it is used whenever a result $X \geq 3$ is to be given. A modification of Algorithm P can be applied in this case, as shown in steps M8–M9 of the algorithm below. Proof that this method is valid is left to the reader.

Now we can give the procedure in its entirety:

Algorithm M (*Marsaglia-MacLaren method for normal deviates*). This algorithm uses several auxiliary tables, constructed as explained in the text, and examples appear in Tables 1 and 2. We assume that a binary computer is being used; the procedure for a decimal computer is similar.

M1. [Get U.] Generate a uniform random number $U = .b_0 b_1 b_2 \ldots b_t$. (Here the b's are the bits in the binary representation of U. For reasonable accuracy, t should be at least 24.) Set $\psi \leftarrow b_0$. (Later, ψ will be used to determine the sign of the result.)

M2. [Large rectangle?] If $b_1 b_2 b_3 b_4 < 10$, where "$b_1 b_2 b_3 b_4$" denotes the binary integer $8b_1 + 4b_2 + 2b_3 + b_4$, set

$$X \leftarrow A[b_1 b_2 b_3 b_4] + .00 b_5 b_6 \ldots b_t$$

and go to M10. Otherwise, if $b_1 b_2 b_3 b_4 b_5 b_6 < 52$, set

$$X \leftarrow B[b_1 b_2 b_3 b_4 b_5 b_6] + .00 b_7 b_8 \ldots b_t$$

Table 2

EXAMPLE OF TABLES USED WITH ALGORITHM M*

j	$S[j]$	$P[j]$	$Q[j]$	$D[j]$	$E[j]$
1	0	0.885	0.881	0.51	16
2	$\frac{1}{4}$	0.895	0.885	0.79	8
3	$\frac{1}{2}$	0.910	0.897	0.90	5.33
4	$\frac{3}{4}$	0.929	0.914	0.98	4
5	1	0.945	0.930	0.99	3.08
6	$\frac{5}{4}$	0.960	0.947	0.99	2.44
7	$\frac{3}{2}$	0.971	0.960	0.98	2.00
8	$\frac{7}{4}$	0.982	0.974	0.96	1.67
9	2	0.987	0.982	0.95	1.43
10	$\frac{9}{4}$	0.991	0.989	0.93	1.23
11	$\frac{5}{2}$	0.994	0.992	0.94	1.08
12	$\frac{11}{4}$	0.997	0.996	0.94	0.95
13	3	1.000			

* In practice, the data in the P, Q, D, E tables would be given with much greater precision.

and go to M10. Otherwise, if $b_1b_2b_3b_4b_5b_6b_7b_8 < 225$, set

$$X \leftarrow C[b_1b_2b_3b_4b_5b_6b_7b_8] + .00b_9b_{10} \ldots b_t$$

and go to M10.

M3. [Wedge or tail?] Find the *smallest* value of j, $1 \leq j \leq 13$, for which $.b_1b_2 \ldots b_t < P[j]$. If $j = 13$, go to step M8.

M4. [Skinny rectangle?] If $.b_1b_2 \ldots b_t < Q[j]$, generate a new uniform random number U, set $X \leftarrow S[j] + \frac{1}{4}U$, and go to M10.

M5. [Get $U \leq V$.] Generate two new uniform deviates, U, V; if $U > V$, exchange $U \leftrightarrow V$. (We are now performing Algorithm L.) Set $X \leftarrow S[j] + \frac{1}{4}U$.

M6. [Easy case?] If $V \leq D[j]$, go to M10.

M7. [Another try?] If $V > U + E[j]$ $(e^{-(X^2 - S[j+1]^2)/2} - 1)$, go back to step M5; otherwise go to M10. (This step is executed with low probability.)

M8. [Get $U^2 + V^2 < 1$.] Generate two new independent uniform deviates, U, V; set $W \leftarrow U^2 + V^2$. If $W \geq 1$, repeat this step.

M9. [Compute $X \geq 3$.] Set $T \leftarrow \sqrt{(9 - 2\ln W)/W}$. Set $X \leftarrow U \times T$. If $X \geq 3$, go to M10. Otherwise set $X \leftarrow V \times T$. If $X \geq 3$, go to M10. Otherwise go back to step M8. (The latter will occur about half the time this step is encountered.)

M10. [Attach sign.] If $\psi = 1$, set $X \leftarrow -X$. ∎

This algorithm is a very pretty example of mathematical theory intimately interwoven with programming ingenuity. It is a fine illustration of the art of computer programming.

Tables A, B, and C have already been described; the other tables required by Algorithm M are constructed as follows:

$$S[j] = (j - 1)/4, \qquad 1 \le j \le 13;$$
$$P[j] = p_1 + p_2 + \cdots + p_{12} + (p_{13} + p_{25}) + \cdots + (p_{12+j} + p_{24+j}),$$
$$1 \le j \le 12; \ P[13] = 1; \qquad (21)$$
$$Q[j] = P[j] - p_{24+j}, \qquad 1 \le j \le 12;$$
$$D[j] = a_j/b_j, \qquad 1 \le j \le 12;$$
$$E[j] = \sqrt{\frac{2}{\pi}}\, e^{-(j/4)^2/2}/b_j p_{j+24}, \qquad 1 \le j \le 12.$$

[The quantities a_j, b_j, p_{j+24} are defined in (19) and (20).]

Table 2 shows the values for our example to only a few significant figures, but in a computer program, these quantities would all appear with full-word precision. Algorithm M requires a total of 101 auxiliary table entries in all.

This method is extremely fast, since 88 percent of the time only the steps M1, M2, and M10 will need to be executed, and the other steps aren't terribly slow either. We have broken the range from 0 to 3 into 12 parts in Fig. 9; if it were broken into more parts, say 48 parts, more table entries would be required, but the computation will stay in steps M1, M2, M10 over 97 percent of the time! Complete tables for both binary and decimal computers are given in the article by Marsaglia, MacLaren, and Bray, *CACM* **7** (1964), 4–10; an additional trick, overlapping parts of Tables A, B, C and S, is employed there in order to save storage space.

(3) Teichroew's method. We can also calculate normal deviates as follows: Generate 12 independent random variables $U_1, U_2, \ldots, U_{12}$, uniformly distributed between zero and one. Let $R = (U_1 + U_2 + \cdots + U_{12} - 6)/4$. Evaluate

$$X = ((((a_9 R^2 + a_7)R^2 + a_5)R^2 + a_3)R^2 + a_1)R, \qquad (22)$$

where

$$a_1 = 3.94984\ 6138, \qquad a_3 = 0.25240\ 8784, \qquad a_5 = 0.07654\ 2912,$$
$$a_7 = 0.00835\ 5968, \qquad a_9 = 0.02989\ 9776. \qquad (23)$$

Then X is a reasonable approximation to a normal deviate. Extremely large values of X are never obtained, but the probability of getting values larger than this method rightfully gives is less than $\frac{1}{50000}$.

This method is based on the fact that R has *approximately* a normal distribution with mean zero and standard deviation $\frac{1}{4}$. Let $F_1(x)$ be the exact distribution of R, and let $F(x)$ be the normal distribution function given in

Eq. (10). We want to set $X = F^{-1}(F_1(R))$; since $F_1(R)$ is uniformly distributed, X will then be normally distributed. The formula in (22) is a polynomial approximation to the function $F^{-1}(F_1(R))$ in the range $|R| \leq 1$.

(4) Comparison of the methods. We have given three methods for generating normal deviates. The polar method is rather slow, but it has essentially perfect accuracy, and it is very easy to write a program for the polar method if we assume square root and logarithm subroutines are available. Teichroew's method is also easy to program, and it requires no other subroutines; therefore it takes considerably less total memory space. Teichroew's method is only approximate, although in most applications its accuracy (an error bounded by 2×10^{-4} when $|R| \leq 1$) is quite satisfactory. Marsaglia's method is considerably faster than either of the others, and like the polar method it gives essentially perfect accuracy. It requires square root, logarithm, and exponential subroutines, and an auxiliary table of 100–400 constants, so its memory space requirement is rather high; yet its speed more than compensates for this on a large computer. A program for Marsaglia's method is considerably more difficult to prepare, but a general-purpose subroutine based on Algorithm M will be a valuable part of any subroutine library. Many applications of normal deviates call for a large quantity of random numbers, so speed of generation is important.

For more information about Teichroew's method, as well as a survey of several other methods now known to be inferior to those discussed here, see the paper by Mervin E. Muller, *JACM* **6** (1959), 376–383.

(5) Variations of the normal distribution. So far we have considered the normal distribution with mean zero and standard deviation one. If X has this distribution, then

$$Y = \mu + \sigma X \tag{24}$$

has the normal distribution with mean μ and standard deviation σ. Furthermore, if X_1 and X_2 are independent normal deviates with mean zero and standard deviation one, and if

$$Y_1 = \mu_1 + \sigma_1 X_1, \qquad Y_2 = \mu_2 + \sigma_2(\rho X_1 + \sqrt{1 - \rho^2}\, X_2), \tag{25}$$

then Y_1 and Y_2 are *dependent* random variables, normally distributed with means μ_1, μ_2 and standard deviations σ_1, σ_2, and with correlation coefficient ρ. (For a generalization to n variables, see exercise 13.)

D. The exponential distribution. Another important type of random number is the *exponential deviate.* This type of number occurs in "arrival time" situations; for example, if a radioactive substance emits alpha particles at a rate so that one particle is emitted every μ sec, on the average, then the amount of time between two successive emissions has the exponential distribution with mean μ.

This distribution is defined by the formula

$$F(x) = 1 - e^{-x/\mu}, \qquad x \geq 0. \tag{26}$$

It follows from this formula that if X has the exponential distribution with mean 1, μX has the exponential distribution with mean μ; therefore it suffices to consider the case $\mu = 1$. Three methods are generally used:

(1) *Logarithm method.* Clearly, if $y = F(x) = 1 - e^{-x}$, $x = F^{-1}(y) = -\ln(1-y)$. Therefore by Eq. (6), $-\ln(1-U)$ has the exponential distribution. Since $1 - U$ is uniformly distributed, when U is, we conclude that

$$X = -\ln U \tag{27}$$

is exponentially distributed with mean one. (The case $U = 0$ must be avoided in the computer program.)

(2) *Random minimization method.* The following algorithm (due to George Marsaglia) computes an exponentially distributed variable without using a logarithm subroutine.

Algorithm E (*Exponential distribution with mean* 1.) We use a table of constants $P[j]$, $Q[j]$, for $j \geq 1$, defined as follows:

$$P[j] = 1 - \frac{1}{e^j}, \qquad Q[j] = \frac{1}{e-1}\left(\frac{1}{1!} + \frac{1}{2!} + \cdots + \frac{1}{j!}\right). \tag{28}$$

The tables extend until the values are equal to the largest fraction which can be held in a computer word.

E1. [Begin fraction part.] Set $j \leftarrow 1$. Generate independent uniform deviates, U_0 and U_1, and set $X \leftarrow U_1$.

E2. [Minimization done?] If $U_0 < Q[j]$, go to E4.

E3. [Minimize.] Set $j \leftarrow j + 1$. Generate a uniform deviate, U_j; and if $X > U_j$, set $X \leftarrow U_j$. Go back to step E2.

E4. [Begin integer part.] (We have now computed the fractional part of our final answer, X, and we will add an appropriate integer to this quantity to complete the calculation.) Generate a new uniform deviate, U, and set $j \leftarrow 1$.

E5. [Correction done?] If $U < P[j]$, the algorithm terminates.

E6. [Correct by 1.] Set $j \leftarrow j + 1$, $X \leftarrow X + 1$, and return to step E5. ∎

To show that this method is valid, we first analyze the distribution of X at the beginning of step E4. If n is the final value of j, we have $X = \min(U_1, U_2, \ldots, U_n)$ for independent uniform deviates $U_1, U_2, \ldots, U_n$; so the probability $X \leq x$ is $p_n(x) = 1 - (1-x)^n$. The probability that n is the final value is $Q[n] - Q[n-1] = 1/(e-1)n!$; so the total probability that

$X \leq x$ is

$$\sum_{n \geq 1} \frac{p_n(x)}{(e-1)n!} = \frac{e}{e-1}(1 - e^{-x}), \qquad 0 \leq x \leq 1.$$

Similarly, we find by considering steps E4–E6 that $\lfloor X \rfloor \leq m$ with probability $P[m+1]$; and the overall probability that $m \leq X \leq m + x$ is

$$(P[m+1] - P[m])\left(\frac{e}{e-1}(1 - e^{-x})\right) = e^{-m} - e^{-(m+x)}, \qquad 0 \leq x \leq 1.$$

This proves X has the distribution $F(x) = 1 - e^{-x}$, for $0 \leq x < \infty$.

This algorithm is reasonably fast; on the average, steps E2 and E5 are executed only 1.582 times, and only 2.582 uniform deviates are calculated per exponential deviate. It may well be faster than calculating the logarithm of a single uniform deviate. An interesting modification of this method has been proposed by M. Sibuya [*Ann. Inst. Stat. Math.* **13** (1962), 231–237].

(*3*) *Rectangle-wedge-tail method.* Just as in the case of the normal distribution (and many others), there is an extremely fast method available for the exponential distribution, based on a decomposition of the frequency function. For the details of this case, see *CACM* **7** (1964), 298–300.

E. Other continuous distributions. We list here, briefly, some other distributions which arise reasonably often in practice, and methods to generate them in terms of the techniques we already know.

(*1*) *The chi-square distribution* with ν degrees of freedom, also called the gamma distribution of order $\nu/2$. We have

$$F(x) = \frac{1}{2^{\nu/2}\Gamma(\nu/2)} \int_0^x t^{\nu/2-1} e^{-t/2} \, dt, \qquad x \geq 0. \tag{29}$$

If $\nu = 2k$ where k is an integer, set $X = 2(Y_1 + Y_2 + \cdots + Y_k)$, where the Y's are independent random variables with the exponential distribution, each with mean 1. If $\nu = 2k + 1$, set $X = 2(Y_1 + \cdots + Y_k) + Z^2$, where the Y's are as before, and Z is an independent random variable with the normal distribution (mean zero, variance one). For proof, see exercise 16. Note that if $Y_1, \ldots, Y_k$ are computed with the logarithm method, $Y_1 + \cdots + Y_k = -\ln(U_1 \ldots U_k)$ requires only one logarithm to be computed.

(*2*) *The beta distribution* with ν_1 and ν_2 degrees of freedom, defined as follows:

$$F(x) = \frac{\Gamma((\nu_1 + \nu_2)/2)}{\Gamma(\nu_1/2)\Gamma(\nu_2/2)} \int_0^x t^{\nu_1/2-1}(1 - t)^{\nu_2/2-1} \, dt, \qquad 0 \leq x \leq 1. \tag{30}$$

Let Y_1 and Y_2 be independent, having the chi-square distribution with ν_1, ν_2 degrees of freedom, respectively; then set $X \leftarrow Y_1/(Y_1 + Y_2)$. Another method, which is valid also for noninteger ν_1 and ν_2, is to set $Y_1 \leftarrow U_1^{2/\nu_1}$, $Y_2 \leftarrow U_2^{2/\nu_2}$, repeating this process if necessary until finding $Y_1 + Y_2 \leq 1$; finally $X \leftarrow Y_1/(Y_1 + Y_2)$. [See M. D. Jöhnk, *Metrika* **8** (1964), 5–15.] Alternatively, if $\nu_1 = 2k_1$ and $\nu_2 = 2k_2$ are both even integers, we may set X to the k_1-th smallest of $k_1 + k_2 - 1$ independent uniform deviates.

(3) *The F-distribution* (variance-ratio distribution) with ν_1 and ν_2 degrees of freedom, defined by

$$F(x) = \frac{\nu_1^{\nu_1/2}\nu_2^{\nu_2/2}\Gamma((\nu_1+\nu_2)/2)}{\Gamma(\nu_1/2)\Gamma(\nu_2/2)} \int_0^x t^{\nu_1/2-1}(\nu_2+\nu_1 t)^{-\nu_1/2-\nu_2/2}\,dt, \qquad (31)$$

where $x \geq 0$. Let Y_1 and Y_2 be independent, having the chi-square distribution with ν_1, ν_2 degrees of freedom, respectively; set $X \leftarrow Y_1\nu_2/Y_2\nu_1$. Or set $X \leftarrow \nu_2 Y/\nu_1(1-Y)$, where Y has the beta distribution (30).

(4) *The t-distribution* with ν degrees of freedom, defined by

$$F(x) = \frac{\Gamma((\nu+1)/2)}{\sqrt{\pi\nu}\,\Gamma(\nu/2)} \int_{-\infty}^x (1+t^2/\nu)^{-(\nu+1)/2}\,dt. \qquad (32)$$

Let Y_1 be a normal deviate (mean 0, variance 1) and let Y_2 be independent of Y_1, having the chi-square distribution with ν degrees of freedom; set $X \leftarrow Y_1/\sqrt{Y_2/\nu}$.

(5) *Random point on n-dimensional sphere with radius one.* Let $X_1, X_2, \ldots, X_n$ be independent normal deviates (mean 0, variance 1); the desired point on the unit sphere is

$$(X_1/r, X_2/r, \ldots, X_n/r), \qquad \text{where} \qquad r = \sqrt{X_1^2 + X_2^2 + \cdots + X_n^2}. \qquad (33)$$

Note that if the X's are calculated using the polar method, Algorithm P, we compute two independent X's each time, and $X_1^2 + X_2^2 = -2\ln S$ (in the notation of that algorithm); this saves a little of the time needed to evaluate r. The validity of this method comes from the fact that the distribution function for the point $(X_1, \ldots, X_n)$ has a density which depends only on the distance from the origin, so when it is projected onto the unit sphere it has the uniform distribution. This method was first suggested by G. W. Brown.

F. Important integer-valued distributions. A probability distribution which consists only of integer values can essentially be handled by the techniques described at the beginning of this section; but some of these distributions are so important in practice, they deserve special mention here.

(1) *The geometric distribution.* If some event occurs with probability p, the number N of independent trials needed until the event first occurs (or between occurrences of the event) has the geometric distribution. We have $N = 1$ with probability p, $N = 2$ with probability $(1-p)p$, $\ldots$, $N = n$ with probability $(1-p)^{n-1}p$. (This is essentially the same situation as we have already considered in the gap test, in Section 3.3.2; it is also directly related to the number of times certain loops in the algorithms of this section are executed, e.g., steps P1–P3 of the polar method.)

A convenient way to generate a variable with this distribution when p is small is to set

$$N = \lceil \ln U/\ln(1-p) \rceil. \qquad (34)$$

To check this formula, we observe that $\lceil \ln U / \ln (1 - p) \rceil = n$ if and only if $n - 1 < \ln U / \ln (1 - p) \leq n$, that is, $(1 - p)^{n-1} > U \geq (1 - p)^n$, and this happens with probability $p(1 - p)^{n-1}$ as required.

The special case $p = \frac{1}{2}$ can be handled more easily on a binary computer, since formula (34) becomes $N = \lceil -\log_2 U \rceil$, that is, N is one more than the number of leading zero bits in the binary representation of U.

(2) *The binomial distribution* (t, p). If some event occurs with probability p and if we carry out t independent trials, the total number N of occurrences equals n with probability $\binom{t}{n} p^n (1 - p)^{t-n}$. (See Section 1.2.10.) No direct method analogous to (34) is available for this distribution. However, we could exploit the fact that if N_1 has the binomial distribution (t_1, p) and if, independently, N_2 has the binomial distribution (t_2, p), then $N_1 + N_2$ has the binomial distribution for $(t_1 + t_2, p)$. When t is large, the binomial distribution is approximately equal to the normal distribution with mean tp and standard deviation $\sqrt{tp(1 - p)}$. See also the technique in exercise 25.

(3) *The Poisson distribution* with mean μ. This distribution is related to the exponential distribution as the binomial distribution is related to the geometric: it represents the number of occurrences, per unit time, of an event which can occur at any instant of time; for example, the number of alpha particles emitted by a radioactive substance in a single second has a Poisson distribution. The probability that $N = n$ is

$$e^{-\mu} \mu^n / n!, \qquad n \geq 0. \tag{35}$$

If N_1 and N_2 are independent Poisson deviates with means μ_1 and μ_2, then $N_1 + N_2 = n$ with probability

$$\sum_{0 \leq k \leq n} \frac{e^{-\mu_1} \mu_1^k}{k!} \frac{e^{-\mu_2} \mu_2^{n-k}}{(n - k)!} = \frac{e^{-(\mu_1 + \mu_2)} (\mu_1 + \mu_2)^n}{n!}$$

so $N_1 + N_2$ has the Poisson distribution with mean $(\mu_1 + \mu_2)$.

Suppose that we wish to write a general-purpose subroutine which generates Poisson deviates with mean μ, where μ is input to the subroutine.

Algorithm Q (*Poisson distribution for variable μ*).

Q1. [Calculate exponental.] Set $p \leftarrow e^{-\mu}$, and set $N \leftarrow 0$, $q \leftarrow 1$. (Although $e^{-\mu}$ is generally calculated in floating-point arithmetic, using a standard subroutine, it is perhaps wise to use fixed-point arithmetic with suitable scaling and rounding for subsequent operations on the quantities p and q.)

Q2. [Get uniform variable.] Generate a random variable, U, uniformly distributed between zero and one.

Q3. [Multiply.] Set $q \leftarrow qU$.

Q4. [Test if $< e^{-\mu}$.] If $q \geq p$, set $N \leftarrow N + 1$ and return to step Q2. Otherwise the algorithm terminates with output N. ∎

To prove the validity of this method, we observe that independent uniform deviates will satisfy the conditions

$$U_1 \geq p, \quad U_1 U_2 \geq p, \quad \ldots, \quad U_1 U_2 \ldots U_n \geq p, \quad U_1 U_2 \ldots U_{n+1} < p$$

with probability

$$p \left(\ln \frac{1}{p} \right)^n \Big/ n!, \quad \text{when} \quad 0 < p \leq 1.$$

This follows by induction on n, by integration of

$$\int_p^1 (p/u_1)(\ln (u_1/p))^{n-1} \, du_1/(n-1)!.$$

Another method for generating the Poisson distribution, due to P. Kribs, is based on the summation property mentioned earlier. It requires more detailed programming than Algorithm Q, but it has the potential of running somewhat faster if it is skillfully coded:

Algorithm K (*Poisson distribution for variable μ.*)   An auxiliary table $M[1] < M[2] < \cdots < M[n]$ is used, as described below.

K1. [Initialize.] Set $m \leftarrow \mu$, $j \leftarrow n$, $N \leftarrow 0$.

K2. [$m \geq M[j]$?] If $m < M[j]$, go to step K5.

K3. [Generate mean $M[j]$.] Let X be a random integer having the Poisson distribution with mean $M[j]$. (This is to be generated efficiently by using tables especially set up for this case, according to the general method for integer distributions found in exercise 21.)

K4. [Update N, m.] Set $m \leftarrow m - M[j]$, $N \leftarrow N + X$, and return to step K2.

K5. [Decrease j.] Decrease j by 1. If $j > 0$, return to step K2, otherwise the algorithm terminates. ∎

To use this algorithm, we must provide special routines for particular values of μ, and these values are given in the table $M[1], M[2], \ldots, M[n]$. For example, we might have $n = 10$, and

$j =$	1	2	3	4	5	6	7	8	9	10	
$M[j] =$	2^{-15}	2^{-12}	2^{-9}	2^{-6}	2^{-3}	2^{-1}	1	2	4	8	(36)

This method is inefficient for large values of μ, say $\mu \geq 50$. When $\mu < M[1] = 2^{-15}$, the above algorithm returns the value $N = 0$, since $N > 0$ with probability $1 - e^{-\mu}$ and this is less than $\frac{1}{32000}$. The Poisson distribution for very small values of μ is extremely easy to generate, because N will, for all practical purposes, always be quite small. Only the distributions for $M[j] = 4$ and 8 in the above list will require much of a table in step K3.

For large variable μ, J. H. Ahrens has developed an efficient but rather complicated method of order $\sqrt{\mu}$. His procedure partitions the Poisson probabilities into two distributions, one of which looks like an isosceles triangle.

EXERCISES

1. [*10*] If α and β are real numbers with $\alpha < \beta$, how would you generate a random real number uniformly distributed between α and β?

2. [*M16*] Assuming that mU is a random integer between 0 and $m - 1$, what is the *exact* probability that $\lfloor kU \rfloor = r$, if $0 \le r < k$? Compare this with the desired probability $1/k$.

▶ **3.** [*14*] Discuss treating U as an integer and *dividing* by k to get a random integer between 0 and $k - 1$, instead of multiplying as suggested in the text. Thus (1) would be changed to

```
ENTA  0
LDX   U
DIV   K
```

with the result appearing in register X. Is this a good method?

4. [*M20*] Prove the two relations in (7).

▶ **5.** [*21*] Suggest an efficient method to compute a random variable with the distribution $px + qx^2 + rx^3$, where $p \ge 0$, $q \ge 0$, $r \ge 0$, and $p + q + r = 1$.

▶ **6.** [*HM21*] A quantity X is computed by the following method:

"*Step* 1. Generate two independent uniform deviates U, V.

Step 2. If $U^2 + V^2 \ge 1$, return to step 1; otherwise set $X \leftarrow U$."

What is the distribution function of X? How many times will step 1 be performed? (Give the mean and standard deviation.)

7. [*M18*] Explain why, in Marsaglia's method for normal deviates, the decision to make p_j a multiple of $\frac{1}{256}$ implies that we should take $p_j = \lfloor 64f(j/4) \rfloor / 256$, $1 \le j \le 12$.

8. [*10*] Why are there "skinny" rectangles $f_{13}, \ldots, f_{24}$ in Marsaglia's method, as well as the larger rectangles $f_1, \ldots, f_{12}$? (Why is this better than putting (f_1, f_{13}), (f_2, f_{14}), ... together into single large rectangles?)

9. [*HM10*] Why is the curve $f(x)$ of Fig. 9 concave downward for $x < 1$, concave upward for $x > 1$?

10. [*HM21*] Derive the formulas for a_j, b_j in Eqs. (20). Also show that $E[j] = 16/j$, if $1 \le j \le 4$; $E[j] = 1/(e^{j/16 - 1/32} - 1)$, if $5 \le j \le 12$.

▶ **11.** [*HM27*] Prove that steps M8–M9 of Algorithm M generate a random variable with the appropriate tail of the normal distribution; i.e., the probability that $X \le x$ should be

$$\int_3^x e^{-t^2/2} \, dt \Big/ \int_3^\infty e^{-t^2/2} \, dt,$$

when $x \ge 3$.

12. [*HM46*] Teichroew's polynomial, Eq. (22), is a truncated sum of Chebyshev polynomials and it may not be the best possible polynomial approximation of degree 9 to the function $F^{-1}(F_1(R))$, in this application. Is there a better polynomial to use?

13. [*HM25*] Given a set of n independent normal deviates, $X_1, X_2, \ldots, X_n$, with mean 0 and variance 1, show how to find constants b_j and a_{ij}, $1 \le j \le i \le n$, so that if

$$Y_1 = b_1 + a_{11}X_1, \qquad Y_2 = b_2 + a_{21}X_1 + a_{22}X_2, \qquad \dots,$$
$$Y_n = b_n + a_{n1}X_1 + a_{n2}X_2 + \cdots + a_{nn}X_n,$$

then $Y_1, Y_2, \dots, Y_n$ are dependent, normally distributed variables, Y_j has mean μ_j, and the Y's have a given covariance matrix (c_{ij}). (The covariance, c_{ij}, of Y_i and Y_j is defined to be the average value of $(Y_i - \mu_i)(Y_j - \mu_j)$. In particular, c_{jj} is the variance of Y_j, the square of its standard deviation. Not all matrices (c_{ij}) can be covariance matrices, and your construction is, of course, only supposed to work whenever a solution to the given conditions is possible.)

14. [*M20*] If X is a random variable with continuous distribution $F(x)$, and if c is a constant, what is the distribution of cX?

15. [*HM21*] If X_1 and X_2 are independent random variables with the respective distributions $F_1(x)$ and $F_2(x)$, and the densities $f_1(x) = F_1'(x)$, $f_2(x) = F_2'(x)$, what are the distribution and density functions of the quantity $X_1 + X_2$?

16. [*HM25*] (a) Show that the chi-square distribution with one degree of freedom is the distribution of X^2, if X has the normal distribution with mean zero and variance one. (b) Show that the chi-square distribution with two degrees of freedom is the exponential distribution with mean 2. (c) Show that if X_1 and X_2 are independent random variables, where X_1 has the chi-square distribution with ν_1 degrees of freedom and X_2 has the chi-square distribution with ν_2 degrees of freedom, then $X_1 + X_2$ has the chi-square distribution with $\nu_1 + \nu_2$ degrees of freedom.

▶ **17.** [*M24*] What is the *distribution function* $F(x)$ for the geometric distribution with probability p? What is the *generating function* $G(z)$? What are the mean and standard deviation of this distribution?

18. [*M24*] Suggest a method to compute a random integer N for which N takes the value n with probability $np^2(1-p)^{n-1}$, $n \geq 0$. (The case of particular interest is when p is rather small.)

19. [*M22*] (a) Find the average number of times step E2 of Algorithm E is executed, and also find the standard deviation. (b) Same question for step E6.

▶ **20.** [*23*] Suppose we want to compute a random variable X with the following finite distribution:

Value of $X =$	x_1	x_2	x_3	x_4	x_5	x_6
With probability $=$	$\frac{90}{512}$	$\frac{81}{512}$	$\frac{131}{512}$	$\frac{10}{512}$	$\frac{32}{512}$	$\frac{168}{512}$

Show how the value of X can be efficiently obtained from a random nine-bit binary integer, $b_1b_2b_3b_4b_5b_6b_7b_8b_9$, using a method analogous to (16) and using auxiliary tables containing less than 35 entries in all.

▶ **21.** [*24*] The situation in the preceding exercise is idealized in the sense that we usually will not have all probabilities an exact multiple of a convenient number like 512. Suppose the probabilities which appear in that exercise are modified so that they are actually equal to the following:

Value of $X =$	x_1	x_2	x_3	x_4	x_5	x_6
With probability $=$	$\frac{89}{512} + \epsilon_1$	$\frac{80}{512} + \epsilon_2$	$\frac{131}{512} + \epsilon_3$	$\frac{10}{512} + \epsilon_4$	$\frac{32}{512} + \epsilon_5$	$\frac{168}{512} + \epsilon_6$

Here $0 \leq \epsilon_j < \frac{1}{512}$, and $\epsilon_1 + \epsilon_2 + \epsilon_3 + \epsilon_4 + \epsilon_5 + \epsilon_6 = \frac{2}{512}$. Show how to make this choice efficiently, when given a uniform binary deviate $U = .b_1b_2 \cdots b_t$, where

t is relatively large. As in exercise 20, try to do this using auxiliary tables containing less than 35 entries in all.

22. [*HM46*] Can the exact Poisson distribution for large μ be obtained by generating an appropriate normal deviate, converting it to an integer in some convenient way, and applying a (possibly complicated) correction a small percent of the time? Can the binomial distribution for large t be done in a similar way?

23. [*HM23*] (J. von Neumann.) Are the following two ways to generate a random quantity X equivalent (i.e., does the quantity X have the same distribution)?

> *Method 1*: Set $X \leftarrow \sin\left((\pi/2)U\right)$, where U is uniform.

> *Method 2*: Generate two uniform deviates, U, V, and if $U^2 + V^2 \geq 1$, repeat until $U^2 + V^2 < 1$. Then set $X \leftarrow |U^2 - V^2|/(U^2 + V^2)$.

24. [*HM40*] (S. Ulam, J. von Neumann.) Let V_0 be a randomly selected real number between 0 and 1, and define the sequence $\langle V_n \rangle$ by the rule $V_{n+1} = 4V_n(1 - V_n)$. Now if this computation is done with perfect accuracy, the result should be a sequence with the distribution $\sin^2 \pi U$, where U is uniform, i.e., with distribution function

$$F(x) = \frac{1}{\sqrt{2\pi}} \int_0^x \frac{dx}{\sqrt{x(1-x)}}.$$

For if we write $V_n = \sin^2 \pi U_n$, we find that $U_{n+1} = (2U_n) \bmod 1$; and by the fact that almost all real numbers have a random binary expansion (see Section 3.5), this sequence U_n is equidistributed. But if the computation of V_n is done with only finite accuracy, the above argument breaks down because we soon are dealing with noise from the roundoff error. (*Reference:* von Neumann's *Collected Works*, Vol. V, pp. 768–770.)

Analyze the sequence $\langle V_n \rangle$ defined above when only finite accuracy is present, both empirically (for various different choices of V_0) and theoretically. Does the sequence have a distribution resembling the expected distribution? Is it of any possible use for random-number generation?

25. [*M25*] Let X_1, X_2, ..., X_5 be binary words each of whose bits is independently 0 or 1 with probability $\frac{1}{2}$. What is the probability that a given bit position of $X_1 \vee (X_2 \wedge (X_3 \vee (X_4 \wedge X_5)))$ contains a 1? Generalize.

3.4.2. Random Sampling and Shuffling

Many data processing applications call for an unbiased choice of n records at random, from a file containing N records. This problem arises, for example, in quality control or other statistical calculations where sampling is needed. Usually N is very large, so that it is impossible to contain all the data in memory at once; therefore we seek an efficient procedure for selecting n records by deciding whether to accept or to reject each record as it comes along.

Several methods have been devised for this problem. The most obvious approach is to select each record with probability n/N; this may sometimes be appropriate, but it gives only an *average* of n records in the sample; the standard deviation is $\sqrt{n(1 - (n/N))}$, and it is possible that the sample will be either too large for the desired application, or too small to give the necessary results.

A simple modification of the "obvious" procedure gives the desired result:

The $(t + 1)$st record should be selected with probability $(n - m)/(N - t)$, if m items have already been selected. This is the appropriate probability, since of all the possible ways to choose n things from N such that m values occur in the first t, exactly

$$\binom{N - t - 1}{n - m - 1} \Big/ \binom{N - t}{n - m} = \frac{n - m}{N - t} \tag{1}$$

of these select the $(t + 1)$st element.

The idea developed in the preceding paragraph may be formulated into the following algorithm:

Algorithm S (*Selection sampling technique*). To select n records at random from a set of N, where $0 < n \leq N$.

S1. [Initialize.] Set $t \leftarrow 0$, $m \leftarrow 0$.

S2. [Generate U.] Generate a random number U, uniformly distributed between zero and one.

S3. [Test.] If $(N - t)U \geq n - m$, go to step S5.

S4. [Select.] Select the next record for the sample, and increase m and t by 1. If $m < n$, go to step S2; otherwise the sample is complete and the algorithm terminates.

S5. [Skip.] Skip the next record (do not include it in the sample), increase t by 1, and go to step S2. ∎

This algorithm, at first glance, may appear to be unreliable and, in fact, incorrect; but a careful analysis (see the exercises below) shows that it is completely trustworthy. It is not difficult to verify that

a) *Exactly* n elements will be selected.

b) The sample is completely unbiased; in particular, the probability that any given element is selected, e.g., the last element of the file, is n/N.

Statement (b) is true in spite of the fact that we are *not* selecting the $(t + 1)$st item with probability n/N, we select it with the probability in Eq. (1)! This has caused some confusion in the published literature. Can the reader explain this seeming contradiction?

(*Note:* When using Algorithm S, care should be taken to use a different source of random numbers U each time the program is run, to avoid connections between the samples obtained on different days. This can be done, for example, by choosing a different value of X_0 for the linear congruential method each time; X_0 could be set to the current date, or to the last X value generated on the previous run of the program.)

We will usually not have to pass over all N records; in fact, since (b) above says the last record is selected with probability n/N, we will terminate the algorithm *before* considering the last record exactly $(1 - n/N)$ of the time. The average number of records considered when $n = 2$ is about $\frac{2}{3}N$, and the general formulas are given in exercises 5 and 6.

Algorithm S can be used only when the value of N is known in advance. On the other hand, suppose that we want to select n items at random from a file, but we do not know how many are present in that file. We could first go through and count the records, then take a second pass to select them; but it is generally better to sample $m \geq n$ of the original items on the first pass, where m is much less than N, so that only m items must be considered on the second pass. The trick is, of course, to do this in such a way that the final result is a truly random sample of the original file.

We will now describe an ingenious technique which makes this unlikely-sounding idea actually work. Assume that we have a random-number generator that will yield at least N distinct values, before any value is repeated; for example, a linear congruential sequence with period length greater than N will do the job. (Period lengths are generally on the order of several billion.) The modification by MacLaren and Marsaglia, Algorithm 3.2.2M, is recommended for this application.

The idea is to compute N random values, and to ascertain which are the largest n of these. The corresponding n records will be the ones which appear in the final sample. During the first pass, we construct a "reservoir" which contains only those m records which are possible candidates, i.e., those records which correspond to a random value that has not been preceded by n larger values. (The first n items will always go into the reservoir.)

Algorithm R (*Reservoir sampling*). To mechanize the process described informally above, we maintain an internal table of n items, of the form (Y, I), where Y is the random value corresponding to the Ith record in the reservoir. Initially this internal table is entirely filled with $(0, 0)$.

R1. [Initialize.] Set $m \leftarrow 0$.

R2. [Generate U.] Let U be a random number uniformly distributed between zero and one. (The algorithm expects U to be *unequal* to any of the other random values previously computed in this step, as explained above.)

R3. [Test.] Let the smallest Y value, of all the (Y, I) entries in the auxiliary table, be Y_0. If $U < Y_0$, proceed to step R6.

R4. [Add to reservoir.] Transfer the next record of the file to the reservoir, and increase m by 1.

R5. [Update table.] Replace the table entry (Y_0, I_0), which now contains Y_0, by (U, m). Go to step R7.

R6. [Skip.] Pass over the next record of the file.

R7. [End of file?] If there are no more records in the file, go on to step R8; otherwise return to step R2.

R8. [Second pass.] Sort the auxiliary table entries (Y, I) on I, and if the resulting values are $I_1 < I_2 < \cdots < I_n$, these are the indices of the records to select from the reservoir for the final sample. ∎

The reader should work out the example of this algorithm which appears in exercise 9.

The (Y, I) table in Algorithm R can be maintained in such a way that the determination of Y_0 in step R3 is very fast, and that the insertion of (U, m) in step R5 is also efficient. For details, see the tree selection sort algorithms in Chapter 5.

The natural question to ask about Algorithm R is, "What is the expected size, m, of the reservoir?" Exercise 11 shows that the average value of m is exactly $n(1 + H_N - H_n)$; this is approximately $n(1 + \ln (N/n))$. So if $N/n = 1000$, the reservoir will contain only about $\frac{1}{125}$ as many items as the original file.

Note that Algorithms S and R can be used to obtain samples for several independent categories simultaneously. For example, if we have a large file of names and addresses of U.S. residents, we could pick random samples of exactly 10 people from each of the 50 states without making 50 passes through the file, and without first sorting the file by state.

Algorithms S and R and a number of other sampling techniques are discussed in a paper by C. T. Fan, Mervin E. Muller, and Ivan Rezucha, *Journal of the American Statistical Association* **57** (1962), 387–402. Algorithm S was independently discovered by T. G. Jones, *CACM* **5** (1962), 343.

The *sampling problem*, as described here, can be regarded as the computation of a random *combination*, according to the conventional definition of combinations of N things taken n at a time (see Section 1.2.6). Now let us consider the problem of computing a random *permutation* of t objects; we will call this the *shuffling problem*, since shuffling a deck of 52 cards is nothing more than subjecting it to a random permutation.

There are two principal methods available for shuffling. The first procedure, suggested by S. Ulam in the early 1950's, is to have, say, five subroutines, each of which applies a certain permutation to the elements $X_1, X_2, \ldots, X_t$. Assume that a *sequence* of random permutations is to be generated. To get the next permutation of the sequence, we apply three of the five available permutations to the previous permutation. These three are not necessarily different from each other, and they are selected at random.

This method is something like the way people ordinarily shuffle a deck of cards (see exercise 13). Perhaps the reader will be curious to know what kind of permutations a human card shark gives to a deck of cards; to answer this question, the author asked William A. Logan (a computer scientist and card shuffler who is believed to be human), to record the results of three shuffles of a deck of 52 cards. The following permutations resulted:

$\pi_1 = $ (1 23 31) (2 42 20) (3 34 52 48 24 28 39 50 15 40 41 9 17 38 25
            43 47 14 8 13 35 11 22 10 12 36 49 46 44 4 37) (5 19 18 45) (6)
            (7 30) (16 51 27 32) (21 29 33 26)

$\pi_2 = $ (1 5 24 15 36 19) (2 18 27 52 37 44 25 47 50 46 20 7 17 10) (3 28
            6 39 13 23 41 43 11 38 49 12 4 34 8 40 21 14 30 16 45 51 22 9 33
            32 35 31 26) (29 42) (48)

$\pi_3 = $ (1 42 19 40 30) (2 51 10 34 44 48) (3 12 33 49 8 23 20 25 14 7 43 36)
    (4 11 26 16 46 32 24 28 9 31 29 52 21 5 47 13 39 38 17 15 37 45 6
    41 50 27) (18 22 35)

These permutations are shown in cycle form (cf. Section 1.3.3). The most efficient way to apply a permutation to a sequence of numbers is to follow its cycle structure; for example, using MIX, the first permutation above could be coded

```
LDA   X+1
LDX   X+31
STX   X+1
LDX   X+23
STX   X+31
STA   X+23
LDA   X+2
LDX   X+20
```

etc.

We would like to make sure that our set of five special permutations is capable of generating all $t!$ permutations, if applied in the appropriate sequence. It can be shown that if two permutations with the respective forms,

$$\pi_1 = (1\ 2) \text{ (any product of disjoint cycles each of odd length)},$$
$$\pi_2 = (1) \text{ (any cycle of length } t - 1 \text{ excluding the element 1)}, \tag{2}$$

are included in the set, it is possible to generate any permutation of t elements by applying π_1 and π_2 repeatedly in some order. (See exercise 12.) John D. Dixon has shown that almost $\frac{3}{4}$ of all pairs of permutations will have this property [*Math. Z.* **110** (1969), 199–205].

The second method for shuffling is intimately related to Algorithm 3.3.2P. The reader should refer back to that algorithm; it gives a simple correspondence between each of the $t!$ possible permutations and a set of numbers $C[1]$, $C[2]$, . . . , $C[t]$ with $0 \leq C[j] < j$. It is easy to compute such a set of numbers at random, and we can use the correspondence to produce a random permutation.

Algorithm P (*Shuffling*). Let $X_1, X_2, \ldots, X_t$ be a set of t numbers to be shuffled.

P1. [Initialize.] Set $j \leftarrow t$.

P2. [Generate U.] Generate a random number U, uniformly distributed between zero and one.

P3. [Exchange.] Set $k \leftarrow \lfloor jU \rfloor + 1$. (Now k is a random integer between 1 and j). Exchange $X_k \leftrightarrow X_j$.

P4. [Decrease j.] Decrease j by 1. If $j > 1$, return to step P2. ∎

This algorithm was first published by L. E. Moses and R. V. Oakford, in *Tables of Random Permutations* (Stanford University Press, 1963), and by R. Durstenfeld, *CACM* **7** (1964), 420. It is superior to Ulam's method because it truly produces "truly random" permutations and takes less memory space.

EXERCISES

1. [*M12*] Explain Eq. (1).

2. [*20*] Prove that exactly n records are selected in Algorithm S, if $0 < n \le N$.

▶ **3.** [*22*] The $(t+1)$st item in Algorithm S is selected with probability $(n-m)/(N-t)$, *not* n/N, yet the text claims the sample is unbiased so each item should be selected with the *same* probability! How can both of these statements be true?

4. [*M23*] Let $p(m, t)$ be the probability that exactly m items are selected from among the first t in the selection sampling technique. Show directly from Algorithm S that

$$p(m, t) = \binom{t}{m}\binom{N-t}{n-m} \Big/ \binom{N}{n}, \qquad \text{for} \qquad 0 \le t \le N.$$

5. [*M24*] What is the average value of t when Algorithm S terminates? (In other words, how many of the N records have passed before the sample is complete?)

6. [*M24*] What is the standard deviation of the value computed in the previous exercise?

▶ **7.** [*M25*] Prove that any *given* choice of n records from the set of N is obtained by Algorithm S with probability $1/\binom{N}{n}$. Therefore the sample is completely unbiased.

▶ **8.** [*18*] What happens in Algorithm R if the file contains less than n records in all? Suggest a way to detect this condition by making a small modification to the algorithm.

9. [*22*] Suppose that Algorithm R is applied with $n = 3$ to a file containing 20 records, and the following numbers U are the random values generated at step R2:

$$0.53,\ 0.97,\ 0.66,\ 0.30,\ 0.81,\ 0.19,\ 0.09,\ 0.31,\ 0.67,\ 0.62,$$
$$0.04,\ 0.05,\ 0.73,\ 0.54,\ 0.42,\ 0.99,\ 0.40,\ 0.78,\ 0.69,\ 0.80;$$

the twenty records themselves are numbered $1, 2, \ldots, 20$. Which records go into the reservoir? Which records go into the final sample?

10. [*30*] Show how the (Y, I) table in Algorithm R may be arranged, as suggested in the text, so that the searching and insertion processes of steps R3 and R5 are performed efficiently.

▶ **11.** [*M25*] Let p_m be the probability that exactly m elements are put into the reservoir during the first pass of Algorithm R. Determine the generating function $G(z) = \sum_m p_m z^m$, and find the mean and standard deviation. (Use the ideas of Section 1.2.10.)

12. [*M23*] Let π_1, π_2 be the permutations in (2). Prove the following: (a) Some power of π_1 equals the two-cycle (12). (b) Some product of π_1 and π_2 equals the two-cycle $(1\ k)$, for each k with $2 \le k \le t$. (c) All cycles $(j\ k)$ with $j \ne k$ can be obtained as a product of π_1 and π_2. (d) All permutations of t elements can be obtained as a product of π_1 and π_2.

13. [*M23*] (S. W. Golomb.) A common procedure for card shuffling is to divide the deck into two parts as equal as possible, and to "riffle" them together. (See the discussion of card-playing etiquette in Hoyle's rules of card games; we read, "A shuffle of this sort should be made about three times to mix the cards thoroughly.") Consider a deck of $2n - 1$ cards $X_1, X_2, \ldots, X_{2n-1}$; a "perfect shuffle" s divides this deck into $X_1, X_2, \ldots, X_n$ and $X_{n+1}, \ldots, X_{2n-1}$ and perfectly interleaves them, to obtain $X_1, X_{n+1}, X_2, X_{n+2}, \ldots, X_{2n-1}, X_n$. The "cut" operation c^j changes $X_1, X_2, \ldots,$ X_{2n-1} into $X_{j+1}, \ldots, X_{2n-1}, X_1, \ldots, X_j$. Show that by combining perfect shuffles and cuts, at most $(2n-1)(2n-2)$ different arrangements of the deck are possible, if $n > 1$.

*3.5. WHAT IS A RANDOM SEQUENCE?

A. Introductory remarks. We have seen in this chapter how to generate sequences

$$\langle U_n \rangle = U_0, U_1, U_2, \ldots \tag{1}$$

of real numbers between zero and one, that is, $0 \le U_n < 1$, which have been called "random" sequences even though they are completely deterministic in character. To justify this terminology, the claim was made that the numbers "behave as if they are truly random." This statement may be satisfactory for practical purposes (at the present time), but it sidesteps a very important philosophical and theoretical question: Precisely what do we mean by "random behavior"? A quantitative definition of random behavior is needed. It is undesirable to talk about concepts that we do not really understand, especially since many apparently paradoxical statements can be made about random numbers.

The mathematical theory of probability and statistics carefully avoids answering the question; it refrains from making absolute statements, and instead expresses everything in terms of how much *probability* is to be attached to statements involving random sequences of independent events. The axioms of probability theory are set up so that abstract probabilities can be computed readily, but nothing is said about what probability really signifies, or how this concept can be applied meaningfully to the actual world. In the book *Probability, Statistics, and Truth* (Macmillan, 1957), R. von Mises discusses this situation in detail, and presents the view that a proper definition of probability depends on obtaining a proper definition of a random sequence.

Let us paraphrase here two statements about random sequences which have been given recently by other authors.

> *D. H. Lehmer (1951):* "A random sequence is a vague notion embodying the idea of a sequence in which each term is unpredictable to the uninitiated and whose digits pass a certain number of tests, traditional with statisticians and depending somewhat on the uses to which the sequence is to be put."

> *J. N. Franklin (1962):* "The sequence (1) is random if it has every property that is shared by all infinite sequences of independent samples of random variables from the uniform distribution."

Franklin's statement essentially generalizes Lehmer's to say that the sequence must satisfy *all* statistical tests. His definition is not completely precise, and we will see at the close of this section that a reasonable interpretation of his statement leads us to conclude that there is no such thing as a random sequence! The definition is therefore too restrictive, so we turn to Lehmer's statement and attempt to make *it* precise. What we really want is a relatively short list of mathematical properties, each of which is satisfied by our intuitive notion of a random sequence; furthermore, this list is to be complete enough so that we are willing to agree that *any* sequence satisfying these properties is "random." In this section, we will develop what seems to be an adequate definition of randomness according to these criteria, although many interesting questions remain to be answered.

Let u and v be real numbers, $0 \leq u < v \leq 1$. If U is a random variable which is uniformly distributed between 0 and 1, the probability that $u \leq U < v$ is equal to $v - u$. For example, if $u = \frac{1}{3}$ and $v = \frac{2}{3}$, the probability that $\frac{1}{3} \leq U < \frac{2}{3}$ is $\frac{1}{3}$. How can we translate this property of the single number U into a property of the infinite sequence $U_0, U_1, U_2, \ldots$? The obvious answer is to count how many times U_n lies between u and v, and the average number of times should equal $v - u$. Our intuitive idea of probability is based in this way on the frequency of occurrence. More precisely, let $\nu(n)$ be the number of values of j, $0 \leq j < n$, such that $u \leq U_j < v$; we want the ratio $\nu(n)/n$ to approach the value $v - u$ as n approaches infinity:

$$\lim_{n \to \infty} \nu(n)/n = v - u.$$

If this condition holds for all choices of u and v, the sequence is said to be *equidistributed*.

Let $S(n)$ be a statement about the integer n and the sequence $U_1, U_2, \ldots$; for example, $S(n)$ might be the statement considered above, namely "$u \leq U_n < v$". We can generalize the idea used in the preceding paragraph to define "the probability that $S(n)$ is true" with respect to a particular infinite sequence: Let $\nu(n)$ be the number of values of j, $0 \leq j < n$, such that $S(j)$ is true.

Definition A. *We say* $\Pr\big(S(n)\big) = \lambda$, *if* $\lim_{n \to \infty} \nu(n)/n = \lambda$. (Read, "The probability that $S(n)$ is true is λ, if the limit as n tends to infinity of $\nu(n)/n$ is λ.")

In terms of this notation, the sequence $U_0, U_1, \ldots$ is equidistributed if and only if $\Pr(u \leq U_n < v) = v - u$, for all real numbers u, v with $0 \leq u < v \leq 1$.

A sequence may be equidistributed without being random. For example, if $U_0, U_1, \ldots$ and $V_0, V_1, \ldots$ are equidistributed sequences, it is not hard to show that the sequence

$$W_0, W_1, W_2, W_3, \ldots = \tfrac{1}{2}U_0, \tfrac{1}{2} + \tfrac{1}{2}V_0, \tfrac{1}{2}U_1, \tfrac{1}{2} + \tfrac{1}{2}V_1, \ldots \qquad (3)$$

is also equidistributed, since the sequence $\frac{1}{2}U_0, \frac{1}{2}U_1, \ldots$ is equidistributed between 0 and $\frac{1}{2}$, while the alternate terms, $\frac{1}{2} + \frac{1}{2}V_0, \frac{1}{2} + \frac{1}{2}V_1, \ldots$, are equidistributed between $\frac{1}{2}$ and 1. In the sequence of W's, a value less than $\frac{1}{2}$ is always followed by a value greater than or equal to $\frac{1}{2}$, and conversely; hence that sequence is not random by any reasonable definition. A stronger property than equidistribution is needed.

A natural generalization of the equidistribution property, which removes the objection stated in the preceding paragraph, is to consider adjacent pairs of numbers of our sequence. We can require the sequence to satisfy the condition

$$\Pr(u_1 \le U_n < v_1 \quad \text{and} \quad u_2 \le U_{n+1} < v_2) = (v_1 - u_1)(v_2 - u_2) \qquad (4)$$

for any four numbers u_1, v_1, u_2, v_2 with $0 \le u_1 < v_1 \le 1$, $0 \le u_2 < v_2 \le 1$. In general, for any positive integer k we can require our sequence to be k-*distributed* in the following sense:

Definition B. *The sequence (1) is said to be k-distributed if*

$$\Pr(u_1 \le U_n < v_1, \ldots, u_k \le U_{n+k-1} < v_k) = (v_1 - u_1) \ldots (v_k - u_k) \qquad (5)$$

for all choices of real numbers u_j, v_j with $0 \le u_j < v_j \le 1$, for $1 \le j \le k$.

An equidistributed sequence is a 1-distributed sequence. Note that if $k > 1$, a k-distributed sequence is always $(k - 1)$-distributed, since we may set $u_k = 0$ and $v_k = 1$ in Eq. (5). Thus, in particular, any sequence which is known to be 4-distributed must also be 3-distributed, 2-distributed, and equidistributed. We can investigate the largest k for which a given sequence is k-distributed; and this leads us to formulate

Definition C. *A sequence is said to be ∞-distributed if it is k-distributed for all positive integers k.*

So far we have considered "[0, 1) sequences," i.e., sequences of real numbers lying between zero and one. The same ideas apply to integer-valued sequences; let us say a sequence $\langle X_n \rangle = X_0, X_1, X_2, \ldots$ is a "b-ary sequence" if each X_n is one of the integers $0, 1, \ldots, b - 1$. Thus, a 2-ary (binary) sequence is a sequence of zeros and ones.

A "b-ary number" $x_1 x_2 \ldots x_k$ is an ordered set of k integers, where $0 \le x_j < b$ for $1 \le j \le k$.

Definition D. *A b-ary sequence is said to be k-distributed if*

$$\Pr(X_n X_{n+1} \ldots X_{n+k-1} = x_1 x_2 \ldots x_k) = 1/b^k \qquad (6)$$

for all b-ary numbers $x_1 x_2 \ldots x_k$.

It is clear from this definition that if $U_0, U_1, \ldots$ is a k-distributed [0, 1) sequence, then the sequence $\lfloor bU_0 \rfloor, \lfloor bU_1 \rfloor, \ldots$ is a k-distributed b-ary sequence. (For if we set $u_j = x_j/b$, $v_j = (x_j + 1)/b$, $X_n = \lfloor bU_n \rfloor$, Eq. (5) becomes Eq. (6).) Furthermore, when a b-ary sequence is k-distributed, it is also $(k - 1)$-distributed: we add together the probabilities for the b-ary numbers

$x_1 \ldots x_{k-1}0$, $x_1 \ldots x_{k-1}1$, $\ldots$, $x_1 \ldots x_{k-1}(b-1)$ to obtain

$$\Pr(X_n \ldots X_{n+k-2} = x_1 \ldots x_{k-1}) \quad = \quad 1/b^{k-1}.$$

(Probabilities for disjoint events are additive; see exercise 5.) It therefore is natural to speak of an ∞-distributed b-ary sequence, as in Definition C above.

The decimal representation of a positive real number in the b-ary number system may be regarded as a b-ary sequence; for example, π corresponds to the 10-ary sequence 3, 1, 4, 1, 5, 9, 2, 6, 5, 3, 5, 8, 9, ... It has been conjectured that this sequence is ∞-distributed, but nobody has yet been able to prove that it is even 1-distributed.

Let us analyze these concepts a little more closely in the case when k equals a million. A binary sequence which is 1000000-distributed is going to have runs of a million zeros in a row! Similarly, a $[0, 1)$ sequence which is 1000000-distributed is going to have runs of a million consecutive values each of which is less than $\frac{1}{2}$. It is true that this will happen only $(\frac{1}{2})^{1000000}$ of the time, on the average, but the fact is that it *does* happen. Indeed, this phenomenon will occur in any truly random sequence, using our intuitive notion of "truly random." One can easily imagine that such a situation will have a drastic effect if this set of a million "truly random" numbers is being used in a computer-simulation experiment; there would be good reason to complain about the random-number generator! However, if we have a sequence of numbers which never has runs of a million consecutive U's less than $\frac{1}{2}$, the sequence is not random, and it will not be a suitable source of numbers for other conceivable applications which use extremely long blocks of U's as input. In summary, *a truly random sequence will exhibit local nonrandomness;* local nonrandomness is necessary in some applications, but it is disastrous in others. We are forced to conclude that *no sequence of "random" numbers can be adequate for every application.*

In a similar vein, one may argue that there is no way to judge whether a *finite* sequence is random or not; any particular sequence is just as likely as any other one. These facts are definitely stumbling blocks if we are ever to have a useful definition of randomness, but they are not really cause for alarm. It is still possible to give a definition for the randomness of infinite sequences of real numbers in such a way that the corresponding theory (viewed properly) will give us a great deal of insight concerning the ordinary finite sequences of rational numbers which are actually generated on a computer. Furthermore, we shall see later in this section that several plausible methods for defining the randomness of finite sequences are in fact available.

B. ∞-distributed sequences. Let us now undertake a brief study of the theory of ∞-distributed sequences. To describe the theory adequately, we will need to use a little "higher mathematics," so we assume in the remainder of this section that the reader knows the material ordinarily taught in an "advanced calculus" course.

First it is convenient to generalize Definition A, since the limit appearing there does not exist for all sequences. Let us define

$$\overline{\Pr}\,(S(n)) = \limsup_{n\to\infty}\,(\nu(n)/n), \qquad \underline{\Pr}\,(S(n)) = \liminf_{n\to\infty}\,(\nu(n)/n). \qquad (7)$$

Then $\Pr\,(S(n))$, if it exists, is the common value of $\underline{\Pr}\,(S(n))$ and $\overline{\Pr}\,(S(n))$.

We have seen that a k-distributed $[0, 1)$ sequence leads to a k-distributed b-ary sequence, if U is replaced by $\lfloor bU \rfloor$. Our first theorem shows that a converse result is true:

Theorem A. *Let* $\langle U_n\rangle = U_0, U_1, U_2, \ldots$ *be a* $[0, 1)$ *sequence. If the sequence*

$$\langle \lfloor b_j U_n \rfloor\rangle = \lfloor b_j U_0 \rfloor, \lfloor b_j U_1 \rfloor, \lfloor b_j U_2 \rfloor,$$

is a k-distributed b_j-ary *sequence for all* b_j *in an infinite sequence of integers,* $1 < b_1 < b_2 < b_3 < \cdots$, *then the original sequence* $\langle U_n\rangle$ *is* k-distributed.

As an example of this theorem, suppose that $b_j = 2^j$. The sequence $\lfloor 2^j U_0 \rfloor, \lfloor 2^j U_1 \rfloor, \ldots$ is essentially the sequence of the first j bits of the binary representation of $U_0, U_1, \ldots$. If all these integer sequences are k-distributed, in the sense of Definition D, the real-valued sequence $U_0, U_1, \ldots$ must also be k-distributed in the sense of Definition B.

Proof of Theorem A. If the sequence $\lfloor bU_0 \rfloor, \lfloor bU_1 \rfloor, \ldots$ is k-distributed, it follows by the addition of probabilities that Eq. (5) holds whenever each u_j and v_j is a rational number with denominator b. Now let u_j, v_j be any real numbers, and let u_j', v_j' be rational numbers with denominator b such that

$$u_j' \le u_j < u_j' + 1/b, \; v_j' \le v_j < v_j' + 1/b.$$

Let $S(n)$ be the statement that $u_1 \le U_n < v_1, \ldots, u_k \le U_{n+k-1} < v_k$. We have

$$\overline{\Pr}\,(S(n)) \le \Pr\left(u_1' \le U_n < v_1' + \frac{1}{b},\; \cdots,\; u_k' \le U_{n+k-1} < v_k' + \frac{1}{b}\right)$$

$$= \left(v_1' - u_1' + \frac{1}{b}\right)\cdots\left(v_k' - u_k' + \frac{1}{b}\right);$$

$$\underline{\Pr}\,(S(n)) \ge \Pr\left(u_1' + \frac{1}{b} \le U_n < v_1',\; \ldots,\; u_k' + \frac{1}{b} \le U_{n+k-1} < v_k'\right)$$

$$= \left(v_1' - u_1' - \frac{1}{b}\right)\cdots\left(v_k' - u_k' - \frac{1}{b}\right).$$

Now $|(v_j' - u_j' \pm 1/b) - (v_j - u_j)| \le 2/b$; since our inequalities hold for all $b = b_j$, we find that $b_j \to \infty$ as $j \to \infty$, and thus

$$(v_1 - u_1)\ldots(v_k - u_k) \le \underline{\Pr}\,(S(n)) \le \overline{\Pr}\,(S(n)) \le (v_1 - u_1)\ldots(v_k - u_k). \quad \blacksquare$$

The next theorem is our main tool for proving things about k-distributed sequences.

Theorem B. *Let $\langle U_n \rangle$ be a k-distributed $[0, 1)$ sequence, and let $f(x_1, x_2, \ldots, x_k)$ be a Riemann-integrable function of k variables; then*

$$\lim_{n \to \infty} \frac{1}{n} \sum_{0 \leq j < n} f(U_j, U_{j+1}, \ldots, U_{j+k-1})$$

$$= \int_0^1 \cdots \int_0^1 f(x_1, x_2, \ldots, x_k) \, dx_1 \ldots dx_k. \qquad (8)$$

Proof. The definition of a k-distributed sequence states that this result is true in the special case that

$$f(x_1, \ldots, x_k) = \begin{cases} 1, & \text{if} \quad u_1 \leq x_1 < v_1, \ldots, u_k \leq x_k < v_k; \\ 0, & \text{otherwise.} \end{cases} \qquad (9)$$

Therefore Eq. (8) is true whenever $f = a_1 f_1 + a_2 f_2 + \cdots + a_m f_m$ and when each f_j is a function of type (9); in other words, Eq. (8) holds whenever f is a "step-function" obtained by (i) partitioning the unit k-dimensional cube into subcells whose faces are parallel to the coordinate axes, and (ii) assigning a constant value to f on each subcell.

Now let f be any Riemann-integrable function. If ϵ is any positive number, we know (by the definition of Riemann-integrability) that there exist step functions $\underline{f}$ and $\overline{f}$ such that $\underline{f}(x_1, \ldots, x_k) \leq f(x_1, \ldots, x_k) \leq \overline{f}(x_1, \ldots, x_k)$, and such that the difference of the integrals of $\underline{f}$, f, and $\overline{f}$ is less than ϵ. Since Eq. (8) holds for $\underline{f}$ and $\overline{f}$, and since

$$\frac{1}{n} \sum_{0 \leq j < n} \underline{f}(U_j, \ldots, U_{j+k-1}) \leq \frac{1}{n} \sum_{0 \leq j < n} f(U_j, \ldots, U_{j+k-1})$$

$$\leq \frac{1}{n} \sum_{0 \leq j < n} \overline{f}(U_j, \ldots, U_{j+k-1}),$$

we conclude Eq. (8) is true also for f. ∎

As the first application of this theorem, consider the *permutation test* described in Section 3.3.2. Let $(p_1, p_2, \ldots, p_k)$ be any permutation of the numbers $1, 2, \ldots, k$; we want to show that

$$\Pr (U_{n+p_1-1} < U_{n+p_2-1} < \cdots < U_{n+p_k-1}) = 1/k!. \qquad (10)$$

To prove this, assume that the sequence $\langle U_n \rangle$ is k-distributed, and let

$$f(x_1, \ldots, x_k) = \begin{cases} 1, & \text{if} \quad x_{p_1} \leq x_{p_2} \leq \cdots \leq x_{p_k} \\ 0, & \text{otherwise.} \end{cases}$$

We have

$$\Pr \left(U_{n+p_1-1} < U_{n+p_2-1} < \cdots < U_{n+p_k-1} \right)$$

$$= \int_0^1 \cdots \int_0^1 f(x_1, \ldots, x_k)\, dx_1 \ldots dx_k$$

$$= \int_0^1 dx_{p_k} \int_0^{x_{p_k}} \cdots \int_0^{x_{p_3}} dx_{p_2} \int_0^{x_{p_2}} dx_{p_1} = \frac{1}{k!}.$$

Corollary P. *If a* $[0, 1)$ *sequence is k-distributed, it satisfies the permutation test of order k, in the sense of Eq. (10).* ∎

We can also show that the *serial correlation test* is satisfied:

Corollary S. *If a* $[0, 1)$ *sequence is* $(k + 1)$-*distributed, the serial correlation coefficient between* U_n *and* U_{n+k} *tends to zero:*

$$\lim_{n \to \infty} \frac{\frac{1}{n} \sum U_j U_{j+k} - \left(\frac{1}{n} \sum U_j \right)\left(\frac{1}{n} \sum U_{j+k} \right)}{\sqrt{\left(\frac{1}{n} \sum U_j^2 - \left(\frac{1}{n} \sum U_j \right)^2 \right)\left(\frac{1}{n} \sum U_{j+k}^2 - \left(\frac{1}{n} \sum U_{j+k} \right)^2 \right)}} = 0.$$

(All summations here are for $0 \le j < n$.)

Proof. By Theorem B, the quantities

$$\frac{1}{n} \sum U_j U_{j+k}, \quad \frac{1}{n} \sum U_j^2, \quad \frac{1}{n} \sum U_{j+k}^2, \quad \frac{1}{n} \sum U_j, \quad \frac{1}{n} \sum U_{j+k}$$

tend to the respective limits $\frac{1}{4}, \frac{1}{3}, \frac{1}{3}, \frac{1}{2}, \frac{1}{2}$ as $n \to \infty$. ∎

Let us now consider some slightly more general distribution properties of sequences. We have defined the k-distribution property for all adjacent k-tuples; for example, a sequence is 2-distributed if and only if the points

$$(U_0, U_1), \quad (U_1, U_2), \quad (U_2, U_3), \quad (U_3, U_4), \quad (U_4, U_5), \quad \cdots$$

are equidistributed in the unit square. It is quite possible, however, that the alternate pairs of points, $(U_1, U_2), (U_3, U_4), (U_5, U_6), \ldots$ are not themselves equidistributed; if the density of points (U_{2n-1}, U_{2n}) is deficient in some area, the other points (U_{2n}, U_{2n+1}) might compensate for it. For example, the periodic binary sequence

$$\langle X_n \rangle = 0, 0, 0, 1, \quad 0, 0, 0, 1, \quad 1, 1, 0, 1, \quad 1, 1, 0, 1, \quad 0, 0, 0, 1, \quad \ldots, \quad (11)$$

which has a period of length 16, is seen to be 3-distributed; yet the sequence of even-numbered elements $\langle X_{2n} \rangle = 0, 0, 0, 0, 1, 0, 1, 0, \ldots$ has three times as many zeros as ones, while the subsequence of odd-numbered elements $\langle X_{2n+1} \rangle = 0, 1, 0, 1, 1, 1, 1, 1, \ldots$ has three times as many ones as zeros.

If a sequence $\langle U_n \rangle$ is ∞-distributed, the example above shows that it is not at all obvious that the subsequence of alternate terms $\langle U_{2n} \rangle = U_0, U_2, U_4, U_6, \ldots$ is ∞-distributed or even 1-distributed. But we shall see that $\langle U_{2n} \rangle$ is, in fact, ∞-distributed, and much more is also true.

Definition E. *A $[0, 1)$ sequence $\langle U_n \rangle$ is said to be (m, k)-distributed if*

$$\Pr (u_1 \leq U_{mn+j} < v_1, \ u_2 \leq U_{mn+j+1} < v_2, \ \ldots, \ u_k \leq U_{mn+j+k-1} < v_k)$$
$$= (v_1 - u_1) \ldots (v_k - u_k)$$

for all choices of real numbers u_r, v_r with $0 \leq u_r < v_r \leq 1$ for $1 \leq r \leq k$, and for all integers j with $0 \leq j < m$.

Thus a k-distributed sequence is the special case $m = 1$ in Definition E; the case $m = 2$ means that the k-tuples in even positions must have the same density as the k-tuples in odd positions, etc.

Several properties of Definition E are obvious:

$$An\ (m, k)\text{-}distributed\ sequence\ is\ (m, \kappa)\text{-}distributed\ for\ 1 \leq \kappa \leq k. \qquad (12)$$

$$An\ (m, k)\text{-}distributed\ sequence\ is\ (d, k)\text{-}distributed\ for\ all\ divisors \qquad (13)$$
$$d\ of\ m.$$

We can define the concept of an (m, k)-distributed b-ary sequence, as in Definition D; and the proof of Theorem A remains valid for (m, k)-distributed sequences.

The next theorem, which is in many ways rather surprising, shows that the property of being ∞-distributed is very strong indeed, much stronger than we imagined it to be when we first considered the definition of the concept.

Theorem C (*Ivan Niven and H. S. Zuckerman*). *An ∞-distributed sequence is (m, k)-distributed for all positive integers m and k.*

Proof. It suffices to prove the theorem for b-ary sequences, by using the generalization of Theorem A just mentioned. Furthermore, we may assume $m = k$: for, by (12) and (13), the sequence will be (m, k)-distributed if it is (mk, mk)-distributed.

So we will prove that *any ∞-distributed b-ary sequence $X_0, X_1, \ldots$ is (m, m)-distributed for all positive integers m.* Our proof is a simplification of the proof given by Niven and Zuckerman in *Pacific Journal of Mathematics* **1** (1951), 103–109.

The theorem is proved by making use of an important idea which is useful in so many mathematical arguments: "If the sum of m quantities and the sum of their squares are both consistent with the hypothesis that the m quantities are equal, that hypothesis is true." In a strong form, this principle may be stated as follows:

Lemma E. *Given m sequences of numbers* $\langle y_{jn} \rangle = y_{j0}, y_{j1}, y_{j2}, \ldots$ *for* $1 \leq j \leq m$, *suppose that*

$$\lim_{n \to \infty} (y_{1n} + y_{2n} + \cdots + y_{mn}) = m\alpha,$$

$$\lim_{n \to \infty} \sup (y_{1n}^2 + y_{2n}^2 + \cdots + y_{mn}^2) \leq m\alpha^2. \tag{14}$$

Then for each j, $\lim_{n \to \infty} y_{jn}$ *exists and equals* α.

An incredibly simple proof of this lemma is given in exercise 9. ∎

Now to continue our proof of Theorem C. Let $x = x_1 x_2 \ldots x_m$ be a b-ary number, and say that x *occurs* at position p of the sequence if $X_{p-m+1} X_{p-m+2} \ldots X_p = x$. Let $\nu_j(n)$ be the number of occurrences of x at position p when $p < n$ and $p \bmod m = j$. Let $y_{jn} = \nu_j(n)/n$; we wish to prove that

$$\lim_{n \to \infty} y_{jn} = 1/mb^m. \tag{15}$$

First we know that

$$\lim_{n \to \infty} (y_{0n} + y_{1n} + \cdots + y_{(m-1)n}) = 1/b^m, \tag{16}$$

since the sequence is m-distributed. By Lemma E and Eq. (16), the theorem will be proved if we can show that

$$\lim_{n \to \infty} \sup (y_{0n}^2 + y_{1n}^2 + \cdots + y_{(m-1)n}^2) \leq 1/mb^{2m}. \tag{17}$$

This inequality is not obvious yet; some rather delicate maneuvering is necessary before we can prove it. Let q be a multiple of m, and consider

$$C(n) = \sum_{0 \leq j < m} \binom{\nu_j(n) - \nu_j(n-q)}{2}. \tag{18}$$

This is the number of pairs of occurrences of x in positions p_1, p_2 with $n - q \leq p_1 < p_2 < n$ and with $p_2 - p_1$ a multiple of m. Consider now the sum

$$S_N = \sum_{1 \leq n \leq N+q} C(n). \tag{19}$$

Each pair of occurrences of x in positions p_1, p_2 with $p_1 < p_2 < p_1 + q$, $p_2 - p_1$ a multiple of m, and $p_1 \leq N$, is counted $p_1 + q - p_2$ times in the total S_N (namely, when $p_2 < n \leq p_1 + q$); and the pairs of such occurrences with $N < p_1 < p_2 < N + q$ are counted $N + q - p_2$ times.

Let $d_t(n)$ be the number of pairs of occurrences of x in positions p_1, p_2 with $p_1 + t = p_2 < n$. The analysis above shows that

$$\sum_{0<t<q/m} (q - mt)\, d_{mt}(N + q) \geq S_N \geq \sum_{0<t<q/m} (q - mt)\, d_{mt}(N). \qquad (20)$$

Since the original sequence is q-distributed,

$$\lim_{N\to\infty} \frac{1}{N} d_{mt}(N) = 1/b^{2m} \qquad (21)$$

for all $0 < t < q/m$, and therefore by (20)

$$\lim_{N\to\infty} \frac{S_N}{N} = \sum_{0<t<q/m} (q - mt)/b^{2m}$$

$$= q(q - m)/2mb^{2m}. \qquad (22)$$

This fact will prove the theorem, after some manipulation.

By definition,

$$2S_N = \sum_{1\leq n\leq N+q} \sum_{0\leq j<m} \left((\nu_j(n) - \nu_j(n - q))^2 - (\nu_j(n) - \nu_j(n - q)) \right),$$

and we can remove the unsquared terms by applying (16) to get

$$\lim_{N\to\infty} \frac{T_N}{N} = q(q - m)/mb^{2m} + q/b^m, \qquad (23)$$

where

$$T_N = \sum_{1\leq n\leq N+q} \sum_{0\leq j<m} (\nu_j(n) - \nu_j(n - q))^2.$$

Using the inequality

$$\frac{1}{r}\left(\sum_{1\leq j\leq r} a_j \right)^2 \leq \sum_{1\leq j\leq r} a_j^2$$

(cf. exercise 1.2.3–30), we find that

$$\limsup_{N\to\infty} \sum_{0\leq j<m} \frac{1}{N(N + q)} \left(\sum_{1\leq n\leq N+q} (\nu_j(n) - \nu_j(n - q)) \right)^2$$

$$\leq q(q - m)/mb^{2m} + q/b^m. \qquad (24)$$

We also have

$$q\nu_j(N) \leq \sum_{N<n\leq N+q} \nu_j(n) = \sum_{1\leq n\leq N+q} (\nu_j(n) - \nu_j(n - q)) \leq q\nu_j(N + q),$$

and putting this into (24) gives

$$\limsup_{N\to\infty} \sum_{0\leq j<m} (\nu_j(N)/N)^2 \leq (q - m)/qmb^{2m} + 1/qb^m. \qquad (25)$$

This formula has been proved whenever q is a multiple of m, and if we let $q \to \infty$, we obtain (17), completing the proof.

For a possibly simpler proof, see J. W. S. Cassels, *Pacific Journal of Mathematics* **2** (1952), 555–557. ∎

Exercises 29 and 30 illustrate the nontriviality of this theorem, and they also illustrate the fact, indicated in (25), that a q-distributed sequence will have probabilities deviating from the true (m, m)-distribution probabilities by essentially $1/\sqrt{q}$ at most. The full hypothesis of ∞-distribution is necessary for the proof of the theorem.

As a result of Theorem C, we can prove that an ∞-distributed sequence passes the serial test, the "maximum of t" test, and the tests on subsequences mentioned in Section 3.3.2. It is not hard to show that the gap test, the poker test, and the run test are also satisfied (see exercises 12 through 14); the coupon collector's test is considerably more difficult to deal with, but it too is satisfied (see exercises 15 and 16).

The existence of ∞-distributed sequences of a rather simple type is guaranteed by the next theorem.

Theorem F (J. Franklin). *The* $[0, 1)$ *sequence* $U_0, U_1, \ldots,$ *with*

$$U_n = (\theta^n \bmod 1) \tag{26}$$

is ∞-*distributed for almost all real numbers* $\theta > 1$. *That is, the set*

$$\{\theta \mid \theta > 1 \text{ and } (26) \text{ is not } \infty\text{-distributed}\}$$

is of measure zero.

The proofs of this theorem and some generalizations are given in Franklin's paper quoted below. ∎

Franklin has shown that θ must be a transcendental number for (26) to be ∞-distributed; yet even though the sequence (26) is known to be ∞-distributed for *almost all* numbers, we still do not know of any *particular* number for which it is true. The powers $(\pi^n \bmod 1)$ have been laboriously computed for $n \leq 10000$, using multiple-precision arithmetic. The most significant 35 bits of each of these numbers, stored on a disk file, have been successfully used as a source of uniform deviates. We have the interesting situation that, by Theorem F, the probability that the powers $(\pi^n \bmod 1)$ are ∞-distributed is equal to 1, yet because there are uncountably many real numbers, this gives us no information as to whether the sequence is really ∞-distributed or not. It is a fairly safe bet that nobody in our lifetimes will ever *prove* that this particular sequence is *not* ∞-distributed; but it may not be. Because of these considerations, one may legitimately wonder if there is any *explicit* sequence which is ∞-distributed; i.e., *is there an algorithm to compute real numbers U_n for all $n \geq 0$, such that the sequence $\langle U_n \rangle$ is ∞-distributed?* The answer is yes, as shown for example by the author in the article "Construction of a Random Sequence," *BIT* **5** (1965), 246–250. The sequence constructed there consists entirely of rational numbers; in fact, each number U_n has a terminating representation in the binary number system. Another construction of an explicit ∞-distributed sequence, somewhat more complicated than the sequence just cited, follows from Theorem W below.

C. Does ∞-distributed = random? In view of all the above theory about ∞-distributed sequences, we can be sure of one thing: the concept of an ∞-distributed sequence is an important one in mathematics. There is also a good deal of evidence that the following statement is a valid formulation of the intuitive idea of randomness:

Definition R1. *A* $[0, 1)$ *sequence is defined to be "random" if it is an ∞-distributed sequence.*

We have seen that sequences meeting this definition will satisfy all of the statistical tests of Section 3.3.2 and many more.

Let us attempt to criticize this definition objectively. First of all, is every "truly random" sequence ∞-distributed? There are uncountably many sequences $U_0, U_1, \ldots$ of real numbers between zero and one. If a truly random number generator is sampled to give values $U_0, U_1, \ldots$, any of the possible sequences may be considered equally likely, and some of the sequences (indeed, infinitely many of them) are not even equidistributed. On the other hand, using any reasonable definition of probability on this space of all possible sequences leads us to conclude that a random sequence is ∞-distributed *with probability one*. We are therefore led to formalize Franklin's definition of randomness (as given at the beginning of this section) in the following way:

Definition R2. *A* $[0, 1)$ *sequence* $\langle U_n \rangle$ *is defined to be "random" if, whenever P is a property such that $P(\langle V_n \rangle)$ holds with probability one for a sequence $\langle V_n \rangle$ of independent samples of random variables from the uniform distribution, then $P(\langle U_n \rangle)$ is true.*

Is it perhaps possible that Definition R1 is equivalent to Definition R2? Let us try out some possible objections to Definition R1, and see if these criticisms are valid.

In the first place, Definition R1 deals only with limiting properties of the sequence as $n \to \infty$. There are ∞-distributed sequences in which the first million elements are all zero; should such a sequence be considered random?

This objection is not very valid. If ϵ is any positive number, there is no reason why the first million elements of a sequence should not all be less than ϵ; as pointed out earlier in this section, there is no legitimate way to say if a finite sequence is random or not. With probability one, a truly random sequence contains infinitely many runs of a million consecutive elements less than ϵ, so why can't this happen at the beginning of the sequence?

On the other hand, consider Definition R2 and let P be the property that all elements of the sequence are distinct; P is true with probability one, so any sequence with a million zeros is not random by *this* criterion.

Now let P be the property that *no* element of the sequence is equal to zero; again, P is true with probability one, so by Definition R2 any sequence with a zero element is not random. More generally, however, let x_0 be any fixed number between zero and one, and let P be the property that no element of the sequence is equal to x_0; Definition R2 now says that no random sequence may contain the element x_0! We can now prove that *no sequence satisfies the condition of Definition R2.* (For if $U_0, U_1, \ldots$ is such a sequence, take $x_0 = U_0$.)

Therefore if R1 is too weak a definition, R2 is certainly too strong. The "right" definition must be less strict than R2. We have not really shown that R1 is too weak, however, so let us continue to attack it some more. As mentioned above, an ∞-distributed sequence of *rational* numbers has been constructed. (Indeed, this is not so surprising: see exercise 18.) Almost all real numbers are irrational; perhaps we should insist that

$$\Pr(U_n \text{ is rational}) = 0$$

for a random sequence.

Note that the definition of equidistribution says that $Pr(u \leq U_n < v) = v - u$. There is an obvious way to generalize this definition, using measure theory: "If $S \subseteq [0, 1)$ is a set of measure μ, then

$$\Pr(U_n \in S) = \mu, \qquad (27)$$

for all random sequences $\langle U_n \rangle$." In particular, if S is the set of rationals, it has measure zero, so no sequence of rational numbers is equidistributed in this generalized sense. It is reasonable to expect that Theorem B could be extended to Lebesgue integration instead of Riemann integration, if property (27) is stipulated. However, once again we find that definition (27) is too strict, for *no* sequence satisfies that property! If $U_0, U_1, \ldots$ is any sequence, the set $S = \{U_0, U_1, \ldots\}$ is of measure zero, yet $\Pr(U_n \in S) = 1$. Thus, by the force of the same argument we used to exclude rationals from random sequences, we can exclude all random sequences.

So far Definition R1 has proved to be defensible. There are, however, some quite valid objections to it. For example, if we have a random sequence in the intuitive sense, the infinite subsequence

$$U_0, U_1, U_4, U_9, \ldots, U_{n^2}, \ldots \qquad (28)$$

should also be a random sequence. This is not always true for an ∞-distributed sequence. (In fact, if we take any ∞-distributed sequence and set $U_{n^2} \leftarrow 0$ for all n, the counts $\nu_k(n)$ which appear in the test of k-distributivity are changed by at most $\sqrt{n}$, so the limits of the ratios $\nu_k(n)/n$ remain unchanged.) Therefore R1 definitely fails to satisfy this randomness criterion.

Perhaps we should strengthen R1 as follows:

Definition R3. *A* $[0, 1)$ *sequence is said to be "random" if every infinite subsequence is ∞-distributed.*

But once again we find this definition is too strict. If $\langle U_n \rangle$ is any equidistributed sequence, it has a monotonic subsequence with $U_{s_0} < U_{s_1} < U_{s_2} < \cdots$.

The secret is to restrict the subsequences so that they could be defined by a man who does not look at U_n before he decides whether or not it is to be in the subsequence. The following definition now suggests itself:

Definition R4. *A $[0, 1)$ sequence $\langle U_n \rangle$ is said to be "random" if, for every effective algorithm which specifies an infinite sequence of distinct nonnegative integers s_n for $n \geq 0$, the subsequence $U_{s_0}, U_{s_1}, \ldots$ corresponding to this algorithm is ∞-distributed.*

The algorithms referred to in this definition are procedures which compute s_n, given n. This definition tests all infinite "recursively enumerable" subsequences, according to the conventional definition of recursive enumerability (see Chapter 11).

A few comments on this definition are appropriate. Certainly the sequence $\langle \pi^n \bmod 1 \rangle$ will *not* satisfy R4, since it is either not equidistributed or there is an effective algorithm which determines an infinite subsequence s_n with $(\pi^{s_0} \bmod 1) < (\pi^{s_1} \bmod 1) < \cdots$. Similarly, *no explicitly defined sequence can satisfy Definition R4;* this is appropriate, if we agree that no explicitly defined sequence can really be random. It is quite likely, however, that the sequence $\langle \theta^n \bmod 1 \rangle$ will satisfy Definition R4, for almost all real numbers $\theta > 1$; this is no contradiction, since almost all θ are uncomputable by algorithms. The following facts are known, for example: (i) The sequence $\langle \theta^n \bmod 1 \rangle$ satisfies Definition R4 for almost all real $\theta > 1$, if "∞-distributed" is replaced by "1-distributed." This theorem was proved by J. F. Koksma, *Compositio Mathematica* **2** (1935), 250–258. (ii) The particular sequence $\langle \theta^{s(n)} \bmod 1 \rangle$ is ∞-distributed for almost all real $\theta > 1$, if $\langle s(n) \rangle$ is a sequence of integers for which $s(n + 1) - s(n) \to \infty$ as $n \to \infty$. For example, we could have $s(n) = n^2$, or $s(n) = \lfloor n \log n \rfloor$.

Definition R4 is much stronger than Definition R1; but it is still reasonable to claim that Definition R4 is too weak. For example, let $\langle U_n \rangle$ be a truly random sequence, and define the subsequence $\langle U_{s_n} \rangle$ by the following rules: $s_0 = 0$, and (for $n > 0$) s_n is the smallest positive integer for which $U_{s_n - 1}$, $U_{s_n - 2}, \ldots, U_{s_n - n}$ are all less than $\frac{1}{2}$. Thus we are considering the subsequence of values following the first consecutive run of n values less than $\frac{1}{2}$. Suppose that "$U_n < \frac{1}{2}$" corresponds to the value "heads" in the flipping of a coin. Gamblers tend to feel that a long run of "heads" makes the opposite condition, "tails," more probable, assuming a true coin is being used; and the subsequence $\langle U_{s_n} \rangle$ just defined corresponds to a gambling system for a man who places his nth bet on the coin toss following the first run of n consecutive "heads." The gambler may think that $Pr(U_{s_n} \geq \frac{1}{2})$ is more than $\frac{1}{2}$, but of course in a truly random sequence, $\langle U_{s_n} \rangle$ will be completely random. No gambling system will ever be able to beat the odds! Definition R4 says nothing about subsequences formed according to such a gambling system, and so apparently we need something more.

Let us define a "subsequence rule" $\mathfrak{R}$ as an infinite sequence of functions $\langle f_n(x_1, \ldots, x_n) \rangle$ where, for $n \geq 0$, f_n is a function of n variables, and the value of $f_n(x_1, \ldots, x_n)$ is either 0 or 1. Here $x_1, \ldots, x_n$ are elements of some set S. (Thus, in particular, f_0 is a constant function, either 0 or 1.) A subsequence rule $\mathfrak{R}$ defines a subsequence of any infinite sequence $\langle X_n \rangle$ of elements of S as follows: *The nth term X_n is in the subsequence $\langle X_n \rangle \mathfrak{R}$ if and only if* $f_n(X_0, X_1, \ldots, X_{n-1}) = 1$. Note that the subsequence $\langle X_n \rangle \mathfrak{R}$ thus defined is not necessarily infinite, and it may in fact be the null subsequence.

The "gambler's subsequence" which was described above corresponds to the following subsequence rule: "$f_0 = 1$; and for $n > 0$, $f_n(x_1, \ldots, x_n) = 1$ if and only if there is some k, $0 < k \leq n$, such that $x_m < \frac{1}{2}$, $x_{m-1} < \frac{1}{2}, \ldots,$ $x_{m-k+1} < \frac{1}{2}$, when $m = n$ but not when $k \leq m < n$."

A subsequence rule $\mathfrak{R}$ is said to be *computable* if there is an effective algorithm which determines the value of $f_n(x_1, \ldots, x_n)$, when $n, x_1, \ldots, x_n$ are given as input. In a definition of randomness, we should restrict ourselves to computable subsequence rules, or else we obtain a definition like R3 above, which is too strong. Effective algorithms cannot deal nicely with arbitrary real numbers as inputs; for example, if a real number x is specified by an infinite base 10 expansion, there is no algorithm to determine if x is $< \frac{1}{3}$ or not, since all digits of the number $0.333 \cdots$ would have to be examined. Therefore computable subsequence rules do not apply to all $[0, 1)$ sequences, and it is convenient to base our next definition on b-ary sequences:

Definition R5. *A b-ary sequence is said to be "random" if every infinite subsequence defined by a computable subsequence rule is 1-distributed.*

A $[0, 1)$ sequence $\langle U_n \rangle$ is said to be "random" if the b-ary sequence $\langle \lfloor b U_n \rfloor \rangle$ is "random" for all integers $b \geq 2$.

Note that Definition R5 says only "1-distributed," not "∞-distributed." It is interesting to note that this may be done without loss of generality. For, we may define an obviously computable subsequence rule $\mathfrak{R}(a_1 \ldots a_k)$ as follows for any b-ary number $a_1 \ldots a_k$: Let $f_n(x_1, \ldots, x_n) = 1$ if and only if $n \geq k - 1$ and $x_{n-k+1} = a_1, \ldots, x_{n-1} = a_{k-1}, x_n = a_k$. Now if $\langle X_n \rangle$ is a k-distributed b-ary sequence, this rule $\mathfrak{R}(a_1 \ldots a_k)$—which selects the subsequence consisting of those terms just following an occurrence of $a_1 \ldots a_k$—defines an infinite subsequence; and if this subsequence is 1-distributed, each of the $(k + 1)$-tuples $a_1 \ldots a_k a_{k+1}$ for $0 \leq a_{k+1} < b$ occurs with probability $1/b^{k+1}$ in $\langle X_n \rangle$. Thus we can prove that a sequence satisfying Definition R5 is k-distributed for all k, by induction on k. Similarly, by considering the "composition" of subsequence rules—if $\mathfrak{R}_1$ defines an infinite subsequence $\langle X_n \rangle \mathfrak{R}_1$, then we can define $\mathfrak{R}_1 \mathfrak{R}_2$ to be the subsequence rule for which $\langle X_n \rangle \mathfrak{R}_1 \mathfrak{R}_2 = (\langle X_n \rangle \mathfrak{R}_1) \mathfrak{R}_2$—we find that all subsequences considered in Definition R5 are ∞-distributed. (See exercise 32.)

The fact that ∞-distribution comes out of Definition R5 as a very special case is encouraging, and it is a good indication that we may at last have found the definition of randomness which we have been seeking. But alas, there still is a problem! It is not clear that sequences satisfying Definition R4 must satisfy Definition R5; the "computable subsequence rules" we have just specified always enumerate subsequences $\langle X_{s_n} \rangle$ for which $s_0 < s_1 < \cdots$, and this leads to so-called "recursive sequences." These do not include all the recursively enumerable sequences allowed by Definition R4, where $\langle s_n \rangle$ does not have to be monotone; it must only satisfy the condition $s_n \neq s_m$ for $n \neq m$.

To meet this objection, we may combine Definitions R4 and R5 as follows:

Definition R6. *A b-ary sequence $\langle X_n \rangle$ is said to be "random" if, for every effective algorithm which specifies an infinite sequence of distinct nonnegative integers $\langle s_n \rangle$ as a function of n and the values of $X_{s_0}, \ldots, X_{s_{n-1}}$, the subsequence $\langle X_{s_n} \rangle$ corresponding to this algorithm is "random" in the sense of Definition R5.*

A $[0, 1)$ sequence $\langle U_n \rangle$ is said to be "random" if the b-ary sequence $\langle \lfloor bU_n \rfloor \rangle$ is "random" for all integers $b \geq 2$.

The author contends that this definition surely meets all reasonable philosophical requirements for randomness, and so it provides an answer to the principal question posed in this section.

D. Existence of random sequences. We have seen that Definition R3 is too strong, in the sense that no sequences could exist satisfying that definition; and the formulation of Definitions R4, R5, and R6 above was carried out in an attempt to recapture the essential characteristics of Definition R3. In order to show that Definition R6 is not overly restrictive, it is still necessary for us to prove that sequences satisfying all these conditions exist. Intuitively, we feel quite sure such sequences exist, because we believe a truly random sequence exists and satisfies Definition R6; but a proof is really necessary to show the definition is consistent.

An interesting method for constructing sequences satisfying Definition R5 has been given by A. Wald. The construction starts with a very simple 1-distributed sequence:

Lemma T. *Let the sequence of real numbers $\langle V_n \rangle$ be defined in terms of the binary system as follows:*

$$V_0 = 0, \quad V_1 = .1, \quad V_2 = .01, \quad V_3 = .11, \quad V_4 = .001, \quad \cdots$$
$$V_n = .c_r \ldots c_1 1 \quad \text{if} \quad n = 2^r + c_1 2^{r-1} + \cdots + c_r. \tag{29}$$

Let $I_{b_1 \ldots b_r}$ denote the set of all real numbers in $[0, 1)$ whose binary representation begins with $0.b_1 \ldots b_r$; thus

$$I_{b_1 \ldots b_r} = [0.b_1 \ldots b_r, 0.b_1 \ldots b_r + 2^{-r}). \tag{30}$$

Then if $\nu(n)$ denotes the number of V_k in $I_{b_1 \ldots b_r}$ for $0 \leq k < n$, we have

$$|\nu(n)/n - 2^{-r}| \leq 1/n. \tag{31}$$

Proof. Since $\nu(n)$ is the number of k for which $k \bmod 2^r = b_r \ldots b_1$, we have $\nu(n) = t$ or $t + 1$ when $\lfloor n/2^r \rfloor = t$. Hence $|\nu(n) - n/2^r| \leq 1$. ∎

It follows from (31) that the sequence $\langle \lfloor 2^r V_n \rfloor \rangle$ is an equidistributed 2^r-ary sequence; hence by Theorem A, $\langle V_n \rangle$ is an equidistributed $[0, 1)$ sequence. Indeed, it is pretty clear that $\langle V_n \rangle$ is about as equidistributed as a $[0, 1)$ sequence can be! (For further discussion of this and related sequences, see J. C. van der Corput, *Proc. Koninklijke Nederlandse Akademie van Wetenschappen* **38** (1935), 813–821, 1058–1066; J. H. Halton, *Numerische Mathematik* **2** (1960), 84–90, 196.)

Now let $\Re_1, \Re_2, \ldots$ be infinitely many subsequence rules; we would like to find a sequence $\langle U_n \rangle$ for which all the infinite subsequences $\langle U_n \rangle \Re_j$ are equidistributed.

Algorithm W (*Wald sequence*). Given an infinite set of subsequence rules $\Re_1, \Re_2, \ldots$, which define subsequences of $[0, 1)$ sequences of *rational* numbers, this procedure defines a $[0, 1)$ sequence $\langle U_n \rangle$. The computation involves infinitely many auxiliary variables $C[a_1, \ldots, a_r]$ where $r \geq 1$ and where $a_j = 0$ or 1 for $1 \leq j \leq r$. These variables are initially all zero.

W1. [Initialize n.] Set $n \leftarrow 0$.

W2. [Initialize r.] Set $r \leftarrow 1$.

W3. [Test $\Re_r$.] If the element U_n is to be in the subsequence defined by $\Re_r$, based on the values of U_k for $0 \leq k < n$, set $a_r \leftarrow 1$; otherwise set $a_r \leftarrow 0$.

W4. [$B[a_1, \ldots, a_r]$ full?] If $C[a_1, \ldots, a_r] < 3 \cdot 4^{r-1}$, go to W6.

W5. [Increase r.] Set $r \leftarrow r + 1$ and return to W3.

W6. [Set U_n.] Increase $C[a_1, \ldots, a_r]$ by 1 and let k be the new value of $C[a_1, \ldots, a_r]$. Set $U_n \leftarrow V_k$, where V_k is defined in Lemma T above.

W7. [Advance n.] Increase n by 1 and return to W2. ∎

Strictly speaking, this is not an algorithm, since it doesn't terminate; it would of course be easy to modify the procedure to stop when n reaches a given value. The reader will find it easier to grasp the idea of the construction if he tries it by hand, replacing the number $3 \cdot 4^{r-1}$ of step W4 by 2^r during this experiment.

Algorithm W is not meant to be a practical source of random numbers; it is intended to serve only a theoretical purpose:

Theorem W. *Let U_n be the sequence of rational numbers defined by Algorithm W, and let k be a positive integer. If the subsequence $\langle U_n \rangle \Re_k$ is infinite, it is 1-distributed.*

Proof. Let $A[a_1, \ldots, a_r]$ denote the (possibly empty) subsequence of $\langle U_n \rangle$ containing precisely those elements U_n which, for all j, $1 \leq j \leq r$, belong to subsequence $\langle U_n \rangle \Re_j$ if $a_j = 1$ and do not belong to subsequence $\langle U_n \rangle \Re_j$ if $a_j = 0$.

It suffices to prove, for all $r \geq 1$ and all pairs of binary numbers $a_1 \ldots a_r$ and $b_1 \ldots b_r$, that $Pr(U_n \in I_{b_1 \ldots b_r}) = 2^{-r}$ with respect to the subsequence $A[a_1, \ldots, a_r]$, whenever the latter is infinite. [See Eq. (30).] For if $r \geq k$, the infinite sequence $\langle U_n \rangle \Re_k$ is the finite union of the disjoint subsequences $A[a_1, \ldots, a_r]$ for $a_k = 1$ and $a_j = 0$ or 1 for $1 \leq j \leq r$, $j \neq k$; and it follows that $Pr(U_n \in I_{b_1 \ldots b_r}) = 2^{-r}$ with respect to $\langle U_n \rangle \Re_k$. (See exercise 33.) This is enough to show the sequence is 1-distributed, by Theorem A.

Let $B[a_1, \ldots, a_r]$ denote the subsequence of $\langle U_n \rangle$ which consists of the values for those n in which $C[a_1, \ldots, a_r]$ is increased by one in step W6 of the algorithm. By the algorithm, $B[a_1, \ldots, a_r]$ is a finite sequence with at most $3 \cdot 4^{r-1}$ elements. All but a finite number of the members of $A[a_1, \ldots, a_r]$ come from the subsequences $B[a_1, \ldots, a_r, \ldots, a_t]$, where $a_j = 0$ or 1 for $r < j \leq t$.

Now assume that $A[a_1, \ldots, a_r]$ is infinite, and let $A[a_1, \ldots, a_r] = \langle U_{s_n} \rangle$ where $s_0 < s_1 < s_2 < \cdots$. If N is a large integer, with $4^r \leq 4^q < N \leq 4^{q+1}$, it follows that the number of values of $k < N$ for which U_{s_k} is in $I_{b_1 \ldots b_r}$ is (except for finitely many elements at the beginning of the subsequence)

$$\nu(N) = \nu(N_1) + \cdots + \nu(N_m).$$

Here m is the number of subsequences $B[a_1, \ldots, a_t]$ listed above in which U_{s_k} appears for some $k < N$; N_j is the number of values of k with U_{s_k} in the corresponding subsequence; and $\nu(N_j)$ is the number of such values which are also in $I_{b_1 \ldots b_r}$. Therefore by Lemma T,

$$|\nu(N) - 2^{-r}N| = |\nu(N_1) - 2^{-r}N_1 + \cdots + \nu(N_m) - 2^{-r}N_m|$$
$$\leq |\nu(N_1) - 2^{-r}N_1| + \cdots + |\nu(N_m) - 2^{-r}N_m|$$
$$\leq m \leq 1 + 2 + 4 + \cdots + 2^{q-r+1} < 2^{q+1}.$$

The inequality on m follows here from the fact that, by our choice of N, U_{s_N} is in $B[a_1, \ldots, a_t]$ for some $t \leq q + 1$.

Therefore

$$|\nu(N)/N - 2^{-r}| \leq 2^{q+1}/N < 2/\sqrt{N}. \quad \blacksquare$$

To show finally that sequences satisfying Definition R5 exist, we note first that if $\langle U_n \rangle$ is a $[0, 1)$ sequence of rational numbers and if $\Re$ is a computable subsequence rule for a b-ary sequence, we can make $\Re$ into a computable subsequence rule $\Re'$ for $\langle U_n \rangle$ by letting $f_n'(x_1, \ldots, x_n)$ in $\Re'$ equal $f_n(\lfloor bx_1 \rfloor, \ldots, \lfloor bx_n \rfloor)$ in $\Re$. If the sequence $\langle U_n \rangle \Re'$ is equidistributed, so is the sequence $\langle \lfloor bU_n \rfloor \rangle \Re$. Now the set of all computable subsequence rules for b-ary sequences, for all values of b, is countable (since only countably many effective algorithms are possible), so they may be listed in some sequence $\Re_1, \Re_2, \ldots$; therefore Algorithm W defines a $[0, 1)$ sequence which is random in the sense of Definition R5.

This brings us to a somewhat paradoxical situation. As we mentioned earlier, no effective algorithm can define a sequence which satisfies Definition R4, and for the same reason there is no effective algorithm which defines a sequence satisfying Definition R5. A proof of the existence of such random sequences is necessarily nonconstructive; how then can Algorithm W construct such a sequence?

There is no contradiction here; we have merely stumbled on the fact that the set of all algorithms cannot be enumerated by an effective algorithm. In other words, there is no effective algorithm to select the jth computable subsequence rule $\mathfrak{R}_j$; this happens because there is no effective algorithm to determine if a computational method ever terminates. (We shall return to this topic in Chapter 11.) Important large classes of algorithms *can* be systematically enumerated; thus, for example, Algorithm W shows that it is possible to construct, with an effective algorithm, a sequence which satisfies Definition R5 if we restrict consideration to subsequence rules which are "primitive recursive."

By modifying step W6 of Algorithm W, so that it sets $U_n \leftarrow V_{k+t}$ instead of V_k, where t is any nonnegative integer depending on $a_1, \ldots, a_r$, we can show there are *uncountably* many [0, 1) sequences satisfying Definition R5.

The following theorem shows still another way to prove the existence of uncountably many random sequences, using a less direct argument based on measure theory, even if the strong definition R6 is used:

Theorem M. *Let the real number x, $0 \le x < 1$, correspond to the binary sequence $\langle X_n \rangle$ if the binary representation of x is $0.X_0X_1 \cdots$. Under this correspondence, almost all x correspond to binary sequences which are random in the sense of Definition R6. (In other words, the set of all real x which correspond to a binary sequence which is nonrandom by Definition R6 has measure zero.)*

Proof. Let $\mathcal{S}$ be an effective algorithm which determines an infinite sequence of distinct nonnegative integers $\langle s_n \rangle$, where the choice of s_n depends only on n and X_{s_k} for $0 \le k < n$; and let $\mathfrak{R}$ be a computable subsequence rule. Then any binary sequence $\langle X_n \rangle$ leads to a subsequence $\langle X_{s_n} \rangle \mathfrak{R}$, and Definition R6 says this subsequence must either be finite or 1-distributed. It suffices to prove that *for fixed $\mathfrak{R}$ and $\mathcal{S}$ the set $N(\mathfrak{R}, \mathcal{S})$ of all real x corresponding to $\langle X_n \rangle$, such that $\langle X_{s_n} \rangle \mathfrak{R}$ is infinite and not 1-distributed, has measure zero.* For x has a nonrandom binary representation if and only if x is in $\bigcup N(\mathfrak{R}, \mathcal{S})$, taken over the countably many choices of $\mathfrak{R}$ and $\mathcal{S}$.

Therefore let $\mathfrak{R}$ and $\mathcal{S}$ be fixed. Consider the set $T(a_1 a_2 \ldots a_r)$ defined for all binary numbers $a_1 a_2 \ldots a_r$ as the set of all x corresponding to $\langle X_n \rangle$, such that $\langle X_{s_n} \rangle \mathfrak{R}$ has $\ge r$ elements whose first r elements are respectively equal to $a_1, a_2, \ldots, a_r$. Our first result is that

$$T(a_1 a_2 \ldots a_r) \qquad \text{has measure} \qquad \le 2^{-r}. \qquad (32)$$

To prove this, we start by observing that $T(a_1 a_2 \ldots a_r)$ is a measurable set: Each element of $T(a_1 a_2 \ldots a_r)$ is a real number $x = 0.X_0 X_1 \cdots$ for which there exists an integer m such that algorithm $\mathcal{S}$ determines distinct values $s_0, s_1, \ldots, s_m$, and rule $\mathfrak{R}$ determines a subsequence of $X_{s_0}, X_{s_1}, \ldots, X_{s_m}$

such that X_{s_m} is the rth element of this sequence. The set of all real $y = 0.Y_0Y_1\cdots$ such that $Y_{s_k} = X_{s_k}$ for $0 \leq k \leq m$ also belongs to $T(a_1a_2\ldots a_r)$, and this is a measurable set consisting of the finite union of dyadic subintervals $I_{b_1\ldots b_t}$. Since there are only countably many such dyadic intervals, we see that $T(a_1a_2\ldots a_r)$ is a countable union of dyadic intervals, and it is therefore measurable. Furthermore, this argument can be extended to show that the measure of $T(a_1\ldots a_{r-1}0)$ equals the measure of $T(a_1\ldots a_{r-1}1)$, since the latter is a union of dyadic intervals obtained from the former by requiring that $Y_{s_k} = X_{s_k}$ for $0 \leq k < m$ and $Y_{s_m} \neq X_{s_m}$. Now since

$$T(a_1\ldots a_{r-1}0) \cup T(a_1\ldots a_{r-1}1) \subseteq T(a_1\ldots a_{r-1}),$$

the measure of $T(a_1a_2\ldots a_r)$ is at most one-half the measure of $T(a_1\ldots a_{r-1})$. The inequality (32) follows by induction on r.

Now that (32) has been established, the remainder of the proof is essentially to show that the binary representations of almost all real numbers are equidistributed. The next few paragraphs constitute a rather long but not difficult proof of this fact, and they serve to illustrate typical estimation techniques in mathematical analysis.

For $0 < \epsilon < 1$, let $B(r, \epsilon)$ be $\bigcup T(a_1\ldots a_r)$, where the union is taken over all binary numbers $a_1\ldots a_r$ for which the number $\nu(r)$ of zeros among $a_1\ldots a_r$ satisfies

$$|\nu(r) - \tfrac{1}{2}r| \geq 1 + \epsilon r.$$

The number of such binary numbers is $C(r, \epsilon) = \sum \binom{r}{k}$ summed over the values of k with $|k - \tfrac{1}{2}r| \geq 1 + \epsilon r$. Let $r = 2t$ be an even integer; we may crudely estimate this quantity $\sum \binom{r}{k}$ as follows:

If $k > 0$,

$$\binom{2t}{t+k} = \binom{2t}{t}\frac{t}{t+1}\frac{t-1}{t+2}\cdots\frac{t-k+1}{t+k}$$

$$< \binom{2t}{t}\frac{t}{t}\frac{t-1}{t}\cdots\frac{t-k+1}{t}$$

$$\leq \binom{2t}{t}e^{-0/t}e^{-1/t}\cdots e^{-(k-1)/t} = \binom{2t}{t}e^{-k(k-1)/r}.$$

Thus

$$C(r, \epsilon) = 2\sum_{k \geq 1+\epsilon r}\binom{2t}{t+k} \leq 2\binom{2t}{t}\sum_{k \geq 1+\epsilon r}e^{-k(k-1)/r}$$

$$\leq 2\binom{2t}{t}te^{-(1+\epsilon r)\epsilon} < r\binom{r}{t}e^{-\epsilon^2 r}.$$

Similarly, when $r = 2t + 1$ we find that

$$C(r, \epsilon) < r \binom{r}{t} e^{-\epsilon^2 r};$$

hence by (32)

$$B(r, \epsilon) \quad \text{has measure} \quad \leq 2^{-r} C(r, \epsilon) < r e^{-\epsilon^2 r}. \tag{33}$$

The next step is to define

$$B^*(r, \epsilon) = B(r, \epsilon) \cup B(r + 1, \epsilon) \cup B(r + 2, \epsilon) \cup \cdots.$$

The measure of $B^*(r, \epsilon)$ is at most $\sum_{k \geq r} k e^{-\epsilon^2 k}$, and this is the remainder of a convergent series so

$$\lim_{r \to \infty} (\text{measure of } B^*(r, \epsilon)) = 0. \tag{34}$$

Now if x is a real number whose binary expansion $0.X_0 X_1 \cdots$ leads to an infinite sequence $\langle X_{s_n} \rangle \mathfrak{R}$ that is not 1-distributed, and if $\nu(r)$ denotes the number of zeros in the first r elements of the latter sequence, then

$$|\nu(r)/r - \tfrac{1}{2}| \geq 2\epsilon,$$

for some $\epsilon > 0$ and infinitely many r. This means x is in $B^*(r, \epsilon)$ for all r. So finally we find that

$$N(\mathfrak{R}, \mathcal{S}) = \bigcup_{t \geq 2} \bigcap_{r \geq 1} B^*(r, 1/t),$$

and, by (34), $\bigcap_{r \geq 1} B^*(r, 1/t)$ has measure zero for all t. Hence $N(\mathfrak{R}, \mathcal{S})$ has measure zero. ∎

From the existence of *binary* sequences satisfying Definition R6, we can show the existence of $[0, 1)$ sequences which are random in this sense. For details, see exercise 36. The consistency of Definition R6 is thereby established.

E. Random finite sequences. An argument was given above to indicate that it is impossible to define the concept of randomness for finite sequences; any given finite sequence is as likely as any other. Still, nearly everyone would agree that the sequence 011101001 is "more random" than 101010101, and even the latter sequence is "more random" than 000000000. Although it is true that truly random sequences will exhibit locally nonrandom behavior, we would expect such behavior only in a long finite sequence, not in a short one.

Several ways for defining the randomness of a finite sequence can be used, and only a few of the ideas will be sketched here. Let us consider only the case of b-ary sequences.

Given a b-ary sequence $X_1, X_2, \ldots X_N$, we can say that

$$\Pr(\mathcal{S}(n)) \approx p, \quad \text{if} \quad |\nu(N)/N - p| \leq 1/\sqrt{N},$$

where $\nu(n)$ is the quantity appearing in Definition A at the beginning of this section. The above sequence can be called "k-distributed" if

$$\Pr(X_n X_{n+1} \ldots X_{n+k-1} = x_1 x_2 \ldots x_k) \approx 1/b^k$$

for all b-ary numbers $x_1 x_2 \ldots x_k$. (Cf. Definition D; unfortunately a sequence may be k-distributed by this new definition when it is not $(k-1)$-distributed.)

A definition of randomness may now be given analogous to Definition R1, as follows:

Definition Q1. *A b-ary sequence of length N is "random" if it is k-distributed (in the above sense) for all positive integers $k \leq \log_b N$.*

According to this definition, for example, there are 170 nonrandom binary sequences of length 11:

00000001111	10000000111	11000000011	11100000001
00000001110	10000000110	11000000010	11100000000
00000001101	10000000101	11000000001	10100000001
00000001011	10000000011	01000000011	01100000001
00000000111			

plus 01010101010 and all sequences with nine or more zeros, plus all sequences obtained from the preceding ones by interchanging ones and zeros.

Similarly, we can formulate a definition for finite sequences analogous to Definition R6. Let **A** be a set of algorithms, each of which is a combination selection and choice procedure that gives a subsequence $\langle X_{s_n} \rangle \Re$, as in the proof of Theorem M.

Definition Q2. *The b-ary sequence $X_1, X_2, \ldots, X_N$ is (n, ϵ)-random with respect to a set of algorithms* **A***, if for every subsequence $X_{t_1}, X_{t_2}, \ldots, X_{t_m}$ determined by an algorithm of* **A** *we have either $m < n$ or*

$$\left| \frac{1}{m} \nu_a(X_{t_1}, \ldots, X_{t_m}) - \frac{1}{b} \right| \leq \epsilon \qquad \text{for} \qquad 0 \leq a < b.$$

Here $\nu_a(x_1, \ldots, x_m)$ is the number of a's in the sequence $x_1, \ldots, x_m$.

(In other words, every sufficiently long subsequence determined by an algorithm of **A** must be approximately equidistributed.) The basic idea in this case is to let **A** be a set of "simple" algorithms; the number (and the complexity) of the algorithms in **A** can grow as N grows.

As an example of Definition Q2, let us consider binary sequences, and let **A** be just the following four algorithms:

a) Take the whole sequence.
b) Take alternate terms of the sequence, starting with the first.
c) Take the terms of the sequence following a zero.
d) Take the terms of the sequence following a one.

Now a sequence $X_1, \ldots, X_8$ is $(4, \frac{1}{8})$-random if:

by (a), $|\frac{1}{8}(X_1 + \cdots + X_8) - \frac{1}{2}| \leq \frac{1}{8}$, i.e., if there are 3, 4, or 5 ones;

by (b), $|\frac{1}{4}(X_1 + X_3 + X_5 + X_7) - \frac{1}{2}| \leq \frac{1}{8}$, i.e., if there are two ones in odd-numbered positions.

by (c), there are three possibilities depending on how many zeros occupy positions $X_1, \ldots, X_7$: if there are 2 or 3 zeros here, there is no condition to test (since $n = 4$); if there are 4 zeros, they must be followed by two zeros and two ones; and if there are 5 zeros, they must be followed by two or three zeros.

by (d), we get conditions similar to those implied by (c).

It turns out that only the following binary sequences of length 8 are $(4, \frac{1}{8})$-random with respect to these rules:

00001011	00101001	01001110	01101000
00011010	00101100	01011011	01101100
00011011	00110010	01011110	01101101
00100011	00110011	01100010	01110010
00100110	00110110	01100011	01110110
00100111	00111001	01100110	

plus those obtained by interchanging 0 and 1 consistently.

It is clear that we could make the set of algorithms so large that no sequences satisfy the definition, when n and ϵ are reasonably small. A. N. Kolmogorov has proved that an (n, ϵ)-random binary sequence *will* always exist, for any given N, if the number of algorithms in **A** does not exceed

$$\frac{1}{2}e^{2n\epsilon^2(1-\epsilon)}. \tag{35}$$

This result is not nearly strong enough to show that sequences satisfying Definition Q1 will exist, but the latter can be constructed efficiently using the procedure of Rees in exercise 3.2.2–21.

Still another interesting approach to a definition of randomness has been taken by Per Martin-Löf [*Information and Control* **9** (1966), 602–619]. Given a finite b-ary sequence $X_1, \ldots, X_N$, let $l(X_1, \ldots, X_N)$ be the length of the shortest Turing machine program which generates this sequence. (For a definition of Turing machines, see Chapter 11; alternatively, we could use certain classes of effective algorithms, such as those defined in exercise 1.1–8.) Then $l(X_1, \ldots, X_N)$ is a measure of the "patternlessness" of the sequence, and we may equate this idea with randomness. The sequences of length N which maximize $l(X_1, \ldots, X_N)$ may be called random. (From the standpoint of practical random-number generation by computer, this is, of course, the worst definition of "randomness" that can be imagined!)

Essentially the same definition of randomness was independently given by G. Chaitin at about the same time; see *JACM* **16** (1969), 145–159. It is interesting to note that even though this definition makes no reference to equidistribution properties as our other definitions have, Martin-Löf and Chaitin have proved that random sequences of this type also have the expected equidistribution properties. In fact, Martin-Löf has demonstrated that such sequences satisfy *all* computable statistical tests for randomness, in an appropriate sense.

F. Summary, history, and bibliography. We have defined several degrees of randomness that a sequence might possess.

An infinite sequence which is ∞-distributed satisfies a great many useful properties which are expected of random sequences, and there is a rich theory concerning ∞-distributed sequences. (The exercises which follow develop several important properties of ∞-distributed sequences which have not been mentioned in the text.) Definition R1 is therefore an appropriate basis for theoretical studies of randomness.

The concept of an ∞-distributed b-ary sequence was introduced in 1909 by Emile Borel. He essentially defined the concept of an (m, k)-distributed sequence, and showed that the b-ary representations of almost all real numbers are (m, k)-distributed for all m and k. He called such numbers *normal* to base b. An excellent discussion of this topic appears in his well-known book, *Leçons sur la théorie des fonctions* (2nd ed., 1914), 182–216. For later results and further bibliography concerning normal numbers, see W. Schmidt, *Pacific Journal of Mathematics* **10** (1960), 661–672.

The notion of an ∞-distributed sequence of *real* numbers, also called a *completely equidistributed sequence*, was introduced by Joel N. Franklin in his paper "Deterministic Simulation of Random Processes," *Math. Comp.* **17** (1963), 28–59. This important paper solves a number of very interesting problems concerned with k-distributed sequences, and Franklin obtains results about the distribution properties of many special sequences.

We have seen, however, that ∞-distributed sequences need not be sufficiently haphazard to qualify completely as "random." Three definitions, R4, R5, and R6, were formulated above to provide the additional conditions; and Definition R6, in particular, seems to be an appropriate way to define the concept of an infinite random sequence. It is a precise, quantitative statement that apparently coincides with the intuitive idea of true randomness.

Historically, the development of these definitions was primarily influenced by R. von Mises' quest for a good definition of "probability." In *Math. Zeitschrift* **5** (1919), 52–99, von Mises proposed a definition similar in spirit to Definition R5, although stated too strongly (like our Definition R3) so that no sequences satisfying the conditions could possibly exist. Many people noticed this discrepancy, and A. H. Copeland [*Amer. J. Math.* **50** (1928), 535–552] suggested weakening von Mises' definition by substituting what he called "admissible numbers" (or Bernoulli sequences). These are equivalent to ∞-

distributed $[0, 1)$ sequences in which all entries U_n have been replaced by 1 if $U_n < p$ or by 0 if $U_n \geq p$, for a given probability p. Thus Copeland was essentially suggesting a return to Definition R1. Then Abraham Wald showed that it is not necessary to weaken von Mises' definition so drastically, and he proposed substituting a countable set of subsequence rules. In an important paper [*Ergebnisse eines math. Kolloquiums* **8** (Vienna, 1937), 38–72], Wald essentially proved Theorem W, although he made the erroneous assertion that the sequence constructed by Algorithm W also satisfies the stronger condition that $Pr(U_n \in A) =$ measure of A, for all Lebesgue measurable $A \subseteq [0, 1)$. We have observed that no sequence can satisfy this property.

Definition R4 was first formulated by D. W. Loveland [*Zeit. fur Math. Logik und Grundlagen d. Math.* **12** (1966), 279–294], who discovered that Algorithm W can be used to show the existence of binary sequences which satisfy Definition R5 but not R4.

The concept of "computability" was still very much in its infancy when Wald wrote his paper, and A. Church [*Bulletin AMS* **47** (1940), 130–135] showed how the precise notion of "effective algorithm" could be added to Wald's theory to make his definitions completely rigorous. The extension to Definition R6 was due essentially to A. N. Kolmogorov [*Sankhyā* series A, **25** (1963), 369–376], who proposed Definition Q2 for finite sequences in that same paper. Further connections between random sequences and recursive function theory have been explored by D. W. Loveland, *Trans. AMS* **125** (1966), 497–510. See also C.-P. Schnorr [*Z. Wahr. verw. Geb.* **14** (1969), 27–35], who has found strong relations between random sequences and the "species of measure zero" defined by L. E. J. Brouwer in 1919.

The publications of Church and Kolmogorov considered only binary sequences for which $Pr(X_n = 1) = p$ for a given probability p. The discussion which appears in this section is therefore somewhat more general, since a $[0, 1)$ sequence essentially represents all p at once.

In another paper [*Problemy Peredači Informacii* **1** (1965), 3–11], Kolmogorov considered the problem of defining the "information content" of a sequence, and this work led to Martin-Löf's interesting definition of finite random sequences via "patternlessness." (See *IEEE Trans.* **IT-14** (1968), 662–664.) Another definition of randomness, somewhere "between" Definitions Q1 and Q2, had been formulated many years earlier by A. S. Besicovitch [*Math. Zeitschrift* **39** (1934), 146–156].

For further discussion of random sequences, see K. R. Popper, *The Logic of Scientific Discovery* (London, 1959), especially the interesting construction on pp. 162–163, which he first published in 1934.

For more exercises of interest concerning equidistributed $[0, 1)$ sequences, see G. Pólya and G. Szegö, *Aufgaben und Lehrsätze aus der Analysis* (Berlin, 1925), Vol. 1, pp. 67–77. For equidistributed b-ary sequences, see I. Niven, *Trans. Amer. Math. Soc.* **98** (1961), 52–61; B. Zane, *AMM* **71** (1964), 162–164.

EXERCISES

1. [*10*] Can a periodic sequence be equidistributed?

2. [*10*] Consider the periodic binary sequence $0, 0, 1, 1, 0, 0, 1, 1, \ldots$. Is it 1-distributed? Is it 2-distributed? Is it 3-distributed?

3. [*M22*] Construct a periodic ternary sequence which is 3-distributed.

▶ **4.** [*HM22*] Consider the sequence with $U_n = (2^{\lfloor \log_2(n+1) \rfloor}/3) \bmod 1$. What is $\Pr(U_n < \frac{1}{2})$?

5. [*HM14*] Prove that $\Pr(S(n) \text{ and } T(n)) + \Pr(S(n) \text{ or } T(n)) = \Pr(S(n)) + \Pr(T(n))$, for any two statements $S(n)$, $T(n)$, provided at least three of the limits exist. [For example, if a sequence is 2-distributed, we would find that

$$\Pr(u_1 \leq U_n < v_1 \text{ or } u_2 \leq U_{n+1} < v_2)$$
$$= v_1 - u_1 + v_2 - u_2 - (v_1 - u_1)(v_2 - u_2).]$$

6. [*HM23*] Let $S_1(n), S_2(n), \ldots$ be an infinite sequence of statements about mutually disjoint events, i.e., $S_i(n)$ and $S_j(n)$ cannot simultaneously be true if $i \neq j$. Assume that $\Pr(S_j(n))$ exists for each $j \geq 1$. Show that $\underline{\Pr}(S_j(n))$ is true for some $j \geq 1) \geq \sum_{j \geq 1} \Pr(S_j(n))$, and give an example to show that equality need not hold.

7. [*HM27*] As in the previous exercise, let $S_{ij}(n)$, $i \geq 1, j \geq 1$, be infinitely many mutually disjoint events, such that $\Pr(S_{ij}(n))$ exists. Assume that for all $n > 0$, $S_{ij}(n)$ is true for exactly one pair of integers i, j. If $\sum_{i,j \geq 1} \Pr(S_{ij}(n)) = 1$, can it be concluded that, for all $i \geq 1$, $\Pr(S_{ij}(n))$ is true for some $j \geq 1$) exists and equals $\sum_{j \geq 1} \Pr(S_{ij}(n))$?

8. [*M15*] Prove (13).

9. [*HM20*] Prove Lemma E. [*Hint:* Consider $\sum_{1 \leq j \leq m} (y_{jn} - \alpha)^2$.]

▶ **10.** [*HM22*] Where was the fact that q is a multiple of m used in the proof of Theorem C?

▶ **11.** [*HM20*] Use Theorem C to prove that if $\langle U_n \rangle$ is ∞-distributed, so is the sub-sequence $\langle U_{2n} \rangle$.

12. [*HM20*] Show that a k-distributed sequence passes the "maximum of k" test, in the following sense: $\Pr(u \leq \max(U_n, U_{n+1}, \ldots, U_{n+k-1}) < v) = v^k - u^k$.

▶ **13.** [*HM27*] Show that an ∞-distributed $[0, 1)$ sequence passes the "gap test" in the following sense: If $0 \leq \alpha < \beta \leq 1$, and $p = \beta - \alpha$, let $f(0) = 0$, and for $n \geq 1$, let $f(n)$ be the smallest integer $m > f(n-1)$ such that $\alpha \leq U_m < \beta$; then $\Pr(f(n) - f(n-1) = k) = p(1-p)^{k-1}$.

14. [*HM25*] Show that an ∞-distributed sequence passes the "run test" in the following sense: If $f(0) = 1$ and if for $n \geq 1$ $f(n)$ is the smallest integer $m > f(n-1)$ such that $U_{m-1} > U_m$, then

$$\Pr(f(n) - f(n-1) = k) = 2k/(k+1)! - 2(k+1)/(k+2)!.$$

▶ **15.** [*HM30*] Show that an ∞-distributed sequence passes the "coupon-collectors test" when there are only two kinds of coupons, in the following sense: Let $X_1, X_2, \ldots$ be an ∞-distributed binary sequence. Let $f(0) = 0$, and for $n \geq 1$ let $f(n)$ be the smallest integer $m > f(n-1)$ such that $\{X_{f(n-1)+1}, \ldots, X_m\}$ is the set $\{0, 1\}$. Prove that $\Pr(f(n) - f(n-1) = k) = 2^{1-k}$, $k \geq 2$. (Cf. exercise 7.)

16. [*HM38*] Does the coupon-collector's test hold for ∞-distributed sequences when there are more than two kinds of coupons? (Cf. the previous exercise.)

17. [*HM50*] Given that r is a rational number, Franklin has proved that the sequence with $U_n = (r^n \bmod 1)$ is not 2-distributed. But is there any rational number r for which this sequence is equidistributed? In particular, is the sequence equidistributed when $r = \frac{3}{2}$? (Cf. the article by K. Mahler, *Mathematika* **4** (1957), 122–124.)

▶ **18.** [*HM22*] Prove that if $U_0, U_1, \ldots$ is k-distributed, so is the sequence $V_0, V_1, \ldots$, where $V_n = \lfloor nU_n \rfloor / n$.

19. [*HM46*] Consider Definition R4 with "∞-distributed" replaced by "1-distributed". Is there a sequence which satisfies this weaker definition, but which is not ∞-distributed? (Is the weaker definition really weaker?)

20. [*HM50*] Does the sequence with $U_n = (\theta^n \bmod 1)$ satisfy Definition R4 for almost all real numbers $\theta > 1$? [This result would follow if we could either give a negative answer to exercise 19, or if we could show that for any sequence of distinct positive integers $s_0, s_1, s_2, \ldots$ the sequence with $U_n = (\theta^{s_n} \bmod 1)$ is ∞-distributed for almost all $\theta > 1$.]

21. [*HM20*] Let S be a set and let $\mathfrak{M}$ be a collection of subsets of S. Suppose that p is a real-valued function of the sets in $\mathfrak{M}$, for which $p(M)$ denotes the probability that a "randomly" chosen element of S lies in M. Generalize Definitions B and D to obtain a good definition of the concept of a k-distributed sequence $\langle Z_n \rangle$ of elements of S with respect to the probability distribution p.

▶ **22.** [*HM30*] (Hermann Weyl.) Show that the [0, 1) sequence $\langle U_n \rangle$ is k-distributed if and only if

$$\lim_{N \to \infty} \frac{1}{N} \sum_{0 \le n < N} \exp\left(2\pi i(c_1 U_n + \cdots + c_k U_{n+k-1})\right) = 0$$

for every set of integers $c_1, c_2, \ldots, c_k$ not all zero.

23. [*M34*] Show that a b-ary sequence $\langle X_n \rangle$ is k-distributed if and only if all of the sequences $\langle c_1 X_n + c_2 X_{n+1} + \cdots + c_k X_{n+k-1} \rangle$ are *equidistributed*, whenever $c_1, c_2, \ldots, c_k$ are integers with $\gcd(c_1, \ldots, c_k) = 1$.

24. [*M30*] Show that a [0, 1) sequence $\langle U_n \rangle$ is k-distributed if and only if all of the sequences $\langle c_1 U_n + c_2 U_{n+1} + \cdots + c_k U_{n+k-1} \rangle$ are *equidistributed*, whenever $c_1, c_2, \ldots, c_k$ are integers not all zero.

25. [*HM20*] A sequence is called a "white sequence" if all serial correlations are zero, i.e., if the equation in Corollary S is true for *all* $k \ge 1$. (By Corollary S, an ∞-distributed sequence is white.) Show that if a [0, 1) sequence is equidistributed, it is white if and only if

$$\lim_{n \to \infty} \frac{1}{n} \sum_{0 \le j < n} (U_j - \tfrac{1}{2})(U_{j+k} - \tfrac{1}{2}) = 0, \qquad \text{all} \quad k \ge 1.$$

26. [*HM34*] (J. Franklin.) A white sequence, as defined in the previous exercise, can definitely fail to be random. Let $U_0, U_1, \ldots$ be an ∞-distributed sequence; define the sequence $V_0, V_1, \ldots$ as follows:

$$\begin{aligned}
(V_{2n-1}, V_{2n}) &= (U_{2n-1}, U_{2n}) & \text{if} \quad & (U_{2n-1}, U_{2n}) \in G, \\
(V_{2n-1}, V_{2n}) &= (U_{2n}, U_{2n-1}) & \text{if} \quad & (U_{2n-1}, U_{2n}) \notin G,
\end{aligned}$$

where G is the set $\{(x, y) \mid x - \frac{1}{2} \leq y \leq x \text{ or } x + \frac{1}{2} \leq y\}$. Show that (a) $V_0, V_1, \ldots$ is equidistributed and white; (b) $\Pr(V_n > V_{n+1}) = \frac{5}{8}$. (This points out the weakness of the serial correlation test.)

27. [*HM49*] Is the number $\frac{5}{8}$ in the previous exercise the highest possible value for $\Pr(V_n > V_{n+1})$ in an equidistributed, white sequence?

▶ **28.** [*M24*] Use the sequence (11) to construct a $[0, 1)$ sequence which is 3-distributed, for which $\Pr(U_{2n} \geq \frac{1}{2}) = \frac{3}{4}$.

29. [*HM34*] Let $X_0, X_1, \ldots$ be a $(2k)$-distributed binary sequence. Show that

$$\overline{\Pr}(X_{2n} = 0) \leq \frac{1}{2} + \left(\begin{matrix} 2k - 1 \\ k \end{matrix} \right) \Big/ 2^{2k}.$$

▶ **30.** [*M39*] Construct a binary sequence which is $(2k)$-distributed, and for which

$$\Pr(X_{2n} = 0) = \frac{1}{2} + \left(\begin{matrix} 2k - 1 \\ k \end{matrix} \right) \Big/ 2^{2k}.$$

(Therefore the inequality in the previous exercise is the best possible.)

31. [*M30*] Show that $[0, 1)$ sequences exist which satisfy Definition R5, yet $\nu_n/n \geq \frac{1}{2}$ for all $n > 0$, where ν_n is the number of $j < n$ for which $u_n < \frac{1}{2}$. (This might be considered a nonrandom property of the sequence.)

32. [*M24*] Given that $\langle X_n \rangle$ is a "random" b-ary sequence according to Definition R5, and if $\Re$ is a computable subsequence rule which specifies an infinite subsequence $\langle X_n \rangle \Re$, show that the latter subsequence is not only 1-distributed, it is "random" by Definition R5.

33. [*HM24*] Let $\langle U_{r_n} \rangle$ and $\langle U_{s_n} \rangle$ be infinite disjoint subsequences of a sequence $\langle U_n \rangle$. (Thus, $r_0 < r_1 < r_2 < \cdots$ and $s_0 < s_1 < s_2 < \cdots$ are increasing sequences of integers and $r_m \neq s_n$ for any m, n.) Let $\langle U_{t_n} \rangle$ be the combined subsequence, so that $t_0 < t_1 < t_2 < \cdots$ and the set $\{t_n\} = \{r_n\} \cup \{s_n\}$. Show that if $\Pr(U_{r_n} \in A) = \Pr(U_{s_n} \in A) = p$, then $\Pr(U_{t_n} \in A) = p$.

▶ **34.** [*M25*] Define subsequence rules $\Re_1, \Re_2, \Re_3, \ldots$ such that Algorithm W can be used with these rules to give an effective algorithm to construct a $[0, 1)$ sequence satisfying Definition R1.

35. [*M50*] If possible, generalize Algorithm W to obtain a similar construction of sequences that satisfy the more stringent conditions of Definition R6.

36. [*HM30*] Let $\langle X_n \rangle$ be a binary sequence which is "random" according to Definition R6. Show that the $[0, 1)$ sequence $\langle U_n \rangle$ defined in binary notation by the scheme

$$U_0 = 0.X_0$$
$$U_1 = 0.X_1 X_2$$
$$U_2 = 0.X_3 X_4 X_5$$
$$U_3 = 0.X_6 X_7 X_8 X_9$$
$$\cdots$$

is random in the sense of Definition R6.

37. [*M48*] Do there exist sequences which satisfy Definition R4 but not Definition R5?

38. [*M50*] (A. N. Kolmogorov.) Given N, n and ϵ, what is the smallest number of algorithms in a set **A** such that no (n, ϵ)-random binary sequences of length N exist with respect to **A**? (If exact formulas cannot be given, can asymptotic formulas be found? The point of this problem is to discover how close the bound (35) comes to being "best possible.")

39. [*HM45*] (K. F. Roth.) Let U_n be a $[0, 1)$ sequence, and let $\nu_n(u)$ be the number of nonnegative integers $j < n$ such that $0 \le U_j < u$. Prove that there is a positive constant c such that, for any N and for any $[0, 1)$ sequence $\langle U_n \rangle$, we have

$$|\nu_n(u) - un| > c\sqrt{\log N}$$

for some n and u with $0 \le n < N$, $0 \le u < 1$. (In other words, no $[0, 1)$ sequence can be *too* equidistributed.)

3.6. SUMMARY

We have covered a fairly large number of topics in this chapter: how to generate random numbers, how to test them, how to modify them in applications, and how to derive theoretical facts about them. Perhaps the main question in many readers' minds will be, "What is the result of all this theory? What is a simple, virtuous, random-number generator I can use in my programs which I can expect to be a reliable source of random numbers?"

The detailed investigations in this chapter suggest that the following procedure gives the "nicest" and "simplest" random-number generator: At the beginning of the program, set an integer variable X to some value X_0. This variable X is to be used only for the purpose of random-number generation. Whenever a new random number is required by the program, set

$$X \leftarrow (aX + c) \bmod m \tag{1}$$

and use the new value of X as the random value. It is necessary to choose X_0, a, c, and m properly, and to use the random numbers wisely, according to the following principles:

i) The number X_0 may be chosen arbitrarily. If the program is run several times and a different source of random numbers is desired each time, set X_0 to the last value attained by X on the preceding run; or (if more convenient) set X_0 to the current date and time.

ii) The number m should be large. It may conveniently be taken as the computer's word size, since this makes the computation of $(aX + c) \bmod m$ quite efficient. Section 3.2.1.1 discusses the choice of m in more detail. The computation of $(aX + c) \bmod m$ must be done *exactly*, with no roundoff error.

iii) If m is a power of 2 (i.e., if a binary computer is being used), pick a so that $a \bmod 8 = 5$. If m is a power of 10 (i.e., if a decimal computer is being used), choose a so that $a \bmod 200 = 21$. This choice of a together with the choice of c given below ensures that the random-number generator will produce all m different possible values of X before it starts to repeat (see Section 3.2.1.2) and ensures high "potency" (see Section 3.2.1.3).

iv) The multiplier a should be larger than $\sqrt{m}$, preferably larger than $m/100$, but smaller than $m - \sqrt{m}$. The digits in the binary or decimal representation of a should *not* have a simple, regular pattern; for example, see Section 3.2.1.3 and exercise 3.2.1.3–8. The best policy is to take some haphazard constant to be the multiplier, such as

$$a = 3141592621 \tag{2}$$

(which satisfies both of the conditions in (iii)). These considerations will almost always give a reasonably good multiplier; but if the random-number generator is to be used extensively, the multiplier a should also be chosen so that it passes the "spectral test" (Section 3.3.4).

v) The constant c should be an odd number (when m is a power of 2) and also not a multiple of 5 (when m is a power of 10). It is preferable to choose c so that c/m has approximately the value given in Section 3.3.3, Eq. (41).

vi) The least significant (right-hand) digits of X are not very random, so decisions based on the number X should always be primarily influenced by the most significant digits. It is generally best to think of X as a random fraction X/m between 0 and 1, that is, to visualize X with a decimal point at its left, than to regard X as a random integer between 0 and $m - 1$. To compute a random integer between 0 and $k - 1$, one should multiply by k and truncate the result (see the beginning of Section 3.4.2).

These six "rules" are the most important things to know about computer-generated random numbers. Unfortunately, quite a bit of published material in existence at the time this chapter was written recommends the use of generators which violate the suggestions above; many of these articles contain important general contributions to the science of random-number generation, yet the authors did not find out until later that the particular method they recommended was inadequate.

Perhaps further research will show that even the random-number generators recommended here are unsatisfactory; we hope this is not the case, but the history of the subject warns us to be cautious. The most prudent policy for a person to follow is to run each Monte Carlo program at least twice using quite different sources of random numbers, before taking the answers of the program seriously; this not only will give an indication of the stability of the results, it also guards against the chance of trusting in a generator with hidden deficiencies. (Every random-number generator will fail in at least one application.)

Although linear congruential sequences of the type described here are usually an adequate source of random numbers, an important limitation on their randomness has been mentioned in Section 3.3.4 (see Eq. 3.3.4–13). The result of exercise 3.3.4–27 shows that linear congruential sequences are unsuitable for "high resolution" Monte Carlo applications. If more independence is required, the techniques of Section 3.2.2 should be used.

The article "Random Number Generators" by T. E. Hull and A. R. Dobell, *SIAM Review* **4** (1962), 230–254, contains an excellent bibliography of the pre-1962 literature on the subject. The existence of such a complete bibliography makes it unnecessary to list other references to the early literature here, in addition to those mentioned elsewhere in the chapter.

The paper "Uniform Random Generators" by M. Donald MacLaren and George Marsaglia, *JACM* **12** (1965), 83–89, is a particularly noteworthy post-1962 development, since this paper is the first one which really points out the deficiencies of linear congruential generators with improper multipliers. Algorithm 3.2.2M, a good method which considerably improves the randomness of a generator with the computation time only doubled, was also introduced in that article.

The book *Random Number Generators* by Birger Jansson (Stockholm: Almqvist and Wiksell, 1966), is another survey of many of the topics considered in this chapter.

For a detailed study of the use of random numbers in numerical analysis, see J. M. Hammersley and D. C. Handscomb, *Monte Carlo Methods* (London: Methuen, 1964). This book shows that many numerical methods are enhanced by using numbers which are "quasirandom," designed specifically to solve a certain problem (not necessarily satisfying the statistical tests we have discussed).

The exercises below include some interesting types of computer programming projects based on uses of random numbers, and many similar ideas will perhaps occur to the reader. Almost all good computer programs contain at least one random-number generator.

EXERCISES

1. [*21*] Write a `MIX` subroutine with the following characteristics:

Calling sequence: `JMP   RANDI`

Entry conditions: rA = k, a positive integer < 5000

Exit conditions: rA $\leftarrow$ a random integer Y, $1 \leq Y \leq k$, with each integer equally probable; rX = ?; overflow off.

[Use the method suggested in this section, except take $c = 1$.]

▶ 2. [*21*] Some people have been afraid computers will someday take over the world; but they are reassured by the statement that a machine cannot do anything really new, since it is only obeying the commands of its master, the programmer. Lady Lovelace wrote in 1844, "The Analytical Engine has no pretensions to *originate* anything. It can do *whatever we know how to order it* to perform." Her statement has been further elaborated by many philosophers. Discuss this topic, from the standpoint of random-number generators.

3. [*32*] "A dice game." Write a program that simulates a roll of two dice, each of which takes on the values 1, 2, . . . , 6 with equal probability. If the total is 7 or 11 on the first roll, the game is won; a total of 2, 3, or 12 loses; and on any other total, call that total the "point" and continue rolling dice until either a 7 occurs (a loss) or the point occurs again (a win).

Play ten games. The result of each roll of the dice should be printed in the form *mn*, where *m* and *n* are the contents of the two dice, followed by some appropriate comment (like "snake eyes" or "little Joe" or "the hard way," etc.)

4. [*40*] "Solitaire" (patience). Some people spend a lot of valuable time playing card games of solitaire, and perhaps automation will make an important inroad in this area. Write a program that (a) shuffles a simulated deck of cards; (b) plays some common game of solitaire based on the order of the cards in the deck; (c) prints out the result of the game, i.e., how close the program came to winning. Several games should be played. The program might be set up to "cheat" upon request.

5. [*46*] *Creative writing by computer.* A television program entitled "The Thinking Machine," broadcast by the CBS television network on October 26, 1960, featured (among other things) two Western-style playlets which were written by a computer program. Here are the two scripts as they were printed out by the computer:

Saga #1. (The gun is in the right hand; the money is in the left hand; the glass is on the table; the bottle is on the table; the holster is on the robber; the sheriff's gun is in the sheriff's right hand; the sheriff's holster is on the sheriff.)

ROBBER: (The robber is at the window.) Go to door; open door; go thru door; close door; go to corner; put money down at corner; go to table; put gun on table; sit and fidgit; sit at table; pick up glass with right hand (empty); put glass on table; pick up bottle with right hand; pour; put bottle on table; pick up glass with right hand; take a drink from glass; put glass on table; pick up bottle with right hand; sit at table; sit at table; go to corner; go to window; go to table; put bottle on table; sit and fidgit; sit at table; sit and fidgit; go to window; go to table; pick up glass with right hand.

SHERIFF: (The sheriff is at the window.) See robber; (robber sees sheriff); go to door.

ROBBER: Put glass on table; pick up gun with right hand; check gun.

SHERIFF: Wait; open door; see robber; (robber sees sheriff); go thru door.

ROBBER: Go to window; aim; fire; SHERIFF NICKED.

SHERIFF: Go to window; aim; fire; MISSED; go to door; go to window.

ROBBER: Go to door; aim; aim.

SHERIFF: Aim; fire; MISSED.

ROBBER: Fire; SHERIFF NICKED.

SHERIFF: Go to door; aim; fire; MISSED; go thru door; aim.
ROBBER: Aim; fire; MISSED; aim; fire; MISSED.
SHERIFF: Fire; MISSED; go to window; aim; fire; MISSED.
ROBBER: Aim; fire; MISSED; aim; fire; MISSED; aim; fire; SHERIFF NICKED.
SHERIFF: Aim; fire; ROBBER HIT.
ROBBER: Drop gun; robber dies.
SHERIFF: Put gun in holster; go to table; pick up glass with right hand (empty); take glass from right hand with left hand; pick up bottle with right hand; pour; put bottle on table; take glass from left hand with right hand; take a drink from glass; take glass from right hand with left hand; pick up bottle with right hand; pour; put bottle on table; take glass from left hand with right hand; take a drink from glass; put glass on table; go to corner; pick up money with right hand; go to door; go thru door; close door. CURTAIN.

Saga #2. (The gun is in the right hand; the money is in the left hand; the glass is on the table; the bottle is on the table; the holster is on the robber; the sheriff's gun is in the sheriff's right hand; the sheriff's holster is on the sheriff.)

ROBBER: (The robber is at the window.) Go to door; open door; go thru door; close door; go to corner; put money down at corner; go to window; put gun down at window; lean on window and look; lean on window and look; go to corner; count money; go to table; pick up glass with right hand (empty); take glass from right hand with left hand; pick up bottle with right hand; pour; put bottle on table; take glass from left hand with right hand; take a drink from glass; put glass on table; pick up bottle with right hand; pour; go to corner; put bottle down at corner; go to window; pick up gun with right hand; check gun; put gun in holster; go to table; pick up glass with right hand; take a drink from glass; go to window; put glass down at window.
SHERIFF: (The sheriff is at the window.) See robber; (robber sees sheriff); go to door.
ROBBER: Take gun from holster with right hand; check gun; go to door; check gun; put gun down at door.
SHERIFF: Open door; see robber; (robber sees sheriff); go thru door; go to window.
ROBBER: Pick up gun with right hand.
SHERIFF: Go to table.
ROBBER: Aim; fire; MISSED; aim; fire; SHERIFF HIT; blow out barrel; put gun in holster.
SHERIFF: Drop gun; sheriff dies.
ROBBER: Go to corner; pick up money with right hand; go to door; go thru door; close door. CURTAIN.

A careful reading of the above scripts reveals the highly intense drama present here. The computer program was careful to keep track of the locations of each player, the contents of his hands, etc. Actions taken by the players are random, governed by certain probabilities; the probability of a foolish action is increased depending on how much he has had to drink and on how often he has been nicked in a shot. The reader may study the above scripts to deduce further properties of the program.

Of course, even the best scripts are rewritten before they are produced, and this is especially true when an inexperienced writer has prepared the original script. Here are the scripts just as they were actually used in the show:

Saga #1. Music up.
MS Robber peering thru window of shack.
CU Robber's face.
MS Robber entering shack.
CU Robber sees whiskey bottle on table.

CU   Sheriff outside shack.
MS   Robber sees sheriff.
LS   Sheriff in doorway over shoulder of robber, both draw.
MS   Sheriff drawing gun.
LS   Shooting it out. Robber gets shot.
MS   Sheriff picking up money bags.
MS   Robber staggering.
MS   Robber dying. Falls across table, after trying to take last shot at sheriff.
MS   Sheriff walking thru doorway with money.
MS   of robber's body, now still, lying across table top. Camera dollies back. (Laughter)

Saga #2. Music up.
CU   of window. Robber appears.
MS   Robber entering shack with two sacks of money.
MS   Robber puts money bags on barrel.
CU   Robber—sees whiskey on table.
MS   Robber pouring himself a drink at table. Goes to count money. Laughs.
MS   Sheriff outside shack.
MS   thru window.
MS   Robber sees sheriff thru window.
LS   Sheriff entering shack. Draw. Shoot it out.
CU   Sheriff. Writhing from shot.
M/2 shot Sheriff staggering to table for a drink . . . falls dead.
MS   Robber leaves shack with money bags.*

[*Note:* CU = "close up", MS = "medium shot", etc. The above details were kindly furnished to the author by Thomas H. Wolf, producer of the television show, who suggested the idea of a computer-written playlet in the first place, and also by Douglas T. Ross and Harrison R. Morse who designed and checked out the computer program.]

The reader will undoubtedly have many ideas about how he could prepare his own computer program to do creative writing; and that is the point of this exercise.

---

*© 1962 by Columbia Broadcasting System, Inc. All Rights Reserved. Used by permission. For further information, see J. E. Pfeiffer, *The Thinking Machine* (New York: J. B. Lippincott, 1962).

CHAPTER FOUR

ARITHMETIC

Seeing there is nothing (right well beloved Students in the Mathematickes)
that is so troublesome to Mathematicall practise, nor that doth more molest
and hinder Calculators, then the Multiplications, Divisions, square and
cubical Extractions of great numbers, which besides the tedious expence of
time are for the most part subject to many slippery errors. I began therefore
to consider in my minde, by what certaine and ready Art I might
remove those hindrances.

— JOHN NAPIER (1614)

I do hate sums. There is no greater mistake than to call arithmetic an exact
science. There are . . . hidden laws of number which it requires a mind
like mine to perceive. For instance, if you add a sum from the bottom up,
and then again from the top down, the result is always different.

— MRS. LA TOUCHE (19th century)

I cannot conceive that anybody will require multiplications at the rate
of 40,000, or even 4,000 per hour; such a revolutionary change as the
octonary scale should not be imposed upon mankind in general
for the sake of a few individuals.

— F. H. WALES (1936)

The chief purpose of this chapter is to make a careful study of the four basic processes of arithmetic: addition, subtraction, multiplication, and division. Many people regard arithmetic as a trivial thing which children learn and computers do, but we will see that arithmetic is an intriguing topic with many interesting facets. It is important to make a thorough study of efficient methods for calculating with numbers, since arithmetic underlies so many computer applications.

Arithmetic is, in fact, a lively subject which has played an important part in the history of the world, and it still is undergoing rapid development. In this chapter, we will analyze algorithms for doing arithmetic operations on many types of quantities, such as "floating-point" numbers, extremely large numbers, fractions (rational numbers), polynomials, and power series; and we will also discuss related topics such as radix conversion, factoring of numbers, and the evaluation of polynomials.

4.1. POSITIONAL NUMBER SYSTEMS

The way we do arithmetic is intimately related to the way we represent the numbers we deal with, so it is appropriate to begin our study of arithmetic with a discussion of the principal means for representing numbers.

Positional notation using base b (alternatively called *radix b*) is defined by the rule

$$(\ldots a_3 a_2 a_1 a_0 . a_{-1} a_{-2} \ldots)_b$$
$$= \cdots + a_3 b^3 + a_2 b^2 + a_1 b^1 + a_0 + a_{-1} b^{-1} + a_{-2} b^{-2} + \cdots ; \qquad (1)$$

for example, $(520.3)_6 = 5 \cdot 6^2 + 2 \cdot 6^1 + 0 + 3 \cdot 6^{-1} = 192\frac{1}{2}$. Our conventional decimal number system is, of course, the special case when b is ten, and when the a's are chosen from the "decimal digits" 0, 1, 2, 3, 4, 5, 6, 7, 8, 9; in this case the subscript b in (1) may be omitted.

The simplest generalizations of the decimal number system are obtained when we take b to be an integer greater than 1 and when we require the a's to be integers in the range $0 \le a_k < b$. This gives us the standard binary ($b = 2$), ternary ($b = 3$), quaternary ($b = 4$), quinary ($b = 5$), ... number systems. In general, we could take b to be any number, and we could choose the a's from any specified set of numbers; this leads to some interesting situations, as we will see.

The dot which appears between a_0 and a_{-1} in (1) is called the *radix point*. (When $b = 10$, it is also called the decimal point, and when $b = 2$, it is sometimes called the binary point, etc.) Europeans often use a comma instead of a dot to denote the radix point.

The a's in (1) are called the *digits* of the representation. A digit a_k for large k is often said to be "more significant" than the digits a_k for small k; accordingly, the leftmost or "leading" digit is referred to as the *most significant digit*, and the rightmost or "trailing" digit is referred to as the *least significant digit*. In the standard binary system the binary digits are often called *bits*. In the standard hexadecimal system (radix sixteen) the hexadecimal digits zero through fifteen are usually denoted by

0,1,2,3,4,5,6,7,8,9,A,B,C,D,E,F.

The historical development of number representations is a fascinating story, since it parallels the development of civilization itself. We would be going far afield if we were to examine this history in minute detail, but it will be instructive to look at its main features here:

The earliest forms of number representations, still found in primitive cultures, are generally based on groups of fingers, or piles of stones, etc., usually with special conventions about replacing a larger pile or group of, say, five or ten objects by one object of a special kind or in a special place. Such systems lead naturally to the earliest ways of representing numbers in written form, such as the systems of Babylonian, Egyptian, Greek, Chinese, and Roman numerals, but these notations are quite inconvenient for performing arithmetic operations except in the simplest cases.

During the twentieth century, historians of mathematics have made extensive studies of early cuneiform tablets found by archeologists in the Middle East. These studies show that the Babylonian people actually had two distinct systems of number representation: Numbers used in everyday business transactions were written in a notation based on grouping by tens, hundreds, etc.; this notation was inherited from earlier Mesopotamian civilizations, and large numbers were seldom required. When more difficult mathematical problems were considered, however, Babylonian mathematicians made extensive use of a sexagesimal (radix sixty) positional notation which was highly developed at least as early as 1750 B.C. This notation was unique in that it was actually a *floating-point* form of representation with exponents omitted; the proper scale factor or power of sixty was to be supplied by the context, so that, for example, the numbers 2, 120, 7200, $\frac{1}{30}$, etc. were all written in an identical manner. This notation was especially convenient for multiplication and division, using auxiliary tables, since radix-point alignment had no effect on the answer; the same idea is applied today in the use of slide rules. As examples of this Babylonian notation, consider the following excerpts from early tables: The square of 30 is 15 (which may also be read, "The square of $\frac{1}{2}$ is $\frac{1}{4}$"); the reciprocal of 81 = $(1\ 21)_{60}$ is $(44\ 26\ 40)_{60}$; and the square of the latter is $(32\ 55\ 18\ 31\ 6\ 40)_{60}$. The Babylonians had a symbol for zero, but because of their "floating-point" philosophy, it was used only within numbers, not at the right end to denote a scale factor. For the interesting story of early Babylonian mathematics, see O. Neugebauer, *The Exact Sciences in Antiquity* (Princeton University Press, 1952), and B. L. van der Waerden, *Science Awakening*, tr. by A. Dresden (Groningen: P. Noordhoff, 1954).

Fixed-point positional notation was apparently developed first by the Maya Indians in central America 2000 years ago; their radix 20 system was highly developed, especially in connection with astronomical records and calendar dates. But the Spanish conquerors destroyed nearly all of the Maya books on history and science, so we do not know how sophisticated they had become at arithmetic; some special-purpose multiplication tables have been found, but no examples of division are known [cf. J. Eric S. Thompson, *Contributions to Amer. Anthropology and History* 7 (Carnegie Inst. of Washington, 1942), 37–62].

Several centuries before Christ, the Greek people employed an early form of the abacus to do their arithmetical calculations, using sand and/or pebbles on a board which had rows or columns corresponding in a natural way to our decimal system. It is perhaps surprising to us that the same positional notation was never adapted to written forms of numbers, since we are so accustomed to reckoning with the decimal system using pencil and paper; but the greater ease of calculating by abacus (since handwriting was not a common skill, and since abacus calculation makes it unnecessary to memorize addition and multiplication tables) probably made the Greeks feel it would be silly even to suggest that reckoning could be done better on "scratch paper." At the same time Greek astronomers did make use of a sexagesimal positional notation for fractions, which they learned from the Babylonians.

Our decimal notation, which differs from the more ancient forms primarily because of its fixed radix point, together with its symbol for zero to mark an empty position, was developed first in India among the Hindu people. The exact date when this notation first appeared is quite uncertain; about 600 A.D. seems to be a good guess. Hindu science was rather highly developed at that time, particularly in astronomy. The earliest known Hindu manuscripts which show this notation have numbers written backwards (with the most significant digit at the right), but soon it became standard to put the most significant digit at the left.

About 750 A.D., the Hindu principles of decimal arithmetic were brought to Persia, as several important works were translated into Arabic; a picturesque account of this development is given in a Hebrew document, which has been translated into English in *AMM* **15** (1918), 99–108. Not long after this, al-Khowârizmî wrote his Arabic textbook on the subject. (As noted in Chapter 1, our word "algorithm" comes from al-Khowârizmî's name.) His book was translated into Latin and was a strong influence on Leonardo Pisano (Fibonacci), whose book on arithmetic (1202 A.D.) played a major role in the spreading of Hindu-Arabic numerals into Europe. It is interesting to note that the left-to-right order of writing numbers was unchanged during these two transitions from Hindu to Arabic and from Arabic to Latin, although Arabic is written from right to left while Hindu and Latin are written from left to right. A detailed account of the subsequent propagation of decimal numeration and arithmetic into all parts of Europe during the period from 1200 to 1600 A.D. is given by David Eugene Smith in his *History of Mathematics* **1** (Boston: Ginn and Co., 1923), Chapters 6 and 8.

Decimal notation was applied at first only to integer numbers, not to fractions. Arabic astronomers, who required fractions in their star charts and other tables, continued to use the notation of Ptolemy (the famous Greek astronomer) which was based on sexagesimal fractions. This system still survives today, in our trigonometric units of "degrees, minutes, and seconds," and also in our units of time, as a remnant of the original Babylonian sexagesimal notation. Early European mathematicians also used sexagesimal fractions when dealing with noninteger numbers; for example, Fibonacci gave the value

$$1° \ 22' \ 7'' \ 42''' \ 33^{IV} \ 4^{V} \ 40^{VI}$$

as an approximation to the root of the equation $x^3 + 2x^2 + 10x = 20$.

The use of decimal notation also for tenths, hundredths, etc., in a similar way seems to be a comparatively minor change; but, of course, it is hard to break with tradition, and sexagesimal fractions have an advantage over decimal fractions in that numbers such as $\frac{1}{3}$ can be expressed exactly in a simple way. Chinese mathematicians (who never used sexagesimals) were the first people to work with the equivalent of decimal fractions, although their numeral system (lacking zero) was not originally a positional number system in the strict sense. Chinese units of weights and measures were decimal, so that Tsu Chhung-Chih (who died c. 500 A.D.) was able to express an approximation to π as "3 chang, 1 chhih, 4 tshun, 1 fên, 5 li, 9 hao, 2 miao, 7 hu." Here chang, . . . , hu are units of length; 1 hu (the diameter of a silk thread) equals 1/10 miao, etc. The use of such decimal-like fractions was fairly widespread in China after about 1250 A.D. The first known occurrence of decimal fractions in a true positional system is in a 10th century arithmetic text written in Damascus by an obscure mathematician named al-Uqlîdisî ("the Euclidean"). He used the symbol ′ for a decimal point, and expressed some fractions and approximate square and cube roots; but he did not realize that the integer and fraction parts of numbers could be multiplied simultaneously. His work did not become well known, and five more centuries passed before decimal fractions were reinvented by a Persian mathematician, al-Kashî, who died c. 1436. Al-Kashî was a highly skillful calculator, who gave the value of 2π as follows, correct to 16 decimal places:

integer		fractions															
0	6	2	8	3	1	8	5	3	0	7	1	7	9	5	8	6	5

This was by far the best approximation to π known until Ludolph van Ceulen laboriously calculated 35 decimal places during the period 1596–1610. The earliest known example of decimal fractions in Europe occurs in a 15th-century text where, for example, 153.5 is multiplied by 16.25 to get 2494.375; this was referred to as a "Turkish method." In 1525, Christof Rudolff of Germany discovered decimal fractions for himself; his work never became well-known, either. François Viète suggested the idea again in 1579. Finally, an arithmetic text by Simon Stevin of Belgium, who independently hit on the idea of decimal fractions in 1585, became popular. His work, and the discovery of logarithms soon afterwards, made decimal fractions very common in Europe during the 17th century. [See D. E. Smith, *History of Mathematics* **2** (Boston: Ginn and Co., 1925), 228–247, and C. B. Boyer, *History of Mathematics* (New York: Wiley, 1968), for further remarks and references.]

The binary system of notation has its own interesting history. Many primitive tribes in existence today are known to use a binary or "pair" system of counting (making groups of two instead of five or ten), but they do not count in a true radix 2 system, since they do not treat powers of 2 in a special manner. See *The Diffusion of Counting Practices* by Abraham Seidenberg, Univ. Calif. Publ. in Math. **3** (1960), 215–300, for interesting details about primitive number systems. Another "primitive" example of an essentially binary system is the conventional musical notation for expressing rhythms and durations of time.

The Rhind papyrus, which is one of the first nontrivial mathematical documents known (Egypt, c. 1650 B.C.), uses a decimally oriented scheme of notation for numbers, but it shows how to perform multiplication operations by successive doubling and adding. This device is inherently based on the binary representation of the multiplier, although the binary system was not specifically pointed out.

Nondecimal number systems were discussed in Europe during the seventeenth century. For many years astronomers had occasionally used sexagesimal arithmetic both for the integer and the fractional parts of numbers, primarily when performing multiplication [see John Wallis, *Treatise of Algebra* (Oxford, 1685), 18–22, 30]. The fact that *any* positive number could serve as radix was apparently first stated in print by Blaise Pascal in *De numeris multiplicibus*, which was written about 1658 [see Pascal's *Œuvres Complètes* (Paris: Éditions de Seuil, 1963), 84–89]. Pascal wrote, "Denaria enim ex institute hominum, non ex necessitate naturae ut vulgus arbitratur, et sane satis inepte, posita est"; i.e., "The decimal system has been established, somewhat foolishly to be sure, according to man's custom, not from a natural necessity as most people would think." He stated that the duodecimal (radix twelve) system would be a welcome change, and he gave a rule for testing a duodecimal number for divisibility by 9. Erhard Wiegel proposed the quaternary (radix four) system in a number of publications beginning in 1673. A detailed discussion of radix twelve arithmetic was given by Joshua Jordaine, *Duodecimal Arithmetick* (London, 1687).

Although decimal notation was almost exclusively used for arithmetic during that era, other systems of weights and measures were rarely if ever based on multiples of 10, and many business transactions required a good deal of skill in adding quantities such as pounds, shillings, and pence. For centuries, merchants had therefore learned to compute sums and differences of quantities expressed in peculiar units of currency, weights, and measures; and this was actually arithmetic in a nondecimal number system. The common units of liquid measure in England, dating from the 13th century or earlier, are particularly noteworthy:

2 gills = 1 chopin	2 demibushels = 1 bushel or firkin
2 chopins = 1 pint	2 firkins = 1 kilderkin
2 pints = 1 quart	2 kilderkins = 1 barrel
2 quarts = 1 pottle	2 barrels = 1 hogshead
2 pottles = 1 gallon	2 hogsheads = 1 pipe
2 gallons = 1 peck	2 pipes = 1 tun
2 pecks = 1 demibushel	

Quantities of liquid expressed in gallons, pottles, quarts, pints, etc. were essentially written in binary notation. Perhaps the true inventors of binary arithmetic were English wine merchants!

The first known appearance of binary notation was about 1605 in some unpublished manuscripts of Thomas Harriot (1560–1621). Harriot was a creative man, who came to America as a representative of Sir Walter Raleigh; he invented (among other things) a notation like that now used for "less than" and "greater than" relations; but for some reason he chose not to publish many of his discoveries   Excerpts from his notes on binary arithmetic have been reproduced by John W. Shirley, *Amer. J. Physics* **19** (1951), 452–454. The first published discussion of the binary system was given in a comparatively little-known work by a Spanish bishop, Juan Caramuel Lobkowitz, *Mathesis biceps* **1** (Campaniæ, 1670), 45–48; Caramuel discussed the representation of numbers in radices 2, 3, 4, 5, 6, 7, 8, 9, 10, 12, and 60 at some length, but gave no examples of arithmetic operations in nondecimal systems (except for the trivial operation of adding unity).

Ultimately, an article by G. W. Leibnitz [*Memoires de l'Academie Royale des Sciences* (Paris, 1703), 110–116], which illustrated binary addition, subtraction, multiplication, and division, really brought binary notation into the limelight, and this article is usually referred to as the birth of radix 2 arithmetic. Leibnitz later referred to the binary system quite frequently [cf. W. Ahrens, *Mathematische Unterhaltungen und Spiele* **1** (Leipzig: Teubner, 1910), 27–28]. He did not recommend it for practical calculations, but he stressed its importance in number-theoretical investigations, since patterns in number sequences are often more apparent in binary notation than they are in decimal; he also saw a mystical significance in the fact that everything is expressible in terms of zero and one [see Laplace, *Théorie Analytique des Probabilités*, 3rd ed. (Paris, 1820), cix].

It is interesting to note that the important concept of negative powers to the right of the radix point was not yet well understood at that time. Leibnitz asked James Bernoulli to calculate π in the binary system, and Bernoulli "solved" the problem by taking a 35-digit approximation to π, multiplying it by 10^{35}, and then expressing this integer in the binary system as his answer. On a smaller scale this would be like saying that $\pi \approx 3.14$, and $(314)_{10} = (100111010)_2$; hence π in binary is 100111010! [See Leibnitz, *Math. Schriften* (ed. Gehrhardt), **3** (Halle: 1855), 97; two of the 118 bits in the answer are incorrect, due to computational errors.] The motive for Bernoulli's calculation was apparently to see whether any simple pattern could be observed in this representation of π.

Charles XII of Sweden, whose talent for mathematics perhaps exceeded that of all other kings in the history of the world, hit on the idea of radix-8 arithmetic about 1717. This was probably his own invention, although he had met Leibnitz briefly in 1707. Charles felt radix 8 or 64 would be more convenient for calculation than the decimal system, and he considered introducing octal arithmetic into Sweden; but he died in battle before carrying out such a change. [See *The Works of Voltaire* **21** (Paris: E. R. DuMont, 1901), 49; E. Swedenborg, *Gentleman's Magazine* **24** (1754), 423–424.]

About 140 years later, a prominent Swedish-American civil engineer named John W. Nystrom decided to carry Charles XII's plans a step further, and he devised a complete system of numeration, weights, and measures based on hexadecimal (radix 16) arithmetic. He wrote, "I am not afraid, or do not hesitate, to advocate a binary system of arithmetic and metrology. I know I have nature on my side; if I do not succeed to impress upon you its utility and great importance to mankind, it will reflect that much less credit upon our generation, upon our scientific men and philosophers." Nystrom devised special means for pronouncing hexadecimal numbers; e.g., $(B0160)_{16}$ was to be read "vybong, bysanton." His entire system was called the Tonal System, and it is described in *J. Franklin Inst.* **46** (1863), 263–275, 337–348, 402–407. A similar system, but using radix 8, was proposed about the same time by Alfred B. Taylor [*Proc. Amer. Pharmaceutical Assoc.* **8** (1859), 115–216; *Proc. Amer. Philosophical Soc.* **24** (1887), 296–366]. Increased use of the French (metric) system of weights and measures led to extensive debate about the merits of decimal arithmetic during that era.

The binary system was well-known as a curiosity ever since Leibnitz's time, and about 20 early references to it have been compiled by R. C. Archibald [*AMM* **25** (1918), 139–142]. It was applied chiefly to the calculation of powers, as explained in Section 4.6.3, and to the analysis of certain games and puzzles. G. Peano [*Atti della R. Accademia delle Scienze di Torino* **34** (1898), 47–55] used binary notation as the basis of a "logical" character set of 256 symbols. Joseph Bowden [*Special Topics in Theoretical Arithmetic* (Garden City: 1936), 49] gave a system of nomenclature for hexadecimal numbers.

Increased interest in mechanical devices for doing arithmetic, especially for multiplication, led several people to consider the binary system for this purpose. A particularly delightful account of this activity is given in the article "Binary Calculation" by E. William Phillips [*Journal of the Institute of Actuaries* **67** (1936), 187–221] together with a record of the discussion which followed a lecture he gave on the subject. Phillips begins by saying, "The ultimate aim [of this paper] is to persuade the whole civilized world to abandon decimal numeration and to use octonal [i.e., radix 8] numeration in its place."

Modern readers of Phillips's article will perhaps be surprised to discover that a radix-8 number system was properly referred to as "octonary" or "octonal," according to all dictionaries of the English language at that time, just as the radix-10 number system is properly called either "denary" or "decimal"; the word "octal" did not appear in English language dictionaries until 1961, and it apparently originated as a term for the "base" of a certain class of vacuum tubes. The word "hexadecimal," which has crept into our language even more recently, is a mixture of Greek and Latin stems; more proper terms would be "senidenary" or "sedecimal" or even "sexadecimal," but the latter is perhaps too risqué for computer programmers.

The comment by Mr. Wales which is quoted at the beginning of this chapter has been taken from the discussion printed with Mr. Phillips's paper. Another man who attended the same lecture pointed out a disadvantage of the octal system for business purposes: "5% becomes 3.1463 per 64, which sounds rather horrible."

Some calculating machines based on binary notation were developed in France during the early 1930's. See L. Couffignal and R. Valtat, *Comptes Rendus* **197** (Paris, 1933). 877; **202** (1936), 1745–1747, 1970–1972.

The first vacuum-tube computer circuits were designed in 1937 by John V. Atanasoff, and the first relay computer circuits were designed independently in the same year by George R. Stibitz. Both men used the binary system for arithmetic in these planned computers, although Stibitz developed excess-3 binary-coded-decimal notation soon afterwards. At about the same time in Germany, Konrad Zuse built a mechanical floating-point binary computer; he later replaced its mechanical logic by relay circuitry which was operative in 1941.

The first American high-speed computers, built in the early 1940's. used decimal arithmetic. But in 1946, an important memorandum by A. W. Burks, H. H. Goldstine, and J. von Neumann, in connection with the design of the first stored-program computers, gave detailed reasons for the decision to make a radical departure from tradition and to use base-two notation [see John von Neumann, *Collected Works*, Vol. 5, 41–65]. Since then binary computers have become commonplace. After a dozen years of experience with binary machines, a discussion of the relative advantages and disadvantages of binary notation was given by W. Buchholz in his paper "Fingers or Fists?" [*CACM* **2** (December, 1959), 3–11].

The MIX computer used in this book has been defined so that it can be either binary or decimal. It is interesting to note that nearly all MIX programs can be expressed without knowing whether binary or decimal notation is being used—even when we are doing calculations involving multiple-precision arithmetic. Thus we find that the choice of radix does not significantly influence computer programming. (Noteworthy exceptions to this statement, however, are the "Boolean" algorithms discussed in Chapter 7; see also Algorithm 4.5.2B.)

There are several different methods for representing *negative* numbers in a computer, and this sometimes influences the way arithmetic is done. In order to understand these other notations, let us first consider MIX as if it were a decimal computer; then each word contains 10 digits and a sign, e.g.,

$$- \ 12345 \ 67890. \tag{2}$$

This is called the *signed-magnitude* representation. Such a representation corresponds to common notational conventions, so it is preferred by many programmers. A potential disadvantage is that minus zero and plus zero can both be represented, while they usually should mean the same number; this possibility requires some care in practice.

Most mechanical calculators that do decimal arithmetic use another system called *ten's complement* notation. If we subtract 1 from 00000 00000, we get 99999 99999 in this notation; in other words, no explicit sign is attached to the number, and calculation is done *modulo* 10^{10}. The number $- \ 12345 \ 67890$ would appear as

$$87654 \ 32110 \tag{3}$$

in ten's complement notation. It is conventional to regard any number whose leading digit is 5, 6, 7, 8, or 9 as a negative value in this notation, although with respect to addition and subtraction there is no harm in regarding (3) as the number $+87654 \ 32110$ if it is convenient to do so. There is no problem of "minus zero" with ten's complement notation. The major difference between signed magnitude and ten's complement notations in practice is that shifting

right does not divide the magnitude by ten; for example, the number $-11 =$
$\ldots 99989$, shifted right one, gives $\ldots 99998 = -2$ (assuming that a shift to
the right inserts "9" as the leading digit when the number shifted is negative).
In general, x shifted right one digit in ten's complement notation will give
$\lfloor x/10 \rfloor$, whether x is positive or negative. A possible disadvantage of the ten's
complement system is the fact that it is not symmetric about zero; the largest
negative number representable in p digits is $500 \ldots 0$, and it is not the negative
of any p-digit positive number. Thus it is possible that changing x to $-x$ will
cause overflow.

Another notation which has been used since the earliest days of high-speed
computers is called *nines' complement* representation. In this case the number
$- 12345\ 67890$ would appear as

$$87654\ 32109. \tag{4}$$

Each digit of a negative number $-x$ is equal to 9 minus the corresponding digit
of x. It is not difficult to see that the nines' complement notation for a negative
number is always one less than the corresponding ten's complement notation;
addition and subtraction are done modulo $10^{10} - 1$, which means that a carry
off the left end is to be added at the right end. (Cf. Section 3.2.1.1.) Again there
is a potential problem with minus zero, since $99999\ 99999$ and $00000\ 00000$
denote the same value.

The ideas just explained for base-10 arithmetic apply in a similar way to
base-2 arithmetic, where we have *signed magnitude, two's complement*, and *ones'
complement* notations. The MIX computer, as used in the examples of this
chapter, deals only with signed-magnitude arithmetic; alternative procedures
for complement notations are discussed in the accompanying text when it is
important to do so.

Most computer manuals tell us that the machine's circuitry assumes that
the radix point is situated in a particular place within each computer word.
This advice should usually be disregarded; it is better to learn the rules concerning where the radix point will appear in the result of an instruction if we
assume it lies in a certain place beforehand. For example, in the case of MIX we
could regard our operands either as integers with the radix point at the extreme
right, or as fractions with the radix point at the extreme left, or as some mixture
of these two extremes; the rules for the appearance of the radix point in each
result are straightforward.

It is easy to see that there is a simple relation between radix b and radix b^k:

$$(\ldots a_3 a_2 a_1 a_0 . a_{-1} a_{-2} \ldots)_b = (\ldots A_3 A_2 A_1 A_0 . A_{-1} A_{-2} \ldots)_{b^k}, \tag{5}$$

where

$$A_j = (a_{kj+k-1} \ldots a_{kj+1} a_{kj})_b;$$

see exercise 8. Thus we obtain the simple techniques for converting at sight
between, say, binary and octal notation.

There are many interesting variations on positional number systems besides the standard b-ary systems discussed so far. For example, we might have numbers in base (-10), so that

$$(\ldots a_3 a_2 a_1 a_0 . a_{-1} a_{-2} \ldots)_{-10}$$
$$= \cdots + a_3(-10)^3 + a_2(-10)^2 + a_1(-10)^1 + a_0 + \cdots$$
$$= \cdots - 1000a_3 + 100a_2 - 10a_1 + a_0 - \tfrac{1}{10}a_{-1} + \tfrac{1}{100}a_{-2} - \cdots.$$

Here the individual digits satisfy $0 \le a_k \le 9$ just as in the decimal system. The number 12345 67890 would appear in the "negadecimal" system as

$$(1\ 93755\ 73910)_{-10},\qquad\qquad(6)$$

since the latter represents $10305070900 - 9070503010$. It is interesting to note that the negative of this number, $-12345\ 67890$, would be written

$$(28466\ 48290)_{-10},\qquad\qquad(7)$$

and, in fact, *every real number whether positive or negative can be represented without a sign* in the -10 system.

Negative-base systems were apparently first mentioned in the literature by Z. Pawlak and A. Wakulicz [*Bulletin de l'Academie Polonaise des Sciences, Classe III*, **5** (1957), 233–236; Série des sciences techniques **7** (1959), 713–721], and by L. Wadel [*IRE Transactions* **EC-6** (1957), 123]. For further references see IEEE *Transactions* **EC-12** (1963), 274–276; *Computer Design* **6** (May, 1967), 52–63. (There is evidence that the idea of negative bases occurred independently to quite a few people, because of the growing interest in hardware design.)

G. F. Songster investigated base -2 in his master's thesis (U. Penn., 1956) at the suggestion of G. W. Patterson. D. E. Knuth had also discussed negative-base systems in 1955 in a short, typewritten paper submitted to a "science talent search" contest for high-school seniors, together with a further generalization to complex-valued bases. The fact that all numbers can be represented with negative radix had been noted in another context several years earlier by N. G. de Bruijn [*Publ. Math. Debrecen* **1** (1950), 232–242, esp. p. 240], although he did not apply this idea to arithmetic.

The base $2i$ gives an interesting system called the "quater-imaginary" number system (by analogy with "quaternary") since *every complex number can be represented with the digits 0, 1, 2, and 3 without a sign* in this system. [See *CACM* **3** (1960), 245–247.] For example,

$$(11210.31)_{2i} = 1 \cdot 16 + 1 \cdot (-8i) + 2 \cdot (-4) + 1 \cdot (2i) + 3 \cdot (-\tfrac{1}{2}i) + 1(-\tfrac{1}{4})$$
$$= 7\tfrac{3}{4} - 7\tfrac{1}{2}i.$$

Here the number $(a_{2n} \ldots a_1 a_0 . a_{-1} \ldots a_{-2k})_{2i}$ is equal to

$$(a_{2n} \ldots a_2 a_0 . a_{-2} \ldots a_{-2k})_{-4} + 2i(a_{2n-1} \ldots a_3 a_1 . a_{-1} \ldots a_{-2k+1})_{-4},$$

so conversion to and from quater-imaginary notation reduces to conversion to and from negative quaternary representation of the real and imaginary parts. The interesting property of this system is that it allows multiplication and division of complex numbers to be done in a fairly unified manner without treating real and imaginary parts separately. For example, we can multiply two numbers in this system much as we do with any base, merely using a different "carry" rule: whenever a digit exceeds 4 we subtract 4 and "carry" a -1 two columns to the left; when a digit is negative, we add four to it and "carry" $+1$ two columns to the left. A study of the following example shows this peculiar carry rule at work:

$$
\begin{array}{llll}
& 1\ 2\ 2\ 3\ 1 & \quad [9 - 10i] \\
& 1\ 2\ 2\ 3\ 1 & \quad [9 - 10i] \\
\hline
& 1\ 2\ 2\ 3\ 1 \\
1\ 0\ 3\ 2\ 0\ 2\ 1\ 3 \\
\quad\ 1\ 3\ 0\ 2\ 2 \\
\quad 1\ 3\ 0\ 2\ 2 \\
1\ 2\ 2\ 3\ 1 \\
\hline
0\ 2\ 1\ 3\ 3\ 3\ 1\ 2\ 1 & \quad [-19 - 180i]
\end{array}
$$

A similar system which uses just the digits 0 and 1 may be based on $\sqrt{2}i$, but this requires an infinite nonrepeating expansion for the simple number "i" itself.

A "binary" complex number system may also be obtained by using the base $i - 1$, as suggested by W. Penney [*JACM* **12** (1965), 247–248]:

$$(\ldots a_4 a_3 a_2 a_1 a_0 . a_{-1} \ldots)_{i-1}$$
$$= \cdots -4a_4 + (2 + 2i)a_3 - 2ia_2 + (i - 1)a_1 + a_0 - \tfrac{1}{2}(i + 1)a_{-1} + \cdots.$$

In this system, only the digits 0 and 1 are needed. One way to show that every complex number has such a representation is to consider the interesting set S shown in Fig. 1; this set is, by definition, all points which can be written as $\sum_{k \geq 1} a_k (i - 1)^{-k}$, for an infinite sequence $a_1, a_2, a_3, \ldots$ of zeros and ones. Figure 1 shows how S can be decomposed into 256 pieces congruent to $\frac{1}{16}S$; note that if the diagram of S is rotated clockwise by 135°, we obtain two adjacent sets congruent to $(1/\sqrt{2})S$ (since $(i - 1)S = S \cup (S + 1)$). For details of a proof that S contains all complex numbers that are of sufficiently small magnitude, see exercise 18. (Actually, the boundary of S contains many jagged edges; these corners have been rounded off in Figure 1.)

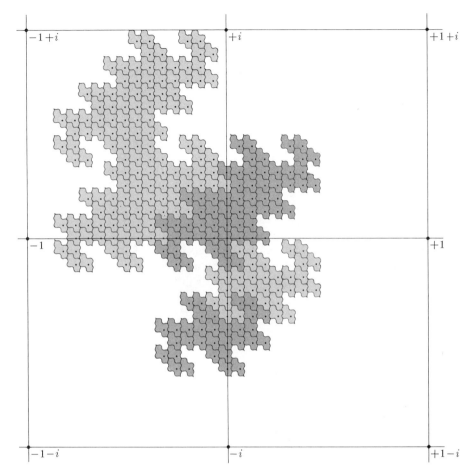

Fig. 1. The set S.

Perhaps the prettiest number system of all is the *balanced ternary* notation, which consists of base 3 representation using the "trits" -1, 0, $+1$ instead of 0, 1, and 2. If we use the symbol $\bar{1}$ to stand for -1, we have the following examples of balanced ternary numbers:

Balanced ternary	Decimal
$1\,0\,\bar{1}$	8
$1\,1\,\bar{1}\,0.\bar{1}\,\bar{1}$	$32\frac{5}{9}$
$\bar{1}\,\bar{1}\,1\,0.1\,1$	$-32\frac{5}{9}$
$\bar{1}\,\bar{1}\,1\,0$	-33
$0.1\,1\,1\,1\,1\ldots$	$\frac{1}{2}$

The balanced ternary number system has many pleasant properties:

a) The negative of a number is obtained by interchanging 1 and $\bar{1}$.

b) The sign of a number is given by its most significant nonzero "trit," and more generally we can compare any two numbers by reading them from left to right and using lexicographic order, as in the decimal system.

c) The operation of rounding to the nearest integer is identical to truncation (i.e., deleting everything to the right of the decimal point).

Addition in the balanced ternary system is quite simple, using the table

$$
\begin{array}{ccccccccccccccccccccccccccc}
\bar{1}&\bar{1}&\bar{1}&\bar{1}&\bar{1}&\bar{1}&\bar{1}&\bar{1}&\bar{1}&0&0&0&0&0&0&0&0&0&1&1&1&1&1&1&1&1&1\\
\bar{1}&\bar{1}&\bar{1}&0&0&0&1&1&1&\bar{1}&\bar{1}&\bar{1}&0&0&0&1&1&1&\bar{1}&\bar{1}&\bar{1}&0&0&0&1&1&1\\
\bar{1}&0&1&\bar{1}&0&1&\bar{1}&0&1&\bar{1}&0&1&\bar{1}&0&1&\bar{1}&0&1&\bar{1}&0&1&\bar{1}&0&1&\bar{1}&0&1\\
\hline
\bar{1}0&\bar{1}1&\bar{1}&\bar{1}1&\bar{1}&0&\bar{1}&0&1&\bar{1}1&\bar{1}&0&\bar{1}&0&1&0&1&1\bar{1}&\bar{1}&0&1&0&1&1\bar{1}&1&1\bar{1}&10
\end{array}
$$

(The three inputs to the addition are the digits of the numbers to be added and the carry digits.) Subtraction is negation followed by addition; and multiplication also reduces to negation and addition, as in the following example:

$$
\begin{array}{r}
1\,\bar{1}\,0\,\bar{1} \quad [17]\\
1\,\bar{1}\,0\,\bar{1} \quad [17]\\
\hline
\bar{1}\,1\,0\,1\\
\bar{1}\,1\,0\,1\,0\\
1\,\bar{1}\,0\,\bar{1}\\
\hline
0\,1\,1\,\bar{1}\,\bar{1}\,0\,1 \quad [289]
\end{array}
$$

For division, see exercise 4.3.1–31.

One way to find the representation of a number in the balanced ternary system is to start by representing it in the ternary notation; for example,

$$208.3 = (21201.022002200220\ldots)_3.$$

(A very simple pencil-and-paper method for converting to ternary notation is given in exercise 4.4–12.) Now add the infinite number $\ldots 11111.11111 \ldots$ in ternary notation; we obtain, in the above example,

$$(\ldots 11111210012.210121012101 \ldots)_3.$$

Finally, subtract $\ldots 11111.11111 \ldots$ by decrementing each digit; we get

$$208.3 = (10\bar{1}\bar{1}01.10\bar{1}010\bar{1}0\ldots)_3. \tag{8}$$

This process may clearly be made rigorous if we replace the artificial infinite number $\ldots 11111.11111 \ldots$ by a number with suitably many ones.

Representation of numbers in the balanced ternary system is implicitly present in a famous mathematical puzzle, which is commonly called "Bachet's prob-

lem of weights," although it was already stated by Fibonacci four centuries before
Bachet wrote his book. [See W. Ahrens, *Mathematische Unterhaltungen und
Spiele* 1 (Leipzig: Teubner, 1910), Section 3.4.] Positional number systems with
negative digits were invented by Sir John Leslie [*The Philosophy of Arithmetic*
(Edinburgh, 1817); see pp. 33–34, 54, 64–65, 117, 150], and independently by A.
Cauchy [*Comptes Rendus* 11 (Paris, 1840), 789–798], who pointed out that negative digits make it unnecessary for a person to memorize the multiplication table
past 5×5. The first true appearance of "pure" balanced ternary notation was
in an article by Léon Lalanne [*Comptes Rendus* 11 (Paris, 1840), 903–905], who
was a designer of mechanical devices for arithmetic. The system was mentioned
only rarely for 100 years after Lalanne's paper, until the development of the
first electronic computers at the Moore School of Electrical Engineering in
1945–1946; at that time it was given serious consideration along with the binary
system as a possible replacement for the decimal system. The complexity of
arithmetic circuitry for balanced ternary arithmetic is not much greater than it
is for the binary system, and a given number requires only $\ln 2/\ln 3 \approx 63\%$
as many digit positions for its representation. Discussions of the balanced
ternary number system appear in *AMM* **57** (1950), 90–93; and in *High-Speed
Computing Devices*, Engineering Research Associates (McGraw-Hill, 1950),
287–289. So far no substantial application of balanced ternary notation has
been made, but perhaps its symmetric properties and simple arithmetic will
prove to be quite important some day (when the "flip-flop" is replaced by a
"flip-flap-flop").

Another important generalization of the simple positional notation is a
mixed radix system. If we have a sequence of numbers $\langle b_k \rangle$ (where k may be
negative), we define

$$\left[\begin{matrix} \ldots, a_3, a_2, a_1, a_0; & a_{-1}, a_{-2}, \ldots \\ \ldots, b_3, b_2, b_1, b_0; & b_{-1}, b_{-2}, \ldots \end{matrix} \right]$$
$$= \cdots + a_3 b_2 b_1 b_0 + a_2 b_1 b_0 + a_1 b_0 + a_0 + a_{-1}/b_{-1} + a_{-2}/b_{-1} b_{-2} + \cdots . \tag{9}$$

In the simplest mixed-radix systems, we work only with integers; we let
$b_0, b_1, b_2, \ldots$ be integers greater than one, and deal only with numbers that
have no radix point, where a_k is required to lie in the range $0 \le a_k < b_k$.

One of the most important mixed radix systems is the *factorial number
system*, where $b_k = k + 2$. Using this system, we can represent every nonnegative integer uniquely in the form

$$c_n n! + c_{n-1}(n-1)! + \cdots + c_2 2! + c_1 \tag{10}$$

where $0 \le c_k \le k$.

Mixed-radix systems are familiar in everyday life, when we deal with units
of measure. For example, the quantity "3 weeks, 2 days, 9 hours, 22 minutes,

57 seconds, and 492 milliseconds" is equal to

$$\begin{bmatrix} 3, 2, & 9, 22, 57; & 492 \\ 7, 24, 60, 60; & 1000 \end{bmatrix} \text{seconds}.$$

The quantity "10 pounds, 6 shillings, and thruppence ha'penny" was once $\begin{bmatrix} 10, & 6, & 3; & 1 \\ & 20, & 12; & 2 \end{bmatrix}$ pence in English currency, before England changed to a pure decimal monetary system.

It is possible to add and subtract mixed radix numbers by using a straight-forward generalization of the usual addition and subtraction algorithms, pro-vided of course that the same mixed radix system is being used for both operands (see exercise 4.3.1–9). Similarly, we can easily multiply or divide a mixed-radix number by small integer constants, using simple extensions of the familiar pencil-and-paper methods.

Mixed-radix systems were first discussed generally by Georg Cantor [*Zeit-schrift für Mathematik und Physik* **14** (1869), 121–128]. For further information about mixed-radix systems, see exercises 26 and 29.

Besides the systems described in this section, there are several other ways to represent numbers, which are mentioned elsewhere in this series: the binomial number system (exercise 1.2.6–56); the Fibonacci number system (exercise 1.2.8–34); the phi number system (exercise 1.2.8–35); modular representations (Section 4.3.2); Gray code (Section 7.2.1); and roman numerals (Section 9.1).

Some questions concerning *irrational* radices have been investigated by W. Parry, *Acta Mathematica*, Acad. Sci. Hung., **11** (1960), 401–416.

EXERCISES

1. [*15*] Express the numbers $-10, -9, -8, \ldots, 8, 9, 10$ in the base -2 number system.

▶ **2.** [*24*] Consider the following four number systems: (a) binary (signed magnitude); (b) negative binary (radix -2); (c) balanced ternary; and (d) radix $b = \frac{1}{10}$. Use each of these number systems to express each of the following three numbers: (i) -49; (ii) $-3\frac{1}{7}$ (show the repeating cycle); (iii) π (to a few significant figures).

3. [*20*] Express $-49 + i$ in the quater-imaginary system.

4. [*15*] Assume that we have a MIX program in which location A contains a number for which the radix point lies between bytes 3 and 4, while location B contains a number whose radix point lies between bytes 2 and 3. (The leftmost byte is number 1.) Where will the radix point be, in registers A and X, after the instructions

a)   LDA   A
      MUL   B   ?

b)   LDA   A
      SRAX 5   ?
      DIV   B

5. [*00*] Explain why a negative integer in nines' complement notation has a repre-sentation in ten's complement notation that is always one less, if the representations are regarded as positive.

6. [*16*] What are the largest and smallest p-bit integers that can be represented in (a) signed-magnitude binary notation, (b) two's complement notation, (c) ones' complement notation?

7. [*M20*] The text defines ten's complement notation only for integers represented in a single computer word. Is there a way to define a ten's complement notation *for all real numbers*, having "infinite precision," analogous to the text's definition? Is there a similar way to define a nines' complement notation for all real numbers?

8. [*M10*] Prove Eq. (5).

▶ **9.** [*15*] Change the following *octal* numbers to *hexadecimal* notation, using the hexadecimal digits 0, 1, . . . , F: *12; 5655; 2550276; 76545336; 3726755.*

10. [*M22*] Generalize Eq. (5) to mixed radix notation.

11. [*22*] Give an algorithm which computes the sum of $(a_n \ldots a_1 a_0)_{-2}$ and $(b_n \ldots b_1 b_0)_{-2}$, using the -2 number system, obtaining the answer $(c_{n+2} \ldots c_1 c_0)_{-2}$.

12. [*23*] Give algorithms to convert (a) the binary signed magnitude number $\pm(a_n \ldots a_0)_2$ to its negative binary form $(b_{n+1} \ldots b_0)_{-2}$; and (b) the negative binary number $(b_{n+1} \ldots b_0)_{-2}$ to its signed magnitude form $\pm(a_{n+1} \ldots a_0)_2$.

▶ **13.** [*M21*] In the decimal system there are some numbers with two infinite decimal expansions, e.g., $2.3599999 \ldots = 2.3600000 \ldots$ Does the *negative decimal* (base -10) system have unique expansions, or are there real numbers with two different infinite expansions in the base also?

14. [*14*] Multiply $(11321)_{2i}$ by itself in the quater-imaginary system using the method illustrated in the text.

15. [*M24*] What are the sets S, analogous to Fig. 1, for the negative decimal and for the quater-imaginary number systems? In other words, what are the sets

$$\left\{ \sum_{k \geq 1} a_k (-10)^{-k} \, \middle| \, 0 \leq a_k \leq 9, \, a_k \text{ an integer, for all } k \right\}$$

and

$$\left\{ \sum_{k \geq 1} a_k (2i)^{-k} \, \middle| \, 0 \leq a_k \leq 3, \, a_k \text{ an integer, for all } k \right\} ?$$

16. [*M24*] Design an algorithm to add 1 to $(a_n \ldots a_1 a_0)_{i-1}$ in the $i - 1$ number system.

17. [*M30*] It may seem peculiar that the number $i - 1$ has been given as a number-system base, instead of the similar but simpler number $i + 1$. Can every complex number $a + bi$, where a and b are integers, be represented in a positional number system to base $i + 1$, using only the digits 0 and 1?

18. [*HM32*] Show that the set S of Fig. 1 is a closed set which contains a neighborhood of the origin. (Consequently, every complex number has a "binary" representation to base $i - 1$.)

19. [*HM42*] Study the set S of Fig. 1 in more detail; e.g., analyze its boundary.

20. [*M22*] Show that any real number (positive, negative, or zero) can be expressed in a decimal system using the digits -1, 0, 1, 2, 3, 4, 5, 6, 7, 8. (*Not* 9.)

▶ **21.** [*M22*] (C. E. Shannon.) Can every real number (positive, negative, or zero) be expressed in a "balanced-decimal" system, i.e., in the form $\sum_{k \leq n} a_k 10^k$, for some integer n and some sequence a_n, a_{n-1}, a_{n-2}, $\ldots$, where each a_k is one of the ten numbers $\{-4\frac{1}{2}, -3\frac{1}{2}, -2\frac{1}{2}, -1\frac{1}{2}, -\frac{1}{2}, \frac{1}{2}, 1\frac{1}{2}, 2\frac{1}{2}, 3\frac{1}{2}, 4\frac{1}{2}\}$? (Note that zero is not one of the allowed digits, but we implicitly assume that a_{n+1}, a_{n+2}, $\ldots$ are zero.) Find all representations of zero in this number system, and find all representations of unity.

22. [*HM25*] Let $\alpha = -\sum_{m \geq 1} 10^{-m^2}$. Given $\epsilon > 0$ and any real number x, prove that there is a "decimal" representation such that $0 < |x - \sum_{0 \leq k \leq n} a_k 10^k| < \epsilon$, where each a_k is allowed to be only one of the three values 0, 1, or α. (Note that no negative powers of 10 are used in this representation!)

23. [*HM30*] Find all sets D of ten or less nonnegative real numbers, such that (a) $0 \in D$, and (b) all positive real numbers can be expressed in a "decimal" system as $\sum_{k \leq n} a_k 10^k$ with each $a_k \in D$.

24. [*HM50*] Find all sets D of ten or less real numbers, such that *every* nonnegative real number can be expressed in the form $\sum_{k \leq n} a_k 10^k$, for some n, with each $a_k \in D$. (Cf. exercises 20–23.)

25. [*M25*] (S. A. Cook.) Let b, u, and v be positive integers, where $b \geq 2$ and $0 < v < b^m$. Show that the base b representation of u/v does not contain a run of m consecutive digits equal to $b - 1$ to the right of the radix point. (By convention, no runs of infinitely many $(b - 1)$'s are permitted in the standard base b representation.)

▶ **26.** [*HM30*] (N. S. Mendelsohn.) Let $\langle \beta_n \rangle$ be a sequence of real numbers defined for all integers n, $-\infty < n < \infty$, such that

$$\beta_n > \beta_{n+1}; \qquad \lim_{n \to \infty} \beta_n = \infty; \qquad \lim_{n \to -\infty} \beta_n = 0.$$

Let $\langle c_n \rangle$ be an arbitrary sequence of positive integers, defined for all integers n, $-\infty < n < \infty$. Let us say that a number x has a "generalized representation" if there is an integer n and a sequence of integers a_n, a_{n-1}, a_{n-2}, $\ldots$ such that $x = \sum_{k \leq n} a_k \beta_k$, where $a_n \neq 0$, $0 \leq a_k \leq c_k$, and $a_k < c_k$ for infinitely many k.

Show that every positive real number x has exactly one generalized representation if and only if $\beta_{n+1} = \sum_{k \leq n} c_k \beta_k$ for all n. (Consequently, the mixed radix systems with integer bases have this property; and the mixed radix systems with $\beta_1 = (c_0 + 1)\beta_0$, $\beta_2 = (c_1 + 1)(c_0 + 1)\beta_0$, $\ldots$, $\beta_{-1} = \beta_0/(c_{-1} + 1)$, $\ldots$ are the most general number systems of this type.)

27. [*M21*] Show that every nonzero integer has a unique "reversing binary representation" $2^{e_0} - 2^{e_1} + \cdots + (-1)^k 2^{e_k}$ where $e_0 < e_1 < \cdots < e_k$.

▶ **28.** [*M24*] Show that every nonzero complex number of the form $a + bi$ where a and b are integers has a unique "revolving binary representation"

$$(1 + i)^{e_0} + i(1 + i)^{e_1} - (1 + i)^{e_2} - i(1 + i)^{e_3} + \cdots + i^k (1 + i)^{e_k},$$

where $e_0 < e_1 < \cdots < e_k$. (Cf. exercise 27.)

29. [*M35*] (N. G. de Bruijn.) Let S_0, S_1, S_2, $\ldots$ be sets of nonnegative integers; we will say that the collection S_0, S_1, S_2, $\ldots$ has Property B if every nonnegative integer n can be written in the form

$$n = s_0 + s_1 + s_2 + \cdots, \qquad s_j \in S_j$$

in exactly one way. (Property B implies that $0 \in S_j$ for all j, since $n = 0$ can only be represented as $0 + 0 + 0 + \cdots$.) Any mixed radix number system with radices b_0, b_1, b_2, ... provides an example of sets satisfying Property B, if we let $S_j = \{0, q_j, \ldots, (b_j - 1)q_j\}$, where $q_j = b_0 b_1 \cdots b_{j-1}$; here the representation of $n = s_0 + s_1 + s_2 + \cdots$ corresponds in an obvious manner to its mixed radix representation (9). Furthermore, if S_0, S_1, S_2, ... has Property B, and if A_0, A_1, A_2, ... is any partition of the nonnegative integers (so that $A_0 \cup A_1 \cup A_2 \cup \cdots = \{0, 1, 2, \ldots\}$ and $A_i \cap A_j = \emptyset$ for $i \neq j$; some A_j's may be empty), then the "collapsed" collection T_0, T_1, T_2, ... also has Property B, where T_j is the set of all sums $\sum_{i \in A_j} s_i$ taken over all possible choices of $s_i \in S_i$.

Prove that *any* collection T_0, T_1, T_2, ... which satisfies Property B may be obtained by collapsing some collection S_0, S_1, S_2, ... that corresponds to a mixed radix number system.

30. (*M39*) (N. G. de Bruijn.) The radix -2 number system shows us that every integer (positive, negative, or zero) has a unique representation of the form

$$(-2)^{e_1} + (-2)^{e_2} + \cdots + (-2)^{e_t}, \qquad e_1 > e_2 > \cdots > e_t \geq 0, \qquad t \geq 0.$$

The purpose of this exercise is to explore generalizations of this phenomenon.

a) Let b_0, b_1, b_2, ... be a sequence of integers such that every integer n has a unique representation of the form

$$n = b_{e_1} + b_{e_2} + \cdots + b_{e_t}, \qquad e_1 > e_2 > \cdots > e_t \geq 0, \qquad t \geq 0.$$

(Such a sequence $\langle b_n \rangle$ is called a "binary basis.") Show that there is an index j such that b_j is odd, but b_k is even for all $k \neq j$.

b) Prove that a binary basis $\langle b_n \rangle$ can always be rearranged into the form d_0, $2d_1$, $4d_2$, ... $= \langle 2^n d_n \rangle$, where each d_k is odd.

c) If each of d_0, d_1, d_2, ... in (b) is ± 1, prove that $\langle b_n \rangle$ is a binary basis if and only if there are infinitely many $+1$'s and infinitely many -1's.

d) Prove that 7, $-13 \cdot 2$, $7 \cdot 2^2$, $-13 \cdot 2^3$, ..., $7 \cdot 2^{2k}$, $-13 \cdot 2^{2k+1}$, ... is a binary basis, and find the representation of $n = 1$.

▶ **31.** [*M35*] A generalization of two's complement arithmetic, called "2-adic numbers," was invented about 1900 by K. Hensel. (Hensel in fact treated *p-adic numbers*, for any prime p.) A 2-adic number may be regarded as a binary number

$$u = (\ldots u_3 u_2 u_1 u_0 . u_{-1} \ldots u_{-n})_2,$$

whose representation extends infinitely far to the left, but only finitely many places to the right, of the binary point. Addition, subtraction, and multiplication of 2-adic numbers is done according to the ordinary procedures of arithmetic, which can in principle be extended indefinitely to the left. For example,

$$7 = (\ldots 000000000000111)_2$$
$$-7 = (\ldots 111111111111001)_2$$
$$\tfrac{7}{4} = (\ldots 000000000000001.11)_2$$

$$\tfrac{1}{7} = (\ldots 110110110110111)_2$$
$$-\tfrac{1}{7} = (\ldots 001001001001001)_2$$
$$\tfrac{1}{10} = (\ldots 110011001100110.1)_2$$
$$\sqrt{-7} = (\ldots 100000010110101)_2 \text{ or } (\ldots 011111101001011)_2.$$

Here 7 is the ordinary binary integer seven, while -7 is its two's complement (extending infinitely to the left); it is easy to verify that the ordinary procedure for addition of binary numbers will give $-7 + 7 = (\ldots 00000)_2 = 0$, when the procedure is continued indefinitely. The values of $\frac{1}{7}$ and $-\frac{1}{7}$ are the unique 2-adic numbers which, when formally multiplied by 7, give 1 and -1, respectively. The values of $\frac{7}{4}$ and $\frac{1}{10}$ are examples of 2-adic numbers which are not 2-adic "integers," since they have nonzero bits to the right of the binary point. The two values of $\sqrt{-7}$, which are negatives of each other, are the two 2-adic numbers which, when formally squared, yield the value $(\ldots 111111111111001)_2$.

a) Prove that any 2-adic number u can be divided by any nonzero 2-adic number v to obtain a unique 2-adic number w satisfying $u = vw$. (Hence the set of 2-adic numbers forms a "field"; cf. Section 4.6.1.)

b) Prove that the 2-adic representation of the rational number $-1/(2n+1)$ may be obtained as follows, when n is a positive integer: First find the ordinary binary expansion of $1/(2n+1)$, which has the periodic form $(0.\alpha\alpha\alpha\ldots)_2$ for some string α of 0's and 1's. Then $-1/(2n+1)$ is the 2-adic number $(\ldots \alpha\alpha\alpha)_2$.

c) Prove that the representation of a 2-adic number u is periodic (that is, $u_{N+\lambda} = u_N$ for all large N, for some $\lambda \geq 1$) if and only if u is rational (that is, $u = m/n$, for some integers m and n).

d) Prove that, when n is an integer, $\sqrt{n}$ is a 2-adic number if and only if $n \bmod 2^{2k+3} = 2^{2k}$ for some nonnegative integer k. (Thus, either $n \bmod 8 = 1$, or $n \bmod 32 = 4$, etc.)

4.2. FLOATING–POINT ARITHMETIC

4.2.1. Single–Precision Calculations

In this section, we shall study the basic principles of doing arithmetic on "floating-point" numbers, by analyzing the internal mechanism of such calculations. Perhaps many readers will have little interest in this subject, since their computers either have built-in floating-point instructions or their computer manufacturer has supplied suitable subroutines. But, in fact, the material of this section should not merely be the concern of computer-design engineers or of a small clique of people who write library subroutines for new machines; *every* well-rounded programmer ought to have a knowledge of what goes on during the elementary steps of floating-point arithmetic. This subject is not at all as trivial as most people think; it involves a surprising amount of interesting information.

A. Floating-point notation. We have discussed "fixed-point" notation for numbers in Section 4.1; in such a case the programmer knows where the radix point is assumed to lie in the numbers he manipulates. For many purposes it is considerably more convenient to let the position of the radix point be dynamically variable or "floating" as a program is running, and to carry with each number an indication of the corresponding radix point position. This idea has been used for many years in scientific calculations, especially for expressing very large numbers like Avogadro's number $N = 6.02250 \times 10^{23}$, or very small numbers like Planck's constant $\hbar = 1.0545 \times 10^{-27}$ erg sec.

In this section, we shall work with *base b, excess q, floating-point numbers with p digits:* Such a number is represented as two quantities (e, f), and this notation stands for

$$(e, f) = f \times b^{e-q}. \tag{1}$$

Here e is an integer having a specified range, and f is a signed fraction. We will adopt the convention that

$$|f| < 1;$$

in other words, the radix point appears at the left of the positional representation of f. More precisely, the stipulation that we have p digit numbers means that $b^p f$ is an integer, and that

$$-b^p < b^p f < b^p. \tag{2}$$

The term "floating binary" implies that $b = 2$, "floating decimal" implies $b = 10$, etc. Using excess 50 floating-decimal numbers with 8 digits, we can write for example

$$\text{Avogadro's number } N = (74, +.60225000);$$
$$\text{Planck's constant } \quad \hbar = (24, +.10545000). \tag{3}$$

The two components e and f of a floating-point number are called the *exponent* and the *fraction* parts, respectively. (Other names are occasionally used for this purpose, notably "characteristic" and "mantissa"; but it is an abuse of terminology to call the fraction part a mantissa, since this concept has quite a different meaning in connection with logarithms, and furthermore the English word mantissa means "a worthless addition.")

In this section, we shall concentrate almost entirely on signed-magnitude representation for the fraction parts f, since complement notations for floating-point numbers apparently do not possess as many desirable properties (see Section 4.2.2).

A floating-point number (e, f) is said to be *normalized* if the most significant digit of the representation of f is nonzero, so that

$$1/b \le |f| < 1; \tag{4}$$

or if $f = 0$ and e has its smallest possible value. It is possible to tell which of two normalized floating-point numbers has a greater magnitude by comparing the exponent parts first, and then testing the fraction parts only if the exponents are equal.

The MIX computer assumes that its floating-point numbers have the form

$$\boxed{\pm} \; \boxed{e} \; \boxed{f} \; \boxed{f} \; \boxed{f} \; \boxed{f} \; . \tag{5}$$

Here we have base b excess q floating-point notation with four "digits" of precision, where b is the byte size (e.g., $b = 64$ or $b = 100$), and q is equal to $\lfloor \frac{1}{2}b \rfloor$. The fraction part is $\pm ffff$, and e is the exponent, which lies in the range $0 \le e < b$. This internal representation is typical of the conventions in most existing computers, although b is a much larger base than normal.

B. Normalized calculations. Most floating-point routines now in use deal almost entirely with normalized numbers: inputs to the routines are assumed to be normalized, and the outputs are always normalized.

Let us now study the operations of floating-point arithmetic in detail. At the same time we can consider the construction of subroutines for these operations (assuming that we have a computer without built-in floating-point hardware).

Machine-language subroutines for floating-point arithmetic are usually written in a very machine-dependent manner, using many of the wildest idiosyncrasies of the computer, and so we find that floating-point addition subroutines for two different machines usually bear little superficial resemblance to each other. Yet a careful study of numerous subroutines for both binary and decimal computers reveals that these programs actually have quite a lot in common, and it is possible to discuss the topics in a machine-independent way.

The first (and by far the most difficult!) algorithm we shall discuss in this section is a procedure for floating-point addition:

$$(e_u, f_u) \oplus (e_v, f_v) = (e_w, f_w). \tag{6}$$

Note: Since floating-point arithmetic is inherently approximate, not exact, we will use the symbols

$$\oplus, \; \ominus, \; \otimes, \; \oslash$$

to denote floating-point addition, subtraction, multiplication, and division, respectively, in order to distinguish the approximate operations from the true ones.

The basic idea involved in floating-point addition is fairly simple: Assuming that $e_u \ge e_v$, we take $e_w = e_u$, $f_w = f_u + f_v/b^{e_u - e_v}$ (thereby aligning the radix points for a meaningful addition), and normalize the result. Several situations can arise which make this process nontrivial, and the following algorithm explains the method more precisely.

Algorithm A (*Floating-point addition*). Given base b, excess q, p-digit, normalized floating-point numbers $u = (e_u, f_u)$ and $v = (e_v, f_v)$, we will form the

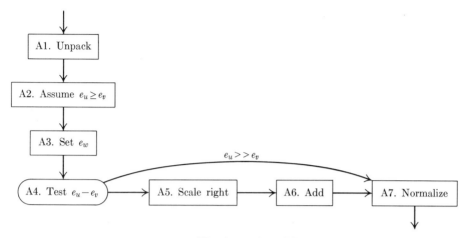

Fig. 2. Floating-point addition.

sum $w = u \oplus v$. The same algorithm may be used for floating-point subtraction, if $-v$ is substituted for v. The base b is assumed to be even.

A1. [Unpack.] Separate the exponent and fraction parts of the representations of u and v.

A2. [Assume $e_u \geq e_v$.] If $e_u < e_v$, interchange u and v. (In many cases, it is best to combine step A2 with step A1 or with some of the later steps.)

A3. [Set e_w.] Set $e_w \leftarrow e_u$.

A4. [Test $e_u - e_v$.] If $e_u - e_v \geq p + 2$ (large difference in exponents), set $f_w \leftarrow f_u$ and go to step A7. (Alternatively, since we are assuming u is normalized, we could terminate the algorithm, but it is often useful to be able to normalize a possibly unnormalized number by adding zero to it.)

A5. [Scale right.] Shift f_v to the right $e_u - e_v$ places, i.e., divide it by $b^{e_u - e_v}$. *Note:* This will be a shift of up to $p + 1$ places, and the next step (which adds f_u to f_v) thereby requires an accumulator capable of holding $2p + 1$ base b digits to the right of the radix point. If such a large accumulator is not available, it is possible to shorten the requirement to $p + 2$ places, if proper precautions are taken; the details are given in exercise 5.

A6. [Add.] Set $f_w \leftarrow f_u + f_v$.

A7. [Normalize.] (At this point (e_w, f_w) represents the sum of u and v, but f_w may have more than p digits, and it may be greater than unity or less than $1/b$.) Perform Algorithm N below, to normalize and round (e_w, f_w) into the final answer. ∎

Algorithm N (*Normalization*). A "raw exponent" e and a "raw fraction" f are converted to normalized form, rounding if necessary to p digits. This algorithm assumes that $|f| < b$ and that b is even.

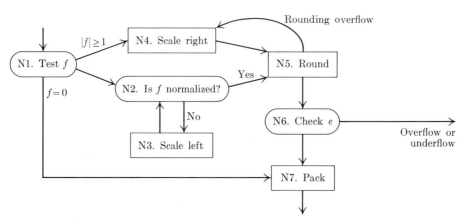

Fig. 3. Normalization of (e, f).

N1. [Test f.] If $|f| \geq 1$ ("fraction overflow"), go to step N4. If $f = 0$, set e_w to its lowest possible value and go to step N7.

N2. [Is f normalized?] If $|f| \geq 1/b$, go to step N5.

N3. [Scale left.] Shift f to the left by one digit position (i.e., multiply it by b), and decrease e by 1. Return to step N2.

N4. [Scale right.] Shift f to the right by one digit position (i.e., divide it by b), and increase e by 1.

N5. [Round.] Round f to p places. (We take this to mean that $f \leftarrow b^{-p}\lfloor b^p f + \frac{1}{2} \rfloor$ when $f > 0$, $f \leftarrow b^{-p}\lceil b^p f - \frac{1}{2} \rceil$ when $f < 0$; other rounding rules can be used, but this common definition appears to lend itself best to the theory which follows later in this chapter.) It is important to note that this rounding operation can make $|f| = 1$ ("rounding overflow"); in such a case, return to step N4.

N6. [Check e.] If e is too large, i.e., larger than its allowed range, an *exponent overflow* condition is sensed. If e is too small, an *exponent underflow* condition is sensed. (See the discussion below; these situations are usually error conditions, in the sense that the result cannot be expressed as a normalized floating-point number in the required range.)

N7. [Pack.] Put e and f together into the desired output representation. ∎

Some simple examples of floating-point addition are given in exercise 4.

The following MIX subroutines, for addition and subtraction of numbers having the form (5), show how Algorithms A and N can be expressed as computer programs. The subroutines below are designed to take one input u from symbolic location ACC, and the other input v comes from register A upon entrance to the subroutine. The output w appears both in register A and location ACC.

Thus, a fixed-point coding sequence

$$\text{LDA A;} \quad \text{ADD B;} \quad \text{SUB C;} \quad \text{STA D;} \tag{7}$$

would correspond to the floating-point coding sequence

$$\text{LDA A, STA ACC;} \quad \text{LDA B, JMP FADD;} \quad \text{LDA C, JMP FSUB;} \quad \text{STA D.} \tag{8}$$

Program A (*Addition, subtraction, and normalization*). The following program is a subroutine for Algorithm A, and it is also designed so that the normalization portion can be used by other subroutines which appear later in this section. In this program and in many other programs throughout this chapter, OFLO stands for a subroutine which prints out a message to the effect that MIX's overflow toggle was unexpectedly found to be "on."

01	EXP	EQU	1:1	Definition of exponent field.
02	FSUB	STA	TEMP	Floating-point subtraction subroutine:
03		LDAN	TEMP	Change sign of operand.
04	FADD	STJ	EXITF	Floating-point addition subroutine:
05		JOV	OFLO	Ensure overflow is off.
06		STA	TEMP	TEMP $\leftarrow v$.
07		LDX	ACC	rX $\leftarrow u$.
08		CMPA	ACC(EXP)	Steps A1, A2, A3 are combined here:
09		JLE	1F	Jump if $e_v \leq e_u$.
10		STA	FU(0:4)	FU $\leftarrow \pm f f f f 0$ (u, v interchanged).
11		LD2	TEMP(EXP)	rI2 $\leftarrow e_w$.
12		STX	FV(0:4)	
13		LD1N	ACC(EXP)	rI1 $\leftarrow -e_v$.
14		JMP	4F	
15	1H	STX	FU(0:4)	FU $\leftarrow \pm f f f f 0$.
16		LD2	ACC(EXP)	rI2 $\leftarrow e_w$.
17		JE	3F	Jump if $e_u = e_v$ (saves time).
18		STA	FV(0:4)	
19		LD1N	TEMP(EXP)	rI1 $\leftarrow -e_v$.
20	4H	INC1	0,2	rI1 $\leftarrow e_u - e_v$. (Step A4 unnecessary.)
21	5H	LDA	FV	A5. *Scale right.*
22		LDX	FV(0:0)	Set rX to zero with sign of v.
23		SRAX	0,1	Shift right $e_u - e_v$ places.
24		JMP	6F	
25	3H	SLA	1	Align radix point.
26		ENTX	0	Clear rX.
27	6H	ADD	FU	A6. *Add.*
28		JOV	N4	A7. *Normalize.* Jump if fraction overflow.
29		CMPA	=0=(1:1)	
30		JNE	N5	Jump if f is normalized.

31		JAP	1F	Otherwise, check if rA, rX
32		JAZ	8F	have opposite signs.
33		JXNP	N2	
34		INCA	1	Adjust for case of opposite signs.
35		JMP	2F	
36	1H	JXNN	N2	
37		DECA	1	Adjust for case of opposite signs.
38	2H	STX	TEMP	Complement register X.
39		SRAX	5	
40		LDA	TEMP	
41		STA	HALF(0:0)	Store sign of rA.
42		SUB	HALF	(This assumes that the
43		SUB	HALF	byte size is even.)
44		SRC	4	Jump into normalization routine.
45		JMP	N3A	
46	HALF	CON	1//2	One half the word size.
47	FU	CON	0	Fraction part f_u.
48	FV	CON	0	Fraction part f_v.
49	*			
50	8H	LDA	FV(0:0)	Set rA to proper sign (see below).
51	NORM	JXNZ	N2	*N1. Test f.*
52	NORM1	JANZ	N2	Assume rI2 ≡ e, rAX ≡ f.
53	ZERO	STZ	ACC	If $f = 0$, set $e \leftarrow 0$.
54		ENN2	1	Exit from the subroutine.
55		JMP	EXITF	
56	N3	SLAX	1	*N3. Scale left.*
57	N3A	DEC2	1	Decrease e by 1.
58	N2	CMPA	=0=(1:1)	*N2. Is f normalized?*
59		JE	N3	If leading byte zero, to N3.
60	N5	STA	TEMP(0:1)	*N5. Round.*
61		ADD	TEMP(0:1)	Add trailing byte of rA to rA.
62		JNOV	N6	If no rounding overflow, go to N6.
63	N4	ENTX	1	*N4. Scale right.*
64		SRC	1	Shift right and insert "1" at left.
65		INC2	1	Increase e by 1.
66		STA	TEMP(0:1)	Round.
67		ADD	TEMP(0:1)	
68	N6	J2N	EXPUN	*N6. Check e.* Underflow if $e < 0$.
69		ENTX	0,2	*N7. Pack.* rX $\leftarrow e$.
70		SRC	1	Insert e into 1:1 field.
71		STA	ACC	Store answer in ACC.
72	BYTE	EQU	1(4:4)	Byte size.
73		DEC2	BYTE	
74	EXITF	J2N	*	Exit, unless e greater than byte size.
75	EXPOV	HLT	2	Exponent overflow detected.
76	EXPUN	HLT	1	Exponent underflow detected.
77	ACC	CON	0	Floating-point accumulator. ∎

The above program is slightly more general than Algorithm A, since it does not assume the input operands are normalized. A true result rounded to four bytes is given in all cases when the inputs are normalized, and a true result is also given in almost all other cases as well (see exercise 8). The rather long section of code from lines 29 to 46 is a consequence of the fact that MIX has only a 5-byte accumulator for adding signed numbers, while in general $2p + 1 = 9$ places of accuracy are required. This section of the code could be omitted without much loss of accuracy if normalized operands are assumed, and the entire program could in fact be shortened to about half its present length if we were willing to sacrifice a little bit more of the accuracy, but we shall see in the next section that it is better to obtain the best possible accuracy in all cases.

The running time for floating-point addition and subtraction depends on several factors which are analyzed in Section 4.2.4.

Now let us consider multiplication and division, which are simpler than addition, and which are somewhat similar to each other.

Algorithm M (*Floating-point multiplication or division*). Given base b, excess q, p-digit, normalized floating-point numbers $u = (e_u, f_u)$ and $v = (e_v, f_v)$, we form the product $w = u \otimes v$ or the quotient $w = u \oslash v$. Let b be even.

M1. [Unpack.] Separate the exponent and fraction parts of the representations of u and v. (Sometimes it is convenient, but not necessary, to test the operands for zero during this step.)

M2. [Operate.] Set

$$e_w \leftarrow e_u + e_v - q, \qquad f_w \leftarrow f_u f_v \qquad\qquad \text{for multiplication;}$$

$$e_w \leftarrow e_u - e_v + q + 1, \qquad f_w \leftarrow \left(\frac{1}{b} f_u\right) \Big/ f_v \qquad \text{for division.} \tag{9}$$

(Since the input numbers are assumed to be normalized, it follows that either $f_w = 0$, or $1/b^2 \le |f_w| < 1$, or a division-by-zero error has occurred.) If necessary, f_w may be truncated to $p + 2$ digits at this point.

M3. [Normalize.] Perform Algorithm N on (e_w, f_w) to normalize, round, and pack the result. (*Note:* Normalization is simpler in this case, since scaling left occurs at most once, and "rounding overflow" cannot occur after division.) ∎

The following MIX subroutines, which use the same conventions as Program A above, illustrate the machine considerations necessary in connection with Algorithm M.

Program M (*Multiplication and division*).

```
01   Q      EQU   BYTE/2      q is half the byte size.
02   FMUL   STJ   EXITF       Floating-point multiplication subroutine:
03          JOV   OFLO        Ensure overflow is off.
```

04		STA	TEMP	TEMP $\leftarrow v$.
05		LDX	ACC	rX $\leftarrow u$.
06		STX	FU(0:4)	FU $\leftarrow \pm ffff0$.
07		LD1	TEMP(EXP)	
08		LD2	ACC(EXP)	
09		INC2	-Q,1	rI2 $\leftarrow e_u + e_v - q$.
10		SLA	1	
11		MUL	FU	Multiply f_u times f_v.
12		JMP	NORM	Normalize, round, and exit.
13	*			
14	FDIV	STJ	EXITF	Floating-point division subroutine:
15		JOV	OFLO	Ensure overflow is off.
16		STA	TEMP	TEMP $\leftarrow v$.
17		STA	FV(0:4)	FV $\leftarrow \pm ffff0$.
18		LD1	TEMP(EXP)	
19		LD2	ACC(EXP)	
20		DEC2	-Q,1	rI2 $\leftarrow e_u - e_v + q$.
21		ENTX	0	
22		LDA	ACC	
23		SLA	1	rA $\leftarrow f_u$.
24		CMPA	FV(1:5)	
25		JL	*+3	Jump if $|f_u| < |f_v|$.
26		SRA	1	Otherwise, scale f_u right
27		INC2	1	and increase rI2 by 1.
28		DIV	FV	Divide.
29		JNOV	NORM1	Normalize, round, and exit.
30	DVZRO	HLT	3	Unnormalized or zero divisor. ∎

The most noteworthy feature of this program is the provision for division in lines 24–27, which is made in order to ensure enough accuracy to round the answer. If $|f_u| < |f_v|$, straightforward application of Algorithm M would leave a result of the form "$\pm 0ffff$" in register A, and this would not allow a proper rounding without a careful analysis of the remainder (which appears in register X). So the program computes $f_w \leftarrow f_u/f_v$ in this case, ensuring that f_w is either zero or normalized in all cases.

We occasionally need to convert between fixed- and floating-point representations. A "fix-to-float" routine is easily obtained with the help of the normalization algorithm above; for example, in MIX, the following subroutine converts an integer to floating-point form:

01	FLOT	STJ	EXITF	Assume that rA = u, an integer.	
02		JOV	OFLO	Ensure overflow is off.	
03		ENT2	Q+5	Set raw exponent.	(10)
04		ENTX	0		
05		JMP	NORM	Normalize, round, and exit. ∎	

A "float-to-fix" subroutine is the subject of exercise 14.

The debugging of floating-point subroutines is usually a difficult job, since there are so many cases to consider. Here is a list of common pitfalls which often trap a programmer or machine designer who is preparing floating-point routines:

1) *Losing the sign.* On many machines (not MIX), shift instructions between registers will affect the sign, and the shifting operations used in normalizing and scaling numbers must be carefully analyzed. The sign is also lost frequently when minus zero is present. (For example, see lines 32 and 50 of Program A, which are used to set the sign of rA to the sign of v when register A is zero after the addition. See also exercise 6.)

2) *Failure to treat exponent underflow or overflow properly.* The size of e_w should not be checked until *after* the rounding and normalization, because preliminary tests may give an erroneous indication. Exponent underflow and overflow can occur on floating-point addition and subtraction, not only during multiplication and division; and even though this is a rather rare occurrence, it must be tested each time.

It has unfortunately become customary in many instances to ignore exponent underflow and simply to set underflowed results to zero with no indication of error. This causes a serious loss of accuracy in most cases (indeed, it is the loss of *all* the significant digits), so the programmer really ought to be told when underflow has occurred. Setting the result to zero is appropriate only in certain cases when the result is to be added to a significantly larger quantity. When exponent underflow is not detected, we find mysterious situations in which $(u \otimes v) \otimes w$ is zero, but $(u \otimes w) \otimes v$ is not, since $u \otimes v$ results in exponent underflow but $(u \otimes w) \otimes v$ can be calculated without any exponents falling out of range. Similarly, we can find positive numbers a, b, c, d, and y such that

$$(a \otimes y \oplus b) \oslash (c \otimes y \oplus d) = \tfrac{2}{3},$$
$$(a \oplus b \oslash y) \oslash (c \oplus d \oslash y) = 1 \tag{11}$$

if exponent underflow is not detected. (See exercise 9.) Even though floating-point routines are not precisely accurate, such a disparity as (11) is quite unexpected when a, b, c, d, and y are all *positive!* Exponent underflow is usually not anticipated by a programmer, so he should be told about it (and possibly be given the option of deciding what to do about it).

3) *Inserted garbage.* When scaling to the left it is important to keep from introducing anything but zeros at the right. For example, note the "ENTX 0" and "LDX FV(0:0)" instructions in Program A, lines 22 and 26, and the all-too-easily-forgotten "ENTX 0" instruction in line 04 of the FLOT subroutine (10). (But it is not necessary to clear register X after line 28 in the division subroutine, since register A is either normalized at that point or both A and X are zero.)

4) *Unforeseen rounding overflow.* When a number like .999999997 is rounded to 8 digits, a carry will occur to the left of the decimal point, and the result

must be scaled to the right. Many people have mistakenly concluded that rounding overflow is impossible during multiplication, since they look at the maximum value of $|f_u f_v|$ which is $1 - 2b^{-p} + b^{-2p}$, and this cannot round up to 1. The fallacy in this reasoning is exhibited in exercise 11. Curiously, the phenomenon of rounding overflow *is* impossible during floating-point division (see exercise 12).

(*Note:* There is one school of thought which says it is harmless to "round" .999999997 to .999999990 instead of to 1.0000000, since the latter result represents all numbers in the interval $[1.0000000 - 5 \times 10^{-8}, 1.0000000 + 5 \times 10^{-8})$, while the former result represents all numbers in the much smaller interval $[.99999999 - 5 \times 10^{-9}, .99999999 + 5 \times 10^{-9})$. Even though the latter interval does not contain the original number .999999997, each number of the second interval is contained in the first, so subsequent calculations with the second interval are no less accurate than with the first. But this argument is incompatible with the mathematical philosophy of floating-point arithmetic which is expressed in Section 4.2.2.)

5) *Rounding before normalizing.* . Inaccuracies are caused by premature rounding in the wrong digit position. This error is obvious when rounding is being done to the left of the appropriate position; and it is also dangerous in the less obvious cases where rounding first is done too far to the right, then is followed by rounding in the true position. For example, repeated rounding of the number 1.444445 to one less digit at each step will yield the sequence 1.44445, 1.4445, 1.445, 1.45, 1.5, 2! For this reason it is a mistake to round during the "scaling-right" operation in step A5. (Yet the special case of rounding in step N5, then rounding again after rounding overflow has occurred, is harmless, because rounding overflow always yields ± 1.000000 and this is unaffected by the subsequent rounding process.)

6) *Failure to retain enough precision in intermediate calculations.* Detailed analyses of the accuracy of floating-point arithmetic, made in the next section, suggest strongly that normalizing floating-point routines should always deliver a properly rounded result to the maximum possible accuracy. There should be no exceptions to this dictum, even in cases which occur with extremely low probability; the appropriate number of significant digits should be retained throughout the computations, as stated in Algorithms A and M.

C. Floating-point hardware. Nearly every large computer intended for scientific calculations includes floating-point arithmetic as part of its repertoire of built-in operations. Unfortunately, the design of such hardware usually includes some anomalies which result in dismally poor behavior in certain circumstances, and we hope that future computer designers will pay more attention to providing the proper behavior than they have in the past.

The MIX computer, which is being used as an example of a "typical" machine in this series of books, has an optional "floating-point attachment" (available at extra cost) which includes the following six operations:

• FADD, FSUB, FMUL, FDIV, FLOT, FCMP (C $= 1, 2, 3, 4, 5, 56$, respectively; F $= 6$). The contents of rA after the operation "FADD V" are the same as the contents of rA after the operations

```
STA   ACC
LDA   V
JMP   FADD
```

where FADD is the subroutine which appears earlier in this section, except that both operands are automatically normalized before entry to the subroutine, if they are not already in normalized form. Similar remarks apply to FSUB, FMUL, and FDIV. The contents of rA after the operation "FLOT" are the contents after "JMP FLOT" in the subroutine (10) above. The contents of rA are unchanged by the operation "FCMP V"; this instruction sets the comparison indicator to less, equal, or greater, depending on whether the contents of rA are "definitely less than," "approximately equal to," or "definitely greater than" V; this subject is discussed in the next section, and the precise action is defined by the subroutine FCMP of exercise 4.2.2–17 with EPSILON in location 0. No register other than rA is affected by any of the floating-point operations. If exponent overflow or underflow occurs, the overflow toggle is turned on and the exponent of the answer is given modulo the byte size. Division by zero is undefined.

Sometimes it is helpful to use floating-point operators in a nonstandard manner. For example, if the operation FLOT had not been included as part of MIX's floating-point attachment, we could easily achieve its effect on 4-byte numbers by writing

$$
\begin{array}{lll}
\text{FLOT} & \text{STJ} & \text{9F} \\
& \text{SLA} & 1 \\
& \text{ENTX} & \text{Q+4} \\
& \text{SRC} & 1 \\
& \text{FADD} & \text{=0=} \\
\text{9H} & \text{JMP} & * \\
\end{array} \qquad (12)
$$

This routine is not strictly equivalent to the FLOT operator, since it assumes that the 1 : 1 byte of rA is zero, and it destroys rX. The handling of more general situations is a little tricky, because rounding overflow can occur even during a FLOT operation.

Similarly, if we want to round a number u from floating-point form to the nearest fixed-point integer, and if we know that number is nonnegative and will fit in at most three bytes, we may write

```
FADD   FUDGE
```

where FUDGE contains the constant

+	Q+4	1	0	0	0	;

the result in rA will be

$$\boxed{\ +\ |\ \text{Q+4}\ |\ \ 1\ \ |\ \ \ \text{round}\ (u)\ \ \ }\ . \tag{13}$$

Some machines have floating-point hardware which suppresses automatic rounding and gives additional information about the less-significant digits that would ordinarily be rounded off. Such facilities are of use in extending the precision of floating-point calculations, although they slow down single-precision arithmetic slightly when rounding is desired.

D. History and bibliography. The origins of floating-point notation can be traced back to the Babylonians (c. 1800 B.C.), who made use of radix 60 floating-point arithmetic but did not have a notation for the exponents. The appropriate exponent was always somehow "understood" by the man doing the calculations. At least one case has been found in which the wrong answer was given, because addition was performed with improper alignment of the operands, but such examples are very rare; see O. Neugebauer, *The Exact Sciences in Antiquity* (Princeton University Press, 1952), 26–27. Another early contribution to floating-point notation is due to the Greek mathematician Apollonius (3rd century B.C.), who apparently was the first to explain how to simplify multiplication by collecting powers of 10 separately from their coefficients, at least in simple cases. For a discussion of Apollonius's method, see Pappus, *Mathematical Collections* (4th century A.D.). After the Babylonian civilization died out, the first significant uses of floating-point notation for products and quotients did not emerge until much later, about the time logarithms were invented (1600) and shortly afterwards when Oughtred invented the slide rule (1630). The modern notation "x^n" for exponents was being introduced at about the same time; separate symbols for x squared, x cubed, etc. had been in use before this.

Floating-point arithmetic was incorporated into the design of some of the earliest computers, appearing first in the Bell Laboratories' Model V and then in the Harvard Mark II. Both of these machines were relay calculators which were designed in 1944; George Stibitz had already proposed floating-point hardware for Bell Laboratories' computers in 1939. [See *Math. Comp.* **3** (1948), 1–13; *Proc. Symp. on Large-Scale Digital Calculating Machinery* (Harvard, 1947), 41–68, 69–79; *Datamation* **13** (April, 1967), 35–44, (May, 1967), 45–49.] Floating-point arithmetic was also used about the same time in a German computer built by K. Zuse; this machine included conventions for dealing with two special quantities, "∞" and "undefined." [See *Zeitschrift für angewandte Mathematik und Physik* **1** (1950), 345–346.]

The use of floating-binary arithmetic was seriously considered in 1944–1946 by researchers at the Moore School in their plans for the first *electronic* digital computers, but it turned out to be much harder to implement floating-point circuitry with tubes than with relays. The group realized that scaling is a problem in programming, but felt that it is only a very small part of a total programming job and is usually worth the time and trouble it takes, since it tends to keep a programmer aware of the numerical accuracy he is getting. Furthermore, they argued that floating-point representation takes up valuable memory space, since the exponents must be stored, and that it becomes difficult to adapt floating-point arithmetic to multiple precision calculations.   [See von Neumann's *Collected Works*, **5** (New York: Macmillan, 1963), 43, 73–74.] At this time, of course, they were designing the first stored-program computer and the second electronic computer, and their choice was to be either fixed point *or* floating point, not both.   They anticipated the coding of floating-binary routines, and in fact "shift-left" and "shift-right" instructions were put into their machine primarily to make such routines more efficient. The first machine to have both kinds of arithmetic in its hardware was apparently a computer developed at General Electric Company [see *Proc. 2nd Symp. on Large-Scale Digital Calculating Machinery* (Harvard, 1948), 65–69].

Floating-point subroutines and interpretive systems for early machines were coded by D. J. Wheeler and others, and the first publication of such routines was in *The Preparation of Programs for an Electronic Digital Computer* by Wilkes, Wheeler, and Gill (Reading, Mass.: Addison-Wesley, 1951), subroutines A1–A11, pp. 35–37, 105–117.   It is interesting to note that floating *decimal* subroutines are described here, although a binary computer was being used; in other words, the numbers were represented as $10^e f$, not $2^e f$, and therefore the scaling operations required multiplication or division by 10.   The concept of floating-binary arithmetic was not unknown at the time, since it had been mentioned by von Neumann in a widely-circulated document; this use of the more cumbersome floating-decimal arithmetic indicates strong ties with the scientific power-of-10 notation and an unwillingness to surrender completely to the binary scale.

Most published references to the details of floating-point arithmetic routines are scattered in "technical memorandums" distributed by various computer manufacturers, but there have been occasional appearances of these routines in the open literature.   Besides the reference above, see R. H. Stark and D. B. MacMillan, *Math. Comp.* **5** (1951), 86–92, where a plugboard-wired program is described; D. McCracken, *Digital Computer Programming* (New York: Wiley, 1957), 121–131; J. W. Carr III, *CACM* **2** (May, 1959), 10–15; W. G. Wadey, *JACM* **7** (1960), 129–139; D. E. Knuth, *JACM* **8** (1961), 119–128; O. Kesner, *CACM* **5** (1962), 269–271; F. P. Brooks and K. E. Iverson, *Automatic Data Processing* (New York: Wiley, 1963), 184–199.   For a discussion of floating-point arithmetic from a computer designer's standpoint, see "Floating-point operation" by S. G. Campbell, in *Planning a Computer System*, ed. by W. Buchholz (New York: McGraw-Hill, 1962), 92–121.   Additional references are given in Section 4.2.2.

EXERCISES

   1. [*10*] How would Avogadro's number and Planck's constant be represented in base 100, excess 50, four-digit floating-point notation? (This would be the representation used by MIX, as in (5), if the byte size is 100.)

2. *[12]* Assume that the exponent e is constrained to lie in the range $0 \le e \le E$; what are the largest and smallest positive values which can be written as base b, excess q, p-digit floating-point numbers? What are the largest and smallest positive values which can be written as *normalized* floating-point numbers with these specifications?

3. *[20]* Show that if we are using normalized floating-binary arithmetic, there is a way to increase the precision slightly without loss of memory space: A p-bit fraction part can be represented using only $p - 1$ bit positions of a computer word, if the range of exponent values is decreased very slightly.

▶ **4.** *[15]* Assume that $b = 10$, $p = 8$. What result does Algorithm A give for $(50, +.98765432) \oplus (49, +.33333333)$? For $(53, -.99987654) \oplus (54, +.10000000)$? For $(45, -.50000001) \oplus (54, +.10000000)$?

5. *[M23]* Prove that the following operation can be placed between steps A5 and A6 of Algorithm A, without changing the result: If f_u and f_v have the same sign, replace f_v by $\text{sign}\,(f_v)b^{-p-2}\lfloor b^{p+2}|f_v|\rfloor$; if f_u and f_v have opposite signs, replace f_v by $\text{sign}\,(f_v)b^{-p-2}\lceil b^{p+2}|f_v|\rceil$. (The effect of this operation is to "truncate" f_v to $p + 2$ digits, to minimize the length of register which is needed for the addition in step A6.)

6. *[22]* Would it be a good idea to replace lines 38–40 of Program A by the single instruction "SLC 5"?

7. *[M21]* What changes should be made to Algorithm A so that it delivers a properly rounded, normalized answer even when the inputs are not normalized? (The term "properly rounded" means that the result has the maximum possible accuracy in p digits, assuming that the inputs u and v are exactly equal to $f_u \times b^{e_u - q}$ and $f_v \times b^{e_v - q}$, although possibly unnormalized. In particular, u or v might have a fraction part equal to zero with a very large exponent, and such an operand would be treated as zero in the context of this exercise.)

8. *[M25]* Give examples of input values for which the FADD subroutine in Program A does not yield a "properly rounded, normalized answer" in the sense of exercise 7.

9. *[M24]* (W. Kahan.) Assume that the occurrence of exponent underflow causes the result to be replaced by zero, with no error indication given. Using excess zero, eight-digit floating decimal numbers with e in the range $-50 \le e < 50$, find positive values of a, b, c, d, and y such that (11) holds.

10. *[M15]* Give an example of normalized eight-digit floating-decimal numbers u and v for which rounding overflow occurs in addition.

▶ **11.** *[M20]* Give an example of normalized eight-digit floating-decimal numbers u and v for which rounding overflow occurs in multiplication.

12. *[M25]* Prove that rounding overflow cannot occur during the normalization phase of floating-point division.

▶ **13.** *[M23]* Mr. J. H. Quick modified the trick of (13) in order to *truncate* a positive floating-point number u to the fixed-point integer $\lfloor u \rfloor$, assuming that $0 < u < b^3$, as follows: FSUB ONEHALF; FADD FUDGE; SUB FUDGE. (Here ONEHALF is the floating-point representation of $\frac{1}{2}$.) His idea was based on the fact that $u - \frac{1}{2}$ rounds to $\lfloor u \rfloor$. But he later found that this idea does not always work. What went wrong? Is there another trick he could use instead?

14. [*23*] Write a `MIX` subroutine which begins with an arbitrary floating-point number in register A, not necessarily normalized, and which converts it to the nearest fixed-point integer (or determines that it is too large in absolute value to make such a conversion possible).

▶ **15.** [*28*] Write a `MIX` subroutine, to be used in connection with the other subroutines of this section, that calculates u mod 1, that is, $u - \lfloor u \rfloor$, given a floating-point number u. A properly rounded result (in the sense of exercise 7) should be given; therefore when u is a very small negative number, u mod 1 will be rounded so that the result is unity (even though u mod 1 has been defined to be always *less* than unity, as a real number).

16. [*HM21*] (Robert L. Smith.) Design an algorithm to compute the real and imaginary parts of the complex number $(a + bi)/(c + di)$, given real floating-point values a, b, c, and d. Avoid the computation of $c^2 + d^2$, since it would cause floating-point overflow even when $|c|$ or $|d|$ is approximately the square root of the maximum allowable floating-point value.

17. [*40*] (John Cocke.) Explore the idea of extending the range of floating-point numbers by defining a single-word representation in which the precision of the fraction decreases as the magnitude of the exponent increases.

4.2.2. Accuracy of Floating-Point Arithmetic

Floating-point computation is by nature inexact, and it is not difficult to misuse it so that the computed answers consist almost entirely of "noise." One of the principal problems of numerical analysis is to determine how accurate the results of certain numerical methods will be; a "credibility-gap" problem is involved here: we don't know how much of the computer's answers to believe. Novice computer users solve this problem by implicitly trusting in the computer as an infallible authority; they tend to believe all digits of a printed answer are significant. Disillusioned computer users have just the opposite approach, they are constantly afraid their answers are almost meaningless. Many a serious mathematician has attempted to give rigorous analyses of a sequence of floating-point operations, but has found the task to be so formidable he has tried to content himself with plausibility arguments instead.

A thorough examination of error analysis techniques is, of course, beyond the scope of this book, but we will in this section study some of the characteristics of floating-point arithmetic errors. Our goal is to discover how to perform floating-point arithmetic in such a way that reasonable analyses of error propagation are facilitated as much as possible.

A rough (but reasonably useful) way to express the behavior of floating-point arithmetic can be based on the concept of "significant figures" or *relative error*. If we are representing an exact real number x inside a computer by using the approximation $\hat{x} = x(1 + \epsilon)$, the quantity $\epsilon = (\hat{x} - x)/x$ is called the relative error of approximation. Roughly speaking, the operations of floating-point multiplication and division do not magnify the relative error by very much; but floating-point subtraction of nearly equal quantities (and floating-point addition, $u \oplus v$, where u is nearly equal to $-v$) can very greatly increase the relative error. So we have a general rule of thumb, that a substantial loss of accuracy is expected from such additions and subtractions, but not from multiplications and divisions.

One of the consequences of the possible unreliability of floating-point addition is that the associative law breaks down:

$$(u \oplus v) \oplus w \neq u \oplus (v \oplus w), \quad \text{for certain } u, v, w. \tag{1}$$

For example,

$$(11111113. \oplus -11111111.) \oplus 7.5111111 = 2.0000000 \oplus 7.5111111$$
$$= 9.5111111;$$
$$11111113. \oplus (-11111111. \oplus 7.5111111) = 11111113. \oplus -11111103.$$
$$= 10.000000.$$

(All examples in this section are given in eight-digit floating-decimal arithmetic, with exponents indicated by an explicit indication of the decimal-point position. Recall that, as in Section 4.2.1, the symbols $\oplus$, $\ominus$, $\otimes$, $\oslash$ are being used to stand for floating-point operations which correspond to the exact operations $+, -, \times, /$.)

In view of the failure of the associative law, the comment of Mrs. La Touche which appears at the beginning of this chapter [taken from *Math. Gazette* **12** (1924), 95] makes a good deal of sense with respect to floating-point arithmetic. Mathematical notations like "$a_1 + a_2 + a_3$" or "$\sum_{1 \leq k \leq n} a_k$" are inherently based upon the assumption of associativity, so a programmer must be especially careful that he does not implicitly assume the validity of the associative law.

A. An axiomatic approach. Although the associative law is not valid, the commutative law

$$u \oplus v = v \oplus u \tag{2}$$

does hold, and this law can be a valuable conceptual asset in programming and in the analysis of programs. This example suggests that we should look for important laws which are satisfied by $\oplus$, $\ominus$, $\otimes$, and $\oslash$; furthermore, it is not unreasonable to say that *floating-point routines should be designed to preserve as many of the ordinary mathematical laws as possible.*

Let us now consider some of the other basic laws which are valid for normalized floating-point operations as described in the previous section. First we have

$$u \ominus v = u \oplus -v; \tag{3}$$

$$-(u \oplus v) = -u \oplus -v; \tag{4}$$

$$u \oplus v = 0 \quad \text{if and only if} \quad v = -u; \tag{5}$$

$$u \oplus 0 = u \tag{6}$$

From these laws we can derive further identities; for example,

$$u \ominus v = -(v \ominus u). \tag{7}$$

(See exercise 1.) The above laws would *not* strictly be true if two's complement notation were used for fraction parts in floating-binary arithmetic instead of the signed-magnitude representation; see exercise 11. From this standpoint the signed-magnitude representation of floating-point numbers has a theoretical advantage.

Identities (2) to (6) are immediate from the algorithms in Section 4.2.1. The following rule is slightly less obvious:

$$\text{if} \quad u \leq v \quad \text{then} \quad u \oplus w \leq v \oplus w. \tag{8}$$

To prove this rule, let us define

$$\begin{aligned}
\text{round}(x, p) &= \text{``}x \text{ rounded to } p \text{ digits''} \\
&= \begin{cases}
b^{e-p} \lfloor b^{p-e} x + \tfrac{1}{2} \rfloor, & \text{if} \quad b^{e-1} \leq x < b^e; \\
0, & \text{if} \quad x = 0; \\
b^{e-p} \lceil b^{p-e} x - \tfrac{1}{2} \rceil, & \text{if} \quad b^{e-1} \leq -x < b^e;
\end{cases}
\end{aligned} \tag{9}$$

if x is a real number and p is a positive integer. (Compare with Algorithm 4.2.1N, step N5.) Note that

$$\text{round}(-x, p) = -\text{round}(x, p); \qquad \text{round}(bx, p) = b\,\text{round}(x, p); \tag{10}$$

and, furthermore, the definitions of Section 4.2.1 satisfy the important relationships

$$u \oplus v = \text{round}(u + v, p), \tag{11}$$

$$u \ominus v = \text{round}(u - v, p), \tag{12}$$

$$u \otimes v = \text{round}(u \times v, p), \tag{13}$$

$$u \oslash v = \text{round}(u/v, p), \tag{14}$$

provided that no exponent overflow or underflow occurs, i.e., provided that $u + v$, $u - v$, $u \times v$, and u/v are in a proper range. This observation leads to a simple proof of all the identities above, including (8), which is a consequence of the fact that $\text{round}(x, p)$ is nondecreasing as x increases.

We may now write down several more identities which follow from the above equations:

$$u \otimes v = v \otimes u, \qquad (-u) \otimes v = -(u \otimes v), \qquad 1 \otimes v = v;$$

$$u \otimes v = 0 \quad \text{if and only if} \quad u = 0 \quad \text{or} \quad v = 0;$$

$$(-u) \oslash v = u \oslash (-v) = -(u \oslash v), \qquad 0 \oslash v = 0,$$

$$u \oslash 1 = u, \qquad u \oslash u = 1;$$

if $u \leq v$ and $w > 0$ then $u \otimes w \leq v \otimes w$, $u \oslash w \leq v \oslash w$, and $w \oslash u \geq w \oslash v$. So there is a good deal of regularity present in spite of the inexactness of the floating-point operations, if the operations are defined according to a consistent set of conventions.

Several familiar rules of algebra are still, of course, conspicuously lacking from the collection of identities above; the associative law for floating-point multiplication is not strictly true, as shown in exercise 3, and the distributive law between $\otimes$ and $\oplus$ can fail rather badly: Let $u = 20000.000$, $v = -6.0000000$, and $w = 6.0000003$; then

$$(u \otimes v) \oplus (u \otimes w) = -120000.00 \oplus 120000.01 = .010000000$$
$$u \otimes (v \oplus w) = 20000.000 \otimes .00000030000000 = .0060000000$$

so

$$u \otimes (v \oplus w) \neq (u \otimes v) \oplus (u \otimes w). \tag{15}$$

Similarly, it is not difficult to find examples where $2(u^2 \oplus v^2) < (u \oplus v)^2$ in floating-point arithmetic; a programmer calculating the standard deviation of some observations by the formula

$$\sigma = \frac{1}{n} \sqrt{n \sum_{1 \leq k \leq n} x_k^2 - \left(\sum_{1 \leq k \leq n} x_k \right)^2}$$

may find himself taking the square root of a negative number!

Even when algebraic laws do not hold exactly, we can use these techniques to determine how close a law comes to being valid. The definition of round(x, p) implies that

$$\text{round}(x, p) = x(1 + \delta_p(x)), \tag{16}$$

where

$$|\delta_p(x)| \leq \tfrac{1}{2} b^{1-p}. \tag{17}$$

Therefore we can always write

$$a \oplus b = (a + b)(1 + \delta_p(a + b)), \tag{18}$$
$$a \ominus b = (a - b)(1 + \delta_p(a - b)), \tag{19}$$
$$a \otimes b = (a \times b)(1 + \delta_p(a \times b)), \tag{20}$$
$$a \oslash b = (a/b)(1 + \delta_p(a/b)). \tag{21}$$

Here we can estimate the relative error of normalized floating-point calculations in a very simple way. Formulas (18) to (21) are the main tools for error estimation in normalized floating-point arithmetic.

As an example of typical error-estimation procedures, let us consider the associative law for multiplication. Exercise 3 shows that $(u \otimes v) \otimes w$ is not in general equal to $u \otimes (v \otimes w)$; but the situation in this case is much better than it was with respect to the associative law of addition (1) and the distributive law (15). In fact, we have by (17), (20),

$$(u \otimes v) \otimes w = ((uv)(1 + \delta_1)) \otimes w = uvw(1 + \delta_1)(1 + \delta_2),$$
$$u \otimes (v \otimes w) = u \otimes ((vw)(1 + \delta_3)) = uvw(1 + \delta_3)(1 + \delta_4)$$

for some δ_1, δ_2, δ_3, δ_4, provided that no exponent underflow or overflow occurs, where $|\delta_j| \leq \frac{1}{2}b^{1-p}$ for each j. Hence

$$\frac{(u \otimes v) \otimes w}{u \otimes (v \otimes w)} = \frac{(1 + \delta_1)(1 + \delta_2)}{(1 + \delta_3)(1 + \delta_4)} = 1 + \delta,$$

where

$$|\delta| \leq 2b^{1-p}/(1 - \tfrac{1}{2}b^{1-p})^2. \tag{22}$$

We have thereby established the fact that $(u \otimes v) \otimes w$ is *approximately equal to* $u \otimes (v \otimes w)$, except when exponent overflow or underflow is a problem. It is worth while studying this intuitive idea of being "approximately equal" in more detail; can we make such a statement more precise in a reasonable way?

A programmer using floating-point arithmetic almost never wants to test if $u = v$ (or at least he hardly ever should try to do so), because this is an extremely improbable occurrence. For example, if a recurrence relation

$$x_{n+1} = f(x_n)$$

is being used, where the theory in some textbook says x_n approaches a limit as $n \to \infty$, it is usually a mistake to wait until $x_{n+1} = x_n$ for some n, since the sequence x_n might be periodic with a longer period due to the rounding of intermediate results. The proper procedure is to wait until $|x_{n+1} - x_n| < \delta$, for some suitably chosen number δ; but since we don't necessarily know the order of magnitude of x_n in advance, it is even more proper to wait until

$$|x_{n+1} - x_n| \leq \epsilon |x_n|; \tag{23}$$

now ϵ is a number that is much easier to select. This relation (23) is another way of saying that x_{n+1} and x_n are approximately equal; and this discussion indicates that a relation of "approximately equal" would be more useful than the traditional relation of equality, when floating-point computations are considered, if we could only define an approximation relation which is suitable.

In other words, the fact that strict equality of floating-point values is of little importance implies that we ought to have a new operation, *floating-point comparison*, which is intended to help assess the relative values of two floating-point quantities. The following definitions seem to be appropriate, for base b, excess q, floating-point numbers $u = (e_u, f_u)$ and $v = (e_v, f_v)$:

$$u \prec v \quad (\epsilon) \quad \text{if and only if} \quad v - u > \epsilon \max(b^{e_u - q}, b^{e_v - q}); \tag{24}$$

$$u \sim v \quad (\epsilon) \quad \text{if and only if} \quad |v - u| \leq \epsilon \max(b^{e_u - q}, b^{e_v - q}); \tag{25}$$

$$u \succ v \quad (\epsilon) \quad \text{if and only if} \quad u - v > \epsilon \max(b^{e_u - q}, b^{e_v - q}); \tag{26}$$

$$u \approx v \quad (\epsilon) \quad \text{if and only if} \quad |v - u| \leq \epsilon \min(b^{e_u - q}, b^{e_v - q}). \tag{27}$$

These definitions imply that exactly one of the conditions $u \prec v$ (definitely less than), $u \sim v$ (approximately equal to), or $u > v$ (definitely greater than) must always hold for any given pair of values u and v. The relation $u \approx v$ is somewhat stronger than $u \sim v$, and it might be read "u is essentially equal to v." All of the relations are given in terms of a positive real number ϵ which measures the degree of approximation being considered.

One way to view the above definitions is to associate a set $S(u) = \{x \mid |x - u| \le \epsilon b^{e_u - q}\}$ with each floating-point number u; $S(u)$ represents a set of values near u based on the exponent of u's floating-point representation. In these terms, we have $u < v$ if and only if $S(u) < v$ and $u < S(v)$; $u \sim v$ if and only if $u \in S(v)$ or $v \in S(u)$; $u > v$ if and only if $u > S(v)$ and $S(u) > v$; $u \approx v$ if and only if $u \in S(v)$ and $v \in S(u)$. (Here we are assuming that the parameter ϵ, which measures the degree of approximation, is a constant; a more complete notation would indicate the dependence of $S(u)$ upon ϵ.)

Here are some simple consequences of the above definitions:

$$\text{if} \quad u < v \quad (\epsilon) \quad \text{then} \quad v > u \quad (\epsilon); \tag{28}$$

$$\text{if} \quad u \approx v \quad (\epsilon) \quad \text{then} \quad u \sim v \quad (\epsilon); \tag{29}$$

$$u \approx u \quad (\epsilon); \tag{30}$$

$$\text{if} \quad u < v \quad (\epsilon) \quad \text{then} \quad u < v; \tag{31}$$

$$\text{if} \quad u < v \quad (\epsilon_1) \quad \text{and} \quad \epsilon_1 \ge \epsilon_2 \quad \text{then} \quad u < v \quad (\epsilon_2); \tag{32}$$

$$\text{if} \quad u \sim v \quad (\epsilon_1) \quad \text{and} \quad \epsilon_1 \le \epsilon_2 \quad \text{then} \quad u \sim v \quad (\epsilon_2); \tag{33}$$

$$\text{if} \quad u \approx v \quad (\epsilon_1) \quad \text{and} \quad \epsilon_1 \le \epsilon_2 \quad \text{then} \quad u \approx v \quad (\epsilon_2); \tag{34}$$

$$\text{if} \quad u < v \quad (\epsilon_1) \quad \text{and} \quad v < w \quad (\epsilon_2) \quad \text{then} \quad u < w \quad (\epsilon_1 + \epsilon_2); \tag{35}$$

$$\text{if} \quad u \approx v \quad (\epsilon_1) \quad \text{and} \quad v \approx w \quad (\epsilon_2) \quad \text{then} \quad u \sim w \quad (\epsilon_1 + \epsilon_2). \tag{36}$$

Moreover, we can prove without difficulty that

$$|u - v| \le \epsilon|u| \quad \text{and} \quad |u - v| \le \epsilon|v| \quad \text{implies} \quad u \approx v \quad (\epsilon); \tag{37}$$

$$|u - v| \le \epsilon|u| \quad \text{or} \quad |u - v| \le \epsilon|v| \quad \text{implies} \quad u \sim v \quad (\epsilon); \tag{38}$$

and conversely, for *normalized* numbers u and v, when $\epsilon < 1$,

$$u \approx v \quad (\epsilon) \quad \text{implies} \quad |u - v| \le b\epsilon|u| \quad \text{and} \quad |u - v| \le b\epsilon|v|; \tag{39}$$

$$u \sim v \quad (\epsilon) \quad \text{implies} \quad |u - v| \le b\epsilon|u| \quad \text{or} \quad |u - v| \le b\epsilon|v|. \tag{40}$$

As an example of all of these relations, we have [as in (22)]

$$|(u \otimes v) \otimes w - u \otimes (v \otimes w)| = |u \otimes (v \otimes w)| \left| \frac{(1 + \delta_1)(1 + \delta_2)}{(1 + \delta_3)(1 + \delta_4)} - 1 \right|$$

$$\le \frac{2b^{1-p}}{(1 - \frac{1}{2}b^{1-p})^2} |u \otimes (v \otimes w)|.$$

The same inequality is valid with $(u \otimes v) \otimes w$ and $u \otimes (v \otimes w)$ inter-changed. Hence by (27),

$$(u \otimes v) \otimes w \approx u \otimes (v \otimes w) \qquad (\epsilon) \qquad (41)$$

if $\epsilon \geq 2b^{1-p}/(1 - b^{1-p})^2$. For example, if $b = 10$ and $p = 8$, we may take $\epsilon = 0.00000021$. Thus we have a pretty good "associative law."

The relations $<$, $\sim$, $>$, and $\approx$ are useful within numerical algorithms, and it is therefore a good idea to provide routines for comparing floating-point numbers as well as for doing arithmetic on them.

Let us now shift our attention back to the question of finding *exact* relations which are satisfied by the floating-point operations. It is interesting to note that floating-point addition and subtraction are not completely intractable from an axiomatic standpoint, since they do satisfy the nontrivial identities stated in Theorems A and B:

Theorem A. *Let u and v be normalized floating point numbers. Then*

$$((u \oplus v) \ominus u) + ((u \oplus v) \ominus ((u \oplus v) \ominus u)) = u \oplus v, \quad (42)$$

provided that no exponent overflow or underflow occurs.

Note: This rather cumbersome looking identity can be rewritten in a simpler manner as follows: Let

$$u' = (u \oplus v) \ominus v, \qquad v' = (u \oplus v) \ominus u; \qquad (43)$$
$$u'' = (u \oplus v) \ominus v', \qquad v'' = (u \oplus v) \ominus u'.$$

Intuitively, u' and u'' should be approximations to u, and v' and v'' should be approximations to v. Theorem A tells us that

$$u \oplus v = u' + v'' = u'' + v'. \qquad (44)$$

This is a stronger statement than the identity

$$u \oplus v = u' \oplus v'' = u'' \oplus v', \qquad (45)$$

which is another consequence of Theorem A (see exercise 12).

Proof. By symmetry of the hypotheses, we may prove Theorem A completely by proving only that (44) holds when $u \geq |v|$. In the following proof it is convenient to use the abbreviations

$$d = e_u - e_v \geq 0, \qquad w = u \oplus v \geq 0, \qquad (46)$$

and to work with integers instead of fractions; if we have a small letter x denoting the normalized floating-point quantity (e_x, f_x), the corresponding capital letter X denotes $b^{p+e_x-e_v}f_x$. In particular, we have $U = b^{p+d}f_u$, $V = b^p f_v$; these quantities as well as U', V', U'', V'', and W are integers whose decimal points are aligned so that (44) is equivalent to

$$W = U' + V'' = U'' + V'. \tag{47}$$

The proof now divides into a rather tedious list of special cases.

Case 1, $e_w = e_u$. (See Fig. 4(i).) Here $U + V = W + R$, where

$$R \equiv V \ (\text{modulo } b^d), \qquad -\tfrac{1}{2}b^d \le R < \tfrac{1}{2}b^d.$$

Now $U' = \text{round}(W - V, p) = \text{round}(U - R, p)$. There are two subcases. *Case (1a)*, $R = -\tfrac{1}{2}b^d$. Then $U' = U + b^d$, $V' = V - R$, $U'' = U$, $V'' = V - R - b^d = V - \tfrac{1}{2}b^d$. *Case (1b)*, $R \ne -\tfrac{1}{2}b^d$. Then $U' = U$, $V' = V - R$, $U'' = U$, $V'' = V - R$.

Case 2, $e_w = e_u + 1$. (See Fig. 4(ii).) It follows that $V > 0$ and $d \le p$; and $U + V = W + R$ where

$$R \equiv V + b^d U_0 \ (\text{modulo } b^{d+1}), \qquad -\tfrac{1}{2}b^{d+1} \le R < \tfrac{1}{2}b^{d+1},$$

and U_0 is the least significant digit of f_u. Again there are subcases: *Case (2a)*, $U - R \ge b^{d+p} - \tfrac{1}{2}b^d$. Then since $U - R \le b^{d+p} - b^d + \tfrac{1}{2}b^{d+1}$ we must have $U' = b^{d+p} = U - R + Q$ where

$$Q \equiv V \ (\text{modulo } b^{d+1}), \qquad -\tfrac{1}{2}b^{d+1} < Q \le \tfrac{1}{2}b^{d+1}.$$

Hence $V'' = V - Q$. *Case (2b)*, $b^{d+p} - \tfrac{1}{2}b^d > U - R \ge b^{d+p-1} - \tfrac{1}{2}b^{d-1}$. Then $U' = U - R + Q$, where

$$Q \equiv V \ (\text{modulo } b^d), \qquad -\tfrac{1}{2}b^d < Q \le \tfrac{1}{2}b^d, \qquad \text{and} \qquad V'' = V - Q.$$

Case (2c), $b^{d+p-1} - \tfrac{1}{2}b^{d-1} > U - R$. This is impossible, since it obviously cannot occur when $d = 0$; and when $d > 0$ it implies $R > 0$, so $U + V > W$, and $U - R \ge W - V - R + 1 > b^{d+p} - (b^p - 1) - \tfrac{1}{2}b^{d+1} + 1 \ge b^{d+p-1}$, a contradiction.

Finally to complete the analysis of Case 2, we need to calculate $V' = \text{round}(V - R, p)$. Here $V - R$ has at most $p + 1$ digits, and the least significant d of them are zero, so if $d \ne 0$, we have $V' = V - R$, $U'' = U$. When

$U = uuuu\,uuuu00$ $V = \quad vvvvvvvv$ <hr> $W = wwwwwwww00$ (i)	$U = \quad uuuuuuuu00$ $V = \qquad vvvvvvvv$ <hr> $W = wwwwwwww000$ (ii)	$U = uuuuuuuu00$ $V = \qquad vvvvvvvv$ <hr> $W = \quad wwwwwwww0$ (iii)

Fig. 4. Possible alignments in floating-point addition.

$d = 0$ we must have $V' = V - R$ except when $V' = b^{p+d}$, and the latter case is an unusual situation in which W attains its maximum value $2b^{p+d}$; here $U'' = b^{p+d}$ and $b > 3$.

Case 3, $e_w < e_u$. (See Fig. 4(iii).) Here $V < 0$. *Case (3a), $d \le 1$.* Then $U + V = W$, so $U' = U'' = U$, $V' = V'' = V$. *Case (3b), $d > 1$.* Then $e_w = e_u - 1$, and $U + V = W + R$, where

$$R \equiv V \text{ (modulo } b^{d-1}), \qquad -\tfrac{1}{2}b^{d-1} \le R < \tfrac{1}{2}b^{d-1}.$$

This is similar to Case 1, but simpler because R has a smaller range. We have $U' = U$, $V' = V - R$, $U'' = U$, $V'' = V - R$. ∎

Theorem A exhibits a regularity property of floating-point addition, but it doesn't seem to be an especially useful result. The following theorem is much more significant:

Theorem B. *Under the hypotheses of Theorem A and (43),*

$$u + v = (u \oplus v) + ((u \ominus u') \oplus (v \ominus v'')). \tag{48}$$

Proof. By considering each case which arises in the proof of Theorem A, we invariably find that

$$u \ominus u' = u - u', \qquad v \ominus v' = v - v',$$
$$u \ominus u'' = u - u'', \qquad v \ominus v'' = v - v'',$$
$$((u \ominus u') \oplus (v \ominus v'')) = ((u - u') + (v - v''))$$
$$= ((u - u'') + (v - v'))$$
$$= ((u \ominus u'') \oplus (v \ominus v')),$$

since each of these quantities can be expressed exactly as a p-digit floating-point number with no rounding required. For example, in Case 2 we have $U - U' = R - Q \equiv 0$ (modulo b^d); and in every case $|R| < b^p$. The theorem now follows by combining the above identities with Theorem A. ∎

Theorem B gives *an explicit formula for the difference* between $u + v$ and $u \oplus v$, in terms of quantities that can be calculated directly using single precision floating-point arithmetic. Therefore it is often possible to increase the accuracy of single-precision floating-point calculations by accumulating the "correction terms" $(u \ominus u') \oplus (v \ominus v'')$.

A reader who has painstakingly checked through all details of the proof of Theorems A and B will realize the immense simplification that has been afforded by the simple rule

$$u \oplus v = \text{round}(u + v, p).$$

If our floating-point addition routine would fail to give this result even in a few rare cases, the whole proof would become enormously more complicated and perhaps it would even break down completely.

Theorem B would be false if we used truncation arithmetic instead of rounding, i.e., if we substituted $\mathrm{trunc}(x, p)$ for $\mathrm{round}(x, p)$ in Eqs. (11) through (15); this function $\mathrm{trunc}(x, p)$ is like $\mathrm{round}(x, p)$ when $x > 0$ except that the quantity $\frac{1}{2}$ which appears in (9) is replaced by zero. An exception to Theorem B would then occur for cases such as

$$(20, +.10000001) \oplus (10, -.10000001) = (20, +.10000000),$$

when the difference between $u + v$ and $u \oplus v$ could not be expressed exactly as a floating-point number. If truncation is made in a different manner, by employing truncation in the middle of Algorithm 4.2.1A indiscriminately without respect to the signs of the numbers, it appears that Theorems A and B would be valid, but the resulting $\oplus$ operation is much less tractable to mathematical analysis.

Many people feel that, since floating-point arithmetic is inexact by nature, there is no harm in making it just a little bit less exact in certain rather rare cases, if it is convenient to do so. This policy saves a few cents in the design of computer hardware, or a small percentage of the average running time of a subroutine. But the above discussion shows that such a policy is mistaken. Even though the speed of the FADD subroutine, Program 4.2.1A, could be increased by about five per cent if we took the liberty of rounding incorrectly in a few cases, we are much better off leaving it as it is. The reason is not to glorify "bit chasing," nor to give fantastically superior answers in the average program; a more important, fundamental issue is at stake here: *Numerical subroutines should deliver results which satisfy simple, useful mathematical laws whenever possible.* The crucial formula $u \oplus v = \mathrm{round}(u + v, p)$, for example, is a "regularity" property which makes a great deal of difference between whether mathematical analysis of computational algorithms is worth doing or worth avoiding! Without any underlying symmetry properties, the job of proving interesting results becomes extremely unpleasant. The enjoyment of the tools one works with is, of course, an essential ingredient of successful work.

B. Unnormalized floating-point arithmetic. The policy of normalizing all floating-point numbers may be construed in two ways: We may look on it favorably by saying it is an attempt to get the maximum possible accuracy obtainable with a given degree of precision, or we may consider it to be potentially dangerous in that it tends to imply the results are more accurate than they really are. When we normalize the result of $(1, +.31428571) \ominus (1, +.31415927)$ to $(-2, +.12644000)$, we are suppressing information about the possibly greater inaccuracy of the latter quantity. Such information would be retained if the answer were left as $(1, +.00012644)$.

The input data to a problem is frequently not known as precisely as the floating-point representation allows. For example, the values of Avogadro's number and Planck's constant are not known to eight significant digits, and it might be more appropriate to denote them, respectively, by

$$(27, +.00060225) \qquad \text{and} \qquad (-23, +.00010545)$$

instead of by $(24, +.60225000)$ and $(-26, +.10545000)$. It would be nice if we could give our input data for each problem in an unnormalized form which expresses how much precision is assumed, and if the output would indicate just how much precision is known in the answer. Unfortunately, this is a terribly difficult problem, although the use of unnormalized arithmetic can help to give some indication. For example, we can say with a fair degree of certainty that the product of Avogadro's number by Planck's constant is $(0, +.00063507)$, and their sum is $(27, +.00060225)$. (The purpose of this example is not to suggest that any important physical significance should be attached to the sum and product of these fundamental constants; the point is that it is possible to preserve a little of the information about precision in the result of calculation with imprecise quantities, when the original operands are independent of each other.)

The rules for unnormalized arithmetic are simply this: let l_u be the number of leading zeros in the fraction part of $u = (e_u, f_u)$, so that l_u is the largest integer $\leq p$ with $|f_u| < b^{-l_u}$. Then addition and subtraction are performed just as in Algorithm 4.2.1A, except that all scaling to the left is suppressed. Multiplication and division are performed as in Algorithm 4.2.1M, except that the answer is scaled right or left so that precisely max (l_u, l_v) leading zeros appear. Essentially the same rules have been used in manual calculation for many years.

It follows that, for unnormalized computations,

$$e_u \oplus v, \; e_u \ominus v = \text{max } (e_u, e_v) + (0 \text{ or } 1) \tag{49}$$

$$e_u \otimes v = e_u + e_v - q - \text{min } (l_u, l_v) - (0 \text{ or } 1) \tag{50}$$

$$e_u \oslash v = e_u - e_v + q - l_u + l_v + \text{max } (l_u, l_v) + (0 \text{ or } 1). \tag{51}$$

When the result of a calculation is zero, an unnormalized zero (often called an "order of magnitude zero") is given as the answer; this indicates that the answer may not truly be zero, we just don't know any of its significant digits!

When unnormalized floating-point arithmetic is used, the formulas (18) to (21) which we have used for error analysis are no longer valid. Instead, let us define

$$\delta_u = \tfrac{1}{2} b^{e_u - q - p} \qquad \text{if} \qquad u = (e_u, f_u). \tag{52}$$

This quantity depends on the representation of u, not just on the value $b^{e_u - q} f_u$.

We now have the inequalities

$$|u \oplus v - (u + v)| \le \delta_{u \oplus v}, \tag{53}$$

$$|u \ominus v - (u - v)| \le \delta_{u \ominus v}, \tag{54}$$

$$|u \otimes v - (u \times v)| \le \delta_{u \otimes v}, \tag{55}$$

$$|u \oslash v - (u / v)| \le \delta_{u \oslash v}, \tag{56}$$

in place of (18) through (21). These inequalities are simple consequences of the rounding rule, and they apply to normalized as well as unnormalized arithmetic; the main difference between the two types of error analysis is the definition of the exponent of the result of each operation (Eqs. (49) to (51)).

The relations $\prec$, $\sim$, $\succ$, and $\approx$ which we have defined above in (24), (25), (26), and (27) are valid and meaningful for unnormalized numbers as well as for normalized numbers. As an example of the use of these relations, let us prove an approximate associative law for unnormalized addition (analogous to (41)):

$$(u \oplus v) \oplus w \approx u \oplus (v \oplus w) \qquad (\epsilon), \tag{57}$$

for suitable ϵ. We have

$$
\begin{aligned}
|(u \oplus v) \oplus w - (u + v + w)| &\le |(u \oplus v) \oplus w - ((u \oplus v) + w)| \\
&\quad + |u \oplus v - (u + v)| \\
&\le \delta_{(u \oplus v) \oplus w} + \delta_{u \oplus v} \\
&\le 2\delta_{(u \oplus v) \oplus w}.
\end{aligned}
$$

A similar formula holds for $|u \oplus (v \oplus w) - (u + v + w)|$. Now since $e_{(u \oplus v) \oplus w} = \max(e_u, e_v, e_w) + (0, 1, \text{or } 2)$, we have $\delta_{(u \oplus v) \oplus w} \le b^2 \delta_{u \oplus (v \oplus w)}$. Therefore we find that (57) is valid when $\epsilon \ge 2b^{2-p}$; unnormalized addition is not as erratic as normalized addition with respect to the associative law.

It should be emphasized that unnormalized arithmetic is by no means a panacea; there are examples where it indicates greater accuracy than is present (e.g., adding up a great many small quantities of about the same magnitude, or finding the nth power of a number for large n), and there are many more examples when it indicates poor accuracy while normalized arithmetic does, in fact, produce fairly accurate results. There is an important reason why no straightforward one-operation-at-a-time method of error analysis is really reliable, namely the fact that operands are usually not independent of each other. This means that errors tend to cancel or reinforce each other in unusual ways. For example, suppose that x is approximately $\frac{1}{2}$, and suppose that we have an approximation $y = x + \delta$ with absolute error δ. If we now wish to compute $x(1 - x)$, we can form $y(1 - y)$; if $x = \frac{1}{2} + \epsilon$ we find $y(1 - y) = x(1 - x) - 2\epsilon\delta - \delta^2$: The error has decreased by a factor of 2ϵ. This is just one case where multiplication of imprecise quantities can lead to a quite accurate result when the operands are not independent of each other. A more obvious example is the computation of $x \ominus x$, which can be obtained with perfect accuracy regardless of how bad an approximation to x we begin with.

The extra information which unnormalized arithmetic gives us can often be more important than the information it destroys during an extended calculation, but (as usual) it is necessary to be careful when we are using it. Examples of the proper use of unnormalized arithmetic are discussed by R. L. Ashenhurst and N. Metropolis, in *Computers and Computing*, *AMM* Slaught Memorial Papers, **10** (February, 1965), 47–59; and by R. L. Ashenhurst, in *Error in Digital Computation*, vol. II, ed. by L. B. Rall (New York: Wiley, 1965), 3–37. Appropriate methods for computing standard mathematical functions with both input and output in unnormalized form are given by R. L. Ashenhurst in *JACM* **11** (1964), 168–187.

Another approach to the problem of error determination is the so-called "interval" or "range" arithmetic, in which upper and lower bounds on each number are maintained during the calculations. Thus, for example, if we know that $u_0 \leq u \leq u_1$ and $v_0 \leq v \leq v_1$, then $u_0 + v_0 \leq u + v \leq u_1 + v_1$, $u_0 - v_1 \leq u - v \leq u_1 - v_0$, and (assuming that u_0 and v_0 are positive) $u_0 v_0 \leq uv \leq u_1 v_1$, $u_0/v_1 \leq u/v \leq u_1/v_0$. Similar rules for the other cases are straightforward; a problem arises, of course, when $v_0 < 0 < v_1$ and we are trying to divide by v. The calculations on u_0, u_1, v_0, and v_1 are to be performed with suitable rounding (down for lower bounds, up for upper bounds). Such a calculation takes roughly only twice as long as the same calculation would take using ordinary arithmetic, and since it provides a rigorous error estimate, it can be quite valuable. However, due to the dependence of intermediate values upon each other, the final estimates are often too pessimistic; there are also some problems concerned with iterative numerical methods. For a discussion of interval arithmetic, and some modifications, see A. Gibb, *CACM* **4** (1961), 319–320; B. A. Chartres, *JACM* **13** (1966), 386–403; and the book *Interval Analysis* by Ramon E. Moore (Englewood Cliffs: Prentice Hall, 1966).

C. History and bibliography. The first analysis of floating-point arithmetic was given by F. L. Bauer and K. Samelson, "Optimale Rechengenauigkeit bei Rechenanlagen mit gleitendem Komma," *Zeitschrift für angewandte Math. und Physik* **4** (1953), 312–316. The next publication was not until over five years later: J. W. Carr III, "Error analysis in floating-point arithmetic," *CACM* **2** (May, 1959), 10–15. See also P. C. Fischer, *Proc. ACM 13th Nat. Meeting* (Urbana, Illinois, 1958), paper 39. The book *Rounding Errors in Algebraic Processes* (Englewood Cliffs: Prentice-Hall, 1963), by J. H. Wilkinson, shows how to apply error analysis of the individual arithmetic operations to the error analysis of large-scale problems; see also his treatise on *The Algebraic Eigenvalue Problem* (Oxford: Clarendon Press, 1965).

The relations $<$, $\sim$, $>$, $\approx$ introduced in this section are similar to ideas published by A. van Wijngaarden, "Numerical analysis as an independent science," *BIT* **6** (1966), 66–81. Theorems A and B above were inspired by some related work of Ole Møller, *BIT* **5** (1965), 37–50, 251–255; see also W. Kahan, *CACM* **8** (1965), 40.

Unnormalized floating-point arithmetic was recommended by F. L. Bauer and K. Samelson in the article cited above, and it was independently used by J. W. Carr III at the University of Michigan in 1953. Several years later, the MANIAC III computer was designed to include both kinds of arithmetic in its hardware; see R. L. Ashenhurst and N. Metropolis, *JACM* **6** (1959), 415–428; *IEEE Transactions on Electronic Computers* **EC-12** (1963), 896–901; R. L. Ashenhurst, *Proc. Spring Joint Computer Conf.* **21** (1962), 195–202. See also H. L. Gray and C. Harrison, Jr., *Proc. Eastern Joint Computer Conf.* **16** (1959), 244–248, and W. G. Wadey, *JACM* **7** (1960), 129–139, for further early discussions of unnormalized arithmetic.

EXERCISES

(In these problems, normalized floating-point arithmetic is assumed unless the contrary is specified.)

1. [*M18*] Prove that identity (7) is a consequence of (2) through (6).

2. [*M20*] Use identities (2) through (8) to prove that $(u \oplus x) \oplus (v \oplus y) \geq u \oplus v$ whenever $x \geq 0$ and $y \geq 0$.

3. [*M20*] Find eight-digit floating-decimal numbers u, v, and w such that

$$u \otimes (v \otimes w) \neq (u \otimes v) \otimes w,$$

and no exponent overflow or underflow occurs during any of these computations.

4. [*10*] Is it possible to have floating-point numbers u, v, and w for which exponent overflow occurs during the calculation of $u \otimes (v \otimes w)$ but not during the calculation of $(u \otimes v) \otimes w$?

5. [*M20*] Is $u \oslash v = u \otimes (1 \oslash v)$ an identity, for all floating-point numbers u and $v \neq 0$ such that no exponent overflow or underflow occurs?

6. [*M22*] Are either of the following two identities valid for all floating-point numbers u? (a) $0 \ominus (0 \ominus u) = u$. (b) $1 \oslash (1 \oslash u) = u$.

7. [*M20*] Given that $\delta_p(x)$ is defined by Eq. (16), prove the inequality (17).

▶ **8.** [*20*] Let $\epsilon = 0.0001$; which of the relations $u < v$ (ϵ), $u \sim v$ (ϵ), $u > v$ (ϵ), $u \approx v$ (ϵ) hold for the following pairs of base 10, excess 0, eight-digit floating-point numbers?

a) $u = (1, +.31415927)$, $v = (1, +.31416000)$;
b) $u = (0, +.99997000)$, $v = (1, +.10000039)$;
c) $u = (24, +.60225200)$, $v = (27, +.00060225)$;
d) $u = (24, +.60225200)$, $v = (31, +.00000006)$;
e) $u = (24, +.60225200)$, $v = (32, +.00000000)$.

9. [*M22*] Prove (36) and explain why the conclusion cannot be strengthened to $u \approx w$ $(\epsilon_1 + \epsilon_2)$.

▶ **10.** [*M25*] (W. Kahan.)   A certain computer performs floating-point arithmetic without proper rounding, and, in fact, its floating-point multiplication routine ignores all but the first p most significant digits of the $2p$-digit product $f_u f_v$. (Thus when $f_u f_v < 1/b$, the least-significant digit of $u \bigotimes v$ always comes out to be zero, due to subsequent normalization.) Show that this causes the monotonicity of multiplication to fail; i.e., there are positive normalized floating-point numbers u, v, w such that $u < v$ but $u \bigotimes w > v \bigotimes w$.

▶ **11.** [*M28*] Instead of using signed-magnitude notation for the fraction parts of a floating-binary number, we could use two's complement notation (see Section 4.1) as follows: The fraction part f of a positive number lies in the range $(0.100 \ldots 0)_2 = \frac{1}{2} \le f \le 1 - 2^{-p} = (0.111 \ldots 1)_2$, as presently, but the fraction part f of a *negative* number lies in the range $(1.000 \ldots 0)_2 = -1 \le f \le -\frac{1}{2} - 2^{-p} = (1.011 \ldots 1)_2$. Floating-point addition and subtraction may be done by using a straightforward generalization of Algorithm 4.2.1A: We may carry sufficient accuracy during the calculations to obtain a true sum or difference, then after normalizing the fraction so that its first p bits have the appropriate form, we "round" the result by adding one in the $(p + 1)$st bit and discarding all but the first p bits, renormalizing in case of rounding overflow. For example, $(2, 0.11111111) \ominus (6, 0.10000000)$ would be calculated first as $(6, 1.100011111111)$, normalized to $(5, 1.00011111111)$, and then rounded to $(5, 1.00100000)$. Taking the numbers in the opposite order, we find that

$$(6, 0.10000000) \ominus (2, 0.11111111) = (5, 0.111000000),$$

which is the negative of the preceding answer, so (7) holds in this case.

Find two's complement floating-binary numbers u and v for which (7) is *not* true, and for which no exponent overflow or underflow occurs during the calculation.

12. [*M15*] Why does (45) follow from (44)?

▶ **13.** [*M25*] Some programming languages (and even some computers) make use of floating-point arithmetic only, with no provision for exact calculations with integers. If operations on integers are desired, we can, of course, represent an integer as a floating-point number, and when the floating-point operations satisfy the basic definitions (11) to (14) of this section, we know that all floating-point operations will be exact, provided that the operands and the answer can each be represented exactly with p significant digits. Therefore so long as we know the numbers aren't too large, we can add, subtract, or multiply integers with no inaccuracy due to rounding errors.

But suppose that a programmer wants to determine if m is an exact multiple of n, when m and $n \ne 0$ are integers. Suppose further that a subroutine is available, as in exercise 4.2.1–15, which determines round $(u \bmod 1, p) = u \widehat{\bmod} 1$ for any floating-point quantity u. One good way to determine whether or not m is a multiple of n would perhaps be to test whether or not $((m \oslash n) \widehat{\bmod} 1) = 0$, using the assumed subroutine; but perhaps the rounding errors in the floating-point calculations will invalidate this test.

Find suitable conditions on the range of integer values $n \ne 0$ and m, such that m is a multiple of n if and only if $(m \oslash n) \widehat{\bmod} 1 = 0$. In other words, show that if m and n are not too large, this test is valid.

14. [*M27*] Find a suitable ϵ such that $(u \bigotimes v) \bigotimes w \approx u \bigotimes (v \bigotimes w) (\epsilon)$, when *unnormalized* multiplication is being used. (This generalizes (41), since unnormalized multiplication is exactly the same as normalized multiplication when the input operands u, v, and w are normalized.)

▶ **15.** [*M24*] (H. Björk.) Does the computed midpoint of an interval always lie between the endpoints? (In other words, does $u \leq v$ imply that $u \leq (u \oplus v) \oslash 2 \leq v$?)

16. [*HM23*] Assume that u and v are real numbers independently and uniformly distributed in the intervals $0 < u_0 - \delta \leq u \leq u_0 + \delta$, $0 < v_0 - \epsilon \leq v \leq v_0 + \epsilon$. (a) What is the mean value of uv? (b) What is the mean value of u/v? [The answer to this problem may have a bearing on the proper way to round the result of a multiplication or division operation.]

17. [*28*] Write a MIX subroutine, FCMP, which compares the floating-point number u in location ACC with the floating-point number v in register A, and which sets the comparison indicator to less, equal, or greater, according as $u < v$, $u \sim v$, or $u > v$ (ϵ); here ϵ is stored in location EPSILON as a nonnegative fixed-point quantity with the decimal point assumed at the left of the word.

18. [*M40*] In unnormalized arithmetic is there a suitable number ϵ such that

$$u \otimes (v \oplus w) \approx (u \otimes v) \oplus (u \otimes w) \qquad (\epsilon)?$$

*4.2.3. Double-Precision Calculations

Up to now we have considered "single-precision" floating-point arithmetic, which essentially means that the floating-point values we have dealt with can be stored in a single machine word. When single-precision floating-point arithmetic does not yield sufficient accuracy for a given application, the precision can be increased by suitable programming techniques which use two or more words of memory to represent each number. Although we shall discuss the general question of high-precision calculations in Section 4.3, it is appropriate to give a separate discussion of double-precision here; special techniques apply to double precision which are comparatively inappropriate for higher precisions, and double precision is a reasonably important topic in its own right, since it is the first step beyond single precision and is applicable to many problems which do not require extremely high precision.

Double-precision calculations are almost always required for floating-point, rather than fixed-point arithmetic, except perhaps in statistical work where fixed-point double-precision arithmetic is commonly used to calculate sums of squares and cross products; since fixed-point versions of double-precision arithmetic are simpler than floating-point versions, we will confine our discussion here to the latter.

Double precision is quite frequently desired not only to extend the precision of the fraction parts of floating-point numbers, but also to increase the range of the exponent part. Thus we shall deal in this section with the following two-word format for double-precision floating-point numbers in the MIX computer:

$$\boxed{\pm} \; \boxed{e} \; \boxed{e} \; \boxed{f} \; \boxed{f} \; \boxed{f} \qquad \boxed{} \; \boxed{} \; \boxed{f} \; \boxed{f} \; \boxed{f} \; \boxed{f} \; \boxed{f} \, . \qquad (1)$$

Here two bytes are used for the exponent and eight bytes are used for the fraction. The exponent is "excess $b^2/2$," where b is the byte size. The sign will appear in the most significant word; it is convenient to ignore the sign of the other word completely.

Another common alternative for extended-precision floating-point arithmetic is to use three computer words, one for the exponent and two for the fraction. Such an arrangement has been called "triple precision," although the techniques of this section apply, since the fraction part is only double precision.

Our discussion of double-precision arithmetic will be rather machine-oriented, because it is only by studying the problems involved in coding these routines that a person can properly appreciate the subject. A careful study of the MIX programs below is therefore essential to the understanding of this section.

It is customary to depart from the idealistic goals of accuracy stated in the preceding two sections and to refrain from rounding double-precision results. Although there is ample reason to squeeze out every possible drop of accuracy in the single-precision case, the situation is different with respect to double precision because of the following considerations: (a) The extra programming required to ensure true rounding in all cases is considerable, so that the routines would take, say, twice as much space and half again as much more time. Speed is a vital concern to most double-precision users. (b) Although the difference between seven- and eight-place accuracy can be noticeable, the difference between 15- and 16-place accuracy is of comparatively little importance. (c) Double-precision routines are not a "staple food" that everybody who wants to employ floating-point arithmetic must use; they are tools which do not have to be so delicately refined, since it is not necessary for a programmer to "enjoy" using them. For these practical reasons, we sacrifice the mathematical aesthetics of a "clean" arithmetical procedure such as we have been discussing in the previous sections, and we turn to a computer programmer's aesthetics of providing a compact and efficient (though perhaps somewhat unclean numerically) set of routines which provide substantially accurate results for the vast majority of cases. In these routines we do not round unless it is convenient to do so, and in fact, we sometimes permit ourselves to be a little bit wrong in the least-significant digit position. It is still possible to give rigorous bounds on the numerical accuracy of the results, but these bounds are not as tight as theoretically possible with the given amount of precision, and the routines do not satisfy many of the simple mathematical laws of the previous section.

Let us now consider addition and subtraction operations from this standpoint. Double-precision addition differs from the single-precision case in the following respects: The addition is performed by separately adding together the least-significant halves and the most-significant halves, taking care of carries appropriately. But since we are doing signed-magnitude arithmetic, it is possible to add the least-significant halves and to get the wrong sign (namely, when the signs of the operands are opposite, and the least-significant half of the smaller operand is bigger than the least-significant half of the larger operand);

so it is helpful to know what sign to expect. Therefore in step A2 (cf. Algorithm 4.2.1A), we not only assume that $e_u \geq e_v$, we also assume that $|u| \geq |v|$. This means we can be sure that the final sign will be the sign of u. In other respects, the double-precision addition is very much like the single-precision routine, only everything is done twice. A comparison of the following program with Program 4.2.1A will be instructive.

Program A (*Double-precision addition*). The subroutine DFADD adds a double-precision floating-point number v, having the form (1), to a double-precision floating-point number u, assuming that v is initially in rAX (i.e., registers A and X), and that u is initially stored in locations ACC and ACCX. The answer appears both in rAX and in (ACC, ACCX). The subroutine DFSUB subtracts v from u under the same conventions. Both input operands are assumed to be normalized, and the answer is normalized. The last portion of this program is a double-precision normalization procedure which is used by other subroutines of this section.

01	EXPD	EQU	1:2	Double-precision exponent field
02	DFSUB	STA	TEMP	Double-precision subtraction:
03		LDAN	TEMP	Change sign of v.
04	DFADD	STJ	EXITDF	Double-precision addition:
05		CMPA	ACC(1:5)	Compare $|u|$ with $|v|$.
06		JG	1F	
07		JL	2F	
08		CMPX	ACCX(1:5)	
09		JLE	2F	
10	1H	STA	ARG	If $|u| < |v|$, interchange $u \leftrightarrow v$.
11		STX	ARGX	
12		LDA	ACC	
13		LDX	ACCX	
14		ENT1	ACC	(ACC and ACCX are in consecutive
15		MOVE	ARG(2)	locations.)
16	2H	STA	TEMP	Now ACC has the sign of the answer.
17		LD1N	TEMP(EXPD)	$rI1 \leftarrow -e_v$.
18		LD2	ACC(EXPD)	$rI2 \leftarrow e_u$.
19		INC1	-8,2	$rI1 \leftarrow e_u - e_v - 8$.
20		J1NN	9F	If $e_u - e_v \geq 8$, ACC is the answer.
21		SLAX	2	Remove exponent.
22		SRAX	9,1	Scale right.
23		SLA	1	
24		SRAX	1	
25		STA	ARG	$0\ v_1\ v_2\ v_3\ v_4$
26		STX	ARGX	$0\ v_5\ v_6\ v_7\ v_8$
27		STA	ARGX(0:0)	Store true sign in both halves.
28		ENTA	1	
29		STA	EXPO	EXPO $\leftarrow 1$ (see below).

30		LDA	ACC	
31		LDX	ACCX	
32		SLAX	1	
33		SLA	1	
34		SRAX	1	
35		STA	ACC	$0\ u_1\ u_2\ u_3\ u_4$
36		STX	TRICK(2:5)	$1\ u_5\ u_6\ u_7\ u_8$
37		STA	TRICK(0:0)	(See comments in text.)
38		STA	1F(0:0)	
39		LDA	ARGX	
40		ADD	TRICK	Add least significant halves, with
41		SRAX	4	true sign of result.
42	1H	DECA	1	Recover from inserted 1 in TRICK.
43		ADD	ACC	Add most significant halves.
44		ADD	ARG	(Overflow cannot occur.)
45	DNORM	JANZ	1F	Normalization routine:
46		JXNZ	1F	f_w in rAX, $e_w = $ EXPO $+$ rI2.
47	DZERO	STZ	ACC	If $f_w = 0$, set $e_w \leftarrow 0$.
48		STZ	ACCX	
49		JMP	9F	
50	2H	SLAX	1	Normalize to left.
51		DEC2	1	
52	1H	CMPA	=0=(1:1)	
53		JE	2B	
54		SRAX	2	(Rounding omitted.)
55		STX	ACCX	
56		STA	ACC	
57		LDA	EXPO	Compute final exponent.
58		INCA	0,2	
59		JAN	EXPUND	Is it negative?
60		STA	ACC(EXPD)	
61		CMPA	=1(3:3)=	Is it more than two bytes?
62		JL	8F	
63	EXPOVD	HLT	20	
64	EXPUND	HLT	10	
65	9H	LDX	ACCX	Bring answer into accumulator.
66	8H	LDA	ACC	
67	EXITDF	JMP	*	Exit from subroutine.
68	TRICK	CON	1(1:1)	
69	ARG	CON	0	
70	ARGX	CON	0	
71	ACC	CON	0	Floating-point accumulator
72	ACCX	CON	0	
73	EXPO	CON	0	Part of "raw exponent" ▮

When the least-significant halves are added together in this program, an extra digit "1" is inserted at the left of the word which is known to have the correct sign. After the addition, this byte can be 0, 1, or 2, depending on the circumstances, and all three cases are handled simultaneously in this way. (Compare this with the rather cumbersome method of complementation which is used in Program 4.2.1A.)

It is worth noting that register A can be zero after the instruction on line 44 has been performed; and, because of the way MIX defines the sign of a zero result, the accumulator contains the correct sign which is to be attached to the result if register X is nonzero. If lines 43 and 44 were interchanged, the program would be incorrect (although both instructions are "ADD")!

·Now let us consider double-precision multiplication. The product has four components:

$$
\begin{array}{llllllllll}
u & u & u & u & u & u & u & u & 0 & 0 & \quad = u_m + \epsilon u_l \\
v & v & v & v & v & v & v & v & 0 & 0 & \quad = v_m + \epsilon v_l \\
\hline
x & x & x & x & x & x & 0 & 0 & 0 & 0 & \quad \epsilon^2 u_l \times v_l \\
\end{array}
$$

$$
\begin{array}{l}
x\ x\ x\ x\ |\ x \quad x\ x\ x\ 0\ 0 \qquad \epsilon u_m \times v_l \qquad (2) \\
x\ x\ x\ x\ |\ x \quad x\ x\ x\ 0\ 0 \qquad \epsilon u_l \times v_m \\
x\ x\ x\ x\ x \quad x\ x\ x\ x\ |\ x \qquad u_m \times v_m \\
\hline
w\ w\ w\ w\ w \quad w\ w\ w\ w\ |\ w \quad w\ w\ w\ w\ w \quad w\ 0\ 0\ 0\ 0.
\end{array}
$$

Since we need only the leftmost eight bytes, it is convenient to work only with the digits to the left of the vertical line in (2), and this means in particular that we need not even compute the product of the two least significant halves.

Program M (*Double-precision multiplication*). The input and output conventions for this subroutine are the same as for Program A.

01	ABS	EQU	1:5	Field definition for absolute value
02	BYTE	EQU	1(4:4)	Byte size
03	QQ	EQU	BYTE*BYTE/2	Excess of double precision exponent
04	DFMUL	STJ	EXITDF	Double precision multiplication:
05		STA	TEMP	
06		SLAX	2	Remove exponent.
07		STA	ARG	v_m
08		STX	ARGX	v_l
09		LDA	TEMP(EXPD)	
10		ADD	ACC(EXPD)	
11		STA	EXPO	EXPO $\leftarrow e_u + e_v$.
12		ENT2	-QQ	rI2 $\leftarrow$ -QQ.
13		LDA	ACC	
14		LDX	ACCX	
15		SLAX	2	Remove exponent.
16		STA	ACC	u_m
17		STX	ACCX	u_l
18		MUL	ARGX	$u_m \times v_l$
19		SRA	1	
20		STA	TEMP	$0\ x\ x\ x\ x$

21	LDA	ARG(ABS)	
22	MUL	ACCX(ABS)	$v_m \times u_l$
23	SRA	1	
24	ADD	TEMP(ABS)	(No overflow possible)
25	STA	TEMP	
26	LDA	ARG	
27	MUL	ACC	$v_m \times u_m$
28	STA	TEMP(0:0)	True sign of result.
29	STA	ACC	Now prepare to add all the
30	STX	ACCX	partial products together.
31	LDA	ACCX(0:4)	$0\ x\ x\ x\ x$
32	ADD	TEMP	(No overflow possible)
33	SRAX	4	
34	ADD	ACC	(No overflow possible)
35	JMP	DNORM	Normalize and exit. ▌

Note the careful treatment of signs in this program, and note also the fact that the range of exponents makes it impossible to compute the final exponent using an index register. Program M is perhaps a little too slipshod in accuracy, since it throws away all the information to the right of the vertical line in (2) and this can make the least significant byte as much as 2 in error. A little more accuracy can be achieved as discussed in exercise 4.

Double-precision floating division is the most difficult routine, or at least the most frightening prospect we have encountered so far in this chapter. Actually, it is not terribly complicated; let us write the numbers to be divided in the form

$$\frac{u_m + \epsilon u_l}{v_m + \epsilon v_l},$$

where ϵ is the reciprocal of the word size of the computer, and where v_m is assumed to be normalized. The fraction can now be expanded as follows:

$$\frac{u_m + \epsilon u_l}{v_m + \epsilon v_l} = \frac{u_m + \epsilon u_l}{v_m} \left(\frac{1}{1 + \epsilon(v_l/v_m)} \right)$$

$$= \frac{u_m + \epsilon u_l}{v_m} \left(1 - \epsilon \left(\frac{v_l}{v_m} \right) + \epsilon^2 \left(\frac{v_l}{v_m} \right)^2 - \cdots \right). \qquad (3)$$

Since $0 \leq |v_l| < 1$ and $1/b \leq |v_m| < 1$, we have $|v_l/v_m| < b$, and the error from dropping terms in ϵ^2 can be disregarded. Our method therefore is to compute $w_m + \epsilon w_l = (u_m + \epsilon u_l)/v_m$, and then to subtract ϵ times $w_m v_l/v_m$ from the result.

In the following program, lines 27–32 do the lower half of a double-precision addition, using another method for forcing the appropriate sign as an alternative to the "TRICK" of Program A.

Program D (*Double-precision division*). This program adheres to the same conventions as Programs A and M.

01	DFDIV	STJ	EXITDF	Double precision division:
02		JOV	OFLO	Ensure overflow is off.
03		STA	TEMP	
04		SLAX	2	Remove exponent.
05		STA	ARG	v_m
06		STX	ARGX	v_l
07		LDA	ACC(EXPD)	
08		SUB	TEMP(EXPD)	
09		STA	EXPO	EXPO $\leftarrow e_u - e_v$.
10		ENT2	QQ+1	rI2 $\leftarrow$ QQ $+$ 1.
11		LDA	ACC	
12		LDX	ACCX	
13		SLAX	2	Remove exponent.
14		SRAX	1	(Cf. Algorithm 4.2.1M)
15		DIV	ARG	If overflow, it is detected below.
16		STA	ACC	w_m
17		SLAX	5	Use remainder in further division.
18		DIV	ARG	
19		STA	ACCX	$\pm\, w_l$
20		LDA	ARGX(1:4)	
21		ENTX	0	
22		DIV	ARG(ABS)	rA $\leftarrow \lfloor\lfloor b^4 v_l / v_m\rfloor\rfloor / b^5$
23		JOV	DVZROD	Did division cause overflow?
24		MUL	ACC(ABS)	$\lvert w_m v_l / b v_m\rvert$
25		SRAX	4	Multiply by b, and save
26		SLC	5	leading byte in rX.
27		SUB	ACCX(ABS)	Subtract $\lvert w_l\rvert$.
28		DECA	1	Force minus sign.
29		SUB	WM1	
30		JOV	*+2	If no overflow, carry one more
31		INCX	1	to upper half.
32		SLC	5	(Now rA $\le$ 0)
33		ADD	ACC(ABS)	Subtract $\lvert$rA$\rvert$ from $\lvert w_m\rvert$.
34		STA	ACC(ABS)	(Here rA $\ge$ 0)
35		LDA	ACC	w_m with correct sign
36		JMP	DNORM	Normalize and exit.
37	DVZROD	HLT	30	Unnormalized or zero divisor
38	1H	EQU	1(1:1)	
39	WM1	CON	1B−1,BYTE−1(1:1)	Word size minus one █

Here are the approximate average computation times for these double-precision subroutines, compared to the single-precision subroutines which appear in Section 4.2.1:

	Single precision	Double precision
Addition	46	97
Subtraction	50	101
Multiplication	47	112
Division	50.5	127.5

For extension of the methods of this section to triple-precision floating point fraction parts, see Y. Ikebe, *CACM* **8** (1965), 175–177.

EXERCISES

1. [*16*] Try the double-precision division technique by hand, with $\epsilon = \frac{1}{1000}$, when dividing 180000 by 314159. (Thus, let $(u_m, u_l) = (.180, .000)$ and $(v_m, v_l) = (.314, .159)$, and find the quotient using the method suggested in the text following (3).)

2. [*20*] Would it be a good idea to insert the instruction "ENTX 0" between lines 40 and 41 of Program A, and/or between lines 32 and 33 of Program M, in order to keep unwanted information left over in register X from interfering with the accuracy of the results?

3. [*M20*] Explain why overflow cannot occur during Program M.

4. [*22*] How should Program M be changed so that extra accuracy is achieved, essentially by moving the vertical line in (2) over to the right one position? Specify all changes that are required, and determine the difference in execution time caused by these changes.

▶ **5.** [*24*] How should Program A be changed so that extra accuracy is achieved, essentially by working with a nine-byte accumulator instead of an eight-byte accumulator to the right of the decimal point? Specify all changes that are required, and determine the difference in execution time caused by these changes.

6. [*23*] Assume that the double-precision subroutines of this section and the single-precision subroutines of Section 4.2.1 are being used in the same main program. Write a subroutine which converts a single-precision floating-point number into double-precision form (1), and write another subroutine which converts a double-precision floating-point number into single-precision form (or which reports exponent overflow or underflow if the conversion is impossible).

▶ **7.** [*M30*] Estimate the accuracy of the double-precision subroutines in this section, by finding suitable constants δ_1, δ_2, and δ_3 such that the relative errors

$$|((u \oplus v) - (u + v))/(u + v)|, \qquad |((u \otimes v) - (u \times v))/(u \times v)|,$$
$$|((u \oslash v) - (u/v))/(u/v)|$$

are respectively bounded by δ_1, δ_2, and δ_3.

8. [*M28*] Estimate the accuracy of the "improved" double-precision subroutines of exercises 4 and 5, in the sense of exercise 7.

4.2.4. Statistical Distribution

In order to analyze the average behavior of floating-point arithmetic algorithms (and in particular to determine their average running time), we need some statistical information that allows us to determine how often various cases arise. The purpose of this section is to discuss the empirical and theoretical properties of the distribution of floating-point numbers.

A. Addition and subtraction routines. The execution time for a floating-point addition or subtraction depends largely on the initial difference of exponents, and also on the number of normalization steps required (to the left or to the right). No way is known to give a good theoretical model which tells what characteristics to expect, but extensive empirical investigations have been made by D. W. Sweeney [*IBM Systems J.* 4 (1965), 31–42].

By means of a special tracing routine, Sweeney ran six "typical" large-scale numerical programs, selected from several different computing laboratories, and examined each floating addition or subtraction operation very carefully. Over one million floating-point addition-subtractions were involved in gathering this data. About one out of every nine instructions executed by the tested programs was either FADD or FSUB.

Let us consider subtraction to be addition preceded by negating the second operand; therefore we may give all the statistics as if we were merely doing addition. Sweeney's results can be summarized as follows:

One of the two operands to be added was found to be equal to zero about 8 percent of the time, and this was usually the accumulator (ACC). The other 92 percent of the cases split about equally between operands of the same or of opposite signs, and about equally between cases where $|u| \leq |v|$ or $|v| \leq |u|$. The computed answer was zero about 1.5 percent of the time.

After interchanging operands if necessary so that $e_u \geq e_v$, the difference between exponents had a behavior approximately as the probabilities given in this table, for various radices b:

$e_u - e_v$	$b = 2$	$b = 10$	$b = 16$	$b = 64$
0	0.33	0.47	0.47	0.56
1	0.12	0.23	0.26	0.27
2	0.09	0.11	0.11	0.04
3	0.07	0.03	0.02	0.02
4	0.07	0.01	0.01	0.01
5	0.04	0.01	0.01	0.01
over 5	0.28	0.14	0.12	0.09

(The last line includes essentially all of the cases when $v = 0$; and, as we have seen, this amounts to 8 percent of the total.)

When u and v have the same sign and are normalized, then $u + v$ either requires one shift to the *right* (for fraction overflow), or no normalization shifts whatever. When u and v have opposite signs, we have zero or more *left* shifts during the normalization. The following table gives the observed number of shifts required:

	$b = 2$	$b = 10$	$b = 16$	$b = 64$
Shift right 1	0.20	0.07	0.05	0.02
No shift	0.60	0.80	0.82	0.87
Shift left 1	0.07	0.08	0.07	0.06
Shift left 2	0.03	0.01	0.01	0.01
Shift left 3	0.02	0.00	0.01	0.01
Shift left 4	0.02	0.01	0.01	0.01
Shift left >4	0.06	0.03	0.03	0.02

The last line includes all cases where the result was zero. The average number of left shifts, not counting the cases where fraction overflow occurred, was about 0.9 when $b = 2$; 0.2 when $b = 10$ or 16; and 0.1 when $b = 64$.

B. The fraction parts. Further analysis of floating-point routines can be based on the *statistical distribution of the fraction parts* of randomly chosen normalized floating-point numbers. In this case the facts are quite surprising, and there is an interesting theory which accounts for the unusual phenomena which are observed.

For convenience let us temporarily assume we are dealing with floating-decimal (i.e., base 10) arithmetic; modifications of the following discussion to any other positive integer base b will be very straightforward. Let us assume we are given a "random" positive normalized number $(e, f) = 10^e \cdot f$. Since f is normalized, we know that its leading digit is 1, 2, 3, 4, 5, 6, 7, 8, or 9, and it seems natural to assume that each of these leading digits will occur about one-ninth of the time. But, in fact, the behavior in practice is quite different. For example, the leading digit tends to be equal to 1 over 30 percent of the time!

One way to test the assertion just made is to take a table of physical constants (e.g., the speed of light, the force of gravity) from some standard reference. If we look at the *Handbook of Mathematical Functions* (U.S. Dept. of Commerce, 1964), for example, we find that 8 of the 28 different physical constants given in Table 2.3, roughly 29 percent, have leading digit equal to 1. We might also try looking at census reports, or a Farmer's Almanack, etc.

The pages in well-used tables of logarithms tend to get quite dirty in the front, while the last pages stay relatively clean and neat. This phenomenon was apparently first mentioned by the American astronomer Simon Newcomb [*Amer. J. Math.* **4** (1881), 39–40], who gave good grounds for believing that the leading digit d occurs with probability $\log_{10}(1 + 1/d)$. The same distribution was discovered empirically, many years later, by F. Benford [*Proc. Amer.*

Philosophical Soc. **78** (1938), 551], who was unaware of Newcomb's earlier note. We shall see that this leading-digit law is a natural consequence of the way we write numbers in floating-point notation.

If we take any positive number u, its leading digits are determined by the value $(\log_{10} u) \bmod 1$: The leading digit is less than d if and only if

$$(\log_{10} u) \bmod 1 < \log_{10} d, \tag{1}$$

since $10 f_u = 10^{(\log_{10} u) \bmod 1}$.

Now if we have any "random" positive number U, chosen from some reasonable distribution such as might occur in nature, we might expect that $(\log_{10} U) \bmod 1$ would be uniformly distributed between zero and one, at least to a very good approximation. (Similarly, we expect $U \bmod 1$, $U^2 \bmod 1$, $\sqrt{U + \pi} \bmod 1$, etc., to be uniformly distributed. We expect a roulette wheel to be unbiased, for essentially the same reason.) Therefore by (1) the leading digit will be 1 with probability $\log_{10} 2 \approx 30.103$ percent; it will be 2 with probability $\log_{10} 3 - \log_{10} 2 \approx 17.609$ percent; and, in general, if r is any real value between 1 and 10, we ought to have $10 f_U \leq r$ approximately $\log_{10} r$ of the time.

Another way to explain this law is to say that a random value U should appear at a random point on a slide rule (that is, any position on a slide rule is equally probable). For the distance from the left end of a slide rule to the position of U is proportional to $(\log_{10} U) \bmod 1$. The analogy between slide rules and floating-point calculation is very close when multiplication and division are being considered.

The fact that leading digits tend to be small is important to keep in mind; it makes the most obvious techniques of "average error" estimation for floating-point calculations invalid. The relative error is usually a little more than expected.

Of course, it may justly be said that the heuristic argument above does not prove the stated law. It merely shows us a plausible reason why the leading digits behave the way they do. Another approach to the analysis of leading digits has been suggested by R. S. Pinkham and R. Hamming [*Ann. Math. Stat.* **32** (1961), 1223–1230]: Let $p(r)$ be the probability that $10 f_U \leq r$, where $1 \leq r \leq 10$ and f_U is the normalized fraction part of a random normalized floating-point number U. If we think of random quantities in the real world, we observe that they are measured in terms of arbitrary units; and if we were to change the definition of a meter or a gram, many of the fundamental physical constants would have different values. Suppose then that all of the numbers in the universe are suddenly multiplied by a constant factor c; our universe of random floating-point quantities should be essentially unchanged by this transformation, so $p(r)$ should not be affected.

Multiplying everything by c changes $(\log_{10} U) \bmod 1$ to $(\log_{10} U + \log_{10} c) \bmod 1$. It is now time to set up the formulas which describe the desired behavior; we may assume that $1 \le c \le 10$. By definition,

$$p(r) = \text{probability } ((\log_{10} U) \bmod 1 \le \log_{10} r).$$

By our assumption, we should also have

$$p(r) = \text{probability } ((\log_{10} U + \log_{10} c) \bmod 1 \le \log_{10} r)$$

$$= \begin{cases} \text{probability } ((\log_{10} U) \bmod 1 \le \log_{10} r - \log_{10} c \\ \qquad\text{or}\qquad (\log_{10} U) \bmod 1 \ge 1 - \log_{10} c), \qquad \text{if} \qquad c \le r; \\ \text{probability } (1 - \log_{10} c \le (\log_{10} U) \bmod 1 \le 1 + \log_{10} r - \log_{10} c), \\ \qquad\qquad\qquad\qquad\qquad\qquad\qquad\qquad\qquad \text{if} \qquad c \ge r; \end{cases}$$

$$= \begin{cases} p(r/c) + 1 - p(10/c), & \text{if} \quad c \le r; \\ p(10r/c) - p(10/c), & \text{if} \quad c \ge r. \end{cases} \tag{2}$$

Let us now extend the function $p(r)$ to values outside the range $1 \le r \le 10$, by defining $p(10^n r) = p(r) + n$; then if we replace $10/c$ by d, the last equation of (2) may be written

$$p(rd) = p(r) + p(d). \tag{3}$$

If our assumption about invariance of the distribution under multiplication by a constant factor is valid, then Eq. (3) must hold for all $r > 0$ and $1 \le d \le 10$. The facts that $p(1) = 0$, $p(10) = 1$ now imply that

$$1 = p(10) = p((\sqrt[n]{10})^n) = p(\sqrt[n]{10}) + p((\sqrt[n]{10})^{n-1}) = \cdots = np(\sqrt[n]{10});$$

hence we deduce that $p(10^{m/n}) = m/n$ for all positive integers m and n. If we now decide to require that p is continuous, we are forced to conclude that $p(r) = \log_{10} r$, and this is the desired law.

   Although this argument may be more convincing than the first one, it doesn't really hold up under scrutiny. We are assuming that there is some underlying distribution of numbers $F(u)$ such that a given positive number U is $\le u$ with probability $F(u)$; and that

$$p(r) = \sum_m (F(10^m r) - F(10^m)) \tag{4}$$

summed over all values $-\infty < m < \infty$. Our argument concluded that

$$p(r) = \log_{10} r.$$

Using the same argument, we can "prove" that

$$\sum_m (F(b^m r) - F(b^m)) = \log_b r, \tag{5}$$

for each integer $b \geq 2$, when $1 \leq r \leq b$. But there *is* no distribution function F which satisfies this equation for all such b and r!

One way out of the difficulty is to regard the logarithm law $p(r) = \log_{10} r$ as only a very close *approximation* to the true distribution. The true distribution itself may perhaps be changing as the universe expands, becoming a better and better approximation as time goes on; and if we replace 10 by an arbitrary base b, the approximation appears to be less accurate (at any given time) as b gets larger. Another rather appealing way to resolve the dilemma, by abandoning the traditional idea of a distribution function, has been suggested by R. A. Raimi, *AMM* **76** (1969), 342–348.

The hedging in the last paragraph is probably a very unsatisfactory explanation, and so the following further calculation (which sticks to rigorous mathematics and avoids any intuitive, yet paradoxical, notions of probability) should be welcome: Let us consider the distribution of the leading digits of the *positive integers*, instead of the distribution for some imagined set of real numbers. The investigation of this topic is quite interesting, not only because it sheds some light on the probability distributions of floating-point data, but also because it makes a very instructive example of how to combine the methods of discrete mathematics with the methods of infinitesimal calculus.

In the following discussion, let r be a fixed real number, $1 \leq r \leq 10$; we will attempt to make a reasonable attempt at defining $p(r)$, the "probability" that the representation $10^{e_N} \cdot f_N$ of a "random" positive integer N has $10 f_N < r$.

To start, let us try to find the probability using a limiting method like the definition of "Pr" in Section 3.5. One nice way to rephrase that definition is to define

$$P_0(n) = \begin{cases} 1, & \text{if } n = 10^e \cdot f \text{ where } 10f < r; \\ & \qquad \text{i.e., if } (\log_{10} n) \bmod 1 < \log_{10} r; \quad (6) \\ 0, & \text{otherwise.} \end{cases}$$

Now $P_0(1), P_0(2), \ldots$ is an infinite sequence of zeros and ones, with ones to represent the cases which contribute to the probability. We can try to "average out" this sequence, by defining

$$P_1(n) = \frac{1}{n} \sum_{1 \leq k \leq n} P_0(k). \qquad (7)$$

It is natural to let $\lim_{n \to \infty} P_1(n)$ be the "probability" $p(r)$ we are after, and that is just what we did in Section 3.5.

But in this case the limit does not exist: For example, let us consider the subsequence

$$P_1(s), P_1(10s), P_1(100s), \ldots, P_1(10^n s), \ldots,$$

where s is a real number, $1 \leq s \leq 10$. If $s \leq r$, we find that

$P_1(10^n s)$

$$= \frac{1}{10^n s} (\lceil r \rceil - 1 + \lceil 10r \rceil - 10 + \cdots + \lceil 10^{n-1}r \rceil - 10^{n-1} + \lfloor 10^n s \rfloor$$
$$+ 1 - 10^n)$$

$$= \frac{1}{10^n s} (r(1 + 10 + \cdots + 10^{n-1}) + O(n) + \lfloor 10^n s \rfloor - 1 - 10 - \cdots - 10^n)$$

$$= \frac{1}{10^n s} (\tfrac{1}{9}(10^n r - 10^{n+1}) + \lfloor 10^n s \rfloor + O(n)), \tag{8}$$

where $r = r_0 . r_1 r_2 \ldots$ in decimal notation. As $n \to \infty$, $P_1(10^n s)$ therefore approaches the limiting value $1 + (r - 10)/9s$. The above calculation for the case $s \le r$ can be modified so that it is valid for $s > r$ if we replace $\lfloor 10^n s \rfloor$ $+ 1$ by $\lceil 10^n r \rceil$; so when $s \ge r$, we find that the limiting value is $10(r - 1)/9s$. [See J. Franel, *Naturforschende Gesellschaft, Vierteljahrsschrift* **62** (Zürich, 1917), 286–295.]

Therefore $P_1(n)$ has subsequences whose limit goes from $(r - 1)/9$ up to $10(r - 1)/9r$ and down again to $(r - 1)/9$, as s goes from 1 to r to 10. We see that $P_1(n)$ has no limit, and $P_1(n)$ is not a particularly good approximation to our conjectured answer $\log_{10} r$ either!

Since $P_1(n)$ doesn't approach a limit, we can try to use the same idea as (7) once again, to "average out" this anomalous behavior. In general, let

$$P_{m+1}(n) = \frac{1}{n} \sum_{1 \le k \le n} P_m(k). \tag{9}$$

Then $P_{m+1}(n)$ will tend to be a more well-behaved sequence than $P_m(n)$. Let us try to examine the behavior of $P_{m+1}(n)$, for large n. Our experience above with the special case $m = 0$ indicates that it might be worth while to consider the subsequence $P_{m+1}(10^n s)$; and the following results can now be derived:

Lemma Q. *For any integer $m \ge 1$ and any real number $\epsilon > 0$, there are functions $Q_m(s)$, $R_m(s)$ and an integer $N_m(\epsilon)$, such that whenever $n > N_m(\epsilon)$ and $1 \le s \le 10$, we have*

$$|P_m(10^n s) - Q_m(s)| < \epsilon, \qquad \text{if} \qquad s \le r;$$
$$|P_m(10^n s) - (Q_m(s) + R_m(s))| < \epsilon, \qquad \text{if} \qquad s > r. \tag{10}$$

Furthermore the functions $Q_m(s)$, $R_m(s)$ satisfy the relations

$$Q_m(s) = \frac{1}{s} \left(\frac{1}{9} \int_1^{10} Q_{m-1}(t) \, dt + \int_1^s Q_{m-1}(t) \, dt + \frac{1}{9} \int_r^{10} R_{m-1}(t) \, dt \right);$$

$$R_m(s) = \frac{1}{s} \int_r^s R_{m-1}(t) \, dt; \tag{11}$$

$$Q_0(s) = 1, \qquad R_0(s) = -1.$$

Proof. Consider the functions $Q_m(s)$, $R_m(s)$ defined by (11), and let

$$S_m(t) = \begin{cases} Q_m(t), & t \leq r, \\ Q_m(t) + R_m(t), & t > r. \end{cases} \tag{12}$$

We will prove the lemma by induction on m.

First, let $m = 1$; then $Q_1(s) = \dfrac{1}{s}(1 + (s - 1) + (r - 10)/9) = 1 + (r - 10)/9s$, and $R_1(s) = (r - s)/s$. From (8) we find that

$$|P_1(10^n s) - S_1(s)| = O(n)/10^n;$$

this establishes the lemma when $m = 1$.

Now for $m > 1$, we have

$$P_m(10^n s) = \frac{1}{s}\left(\sum_{0 \leq j < n} \frac{1}{10^{n-j}} \sum_{10^j \leq k < 10^{j+1}} \frac{1}{10^j} P_{m-1}(k) \right.$$

$$\left. + \sum_{10^n \leq k \leq 10^n s} \frac{1}{10^n} P_{m-1}(k) \right),$$

and we want to approximate this quantity. The difference

$$\left| \sum_{10^j \leq k \leq 10^j q} \frac{1}{10^j} P_{m-1}(k) - \sum_{10^j \leq k \leq 10^j q} \frac{1}{10^j} S_{m-1}\left(\frac{k}{10^j} \right) \right| \tag{13}$$

is less than $(q - 1)\epsilon$ when $1 \leq q \leq 10$ and $j > N_{m-1}(\epsilon)$; and since $S_{m-1}(t)$ is continuous, it is a Riemann-integrable function, and the difference

$$\left| \sum_{10^j \leq k \leq 10^j q} \frac{1}{10^j} S_{m-1}\left(\frac{k}{10^j} \right) - \int_1^q S_{m-1}(t)\, dt \right| \tag{14}$$

is less than ϵ for all j greater than some number N, independent of q, by the definition of integration. We may choose N to be $> N_{m-1}(\epsilon)$. Therefore for $n > N$, the difference

$$\left| P_m(10^n s) - \frac{1}{s}\left(\sum_{0 \leq j < n} \frac{1}{10^{n-j}} \int_1^{10} S_{m-1}(t)\, dt + \int_1^s S_{m-1}(t)\, dt \right) \right| \tag{15}$$

is bounded by

$$\sum_{0 \leq j \leq N} \frac{M}{10^{n-j}} + \sum_{N < j < n} \frac{10\epsilon}{10^{n-j}} + 10\epsilon,$$

if M is an upper bound for (13) + (14) which is valid for all j. Finally, the sum

$\sum_{0 \leq j < n} (1/10^{n-j})$, which appears in (15), is equal to $(1 - 1/10^n)/9$; so

$$\left| P_m(10^n s) - \frac{1}{s} \left(\frac{1}{9} \int_1^{10} S_{m-1}(t) \, dt + \int_1^s S_{m-1}(t) \, dt \right) \right|$$

can be made smaller than $\frac{10}{9}(10\epsilon)$ if n is taken large enough. Comparing this with (10) and (11) completes the proof. ∎

The gist of Lemma Q is that we know

$$\lim_{n \to \infty} P_m(10^n s) = S_m(s). \tag{16}$$

Also, since $S_m(s)$ is not constant as s varies, the limit

$$\lim_{n \to \infty} P_m(n),$$

which would be our desired "probability," does not exist for any m. The situation is shown in Fig. 5, which shows the values of $S_m(s)$ when m is small and $r = 2$.

Even though $S_m(s)$ is not a constant, so that we do not have a definite limit for $P_m(n)$, note that already for $m = 3$ in Fig. 5 the value of $S_m(s)$ stays very close to $\log_{10} 2 = 0.30103 \ldots$. Therefore we have good reason to suspect that $S_m(s)$ is very close to $\log_{10} r$ for all large m, and, in fact, that the sequence of functions $\langle S_m(s) \rangle$ converges uniformly to the constant function $\log_{10} r$.

It is interesting to prove this conjecture by explicitly calculating $Q_m(s)$ and $R_m(s)$ for all m, as in the proof of the following theorem:

Theorem F. *Given $\epsilon > 0$, there exists a number N such that*

$$|P_m(n) - \log_{10} r| < \epsilon \tag{17}$$

whenever $m, n > N$.

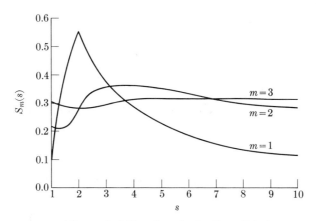

Fig. 5. The probability that the leading digit is 1.

Proof. In view of Lemma Q, we can prove this result if we can show there is a number M depending on ϵ such that, for $1 \le s \le 10$,

$$|Q_m(s) - \log_{10} r| < \epsilon \qquad \text{and} \qquad |R_m(s)| < \epsilon \qquad (18)$$

for all $m > M$.

It is not difficult to solve the recurrence formula (11) for R_m: We have $R_0(s) = -1$, $R_1(s) = -1 + r/s$, $R_2(s) = -1 + (r/s)(1 + \ln(s/r))$, and in general

$$R_m(s) = -1 + \frac{r}{s}\left(1 + \frac{1}{1!}\ln\left(\frac{s}{r}\right) + \frac{1}{2!}\left(\ln\left(\frac{s}{r}\right)\right)^2 + \cdots\right.$$
$$\left. + \frac{1}{(m-1)!}\left(\ln\left(\frac{s}{r}\right)\right)^{m-1}\right). \qquad (19)$$

For the stated range of s, this converges uniformly to

$$-1 + (r/s)\exp\left(\ln(s/r)\right) = 0.$$

The recurrence (11) for Q_m takes the form

$$Q_m(s) = \frac{1}{s}\left(c_m + 1 + \int_1^s Q_{m-1}(t)\,dt\right), \qquad (20)$$

where

$$c_m = \frac{1}{9}\left(\int_1^{10} Q_{m-1}(t)\,dt + \int_r^{10} R_{m-1}(t)\,dt\right) - 1. \qquad (21)$$

The solution to the recurrence (20) is, once again, easily found by trying out the first few cases and guessing at a formula which can be proved by induction; we find that

$$Q_m(s) = 1 + \frac{1}{s}\left(c_m + \frac{1}{1!}c_{m-1}\ln s + \frac{1}{2!}(\ln s)^2 + \cdots\right.$$
$$\left. + \frac{1}{(m-1)!}(\ln s)^{m-1}\right). \qquad (22)$$

It remains for us to calculate the coefficients c_m, which by (19), (21), and (22) satisfy the relations

$$c_1 = (r - 10)/9$$

$$c_{m+1} = \frac{1}{9}\left(c_m \ln 10 + \frac{1}{2!}c_{m-1}(\ln 10)^2 + \cdots + \frac{1}{m!}c_1(\ln 10)^m\right.$$
$$\left. + r\left(1 + \frac{1}{1!}\ln\frac{10}{r} + \cdots + \frac{1}{m!}\left(\ln\frac{10}{r}\right)^m\right) - 10\right). \qquad (23)$$

This sequence appears to be very complicated, but actually it is not difficult to analyze it by means of generating functions. Let

$$C(z) = c_1 z + c_2 z^2 + c_3 z^3 + \cdots;$$

then since $10^z = 1 + \ln 10 + (1/2!)(\ln 10)^2 + \cdots$, we deduce that

$$
\begin{aligned}
c_{m+1} &= \frac{1}{10} c_{m+1} + \frac{9}{10} c_{m+1} \\
&= \frac{1}{10} \left(c_{m+1} + c_m \ln 10 + \cdots + \frac{1}{m!} c_1 (\ln 10)^m \right) \\
&\qquad\qquad + \frac{r}{10} \left(1 + \cdots + \frac{1}{m!} \left(\ln \frac{10}{r} \right)^m \right) - 1
\end{aligned}
$$

is the coefficient of z^{m+1} in the function

$$\tfrac{1}{10} C(z) 10^z + \frac{rz}{10} \left(\frac{10}{r} \right)^z \left(\frac{1}{1-z} \right) - \frac{1}{1-z}. \tag{24}$$

This condition holds for all values of m, so (24) must equal $C(z)$, and we obtain the explicit formula

$$C(z) = \frac{-z}{1-z} \left(\frac{(10/r)^{z-1} - 1}{10^{z-1} - 1} \right). \tag{25}$$

We want to study the asymptotic properties of the coefficients of $C(z)$, to complete our analysis. The large parenthesized quantity in (25) approaches $\ln(10/r)/\ln 10 = 1 - \log_{10} r$ as $z \to 1$, so we see that

$$C(z) + \frac{1 - \log_{10} r}{1-z} = R(z) \tag{26}$$

is an analytic function of the complex variable z in the circle

$$|z| < \left| 1 + \frac{2\pi i}{\ln 10} \right|.$$

In particular, $R(z)$ converges for $z = 1$, so its coefficients approach zero. This proves that the coefficients of $C(z)$ behave like those of $(\log_{10} r - 1)/(1 - z)$, that is,

$$\lim_{m \to \infty} c_m = \log_{10} r - 1.$$

Finally, we may combine this with (22), to show that $Q_m(s)$ approaches

$$1 + \frac{\log_{10} r - 1}{s} \left(1 + \ln s + \frac{1}{2!} (\ln s)^2 + \cdots \right) = \log_{10} r$$

uniformly for $1 \le s \le 10$. ∎

Therefore we have established the logarithmic law for integers by direct calculation, at the same time seeing that it is an extremely good approximation to the average behavior although it is never precisely achieved. Similar results for other distributions have been reported by W. H. Furry and H. Hurwitz in *Nature* **155** (Jan. 13, 1945), 52–53. The proofs of Lemma Q and Theorem F that have been given here are simplifications and amplifications of methods due to B. J. Flehinger, *AMM* **73** (1966), 1056–1061. Another interesting treatment of floating-point distribution has been given by Alan G. Konheim, *Math. Comp.* **19** (1965), 143–144.

EXERCISES

1. [*13*] If u and v are floating-decimal numbers *with the same sign*, what is the approximate probability that fraction overflow occurs during the calculation of $u \oplus v$, according to Sweeney's tables?

2. [*40*] Make further tests of floating-point addition and subtraction, to confirm or improve on the accuracy of Sweeney's tables.

3. [*15*] What is the probability that the two leading digits of a floating-decimal number are "23", according to the logarithm law?

4. [*18*] The text points out that the front pages of a well-used table of logarithms get dirtier than the back pages do. What if we had an *antilogarithm* table instead, i.e., a table giving the value of x when $\log_{10} x$ is given. Which pages of such a table would be the dirtiest?

▶ **5.** [*M20*] Suppose that U is a real number uniformly distributed in the interval $0 < U < 1$. What is the distribution of the leading digits of U?

6. [*22*] If we have binary computer words containing $n + 1$ bits, we might use p bits for the fraction part of floating-binary numbers, one bit for the sign, and $n - p$ bits for the exponent. This means that the range of values representable, i.e., the ratio of the largest to the smallest positive normalized value, is essentially $2^{2^{n-p}}$. The same computer word could be used to represent floating *hexadecimal* numbers, i.e., floating-point numbers with base 16, with $p + 2$ bits for the fraction part $((p + 2)/4$ hexadecimal digits) and $n - p - 2$ bits for the exponent; then the range of values would be $16^{2^{n-p-2}} = 2^{2^{n-p}}$, the same as before, and with more bits in the fraction part. This may sound as if we are getting something for nothing, but the normalization condition for base 16 is weaker in that there may be up to three leading zero bits in the fraction part; thus not all of the $p + 2$ bits are "significant bits."

On the basis of the logarithm law, what are the probabilities that the fraction part of a positive normalized base 16 floating-point number has exactly 0, 1, 2, and 3 leading zero bits? Discuss the desirability of hexadecimal versus binary, on the grounds of the topics discussed in this section.

7. [*HM28*] Prove that there is no distribution function $F(u)$ which satisfies (5) for each integer $b \geq 2$, and for all real values r in the range $1 \leq r \leq b$.

8. [*M23*] Does (10) hold when $m = 0$ for suitable $N_0(\epsilon)$?

9. [*HM24*] Let $\langle x_n \rangle$ be a bounded sequence of real numbers such that $\lim_{n \to \infty} x_{\lfloor 10^n s \rfloor} = q(s)$ exists for $1 \leq s \leq 10$, where $q(s)$ is a continuous function of s. Does it follow that $\lim \inf_{n \to \infty} x_n = \inf_{1 \leq s \leq 10} q(s)$?

▶ **10.** [*HM28*] The argument in the text shows that $c_m = \log_{10} r - 1 + \epsilon_m$, where ϵ_m approaches zero as $m \to \infty$. Obtain the next term in the asymptotic expansion of c_m.

11. [*M15*] Show that if U is a random variable which is distributed according to the logarithmic law, then so is $1/U$.

12. [*M20*] Assume that U and V are random, normalized, positive floating-point numbers, whose fraction parts are independently distributed according to the logarithmic law. What is the distribution of the fraction part of $U \otimes V$? (Ignore rounding of the result.)

▶ **13.** [*M20*] The floating-point multiplication routine, Algorithm 4.2.1M, requires zero or one left shifts during normalization, depending on whether $f_u f_v \geq 1/b$ or not. Assuming that the input operands are independently distributed according to the logarithmic law, what is the probability that no left shift is needed for normalization of the result?

▶ **14.** [*HM30*] Let U and V satisfy the assumptions of exercise 12, and let p_k be the probability that the difference in their exponents is k. Give an equation for the probability that "fraction overflow" occurs during the floating-point addition of $U \oplus V$, in terms of the base b and the quantities $p_0, p_1, p_2, \ldots$. Compare this result with the observed behavior in Sweeney's table.

15. [*HM28*] Let U, V, p_0, p_1, $\ldots$ be as in exercise 14, and assume that base 10 arithmetic is being used. Show that regardless of the values of p_0, p_1, p_2, the sum $U \oplus V$ will *not* obey the logarithmic law exactly, and in fact the probability that $U \oplus V$ has leading digit 1 is always strictly *less* than $\log_{10} 2$.

4.3. MULTIPLE-PRECISION ARITHMETIC

4.3.1. The Classical Algorithms

Let us now consider operations on numbers which have arbitrarily high precision. For simplicity in exposition, let us assume that we are working with integers instead of numbers which have an embedded radix point. In this section we shall discuss algorithms for:

a) addition or subtraction of n-place integers, giving an n-place answer and a carry;

b) multiplication of an n-place integer by an m-place integer, giving an $(m + n)$-place answer;

c) division of an $(m + n)$-place integer by an n-place integer, giving an $(m + 1)$-place quotient and an n-place remainder.

These may be called "the classical algorithms," since the word "algorithm" was used only in connection with these processes for several centuries. (By the term "n-place integer," we mean any integer less than b^n, where b is the radix of the conventional positional notation in which the numbers are expressed; such numbers can be written using at most n "places" in this notation.)

It is a straightforward matter to apply the classical algorithms for integers to problems involving numbers with embedded radix points, or rational numbers, or extended-precision floating-point numbers, in the same way as the arithmetic operations defined for integers in MIX are applied to these more general problems.

In this section we shall study algorithms which do operations (a), (b), and (c) above for integers expressed in radix b notation, where b is any given integer ≥ 2. Thus the algorithms are quite general definitions of arithmetic processes, and as such they are unrelated to any particular computer. But the discussion in this section will also be somewhat machine-oriented, since we are chiefly concerned with efficient methods for doing high-precision calculations by computer. Although our examples are based on the mythical MIX computer, essentially the same considerations apply to nearly every other machine. For convenience, let us assume first that we have a computer (like MIX) which uses the signed-magnitude representation for numbers; suitable modifications for complement notations are discussed near the end of this section.

The most important fact to understand about extended-precision numbers is that they may be regarded as numbers written in radix b notation, where b is the computer's word size. For example, an integer which fills 10 words on a computer whose word size is $b = 10^{10}$ has 100 decimal digits; but we will consider it to be a 10-place number to the base 10^{10}. This assumption is justified for the same reason that we may convert, say, from binary to octal notation, simply by grouping the bits together. (See Eq. 4.1–5.)

In these terms, we are given the following primitive operations to work with:

a_0) addition or subtraction of one-place integers, giving a one-place answer and a carry;

b_0) multiplication of a one-place integer by another one-place integer, giving a two-place answer;

c_0) division of a two-place integer by a one-place integer, provided that the quotient is a one-place integer, and yielding also a one-place remainder.

By adjusting the word size, if necessary, nearly all computers will have these three operations available, and so we will construct our algorithms (a), (b), and (c) mentioned above in terms of the primitive operations (a_0), (b_0), and (c_0).

Since we are visualizing extended-precision integers as base b numbers, it is sometimes helpful to think of the situation when $b = 10$, and to imagine that we are doing the arithmetic by hand. Then operation (a_0) is analogous to memorizing the addition table; (b_0) is analogous to memorizing the multiplication table; and (c_0) is essentially memorizing the multiplication table in reverse. The more complicated operations (a), (b), (c) on high-precision numbers can now be done using the simple addition, subtraction, multiplication, and long division procedures we are taught in elementary school. In fact, most of the algorithms we shall discuss in this section are essentially only mechanizations of familiar pencil-and-paper operations. Of course, we must state the algorithms much more precisely than they have ever been stated in the fifth grade, and we should also attempt to minimize computer memory and running time requirements.

To avoid a tedious discussion and cumbersome notations, let us assume that all numbers we deal with are *nonnegative*. The additional work of computing the signs, etc., is quite straightforward, and the reader will find it easy to fill in any details of this sort.

First comes addition, which of course is very simple, but it is worth studying since the same ideas occur in the other algorithms also:

Algorithm A (*Addition of nonnegative integers*). Given nonnegative n-place integers $u_1 u_2 \ldots u_n$ and $v_1 v_2 \ldots v_n$ with radix b, this algorithm forms their sum, $(w_0 w_1 w_2 \ldots w_n)_b$. (Here w_0 is the "carry," and it will always be equal to 0 or 1.)

A1. [Initialize.] Set $j \leftarrow n$, $k \leftarrow 0$. (The variable j will run through the various digit positions, and the variable k keeps track of carries at each step.)

A2. [Add digits.] Set $w_j \leftarrow (u_j + v_j + k) \bmod b$, and $k \leftarrow \lfloor (u_j + v_j + k)/b \rfloor$. (In other words, k is set to 1 or 0, depending on whether a "carry" occurred or not, i.e., whether $u_j + v_j + k \geq b$ or not. At most one carry is possible during the two additions, since we always have

$$u_j + v_j + k \leq (b - 1) + (b - 1) + 1 < 2b.)$$

A3. [Loop on j.] Decrease j by one. Now if $j > 0$, go back to step A2; otherwise set $w_0 \leftarrow k$ and terminate the algorithm. ∎

For a formal proof that Algorithm A is valid, see exercise 4.

A MIX program for this addition process might take the following form:

Program A (*Addition of nonnegative integers*). Let $\mathrm{LOC}(u_j) \equiv \mathsf{U} + j$, $\mathrm{LOC}(v_j) \equiv \mathsf{V} + j$, $\mathrm{LOC}(w_j) \equiv \mathsf{W} + j$, $\mathrm{rI1} \equiv j$, $\mathrm{rA} \equiv k$, word size $\equiv b$, $\mathsf{N} \equiv n$.

01		ENT1	N	1	A1. Initialize.
02		JOV	OFLO	1	Ensure overflow is off.
03	1H	ENTA	0	$N + 1 - K$	$k \leftarrow 0$.
04		J1Z	3F	$N + 1 - K$	To A3 if $j = 0$.
05	2H	ADD	U,1	N	A2. Add digits.
06		ADD	V,1	N	
07		STA	W,1	N	
08		DEC1	1	N	A3. Loop on j.
09		JNOV	1B	N	If no overflow, set $k \leftarrow 0$.
10		ENTA	1	K	Otherwise, set $k \leftarrow 1$.
11		J1P	2B	K	To A2 if $j \neq 0$.
12	3H	STA	W	1	Store final carry in w_0. ∎

The running time for this program is $10N + 6$ cycles, independent of the number of carries, K. The quantity K is analyzed in detail at the close of this section.

Many modifications of Algorithm A are possible, and only a few of these are mentioned in the exercises below. A chapter on generalizations of this algorithm might be entitled, "How to design adding circuits for a digital computer."

The problem of subtraction is similar to addition, but the differences are worth noting:

Algorithm S (*Subtraction of nonnegative integers*). Given nonnegative n-place integers $u_1 u_2 \ldots u_n \geq v_1 v_2 \ldots v_n$ with radix b, this algorithm forms their nonnegative difference, $(w_1 w_2 \ldots w_n)_b$.

S1. [Initialize.]   Set $j \leftarrow n$, $k \leftarrow 0$.

S2. [Subtract digits.]   Set $w_j \leftarrow (u_j - v_j + k) \bmod b$, and $k \leftarrow \lfloor (u_j - v_j + k)/b \rfloor$. (In other words, k is set to -1 or 0, depending on whether a "borrow" occurred or not, i.e., whether $u_j - v_j + k < 0$ or not. In the calculation of w_j note that we must have $-b = 0 - (b-1) + (-1) \leq u_j - v_j + k \leq (b-1) - 0 + 0 < b$; hence $0 \leq u_j - v_j + k + b < 2b$, and this suggests the method of computer implementation explained below.)

S3. [Loop on j.]   Decrease j by one. Now if $j > 0$, go back to step S2; otherwise terminate the algorithm. (When the algorithm terminates, we should have $k = 0$; the condition $k = -1$ will occur if and only if $v_1 \ldots v_n > u_1 \ldots u_n$, and this is contrary to the given assumptions. See exercise 12.) ∎

In a MIX program to implement subtraction, it is most convenient to retain the value $1 + k$ instead of k throughout the algorithm, and we can then calculate $u_j - v_j + (1 + k) + (b - 1)$ in step S2. (Recall that b is the word size.) This is illustrated in the following code:

Program S (*Subtraction of nonnegative integers*). This program is analogous to Program A; we have $\text{rI1} \equiv j$, $\text{rA} \equiv 1 + k$. Here, as in other programs of this section, location WM1 contains the constant $b - 1$, the largest possible value that can be stored in a MIX word; cf. Program 4.2.3D, lines 38–39.

01		ENT1	N	1	*S1. Initialize.*
02		JOV	OFLO	1	Ensure overflow is off.
03	1H	J1Z	DONE	$K + 1$	Terminate if $j = 0$.
04		ENTA	1	K	Set $k \leftarrow 0$.
05	2H	ADD	U,1	N	*S2. Subtract digits.*
06		SUB	V,1	N	Compute $u_j - v_j + k + b$.
07		ADD	WM1	N	
08		STA	W,1	N	(May be minus zero.)
09		DEC1	1	N	*S3. Loop on j.*
10		JOV	1B	N	If overflow, set $k \leftarrow 0$.
11		ENTA	0	$N - K$	Otherwise, set $k \leftarrow -1$.
12		J1P	2B	$N - K$	Back to S2.
13		HLT	5		(Error, $v > u$) ∎

The running time for this program is $12N + 3$ cycles, which is slightly longer than that for Program A.

The reader may wonder if it would not be worth while to have a combined addition-subtraction routine in place of the two algorithms A and S. Study of the computer programs shows that it is generally better to use two different routines, so that the inner loop of the computation can be performed as rapidly as possible.

Our next problem is multiplication, and here we carry the ideas used in Algorithm A a little further:

Algorithm M (*Multiplication of nonnegative integers*). Given nonnegative integers $u_1 u_2 \ldots u_n$ and $v_1 v_2 \ldots v_m$ with radix b, this algorithm forms their product $(w_1 w_2 \ldots w_{m+n})_b$. (The conventional pencil-and-paper algorithm for this process is based on forming the partial products $(u_1 u_2 \ldots u_n) \times v_j$ first, for $1 \leq j \leq m$, and then adding these products together with an appropriate scale factor; but in a computer it is best to do the addition concurrently with the multiplication, as described in this algorithm.)

M1. [Initialize.]   Set $w_{m+1}, w_{m+2}, \ldots, w_{m+n}$ all to zero. Set $j \leftarrow m$. (If $w_{m+1}, \ldots, w_{m+n}$ were not cleared to zero in this step, we would have a more general algorithm which sets

$$(w_1 \ldots w_{m+n}) \leftarrow (u_1 \ldots u_n) \times (v_1 \ldots v_m) + (w_{m+1} \ldots w_{m+n}).)$$

M2. [Zero multiplier?] If $v_j = 0$, set $w_j \leftarrow 0$ and go to step M6. (This test saves a good deal of time if there is a reasonable chance that v_j is zero, but otherwise it may be omitted without affecting the validity of the algorithm.)

M3. [Initialize i.]   Set $i \leftarrow n$, $k \leftarrow 0$.

M4. [Multiply and add.] Set $t \leftarrow u_i \times v_j + w_{i+j} + k$; then set $w_{i+j} \leftarrow t \bmod b$, $k \leftarrow \lfloor t/b \rfloor$. (Here the "carry" k will always be in the range $0 \leq k < b$; see below.)

M5. [Loop on i.] Decrease i by one. Now if $i > 0$, go back to step M4; otherwise set $w_j \leftarrow k$.

M6. [Loop on j.] Decrease j by one. Now if $j > 0$, go back to step M2; otherwise the algorithm terminates. ∎

Algorithm M is illustrated in Table 1, assuming that $b = 10$, by showing the states of the computation at the beginning of steps M5 and M6. A proof of Algorithm M appears in the answer to exercise 14.

The two inequalities

$$0 \leq t < b^2, \qquad 0 \leq k < b \tag{1}$$

are crucial for an efficient implementation of this algorithm, since they point out how large a register is needed for the computations. These inequalities may

Table 1

MULTIPLICATION OF 914 BY 84.

Step	i	j	u_i	v_j	t	w_1	w_2	w_3	w_4	w_5
M5	3	2	4	4	16	x	x	0	0	6
M5	2	2	1	4	05	x	x	0	5	6
M5	1	2	9	4	36	x	x	6	5	6
M6	0	2	x	4	36	x	3	6	5	6
M5	3	1	4	8	37	x	3	6	7	6
M5	2	1	1	8	17	x	3	7	7	6
M5	1	1	9	8	76	x	6	7	7	6
M6	0	1	x	8	76	7	6	7	7	6

be proved by induction as the algorithm proceeds, for if we have $k < b$ at the start of step M4, we have

$$u_i \times v_j + w_{i+j} + k \le (b-1) \times (b-1) + (b-1) + (b-1) = b^2 - 1 < b^2.$$

The following MIX program shows the considerations which are necessary when Algorithm M is implemented on a computer. The coding for step M4 would be a little simpler if our computer had a "multiply-and-add" instruction, or if it had a double-length accumulator for addition.

Program M. (*Multiplication of nonnegative integers*). This program is analogous to Program A. $rI1 \equiv i$, $rI2 \equiv j$, $rI3 \equiv i+j$, CONTENTS(CARRY) $\equiv k$.

01		ENT1	N	1	M1. Initialize.
02		JOV	OFLO	1	Ensure overflow is off.
03		STZ	W+M,1	N	$w_{m+i} \leftarrow 0$.
04		DEC1	1	N	
05		J1P	*-2	N	Repeat for $n \ge i > 0$.
06		ENT2	M	1	$j \leftarrow m$.
07	1H	LDX	V,2	M	M2. Zero multiplier?
08		JXZ	8F	M	If $v_j = 0$, set $w_j \leftarrow 0$ and go to M6.
09		ENT1	N	$M - Z$	M3. Initialize i.
10		ENT3	N,2	$M - Z$	$i \leftarrow n$, $(i+j) \leftarrow n+j$.
11		ENTX	0	$M - Z$	$k \leftarrow 0$.
12	2H	STX	CARRY	$(M - Z)N$	M4. Multiply and add.
13		LDA	U,1	$(M - Z)N$	u_i
14		MUL	V,2	$(M - Z)N$	$\times v_j$
15		SLC	5	$(M - Z)N$	Interchange $rA \leftrightarrow rX$.
16		ADD	W,3	$(M - Z)N$	Add w_{i+j} to lower half.
17		JNOV	*+2	$(M - Z)N$	Did overflow occur?
18		INCX	1	K	If so, carry one into rX.
19		ADD	CARRY	$(M - Z)N$	Add k to lower half.
20		JNOV	*+2	$(M - Z)N$	Did overflow occur?
21		INCX	1	K'	If so, carry one into rX.
22		STA	W,3	$(M - Z)N$	$w_{i+j} \leftarrow t \bmod b$.

23		DEC1	1	$(M-Z)N$	$M5$. *Loop on* i.
24		DEC3	1	$(M-Z)N$	Decrease i and $(i+j)$ by one.
25		J1P	2B	$(M-Z)N$	Back to M4 if $i > 0$; rX $= \lfloor t/b \rfloor$.
26	8H	STX	W,2	M	Set $w_j \leftarrow k$.
27		DEC2	1	M	$M6$. *Loop on* j.
28		J2P	1B	M	Repeat until $j = 0$. ∎

The execution time of Program M depends on the number of places, M, in the multiplier; the number of places, N, in the multiplicand; the number of zeros, Z, in the multiplier; and the number of carries, K and K' which occur during the addition to the lower half of the product in the computation of t. If we approximate both K and K' by the reasonable (although somewhat pessimistic) values $\frac{1}{2}(M-Z)N$, we find that the total running time comes to $28MN + 10M + 4N + 3 - Z(28N + 3)$ cycles. If step M2 were deleted, the running time would be $28MN + 7M + 4N + 3$ cycles, so this step is not advantageous unless the density of zero positions within the multiplier is $Z/M > 3/(28N + 3)$. If the multiplier is chosen completely at random, this ratio Z/M is expected to be only about $1/b$, which is extremely small; so step M2 is generally *not* worth while.

Algorithm M is not the fastest way to multiply when m and n are large, although it has the advantage of simplicity. Speedier methods are discussed in Section 4.3.3; even when $m = n = 4$, it is possible to multiply numbers in a little less time than is required by Algorithm M.

The final algorithm of concern to us in this section is long division, in which we want to divide $(n + m)$-place integers by n-place integers. Here the ordinary pencil-and-paper method involves a certain amount of guesswork and ingenuity on the part of the man doing the division, and so we either have to eliminate this from the algorithm or have to develop some theory to explain it more carefully.

A moment's reflection about the ordinary process of long division shows that the general problem breaks down into simpler steps, each of which is the division of an $(n + 1)$-place number u by the n-place divisor v, where $0 \leq u/v < b$; the remainder after each step is less than v, so it may be used in the succeeding step. For example, if we are asked to divide 3142 by 47, we first divide 314 by 47, getting 6 and a remainder of 32; then we divide 322 by 47, getting 6 and a remainder of 40; thus we have a quotient of 66 and a remainder of 40. It is clear that this same idea works in general, and so our search for an appropriate division algorithm reduces to the following problem (Fig. 6):

Let $u = u_0 u_1 \ldots u_n$ and $v = v_1 v_2 \ldots v_n$ be nonnegative integers in radix b notation, such that $u/v < b$. Find an algorithm to determine $q = \lfloor u/v \rfloor$.

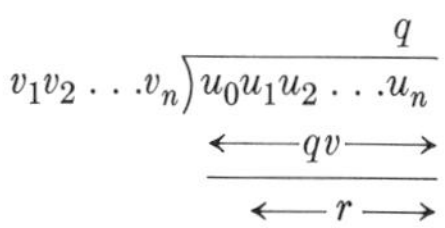

Fig. 6. Wanted: a way to determine q rapidly.

$$v_1 v_2 \ldots v_n \overline{\smash{\big)}\, u_0 u_1 u_2 \ldots u_n}^{\textstyle q}$$
$$\longleftarrow qv \longrightarrow$$
$$\longleftarrow r \longrightarrow$$

We may observe that the condition $u/v < b$ is equivalent to the condition that $u/b < v$; i.e., $\lfloor u/b \rfloor < v$; and this is the condition that $u_0 u_1 \ldots u_{n-1} < v_1 v_2 \ldots v_n$. Furthermore, if we write $r = u - qv$, then q is the unique integer such that $0 \leq r < v$.

The most obvious approach to this problem is to make a guess about q, based on the most significant digits of u and v. It isn't obvious that such a method will be reliable enough, but it is worth investigating; let us therefore set

$$\hat{q} = \min\left(\left\lfloor \frac{u_0 b + u_1}{v_1} \right\rfloor, \ b - 1\right). \tag{2}$$

This formula says $\hat{q}$ is obtained by dividing the two leading digits of u by the leading digit of v; and if the result is b or more we can replace it by $(b - 1)$.

It is a remarkable fact, which we will now investigate, that this value $\hat{q}$ is always a very good approximation to the desired answer q, so long as v_1 is reasonably large. In order to analyze how close $\hat{q}$ comes to q, we will first prove that $\hat{q}$ is never too small.

Theorem A. *In the notation above, $\hat{q} \geq q$.*

Proof. Since $q \leq b - 1$, the theorem is certainly true if $\hat{q} = b - 1$. Suppose therefore that $\hat{q} < b - 1$; it follows that $\hat{q} = \lfloor (u_0 b + u_1)/v_1 \rfloor$; hence $\hat{q} v_1 \geq u_0 b + u_1 - v_1 + 1$. Therefore

$$u - \hat{q}v \leq u - \hat{q}v_1 b^{n-1} \leq u_0 b^n + \cdots + u_n$$
$$- (u_0 b^n + u_1 b^{n-1} - v_1 b^{n-1} + b^{n-1})$$
$$= u_2 b^{n-2} + \cdots + u_n - b^{n-1} + v_1 b^{n-1} < v_1 b^{n-1} \leq v.$$

Since $u - \hat{q}v < v$, we must have $\hat{q} \geq q$. ∎

We will now prove that $\hat{q}$ cannot be much larger than q in practical situations. Assume that $\hat{q} \geq q + 3$. We have

$$\hat{q} \leq \frac{u_0 b + u_1}{v_1} = \frac{u_0 b^n + u_1 b^{n-1}}{v_1 b^{n-1}} \leq \frac{u}{v_1 b^{n-1}} < \frac{u}{v - b^{n-1}}.$$

(The case $v = b^{n-1}$ is impossible, for if $v = (1\,0\,0 \cdots 0)_b$ then $q = \hat{q}$.) Furthermore, since $q > (u/v) - 1$,

$$3 \leq \hat{q} - q < \frac{u}{v - b^{n-1}} - \frac{u}{v} + 1 = \frac{u}{v}\left(\frac{b^{n-1}}{v - b^{n-1}}\right) + 1.$$

Therefore

$$\frac{u}{v} > 2\left(\frac{v - b^{n-1}}{b^{n-1}}\right) \geq 2(v_1 - 1).$$

Finally, since $b - 4 \geq \hat{q} - 3 \geq q = \lfloor u/v \rfloor \geq 2(v_1 - 1)$, we have $v_1 < \lfloor b/2 \rfloor$. This proves Theorem B:

Theorem B. *If $v_1 \geq \lfloor b/2 \rfloor$, then $\hat{q} - 2 \leq q \leq \hat{q}$.* ∎

The most important part of this theorem is that *the conclusion is independent of b*; no matter how large b is, the trial quotient $\hat{q}$ will never be more than 2 in error!

The condition that $v_1 \geq \lfloor b/2 \rfloor$ is very much like a normalization condition (in fact, it is exactly the condition of normalization in a binary computer). One simple way to ensure that v_1 is sufficiently large is to multiply *both* u and v by $\lfloor b/(v_1 + 1) \rfloor$; this does not change the value of u/v, and in exercise 23 it is proved that this will always make the new value of v_1 large enough. (*Note:* For another way to normalize the divisor, see exercise 28.)

Now that we have armed ourselves with all of these facts, we are in a position to write the desired long division algorithm. This algorithm uses a slightly improved choice of $\hat{q}$ in step D3 which guarantees that $q = \hat{q}$ or $\hat{q} - 1$; in fact, the improved choice of $\hat{q}$ made here is almost always accurate.

Algorithm D (*Division of nonnegative integers*). Given nonnegative integers $u = u_1 u_2 \ldots u_{m+n}$ and $v = v_1 v_2 \ldots v_n$ with radix b, where $v_1 \neq 0$ and $n > 1$, we form the quotient $\lfloor u/v \rfloor = (q_0 q_1 \ldots q_m)_b$ and the remainder $u \bmod v = (r_1 r_2 \ldots r_n)_b$. (This notation is slightly different from that used in the above proofs. When $n = 1$, the simpler algorithm of exercise 16 should be used.)

D1. [Normalize.] Set $d \leftarrow \lfloor b/(v_1 + 1) \rfloor$. Set $u_0 u_1 u_2 \ldots u_{m+n}$ equal to $u_1 u_2 \ldots u_{m+n}$ times d. Set $v_1 v_2 \ldots v_n$ equal to $v_1 v_2 \ldots v_n$ times d. (Note the introduction of the new digit position u_0 at the left of u_1; if $d = 1$, all we need to do in this step is to set $u_0 \leftarrow 0$. On a binary computer it may be preferable to choose d to be a power of 2 instead of using the value suggested here; any value of d which results in $v_1 \geq \lfloor b/2 \rfloor$ will suffice here.)

D2. [Initialize j.] Set $j \leftarrow 0$. (The loop on j, steps D2 through D7, will be essentially a division of $u_j u_{j+1} \ldots u_{j+n}$ by $v_1 v_2 \ldots v_n$ to get a single quotient digit q_j; cf. Fig. 6.)

D3. [Calculate $\hat{q}$.] If $u_j = v_1$, set $\hat{q} \leftarrow b - 1$; otherwise set $\hat{q} \leftarrow \lfloor (u_j b + u_{j+1})/v_1 \rfloor$. Now test if $v_2 \hat{q} > (u_j b + u_{j+1} - \hat{q} v_1) b + u_{j+2}$; if so, decrease $\hat{q}$ by 1 and repeat this test. (The latter test determines at high speed most of the cases in which the trial value $\hat{q}$ is one too large, and it eliminates *all* cases where $\hat{q}$ is two too large; see exercises 19, 20, 21.)

D4. [Multiply and subtract.] Replace $u_j u_{j+1} \ldots u_{j+n}$ by $u_j u_{j+1} \ldots u_{j+n}$ minus ($\hat{q}$ times $v_1 v_2 \ldots v_n$). This step (analogous to steps M3 to M5 of Algorithm M) consists of a simple multiplication by a one-place number, combined with a subtraction. The digits $u_j u_{j+1} \ldots u_{j+n}$ should be kept positive; if the result of this step is actually negative, $u_j u_{j+1} \ldots u_{j+n}$ should be left as the true value plus b^{n+1}, i.e., as the b's complement of the true value, and a "borrow" to the left should be remembered.

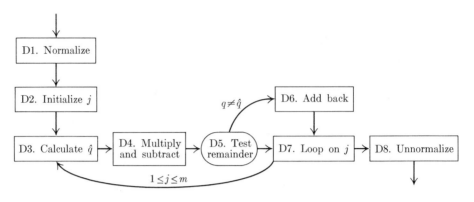

Fig. 7. Long division.

D5. [Test remainder.] Set $q_j \leftarrow \hat{q}$. If the result of step D4 was negative, go to step D6; otherwise go on to step D7.

D6. [Add back.] (The probability that this step is necessary is very small, on the order of only $3/b$.) Decrease q_j by 1, and add $0v_1v_2 \ldots v_n$ to $u_j u_{j+1} u_{j+2} \ldots u_{j+n}$. (A carry will occur to the left of u_j, and it should be ignored since it cancels with the "borrow" which occurred in D4.)

D7. [Loop on j.] Increase j by one. Now if $j \leq m$, go back to D3.

D8. [Unnormalize.] Now $q_0 q_1 \ldots q_m$ is the desired quotient, and the desired remainder may be obtained by dividing $u_{m+1} \ldots u_{m+n}$ by d. ∎

The representation of Algorithm D as a MIX program has several points of interest:

Program D (*Division of nonnegative integers*). The conventions of this program are analogous to Program A; rI1 $\equiv i$, rI2 $\equiv j - m$, rI3 $\equiv i + j$. Steps D1 and D8 have been left as exercises.

001	D1	JOV	OFLO	1	*D1. Normalize.*
...					(See exercise 25)
039	D2	ENN2	M	1	*D2. Initialize j.*
040		STZ	V	1	Set $v_0 \leftarrow 0$, for convenience in D4.
041	D3	LDA	U+M,2(1:5)	$M+1$	*D3. Calculate $\hat{q}$.*
042		LDX	U+M+1,2	$M+1$	rAX $\leftarrow u_j b + u_{j+1}$.
043		DIV	V+1	$M+1$	rA $\leftarrow \lfloor \text{rAX}/v_1 \rfloor$.
044		JOV	1F	$M+1$	Jump if quotient $= b$.
045		STA	QHAT	$M+1$	$\hat{q} \leftarrow$ rA.
046		STX	RHAT	$M+1$	$\hat{r} \leftarrow u_j b + u_{j+1} - \hat{q} v_1$
047		JMP	2F	$M+1$	$= (u_j b + u_{j+1}) \bmod v_1$.
048	1H	LDX	WM1		rX $\leftarrow b - 1$.
049		LDA	U+M+1,2		rA $\leftarrow u_{j+1}$. (Here $u_j = v_1$.)
050		JMP	4F		

051	3H	LDX	QHAT	E	
052		DECX	1	E	Decrease $\hat{q}$ by one.
053		LDA	RHAT	E	Adjust $\hat{r}$ accordingly:
054	4H	STX	QHAT	E	$\hat{q} \leftarrow \text{rX}.$
055		ADD	V+1	E	$\text{rA} \leftarrow \text{rA} + v_1.$
056		JOV	D4	E	(If $\hat{r}$ will be $\geq b$, $v_2\hat{q}$ will be $< \hat{r}b$.)
057		STA	RHAT	E	$\hat{r} \leftarrow \text{rA}.$
058	2H	LDA	QHAT	$M + E + 1$	
059		MUL	V+2	$M + E + 1$	
060		CMPA	RHAT	$M + E + 1$	Test if $v_2\hat{q}$ is $\leq \hat{r}b + u_{j+2}$.
061		JL	D4	$M + E + 1$	
062		JG	3B	E	
063		CMPX	U+M+2,2		
064		JG	3B		If not, $\hat{q}$ is too large.
065	D4	ENTX	1	$M + 1$	$D4.$ *Multiply and subtract.*
066		ENT1	N	$M + 1$	$i \leftarrow n.$
067		ENT3	M+N,2	$M + 1$	$(i + j) \leftarrow n + j.$
068	2H	STX	CARRY	$(N + 1)(M + 1)$	(Here $1 - b < \text{rX} \leq +1$.)
069		LDAN	V,1	$(N + 1)(M + 1)$	
070		MUL	QHAT	$(N + 1)(M + 1)$	$\text{rAX} \leftarrow -\hat{q}v_i.$
071		SLC	5	$(N + 1)(M + 1)$	Interchange $\text{rA} \leftrightarrow \text{rX}.$
072		ADD	CARRY	$(N + 1)(M + 1)$	Add the contribution from the
073		JNOV	*+2	$(N + 1)(M + 1)$	digit to the right, plus 1.
074		DECX	1	K	If sum is $\leq -b$, carry $-1.$
075		ADD	U,3	$(N + 1)(M + 1)$	Add $u_{i+j}.$
076		ADD	WM1	$(N + 1)(M + 1)$	Add $b - 1$ to force $+$ sign.
077		JNOV	*+2	$(N + 1)(M + 1)$	If no overflow, carry $-1.$
078		INCX	1	K'	$\text{rX} \equiv \text{carry} +1.$
079		STA	U,3	$(N + 1)(M + 1)$	$u_{i+j} \leftarrow \text{rA}$ (may be minus zero).
080		DEC1	1	$(N + 1)(M + 1)$	
081		DEC3	1	$(N + 1)(M + 1)$	
082		J1NN	2B	$(N + 1)(M + 1)$	Repeat for $n \geq i \geq 0.$
083	D5	LDA	QHAT	$M + 1$	$D5.$ *Test remainder.*
084		STA	Q+M,2	$M + 1$	Set $q_j \leftarrow \hat{q}.$
085		JXP	D7	$M + 1$	(Here $\text{rX} = 0$ or 1, since $v_0 = 0$.)
086	D6	DECA	1		$D6.$ *Add back.*
087		STA	Q+M,2		Set $q_j \leftarrow \hat{q}_j - 1.$
088		ENT1	N		$i \leftarrow n.$
089		ENT3	M+N,2		$(i + j) \leftarrow n + j.$
090	1H	ENTA	0		(This is essentially Program A.)
091	2H	ADD	U,3		
092		ADD	V,1		
093		STA	U,3		
094		DEC1	1		
095		DEC3	1		
096		JNOV	1B		
097		ENTA	1		
098		J1P	2B		(Not necessary to add to u_j.)

099	D7	INC2	1	$M + 1$
100		J2NP	D3	$M + 1$
101	D8	...		

099 D7 INC2 1 $M + 1$ D7. *Loop on j.*
100   J2NP D3 $M + 1$ Repeat for $0 \leq j \leq m$.
101 D8 ...   (See exercise 26) ▮

Note how easily the rather complex appearing calculations and decisions of step D3 can be handled inside the machine. Note also that the program for step D4 is analogous to Program M, except that the ideas of Program S have also been incorporated. In step D6, use has been made of the fact that $v_0 = 0$, and that u_j is not needed in the subsequent calculations; a strict interpretation of Algorithm D would require line 098 to be "J1NN 2B".

The running time for Program D can be estimated by considering the quantities M, N, E, K, and K' shown in the program. (These quantities ignore several situations which can only occur with very low probability; for example, we may assume that lines 048–050, 063–064, and step D6 are never executed.) Here $M + 1$ is the number of words in the quotient; N is the number of words in the divisor; E is the number of times q is adjusted downwards in step D3; K and K' are the number of times certain "carry" adjustments are made during the multiply-subtract loop. If we assume that $K + K'$ is approximately $(N + 1)(M + 1)$, and that E is approximately $\frac{1}{2}M$, we get a total running time of approximately

$$30MN + 30N + 91M + 113$$

cycles, plus $67N + 235M + 4$ more if $d > 1$. (The program segments of exercises 25 and 26 are included in these totals.) When M and N are large, this is only about seven percent longer than the time Program M takes to multiply the quotient by the divisor.

Further commentary on Algorithm D appears in the exercises at the close of this section.

It is possible to debug programs for multiple-precision arithmetic by using the multiplication and addition routines to check the result of the division routine, etc. The following type of test data is occasionally useful:

$$(t^m - 1)(t^n - 1) = t^{m+n} - t^n - t^m + 1.$$

If $m < n$, this number has the radix t expansion

$$\underbrace{(t - 1) \ \cdots \ (t - 1)}_{m - 1 \text{ places}} \ (t - 2) \ \underbrace{(t - 1) \ \cdots \ (t - 1)}_{n - m \text{ places}} \ \underbrace{0 \ \cdots \ 0}_{m - 1 \text{ places}} \ 1;$$

for example, $(10^3 - 1)(10^5 - 1) = 99899001$. In the case of Program D, it is also necessary to find some test cases which cause the rarely executed parts of the program to be used; some portions of that program would probably never get tested even if a million random test cases were tried.

Now that we have seen how to operate with signed-magnitude numbers, let us consider what approach should be taken to the same problems when a computer with complement notation is being used. For two's complement and ones' complement notations, it is best to let the radix b be *one-half* the word size; thus for a 32-bit computer word we would use $b = 2^{31}$ in the above algorithms. The sign bit of all but the most significant word of a multiple-precision number will be zero, so that no anomalous sign correction takes place during the computer's multiplication and division operations. In fact, the basic meaning of complement notation requires that we consider all but the most significant word to be nonnegative: For example, assuming a 10-bit word, the two's complement number

$$1101111110 \quad 111111010 \quad 011101011$$

(where the sign is given only for the most significant word) is properly thought of as

$$-2^{27} + (101111110)_2 \cdot 2^{18} + (111111010)_2 \cdot 2^9 + (011101011)_2.$$

Addition of signed numbers is slightly easier when complement notations are being used, since the routine for adding n-place nonnegative integers can be used for arbitrary n-place integers; the sign appears only in the first word, so the less significant words may be added together irrespective of the actual sign. (Special attention must be given to the leftmost carry when ones' complement notation is being used, however; it must be added into the least significant word, and possibly propagated further to the left.) Similarly, we find that subtraction of signed numbers is slightly simpler with complement notation. On the other hand, multiplication and division seem to be done most easily by working with nonnegative quantities and doing suitable complementation operations beforehand to make sure both operands are nonnegative; it may be possible to avoid this complementation by devising some tricks for working directly with negative numbers in a complement notation, and it is not hard to see how this could be done in double-precision multiplication, but care should be taken not to slow down the inner loops of the subroutines when high precision is required. Note that the product of two m-place numbers in two's complement notation may require $2m + 1$ places: the square of $-b^m$ is b^{2m}.

Let us now turn to an analysis of the quantity K that arises in Program A, i.e., the number of carries that occur when n-place numbers are being added together. This quantity K plays no part in the total running time of Program A, but it does affect the running time of the counterparts of Program A that deal with complement notations, and its analysis is interesting in itself as a significant application of generating functions.

Suppose now that u and v are independent random n-place integers uniformly distributed in the range $0 \leq u, v < b^n$. Let p_{nk} be the probability that exactly k carries occur in the addition of u to v, *and* that one of these carries occurred in the most significant position (so that $u + v \geq b^n$). Similarly, let q_{nk} be the probability that exactly k carries occur, but there is no carry in the most significant position. Then it is not hard to see that

$$p_{0k} = 0, \qquad q_{0k} = \delta_{0k}, \qquad \text{for all } k;$$

$$p_{(n+1)(k+1)} = \frac{b+1}{2b} p_{nk} + \frac{b-1}{2b} q_{nk}, \tag{3}$$

$$q_{(n+1)k} = \frac{b-1}{2b} p_{nk} + \frac{b+1}{2b} q_{nk};$$

this happens because $(b-1)/2b$ is the probability that $u_1 + v_1 \geq b$ and $(b+1)/2b$ is the probability that $u_1 + v_1 + 1 \geq b$, when u_1 and v_1 are independently and uniformly distributed integers in the range $0 \leq u_1, v_1 < b$.

To obtain further information about these quantities p_{nk} and q_{nk}, we may set up the generating functions

$$P(z, t) = \sum_{k,n} p_{nk} z^k t^n, \qquad Q(z, t) = \sum_{k,n} q_{nk} z^k t^n; \tag{4}$$

from (3) we have the basic relations

$$P(z, t) = zt \left(\frac{b+1}{2b} P(z, t) + \frac{b-1}{2b} Q(z, t) \right),$$

$$Q(z, t) = 1 + t \left(\frac{b-1}{2b} P(z, t) + \frac{b+1}{2b} Q(z, t) \right).$$

These two equations are readily solved for $P(z, t)$ and $Q(z, t)$; and if we let

$$G(z, t) = P(z, t) + Q(z, t) = \sum_n G_n(z) t^n,$$

where $G_n(z)$ is the generating function for the total number of carries when n-place numbers are added, we find that

$$G(z, t) = (b - zt)/p(z, t), \quad \text{where} \quad p(z, t) = b - \tfrac{1}{2}(1 + b)(1 + z)t + zt^2. \tag{5}$$

Note that $G(1, t) = 1/(1 - t)$, and this checks with the fact that $G_n(1)$ must equal 1 (it is the sum of all the possible probabilities). Taking partial derivatives of (5) with respect to z, we find that

$$\frac{\partial G}{\partial z} = \sum_n G_n'(z) t^n = \frac{-t}{p(z, t)} + \frac{t(b - zt)(b + 1 - 2t)}{2p(z, t)^2};$$

$$\frac{\partial^2 G}{\partial z^2} = \sum_n G_n''(z) t^n = \frac{-t^2(b + 1 - 2t)}{p(z, t)^2} + \frac{t^2(b - zt)(b + 1 - 2t)}{p(z, t)^3}.$$

Now let us put $z = 1$ and expand in partial fractions:

$$\sum_n G_n'(1)t^n = \frac{t}{2}\left(\frac{1}{(1-t)^2} - \frac{1}{(b-1)(1-t)} + \frac{1}{(b-1)(b-t)}\right),$$

$$\sum_n G_n''(1)t^n = \frac{t^2}{2}\left(\frac{1}{(1-t)^3} - \frac{1}{(b-1)^2(1-t)} + \frac{1}{(b-1)^2(b-t)}\right.$$
$$\left. + \frac{1}{(b-1)(b-t)^2}\right).$$

It follows that the average number of carries, i.e., the mean value of K, is

$$G_n'(1) = \frac{1}{2}\left(n - \frac{1}{b-1}\left(1 - \left(\frac{1}{b}\right)^n\right)\right); \tag{6}$$

the variance is

$$G_n''(1) + G_n'(1) - G_n'(1)^2$$
$$= \frac{1}{4}\left(n + \frac{2n}{b-1} - \frac{2b+1}{(b-1)^2} + \frac{2b+2}{(b-1)^2}\left(\frac{1}{b}\right)^n - \frac{1}{(b-1)^2}\left(\frac{1}{b}\right)^{2n}\right). \tag{7}$$

So the number of carries is just slightly less than $\frac{1}{2}n$ under these assumptions.

History and Bibliography. The early history of the classical algorithms described in this section is left as an interesting project for the reader, and only the history of their implementation on computers will be traced here.

The use of 10^n as an assumed radix when multiplying large numbers on a desk calculator was discussed by D. N. Lehmer and J. P. Ballantine, *AMM* **30** (1923), 67–69.

Double-precision arithmetic on computers was first treated by J. von Neumann and H. H. Goldstine [J. von Neumann, *Collected Works*, vol. 5, p. 142–151]. Theorems A and B above are due to D. A. Pope and M. L. Stein [*CACM* **3** (1960), 652–654]; their article also contains a bibliography of earlier work on double precision routines. Other ways of choosing the trial quotient $\hat{q}$ have been discussed by A. G. Cox and H. A. Luther, *CACM* **4** (1961), 353 [divide by $v_1 + 1$ instead of v_1], and by M. L. Stein, *CACM* **7** (1964), 472–474 [divide by v_1 or $v_1 + 1$ according to the magnitude of v_2]; Krishnamurthy [*CACM* **8** (1965), 179–181] showed that examination of the single-precision remainder in the latter method leads to an improvement over Theorem B. Krishnamurthy and Nandi, *CACM* **10** (1967), 809–813, suggest a way to replace normalization and unnormalization operations of Algorithm D by a calculation of $\hat{q}$ based on several leading digits of the operands.

Several other methods for division have been suggested:

(1) "Fourier division" [J. Fourier, *Analyse des équations déterminées* (Paris, 1831), Sec. 2.21]. This method, which was often used on desk calculators,

essentially obtains each new quotient digit by increasing the precision of the divisor and the dividend at each step. Some rather extensive tests by the author have indicated that this method is certainly inferior to the "divide and correct" technique above, but there may be some applications in which Fourier division is practical. See D. H. Lehmer, "A Cross-Division Process," *AMM* **33** (1926), 198–206; J. V. Uspensky, *Theory of Equations* (New York: McGraw-Hill, 1948), 159–164.

(2) "Newton's method" for evaluating the reciprocal of a number was extensively used in early computers when there was no single-precision division instruction. The idea is to find some initial approximation x_0 to the number $1/v$, then to let $x_{n+1} = 2x_n - vx_n^2$. This method converges rapidly to $1/v$, since $x_n = 1/v - \epsilon$ implies that $x_{n+1} = 1/v - v\epsilon^2$. Convergence to third order, i.e., with ϵ replaced by $O(\epsilon^3)$ at each step, can be obtained using the formula

$$x_{n+1} = x_n + x_n(1 - vx_n) + x_n(1 - vx_n)^2$$
$$= x_n(1 + (1 - vx_n)(1 + (1 - vx_n))),$$

etc.; see P. Rabinowitz, *CACM* **4** (1961), 98. For calculations on extremely large numbers, Newton's second-order method (followed by multiplication by u) can actually be considerably faster than Algorithm D, if we increase the precision of x_n at each step and if we also use the fast multiplication routines of Section 4.3.3. (See Algorithm 4.3.3D for details.)

(3) Division methods have also been based on the evaluation of

$$\frac{u}{v + \epsilon} = \frac{u}{v}\left(1 - \left(\frac{\epsilon}{v}\right) + \left(\frac{\epsilon}{v}\right)^2 - \left(\frac{\epsilon}{v}\right)^3 + \cdots\right).$$

See "Large-number division by calculating machine," by H. H. Laughlin, *AMM* **37** (1930), 287–293. We have used this idea in the double-precision case [Eq. 4.2.3–(3)].

Besides the references just cited, the following articles concerning multiple-precision arithmetic are of interest: High-precision floating-point routines using ones' complement arithmetic are described by A. H. Stroud and D. Secrest, *Comp. J.* **6** (1963), 62–66. Extended-precision subroutines for use in FORTRAN programs are described by B. I. Blum, *CACM* **8** (1965), 318–320; and for use in ALGOL by M. Tienari and V. Suokonautio, *BIT* **6** (1966), 332–338. Arithmetic on integers with *unlimited* precision, making use of linked memory allocation techniques, has been elegantly described by G. E. Collins, *CACM* **9** (1966), 578–589.

We have restricted our discussion in this section to arithmetic techniques for use in computer programming. There are many algorithms for *hardware* implementation of arithmetic operations which are very interesting but which appear to be inapplicable to computer programs for high-precision numbers; for example, see G. W. Reitwiesner, "Binary Arithmetic," *Advances in Com-*

puters **1** (New York: Academic Press, 1960), 231–308; O. L. MacSorley, *Proc. IRE* **49** (1961), 67–91; G. Metze, *IRE Transactions* **EC-11** (1962), 761–764; H. L. Garner, "Number Systems and Arithmetic," *Advances in Computers* **6** (New York: Academic Press, 1965), 131–194. The minimum achievable execution time for hardware addition and multiplication operations has been investigated by S. Winograd, *JACM* **12** (1965), 277–285; **14** (1967), 793–802.

EXERCISES

1. [*42*] Study the early history of the classical algorithms for arithmetic, by looking up the writings of, say, Sun Tsǔ, al Khowârizmî, Fibonacci, and Robert Recorde, and by translating their methods as faithfully as possible into more precise algorithmic notation.

2. [*15*] Generalize Algorithm A so that it does "column addition," i.e., obtains the sum of m nonnegative n-place integers. (Assume that $m \leq b$.)

3. [*21*] Write a MIX program for the algorithm of exercise 2, and estimate its running time as a function of m and n.

4. [*M21*] Give a formal proof of the validity of Algorithm A, using the method of "inductive assertions" as explained in Section 1.2.1.

5. [*21*] Algorithm A adds the two inputs by going from right to left, but sometimes the data is more readily accessible from left to right. Design an algorithm which produces the same answer as Algorithm A, but which generates the digits of the answer from left to right, and goes back to change previous values if a carry occurs to make a previous value incorrect. (*Note:* Early Hindu and Arabic manuscripts were based on addition from left to right in this way; the right-to-left addition algorithm was a refinement due to later Arabic writers, perhaps because Arabic is written from right to left.)

▶ **6.** [*22*] Design an algorithm which adds from left to right (as in exercise 5), but which does not store a digit of the answer until this digit cannot possibly be affected by future carries; there is to be no changing of any answer digit once it has been stored. [*Hint:* Keep track of the number of consecutive $(b-1)$'s which have not yet been stored in the answer.] This sort of algorithm would be appropriate, for example, in a situation where the input and output numbers are to be read and written from left to right on magnetic tapes.

7. [*M26*] Determine the average number of times the algorithm of exercise 5 will find that a carry makes it necessary to go back and change k digits of the partial answer, for $k = 1, 2, \ldots, n$. (Assume that both inputs are independently and uniformly distributed between 0 and $b^n - 1$.)

8. [*M26*] Write a MIX program for the algorithm of exercise 5, and determine its average running time based on the expected number of carries as computed in the text.

▶ **9.** [*21*] Generalize Algorithm A to obtain an algorithm which adds two n-place numbers in a *mixed radix* number system, with bases $b_0, b_1, \ldots$ (from right to left). Thus the least significant digits lie between 0 and $b_0 - 1$, the next digits lie between 0 and $b_1 - 1$, etc.; cf. Eq. 4.1–(9).

10. [*18*] Would Program S work properly if the instructions on lines 06 and 07 were interchanged? If the instructions on lines 05 and 06 were interchanged?

11. [*10*] Design an algorithm which compares two nonnegative n-place integers $u = u_1u_2\ldots u_n$ and $v = v_1v_2\ldots v_n$ with radix b, to determine whether $u < v$, $u = v$, or $u > v$.

12. [*16*] Algorithm S assumes that we know which of the two input operands is the larger; if this information is not known, we could go ahead and perform the subtraction anyway, and we would find that an extra "borrow" is still present at the end of the algorithm. Design another algorithm which could be used (if there is a "borrow" present at the end of Algorithm S) to complement $w_1w_2\ldots w_n$ and therefore to obtain the absolute value of the difference of u and v.

13. [*21*] Write a MIX program which multiplies $(u_1u_2\ldots u_n)_b$ by v, where v is a single-precision number (i.e., $0 \le v < b$), producing the answer $(w_0w_1\ldots w_n)_b$. How much running time is required?

▶ **14.** [*M24*] Give a formal proof of the validity of Algorithm M, using the method of "inductive assertions" as explained in Section 1.2.1.

15. [*M20*] If we wish to form the product of two n-place fractions, $(.u_1u_2\ldots u_n)_b \times (.v_1v_2\ldots v_n)_b$, and to obtain only an n-place approximation $(.w_1w_2\ldots w_n)_b$ to the result, Algorithm M could be used to obtain a $2n$-place answer which is then rounded to the desired approximation. But this involves about twice as much work as is necessary for reasonable accuracy, since the products u_iv_j for $i+j > n+2$ contribute very little to the answer.

Give an estimate of the maximum error that can occur, if these products u_iv_j for $i+j > n+2$ are not computed during the multiplication, but are assumed to be zero.

▶ **16.** [*20*] Design an algorithm which divides a nonnegative n-place integer $u_1u_2\ldots u_n$ by v, where v is a single precision number (i.e., $0 < v < b$), producing the quotient $w_1w_2\ldots w_n$ and remainder r.

17. [*M20*] In the notation of Fig. 6, assume that $v_1 \ge \lfloor b/2 \rfloor$; show that if $u_0 = v_1$, we must have $q = b - 1$ or $b - 2$.

18. [*M20*] In the notation of Fig. 6, show that if $q' = \lfloor(u_0b + u_1)/(v_1 + 1)\rfloor$, then $q' \le q$.

▶ **19.** [*M21*] In the notation of Fig. 6, let $\hat{q}$ be an approximation to q, and let $\hat{r} = u_0b + u_1 - \hat{q}v_1$. Assume that $v_1 > 0$. Show that if $v_2\hat{q} > b\hat{r} + u_2$, then $q < \hat{q}$. [*Hint:* Strengthen the proof of Theorem A by examining the influence of v_2.]

20. [*M22*] Using the notation and assumptions of exercise 19, show that if $v_2\hat{q} \le b\hat{r} + u_2$, then $\hat{q} = q$ or $q = \hat{q} - 1$.

▶ **21.** [*M23*] Show that if $v_1 \ge \lfloor b/2 \rfloor$, and if $v_2\hat{q} \le b\hat{r} + u_2$ but $\hat{q} \ne q$ in the notation of exercises 19 and 20, then $u - qv \ge (1 - 3/b)v$. (The latter event occurs with approximate probability $3/b$, so that when b is the word size of a computer we must have $q_j = \hat{q}$ in Algorithm D except in very rare circumstances.)

▶ **22.** [*24*] Find an example of a four-digit number divided by a three-digit number, using Algorithm D when the radix b is 10, for which step D6 is necessary.

23. [*M23*] Given that v and b are integers, and that $1 \le v < b$, prove that $\lfloor b/2 \rfloor \le v \lfloor b/(v+1) \rfloor < (v+1)\lfloor b/(v+1) \rfloor \le b$.

24. [*M20*] Using the law of the distribution of leading digits explained in Section 4.2.4, give an approximate formula for the probability that $d = 1$ in Algorithm D. (When $d = 1$, it is, of course, possible to omit most of the calculation in steps D1 and D8.)

25. [*26*] Write a MIX routine for step D1, which is needed to complete Program D.

26. [*21*] Write a MIX routine for step D8, which is needed to complete Program D.

27. [*M20*] Prove that at the beginning of step D8 in Algorithm D, the number $u_{m+1}u_{m+2} \dots u_{m+n}$ is always an exact multiple of d.

28. [*M30*] (A. Svoboda, *Stroje na Zpracování Informací* **9** (1963), 25–32.) Let $v = (v_1 v_2 \dots v_n)_b$ be any radix b integer, where $v_1 \ne 0$. Perform the following operations:

N1. If $v_1 < b/2$, multiply v by $\lfloor (b+1)/(v_1+1) \rfloor$. Let the result of this step be $(v_0 v_1 v_2 \dots v_n)_b$.

N2. If $v_0 = 0$, set $v \leftarrow v + (1/b)\lfloor b(b-v_1)/(v_1+1) \rfloor v$; let the result of this step be $(v_0 v_1 v_2 \dots v_n.v_{n+1} \dots)_b$. Repeat step N2 until $v_0 \ne 0$.

Prove that step N2 will be performed at most three times, and that we must always have $v_0 = 1$, $v_1 = 0$ at the end of the calculations.

[*Note:* If u and v are both multiplied by the above constants, we do not change the value of the quotient u/v, and the divisor has been converted into the form $(1\ 0\ v_2 \dots v_n.v_{n+1}v_{n+2}v_{n+3})_b$. This form of the divisor may be very convenient because, in the notation of Algorithm D, we may simply take $\hat{q} = u_j$ as a trial divisor at the beginning of step D3, or $\hat{q} = b - 1$ when $(u_{j-1}, u_j) = (1, 0)$.]

29. [*15*] Prove or disprove: At the beginning of step D7 of Algorithm D, we always have $u_j = 0$.

▶ **30.** [*22*] If memory space is limited, it may be desirable to use the same storage locations for both input and output during the performance of some of the algorithms in this section. Is it possible to have $w_1, \dots, w_n$ stored in the same respective locations as $u_1, \dots, u_n$ or $v_1, \dots, v_n$ during Algorithm A or S? Is it possible to have $q_0, \dots, q_m$ occupy the same locations as $u_0, \dots, u_m$ in Algorithm D? Is there any permissible overlap of memory locations between input and output in Algorithm M?

31. [*28*] Assume that $b = 3$ and that $u = (u_1 \dots u_{m+n})_3$, $v = (v_1 \dots v_n)_3$ are integers in *balanced ternary* notation (cf. Section 4.1), $v_1 \ne 0$. Design a long-division algorithm which divides u by v, obtaining a remainder whose absolute value does not exceed $\frac{1}{2}|v|$. Try to find an algorithm which would be efficient if incorporated into the arithmetic circuitry of a balanced ternary computer.

32. [*M40*] Assume that $b = 2i$ and that u and v are complex numbers expressed in the quater-imaginary number system. Design algorithms which divide u by v, perhaps obtaining a suitable remainder of some sort, and compare their efficiency. *References:* M. Nadler, *CACM* **4** (1961), 192–193; Z. Pawlak and A. Wakulicz, *Bull. de l' Acad. Polonaise des Sciences*, Classe III, **5** (1957), 233–236 (see also pp. 803–804); and exercise 4.1–15.

33. [*M40*] Design an algorithm for taking square roots, analogous to Algorithm D and to the pencil-and-paper method for extracting square roots.

34. [*40*] Develop a set of computer subroutines for doing the four arithmetic operations on arbitrary integers, putting no constraint on the size of the integers except for the implicit assumption that the total memory capacity of the computer should not be exceeded. (Use linked memory allocation, so that no time is wasted in finding room to put the results.)

35. [*40*] Develop a set of computer subroutines for "decuple-precision floating-point" arithmetic, using excess 0, base b, nine-place floating-point number representation, where b is the computer word size, and allowing a full word for the exponent. (Thus each floating-point number is represented in 10 words of memory, and all scaling is done by moving full words instead of shifting within the words.)

36. [*M42*] Compute the values of the fundamental constants listed in Appendix B to much higher precision than the 40-place values listed there. (*Note:* The first 100,000 digits of the decimal expansion of π were published by D. Shanks and J. W. Wrench, Jr., in *Math. Comp.* **16** (1962), 76–99. In 1967, Jean Guilloud computed π's first 500,000 digits.)

*4.3.2. Modular Arithmetic

Another interesting alternative is available for doing arithmetic on large integer numbers, based on some simple principles of number theory. The idea is to have several "moduli" $m_1, m_2, \ldots, m_r$ which contain no common factors, and to work indirectly with "residues" $u \bmod m_1$, $u \bmod m_2 \ldots$, $u \bmod m_r$ instead of directly with the number u.

For convenience in notation throughout this section, let

$$u_1 = u \bmod m_1, \qquad u_2 = u \bmod m_2, \qquad \ldots, \qquad u_r = u \bmod m_r. \qquad (1)$$

It is easy to compute $(u_1, u_2, \ldots, u_r)$ from an integer number u by means of division; and—more important—no information is lost in this process, since we can always recompute u from $(u_1, u_2, \ldots, u_r)$ provided that we know u is not too large. For example, if $0 \leq u < v \leq 1000$, it is impossible to have $(u \bmod 7, u \bmod 11, u \bmod 13)$ equal to $(v \bmod 7, v \bmod 11, v \bmod 13)$. This is a consequence of the "Chinese Remainder Theorem" stated below.

Therefore we may regard $(u_1, u_2, \ldots, u_r)$ as a new type of internal computer representation, a "modular representation," of the integer u.

The advantages of a modular representation are that addition, subtraction, and multiplication are very simple:

$$(u_1, \ldots, u_r) + (v_1, \ldots, v_r) = ((u_1 + v_1) \bmod m_1, \ldots, (u_r + v_r) \bmod m_r), \qquad (2)$$

$$(u_1, \ldots, u_r) - (v_1, \ldots, v_r) = ((u_1 - v_1) \bmod m_1, \ldots, (u_r - v_r) \bmod m_r), \qquad (3)$$

$$(u_1, \ldots, u_r) \times (v_1, \ldots, v_r) = ((u_1 \times v_1) \bmod m_1, \ldots, (u_r \times v_r) \bmod m_r). \qquad (4)$$

It is easy to prove these formulas; for example, to prove (4) we must show that $uv \bmod m_j = (u \bmod m_j)(v \bmod m_j) \bmod m_j$ for each modulus m_j. But this is a basic fact of elementary number theory: $x \bmod m_j = y \bmod m_j$ if and only if $x \equiv y$ (modulo m_j); furthermore if $x \equiv x'$ and $y \equiv y'$, then $xy \equiv x'y'$ (modulo m_j); hence $(u \bmod m_j)(v \bmod m_j) \equiv uv$ (modulo m_j).

The disadvantages of a modular representation are that it is comparatively difficult to test whether a number is positive or negative or to test whether or not $(u_1, \ldots, u_r)$ is greater than $(v_1, \ldots, v_r)$. It is also difficult to test whether or not overflow has occurred as the result of an addition, subtraction, or multiplication, and it is even more difficult to perform division. When these operations are required frequently in conjunction with addition, subtraction, and multiplication, the use of modular arithmetic can be justified only if fast means of conversion into and out of the modular representation are available. Therefore conversion between modular and positional notation is one of the principal topics of interest to us in this section.

The processes of addition, subtraction, and multiplication using (2), (3), and (4) are called residue arithmetic or *modular arithmetic*. The range of numbers that can be handled by modular arithmetic is equal to $m = m_1 m_2 \ldots m_r$, the product of the moduli. Therefore we see that the amount of time required to add, subtract, or multiply n-digit numbers using modular arithmetic is essentially proportional to n (not counting the time to convert in and out of modular representation). This is no advantage at all when addition and subtraction are considered, but it can be a considerable advantage with respect to multiplication since the conventional method of the preceding section requires an execution time proportional to n^2.

Moreover, on a computer which allows many operations to take place simultaneously, modular arithmetic can be a significant advantage even for addition and subtraction; the operations with respect to different moduli can all be done at the same time, so we obtain a substantial savings of execution time. The same kind of decrease in execution time could not be achieved by the conventional techniques discussed in the previous section, since carry propagation must be considered. Perhaps some day highly parallel computers will make simultaneous operations commonplace, so that modular arithmetic will be of significant importance in "real-time" calculations when a quick answer to a single problem requiring high precision is needed. (With highly parallel computers, it is often preferable to run k *separate* programs simultaneously, instead of running a *single* program k times as fast, since the latter alternative is more complicated but does not utilize the machine any more efficiently; "real-time" calculations are exceptions which make the inherent parallelism of modular arithmetic more significant.)

Now let us examine the basic fact which underlies the modular representation of numbers:

Theorem C (*Chinese Remainder Theorem*). *Let $m_1, m_2, \ldots, m_r$ be positive integers which are relatively prime in pairs, i.e.,*

$$\gcd (m_j, m_k) = 1 \qquad \text{when} \qquad j \neq k. \tag{5}$$

Let $m = m_1 m_2 \ldots m_r$, and let $a, u_1, u_2, \ldots, u_r$ be integers. Then there is exactly one integer u which satisfies the conditions

$$a \leq u < a + m, \quad \text{and} \quad u \equiv u_j \ (\text{modulo } m_j) \quad \text{for} \quad 1 \leq j \leq r.$$
$$\tag{6}$$

Proof. If $u \equiv v$ (modulo m_j) for $1 \leq j \leq r$, then $u - v$ is a multiple of m_j for all j, so (5) implies that $u - v$ is a multiple of $m = m_1 m_2 \ldots m_r$. This argument shows that there is *at most* one solution of (6). To complete the proof we must only show the existence of *at least* one solution, and this can be done in two simple ways:

METHOD 1 ("Nonconstructive" proof). As u runs through the m distinct values $a \leq u < a + m$, the r-tuples $(u \bmod m_1, \ldots, u \bmod m_r)$ must also run through m distinct values, since (6) has at most one solution. But there are exactly $m_1 m_2 \ldots m_r$ possible r-tuples $(v_1, \ldots, v_r)$ such that $0 \leq v_j < m_j$. Therefore each r-tuple must occur exactly once, and there must be some value of u for which $(u \bmod m_1, \ldots, u \bmod m_r) = (u_1, \ldots, u_r)$.

METHOD 2 ("Constructive" proof). We can find numbers M_j, $1 \leq j \leq r$, such that

$$M_j \equiv 1 \ (\text{modulo } m_j); \quad M_j \equiv 0 \ (\text{modulo } m_k) \quad \text{for} \quad k \neq j. \tag{7}$$

This follows because (5) implies that m_j and m/m_j are relatively prime, so we may take

$$M_j = (m/m_j)^{\varphi(m_j)} \tag{8}$$

by Euler's theorem (exercise 1.2.4–28). Now the number

$$u = a + ((u_1 M_1 + u_2 M_2 + \cdots + u_r M_r - a) \bmod m) \tag{9}$$

satisfies all the conditions of (6). ∎

A very special case of this theorem was stated by the Chinese mathematician Sun-Tsŭ, who gave a rule called tái-yen ("great generalization"); the date of his writing is very uncertain, it may have been as early as 200 B.C. or as late as 200 A.D. Curiously, the Greek mathematician Nichomachus gave exactly the *same* special case of the theorem at about the same time (100 A.D.). Theorem C was apparently first stated and proved in its proper generality by L. Euler in 1734, although a description of most of the necessary principles was given in China by Chhin Chiu-Shao in his *Shu Shu Chiu Chang* (1247).

As a consequence of Theorem C, we may use modular representation for numbers in any consecutive interval of $m = m_1 m_2 \ldots m_r$ integers. For example, we could take $a = 0$ in (6), and work only with nonnegative integers u less than m. On the other hand, when addition and subtraction are being done, as well as multiplication, it is usually most convenient to assume that all the moduli $m_1, m_2, \ldots, m_r$ are odd numbers, so that $m = m_1 m_2 \ldots m_r$ is odd, and to work with integers in the range

$$-\frac{m}{2} < u < \frac{m}{2}, \tag{10}$$

which is completely symmetrical about zero.

To perform the basic operations indicated in (2), (3), and (4), we need to compute $(u_j + v_j) \bmod m_j$, $(u_j - v_j) \bmod m_j$, and $u_j v_j \bmod m_j$, when $0 \leq u_j, v_j < m_j$. If m_j is a single-precision number, it is most convenient to form $u_j v_j \bmod m_j$ by doing a multiplication and then a division operation. For addition and subtraction, the situation is a little simpler, since no division is necessary; the following formulas may conveniently be used;

$$(u_j + v_j) \bmod m_j = \begin{cases} u_j + v_j, & \text{if} \quad u_j + v_j < m_j; \\ u_j + v_j - m_j, & \text{if} \quad u_j + v_j \geq m_j. \end{cases} \tag{11}$$

$$(u_j - v_j) \bmod m_j = \begin{cases} u_j - v_j, & \text{if} \quad u_j - v_j \geq 0; \\ u_j - v_j + m_j, & \text{if} \quad u_j - v_j < 0. \end{cases} \tag{12}$$

(Cf. Section 3.2.1.1.) In this case, since we want m to be as large as possible, it is easiest to let m_1 be the largest odd number that fits in a computer word, to let m_2 be the largest odd number $< m_1$ that is relatively prime to m_1, to let m_3 be the largest odd number $< m_2$ that is relatively prime to both m_1 and m_2, and so on until enough m_j's have been found to give the desired range m. Efficient ways to determine whether or not two integers are relatively prime are discussed in Section 4.5.2.

As a simple example, suppose that we have a decimal computer with a word size of only 100. Then the procedure described in the previous paragraph would give

$$m_1 = 99, \quad m_2 = 97, \quad m_3 = 95, \quad m_4 = 91, \quad m_5 = 89, \quad m_6 = 83, \tag{13}$$

and so on.

On binary computers it is often desirable to choose the m_j in a different way, by selecting

$$m_j = 2^{e_j} - 1. \tag{14}$$

In other words, each modulus is one less than a power of 2. Such a choice of m_j often makes the basic arithmetic operations simpler, because it is relatively easy to work modulo $2^{e_j} - 1$, as in ones' complement arithmetic. When the moduli are chosen according to this strategy, it is helpful to relax the condition $0 \leq u_j < m_j$ slightly, so that we require only

$$0 \leq u_j < 2^{e_j}, \qquad u_j \equiv u \pmod{2^{e_j} - 1}. \tag{15}$$

Thus, the value $u_j = m_j = 2^{e_j} - 1$ is allowed as an optional alternative to $u_j = 0$, since this does not affect the validity of Theorem C, and it means we are allowing u_j to be any e_j-bit binary number. Under this assumption, the operations of addition and multiplication modulo m_j become the following:

$$u_j \oplus v_j = \begin{cases} u_j + v_j, & \text{if} \quad u_j + v_j < 2^{e_j}; \\ ((u_j + v_j) \bmod 2^{e_j}) + 1, & \text{if} \quad u_j + v_j \geq 2^{e_j}. \end{cases} \tag{16}$$

$$u_j \otimes v_j = (u_j v_j \bmod 2^{e_j}) \oplus \lfloor u_j v_j / 2^{e_j} \rfloor. \tag{17}$$

[Here $\oplus$ and $\otimes$ refer to the operations to be done on the individual components of $(u_1, \ldots, u_r)$ and $(v_1, \ldots, v_r)$ when adding or multiplying, respectively, using the convention (15).] Equation (12) may be used for subtraction. Clearly, these operations can be readily performed even when m_j is larger than the computer's word size; it is a simple matter to compute the remainder of a positive number modulo a power of 2, or to divide a number by a power of 2. In (17) we have the sum of the "upper half" and the "lower half" of the product, as discussed in exercise 3.2.1.1–8.

If moduli of the form $2^{e_j} - 1$ are to be used, we must know under what conditions the number $2^e - 1$ is relatively prime to the number $2^f - 1$. Fortunately, there is a very simple rule,

$$\gcd(2^e - 1, 2^f - 1) = 2^{\gcd(e,f)} - 1, \tag{18}$$

which states in particular that $2^e - 1$ *and* $2^f - 1$ *are relatively prime if and only if* e *and* f *are relatively prime.* Equation (18) follows from Euclid's algorithm and the identity

$$(2^e - 1) \bmod (2^f - 1) = 2^{e \bmod f} - 1. \tag{19}$$

(See exercise 6.) Thus we could choose for example $m_1 = 2^{35} - 1$, $m_2 = 2^{34} - 1$, $m_3 = 2^{33} - 1$, $m_4 = 2^{31} - 1$, $m_5 = 2^{29} - 1$, if we had a computer with word size 2^{35} and if we wanted to represent numbers up to $m_1 m_2 m_3 m_4 m_5 > 2^{161}$. This range of integers is not big enough to make modular arithmetic faster than the conventional method, and we usually find that modular arithmetic using convention (15) is advantageous only when the m_j are larger than the word size.

As we have already observed, the operations of conversion to and from modular representation are very important. If we are given a number u, its modular representation $(u_1, \ldots, u_r)$ may be obtained by dividing u by $m_1, \ldots, m_r$ and saving the remainders. A possibly more attractive procedure, if $u = (v_m v_{m-1} \ldots v_0)_b$, is to evaluate the polynomial

$$(\ldots (v_m b + v_{m-1})b + \ldots)b + v_0$$

using modular arithmetic. When $b = 2$ and when the modulus m_j has the special form $2^{e_j - 1}$, both of these methods reduce to quite a simple procedure.

Consider the binary representation of u with blocks of e_j bits grouped together,

$$u = a_t A^t + a_{t-1} A^{t-1} + \cdots + a_1 A + a_0, \tag{20}$$

where $A = 2^{e_j}$ and $0 \le a_k < 2^{e_j}$ for $0 \le k \le t$. Then

$$u \equiv a_t + a_{t-1} + \cdots + a_1 + a_0 \quad (\text{modulo } 2^{e_j} - 1), \tag{21}$$

since $A \equiv 1$. Therefore we may obtain u_j by adding the e_j-bit numbers $a_t \oplus \cdots \oplus a_1 \oplus a_0$, modulo $2^{e_j} - 1$, as in Eq. (16). This process is similar to the familiar device of "casting out nines" which is used to determine u mod 9 when u is expressed in the decimal system.

Conversion back from modular form to positional notation is somewhat more difficult. It is interesting in this regard to make a few side remarks about the way computers make us change our viewpoint towards mathematical proofs: Theorem C tells us that the conversion from $(u_1, \ldots, u_r)$ to u is possible, and two proofs are given. The first proof we considered is a classical one which makes use only of very simple concepts, namely the facts that

i) any number which is a multiple of m_1 and of $m_2, \ldots,$ and of m_r, must be a multiple of $m_1 m_2 \ldots m_r$ when the m_j's are pairwise relatively prime; and

ii) if m things are put into m boxes with no two things in the same box, then there must be one in each box.

By traditional notions of mathematical aesthetics, this is no doubt the nicest proof of Theorem C; but from a computational standpoint it is completely worthless! It amounts to saying, "Try $u = a, a + 1, \ldots$ until you find a value for which $u \equiv u_1$ (modulo m_1), $\ldots,$ $u \equiv u_r$ (modulo m_r)."

The second proof of Theorem C is more explicit; it shows how to compute r new constants $M_1, \ldots, M_r$, and to get the solution in terms of these constants by formula (9). This proof uses more complicated concepts (for example, Euler's theorem), but it is much more satisfactory from a computational standpoint, since the constants $M_1, \ldots, M_r$ need to be determined only once. On the other hand, the determination of M_j by Eq. (8) is certainly not trivial, since the evaluation of Euler's φ-function requires, in general, the factorization of m_j into prime powers. Furthermore, M_j is likely to be a terribly large number, even if we compute only the quantity $M_j \bmod m$, which will work just as well as M_j in (9). Since $M_j \bmod m$ is uniquely determined if (7) is to be satisfied (because of the Chinese Remainder Theorem!), we can see that, in any event, Eq. (9) requires a lot of high-precision calculation, and such calculation is just what we wished to avoid by modular arithmetic in the first place.

So we need an even *better* proof of Theorem C if we are going to have a really usable method of conversion from $(u_1, \ldots, u_r)$ to u. Such a method was suggested by H. L. Garner in 1958; it can be carried out using $\binom{r}{2}$ constants c_{ij} for $1 \le i < j \le r$, where

$$c_{ij} m_i \equiv 1 \quad (\text{modulo } m_j). \tag{22}$$

These constants c_{ij} are readily computed using Euclid's algorithm, since Algorithm 4.5.2X determines a, b such that $am_i + bm_j = \gcd(m_i, m_j) = 1$ and we may take $c_{ij} = a$. When the moduli have the special form $2^{e_j} - 1$, a simple method of determining c_{ij} is given in exercise 6.

Once the c_{ij} have been determined satisfying (22), we can set

$$v_1 \leftarrow u_1 \bmod m_1,$$
$$v_2 \leftarrow (u_2 - v_1)c_{12} \bmod m_2,$$
$$v_3 \leftarrow ((u_3 - v_1)c_{13} - v_2)c_{23} \bmod m_3, \qquad (23)$$
$$\ldots$$
$$v_r \leftarrow (\ldots ((u_r - v_1)c_{1r} - v_2)c_{2r} - \cdots - v_{r-1})c_{(r-1)r} \bmod m_r.$$

Then

$$u = v_r m_{r-1} \ldots m_1 + \cdots + v_3 m_2 m_1 + v_2 m_1 + v_1 \qquad (24)$$

is a number satisfying the conditions

$$0 \leq u < m, \qquad u \equiv u_j \quad (\text{modulo } m_j), \qquad 1 \leq j \leq r. \qquad (25)$$

(See exercise 8; another way of rewriting (23) which does not involve as many auxiliary constants is given in exercise 7.) Equation (24) is a *mixed radix representation* of u, which may be converted to binary or decimal notation using the methods of Section 4.4. If $0 \leq u < m$ is not the desired range, an appropriate multiple of m can be added or subtracted after the conversion process.

The advantage of the computation shown in (23) is that the calculation of v_j can be done using only arithmetic mod m_j, which is already built into the modular arithmetic algorithms. (Note that the computation (23) cannot be done in parallel with all the v_j computed at the same time; v_{j-1} must be calculated before v_j. A method which allows more parallelism, but not less total computation, is discussed in an interesting article by A. S. Fraenkel, *Proc. ACM Nat. Conf.* **19** (Philadelphia, 1965), E1.4.)

It is important to observe that the mixed radix representation (24) is sufficient to compare the magnitudes of two modular numbers. For if we know that $0 \leq u < m$ and $0 \leq u' < m$, then we can tell if $u < u'$ by first doing the conversion to $v_1, \ldots, v_r$ and $v_1', \ldots, v_r'$, then testing if $v_r < v_r'$, or if $v_r = v_r'$ and $v_{r-1} < v_{r-1}'$, etc. It is not necessary to convert all the way to binary or decimal notation if we only want to know whether $(u_1, \ldots, u_r)$ is less than $(u_1', \ldots, u_r')$.

The operation of comparing two numbers, or of deciding if a modular number is negative, is intuitively very simple, so we would expect to find a much easier method for making this test than the conversion to mixed radix form. But the following theorem shows that there is little hope of finding a substantially easier method, since the range of a modular number depends essentially on all the residues $(u_1, \ldots, u_r)$ at once:

Theorem S. (*Nicholas Szabó, 1961*). *In terms of the notation above, assume that $m_1 < \sqrt{m}$, and let L be any value in the range*

$$m_1 \leq L \leq m - m_1. \tag{26}$$

Let g be any function such that the set $\{g(0), g(1), \ldots, g(m_1 - 1)\}$ contains less than m_1 values. Then there are numbers u and v such that

$$g(u \bmod m_1) = g(v \bmod m_1), \qquad u \bmod m_j = v \bmod m_j \quad \text{for} \quad 2 \leq j \leq r; \tag{27}$$

$$0 \leq u < L \leq v < m. \tag{28}$$

Proof. By hypothesis, there must exist numbers $u \neq v$ satisfying (27), since g must take on the same value for two different residues. Let (u, v) be a pair of values with $0 \leq u < v < m$ satisfying (27), for which u is a minimum. Since $u' = u - m_1$ and $v' = v - m_1$ also satisfy (27), we must have $u' < 0$ by the minimality of u. Hence $u < m_1 \leq L$; and if (28) does not hold, we must have $v < L$. But $v > u$, and $v - u$ is a multiple of $m_2 \ldots m_r = m/m_1$, so $v \geq v - u \geq m/m_1 > m_1$. Therefore, if (28) does not hold for (u, v), it will be satisfied for the pair $(u'', v'') = (v - m_1, u + m - m_1)$. ∎

Of course, a similar result can be proved for any m_j in place of m_1; and we could also replace (28) by the condition "$a \leq u < a + L \leq v < a + m$" with only minor changes in the proof. Therefore Theorem S shows that many simple functions cannot be used to determine the range of a modular number.

Let us now reiterate the main points of the discussion in this section: Modular arithmetic can be a significant advantage for applications in which the predominant calculations involve exact multiplication (or raising to a power) of large integers, combined with addition and subtraction, but where there is very little need to divide or compare numbers, *nor to test whether intermediate results "overflow" out of range.* (It is important not to forget the latter restriction; methods are available to test for overflow, as in exercise 12, but they are in general so complicated that they nullify the advantages of modular arithmetic.) Several applications for modular computations have been discussed by H. Takahasi and Y. Ishibashi, *Information Processing in Japan* **1** (1961), 28–42.

An example of such an application is the exact solution of linear equations with rational coefficients. For various reasons it is desirable in this case to assume that the moduli $m_1, m_2, \ldots, m_r$ are all large prime numbers; the linear equations can be solved independently modulo each m_j. A detailed discussion of this procedure has been given by I. Borosh and A. S. Fraenkel [*Math. Comp.* **20** (1966), 107–112]. By means of their method, the nine independent solutions of a system of 111 linear equations in 120 unknowns were obtained exactly in less than an hour's running time on the CDC 1604 computer; this is substantially faster than any other known method for obtaining exact solutions. The same procedure is worth while also for solving simultaneous linear equations with floating-point coefficients, when the matrix of coefficients is ill-conditioned. The modular technique (treating the given floating-point coefficients as exact rational numbers) gives a method for obtaining the *true* answers in less time than any other known method can produce reliable *approximate* answers!

The published literature concerning modular arithmetic is mostly oriented towards hardware design, since the carry-free properties of modular arithmetic make it attractive from the standpoint of high-speed operation. The idea was first published by A. Svoboda and M. Valach in the Czechoslovakian journal *Stroje na Zpracování Informací* **3** (1955), 247–295; then independently by H. L. Garner [*IRE Transactions* **EC-8** (1959), 140–147]. The use of moduli of the form $2^{e_j} - 1$ was suggested by A. S. Fraenkel [*JACM* **8** (1961), 87–96], and several advantages of such moduli were demonstrated by A. Schönhage [*Computing* **1** (1966), 182–196]. See the book *Residue Arithmetic and its Applications to Computer Technology* by N. S. Szabó and R. I. Tanaka (New York: McGraw-Hill, 1967), for additional information and a comprehensive bibliography of the subject.

Further discussion of modular arithmetic can be found in part B of Section 4.3.3.

EXERCISES

1. [*20*] Find all integer numbers u which satisfy the conditions $u \bmod 7 = 1$, $u \bmod 11 = 6$, $u \bmod 13 = 5$, and $0 \le u \le 1000$.

2. [*M20*] Would Theorem C still be true if we allowed $a, u_1, u_2, \ldots, u_r$ and u to be arbitrary real numbers (not just integers)?

▶ **3.** [*M26*] (Generalized Chinese Remainder Theorem.)   Let $m_1, m_2, \ldots, m_r$ be positive integers. Let m be the least common multiple of $m_1, m_2, \ldots, m_r$, and let $a, u_1, u_2, \ldots, u_r$ be any integers. Prove that there is exactly one integer u which satisfies the conditions

$$a \le u < a + m, \qquad u \equiv u_j \pmod{m_j}, \qquad 1 \le j \le r,$$

provided that

$$u_i \equiv u_j \pmod{\gcd(m_i, m_j)}, \qquad 1 \le i < j \le r;$$

and there is no such integer u when the latter condition fails to hold.

4. [*20*] Continue the process shown in (13); what would m_7, m_8, m_9, m_{10} be?

▶ **5.** [*M23*] Suppose that the method of (13) is continued until no more m_j can be chosen; does this method give the largest attainable value $m_1 m_2 \ldots m_r$ such that the m_j are odd positive integers less than 100 which are relatively prime in pairs?

6. [*M22*] Let e, f, g be nonnegative integers. (a) Show that $2^e \equiv 2^f$ (modulo $2^g - 1$) if and only if $e \equiv f$ (modulo g). (b) Given that $e \bmod f = d$ and $ce \bmod f = 1$, prove that

$$\left((1 + 2^d + \cdots + 2^{(c-1)d}) \cdot (2^e - 1)\right) \bmod (2^f - 1) = 1.$$

[Thus, we have a comparatively simple formula for the inverse of $2^e - 1$, modulo $2^f - 1$, as required in (22).]

▶ **7.** [*M21*] Show that (23) can be rewritten as follows:

$$v_1 \leftarrow u_1 \bmod m_1,$$
$$v_2 \leftarrow (u_2 - v_1)c_{12} \bmod m_2,$$
$$v_3 \leftarrow (u_3 - (v_1 + m_1 v_2))c_{13}c_{23} \bmod m_3,$$
$$\cdots$$
$$v_r \leftarrow (u_r - (v_1 + m_1(v_2 + m_2(v_3 + \cdots + m_{r-2}v_{r-1}) \ldots)))c_{1r} \ldots c_{(r-1)r} \bmod m_r.$$

If the formulas are rewritten in this way, we see that only $r - 1$ constants $C_j = c_{1j} \ldots c_{(j-1)j} \bmod m_j$ are needed instead of $r(r - 1)/2$ constants c_{ij} as in (23). Discuss the relative merits of this version of the formula as compared to (23), from the standpoint of computer calculation.

8. [*M21*] Prove that the number u defined by (23) and (24) satisfies (25).

9. [*M20*] Show how to go from the values $v_1, \ldots, v_r$ of the mixed radix notation (24) back to the original residues $u_1, \ldots, u_r$, using only arithmetic mod m_j to compute u_j.

10. [*M25*] An integer u which lies in the symmetrical range (10) might be represented by finding the numbers $u_1, \ldots, u_r$ such that $u \equiv u_j$ (modulo m_j) and $-m_j/2 < u_j < m_j/2$, instead of insisting that $0 \leq u_j < m_j$ as in the text. Discuss the modular arithmetic procedures that would be used in this case (including the conversion process, (23)).

11. [*M23*] Assume that all the m_j are odd, and that $u = (u_1, \ldots, u_r)$ is known to be even, where $0 \leq u < m$. Find a reasonably fast method to compute $u/2$ using modular arithmetic.

12. [*M20*] Prove that, if $0 \leq u, v < m$, the modular addition of u and v causes overflow (i.e., is outside the range allowed by the modular representation) if and only if the sum is less than u. (Thus the overflow detection problem is equivalent to the comparison problem.)

▶ **13.** [*M22*] (*Automorphic numbers.*) An n-place decimal number $x > 1$ is called an "automorph" by recreational mathematicians if the last n digits of x^2 are equal to x; i.e., if $x^2 \bmod 10^n = x$. [See *Scientific American* **218** (January, 1968), 125.] For example, 9376 is a 4-digit automorph.

(a) Prove that an n-place number $x > 1$ is an automorph if and only if $x \bmod 5^n = 0$ or 1, and $x \bmod 2^n = 1$ or 0, respectively. [Thus, if $m_1 = 2^n$ and $m_2 = 5^n$, the only two n-place automorphs are the numbers M_1 and M_2 in (7).]

(b) Prove that if x is an n-place automorph, then $(3x^2 - 2x^3) \bmod 10^{2n}$ is a $2n$-place automorph.

(c) Given that $cx^n \equiv 1$ (modulo y^n), what is a simple formula for a number c' such that $c'x^{2n} \equiv 1$ (modulo y^{2n})?

*4.3.3. How Fast Can We Multiply?

The conventional method for multiplication, Algorithm 4.3.1M, requires approximately cmn operations to multiply an m-digit number by an n-digit number, where c is a constant. In this section, let us assume for convenience that $m = n$, and let us consider the following question: *Does every general computer algorithm for multiplying two n-digit numbers require an execution time proportional to n^2, as n increases?*

(In this question, a "general" algorithm means one which accepts, as input, the number n and two arbitrary n-digit numbers in positional notation, and whose output is their product in positional form. Certainly if we were allowed to choose a different algorithm for each value of n, the question would be of no interest, since multiplication could be done for any specific value of n by a "table-lookup" operation in some huge table. The term "computer algorithm" is meant to imply an algorithm which is suitable for implementation on a digital computer such as MIX, and the execution time is to be the time it takes to perform the algorithm on such a computer.)

A. Digital methods. The answer to the above question is, rather surprisingly, "No," and, in fact, it is not very difficult to see why. For convenience, let us assume throughout this section that we are working with integers expressed in binary notation. If we have two $2n$-bit numbers $u = (u_{2n-1} \ldots u_1 u_0)_2$ and $v = (v_{2n-1} \ldots v_1 v_0)_2$, we can write

$$u = 2^n U_1 + U_0, \qquad v = 2^n V_1 + V_0, \tag{1}$$

where $U_1 = (u_{2n-1} \ldots u_n)_2$ is the "most-significant half" of u and $U_0 = (u_{n-1} \ldots u_0)_2$ is the "least-significant half"; and similarly $V_1 = (v_{2n-1} \ldots v_n)_2$, $V_0 = (v_{n-1} \ldots v_0)_2$. Now we have

$$uv = (2^{2n} + 2^n)U_1V_1 + 2^n(U_1 - U_0)(V_0 - V_1) + (2^n + 1)U_0V_0. \tag{2}$$

This formula reduces the problem of multiplying $2n$-bit numbers to three multiplications of n-bit numbers, U_1V_1, $(U_1 - U_0)(V_0 - V_1)$, and U_0V_0, plus some simple shifting and adding operations.

Formula (2) can be used for double-precision multiplication when a quadruple precision result is desired, and it is just a little faster than the traditional method on many machines. It is more important to observe that we can use formula (2) to define a recursive process for multiplication which is significantly faster than the familiar order-n^2 method when n is large: If $T(n)$ is the time required to perform multiplication of n-bit numbers, we have

$$T(2n) \le 3T(n) + cn \tag{3}$$

for some constant c, since the right-hand side of (2) uses just three multiplications plus some additions and shifts. Relation (3) implies by induction that

$$T(2^k) \le c(3^k - 2^k), \qquad k \ge 1, \tag{4}$$

if we choose c to be large enough so that this inequality is valid when $k = 1$; and therefore we have

$$T(n) \le T(2^{\lceil \log_2 n \rceil}) \le c(3^{\lceil \log_2 n \rceil} - 2^{\lceil \log_2 n \rceil})$$
$$\le 3c \cdot 3^{\log_2 n} = 3cn^{\log_2 3}. \tag{5}$$

Relation (5) shows that the running time for multiplication can be reduced from order n^2 to order $n^{\log_2 3} \approx n^{1.585}$, and of course this is a much faster algorithm when n is large.

(A similar but more complicated method for doing multiplication with running time of order $n^{\log_2 3}$ was apparently first suggested by A. Karatsuba, *Doklady Akademiia Nauk SSSR* **145** (1962), 293–294. Curiously, this idea does not seem to have been discovered before 1962; none of the "calculating prodigies" who have become famous for their ability to multiply large numbers mentally have been reported to use any such method, although formula (2) adapted to decimal notation would seem to lead to a reasonably easy way to multiply eight-digit numbers in one's head.)

The running time can be reduced still further, in the limit as n approaches infinity, if we observe that the method just used is essentially the special case $r = 1$ of a more general method that yields

$$T((r + 1)n) \le (2r + 1)T(n) + cn \tag{6}$$

for any fixed r. This more general method can be obtained as follows: Let

$$u = (u_{(r+1)n-1} \ldots u_1 u_0)_2 \qquad \text{and} \qquad v = (v_{(r+1)n-1} \ldots v_1 v_0)_2$$

be broken into $r + 1$ pieces,

$$u = U_r 2^{rn} + \cdots + U_1 2^n + U_0, \qquad v = V_r 2^{rn} + \cdots + V_1 2^n + V_0, \tag{7}$$

where each U_j and each V_j is an n-bit number. Consider the polynomials

$$U(x) = U_r x^r + \cdots + U_1 x + U_0, \qquad V(x) = V_r x^r + \cdots + V_1 x + V_0, \tag{8}$$

and let

$$W(x) = U(x)V(x) = W_{2r} x^{2r} + \cdots + W_1 x + W_0. \tag{9}$$

Since $u = U(2^n)$ and $v = V(2^n)$, we have $uv = W(2^n)$, so we can easily compute uv if we know the coefficients of $W(x)$. The problem is to find a good way to compute the coefficients of $W(x)$ by using only $2r + 1$ multiplications of n-bit numbers plus some further operations which involve only an execution time proportional to n. This can be done by computing

$$U(0)V(0) = W(0), \qquad U(1)V(1) = W(1), \qquad \ldots, \qquad U(2r)V(2r) = W(2r).$$
$$(10)$$

The coefficients of a polynomial of degree $2r$ can be written as a linear combination of the values of that polynomial at $2r + 1$ distinct points; such a linear combination requires an execution time at most proportional to n. (Actually, the products $U(j)V(j)$ are not strictly products of n-bit numbers, but they are products of at most $(n + t)$-bit numbers, where t is a fixed value depending on r. It is easy to design a multiplication routine for $(n + t)$-bit numbers which requires only $T(n) + c_1 n$ operations, since two products of t-bit by n-bit numbers can be done in $c_2 n$ operations when t is fixed.) Therefore we obtain a method of multiplication satisfying (6).

Relation (6) can be used to show that $T(n) \le c_3 n^{\log_{r+1}(2r+1)} < c_3 n^{1+\log_{r+1} 2}$, using a method analogous to the derivation of (5), so we have now proved:

Theorem A. *Given $\epsilon > 0$, there exists a constant $c(\epsilon)$ and a multiplication algorithm such that the number of elementary operations $T(n)$ needed to multiply two n-bit numbers satisfies*

$$T(n) < c(\epsilon)n^{1+\epsilon}. \quad \blacksquare \qquad (11)$$

This theorem is still not the result we are after. It is unsatisfactory for practical purposes in that the method becomes much more complicated as $\epsilon \to 0$ (and therefore as $r \to \infty$), causing $c(\epsilon)$ to grow so rapidly that extremely huge values of n are needed before we have any significant improvement over (5). And it is unsatisfactory for theoretical purposes because it does not make use of the full power of the polynomial method on which it is based. We can obtain a better result if we let r *vary* with n, choosing larger and larger values of r as n increases. This idea is due to A. L. Toom [*Doklady Akademiia Nauk SSSR* **150** (1963), 496–498; tr. into English in *Soviet Mathematics* **3** (1963), 714–716], who used it to show that computer circuitry for multiplication of n-bit numbers can be constructed involving a fairly small number of components as n grows. S. A. Cook [*On the minimum computation time of functions* (Thesis, Harvard University, 1966), 51–77] later showed how Toom's method can be adapted to fast computer programs.

Before we discuss the Toom-Cook algorithm any further, let us study a small example of the transition from $U(x)$ and $V(x)$ to the coefficients of $W(x)$. This example will not demonstrate the efficiency of the method, since the numbers are too small, but it points out some useful simplifications that we can make in the general case. Suppose that we want to multiply $u = 1234$ times $v = 2341$; in binary notation this is $u = (0100\ 1101\ 0010)_2$ times $v = (1001\ 0010\ 0101)_2$. Let $r = 2$; the polynomials $U(x)$, $V(x)$ in (8) are

$$U(x) = 4x^2 + 13x + 2, \qquad V(x) = 9x^2 + 2x + 5.$$

Hence we find, for $W(x) = U(x)V(x)$,

$$
\begin{aligned}
&U(0) = \;\; 2, \quad U(1) = \;\; 19, \quad U(2) = \;\;\; 44, \quad U(3) = \;\;\; 77, \quad U(4) = \;\;\; 118; \\
&V(0) = \;\; 5, \quad V(1) = \;\; 16, \quad V(2) = \;\;\; 45, \quad V(3) = \;\;\; 92, \quad V(4) = \;\;\; 157; \\
&W(0) = 10, \quad W(1) = 304, \quad W(2) = 1980, \quad W(3) = 7084, \quad W(4) = 18526.
\end{aligned}
$$
(12)

Our job now is to compute the five coefficients of $W(x)$ from the latter five values.

There is an attractive little algorithm which can be used to compute the coefficients of a polynomial $W(x) = W_{m-1}x^{m-1} + \cdots + W_1 x + W_0$ when the values $W(0), W(1), \ldots, W(m-1)$ are given: Let us first write

$$
W(x) = \theta_{m-1}x^{\underline{m-1}} + \theta_{m-2}x^{\underline{m-2}} + \cdots + \theta_1 x^{\underline{1}} + \theta_0,
$$
(13)

where $x^{\underline{k}} = x(x-1)\ldots(x-k+1)$, and where the θ_j are unknown as well as the W_j. Now

$$
W(x+1) - W(x) = (m-1)\theta_{m-1}x^{\underline{m-2}} + (m-2)\theta_{m-2}x^{\underline{m-3}} + \cdots + \theta_1,
$$

and by induction we find that for all $k \geq 0$

$$
\begin{aligned}
\frac{1}{k!}\Bigg(W(x+k) &- \binom{k}{1}W(x+k-1) + \binom{k}{2}W(x+k-2) - \cdots \\
&+ (-1)^k W(x) \Bigg) \\
= \binom{m-1}{k}\theta_{m-1}x^{\underline{m-1-k}} &+ \binom{m-2}{k}\theta_{m-2}x^{\underline{m-2-k}} + \cdots + \binom{k}{k}\theta_k.
\end{aligned}
$$
(14)

Denoting the left-hand side of (14) in the customary way as $(1/k!)\,\Delta^k W(x)$, we see that

$$
\frac{1}{k!}\Delta^k W(x) = \frac{1}{k}\left(\frac{1}{(k-1)!}\Delta^{k-1}W(x+1) - \frac{1}{(k-1)!}\Delta^{k-1}W(x)\right)
$$

and $(1/k!)\,\Delta^k W(0) = \theta_k$. So the coefficients θ_j can be evaluated using a very simple method, illustrated here for the polynomial $W(x)$ in (12):

```
  10
            294
 304                1382/2 =  691
          1676                       1023/3 = 341
1980                3428/2 = 1714                     144/4 = 36   (15)
          5104                       1455/3 = 485
7084                6338/2 = 3169
         11442
18526
```

The leftmost column of this tableau is a listing of the given values of $W(0)$, $W(1), \ldots, W(4)$; the kth succeeding column is obtained by computing the difference between successive values of the preceding column and dividing by k. The coefficients θ_j appear at the top of the columns, so that $\theta_0 = 10$, $\theta_1 = 294, \ldots, \theta_4 = 36$, and we have

$$W(x) = 36x^4 + 341x^3 + 691x^2 + 294x^1 + 10$$
$$= (((36(x - 3) + 341)(x - 2) + 691)(x - 1) + 294)x + 10. \quad (16)$$

In general, we can write

$$W(x) = (\ldots ((\theta_{m-1}(x - m + 2) + \theta_{m-2})(x - m + 3) + \theta_{m-3})$$
$$\times (x - m + 4) + \cdots + \theta_1)x + \theta_0,$$

and this formula shows how the coefficients $W_{m-1}, \ldots, W_1, W_0$ can be obtained from the θ's:

$$
\begin{array}{ccccc}
36 & 341 & & & \\
 & -3 \cdot 36 & & & \\
36 & 233 & 691 & & \\
 & -2 \cdot 36 & -2 \cdot 233 & & \quad (17) \\
36 & 161 & 225 & 294 & \\
 & -1 \cdot 36 & -1 \cdot 161 & -1 \cdot 225 & \\
36 & 125 & 64 & 69 & 10
\end{array}
$$

Here the numbers below the horizontal lines successively show the coefficients of the polynomials

$$\theta_{m-1}, \qquad \theta_{m-1}(x - m + 2) + \theta_{m-2},$$
$$\big(\theta_{m-1}(x - m + 2) + \theta_{m-2}\big)(x - m + 3) + \theta_{m-3}, \qquad \text{etc.}$$

From this tableau we have

$$W(x) = 36x^4 + 125x^3 + 64x^2 + 69x + 10,$$

so the answer to our original problem is $1234 \cdot 2341 = W(16)$, where $W(16)$ is obtained by adding and shifting. A generalization of this method is discussed in Section 4.6.4.

The basic Stirling number identity,

$$x^n = \left\{ {n \atop n} \right\} x^{\underline{n}} + \cdots + \left\{ {n \atop 1} \right\} x^{\underline{1}} + \left\{ {n \atop 0} \right\},$$

Eq. 1.2.6–41, shows that if the coefficients of $W(x)$ are nonnegative, so are the numbers θ_j, and in such a case *all of the intermediate results in the above computation are nonnegative.* This further simplifies the Toom-Cook multiplication algorithm, which we will now consider in detail.

Algorithm C (*High-precision multiplication of binary numbers*). Given a positive integer n and two nonnegative n-bit integers u and v, this algorithm forms their $2n$-bit product, w. Four auxiliary stacks are used to hold the long numbers which are manipulated during this algorithm:

Stack U, V: Temporary storage of $U(j)$ and $V(j)$ in step C4.
Stack C: Numbers to be multiplied, and control codes.
Stack W: Storage of $W(j)$.

These stacks may contain either binary numbers or special symbols called code-1, code-2, code-3, and code-4. The algorithm also constructs an auxiliary table of numbers q_k, r_k; this table is maintained in such a manner that it may be stored as a linear list, and all accesses to this table are made in a simple manner so that a single pointer (which traverses the list, moving back and forth) may be used to access the current table entry of interest.

   (Stacks C and W in this algorithm are used to control the recursive mechanism of the multiplication algorithm in a reasonably straightforward manner which is a special case of the general procedures discussed in Chapter 8.)

C1. [Compute q, r tables.] Set stacks U, V, C, and W empty. Set

$$k \leftarrow 1, \qquad q_0 \leftarrow q_1 \leftarrow 16, \qquad r_0 \leftarrow r_1 \leftarrow 4, \qquad Q \leftarrow 4, \qquad R \leftarrow 2.$$

Now if $q_{k-1} + q_k < n$, set

$$k \leftarrow k + 1, \qquad Q \leftarrow Q + R, \qquad R \leftarrow \lfloor \sqrt{Q} \rfloor, \qquad q_k \leftarrow 2^Q, \qquad r_k \leftarrow 2^R,$$

and repeat this operation until $q_{k-1} + q_k \geq n$. (*Note:* The calculation of $R \leftarrow \lfloor \sqrt{Q} \rfloor$ does not require a square root to be taken, since we may simply set $R \leftarrow R + 1$ if $(R + 1)^2 \leq Q$ and leave R unchanged if $(R + 1)^2 > Q$; see exercise 2. In this step we build the sequence

$k =$	0	1	2	3	4	5	6	
$q_k =$	2^4	2^4	2^6	2^8	2^{10}	2^{13}	2^{16}	$\ldots$
$r_k =$	2^2	2^2	2^2	2^2	2^3	2^3	2^4	$\ldots$

The multiplication of 70000-bit numbers would cause this step to terminate with $k = 6$, since $70000 < 2^{13} + 2^{16}$.)

C2. [Put u, v on stack.] Put code-1 on stack C, then place u and v onto stack C as numbers of exactly $q_{k-1} + q_k$ bits each.

C3. [Check recursion level.] Decrease k by 1. If $k = 0$, the top of stack C contains two 32-bit numbers, u and v; set $w \leftarrow uv$ using a built-in routine for multiplying 32-bit numbers, and go to step C10. If $k > 0$, set $r \leftarrow r_k$, $q \leftarrow q_k$, $p \leftarrow q_{k-1} + q_k$, and go on to step C4.

C4. [Break into $r + 1$ parts.] Let the number at the top of stack C be regarded as a list of $r + 1$ numbers with q bits each, $(U_r \ldots U_1 U_0)_{2^q}$. (The top of stack C now contains an $(r + 1)q = (q_k + q_{k+1})$-bit number.) For $j = 0, 1, \ldots, 2r$ compute the p-bit numbers

$$(\cdots (U_r j + U_{r-1}) j + \cdots + U_1) j + U_0 = U(j)$$

and successively put these values onto stack U. (The bottom of stack U now contains $U(0)$, then comes $U(1)$, etc., with $U(2r)$ on top. Note that

$$U(j) \le U(2r) < 2^q((2r)^r + (2r)^{r-1} + \cdots + 1) < 2^{q+1}(2r)^r \le 2^p,$$

by exercise 3.) Then remove $U_r \ldots U_1 U_0$ from stack C.

Now the top of stack C contains another list of $r + 1$ q-bit numbers, $V_r \ldots V_1 V_0$, and the p-bit numbers

$$(\cdots (V_r j + V_{r-1}) j + \cdots + V_1) j + V_0 = V(j)$$

should be put onto stack V in the same way. After this has been done, remove $V_r \ldots V_1 V_0$ from stack C.

C5. [Recurse.] Successively put the following items onto stack C, at the same time emptying stacks U and V:

code-2, $V(2r)$, $U(2r)$, code-3, $V(2r - 1)$, $U(2r - 1)$, ...,

code-3, $V(1)$, $U(1)$, code-3, $V(0)$, $U(0)$.

Put code-4 onto stack W. Go back to step C3.

C6. [Save one product.] (At this point the multiplication algorithm has set w to one of the products $W(j) = U(j)V(j)$.) Put w onto stack W. (This number w contains $2(q_k + q_{k-1})$ bits.) Go back to step C3.

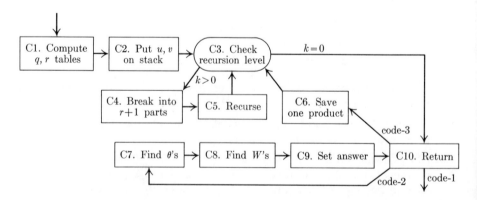

Fig. 8. Toom-Cook algorithm for high-precision multiplication.

C7. [Find θ's.] Set $r \leftarrow r_k$, $q \leftarrow q_k$, $p \leftarrow q_{k-1} + q_k$. (At this point stack W contains ..., code-4, $W(0)$, $W(1)$, ..., $W(2r)$ from bottom to top, where each $W(j)$ is a $2p$-bit number.)

       Now for $j = 1, 2, 3, \ldots, 2r$, perform the following loop: For $t = 2r$, $2r - 1, 2r - 2, \ldots, j$ set $W(t) \leftarrow (W(t) - W(t - 1))/j$. (Here j must increase and t must decrease. The quantity $(W(t) - W(t - 1))/j$ will always be a nonnegative integer which fits in $2p$ bits; cf. (15).)

C8. [Find W's.] For $j = 2r - 1, 2r - 2, \ldots, 1$, perform the following loop: For $t = j, j + 1, \ldots, 2r - 1$ set $W(t) \leftarrow W(t) - jW(t + 1)$. (Here j must decrease and t must increase. The result of this operation will again be a nonnegative $2p$-bit integer; cf. (17).)

C9. [Set answer.] Set w to the $2(q_k + q_{k+1})$-bit integer

$$(\cdots (W(2r)2^q + W(2r - 1))2^q + \cdots)2^q + W(0).$$

Remove $W(2r), \ldots, W(0)$ and code-4 from stack W.

C10. [Return.] Set $k \leftarrow k + 1$. Remove the top of stack C. If it is code-3, go to C6. If it is code-2, put w onto stack W and go to C7. And if it is code-1, terminate the algorithm (w is the answer). ∎

Let us now estimate the running time, $T(n)$, for Algorithm C, in terms of some things we shall call "cycles," i.e., elementary machine operations. Step C1 takes $O(q_k)$ cycles, even if we represent the number q_k internally as a long string of q_k bits followed by some delimiter, since $q_k + q_{k-1} + \cdots + q_0$ will be $O(q_k)$. Step C2 obviously takes $O(q_k)$ cycles.

Now let t_k denote the amount of computation required to get from step C3 to step C10 for a particular value of k (after k has been decreased at the beginning of step C3). Step C3 requires $O(q)$ cycles at most. Step C4 involves r multiplications of a $\log_2 (r + 1)$-bit number by a p-bit number, and r additions of p-bit numbers, all repeated $4r + 2$ times. Thus we need a total of $O(r^2 q \ln r)$ cycles. Step C5 requires moving $4r + 2$ p-bit numbers, so it involves $O(rq)$ cycles. Step C6 requires $O(q)$ cycles, and it is done $2r + 1$ times per iteration. The recursion involved when the algorithm essentially invokes itself (by returning to step C3) requires t_{k-1} cycles each time. Step C7 requires $O(r^2)$ subtractions of p-bit numbers and divisions of $2p$-bit by $(\log_2 r)$-bit numbers, so it requires $O(r^2 q \ln r)$ cycles. Similarly, step C8 requires $O(r^2 q \ln r)$ cycles. Step C9 involves $O(rq)$ cycles, and C10 takes hardly any time at all.

       Summing up we have, for $q = q_k$ and $r = r_k$, $T(n) = O(q_k) + O(q_k) + t_{k-1}$, where

$$
\begin{aligned}
t_k &= O(q) + O(r^2 q \ln r) + O(rq) + (2r + 1)O(q) + O(r^2 q \ln r) \\
&\qquad\qquad + O(r^2 q \ln r) + O(rq) + O(q) + (2r + 1)t_{k-1} \\
&= O(r^2 q \ln r) + (2r + 1)t_{k-1};
\end{aligned}
$$

thus there is a constant c such that

$$t_k \le cr_k^2 q_k \log_2 r_k + (2r_k + 1)t_{k-1}.$$

To complete the estimation of t_k we can prove by brute force that

$$t_k \le Cq_{k+1}2^{2.5\sqrt{\log_2 q_{k+1}}} \qquad (18)$$

for some constant C. Let us choose $C > 20c$, and let us also take C large enough so that (18) is valid for $k \le k_0$, where k_0 will be specified below. Then when $k > k_0$, let $Q_k = \log_2 q_k$, $R_k = \log_2 r_k$; we have by induction

$$t_k \le cq_k r_k^2 \log_2 r_k + (2r_k + 1)Cq_k 2^{2.5\sqrt{Q_k}}$$

$$= Cq_{k+1}2^{2.5\sqrt{\log_2 q_{k+1}}}(\eta_1 + \eta_2),$$

where

$$\eta_1 = \frac{c}{C}R_k 2^{R_k - 2.5\sqrt{Q_{k+1}}} < \frac{1}{20}R_k 2^{-R_k} < 0.05,$$

$$\eta_2 = \left(2 + \frac{1}{r_k}\right)2^{2.5(\sqrt{Q_k} - \sqrt{Q_{k+1}})} \to 2^{-1/4} < 0.85,$$

since

$$\sqrt{Q_{k+1}} - \sqrt{Q_k} = \sqrt{Q_k + \lfloor\sqrt{Q_k}\rfloor} - \sqrt{Q_k} \to \tfrac{1}{2}$$

as $k \to \infty$. It follows that we can find k_0 such that $\eta_2 < 0.95$ for all $k > k_0$, and this completes the proof of (18) by induction.

Finally, therefore, we may compute $T(n)$; since $n > q_{k-1} + q_{k-2}$, we have $q_{k-1} < n$; hence

$$r_{k-1} = 2^{\lfloor\sqrt{\log_2 q_{k-1}}\rfloor} < 2^{\sqrt{\log_2 n}}, \qquad \text{and} \qquad q_k = r_{k-1}q_{k-1} < n2^{\sqrt{\log_2 n}}.$$

Thus

$$t_{k-1} \le Cq_k 2^{2.5\sqrt{\log_2 q_k}} < Cn2^{\sqrt{\log_2 n} + 2.5(\sqrt{\log_2 n} + 1)},$$

and, since $T(n) = O(q_k) + t_{k-1}$, we have finally the following theorem:

Theorem C. *There is a constant c_0 such that the execution time of Algorithm C is less than $c_0 n2^{3.5\sqrt{\log_2 n}}$ cycles.* ∎

This result is noticeably stronger than Theorem A, since $n2^{3.5\sqrt{\log_2 n}} = n^{1+3.5/\sqrt{\log_2 n}}$. By adding a few complications to the algorithm, pushing the ideas to their apparent limits (see exercise 5), we can improve the estimated execution time to

$$T(n) = O(n2^{\sqrt{2\log_2 n}}\log_2 n). \qquad (19)$$

B. A modular method. There is another way to multiply large numbers very rapidly, based on the ideas of modular arithmetic as presented in Section 4.3.2. It is very hard to believe at first that this method can be of advantage, since a multiplication algorithm based on modular arithmetic must include the choice of moduli and the conversion of numbers into and out of modular representation, besides the actual multiplication operation itself. In spite of these formidable difficulties, A. Schönhage discovered that all of these operations can be done very rapidly.

In order to understand the essential mechanism of Schönhage's method, we shall look at a special case. Consider the sequence defined by the rules

$$q_0 = 1, \qquad q_{k+1} = 3q_k - 1, \tag{20}$$

so that $q_k = 3^k - 3^{k-1} - \cdots - 1 = \frac{1}{2}(3^k + 1)$. We will study a procedure that multiplies $(18q_k + 8)$-bit numbers, in terms of a method for multiplying $(18q_{k-1} + 8)$-bit numbers. Thus, if we know how to multiply numbers having $(18q_0 + 8) = 26$ bits, the procedure to be described will show us how to multiply numbers of $(18q_1 + 8) = 44$ bits, then 98 bits, then 260 bits, etc., eventually increasing the number of bits by almost a factor of 3 at each step.

Let $p_k = 18q_k + 8$. When multiplying p_k-bit numbers, the idea is to use the six moduli

$$m_1 = 2^{6q_k-1} - 1, \qquad m_2 = 2^{6q_k+1} - 1, \qquad m_3 = 2^{6q_k+2} - 1,$$
$$m_4 = 2^{6q_k+3} - 1, \qquad m_5 = 2^{6q_k+5} - 1, \qquad m_6 = 2^{6q_k+7} - 1. \tag{21}$$

These moduli are relatively prime, by Eq. 4.3.2–18, since the exponents

$$6q_k - 1, \qquad 6q_k + 1, \qquad 6q_k + 2, \qquad 6q_k + 3, \qquad 6q_k + 5, \qquad 6q_k + 7 \tag{22}$$

are always relatively prime (see exercise 6). The six moduli in (21) are capable of representing numbers up to $m = m_1 m_2 m_3 m_4 m_5 m_6 > 2^{36q_k+16} = 2^{2p_k}$, so there is no chance of overflow in the multiplication of p_k-bit numbers u and v. Thus we may use the following method:

a) Compute $u_1 = u \bmod m_1, \ldots, u_6 = u \bmod m_6$; $v_1 = v \bmod m_1, \ldots, v_6 = v \bmod m_6$.

b) Multiply u_1 by v_1, u_2 by v_2, $\ldots$, u_6 by v_6. These are numbers of at most $6q_k + 7 = 18q_{k-1} + 1 < p_{k-1}$ bits, so the multiplications can be performed by using the assumed p_{k-1}-bit multiplication procedure.

c) Compute $w_1 = u_1 v_1 \bmod m_1$, $w_2 = u_2 v_2 \bmod m_2$, $\ldots$, $w_6 = u_6 v_6 \bmod m_6$.

d) Compute w such that $0 \le w < m$, $w \bmod m_1 = w_1, \ldots, w \bmod m_6 = w_6$.

Let t_k be the amount of time needed for this process. It is not hard to see that operation (a) takes $O(p_k)$ cycles, since the determination of $u \bmod (2^e - 1)$ is quite simple (like "casting-out nines"), as shown in Section 4.3.2. Similarly, operation (c) takes $O(p_k)$ cycles. Operation (b) requires essentially $6t_{k-1}$ cycles. This leaves us with operation (d), which seems to be quite a difficult computation; but Schönhage has found an ingenious way to perform step (d) in $O(p_k \log p_k)$ cycles, and this is the crux of the method. As a consequence, we have

$$t_k = 6t_{k-1} + O(p_k \log p_k).$$

Since $p_k = 3^{k+2} + 17$, we can show that

$$t_k = O(6^k) = O(p_k^{\log_3 6}) = O(p_k^{1.63}).\qquad(23)$$

(See exercise 7.)

So although this method is more complicated than the $O(n^{\log_2 3})$ procedure given at the beginning of the section, it does, in fact, lead to an execution time substantially better than $O(n^2)$ for the multiplication of n-bit numbers. Thus we can improve on the classical method by using either of two completely different approaches.

Let us now analyze operation (d) above. Assume that we are given the positive integers $e_1 < e_2 < \cdots < e_r$, relatively prime in pairs; let

$$m_1 = 2^{e_1} - 1, \quad m_2 = 2^{e_2} - 1, \quad \ldots, \quad m_r = 2^{e_r} - 1.\qquad(24)$$

We are also given numbers $w_1, \ldots, w_r$ such that $0 \le w_j \le m_j$. Our job is *to determine the binary representation of the number w which satisfies the conditions*

$$0 \le w < m_1 m_2 \ldots m_r,$$
$$w \equiv w_1 \pmod{m_1}, \quad \ldots, \quad w \equiv w_r \pmod{m_r}.\qquad(25)$$

The method is based on (23) and (24) of Section 4.3.2; first we compute

$$w'_j = (\cdots((w_j - w'_1)c_{1j} - w'_2)c_{2j} - \cdots - w'_{j-1})c_{(j-1)j} \bmod m_j,\quad(26)$$

for $j = 2, \ldots, r$, where $w'_1 = w_1 \bmod m_1$; then we compute

$$w = (\cdots(w'_r m_{r-1} + w'_{r-1})m_{r-2} + \cdots + w'_2)m_1 + w'_1.\qquad(27)$$

Here c_{ij} is a number such that $c_{ij}m_i \equiv 1 \pmod{m_j}$; these numbers c_{ij} are not given, they must be determined from the e_j's.

The calculation of (26) for all j involves $\binom{r}{2}$ additions modulo m_j, each of which takes $O(e_r)$ cycles, plus $\binom{r}{2}$ multiplications by c_{ij}, modulo m_j. The calculation of w by formula (27) involves r additions and r multiplications by m_j; it is easy to multiply by m_j, since this is just adding, shifting, and subtracting, so it is clear that the evaluation of Eq. (27) takes $O(r^2 e_r)$ cycles. We will soon see that each of the multiplications by c_{ij}, modulo m_j, requires only $O(e_r \log e_r)$ cycles, and therefore *the entire job of conversion can be done in $O(r^2 e_r \log e_r)$ cycles.*

The above observations leave us with the following problem to solve: Given positive integers $e < f$ and a nonnegative integer $u < 2^f$, compute $(cu) \bmod (2^f - 1)$, where c is the number such that $(2^e - 1)c \equiv 1 \pmod{2^f - 1}$; and we must do this in $O(f \log f)$ cycles. The result of exercise 4.3.2–6 gives a formula for c which suggests a procedure that can be used. First we find the least positive integer b such that

$$be \equiv 1 \pmod{f}.\qquad(28)$$

This can be done using Euclid's algorithm in $O((\log f)^3)$ cycles, since Euclid's algorithm applied to e and f requires $O(\log f)$ iterations, and each iteration requires $O((\log f)^2)$ cycles; alternatively, we could be very sloppy here without violating the total time constraint, by simply trying $b = 1, 2$, etc. until (28) is satisfied, and such a process would take $O(f \log f)$ cycles in all. Once b has been found, exercise 4.3.2–6 tells us that

$$c = c[b] = \left(\sum_{0 \le j < b} 2^{ej} \right) \bmod (2^f - 1). \qquad (29)$$

A brute-force multiplication of $(cu) \bmod (2^f - 1)$ would not be good enough to solve the problem, since we do not know how to multiply general f-bit numbers in $O(f \log f)$ cycles. But the special form of c provides a clue: The binary representation of c is composed of bits in a regular pattern, and Eq. (29) shows that the number $c[2b]$ can be obtained in a simple way from $c[b]$. This suggests that we can rapidly multiply a number u by $c[b]$ if we build $c[b]u$ up in $\log_2 b$ steps in a suitably clever manner, such as the following: Let the binary notation for b be

$$b = (b_s \ldots b_2 b_1 b_0)_2;$$

we may calculate the sequences a_k, d_k, u_k, v_k which are defined by the rules

$$
\begin{aligned}
a_0 &= e, & a_k &= 2a_{k-1} \bmod f; \\
d_0 &= b_0 e, & d_k &= (d_{k-1} + b_k a_k) \bmod f; \\
u_0 &= u, & u_k &= (u_{k-1} + 2^{a_{k-1}} u_{k-1}) \bmod (2^f - 1); \\
v_0 &= b_0 u, & v_k &= (v_{k-1} + b_k 2^{d_{k-1}} u_k) \bmod (2^f - 1). \qquad (30)
\end{aligned}
$$

It is easy to prove by induction on k that

$$
\begin{aligned}
a_k &= (2^k e) \bmod f; & d_k &= ((b_k \ldots b_1 b_0)_2 e) \bmod f; \\
u_k &= (c[2^k]u) \bmod (2^f - 1); & v_k &= (c[(b_k \ldots b_1 b_0)_2]u) \bmod (2^f - 1). \qquad (31)
\end{aligned}
$$

Hence the desired result, $(c[b]u) \bmod (2^f - 1)$, is v_s. The calculation of a_k, d_k, u_k, v_k from a_{k-1}, d_{k-1}, u_{k-1}, v_{k-1} takes $O(\log f) + O(\log f) + O(f) + O(f) = O(f)$ cycles, and therefore the entire calculation can be done in $sO(f) = O(f \log f)$ cycles as desired.

The reader will find it instructive to study the ingenious method represented by (30) and (31) very carefully. Similar techniques are discussed in Section 4.6.3.

Schönhage's paper [*Computing* **1** (1966), 182–196] shows that these ideas can be extended to the multiplication of n-bit numbers using $r \approx 2^{\sqrt{2 \log_2 n}}$ moduli, obtaining a method analogous to Algorithm C. We shall not dwell on the details here, since Algorithm C is always superior; in fact, an even better method is next on our agenda!

C. Use of Fourier transforms. The critical problem in high-precision multiplication is the determination of "convolution products" such as

$$u_r v_0 + u_{r-1} v_1 + \cdots + u_0 v_r,$$

and there is an intimate relation between convolutions and finite Fourier transforms. If ω is a K-th root of unity, the one-dimensional Fourier transform of $(u_0, u_1, \ldots, u_{K-1})$ may be defined to be $(\hat{u}_0, \hat{u}_1, \ldots, \hat{u}_{K-1})$, where

$$\hat{u}_s = \sum_{0 \leq t < K} \omega^{st} u_t, \qquad 0 \leq s < K.$$

(Cf. 3.3.4-1.) Letting $(\hat{v}_0, \hat{v}_1, \ldots, \hat{v}_{K-1})$ be defined in the same way, as the transform of $(v_0, v_1, \ldots, v_{K-1})$, it is not difficult to see that $(\hat{u}_0 \hat{v}_0, \hat{u}_1 \hat{v}_1, \ldots, \hat{u}_{K-1} \hat{v}_{K-1})$ is the transform of $(w_0, w_1, \ldots, w_{K-1})$, where

$$w_r = u_r v_0 + u_{r-1} v_1 + \cdots + u_0 v_r + u_{K-1} v_{r+1} + \cdots + u_{r+1} v_{K-1}$$
$$= \sum_{r+s \equiv j (\text{modulo } K)} u_r v_s.$$

When $K \geq 2n$ and $u_{n+1} = u_{n+2} = \cdots = u_{K-1} = v_{n+1} = \cdots = v_{K-1} = 0$, this is just what we need for multiplication; *the transform of a convolution product is the ordinary product of the transforms.* This idea is closely related to Toom's use of polynomials (cf. (9)), with x replaced by a root of unity.

The above property of Fourier transforms was exploited by V. Strassen in 1968, using a sufficiently precise binary representation of the complex number ω, to multiply large numbers faster than was possible under all previously known schemes. In 1970, he and A. Schönhage found an elegant way to modify the method, avoiding all the complications of complex numbers and obtaining a very pretty algorithm capable of multiplying two n-bit numbers in $O(n \log n \log \log n)$ steps. We shall now study their remarkable algorithm.

It is convenient in the first place to replace n by 2^n, and to seek a procedure that multiplies 2^n-bit numbers in $O(2^n n \log n)$ steps; roughly speaking, it will be desirable to reduce the problem of multiplying 2^{2n}-bit numbers to the problem of multiplying 2^n-bit numbers. Furthermore we shall be interested in multiplying numbers mod $(2^{2^n} + 1)$; this is a key idea (cf. exercise 8), and it causes no loss in generality because, for example, we can obtain the exact product of two 32-bit numbers if we multiply them mod $(2^{64} + 1)$.

Arithmetic mod $(2^N + 1)$ is somewhat similar to ones' complement arithmetic, mod $(2^N - 1)$, although it is slightly more complicated; we have already investigated the idea briefly in Section 3.2.1.1. Numbers can be represented as N-bit quantities in binary notation, except for the special value $-1 \equiv 2^N$ which may be denoted by the special symbol $\bar{1}$. Addition mod $(2^N + 1)$ is easily done in $O(N)$ cycles, since a carry off the left end merely means we must subtract 1 at the right; similarly, subtraction mod $(2^N + 1)$ is quite simple. Furthermore, we can multiply by 2^r in $O(N)$ cycles, when $0 \leq r < N$, since

$$2^r \cdot (u_{N-1} \ldots u_1 u_0)_2 \equiv (u_{N-r-1} \ldots u_0 0 \ldots 0)_2 - (0 \ldots 0 u_{N-1} \ldots u_{N-r})_2,$$
$$(\text{modulo } 2^N + 1). \qquad (32)$$

Let us now suppose that $N = 2^n$ and that we wish to obtain the product of u and v, mod $(2^N + 1)$. If u or v is $\bar{1}$ there is no difficulty, so we may assume that

$0 \leq u, v < 2^N$. As in Algorithm C we shall break these N-bit numbers into groups; let $k + \ell = n$, $K = 2^k$, $L = 2^\ell$, and write

$$u = (U_{K-1} \ldots U_1 U_0)_{2^L}, \qquad v = (V_{K-1} \ldots V_1 V_0)_{2^L},$$

regarding u and v as 2^k groups of 2^ℓ-bit numbers. We will select appropriate values for k and ℓ later; it turns out (see exercise 10) that we will need to have

$$k \leq \ell + 1 \tag{33}$$

but no other conditions. The above representation of u and v implies that

$$u \cdot v \equiv W_{K-1} 2^{L(K-1)} + \cdots + W_1 2^L + W_0 \quad (\text{modulo } 2^N + 1), \tag{34}$$

where

$$W_r = (U_r V_0 + U_{r-1} V_1 + \cdots + U_0 V_r) \\ - (U_{K-1} V_{r+1} + \cdots + U_{r+1} V_{K-1}), \qquad 0 \leq r < K.$$

Clearly, $-(K - 1 - r)2^{2L} < W_r < (r + 1)2^{2L}$, so that $|W_r| < 2^{2L+k}$; if we knew the W's, we could compute $uv \bmod (2^N + 1)$ by Eq. (34), in $O(K(2L + k)) = O(N)$ further steps, so our goal is to compute the W_r exactly. The calculation of $(W_0, W_1, \ldots, W_{K-1})$ can be performed in three steps:

(a) Determine $W'_r = W_r \bmod 2^k$,   for $0 \leq r < 2^k$.

(b) Determine $W''_r = W_r \bmod (2^{2L} + 1)$,   for $0 \leq r < 2^k$.

(c) Let $W'''_r = (2^{2L} + 1)((W'_r - W''_r) \bmod 2^k) + W''_r$, and compute

$$W_r = \begin{cases} W'''_r, & \text{if } W'''_r < (r + 1)2^{2L}, \\ W'''_r - 2^k(2^{2L} + 1), & \text{if } W'''_r \geq (r + 1)2^{2L}, \end{cases} \quad \text{for } 0 \leq r < 2^k. \tag{35}$$

(See exercise 9.) This is the second key idea in the construction.

Step (a) does not take much time since k is much smaller than $2L$; we can let $U'_r = U_r \bmod 2^k$, $V'_r = V_r \bmod 2^k$, and "read off" $(U'_r V'_0 + \cdots U'_0 V'_r)$ and $(U'_{K-1} V'_{r+1} + \cdots + U'_{r+1} V'_{K-1})$ from the binary representations of the respective products

$$(U'_{K-1} \ldots U'_1 U'_0)_{2^{3k}} \times (V'_{K-1} \ldots V'_1 V'_0)_{2^{3k}},$$
$$(U'_0 U'_1 \ldots U'_{K-1})_{2^{3k}} \times (V'_0 V'_1 \ldots V'_{K-1})_{2^{3k}}, \tag{36}$$

since each of the desired convolutions is at most $3k$ bits long. The products of the $(3kK)$-bit numbers can be found in $O(N)$ cycles; in fact, the straightforward method at the beginning of this section will compute the products (36) in $O(K^{1.6})$ units of time.

Therefore everything hinges on being able to do step (b) quickly; and the third key idea in the construction is the fact that step (b) can be achieved by using a "fast Fourier transform."

Given a sequence of $K = 2^k$ integers $(a_0, \ldots, a_{K-1})$, and an integer ω such that $\omega^K \equiv 1$ (modulo M), the integer Fourier transform

$$\hat{a}_s = \left(\sum_{0 \le t < K} \omega^{st} a_t \right) \bmod M \tag{37}$$

can be calculated rapidly as follows. (In these formulas the s_j and t_j are either 0 or 1, so that each step represents 2^k computations.)

Step 0. Let $A^{[0]}(t_{k-1}, \ldots, t_0) = a_t$,   where $t = (t_{k-1} \ldots t_0)_2$.

Step 1. Set $A^{[1]}(s_{k-1}, t_{k-2}, \ldots, t_0)$
$$= \big(A^{[0]}(0, t_{k-2}, \ldots, t_0)$$
$$+ \omega^{(s_{k-1}0\ldots0)_2} \cdot A^{[0]}(1, t_{k-2}, \ldots, t_0) \big) \bmod M.$$

Step 2. Set $A^{[2]}(s_{k-1}, s_{k-2}, t_{k-3}, \ldots, t_0)$
$$= \big(A^{[1]}(s_{k-1}, 0, t_{k-3}, \ldots, t_0)$$
$$+ \omega^{(s_{k-2}s_{k-1}0\ldots0)_2} \cdot A^{[1]}(s_{k-1}, 1, t_{k-3}, \ldots, t_0) \big) \bmod M.$$

$\ldots$

Step k. Set $A^{[k]}(s_{k-1}, \ldots, s_1, s_0)$
$$= \big(A^{[k-1]}(s_{k-1}, \ldots, s_1, 0)$$
$$+ \omega^{(s_0 s_1 \cdots s_{k-1})_2} \cdot A^{[k-1]}(s_{k-1}, \ldots, s_1, 1) \big) \bmod M.$$

It is not difficult to prove by induction that we have

$$A^{[j]}(s_{k-1}, \ldots, s_{k-j}, t_{k-j+1}, \ldots, t_0)$$
$$= \sum_{0 \le t_{k-1}, \ldots, t_{k-j} \le 1} \omega^{(s_0 s_1 \cdots s_{k-1})_2 \cdot (t_{k-1} \cdots t_{k-j} 0 \ldots 0)_2} a_t \bmod M,$$

so that

$$A^{[k]}(s_{k-1}, \ldots, s_1, s_0) = \hat{a}_s,   \text{where}   s = (s_0 s_1 \ldots s_{k-1})_2.$$

(Note the reversed order of the binary digits in s. For further discussion of transforms such as this, see Section 4.6.4.)

Now we are ready to do step (b) above, determining $W''_r = W_r$ mod $(2^{2L} + 1)$. Let $\psi = 2^{2(l+1-k)}$, so that $\psi^K = 2^{2L}$. Let

$$(a_0, a_1, \ldots, a_{K-1}) = (U_0, \psi U_1, \ldots, \psi^{K-1} U_{K-1}),$$

and

$$(b_0, b_1, \ldots, b_{K-1}) = (V_0, \psi V_1, \ldots, \psi^{K-1} V_{K-1}).$$

Compute the Fourier transforms $(\hat{a}_0, \hat{a}_1, \ldots, \hat{a}_{K-1})$ and $(\hat{b}_0, \hat{b}_1, \ldots, \hat{b}_{K-1})$, with $\omega = \psi^2$ and $M = 2^{2L} + 1$. Then compute

$$(c_0, c_1, \ldots, c_{K-1}) = (\hat{a}_0 \hat{b}_0, \hat{a}_1 \hat{b}_1, \ldots, \hat{a}_{K-1} \hat{b}_{K-1}) \bmod M,$$

using a high-speed multiplication procedure mod M for each of these K products. Finally compute

$$(\hat{c}_0, \hat{c}_1, \ldots, \hat{c}_{K-1});$$

this is enough to determine $(W_0'', W_1'', \ldots, W_{K-1}'')$ without much more work, since

$$(\hat{c}_0, \hat{c}_1, \ldots, \hat{c}_{K-2}, \hat{c}_{K-1}) \equiv (2^k W_0'', 2^k \psi^{K-1} W_{K-1}'', \ldots, 2^k \psi^2 W_2'', 2^k \psi W_1'') \qquad (38)$$

(modulo M), and the W_r'' may be obtained by an appropriate shifting operation analogous to (32).

This may seem like magic, but it works; a careful study of the above remarks will show that the method is very clever but not completely mysterious! The key reason that (38) holds is that $\psi^K \equiv -1$ (modulo M). In the first place we have

$$\psi^r W_r \equiv \sum_{\substack{0 \le p,q < K \\ p+q \equiv r \text{ (modulo } K)}} (\psi^p U_p)(\psi^q V_q) \text{ (modulo } M),$$

by the definition of W_r in (34). And in the second place we have

$$\sum_{0 \le t < K} \omega^{st} \equiv \begin{cases} K, & \text{if} \quad s \bmod K = 0; \\ 0, & \text{if} \quad s \bmod K \ne 0. \end{cases} \qquad (39)$$

For when $s \bmod K \ne 0$, let $s \bmod K = 2^p q$ where q is odd and $0 \le p < k$; setting $T = 2^{k-1-p}$, we have $\omega^{sT} = \psi^{2sT} = \psi^{Kq} \equiv -1$, hence

$$\sum_{0 \le t < K} \omega^{st} \equiv 2^p \sum_{0 \le t < 2T} \omega^{st} \equiv 2^p \sum_{0 \le t < T} (\omega^{st} + \omega^{s(t+T)}) \equiv 0.$$

This leads to a proof of (38), since

$$\hat{c}_s \equiv \sum_{0 \le t < K} \omega^{st} \hat{a}_t \hat{b}_t \equiv \sum_{0 \le t, t', t'' < K} \omega^{st} \omega^{tt'} a_{t'} \omega^{tt''} b_{t''}$$

$$\equiv \sum_{0 \le t', t'' < K} a_{t'} b_{t''} \sum_{0 \le t < K} \omega^{(s+t'+t'')t} \equiv K \sum_{\substack{0 \le t', t'' < K \\ t'+t''+s \equiv 0 \text{ (modulo } K)}} (\psi^{t'} U_{t'})(\psi^{t''} V_{t''}).$$

The multiplication procedure is nearly complete; it remains for us to specify k and ℓ, and to total up the amount of work involved. Let $M(n)$ denote the time it takes to multiply 2^n-bit numbers mod $(2^{2^n} + 1)$, by the above method, and let $M'(n) = M(n)/2^n$. Step (a) takes $O(N)$ cycles; step (c) takes $KO(L) = O(N)$, and so does the final evaluation of uv using (34). Step (b) involves three Fourier transformations mod $(2^{2L} + 1)$, each of which uses up $O(kN)$ cycles; it involves 2^k multiplications of $a_s b_s$ mod $(2^{2L} + 1)$, requiring $2^k M(\ell + 1)$ cycles; and it involves a few other operations of negligible cost. Thus

$$M(n) = 2^k M(\ell + 1) + O(kN); \qquad M'(n) = 2M'(\ell + 1) + O(k).$$

We get the best reduction of $M'(n)$ when ℓ is chosen to be as low as possible, so we set

$$k = \lceil n/2 \rceil, \qquad \ell = \lfloor n/2 \rfloor.$$

This means there is a constant C such that

$$M'(n) \leq 2M'(\lfloor (n-2)/2 \rfloor + 2) + Cn \qquad \text{for all} \qquad n \geq 3.$$

Iterating this relation yields

$$M'(n) \leq 2^j M'(\lfloor (n-2)/2^j \rfloor + 2) + C(j(n-2) + 2^{j+1} - 2)$$

for $j = 1, 2, \ldots,$; hence when $j \approx \log_2 n$ we have $M'(n) = O(n \log n)$. We have proved the main result of this section:

Theorem S. (A. Schönhage, V. Strassen.) *It is possible to multiply two n-bit numbers in $O(n \log n \log \log n)$ steps.* ∎

An example of the Schönhage-Strassen "fast Fourier multiplication" algorithm will be useful to fix the ideas. Suppose that we want to multiply two 4096-bit numbers, obtaining an 8192-bit product. We will multiply the numbers modulo $(2^{8192} + 1)$, so that $n = 13$, $k = 7$, and $\ell = 6$ in the above discussion. The bits of u and v are each divided into 128 groups of 64 bits each. (In this case the most significant 64 groups are all zero.) The least significant seven bits of each group are inserted into the 2688-bit numbers shown in (36), which are multiplied to give the W'_r of step (a). The 128 groups are also transformed with $\psi = 2$ and $M = 2^{128} + 1$; the operation of step (b) then relies on 128 multiplications $(\hat{a}_0 \hat{b}_0, \ldots, \hat{a}_{127} \hat{b}_{127})$ of 128-bit numbers mod $(2^{128} + 1)$, with simplifications occurring if any of the $\hat{a}$'s or $\hat{b}$'s is $\bar{1}$. Finally, step (c) allows us to determine the W_r which are used to evaluate uv by (34).

Our formulation of step (a) does not turn out to be very efficient in this example; the algorithm needs to be refined somewhat if it is ever to become competitive with Algorithm C for practical ranges of n. Of course, as $n \to \infty$, fast Fourier multiplication becomes vastly superior to Algorithm C.

The word "steps" in Theorem S has been used somewhat loosely; we have implicitly been assuming a "conventional computer" with unlimited random-access memory, which takes one unit of time to read and write any bit. This assumption is quite unrealistic as $n \to \infty$, since we need $O(\log n)$ bits in an instruction or an index register just to be able to distinguish between n memory cells, so the actual time to access memory on a "conventional computer" is really proportional to $\log n$. We often forget this dependence on n because it does not occur on real machines with bounded memory and bounded register size. When $n \to \infty$ the only physically appropriate model seems to be a finite memory with a finite number of arbitrarily long tapes; the fast Fourier transform can be arranged to work with restricted storage access of this kind, hence the $O(n \log n \log \log n)$ time bound applies to the tape model of computers as well as the random-access model.

The difference between these computer models can be clarified by considering another method due to Schönhage and Strassen: If $n = 2^m \cdot m$, so that $m \approx \log_2 n$ and $2^m \approx n/\log_2 n$, it is possible to use the fast Fourier transform over the complex numbers to compute the product of two n-bit numbers by doing $O(m \cdot 2^m)$ multiplications of $6m$-bit numbers. Each of the latter can be broken into $12^2 = 144$ multiplications of $(\frac{1}{2}m)$-bit numbers. Now we can construct a multiplication table containing all products xy with $0 \le x, y < 2^{(1/2)m}$, by repeated addition, in $O(m \cdot 2^{(1/2)m} \cdot 2^{(1/2)m})$ steps; then each of the $O(m \cdot 2^m)$ needed products can be done by table lookup in $O(m)$ steps. The total number of steps for *this* procedure therefore comes to $O(m^2 2^m) = O(n \log n)$; we have gotten rid of the factor $\log \log n$ in Theorem S, but the method really *requires* an unbounded random-access memory since the table lookup cannot be done efficiently with a finite number of tapes. (Of course, a factor of $\log \log n$ is utterly negligible in practice; when n changes from 10^9 to 10^{18}, $\log_2 \log_2 n$ increases by only one.)

Perhaps $O(n \log n \log \log n)$ will turn out to be the fastest achievable multiplication speed, on the tape model, and $O(n \log n)$ on the unlimited random-access model; no such result has yet been proved. The best lower bound known to date is a rather deep theorem proved by S. A. Cook [*On the minimum computation time of functions* (Thesis, Harvard University, 1966), 26–50] in conjunction with S. Aanderaa, that under certain restrictions there is no algorithm which multiplies n-bit numbers with less than

$$O\bigl(n \log n/(\log \log n)^2\bigr) \tag{41}$$

operations. The restrictions under which (41) is a lower bound are rather severe: (a) The $(k+1)$st input symbols of the operands, from right to left, must not be read by the algorithm until after the kth output symbol has been produced; and (b) the internal tables kept by the algorithm must have a "uniform" structure, in an appropriate sense. The latter restriction rules out algorithms which use general List structures for their internal tables, and the first restriction rules out both Algorithm C and Algorithm S. It is still conceivable (though unlikely) that an algorithm which violates (a) or (b) could multiply n-bit numbers in $O(n)$ cycles.

D. Division. Using a fast multiplication routine, we can now show that division can also be accomplished in $O(n \log n \log \log n)$ cycles, for some constant α.

To divide an n-bit number u by an n-bit number v, we may first find an n-bit approximation to $1/v$, then multiply by u to get an approximation $\hat{q}$ to u/v, and, finally, we can make the slight correction necessary to $\hat{q}$ to ensure that $0 \le u - qv < v$ by using another multiplication. From this reasoning, we see that it suffices to have an algorithm which approximates the reciprocal of an n-bit number, in $O(n \log n \log \log n)$ cycles. The following algorithm achieves this, using "Newton's method" as explained at the end of Section 4.3.1:

Algorithm R (*High-precision reciprocal*). Let v have the binary representation $v = (0.v_1 v_2 v_3 \ldots)_2$, where $v_1 = 1$. This algorithm computes an approximation

z to $1/v$, which satisfies

$$|z - 1/v| \leq 2^{-n}.$$

R1. [Initial approximation.] Set $z \leftarrow \frac{1}{4}\lfloor 32/(4v_1 + 2v_2 + v_3)\rfloor$, $k \leftarrow 0$.

R2. [Newtonian iteration.] (At this point we have a number z of the binary form $xx.xx \ldots x$ with $2^k + 1$ places after the radix point, and $z \leq 2$.) Calculate $z^2 = xxx.xx \ldots x$ exactly, using a high-speed multiplication routine. Then calculate $V_k z^2$ exactly, where $V_k = (0.v_1 v_2 \ldots v_{2^{k+1}+3})_2$. Then set $z \leftarrow 2z - V_k z^2 + r$, where $0 \leq r < 2^{-2^{k+1}-1}$ is added if necessary to "round up" z so it is a multiple of $2^{-2^{k+1}-1}$. Finally, set $k \leftarrow k + 1$.

R3. [Test for end.] If $2^k < n$, go back to step R2; otherwise the algorithm terminates. ∎

This algorithm is a modification of a method suggested by S. A. Cook. A similar technique has been used in computer hardware [see Anderson, Earle, Goldschmidt, and Powers, *IBM J. Res. Dev.* **11** (1967), 48–52]. Of course, it is necessary to check the accuracy of Algorithm R quite carefully, because it comes very close to being inaccurate. We will prove by induction that

$$z \leq 2 \quad \text{and} \quad |z - 1/v| \leq 2^{-2^k} \tag{42}$$

at the beginning and end of step R2.

For this purpose, let $\delta_k = 1/v - z_k$, where z_k is the value of z after k iterations of step R2. To start the induction on k, we have $\delta_0 = 1/v - 8/v' + (32/v' - \lfloor 32/v'\rfloor)/4 = \eta_1 + \eta_2$, where $v' = (v_1 v_2 v_3)_2$, $\eta_1 = (v' - 8v)/vv'$ satisfies $-\frac{1}{2} < \eta_1 \leq 0$, and $0 \leq \eta_2 < \frac{1}{4}$. Hence $|\delta_0| < \frac{1}{2}$. Now suppose (42) has been verified for k; then

$$\begin{aligned}
\delta_{k+1} = 1/v - z_{k+1} &= 1/v - z_k - z_k(1 - z_k V_k) - r \\
&= \delta_k - z_k(1 - z_k v) - z_k^2(v - V_k) - r \\
&= \delta_k - (1/v - \delta_k)v\delta_k - z_k^2(v - V_k) - r \\
&= v\delta_k^2 - z_k^2(v - V_k) - r.
\end{aligned}$$

Now

$$0 \leq v\delta_k^2 < \delta_k^2 \leq (2^{-2^k})^2 = 2^{-2^{k+1}},$$

and

$$0 \leq z^2(v - V_k) + r < 4(2^{-2^{k+1}-3}) + 2^{-2^{k+1}-1} = 2^{-2^{k+1}},$$

so $|\delta_{k+1}| \leq 2^{-2^{k+1}}$. We must still verify the first inequality of (42); to show that $z_{k+1} \leq 2$, there are three cases: (a) $V_k = \frac{1}{2}$; then $z_{k+1} = 2$. (b) $V_k \neq \frac{1}{2} = V_{k-1}$; then $z_k = 2$, so $2z_k - z_k^2 V_k \leq 2 - 2^{-2^{k+1}-1}$.   (c) $V_{k-1} \neq \frac{1}{2}$; then $z_{k+1} = 1/v - \delta_{k+1} < 2 - 2^{-2^k-2} + 2^{-2^{k+1}} \leq 2$, since $k > 0$.

The running time of Algorithm R is bounded by

$$2T(4n) + 2T(2n) + 2T(n) + \cdots + O(n)$$

cycles, where $T(n)$ is an upper bound on the time needed to do a multiplication of n-bit numbers. When $T(n) = C\,n \log n \log \log n$, we have $T(4n) + T(2n) + T(n) + \cdots < T(8n)$, so division can be done with a speed comparable to that of multiplication.

E. An even faster multiplication method. It is natural to wonder if multiplication of n-bit numbers can actually be accomplished in just n steps; we have come from n^2 down to $n^{1+\epsilon}$, so perhaps we can squeeze the time down even more. This is still an unsolved problem, as pointed out above, but it is interesting to note that the best possible time, exactly n cycles, *can* be achieved if we leave the domain of conventional computer programming and allow ourselves to build a computer which has an unlimited number of components all acting at once.

A *linear iterative array* of automata is a set of devices $M_1, M_2, M_3, \ldots$ which can each be in a finite set of "states," at each step of the computation. The machines $M_2, M_3, \ldots$ all have *identical* circuitry, and their state at time $t + 1$ is a function of their own state at time t as well as the states of their left and right neighbors at time t. The first machine M_1 is slightly different: its state at time $t + 1$ is a function of its own state and that of M_2, at time t, and also of the *input* at time t. The *output* of a linear iterative array is a function defined on the states of M_1.

Let $u = (u_{n-1} \ldots u_1 u_0)_2$, $v = (v_{n-1} \ldots v_1 v_0)_2$, and $q = (q_{n-1} \ldots q_1 q_0)_2$ be binary numbers, and let $uv + q = w = (w_{2n-1} \ldots w_1 w_0)_2$. It is a remarkable fact that a linear iterative array can be constructed, independent of n, which will output $w_0, w_1, w_2, \ldots$ at times $1, 2, 3, \ldots$, if it is given the inputs (u_0, v_0, q_0), (u_1, v_1, q_1), (u_2, v_2, q_2), $\ldots$ at times $0, 1, 2, \ldots$.

We can state this phenomenon in the language of computer hardware, by saying that it is possible to design a single "integrated circuit module" with the following property: If we wire together sufficiently many of these devices in a straight line, with each module communicating only with its left and right neighbor, the resulting circuitry will produce the $2n$-bit product of n-bit numbers in exactly $2n$ clock pulses.

Here is the basic idea behind this construction: At time 0, M_1 senses (u_0, v_0, q_0) and it therefore is able to output $(u_0 v_0 + q_0) \bmod 2$ at time 1. Then it sees (u_1, v_1, q_1) and it can output $(u_0 v_1 + u_1 v_0 + q_1 + k) \bmod 2$, where k is the "carry" left over from the previous step, at time 2. Next it sees (u_2, v_2, q_2) and outputs $(u_0 v_2 + u_1 v_1 + u_2 v_0 + q_2 + k) \bmod 2$; furthermore, its state records the values of u_2 and v_2 so that machine M_2 will be able to sense these values at time 3, and M_2 will be able to compute $u_2 v_2$ for the benefit of M_1 at time 4. Thus M_1 arranges to start M_2 multiplying the sequence (u_2, v_2), $(u_3, v_3), \ldots$, and M_2 will ultimately give M_3 the job of multiplying (u_4, v_4), (u_5, v_5), etc. Fortunately, things just work out so that no time is lost. The reader will find it interesting to deduce further details from the abbreviated description which follows.

Each automaton has 2^{11} states $(c, x_0, y_0, x_1, y_1, x, y, z_2, z_1, z_0)$, where $0 \le c < 4$ and each of the x's, y's and z's is either 0 or 1. Initially, all devices are in state $(0, 0, 0, 0, 0, 0, 0, 0, 0, 0)$. Suppose that a machine M_j, $j > 1$, is

Table 1

MULTIPLICATION IN A LINEAR ITERATIVE ARRAY

Time	u_j / v_j	q_j	c	x_0	x_1	x	z	c	x_0	x_1	x	z	c	x_0	x_1	x	z
				Module M_1	(y_0 y_1 y)		($z_2z_1z_0$)		**Module M_2**					**Module M_3**			
0	1	1	0	0	0	0	0	0	0	0	0	0	0	0	0	0	0
	1			0	0	0	0		0	0	0	0		0	0	0	0
							0					0					0
1	1	1	1	1	0	0	0	0	0	0	0	0	0	0	0	0	0
	1			1	0	0	1		0	0	0	0		0	0	0	0
							0					0					0
2	1	0	2	1	1	0	1	0	0	0	0	0	0	0	0	0	0
	1			1	1	0	0		0	0	0	0		0	0	0	0
							0					0					0
3	0	1	3	1	1	1	0	0	0	0	0	0	0	0	0	0	0
	0			1	1	1	1		0	0	0	0		0	0	0	0
							1					1					0
4	1	0	3	1	1	0	1	1	1	0	0	0	0	0	0	0	0
	1			1	1	0	0		1	0	0	0		0	0	0	0
							1					1					0
5	0	0	3	1	1	1	0	2	1	0	0	0	0	0	0	0	0
	0			1	1	1	1		1	0	0	0		0	0	0	0
							1					1					0
6	0	0	3	1	1	0	1	3	1	0	1	0	0	0	0	0	0
	0			1	1	0	0		1	0	1	1		0	0	0	0
							0					0					0
7	0	0	3	1	1	0	0	3	1	0	0	0	1	1	0	0	0
	0			1	1	0	0		1	0	0	1		1	0	0	0
							0					0					1
8	0	0	3	1	1	0	0	3	1	0	0	0	2	1	0	0	0
	0			1	1	0	0		1	0	0	1		1	0	0	0
							0					0					0
9	0	0	3	1	1	0	0	3	1	0	0	0	3	1	0	0	0
	0			1	1	0	0		1	0	0	0		1	0	0	0
							0					1					0
10	0	0	3	1	1	0	0	3	1	0	0	0	3	1	0	0	0
	0			1	1	0	0		1	0	0	0		1	0	0	0
							1					0					0
11	0	0	3	1	1	0	0	3	1	0	0	0	3	1	0	0	0
	0			1	1	0	0		1	0	0	0		1	0	0	0
							0					0					0

in state $(c, x_0, y_0, x_1, y_1, x, y, z_2, z_1, z_0)$ at time t, and its left neighbor M_{j-1} is in state $(c^l, x_0^l, y_0^l, x_1^l, y_1^l, x^l, y^l, z_2^l, z_1^l, z_0^l)$ while its right neighbor M_{j+1} is in state $(c^r, x_0^r, y_0^r, x_1^r, y_1^r, x^r, y^r, z_2^r, z_1^r, z_0^r)$. Then M_j will go into state $(c', x_0', y_0', x_1', y_1', x', y', z_2', z_1', z_0')$ at time $t+1$, where

$$
\begin{aligned}
c' &= \min{(c+1, 3)} && \text{if} && c^l = 3, && 0 \text{ otherwise};\\
x_0' &= x^l && \text{if} && c = 0, && x_0 \text{ otherwise};\\
y_0' &= y^l && \text{if} && c = 0, && y_0 \text{ otherwise};\\
x_1' &= x^l && \text{if} && c = 1, && x_1 \text{ otherwise}; && (43)\\
y_1' &= y^l && \text{if} && c = 1, && y_1 \text{ otherwise};\\
x' &= x^l && \text{if} && c \geq 2, && x \text{ otherwise};\\
y' &= y^l && \text{if} && c \geq 2, && y \text{ otherwise};
\end{aligned}
$$

and $(z_2' z_1' z_0')_2$ is the binary notation for

$$
z_0^r + z_1 + z_2^l + \begin{cases} x^l y^l, & \text{if} & c = 0; \\ x_0 y^l + x^l y_0, & \text{if} & c = 1; \\ x_0 y^l + x_1 y_1 + x^l y_0, & \text{if} & c = 2; \\ x_0 y^l + x_1 y + x y_1 + x^l y_0, & \text{if} & c = 3. \end{cases} \qquad (44)
$$

The leftmost machine M_1 behaves in almost the same way as the others; it acts exactly as if there were a machine to its left in state $(3, 0, 0, 0, 0, u, v, q, 0, 0)$ when it is receiving the inputs (u, v, q). The output of the array is the z_0 component of M_1.

Table 1 shows an example of this array acting on the inputs $u = v = (\dots 00010111)_2$, $q = (\dots 00001011)_2$. The output sequence appears in the lower right portion of the states of M_1: $0, 0, 1, 1, 1, 0, 0, 0, 0, 1, 0, \dots$, representing the number $(\dots 01000011100)_2$ from right to left.

This construction is based on a similar one first published by A. J. Atrubin, *IEEE Transactions* **EC–14** (1965), 394–399. S. Winograd [*JACM* **14** (1967), 793–802] has investigated the minimum multiplication time achievable in a logical circuit when n is given and when the inputs are available all at once in coded form; see also C. S. Wallace, *IEEE Trans.* **EC–13** (1964), 14–17.

EXERCISES

1. [*22*] The idea expressed in (2) can be generalized to the decimal system, if the radix 2 is replaced by 10. Using this generalization, calculate 2718 times 3142 (reducing this product of four-digit numbers to three products of two-digit numbers, and reducing each of the latter to products of one-digit numbers).

2. [*M22*] Prove that, in step C1 of Algorithm C, the value of R either stays the same or increases by one when we set $R \leftarrow \lfloor \sqrt{Q} \rfloor$. (Therefore, as observed in that step, we need not calculate a square root.)

3. [*M23*] Prove that the sequences q_k, r_k defined in Algorithm C satisfy the inequality $2^{q_k+1}(2r_k)^{r_k} \leq 2^{q_{k-1}+q_k}$, when $k > 0$.

▶ **4.** [*M28*] (K. Baker.) Show that it is advantageous to evaluate the polynomial $W(x)$ at the points $x = -r, \ldots, 0, \ldots, r$ instead of at the points $x = 0, 1, \ldots, 2r$ as in Algorithm C. The polynomial $U(x)$ can be written $U(x) = U_e(x^2) + xU_o(x^2)$, and similarly $V(x)$ and $W(x)$ can be expanded in this way; show how to exploit this idea, obtaining faster calculations in steps C7 and C8. (*Note:* The method (15), (17) described in the text is a special case of an equally simple method for determining the coefficients of $W(x)$ when its values are given at *arbitrary* points $x_0, x_1, \ldots, x_{2r}$, as shown in Section 4.6.4.)

▶ **5.** [*HM30*] Show that if in step C1 of Algorithm C we set $R \leftarrow \lceil \sqrt{2Q} \rceil + 1$ instead of $R \leftarrow \lfloor \sqrt{Q} \rfloor$, with suitable initial values of q_0, q_1, r_0, and r_1, then (19) can be improved to $t_k \leq q_{k+1}2^{\sqrt{2\log_2 q_{k+1}}}(\log_2 q_{k+1})$. [*Hint:* Cf. the derivation of (37) and the discussion of (40).]

6. [*M23*] Prove that the six numbers in (22) are relatively prime in pairs.

7. [*M23*] Prove (23).

8. [*M27*] Why does the fast Fourier multiplication algorithm bother to work mod $(2^N + 1)$ instead of mod $(2^N - 1)$? It would seem to be much simpler to do everything mod $(2^N - 1)$, avoiding a lot of miscellaneous minus signs in the formulas, since $\omega = 2$ can be used to compute fast Fourier transforms mod $(2^{2^n} - 1)$. What would go wrong?

9. [*M22*] Prove (35).

10. [*M21*] Where is condition (33) used?

▶ **11.** [*M26*] If n is fixed, how many of the automata in the linear iterative array (43), (44) are needed to compute the product of n-bit numbers? (Note that the automaton M_j is only influenced by the component z_0^r of the machine on its right, so we may remove all automata whose z_0 component is always zero whenever the inputs are n-bit numbers.)

12. [*M50*] Improve on the Cook-Aanderaa lower bound (41); is it impossible for a general node-structure automaton (as described in Section 2.6) to multiply n-bit numbers in $O(n)$ cycles?

13. [*M25*] (A. Schönhage.) What is a good upper bound on the time needed to multiply an m-bit number by an n-bit number, when both m and n are very large but n is much larger than m, based on the results proved in this section for $m = n$?

14. [*M40*] Write a program for Algorithm C, incorporating the improvements of exercise 4. Compare it with a program for Algorithm 4.3.1M and with a program based on (2), to see how large n must be before Algorithm C is an improvement.

4.4. RADIX CONVERSION

If men had invented arithmetic by counting with their two fists or their eight fingers, instead of their 10 "digits", we would never have to worry about writing binary-decimal conversion routines. (And we would perhaps never have learned as much about number systems.) In this section, we shall discuss the conversion of numbers from positional notation with one radix into positional notation with another radix; this process is, of course, most important on binary computers when converting decimal input data into binary form, and converting binary answers into decimal form.

A. The four basic methods. Binary-decimal conversion is one of the most machine dependent operations of all, since engineers keep inventing different ways to provide for it in the computer hardware. Therefore we shall discuss only the general principles involved, from which a programmer can select the procedure most well suited to his machine.

We shall assume that only nonnegative numbers enter into the conversion, since the manipulation of signs is easily accounted for.

Let us assume that we are converting from radix b to radix B. (The methods can also be generalized to mixed radix notations, as shown in exercises 1 and 2.) Most radix conversion routines are based on multiplication and division, using one of the following four schemes:

1) Conversion of integers (radix point at the right).

• Method (1a) *Division by B* (using radix b arithmetic). Given an integer number u, we obtain its radix B representation $(U_M \ldots U_1 U_0)_B$ as follows:

$$U_0 = u \bmod B$$
$$U_1 = \lfloor u/B \rfloor \bmod B$$
$$U_2 = \lfloor \lfloor u/B \rfloor /B \rfloor \bmod B$$
$$\cdots$$

etc., stopping when $\lfloor \cdots \lfloor \lfloor u/B \rfloor /B \rfloor \cdots /B \rfloor = 0$.

• Method (1b) *Multiplication by b* (using radix B arithmetic). If u has the radix b representation $(u_m \ldots u_1 u_0)_b$, we can use radix B arithmetic to evaluate the polynomial $u_m b^m + \cdots + u_1 b + u_0 = u$ in the form

$$((\cdots (u_m b + u_{m-1})b + \cdots)b + u_1)b + u_0.$$

2) Conversion of fractions (radix point at the left). Note that it is often impossible to express a terminating radix b fraction $(0.u_{-1} u_{-2} \ldots u_{-m})_b$ *exactly* as a terminating radix B fraction $(0.U_{-1} U_{-2} \ldots U_{-M})_B$. For example, the fraction $\frac{1}{10}$ has the infinite binary representation $(0.0\ 0011\ 0011\ 0011 \ldots)_2$. Therefore methods of rounding the result to M places are sometimes of interest.

• Method (2a) *Multiplication by B* (using radix b arithmetic). Given a fractional number u, we obtain successive digits of its radix B representation $(.U_{-1} U_{-2} \ldots)_B$ as follows:

$$U_{-1} = \lfloor uB \rfloor$$
$$U_{-2} = \lfloor \{uB\} B \rfloor$$
$$U_{-3} = \lfloor \{\{uB\} B\} B \rfloor$$
$$\cdots$$

where $\{x\}$ denotes $x \bmod 1 = x - \lfloor x \rfloor$. If it is desired to round the result to M places, the computation can be stopped after U_{-M} has been calculated, and U_{-M} should be increased by unity if $\{\cdots\{\{uB\}B\}\cdots B\}$ is greater than $\frac{1}{2}$. (Note, however, that this may cause a number of carries to occur, which must be incorporated into the answer using radix B arithmetic. Therefore it would be simpler to add the constant $\frac{1}{2}B^{-M}$ to the original number u before the calculation begins, but this may lead to a terribly incorrect answer when $\frac{1}{2}B^{-M}$ cannot be represented exactly as a radix b number inside the computer. Note further that it is possible for the answer to round up to $(1.00\ldots 0)_B$; see exercise 5.)

- **Method (2b)** *Division by b* (using radix B arithmetic). If u has the radix b representation $(0.u_{-1}u_{-2}\ldots u_{-m})_b$, we can use radix B arithmetic to evaluate $u_{-1}b^{-1} + u_{-2}b^{-2} + \cdots + u_{-m}b^{-m}$ in the form

$$((\cdots (u_{-m}/b + u_{1-m})/b + \cdots + u_{-2})/b + u_{-1})/b.$$

Care should be taken to control errors which might occur due to truncation or rounding in the division by b; these are often negligible, but not always.

To summarize, Methods (1a), (1b), (2a), and (2b) give us two choices for a conversion process, depending on whether our number is an integer or a fraction. And it is certainly possible to convert between integers and fractions by multiplying or dividing by an appropriate power of b or B; therefore there are at least four methods to choose from when trying to do a conversion.

B. Single precision conversion. To illustrate these four methods, let us assume that MIX is a binary computer, and suppose that we want to convert a binary integer u to a decimal integer, that is, $b = 2$ and $B = 10$. Method (1a) could be programmed as follows:

```
      ENT1  0           Set j ← 0.
      LDX   U
      ENTA  0           Set rAX ← u.
1H    DIV   =10=        (rA, rX) ← (⌊rAX/10⌋, rAX mod 10).
      STX   ANSWER,1    U_j ← rX.
      INC1  1           j ← j + 1.
      SRAX  5           rAX ← rA.
      JXP   1B          Repeat until result is zero.  ∎        (1)
```

This requires $18M + 4$ cycles to obtain M digits.

The above method used division by 10; Method (2a) uses *multiplication* by 10, so it might be a little faster. In order to use Method (2a), we must deal with fractions, and this leads to an interesting situation. Let w be the word size of the computer, and assume that $u < 10^n < w$. With a single division we can find q and r, where

$$w(u + 1) = 10^n q + r, \qquad 0 < r \leq 10^n < w. \qquad (2)$$

Now if we apply Method (2a) to the fraction q/w, we will obtain the digits of u from left to right, in n steps, since

$$\left\lfloor 10^n \frac{q}{w} \right\rfloor = \left\lfloor u + 1 - \frac{r}{w} \right\rfloor = u. \tag{3}$$

Here is the corresponding MIX program; this program will work for *all* $u < w$, if $10^n < w < 10^{n+1}$, since it begins with a division by 10^n:

```
        JOV   OFLO          Ensure overflow is off.
        ENTA  0
        LDX   U             rAX ← u.
        DIV   =10ⁿ=
        STA   ANSWER+n      Set rX ← u mod 10ⁿ, Uₙ ← ⌊u/10ⁿ⌋.
        ENT1  n-1           Set j ← n - 1.
        SLAX  5
        INCA  1             rAX ← (u + 1)w.
        DIV   =10ⁿ=         Compute q, r:
        JNOV  1F
        LDA   WM1           If u + 1 = 10ⁿ, q = w - 1.
        JMP   2F
1H      JXNZ  2F            If r = 0,
        DECA  1                  decrease q by 1.
2H      MUL   =10=          Now imagine radix point at left.
        STA   ANSWER,1      Set Uⱼ ← ⌊10x⌋.
        SLAX  5             x ← {10x}.
        DEC1  1             j ← j - 1.
        J1NN  2B            Repeat for n > j ≥ 0.  ∎       (4)
```

This somewhat longer routine requires $16n + 36$ cycles, which in this case is not worth the extra programming effort, since n cannot be large enough to compensate for the long initialization. Note that if $u = 0$, the answer digits are set to minus zero by this program instead of plus zero, since formula (2) requires q to equal -1.

On many binary computers, for which multiplication by 10 is significantly faster than division by 10, this routine will actually be faster than Method (1a). Thus we see that a fractional method can be used for integers, although the use of inexact division makes the numerical analysis nontrivial.

A modification of Method (1a) can be used to avoid division by 10, by replacing it with two multiplications. It is worth mentioning this modification here, because radix conversion is often done by small "satellite" computers which have no division capability. If we let x be an approximation to $\frac{1}{10}$, so that $\frac{1}{10} < x < \frac{1}{10} + 1/w$, it is easy to prove (see exercise 7) that $\lfloor ux \rfloor = \lfloor u/10 \rfloor$ or $\lfloor u/10 \rfloor + 1$, so long as $0 \le u < w$. Therefore, if we compute $u - 10\lfloor ux \rfloor$, we will be able to determine the value of $\lfloor u/10 \rfloor$:

$$\lfloor u/10 \rfloor = \begin{cases} \lfloor ux \rfloor, & \text{if} \quad u - 10\lfloor ux \rfloor \ge 0; \\ \lfloor ux \rfloor - 1, & \text{if} \quad u - 10\lfloor ux \rfloor < 0. \end{cases} \tag{5}$$

This procedure simultaneously determines u mod 10. A MIX program for conversion using this idea appears in exercise 8; it requires about 36 cycles per digit.

The procedure represented by (5) can be used effectively even if the computer has no built-in multiplication instruction, since multiplication by 10 consists of two shifts and one addition $(10 = 2^3 + 2)$, and multiplication by $\frac{1}{10}$ can also be done by combining a few shifting and adding operations judiciously (see exercise 9).

Method (1b) could also be used to convert from binary to decimal, but to do this we need to simulate doubling in a decimal number system. This idea is generally most suitable for incorporation into computer hardware; however, it is possible to program the doubling process for decimal numbers, using binary addition, extraction, and shifting, as shown in Table 1.

Table 1

DOUBLING A BINARY-CODED DECIMAL NUMBER

Operation	General form	Example
1. Given number	$u_1u_2u_3u_4\ \ u_5u_6u_7u_8\ \ u_9\ \ u_{10}u_{11}u_{12}$	0011 0110 1001 $= 3\ 6\ 9$
2. Add 3 to each digit	$v_1v_2v_3v_4\ \ v_5v_6v_7v_8\ \ v_9\ \ v_{10}v_{11}v_{12}$	0110 1001 1100
3. Shift left one	$v_1\ \ v_2v_3v_4v_5\ \ v_6v_7v_8v_9\ \ v_{10}v_{11}v_{12}0$	0 1101 0011 1000
4. Extract low bit	$v_1\ 0\ 0\ 0\ \ v_5\ 0\ 0\ 0\ \ v_9\ \ 0\ \ 0\ \ 0\ \ 0$	0 0001 0001 0000
5. Shift right two	$0\ v_1 0\ 0\ \ \ 0\ v_5 0\ 0\ \ \ 0\ \ v_9\ 0\ \ 0$	0000 0100 0100
6. Shift right one and add	$0\ v_1 v_1 0\ \ \ 0\ v_5 v_5 0\ \ \ 0\ \ v_9\ v_9\ 0$	0000 0110 0110
7. Add result of step 3	$*\ \ *\ *\ *\ *\ \ \ *\ *\ *\ *\ \ \ *\ \ *\ \ *\ \ 0$	0 1101 1001 1110
8. Subtract 6 from each	$y_1\ \ y_2y_3y_4y_5\ \ y_6y_7y_8y_9\ \ y_{10}y_{11}y_{12}0$	0 0111 0011 1000 $= 7\ 3\ 8$

Another related idea is to keep a table of the powers of two in decimal form, and to add the appropriate powers together by simulating decimal addition.

Finally, even Method (2b) can be used for the conversion of binary integers to decimal integers. We can find q as in (2), and then we can simulate the decimal division of q by w, using a "halving" process (exercise 10) which is similar to the doubling process just described, and retaining only the first n digits to the right of the radix point in the answer. In this situation, Method (2b) does not seem to offer advantages over the other three methods already discussed, but we have at least proved the contention made earlier that at least four distinct methods are available for doing the conversion of integers from one radix to another.

Now let us consider decimal-to-binary conversion ($b = 10$, $B = 2$). Method (1a) simulates a decimal division by 2; this is feasible (see exercise 10), but it is primarily suitable for incorporation in hardware instead of programs.

Method (1b) is the most practical method for decimal-to-binary conversion in the great majority of cases. Here it is in MIX code:

	ENT1	M-1	Set $j \leftarrow m - 1$.	
	LDA	INPUT+M	Set $U \leftarrow u_m$.	
1H	MUL	=10=		
	SLAX	5		
	ADD	INPUT,1	$U \leftarrow 10U + u_j$.	
	DEC1	1		
	J1NN	1B	Repeat for $m > j \geq 0$. ∎	(6)

Note again that adding and shifting may be substituted for multiplication by 10.

For the conversion of decimal fractions $(0.u_{-1}u_{-2}\ldots u_{-m})_{10}$, we can use Method (2b), or, more commonly, we can convert the integer $(u_{-1}u_{-2}\ldots u_{-m})_{10}$ by Method (1b) and then divide by 10^m.

C. Hand calculation. Occasionally, it is necessary for computer programmers to convert numbers by hand, and since this is a subject not yet taught in elementary schools, it may be worth while to examine it briefly here. There are very simple pencil-and-paper methods for converting between decimal and octal notations, which are easily learned, so they ought to be more widely known.

Converting octal integers to decimal. The simplest conversion is from octal to decimal; this technique was apparently first published by Walter Soden, *Math.*

Comp. **7** (1953), 273–274. To do the conversion, write down the given octal number; then at the kth step, double the k leading digits using decimal arithmetic, and subtract this from the $k + 1$ leading digits using decimal arithmetic. The process terminates in $n - 1$ steps if the given number has n digits. It is a good idea to insert a radix point to show which digits are being doubled, as shown in the examples given here, in order to prevent embarrassing mistakes.

Example 1. Convert $(5325121)_8$ to decimal.

```
     5.3 2 5 1 2 1
  −  1 0
     ─────────────
     4 3.2 5 1 2 1
  −    8 6
     ─────────────
     3 4 6.5 1 2 1
  −    6 9 2
     ─────────────
     2 7 7 3.1 2 1
  −    5 5 4 6
     ─────────────
     2 2 1 8 5.2 1
  −    4 4 3 7 0
     ─────────────
     1 7 7 4 8 2.1
  −    3 5 4 9 6 4
     ─────────────
     1 4 1 9 8 5 7
```

Answer: $(1419857)_{10}$.

A reasonably good check on the computations may be had by "casting out nines": The sum of the digits of the decimal number must equal the alternating sum and difference of the digits of the octal number, with the rightmost digit of the latter given a plus sign, except for a multiple of nine. In the above example, we have $1 + 4 + 1 + 9 + 8 + 5 + 7 = 35$, and $1 - 2 + 1 - 5 + 2 - 3 + 5 = -1$, and the difference is 36 (a multiple of 9). If this test fails, it can be applied to the $k + 1$ leading digits after the kth step, and the error can be located using a "binary search" procedure; i.e., start by checking the middle result, then use the same procedure on the first or second half of the calculation, depending on whether the middle result is incorrect or correct.

Of course, the "casting-out-nines" process is only about 89 percent reliable; an even better check is to convert the answer back to octal by using an inverse method, which we will now consider.

Converting decimal integers to octal. A similar procedure can be used for the opposite conversion: Write down the given decimal number; at the kth step, double the k leading digits using *octal* arithmetic, and *add* these to the $k + 1$ leading digits using *octal* arithmetic. The process terminates in $n - 1$ steps if the given number has n digits.

Example 2. Convert $(1419857)_{10}$ to octal.

```
1.4 1 9 8 5 7
2
————————————
1 6.1 9 8 5 7          (Note that the nonoctal digits 8 and
3 4                     9 enter into this octal computation.)
————————————           The answer can be checked as dis-
2 1 5.9 8 5 7          cussed above.  This method was
4 3 2                  published by Charles P. Rozier,
————————————           IEEE  Trans.  CE–11  (1962),
2 6 1 3.8 5 7          708–709.
5 4 2 6
————————————
3 3 5 6 6.5 7
6 7 3 5 4
————————————
4 2 5 2 4 1.7
1 0 5 2 5 0 2
————————————
5 3 2 5 1 2 1          Answer: (5325121)₈.
```

The two procedures just given are essentially Method (1b) of the general radix conversion procedures. Doubling and subtracting in decimal notation is like multiplying by $10 - 2 = 8$; doubling and adding in octal notation is like multiplying by $8 + 2 = 10$.

To keep these two methods straight in our minds, it is not hard to remember that we must subtract to go from octal to decimal, since the decimal representation of a number is in general smaller; similarly we must add to go from decimal to octal. The computations are performed using the radix of the answer (otherwise it is easy to see we won't get the right answer)!

Converting fractions. No equally fast method of converting fractions manually is known; the best way seems to be Method (2a), with doubling and adding or subtracting to simplify the multiplications by 10 or by 8. In this case, we reverse the addition-subtraction criterion, adding when we convert to decimal and subtracting when we convert to octal; we also use the radix of the given input number, *not* the radix of the answer, in this computation (see Examples 3 and 4). The process is about twice as hard as the above method for integers.

Example 3. Convert $(.14159)_{10}$ to octal.

```
.1 4 1 5 9
  2 8 3 1 8—
1.1 3 2 7 2
  2 6 5 4 4—
1.0 6 1 7 6
  1 2 3 5 2—
0.4 9 4 0 8
  9 8 8 1 6—
3.9 5 2 6 4
  1 9 0 5 2 8—
7.6 2 1 1 2
  1 2 4 2 2 4—
4.9 6 8 9 6
```

Answer: $(.110374\ldots)_8$.

Example 4. Convert $(.110374)_8$ to decimal.

```
.1 1 0 3 7 4
  2 2 0 7 7 0+
1.3 2 4 7 3 0
  6 5 1 6 6 0+
4.1 2 1 1 6 0
  2 4 2 3 4 0+
1.4 5 4 1 4 0
  1 1 3 0 3 0 0+
5.6 7 1 7 0 0
  1 5 6 3 6 0 0+
8.5 0 2 6 0 0
  1 2 0 5 4 0 0+
6.2 3 3 4 0 0
```

Answer: $(.141586\ldots)_{10}$.

D. Floating-point conversion. When floating-point values are to be converted, it is necessary to deal with both the exponent and the fraction parts simultaneously, since conversion of the exponent will affect the fraction part. Given the number $f \cdot 2^e$ to be converted to decimal, we may express 2^e in the form $F \cdot 10^E$ (usually by means of auxiliary tables), and then convert Ff to decimal. Alternatively, we can multiply e by $\log_{10} 2$ and round this to the nearest integer E; then divide $f \cdot 2^e$ by 10^E and convert the result. Conversely, given the number $F \cdot 10^E$ to be converted to binary, we may convert F and then multiply it by the floating-point number 10^E (again commonly obtained by using tables). Obvious techniques may be used to reduce the maximum size of the auxiliary tables by using several multiplications and/or divisions, although this can cause rounding errors to propagate.

E. Multiple-precision conversion. When converting multiple-precision numbers, it is most convenient to convert blocks of digits at a time, which can be handled by single-precision techniques, and then to combine them using simple multiple-precision techniques. For example, suppose that 10^n is the highest power of 10 less than the computer word size. Then

a) To convert a multiple-precision *integer* from binary to decimal, divide repeatedly by 10^n [thus converting from binary to radix 10^n by Method (1a)]. Single-precision operations will give the n decimal digits for each place of the radix 10^n representation.

b) To convert a multiple-precision *fraction* from binary to decimal, proceed similarly, multiplying by 10^n [i.e., using Method (2a) with $B = 10^n$].

c) To convert a multiple-precision integer from decimal to binary, convert blocks of n digits first; then convert from radix 10^n to binary by using Method (1b).

d) Finally, a multiple-precision fraction may be converted from decimal to binary by a technique similar to (c), using Method (2b).

F. History and Bibliography. Radix conversion techniques implicitly originated in ancient problems dealing with weights, measures, and currencies, when a mixed radix system was generally involved. Auxiliary tables were usually used as an aid to conversion. During the seventeenth century when sexagesimal fractions were being supplanted by decimal fractions, it was necessary to convert between the two systems in order to use existing books of tables; a systematic method to transform fractions from radix 60 to radix 10 and vice versa was given in the 1667 edition of William Oughtred's *Clavis Mathematicae*, Chapter 6, Section 18. (This material was not present in the original 1631 edition of Oughtred's book.) Conversion rules had been given already by al-Kashî of Persia in his *Key to Arithmetic* (c. 1414), where Methods (1a), (1b), and (2a) are clearly explained [*Istoriko-Mat. Issled.* **7** (1954), 126–135], but his work was unknown in Europe. A. M. Legendre [*Théorie des nombres* (Paris, 1798), 229] noted that positive integers may be conveniently converted to binary form by repeatedly dividing by 64.

In 1946, H. H. Goldstine and J. von Neumann gave prominent consideration to radix conversion in their classic memoir, "Planning and coding problems for an electronic computing instrument," because it was necessary to justify the use of binary arithmetic; see John von Neumann, *Collected Works*, **5** (New York: Macmillan, 1963), 127–142. Another early discussion of radix conversion on binary computers was published by F. Koons and S. Lubkin, *Math. Comp.* **3** (1949), 427–431, where a rather unusual method is suggested. The first discussion of floating-point conversion was given somewhat later by F. L. Bauer and K. Samelson [*Zeit. für angewandte Math. und Physik* **4** (1953), 312–316].

The following articles may be useful for further reference: A note by G. T. Lake [*CACM* **5** (1962), 468–469] mentions some hardware techniques for conversion and gives clear examples. A. H. Stroud and D. Secrest [*Comp. J.* **6** (1963), 62–66] have discussed conversion of multiple-precision floating-point numbers. The conversion of *unnormalized* floating-point numbers, preserving the amount of "significance" implied by the representation, has been discussed by H. Kanner [*JACM* **12** (1965), 242–246] and by N. Metropolis and R. L. Ashenhurst [*Math. Comp.* **19** (1965), 435–441]. The latter article gives many interesting details and examples.

EXERCISES

▶ **1.** [*25*] Generalize Method (1b) so that it works with mixed-radix notations, converting

$$a_m b_{m-1} \ldots b_1 b_0 + \cdots + a_1 b_0 + a_0 \quad \text{into} \quad A_M B_{M-1} \ldots B_1 B_0 + \cdots + A_1 B_0 + A_0,$$

where $0 \le a_j < b_j$, $0 \le A_J < B_J$ for $0 \le j < m$, $0 \le J < M$.

Give an example of your generalization by manually converting the quantity "3 days, 9 hours, 12 minutes, and 37 seconds" into long tons, hundredweight, stones, pounds, and ounces. (Let one second equal one ounce. The British system of weights has 1 stone = 14 pounds, 1 hundredweight = 8 stones, 1 long ton = 20 hundredweight.) In other words, let $b_0 = 60$, $b_1 = 60$, $b_2 = 24$, $m = 3$, $B_0 = 16$, $B_1 = 14$, $B_2 = 8$, $B_3 = 20$, $M = 4$; the problem is to find $A_4, \ldots, A_0$ in the proper ranges such that

$$3 b_2 b_1 b_0 + 2 b_1 b_0 + 12 b_0 + 37$$
$$= A_4 B_3 B_2 B_1 B_0 + A_3 B_2 B_1 B_0 + A_2 B_1 B_0 + A_1 B_0 + A_0,$$

using a systematic method which generalizes Method (1b). (All arithmetic is to be done in a mixed-radix system.)

2. [*25*] Generalize Method (1a) so that it works with mixed-radix notations, as in exercise 1, and give an example of your generalization by manually solving the same conversion problem stated in exercise 1.

3. [*25*] (D. Taranto.) The text observes that when fractions are being converted, there is in general no obvious way to decide how many digits to give in the answer. Design a simple generalization of Method (2a) which, given two positive radix b fractions u and ϵ between 0 and 1, converts u to a radix B equivalent U which has just enough places of accuracy to ensure that $|U - u| < \epsilon$. (If $u < \epsilon$, we may take $U = 0$, with zero "places of accuracy.")

4. [*M21*] (a) Prove that every real number which has a terminating *binary* representation also has a terminating *decimal* representation. (b) Find a simple condition on the positive integers b and B, which is satisfied if and only if every real number which has a terminating radix b representation also has a terminating radix B representation.

5. [*M22*] Suppose we convert a nonnegative m-digit radix b fraction to an M-digit radix B fraction by Method (2a), rounding the result as described in the text. Show that it is possible for this rounded result to be equal to unity if and only if $b^m \ge 2B^M$; otherwise the rounded result will be less than unity.

6. [*30*] Discuss using Methods (1a), (1b), (2a), and (2b) when b or B is -2.

7. [*M16*] Given that $\alpha \le x \le \alpha + 1/w$ and $0 \le u \le w$, prove that $\lfloor ux \rfloor = \lfloor \alpha u \rfloor$ or $\lfloor \alpha u \rfloor + 1$.

8. [*24*] Write a `MIX` program analogous to (1) which uses (5) and which does not include any division instructions.

9. [*M27*] Let u be an integer, $0 \le u < 2^{34}$. Assume that the following sequence of operations (equivalent to addition and binary "shift-right" instructions) is performed:

$$v \leftarrow \lfloor \tfrac{1}{2}u \rfloor, \qquad v \leftarrow v + \lfloor \tfrac{1}{2}v \rfloor, \qquad v \leftarrow v + \lfloor 2^{-4}v \rfloor,$$
$$v \leftarrow v + \lfloor 2^{-8}v \rfloor, \qquad v \leftarrow v + \lfloor 2^{-16}v \rfloor, \qquad v \leftarrow \lfloor \tfrac{1}{8}v \rfloor.$$

Prove that $v = \lfloor u/10 \rfloor$ or $\lfloor u/10 \rfloor - 1$.

10. [*22*] The text shows how a binary-coded decimal number can be doubled by using various shifting, extracting, and addition operations on a binary computer. Give an analogous method which computes *half* of a binary-coded decimal number (throwing away the remainder if the number is odd).

11. [*16*] Convert $(57721)_8$ to decimal.

▶ **12.** [*22*] Invent a rapid pencil-and-paper method for converting integers from ternary notation to decimal, and illustrate your method by converting $(1212011210210)_3$ into decimal.

▶ **13.** [*25*] Assume that locations $U + 1$, $U + 2$, ..., $U + m$ contain a multiple-precision fraction $(.u_{-1}u_{-2} \ldots u_{-m})_b$, where b is the word size of MIX. Write a MIX routine which converts this fraction to decimal notation, truncating it to 180 decimal digits. The answer should be printed on two lines, with the digits grouped into 20 blocks of nine each separated by blanks. (Use the CHAR instruction.)

▶ **14.** [*M30*] (A. Schönhage.) The text's method of converting multiple-precision integers requires an execution time of order n^2 to convert an n-digit integer, when n is large. Show that it is possible to convert n-digit decimal integers into binary notation in $O(M(n) \log n)$ cycles, where $M(n)$ is the number of cycles needed to multiply n-bit binary numbers.

15. [*M50*] Can the upper bound on the time to convert large integers, given in exercise 14, be substantially lowered? (Cf. exercise 4.3.3–12.)

16. [*41*] Construct a fast linear iterative array for radix conversion from decimal to binary (cf. Section 4.3.3).

17. [*M40*] Design "ideal" floating-point conversion subroutines, taking p-digit decimal numbers into P-digit binary numbers and vice versa, in both cases producing a true rounded result in the sense of Section 4.2.2.

18. [*HM39*] (David W. Matula.) Let $\mathrm{round}_b(u, p)$ be the function of b, u, and p defined by Eq. 4.2.2–9. Under the assumption that $\log_B b$ is irrational, prove that

$$u = \mathrm{round}_b\big(\mathrm{round}_B (u, P), p\big),$$

for all p-digit base b floating-point numbers u, if and only if $B^{P-1} \geq b^p$. (In other words, an "ideal" input conversion of u into an independent base B, followed by an "ideal" output conversion of this result, will always yield u again if and only if the intermediate precision P is suitably large, as specified by the formula above.)

4.5. RATIONAL ARITHMETIC

4.5.1. Fractions

In many numerical algorithms it is important to know that the answer to a problem is exactly $\tfrac{1}{3}$, not a floating-point number which gets printed as "0.333333574". If arithmetic is done on fractions instead of on approximations to fractions, many computations can be done entirely *without any accumulated rounding errors.* This results in a comfortable feeling of security which is often lacking when floating-point calculations have been made, and it means that the accuracy of the calculation cannot be improved upon.

When fractional arithmetic is desired, the numbers can be represented as pairs of integers, (u/u'), where u and u' are relatively prime to each other and $u' > 0$. The number zero is represented as $(0/1)$. In this form, $(u/u') = (v/v')$ if and only if $u = v$ and $u' = v'$.

Multiplication of fractions is, of course, rather simple; to form $(u/u') \times (v/v') = (w/w')$, we can simply compute uv and $u'v'$. The two products uv and $u'v'$ might not be relatively prime, but if $d = \gcd(uv, u'v')$, the desired answer is $w = uv/d$, $w' = u'v'/d$. (See exercise 2.) Efficient algorithms to compute the greatest common divisor are discussed in Section 4.5.2.

Another way to perform the multiplication is to find $d_1 = \gcd(u, v')$ and $d_2 = \gcd(u', v)$; then the answer is $w = (u/d_1)(v/d_2)$, $w' = (u'/d_2)(v'/d_1)$. (See exercise 3.) This method requires two gcd calculations, but it is not really slower than the former method; the gcd process involves a number of iterations essentially proportional to the logarithm of its inputs, so the total number of iterations needed to evaluate both d_1 and d_2 is essentially the same as the number of iterations during the single calculation of d. Furthermore, each iteration in the evaluation of d_1 and d_2 is potentially faster, because comparatively small numbers are being examined. If u, u', v, v' are single-precision quantities, this method has the advantage that no double-precision numbers appear in the calculation unless it is impossible to represent both of the answers w and w' in single precision form.

Division may be done in a similar manner; see exercise 4.

Addition and subtraction are slightly more complicated. We can set $(u/u') \pm (v/v') = ((uv' \pm u'v)/u'v')$ and reduce this fraction to lowest terms by calculating $d = \gcd(uv' \pm u'v, u'v')$ as in the first multiplication method. But again it is possible to avoid working with such large numbers, if we start by calculating $d_1 = \gcd(u', v')$. If $d_1 = 1$ then $w = uv' \pm u'v$ and $w' = u'v'$ are the desired answers. (According to Theorem 4.5.2C, d_1 will be 1 about 61 percent of the time, if the denominators u' and v' are randomly distributed, so it is wise to single out this case separately.) If $d_1 > 1$, then let $t = u(v'/d_1) \pm v(u'/d_1)$ and calculate $d_2 = \gcd(t, d_1)$; finally the answer is $w = t/d_2$, $w' = (u'/d_1)(v'/d_2)$. (Exercise 6 proves that these values of w and w' are relatively prime to each other.) If single-precision numbers are being used, this method requires only single-precision operations, except that t may be a double-precision number or slightly larger (see exercise 7); since $\gcd(t, d_1) = \gcd(t \bmod d_1, d_1)$, the calculation of d_2 does not require double precision.

For example, to compute $(7/66) + (17/12)$, we form $d_1 = \gcd(66, 12) = 6$; then $t = 7 \cdot 2 + 17 \cdot 11 = 201$, and $d_2 = \gcd(201, 6) = 3$, so the answer is

$$\tfrac{201}{3} / \tfrac{66}{6} \tfrac{12}{3} = 67/44.$$

Experience with fractional calculations shows that in many cases the numbers grow to be quite large; for example, see the remarks following Eq. 3.3.2–22. So if u and u' are intended to be single-precision numbers for each fraction (u/u'), it is important to include tests for overflow in each of the addition, subtraction, multiplication, and division subroutines. For numerical problems in which perfect accuracy is important, a set of subroutines for fractional arithmetic with *arbitrary* precision allowed in numerator and denominator is very useful.

The methods of this section also extend to other number fields besides the rational numbers; for example, we could do arithmetic on quantities of the form $(u + u'\sqrt{5})/u''$, where u, u', u'' are integers, $\gcd(u, u', u'') = 1$, and $u'' > 0$; or on quantities of the form $(u + u'\sqrt[3]{2} + u''\sqrt[3]{4})/u'''$, etc.

To help check out subroutines for rational arithmetic, inversion of matrices with known inverses (e.g., Cauchy matrices, exercise 1.2.3–41) is suggested.

Exact representation of fractions within a computer was first discussed in the literature by P. Henrici, *JACM* **3** (1956), 6–9.

EXERCISES

1. [*15*] Suggest a reasonable computational method for comparing two fractions, to test whether or not $(u/u') < (v/v')$.

2. [*M15*] Prove that if $d = \gcd(u, v)$ then u/d and v/d are relatively prime.

3. [*M20*] Prove that if u and u' are relatively prime, and if v and v' are relatively prime, then $\gcd(uv, u'v') = \gcd(u, v') \gcd(u', v)$.

4. [*21*] Design a division algorithm for fractions, analogous to the second multiplication method of the text. (Note that the sign of v must be considered.)

5. [*10*] Compute $(17/120) + (-27/70)$ by the method recommended in the text.

▶ **6.** [*M23*] Show that if u, u' are relatively prime and if v, v' are relatively prime, then $\gcd(uv' + vu', u'v') = d_1 d_2$ where $d_1 = \gcd(u', v')$ and $d_2 = \gcd(u(v'/d_1) + v(u'/d_1), d_1)$. (Hence if $d_1 = 1$, then $uv' + vu'$ is relatively prime to $u'v'$.)

7. [*M22*] How large can the absolute value of the quantity t become, in the addition-subtraction method recommended in the text, if the input numerators and denominators are less than N in absolute value?

▶ **8.** [*22*] Discuss using $(1/0)$ and $(-1/0)$ as representations for ∞ and $-\infty$, and/or as representations of overflow.

9. [*M23*] If $1 \le u', v' < 2^n$, show that $\lfloor 2^{2n}u/u' \rfloor = \lfloor 2^{2n}v/v' \rfloor$ implies $u/u' = v/v'$.

10. [*41*] Extend the subroutines suggested in exercise 4.3.1–34 so that they deal with "arbitrary" rational numbers.

11. [*M23*] Consider fractions of the form $(u + u'\sqrt{5})/u''$, where u, u', u'' are integers, $\gcd(u, u', u'') = 1$, and $u'' > 0$. Explain how to divide two such fractions and to obtain a quotient having the same form.

4.5.2. The Greatest Common Divisor

If u and v are integers, not both zero, we say that their *greatest common divisor*, gcd (u, v), is the largest integer which evenly divides both u and v. This definition makes sense, because if $u \neq 0$ then no integer greater than $|u|$ can evenly divide u, but the integer 1 does divide both u and v; hence there must be a largest integer that divides them both. When u and v are both zero, every integer evenly divides zero, so the above definition does not apply; it is convenient to set

$$\gcd (0, 0) = 0. \tag{1}$$

The definitions just given obviously imply that

$$\gcd (u, v) = \gcd (v, u), \tag{2}$$

$$\gcd (u, v) = \gcd (-u, v), \tag{3}$$

$$\gcd (u, 0) = |u|. \tag{4}$$

In the previous section, we reduced the problem of expressing a rational number in "lowest terms" to the problem of finding the greatest common divisor of its numerator and denominator. Other applications of the greatest common divisor have been mentioned for example in Sections 3.2.1.2, 3.3.3, 4.3.2, 4.3.3. So the concept of the greatest common divisor is important and worthy of serious study.

The *least common multiple* of two integers u and v, written lcm (u, v), is a related idea which is also important. It is defined to be the smallest positive integer which is a multiple of (i.e., evenly divisible by) both u and v; and lcm $(0, 0) = 0$. The classical method for teaching children how to add fractions $u/u' + v/v'$ is to train them to find the "least common denominator," which is lcm (u', v').

According to the "fundamental theorem of arithmetic" (proved in exercise 1.2.4–21), each positive integer u can be expressed in the form

$$u = 2^{u_2}3^{u_3}5^{u_5}7^{u_7}11^{u_{11}} \cdots = \prod_{p \text{ prime}} p^{u_p}, \tag{5}$$

where the exponents $u_2, u_3, \ldots$ are uniquely determined nonnegative integers, and where all but a finite number of the exponents are zero. From this "canonical factorization" of a positive integer, it is easy to discover one way to compute the greatest common divisor of u and v: By (2), (3), and (4) we may assume that u and v are positive integers, and if both of them have been canonically factored into primes, we have

$$\gcd (u, v) = \prod_{p \text{ prime}} p^{\min(u_p, v_p)}, \tag{6}$$

$$\text{lcm } (u, v) = \prod_{p \text{ prime}} p^{\max(u_p, v_p)}. \tag{7}$$

Thus, for example, the greatest common divisor of $u = 7000 = 2^3 \cdot 5^3 \cdot 7$ and $v = 4400 = 2^4 \cdot 5^2 \cdot 11$ is $2^{\min(3,4)}5^{\min(3,2)}7^{\min(1,0)}11^{\min(0,1)} = 2^3 \cdot 5^2 = 200$. The least common multiple of the same two numbers is $2^4 \cdot 5^3 \cdot 7 \cdot 11 = 154000$.

From formulas (6) and (7) we can easily prove a number of basic identities concerning the gcd and the lcm:

$$\gcd(u, v)w = \gcd(uw, vw), \qquad\qquad \text{if} \quad w \geq 0; \qquad (8)$$

$$\operatorname{lcm}(u, v)w = \operatorname{lcm}(uw, vw), \qquad\qquad \text{if} \quad w \geq 0; \qquad (9)$$

$$u \cdot v = \gcd(u, v) \cdot \operatorname{lcm}(u, v), \qquad \text{if} \quad u, v \geq 0; \quad (10)$$

$$\gcd\big(\operatorname{lcm}(u, v), \operatorname{lcm}(u, w)\big) = \operatorname{lcm}\big(u, \gcd(v, w)\big); \qquad\qquad\qquad (11)$$

$$\operatorname{lcm}\big(\gcd(u, v), \gcd(u, w)\big) = \gcd\big(u, \operatorname{lcm}(v, w)\big). \qquad\qquad\qquad (12)$$

The latter two formulas are "distributive laws" analogous to the familiar identity $uv + uw = u(v + w)$. Equation (10) reduces the calculation of $\gcd(u, v)$ to the calculation of $\operatorname{lcm}(u, v)$, and conversely.

Euclid's algorithm. Although Eq. (6) is useful for theoretical purposes, it is generally no help for calculating a greatest common divisor in practice, because it requires that we first determine the factorization of u and v. There is no known method for finding the prime factors of an integer very rapidly (see Section 4.5.4). But fortunately there is an efficient way to calculate the greatest common divisor of two integers without factoring them, and, in fact, such a method was discovered over 2250 years ago; this is "Euclid's algorithm," which we have already examined in Sections 1.1 and 1.2.1.

Euclid's algorithm, which is found in Book 7, Propositions 1 and 2 of his *Elements* (c. 300 B.C.), and which many scholars conjecture was actually Euclid's rendition of an algorithm due to Eudoxus (c. 375 B.C.), may be called the granddaddy of all algorithms, because it is the oldest nontrivial algorithm which has survived to the present day. (The chief rival for this honor is perhaps the ancient Egyptian method for multiplication, which was based on doubling and adding, and which forms the basis for efficient calculation of nth powers as explained in Section 4.6.3. But the Egyptian manuscripts merely give examples which are not completely systematic, and which were certainly not stated systematically; the Egyptian method is therefore not quite deserving of the name "algorithm." Several ancient Babylonian methods, for doing such things as solving certain sets of quadratic equations in two variables, are also known. Genuine algorithms are involved in this case, not just special solutions to the equations for certain input parameters; even though the Babylonians invariably presented each method in conjunction with an example worked with special parameters, they regularly explained the general procedure in the accompanying text. If by chance the special parameters made one of the factors of a product equal to unity, the multiplication was explicitly performed anyway, whenever the general case required it. [See O. Neugebauer, *The Exact Sciences in Antiquity* (Princeton University Press, 1952), 42.] Many of these Babylonian algorithms predate Euclid by 1500 years, and they are the earliest known instances of written algorithms. They do not have the stature of Euclid's algorithm, however, since they do not involve iteration and since they have been superseded by modern algebraic methods.)

Since Euclid's algorithm is therefore important for historical as well as practical reasons, let us now consider how Euclid himself treated it. Paraphrasing his words into modern terminology, this is essentially what Euclid said:

Proposition. *Given two positive integers, find their greatest common divisor.*

Let A, C be the two given positive integers; it is required to find their greatest common divisor. If C divides A, then C is a common divisor of C and A, since it also divides itself. And it clearly is in fact the greatest, since no greater number than C will divide C.

But if C does not divide A, then continually subtract the lesser of the numbers A, C from the greater, until some number is left which divides the previous one. This will eventually happen, for if unity is left, it will divide the previous number.

Now let E be the positive remainder of A divided by C; let F be the positive remainder of C divided by E; and let F be a divisor of E. Since F divides E and E divides $C - F$, F also divides $C - F$; but it also divides itself, so it divides C. And C divides $A - E$; therefore F also divides $A - E$. But it also divides E; therefore it divides A. Hence it is a common divisor of A and C.

I now claim that it is also the greatest. For if F is not the greatest common divisor of A and C, some larger number will divide them both. Let such a number be G.

Now since G divides C while C divides $A - E$, G divides $A - E$. G also divides the whole of A, so it divides the remainder E. But E divides $C - F$; therefore G also divides $C - F$. G also divides the whole of C, so it divides the remainder F; that is, a greater number divides a smaller one. This is impossible.

Therefore no number greater than F will divide A and C, and so F is their greatest common divisor.

Corollary. This argument makes it evident that any number dividing two numbers divides their greatest common divisor. *Q.E.D.*

Note. Euclid's statements have been simplified here in one nontrivial respect: Greek mathematicians did not regard unity as a "divisor" of another positive integer. Two positive integers were either both equal to unity, or they were relatively prime, or they had a greatest common divisor. In fact, unity was not even considered to be a "number," and zero was of course nonexistent. These rather awkward conventions made it necessary for Euclid to duplicate much of his discussion, and he gave two separate propositions which each are essentially like the one appearing here.

In his discussion, Euclid first suggests subtracting the smaller of the two current numbers from the larger repeatedly, until we get two numbers in which one is a multiple of another; but in the proof he really relies on taking the remainder of one number divided by another (and since he has no concept of zero, he cannot speak of the remainder when one number divides the other). So it is reasonable to say that he imagines each *division* (not the individual subtractions) as a single step of the algorithm, and hence an "authentic" rendition of his algorithm can be phrased as follows:

Algorithm E (*Original Euclidean algorithm*). Given two integers A and C greater than unity, this algorithm finds their greatest common divisor.

E1. [A divisible by C?] If C divides A, the algorithm terminates with C as the answer.

E2. [Replace A by remainder.] If $A \bmod C$ is equal to unity, the given numbers were relatively prime, so the algorithm terminates. Otherwise replace the pair of values (A, C) by $(C, A \bmod C)$ and return to step E1. ▮

The "proof" Euclid gave, which is quoted above, is especially interesting because it is, of course, not really a proof at all! He verifies the result of the algorithm only if step E1 is performed once or thrice. Surely he must have realized that step E1 could take place more than three times, although he made no mention of such a possibility. Not having the notion of a proof by mathematical induction, he could only give a proof for a finite number of cases. (In fact, he often proved only the case $n = 3$ of a theorem which he wanted to establish for general n.) Although Euclid is justly famous for the great advances he made in the art of logical deduction, techniques for giving valid proofs by induction were not discovered until many centuries later, and the crucial ideas for proving the validity of *algorithms* are only now becoming really clear. (See Section 1.2.1 for a complete proof of Euclid's algorithm, together with a discussion of general proof procedures for algorithms.)

It is worth noting that this algorithm for finding the greatest common divisor was chosen by Euclid to be the very first step in his development of the theory of numbers. The same order of presentation is still in use today in modern textbooks. Euclid also gave (Proposition 34) a method to find the least common multiple of two integers u and v, namely to divide u by gcd (u, v) and to multiply by v; this is equivalent to Eq. (10).

If we avoid Euclid's bias against the numbers 0 and 1, we can reformulate Algorithm E in the following way:

Algorithm A (*Modern Euclidean algorithm*). Given nonnegative integers u and v, this algorithm finds their greatest common divisor. (*Note:* The greatest common divisor of *arbitrary* integers u and v may be obtained by applying this algorithm to $|u|$ and $|v|$, because of Eqs. (2) and (3).)

A1. [$v = 0$?] If $v = 0$, the algorithm terminates with u as the answer.

A2. [Take $u \bmod v$.] Set $r \leftarrow u \bmod v$, $u \leftarrow v$, $v \leftarrow r$, and return to A1. (The operations of this step decrease the value of v, but leave $\gcd(u, v)$ unchanged.) ∎

For example, we may calculate $\gcd(40902, 24140)$ as follows:

$$\gcd(40902, 24140) = \gcd(24140, 16762) = \gcd(16762, 7378)$$
$$= \gcd(7378, 2006) = \gcd(2006, 1360) = \gcd(1360, 646)$$
$$= \gcd(646, 68) = \gcd(68, 34) = \gcd(34, 0) = 34.$$

A proof that Algorithm A is valid follows readily from Eq. (4) and the fact that

$$\gcd(u, v) = \gcd(v, u - qv), \tag{13}$$

if q is any integer. Equation (13) holds because any common divisor of u and v is a divisor of both v and $u - qv$, and, conversely, any common divisor of v and $u - qv$ must divide both u and v.

The following MIX program illustrates the fact that Algorithm A can easily be implemented on a computer:

Program A (*Euclid's algorithm*). Assume that u and v are single-precision, non-negative integers, stored respectively in locations U and V; this program puts $\gcd(u, v)$ into rA.

	LDX	U	1	rX ← u.
	JMP	2F	1	
1H	STX	V	T	v ← rX.
	SRAX	5	T	rAX ← rA.
	DIV	V	T	rX ← rAX mod v.
2H	LDA	V	$1+T$	rA ← v.
	JXNZ	1B	$1+T$	Done if rX = 0. ∎

The running time for this program is $19T + 6$ cycles, where T is the number of divisions performed. The discussion in Section 4.5.3 shows that we may take $T = 0.842766 \ln N + 0.06$ as an approximate average value when u and v are independently and uniformly distributed in the range $1 \le u, v \le N$.

A binary method. Since Euclid's patriarchal algorithm has been used for so many centuries, it is a rather surprising fact that it may not always be the best method for finding the greatest common divisor after all! A quite different gcd algorithm, which is primarily suited to binary arithmetic, was discovered by Roland Silver and John Terzian about 1962. (Curiously, they never published their algorithm, and it became another part of the computer "folklore." It was independently discovered and published by J. Stein, *J. Comp. Phys.* **1** (1967), 397–405.) This new algorithm requires no division instruction; it is based solely on the operations of subtraction, testing whether a number is even or odd, and shifting the binary representation of an even number to the right (halving).

The binary gcd algorithm is based on four simple facts about positive integers u and v:

a) If u and v are both even, then gcd $(u, v) = 2$ gcd $(u/2, v/2)$. [See Eq. (8).]

b) If u is even and v is odd, then gcd $(u, v) =$ gcd $(u/2, v)$. [See Eq. (6).]

c) As in Euclid's algorithm, gcd $(u, v) =$ gcd $(u - v, v)$. [See Eqs. (13), (3).]

d) If u and v are both odd, then $u - v$ is even, and $|u - v| <$ max (u, v).

These facts immediately suggest the following algorithm:

Algorithm B (*Binary gcd algorithm*).   Given positive integers u and v, this algorithm finds their greatest common divisor.

B1. [Find power of 2.]   Set $k \leftarrow 0$, and then repeatedly set $k \leftarrow k + 1$, $u \leftarrow u/2$, $v \leftarrow v/2$ zero or more times until u and v are not both even.

B2. [Initialize.]   (Now the original values of u and v have been divided by 2^k, and at least one of their present values is odd.)   If u is odd, set $t \leftarrow -v$ and go to B4.   Otherwise set $t \leftarrow u$.

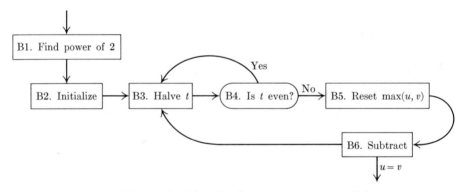

Fig. 9.   Binary algorithm for the greatest common divisor.

B3. [Halve t.]   (At this point, t is even, and nonzero.)   Set $t \leftarrow t/2$.

B4. [Is t even?]   If t is even, go back to B3.

B5. [Reset max (u, v).]   If $t > 0$, set $u \leftarrow t$; otherwise set $v \leftarrow -t$.   (The larger of u and v has been replaced by $|t|$, except perhaps during the first time this step is performed.)

B6. [Subtract.]   Set $t \leftarrow u - v$.   If $t \neq 0$, go back to B3.   Otherwise the algorithm terminates with $u \cdot 2^k$ as the output.   ∎

As an example of Algorithm B, let us consider $u = 40902$, $v = 24140$, the same numbers we have used to try out Euclid's algorithm. Step B1 sets $k \leftarrow 1$, $u \leftarrow 20451$, $v \leftarrow 12070$. Then t is set to -12070, and replaced by -6035; v is replaced by 6035, and the computation proceeds as follows:

u	v	t
20451	6035	$+14416, +7208, +3604, +1802, +901;$
901	6035	$-5134, -2567;$
901	2567	$-1666, -833;$
901	833	$+68, +34, +17;$
17	833	$-816, -408, -204, -102, -51;$
17	51	$-34, -17;$
17	17	$0.$

The answer is $17 \cdot 2^1 = 34$. A few more iterations were necessary here than we needed with Algorithm A, but each iteration was somewhat simpler since no division steps were used.

A MIX program for Algorithm B requires just a little more code than for Algorithm A. In order to make such a program fairly typical of a binary computer representation of Algorithm B, let us assume that MIX is extended to include the following operators:

● SLB (shift left AX binary). C = 6, F = 6. The contents of registers A and X are "shifted left" M binary places; that is, $|rAX| \leftarrow |2^M rAX| \bmod B^{10}$, where B is the byte size. (As with all MIX shift commands, the signs of rA and rX are not affected.)

● SRB (shift right AX binary). C = 6, F = 7. The contents of registers A and X are "shifted right" M binary places; that is, $|rAX| \leftarrow \lfloor |rAX|/2^M \rfloor$.

● JAE, JAO (jump A even, jump A odd). C = 40; F = 6, 7, respectively. A JMP occurs if rA is even or odd, respectively.

● JXE, JXO (jump X even, jump X odd). C = 47; F = 6, 7, respectively. Analogous to JAE, JAO.

Program B (*Binary gcd algorithm*). Assume that u and v are single-precision, positive integers, stored respectively in locations U and V; this program uses Algorithm B to put gcd (u, v) into rA. Register assignments: $t \equiv$ rA, $k \equiv$ rI1.

01	ABS	EQU	1:5	
02	B1	ENT1	0	1	B1. Find power of 2.
03		LDX	U	1	$rX \leftarrow u.$
04		LDAN	V	1	$rA \leftarrow -v.$
05		JMP	1F	1	
06	2H	SRB	1	A	Halve rA, rX.
07		INC1	1	A	$k \leftarrow k + 1.$
08		STX	U	A	$u \leftarrow u/2.$
09		STA	V(ABS)	A	$v \leftarrow v/2.$
10	1H	JXO	B4	$1 + A$	To B4 with $t \leftarrow -v$ if u is odd.
11	B2	JAE	2B	$B + A$	B2. Initialize.
12		LDA	U	B	$t \leftarrow u.$
13	B3	SRB	1	D	B3. Halve t.

14	B4	JAE	B3	$1 - B + D$	B4. Is t even?
15	B5	JAN	1F	C	B5. Reset max (u, v).
16		STA	U	E	If $t > 0$, $u \leftarrow t$.
17		JMP	2F	E	
18	1H	STA	V(ABS)	$C - E$	If $t < 0$, $v \leftarrow -t$.
19	B6	LDA	U	$C - E$	B6. Subtract.
20	2H	SUB	V	C	$t \leftarrow u - v$.
21		JANZ	B3	C	To B3 if $t \neq 0$.
22		LDA	U	1	rA $\leftarrow u$.
23		ENTX	0	1	rX $\leftarrow 0$.
24		SLB	0,1	1	rA $\leftarrow 2^k \cdot$ rA. ∎

The running time of this program is

$$9A + 2B + 8C + 3D - E + 13$$

units, where $A = k$, $B = 1$ if $t \leftarrow u$ in step B2 (otherwise $B = 0$), C is the number of subtraction steps, D is the number of halvings in step B3, and E is the number of times $t > 0$ in step B5. Calculations discussed later in this section imply that we may take $A = \frac{1}{3}$, $B = \frac{1}{3}$, $C = 0.70n - 0.5$, $D = 1.41n - 2.7$, $E = 0.35n - 0.4$ as average values for these quantities, assuming random inputs u and v in the range $1 \leq u, v < 2^n$. The total running time is therefore about $9.5n + 5$ cycles, compared to about $11.1n + 7$ for Program A under the same assumptions. The worst possible running time for u and v in this range occurs when $A = 0$, $B = 0$, $C = n$, $D = 2n - 2$, $E = 1$; this amounts to $14n + 6$ cycles. (The corresponding value for Program A is $26.8n + 19$.)

Thus the greater speed of the iterations in Program B, due to the simplicity of the operations, compensates for the greater number of iterations required. We have found that the binary algorithm is about 15 percent faster than Euclid's algorithm on the MIX computer; of course, the situation may be different on other computers, and in any event both programs are quite efficient. But it appears that not even a procedure as venerable as Euclid's algorithm can withstand progress.

V. C. Harris [*Fibonacci Quarterly* **8** (1970), 102–103] has suggested an interesting cross between Euclid's algorithm and the binary algorithm. If u and v are odd, with $u \geq v > 0$, we can always write $u = qv \pm r$ where $0 \leq r < v$ and r is even; if $r \neq 0$ we set $r \leftarrow r/2$ until r is odd, then set $u \leftarrow v$, $v \leftarrow r$ and repeat the process. In subsequent iterations, $q \geq 3$.

Extensions. We can extend the methods used to calculate gcd (u, v) in order to solve some slightly more difficult problems. For example, assume that we want to compute the greatest common divisor of n integers $u_1, u_2, \ldots, u_n$.

One way to calculate gcd $(u_1, u_2, \ldots, u_n)$, assuming that the u's are all nonnegative, is to extend Euclid's algorithm in the following way: If all u_j are zero, the greatest common divisor is taken to be zero; otherwise if only one u_j is nonzero, it is the greatest common divisor; otherwise replace u_k by $u_k \bmod u_j$ for all $k \neq j$, where u_j is the minimum of the nonzero u's.

The algorithm sketched in the preceding paragraph is a natural generalization of Euclid's method, and it can be justified in a similar manner. But there is a simpler method available, based on the easily verified identity

$$\gcd(u_1, u_2, \ldots, u_n) = \gcd(u_1, \gcd(u_2, \ldots, u_n)). \tag{14}$$

To calculate $\gcd(u_1, u_2, \ldots, u_n)$, we may proceed as follows:

D1. Set $d \leftarrow u_n$, $j \leftarrow n - 1$.

D2. If $d \neq 1$ and $j > 0$, set $d \leftarrow \gcd(u_j, d)$ and $j \leftarrow j - 1$ and repeat this step. Otherwise $d = \gcd(u_1, \ldots, u_n)$.

This method reduces the calculation of $\gcd(u_1, \ldots, u_n)$ to repeated calculations of the greatest common divisor of two numbers at a time. It makes use of the fact that $\gcd(u_1, \ldots, u_j, 1) = 1$; and this will be helpful, since we will already have $\gcd(u_{n-1}, u_n) = 1$ over 60 percent of the time if u_{n-1} and u_n are chosen at random. In most cases, the value of d will rapidly decrease in the first few stages of the calculation, and this makes the remainder of the computation quite fast. Here Euclid's algorithm has an advantage over Algorithm B, in that its running time is primarily governed by the value of $\min(u, v)$, while the running time for Algorithm B is primarily governed by $\max(u, v)$; it would be reasonable to perform one iteration of Euclid's algorithm, replacing u by $u \bmod v$ if u is much larger than v, and then to continue with Algorithm B.

The assertion that $\gcd(u_{n-1}, u_n)$ will be equal to unity more than 60 percent of the time for random inputs is a consequence of the following well-known result of number theory:

Theorem C (E. Cesàro, 1881). *If u and v are integers chosen at random, the probability that $\gcd(u, v) = 1$ is $6/\pi^2$.*

A precise formulation of this theorem, which carefully defines what is meant here by being "chosen at random," appears in exercise 10 with a rigorous proof. Let us content ourselves here with a heuristic argument which shows why the theorem is plausible.

If we assume, without proof, the existence of a well-defined probability p that $\gcd(u, v)$ equals unity, then we can determine the probability that $\gcd(u, v) = d$ for any positive integer d; for the latter event will happen only when u is a multiple of d, v is a multiple of d, and $\gcd(u/d, v/d) = 1$. Thus the probability that $\gcd(u, v) = d$ is equal to $1/d$ times $1/d$ times p, namely p/d^2. Now let us sum these probabilities over all possible values of d; we should get

$$1 = \sum_{d \geq 1} p/d^2 = p(1 + \tfrac{1}{4} + \tfrac{1}{9} + \tfrac{1}{16} + \cdots).$$

Since the sum $1 + \tfrac{1}{4} + \tfrac{1}{9} + \cdots = H_\infty^{(2)}$ is equal to $\pi^2/6$ (cf. Section 1.2.7), we must have $p = 6/\pi^2$ in order to make this equation come out right. ∎

Euclid's algorithm can be extended in another important way: We can calculate integers u' and v' such that

$$uu' + vv' = \gcd(u, v) \tag{15}$$

at the same time $\gcd(u, v)$ is being calculated. This extension of Euclid's algorithm can be described conveniently in vector notation:

Algorithm X (*Extended Euclid's algorithm*). Given nonnegative integers u and v, this algorithm determines a vector (u_1, u_2, u_3) such that $uu_1 + vu_2 = u_3 = \gcd(u, v)$. The computation makes use of auxiliary vectors (v_1, v_2, v_3), $(t_1, t_2 \; t_3)$; all vectors are manipulated in such a way that the relations

$$ut_1 + vt_2 = t_3, \qquad uu_1 + vu_2 = u_3, \qquad uv_1 + vv_2 = v_3 \tag{16}$$

hold throughout the calculation.

X1. [Initialize.] Set $(u_1, u_2, u_3) \leftarrow (1, 0, u)$, $(v_1, v_2, v_3) \leftarrow (0, 1, v)$.

X2. [Is $v_3 = 0$?] If $v_3 = 0$, the algorithm terminates.

X3. [Divide, subtract.] Set $q \leftarrow \lfloor u_3/v_3 \rfloor$, and then set

$$(t_1, t_2, t_3) \leftarrow (u_1, u_2, u_3) - (v_1, v_2, v_3)q,$$
$$(u_1, u_2, u_3) \leftarrow (v_1, v_2, v_3), \qquad (v_1, v_2, v_3) \leftarrow (t_1, t_2, t_3).$$

Return to step X2. ∎

For example, let $u = 40902$, $v = 24140$. At step X2 we have

q	u_1	u_2	u_3	v_1	v_2	v_3
—	1	0	40902	0	1	24140
1	0	1	24140	1	−1	16762
1	1	−1	16762	−1	2	7378
2	−1	2	7378	3	−5	2006
3	3	−5	2006	−10	17	1360
1	−10	17	1360	13	−22	646
2	13	−22	646	−36	61	68
9	−36	61	68	337	−571	34
2	337	−571	34	−710	1203	0

Therefore the solution is $337 \cdot 40902 - 571 \cdot 24140 = 34 = \gcd(40902, 24140)$

The validity of Algorithm X follows from (16) and the fact that the algorithm is identical to Algorithm A with respect to its manipulation of u_3 and v_3. A detailed proof of Algorithm X is discussed in Section 1.2.1. Gordon H. Bradley has observed that we can avoid a good deal of the calculation in Algorithm X by suppressing u_1, v_1, and t_1; then u_1 can be determined afterwards using the relation $uu_1 + vu_2 = u_3$.

Algorithm B, which computes the greatest common divisor using properties of binary notation, does *not* extend in a similar way to an algorithm that solves (15). But there is another binary algorithm which can be used, if q in step X3 is taken to be $2^{\lfloor \log_2(u_3/v_3) \rfloor}$, the largest power of two not exceeding u_3/v_3, instead of $\lfloor u_3/v_3 \rfloor$. This modified algorithm is suitable for incorporation in computer hardware, but it is comparatively difficult to program on most existing machines.

For some instructive extensions to Algorithm X, see exercises 18 and 19 in Section 4.6.1.

The ideas underlying Euclid's algorithm can also be applied to find a *general solution in integers* of any set of linear equations with integer coefficients. For example, suppose that we want to find all integers w, x, y, z which satisfy the two equations

$$10w + 3x + 3y + 8z = 1, \tag{17}$$

$$6w - 7x \qquad - 5z = 2. \tag{18}$$

We can introduce a new variable

$$\lfloor 10/3 \rfloor w + \lfloor 3/3 \rfloor x + \lfloor 3/3 \rfloor y + \lfloor 8/3 \rfloor z = 3w + x + y + 2z = t_1,$$

and use it to eliminate y; Eq. (17) becomes

$$(10 \bmod 3)w + (3 \bmod 3)x + 3t_1 + (8 \bmod 3)z = w + 3t_1 + 2z = 1, \tag{19}$$

and Eq. (18) remains unchanged. The new equation (19) may be used to eliminate w, and (18) becomes

$$6(1 - 3t_1 - 2z) - 7x - 5z = 2;$$

that is,

$$7x + 18t_1 + 17z = 4. \tag{20}$$

Now as before we introduce a new variable

$$x + 2t_1 + 2z = t_2$$

and eliminate x from (20):

$$7t_2 + 4t_1 + 3z = 4. \tag{21}$$

Another new variable can be introduced in the same fashion, in order to eliminate the variable z, which has the smallest coefficient:

$$2t_2 + t_1 + z = t_3.$$

Eliminating z from (21) yields

$$t_2 + t_1 + 3t_3 = 4, \tag{22}$$

and this equation, finally, can be used to eliminate t_2. We are left with two independent variables, t_1 and t_3; substituting back for the original variables, we obtain the general solution

$$
\begin{aligned}
w &= 17 - 5t_1 - 14t_3, \\
x &= 20 - 5t_1 - 17t_3, \\
y &= -55 + 19t_1 + 45t_3, \\
z &= {-8} + t_1 + 7t_3.
\end{aligned}
\tag{23}
$$

In other words, all integer solutions (w, x, y, z) to the original equations (17), (18) are obtained from (23) by letting t_1 and t_3 independently run through all integers.

The general method which has just been illustrated is based on the following procedure: Find a nonzero coefficient c of smallest absolute value in the system of equations. Suppose that this coefficient appears in an equation having the form

$$
cx_0 + c_1 x_1 + \cdots + c_k x_k = d;
\tag{24}
$$

we may assume for simplicity that $c > 0$. If $c = 1$, use this equation to eliminate the variable x_0 from the other equations remaining in the system; then repeat the procedure on the remaining equations. (If no more equations remain, the computation stops, and a general solution in terms of the variables not yet eliminated has essentially been obtained.) If $c > 1$, then if $c_1 \bmod c = \cdots = c_k \bmod c = 0$ we must have $d \bmod c = 0$, otherwise there is no integer solution; divide both sides of (24) by c and eliminate x_0 as in the case $c = 1$. Finally, if $c > 1$ and not all of $c_1 \bmod c, \ldots, c_k \bmod c$ are zero, then introduce a new variable

$$
\lfloor c/c \rfloor x_0 + \lfloor c_1/c \rfloor x_1 + \cdots + \lfloor c_k/c \rfloor x_k = t;
\tag{25}
$$

eliminate the variable x_0 from the other equations, in favor of t, and replace the original equation (24) by

$$
ct + (c_1 \bmod c)x_1 + \cdots + (c_k \bmod c)x_k = d.
\tag{26}
$$

(Cf. (19) and (21) in the above example.)

This process must terminate, since each step either reduces the number of equations or the size of the smallest nonzero coefficient in the system. A study of the above procedure will reveal its intimate connection with Euclid's algorithm. The method is a comparatively simple means of solving linear equations when the variables are required to take on integer values only.

High-precision calculation. If u and v are very large integers, requiring a multiple-precision representation, the binary method (Algorithm B) is a simple and reasonably efficient means of calculating their greatest common divisor.

By contrast, Euclid's algorithm seems much less attractive, since step A2 requires a multiple precision division of u by v. But this difficulty is not really as bad as it seems, since we will prove in Section 4.5.3 that the quotient $\lfloor u/v \rfloor$

is almost always very small; for example, assuming random inputs, the quotient $\lfloor u/v \rfloor$ will be less than 1000 approximately 99.856 percent of the time. Therefore it is almost always possible to find $\lfloor u/v \rfloor$ and $(u \bmod v)$ using single precision calculations, together with the comparatively simple operation of calculating $u - qv$ where q is a single-precision number.

A significant improvement in the speed of Euclid's algorithm when high-precision numbers are involved can be achieved by using a method due to D. H. Lehmer [*AMM* **45** (1938), 227–233]. Working only with the leading digits of large numbers, it is possible to do most of the calculations with single-precision arithmetic, and to make a substantial reduction in the number of multiple-precision operations involved.

Lehmer's method can be illustrated on the eight-digit numbers $u = 27182818$, $v = 10000000$, assuming that we are using a machine with only four-digit words. Let $u' = 2718$, $v' = 1001$, $u'' = 2719$, $v'' = 1000$; then u'/v' and u''/v'' are approximations to u/v, with

$$u'/v' < u/v < u''/v''. \tag{27}$$

The ratio u/v determines the sequence of quotients obtained in Euclid's algorithm; if we carry out Euclid's algorithm simultaneously on the single-precision values (u', v') and (u'', v'') until we get a different quotient, it is not difficult to see that the same sequence of quotients would have appeared to this point if we had worked with the multiple precision numbers (u, v). Thus consider what happens when Euclid's algorithm is applied to (u', v') and to (u'', v''):

u'	v'	q'		u''	v''	q''
2718	1001	2		2719	1000	2
1001	716	1		1000	719	1
716	285	2		719	281	2
285	246	1		281	157	1
146	139	1		157	124	1
139	7	19		124	33	3

After six steps we find that $q' \neq q''$, so the single-precision calculations are suspended; we have gained the knowledge that the calculation would have proceeded as follows if we had been working with the original multiple-precision numbers:

u	v	q	
u_0	v_0	2	
v_0	$u_0 - 2v_0$	1	
$u_0 - 2v_0$	$-u_0 + 3v_0$	2	(28)
$-u_0 + 3v_0$	$3u_0 - 8v_0$	1	
$3u_0 - 8v_0$	$-4u_0 + 11v_0$	1	
$-4u_0 + 11v_0$	$7u_0 - 19v_0$	?	

(The next quotient lies somewhere between 3 and 19.) No matter how many digits are in u and v, the first five steps of Euclid's algorithm would be the same as (28), so long as (27) holds. We can therefore avoid the multiple-precision operations of the first five steps, and replace them all by a multiple-precision calculation of $-4u_0 + 11v_0$ and $7u_0 - 19v_0$. In this case we would obtain $u = 1268728$, $v = 279726$; the calculation could now proceed with $u' = 1268$, $v' = 280$, $u'' = 1269$, $v'' = 279$, etc. With a larger accumulator, more steps could be done by single-precision calculations; our example showed that only five cycles of Euclid's algorithm were combined into one multiple step, but with (say) a word size of 10 digits we could do about twelve cycles at a time. (Results proved in Section 4.5.3 imply that the number of multiple-precision cycles which can be replaced at each iteration is essentially proportional to the logarithm of the word size used in the single-precision calculations.)

Lehmer's method can be formulated as follows:

Algorithm L (*Euclid's algorithm for large numbers*). Let u, v be nonnegative integers, with $u \geq v$, represented in multiple precision. This algorithm computes the greatest common divisor of u and v, making use of auxiliary single-precision p-digit variables u, v, A, B, C, D, T, and q, and auxiliary multiprecision variables t and w.

L1. [Initialize.] If v is small enough to be represented as a single-precision value, calculate $\gcd(u, v)$ by Algorithm A and terminate the computation. Otherwise, let $\hat{u}$ be the p leading digits of u, and let $\hat{v}$ be the p corresponding digits of v; in other words, if radix b notation is being used, $\hat{u} \leftarrow \lfloor u/b^k \rfloor$, $\hat{v} \leftarrow \lfloor v/b^k \rfloor$, where k is as small as possible consistent with the condition $\hat{u} < b^p$.

Set $A \leftarrow 1$, $B \leftarrow 0$, $C \leftarrow 0$, $D \leftarrow 1$. (These variables represent the coefficients in (28), where

$$u = Au_0 + Bv_0, \qquad v = Cu_0 + Dv_0 \tag{29}$$

in the equivalent actions of Algorithm A on multiprecision numbers. We also have

$$u' = \hat{u} + B, \qquad v' = \hat{v} + D, \qquad u'' = \hat{u} + A, \qquad v'' = \hat{v} + C \tag{30}$$

in terms of the notation in the example worked above.)

L2. [Test quotient.] Set $q \leftarrow \lfloor (\hat{u} + A)/(\hat{v} + C) \rfloor$. If $q \neq \lfloor (\hat{u} + B)/(\hat{v} + D) \rfloor$, go to step L4. (This step tests if $q' \neq q''$ in the notation of the above example. Note that single-precision overflow can occur in special circumstances during the computation in this step, but only when $\hat{u} = b^p - 1$ and $A = 1$ or when $\hat{v} = b^p - 1$ and $D = 1$; the conditions

$$0 \leq \hat{u} + A \leq b^p, \qquad 0 \leq \hat{v} + C < b^p, \qquad 0 \leq \hat{u} + B < b^p,$$
$$0 \leq \hat{v} + D \leq b^p \tag{31}$$

will always hold, because of (30). It is possible to have $\hat{v} + C = 0$ or $\hat{v} + D = 0$, but not both simultaneously; therefore division by zero in this step is taken to mean "Go directly to L4.")

L3. [Emulate Euclid.]   Set $T \leftarrow A - qC$, $A \leftarrow C$, $C \leftarrow T$, $T \leftarrow B - qD$, $B \leftarrow D$, $D \leftarrow T$, $T \leftarrow \hat{u} - q\hat{v}$, $\hat{u} \leftarrow \hat{v}$, $\hat{v} \leftarrow T$, and go back to step L2. (These single-precision calculations are the equivalent of multiple-precision operations, as in (28), under the conventions of (29).)

L4. [Multiprecision step.]   If $B = 0$, set $t \leftarrow u \bmod v$, $u \leftarrow v$, $v \leftarrow t$ using multiple-precision division. (This happens only if the single-precision operations cannot simulate any of the multiprecision ones. It implies that Euclid's algorithm requires a very large quotient, and this is an extremely rare occurrence.) Otherwise, set $t \leftarrow Au$, $t \leftarrow t + Bv$, $w \leftarrow Cu$, $w \leftarrow w + Dv$, $u \leftarrow t$, $v \leftarrow w$ (using straightforward multiprecision operations). Go back to step L1. ▮

The values of A, B, C, D remain as single-precision numbers throughout this calculation, because of (31).

Algorithm L requires a somewhat more complicated program than Algorithm B, but with large numbers it will be faster on many computers. It has the advantage that it can be extended, as in Algorithm X (see exercise 17); furthermore, it determines the sequence of quotients obtained in Euclid's algorithm, and this yields the regular continued fraction expansion of a real number (see exercise 4.5.3–18).

Analysis of the binary algorithm. Let us conclude this section by studying the running time of Algorithm B, in order to justify the formulas stated earlier.

An exact determination of the behavior of Algorithm B appears to be exceedingly difficult to derive, but we can begin to study it by means of an approximate model of its behavior. Suppose that u and v are odd numbers, with $u > v$ and

$$\lfloor \log_2 u \rfloor = m, \qquad \lfloor \log_2 v \rfloor = n. \tag{32}$$

(Thus, u is an $(m + 1)$-bit number, and v is an $(n + 1)$-bit number.) Algorithm B forms $u - v$ and shifts this quantity right until obtaining an odd number u' which replaces u. Under random conditions, we would expect to have

$$u' = (u - v)/2$$

about one-half of the time,

$$u' = (u - v)/4$$

about one-fourth of the time,

$$u' = (u - v)/8$$

about one-eighth of the time, and so on. We have

$$\lfloor \log_2 u' \rfloor = m - k - r, \tag{33}$$

where k is the number of places shifted right, and where r is $\lfloor \log_2 u \rfloor - \lfloor \log_2 (u - v) \rfloor$, the number of bits lost at the left during the subtraction of v from u. Note that $r \leq 1$ when $m \geq n + 2$, and $r \geq 1$ when $m = n$. For simplicity, we will assume that $r = 0$ when $m \neq n$ and that $r = 1$ when $m = n$, although this lower bound tends to make u' seem larger than it usually is.

The approximate model we shall use to study Algorithm B is based solely on the values $m = \lfloor \log_2 u \rfloor$ and $n = \lfloor \log_2 v \rfloor$ throughout the course of the algorithm, not on the actual values of u and v. Let us call this approximation a *lattice-point model*, since we will say that we are "at the point (m, n)" when $\lfloor \log_2 u \rfloor = m$ and $\lfloor \log_2 v \rfloor = n$. From point (m, n) the algorithm takes us to (m', n) if $u > v$, or to (m, n') if $u < v$, or terminates if $u = v$. For example, the calculation starting with $u = 20451$, $v = 6035$ which is tabulated after Algorithm B begins at the point $(14, 12)$, then goes to $(9, 12)$, $(9, 11)$, $(9, 9)$, $(4, 9)$, $(4, 5)$, $(4, 4)$, and terminates. In line with the comments of the preceding paragraph, we will make the following assumptions about the probability that we reach a given point just after point (m, n):

Case 1, $m > n$.

Next point	Probability
$(m - 1, n)$	$\frac{1}{2}$
$(m - 2, n)$	$\frac{1}{4}$
$\cdots$	$\cdots$
$(1, n)$	$\frac{1}{2}^{m-1}$
$(0, n)$	$\frac{1}{2}^{m-1}$

Case 2, $m < n$.

Next point	Probability
$(m, n - 1)$	$\frac{1}{2}$
$(m, n - 2)$	$\frac{1}{4}$
$\cdots$	$\cdots$
$(m, 1)$	$\frac{1}{2}^{n-1}$
$(m, 0)$	$\frac{1}{2}^{n-1}$

Case 3, $m = n > 0$.

Next point	Probability
$(m - 2, n)$, $(m, n - 2)$	$\frac{1}{4}$, $\frac{1}{4}$
$(m - 3, n)$, $(m, n - 3)$	$\frac{1}{8}$, $\frac{1}{8}$
$\cdots$	$\cdots$
$(0, n)$, $(m, 0)$	2^{-m}, 2^{-m}
terminate	$\frac{1}{2}^{m-1}$

For example, from point $(5, 3)$ the lattice-point model would go to points $(4, 3)$,

$(3, 3)$, $(2, 3)$, $(1, 3)$, $(0, 3)$ with the respective probabilities $\frac{1}{2}$, $\frac{1}{4}$, $\frac{1}{8}$, $\frac{1}{16}$, $\frac{1}{16}$; from $(4, 4)$ it would go to $(2, 4)$, $(1, 4)$, $(0, 4)$, $(4, 2)$, $(4, 1)$, $(4, 0)$, or would terminate, with the respective probabilities $\frac{1}{4}$, $\frac{1}{8}$, $\frac{1}{16}$, $\frac{1}{4}$, $\frac{1}{8}$, $\frac{1}{16}$, $\frac{1}{8}$. When $m = n = 0$, the formulas above do not apply; the algorithm always terminates in such a case, since $m = n = 0$ implies that $u = v = 1$.

The detailed calculations of exercise 18 show that this lattice-point model is somewhat pessimistic. In fact, when $m > 3$ the actual probability that Algorithm B goes from (m, m) to one of the two points $(m - 2, m)$ or $(m, m - 2)$ is equal to $\frac{1}{8}$, although we have assumed the value $\frac{1}{2}$; the algorithm actually goes from (m, m) to $(m - 3, m)$ or $(m, m - 3)$ with probability $\frac{7}{32}$, not $\frac{1}{4}$; it actually goes from $(m + 1, m)$ to (m, m) with probability $\frac{1}{8}$, not $\frac{1}{2}$. The probabilities in the model are nearly correct when $|m - n|$ is large, but when $|m - n|$ is small the model predicts slower convergence than is actually obtained. In spite of the fact that our model is not a completely faithful representation of Algorithm B, it has one great virtue, namely, that it can be completely analyzed! Furthermore, empirical experiments with Algorithm B show that the behavior predicted by the lattice-point model is analogous to the true behavior.

An analysis of the lattice-point model can be carried out by solving the following rather complicated set of double recurrence relations:

$$A_{mm} = a + \tfrac{1}{2}A_{m(m-2)} + \cdots + \frac{1}{2^{m-1}}A_{m0} + \frac{b}{2^{m-1}}, \qquad \text{if} \quad m \geq 1;$$

$$A_{mn} = c + \tfrac{1}{2}A_{(m-1)n} + \cdots + \frac{1}{2^{m-1}}A_{1n} + \frac{1}{2^{m-1}}A_{0n}, \quad \text{if} \quad m > n \geq 0;$$

$$A_{mn} = A_{nm}, \qquad\qquad\qquad\qquad\qquad\qquad\qquad\quad \text{if} \quad n > m \geq 0.$$

$$\tag{34}$$

The problem is to solve for A_{mn} in terms of m, n, and the parameters a, b, c, and A_{00}. This is an interesting set of recurrence equations, which have an interesting solution.

First we observe that if $0 \leq n < m$,

$$A_{(m+1)n} = c + \sum_{1 \leq k \leq m} 2^{-k}A_{(m+1-k)n} + 2^{-m}A_{0n}$$

$$= c + \tfrac{1}{2}A_{mn} + \sum_{1 \leq k < m} 2^{-k-1}A_{(m-k)n} + 2^{-m}A_{0n}$$

$$= c + \tfrac{1}{2}A_{mn} + \tfrac{1}{2}(A_{mn} - c)$$

$$= \frac{c}{2} + A_{mn}.$$

Hence $A_{(m+k)n} = \tfrac{1}{2}ck + A_{mn}$, by induction on k. In particular, since $A_{10} = c + A_{00}$, we have

$$A_{m0} = \tfrac{1}{2}c(m + 1) + A_{00}, \qquad m > 0. \tag{35}$$

Now let $A_m = A_{mm}$. If $m > 0$, we have

$$A_{(m+1)m} = c + \sum_{1 \le k \le m+1} 2^{-k} A_{(m+1-k)m} + 2^{-m-1} A_{0m}$$

$$= c + \tfrac{1}{2} A_{mm} + \sum_{1 \le k \le m} \left(2^{-k-1}(A_{(m-k)(m+1)} - c/2)\right) + 2^{-m-1} A_{0m}$$

$$= c + \tfrac{1}{2} A_{mm} + \tfrac{1}{2}(A_{(m+1)(m+1)} - a - 2^{-m} b) - \tfrac{1}{4} c(1 - 2^{-m})$$

$$+ 2^{-m-1} \left(\frac{c(m+1)}{2} + A_{00}\right)$$

$$= \tfrac{1}{2}(A_m + A_{m+1}) + \tfrac{3}{4} c - \tfrac{1}{2} a + 2^{-m-1}(c - b + A_{00}) + m2^{-m-2} c.$$

$$(36)$$

Similar maneuvering, as shown in exercise 19, establishes the relation

$$A_{n+1} = \tfrac{3}{4} A_n + \tfrac{1}{4} A_{n-1} + \alpha + 2^{-n-1}\beta + (n+2)2^{-n-1}\gamma, \qquad n \ge 2, \quad (37)$$

where

$$\alpha = \tfrac{1}{4} a + \tfrac{7}{8} c, \ \beta = A_{00} - b - \tfrac{3}{2} c, \quad \text{and} \quad \gamma = \tfrac{1}{2} c.$$

Thus the double recurrence (34) can be transformed into the single recurrence relation in (37). Use of the generating function $G(z) = A_0 + A_1 z + A_2 z^2 + \cdots$ now transforms (37) into the equation

$$(1 - \tfrac{3}{4} z - \tfrac{1}{4} z^2)G(z) = a_0 + a_1 z + a_2 z^2 + \frac{\alpha}{1 - z} + \frac{\beta}{1 - z/2} + \frac{\gamma}{(1 - z/2)^2},$$

$$(38)$$

where a_0, a_1, and a_2 are constants that can be determined by the values of A_0, A_1, and A_2. Since $1 - \tfrac{3}{4} z + \tfrac{1}{4} z^2 = (1 + \tfrac{1}{4} z)(1 - z)$, we can express $G(z)$ by the method of partial fractions in the form

$$G(z) = b_0 + b_1 z + \frac{b_2}{(1 - z)^2} + \frac{b_3}{1 - z} + \frac{b_4}{(1 - z/2)^2} + \frac{b_5}{1 - z/2} + \frac{b_6}{1 + z/4}.$$

Tedious manipulations produce the values of these constants $b_0, \ldots, b_6$, and therefore the coefficients of $G(z)$ are determined. We finally obtain the solution

$$A_{nn} = n(\tfrac{1}{5} a + \tfrac{7}{10} c) + (\tfrac{16}{25} a + \tfrac{2}{5} b - \tfrac{23}{50} c + \tfrac{3}{5} A_{00})$$

$$+ 2^{-n}(-\tfrac{1}{3} cn + \tfrac{2}{3} b - \tfrac{1}{9} c - \tfrac{2}{3} A_{00})$$

$$+ (-\tfrac{1}{4})^n(-\tfrac{16}{25} a - \tfrac{16}{15} b + \tfrac{16}{225} c + \tfrac{16}{15} A_{00}) + \tfrac{1}{2} c \delta_{n0};$$

$$A_{mn} = mc/2 + n(\tfrac{1}{5} a + \tfrac{1}{5} c) + (\tfrac{6}{25} a + \tfrac{2}{5} b + \tfrac{7}{50} c + \tfrac{3}{5} A_{00}) + 2^{-n}(\tfrac{1}{3} c)$$

$$+ (-\tfrac{1}{4})^n(-\tfrac{6}{25} a - \tfrac{2}{5} b + \tfrac{2}{75} c + \tfrac{2}{5} A_{00}), \qquad m > n. \qquad (39)$$

With these elaborate calculations done, we can readily determine the behavior of the lattice-point model. Assume that the inputs u and v to the algorithm are odd, and let $m = \lfloor \log_2 u \rfloor$, $n = \lfloor \log_2 v \rfloor$. The average number of subtraction cycles, namely, the quantity C in the analysis of Program B, the number of times step B6 is executed, is obtained by setting $a = 1$, $b = 0$, $c = 1$, $A_{00} = 1$ in the recurrence (34). By (39) we see that (for $m \geq n$) the lattice model predicts

$$C = \tfrac{1}{2}m + \tfrac{2}{5}n + \tfrac{49}{50} - \tfrac{1}{5}\delta_{mn} \qquad (40)$$

subtraction cycles, plus terms which rapidly go to zero as n approaches infinity.

The average number of times $\gcd(u, v) = 1$ is obtained by setting $a = b = c = 0$, $A_{00} = 1$; this gives the probability that u and v are relatively prime, approximately $\tfrac{3}{5}$. Actually, since u and v are assumed to be odd, they should be relatively prime with probability $8/\pi^2$ (see exercise 13), so this reflects the degree of inaccuracy of the lattice-point model.

The average number of times a path from (m, n) goes through one of the "diagonal" points (m', m') for some $m' \geq 1$ is obtained by setting $a = 1$, $b = c = A_{00} = 0$ in (34); so we find this quantity is approximately

$$\tfrac{1}{5}n + \tfrac{6}{25} + \tfrac{2}{5}\delta_{mn}, \qquad \text{when} \qquad m \geq n.$$

Now we can determine the average number of shifts, the number of times step B3 is performed. (This is the quantity D in Program B.) In any execution of Algorithm B, with u and v both odd, the corresponding path in the lattice model must satisfy the relation

number of shifts + number of diagonal points + $2\lfloor \log_2 \gcd(u, v) \rfloor = m + n$,

since we are assuming that r in (33) is always either 0 or 1. The average value of $\lfloor \log_2 \gcd(u, v) \rfloor$ predicted by the lattice-point model is approximately $\tfrac{4}{5}$ (see exercise 20). Therefore we have, for the total number of shifts,

$$\begin{aligned} D &= m + n - (\tfrac{1}{5}n + \tfrac{6}{25} + \tfrac{2}{5}\delta_{mn}) - \tfrac{4}{5} \\ &= m + \tfrac{4}{5}n - \tfrac{46}{25} - \tfrac{2}{5}\delta_{mn}, \end{aligned} \qquad (41)$$

when $m \geq n$, plus terms which decrease rapidly to zero for large n.

To summarize the most important facts we have derived from the lattice-point model, we have shown that if u and v are odd and if $\lfloor \log_2 u \rfloor = m$, $\lfloor \log_2 v \rfloor = n$, then the quantities C and D which are the critical factors in the running time of Program B will have average values given by

$$C = \tfrac{1}{2}m + \tfrac{2}{5}n + O(1), \qquad D = m + \tfrac{4}{5}n + O(1), \qquad m \geq n. \qquad (42)$$

But the model which we have used to derive (42) is only a pessimistic approximation to the true behavior; Table 1 compares the true average values of C, computed by actually running Algorithm B with all possible inputs, to the values predicted by the lattice-point model, for small m and n. The lattice model is completely accurate when m or n is zero, but it tends to be less accurate when $|m - n|$ is small and min (m, n) is large. When $m = n = 9$, the lattice-point model gives $C = 8.78$, compared to the true value $C = 7.58$.

Table 1

NUMBER OF SUBTRACTIONS IN ALGORITHM B

			n							n			
	0	1	2	3	4	5	0	1	2	3	4	5	
0	1.00	2.00	2.50	3.00	3.50	4.00	1.00	2.00	2.50	3.00	3.50	4.00	0
1	2.00	1.00	2.50	3.00	3.50	4.00	2.00	1.00	3.00	2.75	3.63	3.94	1
2	2.50	2.50	2.25	3.38	3.88	4.38	2.50	3.00	2.00	3.50	3.88	4.25	2
3	3.00	3.00	3.38	3.25	4.22	4.72	3.00	2.75	3.50	2.88	4.13	4.34	3
4	3.50	3.50	3.88	4.22	4.25	5.10	3.50	3.63	3.88	4.13	3.94	4.80	4
5	4.00	4.00	4.38	4.72	5.10	5.19	4.00	3.94	4.25	4.34	4.80	4.60	5
m			Predicted by model						Actual average values				m

Empirical tests of Algorithm B with several million random inputs and with various values of m, n in the range $29 \le m, n \le 37$ show that the actual average behavior of the algorithm is given by

$$C \approx \tfrac{1}{2}m + 0.203n + 1.9 - 0.4(0.6)^{m-n},$$
$$D \approx m + 0.41n - 0.5 - 0.7(0.6)^{m-n}, \qquad m \ge n. \qquad (43)$$

These experiments showed a rather small standard deviation from the observed average values. The coefficients $\tfrac{1}{2}$ and 1 of m in (42) and (43) can be verified rigorously without using the lattice-point approximation (see exercise 21), so the error in the lattice-point model is apparently in the coefficient of n which is too high. Comparison of (42) and (43) makes it tempting at first to conjecture that the true coefficients of n are $\tfrac{1}{5}$ and $\tfrac{2}{5}$, just half of what the lattice model predicts; but the empirical tests showed definitely that the coefficients are not quite as low as this, and, furthermore, the fact that Algorithm B calculates the greatest common divisor makes it unlikely that such a simple coefficient as $\tfrac{1}{5}$ would appear in the analysis. A coefficient such as $2/\pi^2$ would be much more likely, but since no theoretical explanation of (43) has yet been found there is not much point in guessing what the true coefficients are.

The above calculations have been made under the assumption that u and v are odd and in the ranges $2^m < u < 2^{m+1}$, $2^n < v < 2^{n+1}$. If we say instead that u and v are to be *any* integers, independently and uniformly distributed over the ranges

$$1 \le u < 2^N, \qquad 1 \le v < 2^N,$$

then we can calculate the average values of C and D from the data already given; in fact,

$$(2^N - 1)^2 C = N^2 C_{00} + 2N \sum_{1 \le n \le N} (N - n)2^{n-1}C_{n0}$$

$$+ 2 \sum_{1 \le n < m \le N} (N - m)(N - n)2^{m+n-2}C_{mn} + \sum_{1 \le n \le N} (N - n)^2 2^{2n-2}C_{nn}, \tag{44}$$

if C_{mn} denotes the value of C, depending on m and n, which was calculated under our earlier assumptions. (See exercise 22.) The same formula holds for D in terms of D_{mn}. If the indicated sums are carried out using the approximations in (43), we obtain

$$C \approx 0.70N + O(1), \qquad D \approx 1.41N + O(1).$$

(See exercise 23.) This agrees perfectly with the results of further empirical tests, made on several million random inputs for $N \le 30$; the latter tests show that we may take

$$C = 0.70N - 0.5, \qquad D = 1.41N - 2.7 \tag{45}$$

as good estimates of the values, given this distribution of the inputs u and v.

This completes our analysis of the average values of C and D. The other three quantities appearing in the running time of Algorithm B are rather easily analyzed; see exercises 6, 7, and 8.

Thus we know approximately how Algorithm B behaves on the average. Let us now consider a "worst case" analysis: What values of u, v are in some sense the hardest to handle? If we assume as before that

$$\lfloor \log_2 u \rfloor = m \qquad \text{and} \qquad \lfloor \log_2 v \rfloor = n,$$

let us try to find (u, v) which make the algorithm run most slowly. In view of the fact that the subtractions take somewhat longer than the shifts, this question may be phrased by asking which u and v require most subtractions. The answer is somewhat surprising; the maximum value of C is exactly

$$\max(m, n) + 1, \tag{46}$$

although the lattice-point model would predict that substantially higher values of C are possible (see exercise 26). The derivation of the worst case (46) is quite interesting, so it has been left as an amusing problem for the reader to work out by himself (see exercises 27, 28).

EXERCISES

1. [*M21*] How can (8), (9), (10), (11), and (12) be derived easily from (6) and (7)?

2. [*M22*] Given that u divides $v_1 v_2 \ldots v_n$, prove that u divides

$$\gcd(u, v_1) \gcd(u, v_2) \cdots \gcd(u, v_n).$$

3. [*M23*] Show that the number of ordered pairs of positive integers (u, v) such that lcm $(u, v) = n$ is the number of divisors of n^2.

4. [*M21*] Given positive integers u and v, show that there are divisors u' of u and v' of v such that gcd $(u', v') = 1$ and $u'v' = $ lcm (u, v).

▶ **5.** [*26*] Invent an algorithm (analogous to Algorithm B) for calculating the greatest common divisor of two integers based on their *balanced ternary* representation. Demonstrate your algorithm by applying it to the calculation of gcd (40902,24140).

6. [*M22*] Given that u and v are random positive integers, find the mean and standard deviation of the quantity A (the number of right shifts on both u and v during the preparatory phase) which enters into the timing of Program B.

7. [*M20*] Analyze the quantity B which enters into the timing of Program B.

▶ **8.** [*M24*] Show that in Program B, the average value of E is approximately equal to $\frac{1}{2} C_{\text{ave}}$, where C_{ave} is the average value of C.

9. [*18*] Using Algorithm B and hand calculation, find gcd (31408,2718). Also find integers m and n such that $31408m + 2718n = $ gcd (31408,2718), using Algorithm X.

▶ **10.** [*HM24*] Let q_n be the number of ordered pairs of integers (u, v) such that $1 \le u, v \le n$ and gcd $(u, v) = 1$. The object of this exercise is to prove that $\lim_{n \to \infty} q_n/n^2 = 6/\pi^2$, thereby establishing Theorem C.

a) Use the principle of inclusion and exclusion (Section 1.3.3) to show that

$$q_n = n^2 - \sum_{p_1} \lfloor n/p_1 \rfloor^2 + \sum_{p_1 < p_2} \lfloor n/p_1 p_2 \rfloor^2 - \cdots,$$

where the sums are taken over all *prime* numbers p_i.

b) The *Möbius function* $\mu(n)$ is defined by the rules $\mu(1) = 1$, $\mu(p_1 p_2 \ldots p_r) = (-1)^r$ if $p_1, p_2, \ldots, p_r$ are distinct primes, and $\mu(n) = 0$ if n is divisible by the square of a prime. Show that $q_n = \sum_{k \ge 1} \mu(k) \lfloor n/k \rfloor^2$.

c) As a consequence of (b), prove that $\lim_{n \to \infty} q_n/n^2 = \sum_{k \ge 1} \mu(k)/k^2$.

d) Prove that $(\sum_{k \ge 1} \mu(k)/k^2)(\sum_{m \ge 1} 1/m^2) = 1$. *Hint:* When the series are absolutely convergent we have

$$\left(\sum_{k \ge 1} a_k/k^2 \right)\left(\sum_{m \ge 1} b_m/m^2 \right) = \sum_{n \ge 1} \left(\sum_{d \backslash n} a_d b_{n/d} \right) \Big/ n^2.$$

11. [*M22*] What is the probability that gcd $(u, v) \le 3$? (See Theorem C.) What is the *average* value of gcd (u, v)?

12. [*M24*] (E. Cesàro.) If u and v are random positive integers, what is the average number of (positive) divisors they have in common? [*Hint:* See the identity in exercise 10(d), with $a_k = b_m = 1$.]

13. [*HM23*] Given that u and v are random *odd* positive integers, show that they are relatively prime with probability $8/\pi^2$.

14. [*M26*] What are the values of v_1 and v_2 when Algorithm X terminates?

▶ **15.** [*M25*] Design an algorithm to *divide u by v modulo m*, given positive integers u, v, and m, with v relatively prime to m. In other words, your algorithm should find w, in the range $0 \leq w < m$, such that $u \equiv vw$ (modulo m).

16. [*21*] Use the text's method to find a general solution in integers to the following sets of equations:

    a) $3x + 7y + 11z = 1$ b) $3x + 7y + 11z = 1$

      $5x + 7y - 5z = 3$ $5x + 7y - 5z = -3$

▶ **17.** [*M24*] Show how Algorithm L can be extended (as Algorithm A was extended to Algorithm X) to obtain solutions of (15) when u and v are large.

18. [*M37*] Let u and v be odd integers, independently and uniformly distributed in the ranges $2^m < u < 2^{m+1}$, $2^n < v < 2^{n+1}$. What is the *exact* probability that a single "subtract and shift" cycle in Algorithm B, namely, an operation that starts at step B6 and then stops after step B5 is finished, reduces u and v to the ranges $2^{m'} < u < 2^{m'+1}$, $2^{n'} < v < 2^{n'+1}$, as a function of m, n, m', and n'? (This exercise gives more accurate values for the transition probabilities than are used in the text's lattice-point model.)

19. [*M24*] Complete the text's derivation of (38) by establishing (37).

20. [*M26*] Let $\lambda = \lfloor \log_2 \gcd(u, v) \rfloor$. Show that the lattice-point model gives $\lambda = 1$ with probability $\frac{1}{5}$, $\lambda = 2$ with probability $\frac{1}{10}$, $\lambda = 3$ with probability $\frac{1}{20}$, etc., plus correction terms which go rapidly to zero as u, v approach infinity; hence the average value of λ is approximately $\frac{4}{5}$. [*Hint:* Consider the relation between the probability of a path from (m, n) to $(k + 1, k + 1)$ and a corresponding path from $(m - k, n - k)$ to $(1, 1)$.]

21. [*M25*] Let C_{mn}, D_{mn} be the average number of subtraction and shift cycles respectively, in Algorithm B when u and v are odd, $\lfloor \log_2 u \rfloor = m$, $\lfloor \log_2 v \rfloor = n$. Show that for fixed n, $C_{mn} = \frac{1}{2}m + O(1)$ and $D_{mn} = m + O(1)$ as $m \to \infty$.

22. [*M23*] Prove Eq. (44).

23. [*M28*] Show that if $C_{mn} = \alpha m + \beta n + \gamma$ for some constants α, β, and γ, then

$$\sum_{1 \leq n \leq m \leq N} (N - m)(N - n)2^{m+n-2}C_{mn} = 2^{2N}\left(\tfrac{11}{27}(\alpha + \beta)N + O(1)\right),$$

and

$$\sum_{1 \leq n \leq N} (N - n)^2 2^{2n-2}C_{nn} = 2^{2N}\left(\tfrac{5}{27}(\alpha + \beta)N + O(1)\right)$$

as $N \to \infty$.

▶ **24.** [*M30*] If $v = 1$ during Algorithm B, while u is large, it may take fairly long for the algorithm to determine that $\gcd(u, v) = 1$. Perhaps it would be worth while to add a test at the beginning of step B5: "If $t = 1$, the algorithm terminates with 2^k as the answer." Explore the question of whether or not this would be an improvement, by determining the average number of times step B6 is executed with $u = 1$ or $v = 1$, using the lattice-point model.

25. [*M50*] Prove (or disprove) the empirical formulas (43) for the average behavior of the binary gcd algorithm.

26. [*M23*] What is the length of the longest path from (m, n) to $(0, 0)$ in the lattice-point model?

▶ **27.** [*M28*] Find values of u, v with $u \geq v \geq 1$, $\lfloor \log_2 u \rfloor = m$, $\lfloor \log_2 v \rfloor = n$, such that Algorithm B requires $m + 1$ subtraction steps.

28. [*M37*] Prove that the subtraction step B6 of Algorithm B is never executed more than $1 + \lfloor \log_2 \max (u, v) \rfloor$ times.

29. [*M30*] Evaluate the determinant

$$\begin{vmatrix} \gcd (1, 1) & \gcd (1, 2) & \cdots & \gcd (1, n) \\ \gcd (2, 1) & \gcd (2, 2) & \cdots & \gcd (2, n) \\ \vdots & \vdots & & \vdots \\ \gcd (n, 1) & \gcd (n, 2) & \cdots & \gcd (n, n) \end{vmatrix}.$$

30. [*M30*] Show that Euclid's algorithm applied to two n-bit binary numbers requires $O(n^2)$ units of time, as $n \to \infty$. (The same upper bound obviously holds for Algorithm B.)

31. [*M22*] Use Euclid's algorithm to find a simple formula for gcd $(2^m - 1, 2^n - 1)$ when m and n are nonnegative integers.

32. [*M43*] Can the upper bound $O(n^2)$ in exercise 30 be decreased, if another algorithm for calculating the greatest common divisor is used?

33. [*M48*] Analyze V. C. Harris's "binary Euclidean algorithm."

*4.5.3. Analysis of Euclid's algorithm

The execution time of Euclid's algorithm depends on T, the number of times the division step A2 is performed. (See Algorithm 4.5.2A and Program 4.5.2A.) The same quantity T is an important factor in the running time of other algorithms, for example, the evaluation of functions satisfying a reciprocity formula (see Section 3.3.3). A mathematical analysis of this quantity T is interesting and instructive.

Relation to continued fractions. Euclid's algorithm is intimately connected with *continued fractions*, which are expressions of the form

$$\cfrac{b_1}{a_1 + \cfrac{b_2}{a_2 + \cfrac{b_3}{\cdots \cfrac{}{a_{n-1} + \cfrac{b_n}{a_n}}}}} = b_1/(a_1 + b_2/(a_2 + b_3/(\cdots /(a_{n-1} + b_n/a_n) \cdots))). \tag{1}$$

Continued fractions have a beautiful theory which is the subject of several books, e.g., O. Perron, *Die Lehre von den Kettenbrüchen*, 3rd ed. (Stuttgart: Teubner, 1954), 2 vols.; A. Khinchin, *Continued fractions*, tr. by Peter Wynn (Groningen: P. Noordhoff, 1963); H. S. Wall, *Analytic theory of continued fractions* (New York: Van Nostrand, 1948); see also J. Tropfke, *Geschichte der Elementar-Mathematik* **6** (Berlin: Gruyter, 1924), 74–84, for the early history of the subject. It is necessary to limit ourselves to a comparatively brief treatment of that theory here, studying only those aspects which give us more insight into the behavior of Euclid's algorithm.

The continued fractions which are of primary interest to us are those in which all the b's in (1) are equal to unity. For convenience in notation, let us define

$$/x_1, x_2, \ldots, x_n/ = 1/(x_1 + 1/(x_2 + 1/(\cdots + 1/(x_n)\cdots))). \qquad (2)$$

If $n = 0$, the symbol $/x_1, \ldots, x_n/$ is taken to mean 0. Thus, for example,

$$/x_1/ = \frac{1}{x_1}, \qquad /x_1, x_2/ = \frac{1}{x_1 + (1/x_2)} = \frac{x_2}{x_1 x_2 + 1}. \qquad (3)$$

Let us also define the polynomials $Q_n(x_1, x_2, \ldots, x_n)$ of n variables, for $n \geq 0$, by the rule

$$Q_n(x_1, x_2, \ldots, x_n) = \begin{cases} 1, & \text{if } n = 0; \\ x_1, & \text{if } n = 1; \\ x_1 Q_{n-1}(x_2, \ldots, x_n) + Q_{n-2}(x_3, \ldots, x_n), & \text{if } n > 1. \end{cases} \qquad (4)$$

Thus $Q_2(x_1, x_2) = x_1 x_2 + 1$, $Q_3(x_1, x_2, x_3) = x_1 x_2 x_3 + x_1 + x_3$, etc. In general, as noted by L. Euler in the eighteenth century, $Q_n(x_1, x_2, \ldots, x_n)$ is the sum of all terms obtainable by starting with $x_1 x_2 \ldots x_n$ and deleting zero or more nonoverlapping pairs of consecutive variables $x_j x_{j+1}$; there are F_{n+1} such terms. The polynomials defined in (4) are called "continuants."

The basic property of the Q-polynomials is that

$$/x_1, x_2, \ldots, x_n/ = Q_{n-1}(x_2, \ldots, x_n)/Q_n(x_1, x_2, \ldots, x_n), \qquad n \geq 1. \qquad (5)$$

This can be proved by induction, since it implies that

$$x_0 + /x_1, \ldots, x_n/ = Q_{n+1}(x_0, x_1, \ldots, x_n)/Q_n(x_1, \ldots, x_n);$$

hence $/x_0, x_1, \ldots, x_n/$ is the reciprocal of the latter quantity.

The Q-polynomials are symmetrical in the sense that

$$Q_n(x_1, x_2, \ldots, x_n) = Q_n(x_n, \ldots, x_2, x_1). \qquad (6)$$

This follows from Euler's observation above, and as a consequence we have

$$Q_n(x_1, \ldots, x_n) = x_n Q_{n-1}(x_1, \ldots, x_{n-1}) + Q_{n-2}(x_1, \ldots, x_{n-2}). \qquad (7)$$

The Q-polynomials also satisfy the important identity

$$Q_n(x_1, \ldots, x_n)Q_n(x_2, \ldots, x_{n+1}) - Q_{n+1}(x_1, \ldots, x_{n+1})Q_{n-1}(x_2, \ldots, x_n)$$
$$= (-1)^n, \qquad n \geq 1. \qquad (8)$$

(See exercise 4.) The latter equation in connection with (5) implies that

$$/x_1, \ldots, x_n/ = \frac{1}{q_0 q_1} - \frac{1}{q_1 q_2} + \frac{1}{q_2 q_3} - \cdots + \frac{(-1)^{n-1}}{q_{n-1} q_n},$$

$$\text{where} \qquad q_k = Q_k(x_1, \ldots, x_k). \qquad (9)$$

Thus the Q-polynomials are intimately related to continued fractions.

Every real number X in the range $0 \leq X < 1$ has a *regular continued fraction* defined as follows: Let $X_0 = X$, and for all $n \geq 0$ such that $X_n \neq 0$ let

$$A_{n+1} = \lfloor 1/X_n \rfloor, \qquad X_{n+1} = 1/X_n - A_{n+1}. \qquad (10)$$

If $X_n = 0$, the quantities A_{n+1} and X_{n+1} are not defined, and the regular continued fraction for X is $/A_1, \ldots, A_n/$. If $X_n \neq 0$, this definition guarantees that $0 \leq X_{n+1} < 1$, so each of the A's is a positive integer. The definition (10) clearly implies that

$$X = X_0 = \frac{1}{A_1 + X_1} = \frac{1}{A_1 + 1/(A_2 + X_2)} = \cdots ;$$

hence

$$X = /A_1, \ldots, A_{n-1}, A_n + X_n/ \qquad (11)$$

for all $n \geq 1$, whenever X_n is defined. Therefore, if $X_n = 0$, $X = /A_1, \ldots, A_n/$. If $X_n \neq 0$, X lies *between* $/A_1, \ldots, A_n/$ and $/A_1, \ldots, A_n + 1/$, since by (4) and (9) the quantity $q_n = Q_n(A_1, \ldots, A_n + X_n)$ increases monotonically from $Q_n(A_1, \ldots, A_n)$ up to $Q_n(A_1, \ldots, A_n + 1)$ as X_n increases from 0 to 1. In fact,

$$|X - /A_1, \ldots, A_n/| = |/A_1, \ldots, A_n + X_n/ - /A_1, \ldots, A_n/|$$
$$= |/A_1, \ldots, A_n, 1/X_n/ - /A_1, \ldots, A_n/|$$
$$= \left| \frac{Q_n(A_2, \ldots, A_n, 1/X_n)}{Q_{n+1}(A_1, \ldots, A_n, 1/X_n)} - \frac{Q_{n-1}(A_2, \ldots, A_n)}{Q_n(A_1, \ldots, A_n)} \right|$$
$$= 1/Q_n(A_1, \ldots, A_n)Q_{n+1}(A_1, \ldots, A_n, 1/X_n)$$
$$\leq 1/Q_n(A_1, \ldots, A_n)Q_{n+1}(A_1, \ldots, A_n, A_{n+1}) \qquad (12)$$

by (5), (8), and (10). Therefore $/A_1, \ldots, A_n/$ is an extremely close approximation to X. If X is irrational, it is impossible to have $X_n = 0$ for any n, so the regular continued fraction expansion in this case is an *infinite continued fraction* $/A_1, A_2, A_3, \ldots/$. The value of such a continued fraction is defined to be $\lim_{n \to \infty} /A_1, A_2, \ldots, A_n/$, and from the inequality (12) it is clear that this limit equals X.

The regular continued fraction expansion of real numbers has several properties analogous to the representation of numbers in the decimal system. If we use the formulas above to compute the regular continued fraction expansions of some familiar real numbers, we find, for example, that

$$\frac{8}{29} = /3, 1, 1, 1, 2/;$$
$$\sqrt{\frac{8}{29}} = /1, 1, 9, 2, 2, 3, 2, 2, 9, 1, 2, 1, 9, 2, 2, 3, 2, 2, 9, 1, 2, 1, 9, 2, 2, 3, 2, 2, 9, 1, \ldots/;$$
$$\sqrt[3]{2} = 1 + /3, 1, 5, 1, 1, 4, 1, 1, 8, 1, 14, 1, 10, 2, 1, 4, 12, 2, 3, 2, 1, 3, 4, 1, 1, 2, 14, 3, \ldots/;$$
$$\pi = 3 + /7, 15, 1, 292, 1, 1, 1, 2, 1, 3, 1, 14, 2, 1, 1, 2, 2, 2, 2, 1, 84, 2, 1, 1, 15, 3, 13, \ldots/;$$
$$e = 2 + /1, 2, 1, 1, 4, 1, 1, 6, 1, 1, 8, 1, 1, 10, 1, 1, 12, 1, 1, 14, 1, 1, 16, 1, 1, 18, 1, \ldots/;$$
$$\gamma = /1, 1, 2, 1, 2, 1, 4, 3, 13, 5, 1, 1, 8, 1, 2, 4, 1, 1, 40, 1, 11, 3, 7, 1, 7, 1, 1, 5, \ldots/;$$
$$\phi = 1 + /1, 1, 1, 1, 1, 1, 1, 1, 1, \ldots/. \tag{13}$$

The numbers A_1, A_2, $\ldots$ are called the *partial quotients* of X. Note the regular pattern which appears in the partial quotients for $\sqrt{8/29}$, ϕ, and e; the reasons for this behavior are discussed in exercises 12 and 16. There is no apparent pattern in the partial quotients for $\sqrt[3]{2}$, π, or γ.

When X is a rational number, the regular continued fraction corresponds in a natural way to Euclid's algorithm. Let us assume that $X = v/u$, where $u > v \geq 0$. The regular continued fraction process starts with $X_0 = X$; let us define $U_0 = u$, $V_0 = v$. Assuming that $X_n = V_n/U_n \neq 0$, (10) becomes

$$A_{n+1} = \lfloor U_n/V_n \rfloor, \tag{14}$$
$$X_{n+1} = U_n/V_n - A_{n+1} = (U_n \bmod V_n)/V_n.$$

Therefore, if we define

$$U_{n+1} = V_n, \qquad V_{n+1} = U_n \bmod V_n, \tag{15}$$

the condition $X_n = V_n/U_n$ holds throughout the process. Furthermore, (15) is precisely the transformation made on the variables u, v in Euclid's algorithm (see Algorithm 4.5.2A, step A2). For example, since $\frac{8}{29} = /3, 1, 1, 1, 2/$, we know that Euclid's algorithm applied to $u = 29$, $v = 8$ will require exactly five division steps, and the quotients $\lfloor u/v \rfloor$ in step A2 will be successively 3, 1, 1, 1, and 2. Note that the last partial quotient A_n must be 2 or more when $X_n = 0$, $n \geq 1$, since X_{n-1} is less than unity.

From this correspondence with Euclid's algorithm we can see that the regular continued fraction for X terminates at some step with $X_n = 0$ if and only if X is rational; for it is obvious that X_n cannot be zero if X is irrational, and, conversely, we know that Euclid's algorithm always terminates. If the partial quotients obtained during Euclid's algorithm are $A_1, A_2, \ldots, A_n$, then we have, by (5),

$$\frac{v}{u} = \frac{Q_{n-1}(A_2, \ldots, A_n)}{Q_n(A_1, A_2, \ldots, A_n)}. \tag{16}$$

This formula holds also if Euclid's algorithm is applied for $u \le v$, when $A_1 = 0$. Furthermore, because of (8), $Q_{n-1}(A_2, \ldots, A_n)$ and $Q_n(A_1, A_2, \ldots, A_n)$ are relatively prime, and the fraction on the right-hand side of (16) is in lowest terms; therefore

$$u = Q_n(A_1, A_2, \ldots, A_n)d, \qquad v = Q_{n-1}(A_2, \ldots, A_n)d, \qquad (17)$$

where $d = \gcd(u, v)$.

The worst case. We can now apply these observations to determine the behavior of Euclid's algorithm in the "worst case," or in other words to give an upper bound on the number of division steps. The worst case occurs when the inputs are consecutive Fibonacci numbers:

Theorem F (G. Lamé, 1845). *For $n \ge 1$, let u and v be integers with $u > v > 0$ such that Euclid's algorithm applied to u and v requires exactly n division steps, and such that u is as small as possible satisfying these conditions. Then $u = F_{n+2}$, $v = F_{n+1}$.*

Proof. By (17), we must have $u = Q_n(A_1, A_2, \ldots, A_n)d$ where $A_1, A_2, \ldots,$ A_n, d are positive integers and $A_n \ge 2$. Since Q_n is a polynomial with non-negative coefficients, involving all of the variables, the minimum value is achieved only when $A_1 = 1, \ldots, A_{n-1} = 1, A_n = 2, d = 1$. Putting these values in (17) yields the desired result. ∎

(This theorem has the historical claim of being the first practical application of the Fibonacci sequence; since then many other applications of the Fibonacci numbers to algorithms and to the study of algorithms have been discovered.)

As a consequence of Theorem F we have an important corollary:

Corollary. *If $0 \le u, v < N$, the number of division steps required when Algorithm 4.5.2A is applied to u, v is at most $\lceil \log_\phi (\sqrt{5}N) \rceil - 2$.*

Proof. By Theorem F, the maximum number of steps, n, occurs when $u = F_n$ and $v = F_{n+1}$, where n is as large as possible with $F_{n+1} < N$. (The first division step in this case merely interchanges u and v when $n > 1$.) Since $F_{n+1} < N$, we have $\phi^{n+1}/\sqrt{5} < N$ (see Eq. 1.2.8–15), so $n + 1 < \log_\phi (\sqrt{5}N)$. ∎

Note that $\log_\phi (\sqrt{5}N)$ is approximately $2.078 \ln N + 1.672 \approx 4.785 \log_{10} N + 1.672$. See exercises 31 and 35 for extensions of Theorem F.

An approximate model. Now that we know the maximum number of division steps that can occur, let us attempt to find the *average* number. Let $T(m, n)$ be the number of division steps that occur when $u = m, v = n$ are input to Euclid's algorithm. Thus

$$T(m, 0) = 0; \qquad T(m, n) = 1 + T(n, m \bmod n) \qquad \text{if} \qquad n \ge 1. \qquad (18)$$

Let T_n be the average number of division steps when $v = n$ and when u is chosen at random; since only the value of $u \bmod v$ affects the algorithm after the first division step, we may write

$$T_n = \frac{1}{n} \sum_{0 \le k < n} T(k, n).$$ (19)

For example, $T(0, 5) = 1$, $T(1, 5) = 2$, $T(2, 5) = 3$, $T(3, 5) = 4$, $T(4, 5) = 3$, so

$$T_5 = \tfrac{1}{5}(1 + 2 + 3 + 4 + 3) = 2\tfrac{3}{5}.$$

In order to estimate T_n for large n, let us first try an approximation suggested by R. W. Floyd: We might assume that, for $0 \le k < n$, n is essentially "random" modulo k, so that we can set

$$T_n \approx 1 + \frac{1}{n}(T_0 + T_1 + \cdots + T_{n-1}).$$

Then $T_n \approx S_n$, where the sequence $\langle S_n \rangle$ is the solution to the recurrence relation

$$S_0 = 0, \qquad S_n = 1 + \frac{1}{n}(S_0 + S_1 + \cdots + S_{n-1}), \qquad n \ge 1.$$ (20)

(This approximation is analogous to the "lattice-point model" used to investigate Algorithm B in Section 4.5.2.)

The recurrence (20) is readily solved by the use of generating functions. A more direct way to solve it, analogous to our solution of the lattice-point model, is by noting that

$$S_{n+1} = 1 + \frac{1}{n+1}(S_0 + S_1 + \cdots + S_{n-1} + S_n)$$

$$= 1 + \frac{1}{n+1}(n(S_n - 1) + S_n) = S_n + \frac{1}{n+1};$$

hence S_n is $1 + \frac{1}{2} + \cdots + 1/n = H_n$, a harmonic number. Therefore the approximation $T_n \approx S_n$ would tell us that $T_n \approx \ln n + O(1)$.

Comparison of this approximation with tables of the true value of T_n show, however, that $\ln n$ is too large; T_n does not grow this fast. One way to account for the fact that this approximation is too pessimistic is to observe that $n \bmod k \le \frac{1}{3}k$ when $\frac{2}{3}n \le k < n$, $\frac{2}{5}n \le k < \frac{1}{2}n$, $\frac{2}{7}n \le k < \frac{1}{3}n, \ldots$; so $n \bmod k \le \frac{1}{2}k$ in approximately

$$\frac{1}{1 \cdot 3} + \frac{1}{2 \cdot 5} + \frac{1}{3 \cdot 7} + \cdots = 2\left(\frac{1}{2} - \frac{1}{3} + \frac{1}{4} - \frac{1}{5} + \cdots\right)$$

$$= 2(1 - \ln 2) \approx 61\%$$

of the cases $0 \le k < n$. We would ordinarily expect $n \bmod k$ to be less than $\frac{1}{2}k$ about 50 percent of the time; but this calculation demonstrates the fact that $n \bmod k$ actually tends to be smaller than we would expect, on the average. Therefore Euclid's algorithm runs somewhat faster than the model predicts.

A continuous model. The behavior of Euclid's algorithm with $v = N$ is essentially determined by the behavior of the regular continued fraction process when $X = 0/N, 1/N, \ldots, (N-1)/N$. Assuming that N is very large, we are led naturally to a study of regular continued fractions when X is a real number uniformly distributed in $[0, 1)$. Therefore let us define the distribution function

$$F_n(x) = \text{probability that } X_n \leq x, \quad \text{for} \quad 0 \leq x \leq 1, \quad (21)$$

given a uniform distribution of $X = X_0$. By the definition of regular continued fractions, we have $F_0(x) = x$, and

$$F_{n+1}(x) = \sum_{k \geq 1} (\text{probability that } k \leq 1/X_n \leq k + x)$$

$$= \sum_{k \geq 1} (\text{probability that } 1/(k+x) \leq X_n \leq 1/k)$$

$$= \sum_{k \geq 1} \left(F_n(1/k) - F_n(1/(k+x)) \right). \quad (22)$$

If the distributions $F_0(x), F_1(x), \ldots$ defined by these formulas approach a limiting distribution $F_\infty(x) = F(x)$, we will have

$$F(x) = \sum_{k \geq 1} \left(F(1/k) - F(1/(k+x)) \right). \quad (23)$$

One function which satisfies this relation is $F(x) = \log_b (1 + x)$, for any base $b > 1$; see exercise 19. The further condition $F(1) = 1$ implies that we should take $b = 2$. Thus it is reasonable to make a guess that $F(x) = \log_2 (1 + x)$, and that $F_n(x)$ approaches this behavior.

We might conjecture, for example, that $F(\frac{1}{2}) = \log_2 (\frac{3}{2}) \approx 0.58496$; let us see how close $F_n(\frac{1}{2})$ comes to this value for small n. We have $F_0(\frac{1}{2}) = \frac{1}{2}$, and

$$F_1(\tfrac{1}{2}) = \frac{1}{1} - \frac{1}{1 + \frac{1}{2}} + \frac{1}{2} - \frac{1}{2 + \frac{1}{2}} + \cdots$$

$$= 2\left(\frac{1}{2} - \frac{1}{3} + \frac{1}{4} - \frac{1}{5} + \cdots \right) = 2(1 - \ln 2) \approx 0.6137;$$

$$F_2(\tfrac{1}{2}) = \sum_{m \geq 1} \frac{2}{m} \left(\frac{1}{2m+2} - \frac{1}{3m+2} + \frac{1}{4m+2} - \frac{1}{5m+2} + \cdots \right)$$

$$= \sum_{m \geq 1} \frac{2}{m^2} \left(\frac{1}{2} - \frac{1}{3} + \frac{1}{4} - \cdots \right)$$

$$- \sum_{m \geq 1} \frac{4}{m} \left(\frac{1}{2m(2m+2)} - \frac{1}{3m(3m+2)} + \cdots \right)$$

$$= \tfrac{1}{3}\pi^2 (1 - \ln 2) - \sum_{m \geq 1} \frac{4S_m}{m^2}, \quad (24)$$

where $S_m = 1/(4m + 4) - 1/(9m + 6) + 1/(16m + 8) - \cdots$. Using the values of H_x for fractional x found in Table 3 of Appendix B, we find that

$$S_1 = \tfrac{1}{12}, \qquad S_2 = \tfrac{3}{4} - \ln 2, \qquad S_3 = \tfrac{19}{20} - \pi/(2\sqrt{3}),$$

etc.; a numerical evaluation yields $F_2(\tfrac{1}{2}) \approx 0.5748$. Although $F_1(x) = H_x$, it is clear that $F_n(x)$ is difficult to calculate exactly for large n.

The distributions $F_n(x)$ were first studied by K. F. Gauss, who thought of the problem in 1800. His notebook for that year lists various recurrence relations and gives a brief table of values, including the four-place value for $F_2(\tfrac{1}{2})$ which has just been mentioned. After performing these calculations, Gauss wrote, "*Tam complicatae evadunt, ut nulla spes superesse videatur,*" i.e., "They come out so complicated that no hope appears to be left." Twelve years later, Gauss wrote a letter to Laplace in which he posed the problem as one he could not resolve to his satisfaction. He said, "I found by very simple reasoning that, for n infinite, $F_n(x) = \log (1 + x)/\log 2$. But the efforts which I made since then in my inquiries to assign $F_n(x) - \log (1 + x)/\log 2$ for very large but not infinite values of n were fruitless." He never published his "very simple reasoning," and it is not completely clear that he had found a rigorous proof. More than 100 years went by before a proof was finally published, by R. O. Kuz'min [*Atti del Congresso internazionale dei matematici* **6** (Bologna, 1928), 83–89], who showed that

$$F_n(x) = \log_2 (1 + x) + O(e^{-A \sqrt{n}})$$

for some positive constant A. An English account of Kuz'min's work, together with some further results, appears in the book by Khinchin cited earlier in this section. A sharper result was obtained by Paul Lévy [*Bull. Soc. Math. de France* **57** (1929), 178–194]; and since Lévy's interesting work on this problem has apparently not yet appeared in the English language, a fairly detailed and slightly simplified version of Lévy's proof will now be presented.

Given positive integers $a_1, a_2, \ldots, a_n$, the probability that $A_1 = a_1$, $A_2 = a_2, \ldots, A_n = a_n$ is the probability that X lies between

$$/a_1, \ldots, a_n/ \quad \text{and} \quad /a_1, \ldots, a_{n-1}, a_n + 1/ = /a_1, \ldots, a_n, 1/,$$

namely

$$|/a_1, \ldots, a_n/ - /a_1, \ldots, a_n, 1/| = 1/Q_n(a_1, \ldots, a_n)Q_{n+1}(a_1, \ldots, a_n, 1). \tag{25}$$

(Cf. (12).) Let us denote this probability by $p(a_1, a_2, \ldots, a_n)$.

In order to carry out the proof that $F_n(x)$ approaches $\log_2 (1 + x)$, we want to have a more workable formula for $F_n(x)$. It is not difficult to see that $F_n(x)$

is the probability that X lies between $/a_1, \ldots, a_n/$ and $/a_1, \ldots, a_n + x/ = /a_1, \ldots, a_n, 1/x/$, summed over all sets of positive integers $a_1, \ldots, a_n$; thus

$$F_n(x) = \sum_{a_1, \ldots, a_n} |/a_1, \ldots, a_n, 1/x/ - /a_1, \ldots, a_n/|$$

$$= \sum_{a_1, \ldots, a_n} \frac{1}{Q_n(a_1, \ldots, a_n) Q_{n+1}(a_1, \ldots, a_n, 1/x)}$$

$$= \sum_{a_1, \ldots, a_n} p(a_1, \ldots, a_n) \frac{q + q'}{q/x + q'},$$

where $q = Q_n(a_1, \ldots, a_n)$ and $q' = Q_{n-1}(a_1, \ldots, a_{n-1})$. (Properties (4)–(8) of the Q-polynomials will be used freely in this derivation.) By the symmetry condition (6), $q'/q = Q_{n-1}(a_{n-1}, \ldots, a_1)/Q_n(a_n, a_{n-1}, \ldots, a_1) = /a_n, \ldots, a_1/$, so we have

$$F_n(x) = \sum_{a_1, \ldots, a_n} p(a_1, \ldots, a_n) \frac{(1 + /a_n, \ldots, a_1/)x}{1 + /a_n, \ldots, a_1/x}$$

$$= \int_0^1 \frac{(1 + y)x}{1 + yx} \, dG_n(y), \qquad (26)$$

where (see Fig. 10)

$$G_n(y) = \sum_{/a_n, \ldots, a_1/ \le y} p(a_1, \ldots, a_n). \qquad (27)$$

This last conversion from a sum to an integral is a key step in the derivation which can be verified without great difficulty; it is a typical exercise in advanced calculus (see exercise 20).

Now if $0 \le \theta < 1$ and if m is a positive integer,

$$G_{n+1}\left(\frac{1}{m}\right) - G_{n+1}\left(\frac{1}{m + \theta}\right) = \sum p(a_1, \ldots, a_{n+1})$$

summed over all $a_1, \ldots, a_{n+1}$ such that $1/(m + \theta) < /a_{n+1}, \ldots, a_1/ \le 1/m$; that is, $m \le a_{n+1} + /a_n, \ldots, a_1/ < m + \theta$. Therefore $a_{n+1} = m$, and the sum can be rewritten

$$\sum_{/a_n, \ldots, a_1/ < \theta} p(a_1, \ldots, a_n, m) = \sum_{/a_n, \ldots, a_1/ < \theta} \frac{1}{(mq + q')((m + 1)q + q')}$$

$$= \sum_{/a_n, \ldots, a_1/ < \theta} \frac{p(a_1, \ldots, a_n)(1 + q'/q)}{(m + q'/q)(m + 1 + q'/q)}$$

$$= \int_0^\theta \frac{1 + y}{(m + y)(m + 1 + y)} \, dG_n(y). \qquad (28)$$

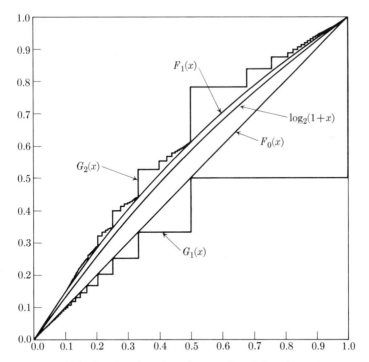

Fig. 10. Distribution functions for continued fraction process.

Let us now define two functions

$$F(y) = G(y) = \log_2 (1 + y). \tag{29}$$

Note that

$$\int_0^\theta \frac{1}{(m + y)(m + 1 + y)}\, dG(y) = \frac{1}{\ln 2} \int_0^1 dy \left(\frac{1}{m + y} - \frac{1}{m + 1 + y} \right)$$

$$= \log_2 (m + \theta) - \log_2 (m + 1 + \theta)$$

$$- \log_2 m + \log_2 (m + 1)$$

$$= G(1/m) - G\big(1/(m + \theta)\big); \tag{30}$$

$$\int_0^1 \frac{(1 + y)x}{1 + yx}\, dG(y) = \frac{1}{\ln 2} \int_0^1 \frac{x\, dy}{1 + yx} = \frac{1}{\ln 2} \int_0^x \frac{du}{1 + u}$$

$$= \log_2 (1 + x) = F(x). \tag{31}$$

In other words, F and G satisfy essentially the same relations we have derived for F_n and G_n.

Now we are ready to prove that $G_n(y) \to G(y)$, and at the same time that $F_n(x) \to F(x)$. We have

$$|F_n(x) - F(x)| = \left| \int_0^1 \frac{(1+y)x}{1+yx} \, d(G_n(y) - G(y)) \right|$$

$$= \left| \int_0^1 (G_n(y) - G(y)) \, d\left(\frac{(1+y)x}{1+yx} \right) \right|,$$

for any fixed x, since integration by parts is a basic property of Riemann-Stieltjes integrals, and since

$$G_n(0) = G(0) = 0, \qquad G_n(1) = G(1) = 1.$$

If we now set

$$\alpha_n = \sup_{0 \le y \le 1} |G_n(y) - G(y)|, \tag{32}$$

we can obtain a good estimate of $|F_n(x) - F(x)|$:

$$|F_n(x) - F(x)| \le \alpha_n \int_0^1 \left| \frac{x-1}{(1+yx)^2} \right| dy = \alpha_n \left| \frac{2x}{1+x} - x \right| \le c\alpha_n, \tag{33}$$

where $c = 3 - 2\sqrt{2} \approx 0.1716$. We also have, for all positive integers m, $G_{n+1}(1/m) = p(a_1, \ldots, a_n, a_{n+1})$ summed over all positive integers $a_1, \ldots, a_{n+1}$ with $/a_{n+1}, a_n, \ldots, a_1/ \le 1/m$, that is, $m \le a_{n+1} + /a_n, \ldots, a_1/$; hence

$$G_{n+1}(1/m) = \text{probability that } A_{n+1} \ge m = F_n(1/m). \tag{34}$$

This equation is illustrated for $n = 0$ and $n = 1$ in Fig. 10.

Finally let $0 \le \theta < 1$; we want to estimate

$$\left| G_{n+1}\left(\frac{1}{m+\theta} \right) - G\left(\frac{1}{m+\theta} \right) \right|$$

$$\le \left| G_{n+1}\left(\frac{1}{m} \right) - G\left(\frac{1}{m} \right) \right|$$

$$+ \left| G_{n+1}\left(\frac{1}{m} \right) - G_{n+1}\left(\frac{1}{m+\theta} \right) - G\left(\frac{1}{m} \right) + G\left(\frac{1}{m+\theta} \right) \right|$$

$$= \left| F_n\left(\frac{1}{m} \right) - F\left(\frac{1}{m} \right) \right| + \left| \int_0^\theta \frac{y+1}{(m+y)(m+1+y)} \, d(G_n(y) - G(y)) \right|$$

$$\le c\alpha_n + \left| \int_0^\theta (G_n(y) - G(y)) \, d\left(\frac{y+1}{(m+y)(m+1+y)} \right) \right.$$

$$\left. + \frac{\theta+1}{(m+\theta)(m+1+\theta)} (G_n(\theta) - G(\theta)) \right|$$

$$\le \alpha_n(c + \psi(m, \theta))$$

where

$$\psi(m, \theta) = \int_0^\theta \left| d\left(\frac{y+1}{(m+y)(m+1+y)} \right) \right| + \frac{\theta+1}{(m+\theta)(m+1+\theta)}.$$

Now by brute-force integration

$$\psi(m, \theta) = \begin{cases} 1/2 - 1/(2+\theta) + 1/(2+\theta) = 1/2, & \text{if} \quad m = 1; \\ 4/(3+\theta) - 2/(2+\theta) - 1/6, & \text{if} \quad m = 2,\ \theta \le \sqrt{2} - 1; \\ 2c - 1/6, & \text{if} \quad m = 2,\ \theta \ge \sqrt{2} - 1; \\ \dfrac{2m}{m+1+\theta} - \dfrac{2m-2}{m+\theta} - \dfrac{1}{m(m+1)} \le \dfrac{3m-2}{m(m+1)(m+2)}, \\ \qquad\qquad\qquad\qquad \text{if} \quad m \ge 3. \end{cases}$$

Hence $|\psi(m, \theta)| \le \frac{1}{2}$ in all cases. We have shown that

$$\left| G_{n+1}\left(\frac{1}{m+\theta} \right) - G\left(\frac{1}{m+\theta} \right) \right| \le \alpha_n(c + \tfrac{1}{2}); \qquad (35)$$

therefore by induction on n we have proved

Theorem L (P. Lévy, 1929). *For all $n \ge 0$ and $0 \le x \le 1$,*

$$|F_n(x) - F(x)| \le (0.68)^n, \qquad |G_n(x) - G(x)| \le (0.68)^n. \quad \blacksquare \qquad (36)$$

In other words, both F_n and G_n converge very rapidly to the distribution $F(x) = G(x) = \log_2(1 + x)$. Perhaps the convergence is even considerably faster than Theorem L indicates; another proof, which demonstrates that the constant 0.68 can at least be lowered to 0.4, has been given by P. Szüsz, *Acta Mathematica* (Acad. Sci. Hung.) **12** (1961), 447–453. Nobody has yet been able to ascertain the exact rate of convergence, so in a sense Gauss's problem has still not been answered.

From continuous to discrete. We have now derived results about the probability distributions for continued fractions when X is a real number uniformly distributed in the interval $[0, 1)$. But since a real number is rational with probability zero (almost all numbers are irrational), these results do not apply directly to Euclid's algorithm. Before we can apply Theorem L to our problem, some technicalities must be overcome. Consider the following observation based on elementary measure theory:

Lemma M. *Let $I_1, I_2, \ldots, J_1, J_2, \ldots$ be pairwise disjoint intervals contained in the interval $[0, 1]$, and let $\mathcal{I} = \bigcup_{k \ge 1} I_k$, $\mathcal{J} = \bigcup_{k \ge 1} J_k$. Assume that $\mathcal{K} = [0, 1] - (\mathcal{I} \cup \mathcal{J})$ has measure zero. Let P_n be the set $\{0/n, 1/n, \ldots, (n-1)/n\}$. Then*

$$\lim_{n \to \infty} \frac{\|\mathcal{I} \cap P_n\|}{n} = \mu(\mathcal{I}). \qquad (37)$$

Here $\mu(\mathcal{I})$ is the Lebesgue measure of $\mathcal{I}$, namely, $\sum_{k \geq 1}$ length (I_k); and $||\mathcal{I} \cap P_n||$ denotes the number of elements of $\mathcal{I} \cap P_n$.)

Proof. Let $\mathcal{I}_N = \bigcup_{1 \leq k \leq N} I_k$, $\mathcal{J}_N = \bigcup_{1 \leq k \leq N} J_k$. Given $\epsilon > 0$, find N large enough so that $\mu(\mathcal{I}_N) + \mu(\mathcal{J}_N) \geq 1 - \epsilon$, and let

$$\mathcal{K}_N = \mathcal{K} \cup \bigcup_{k>N} I_k \cup \bigcup_{k>N} J_k.$$

If I is any interval, having any of the forms (a, b) or $[a, b)$ or $(a, b]$ or $[a, b]$, it is clear that $\mu(I) = b - a$ and

$$n\mu(I) - 1 \leq ||I \cap P_n|| \leq n\mu(I) + 1.$$

Now let $r_n = ||\mathcal{I}_N \cap P_n||$, $s_n = ||\mathcal{J}_N \cap P_n||$, $t_n = ||\mathcal{K}_N \cap P_n||$; we have

$$r_n + s_n + t_n = n;$$

$$n\mu(\mathcal{I}_N) - N \leq r_n \leq n\mu(\mathcal{I}_N) + N;$$

$$n\mu(\mathcal{J}_N) - N \leq s_n \leq n\mu(\mathcal{J}_N) + N.$$

Hence

$$\mu(\mathcal{I}) - \frac{N}{n} - \epsilon \leq \mu(\mathcal{I}_N) - \frac{N}{n} \leq \frac{r_n}{n} \leq \frac{r_n + t_n}{n}$$

$$= 1 - \frac{s_n}{n} \leq 1 - \mu(\mathcal{J}_N) + \frac{N}{n} \leq \mu(\mathcal{I}) + \frac{N}{n} + \epsilon.$$

This holds for all n and for all ϵ; hence $\lim_{n \to \infty} r_n/n = \mu(\mathcal{I})$. ∎

Exercise 25 shows that Lemma M is not trivial, in the sense that some rather restrictive hypotheses are needed to prove (37).

Distribution of partial quotients. Now we can put Theorem L and Lemma M together to derive some solid facts about Euclid's algorithm.

Theorem E. *Let n and k be positive integers, and let $p_k(a, n)$ be the probability that the $(k + 1)$st quotient A_{k+1} in Euclid's algorithm is equal to a, when $v = n$ and u is chosen at random. Then*

$$\lim_{n \to \infty} p_k(a, n) = F_k\left(\frac{1}{a}\right) - F_k\left(\frac{1}{a+1}\right),$$

where $F_k(x)$ is the distribution function (21).

Proof. The set $\mathcal{I}$ of all X in $[0, 1)$ for which $A_{k+1} = a$ is a union of disjoint intervals, and so is the set $\mathcal{J}$ of all X for which $A_{k+1} \neq a$. Lemma M therefore applies, with $\mathcal{K}$ the set of all X for which A_{k+1} is undefined. Furthermore, $F_k(1/a) - F_k(1/(a+1))$ is the probability that $1/(a+1) < X_k \leq 1/a$, which is $\mu(\mathcal{I})$, the probability that $A_{k+1} = a$. ∎

As a consequence of Theorems E and L, we can say that a quotient equal to a occurs with the approximate probability

$$\log_2\left(1 + 1/a\right) - \log_2\left(1 + 1/(a+1)\right) = \log_2\left((a+1)^2/((a+1)^2 - 1)\right).$$

Thus

$$
\begin{aligned}
&\text{a quotient of 1 occurs about } \log_2\left(\tfrac{4}{3}\right) &&= 41.504 \text{ percent of the time;}\\
&\text{a quotient of 2 occurs about } \log_2\left(\tfrac{9}{8}\right) &&= 16.992 \text{ percent of the time;}\\
&\text{a quotient of 3 occurs about } \log_2\left(\tfrac{16}{15}\right) &&= 9.311 \text{ percent of the time;}\\
&\text{a quotient of 4 occurs about } \log_2\left(\tfrac{25}{24}\right) &&= 5.890 \text{ percent of the time.}
\end{aligned}
$$

Actually, if Euclid's algorithm produces the quotients $A_1, A_2, \ldots, A_t$, the nature of the proofs above will guarantee this behavior only for A_k when k is comparatively small with respect to t; the values $A_{t-1}, A_{t-2}, \ldots$ are not covered by this proof. Empirical tests show, however, that the estimates are valid very generally, and we can in fact prove that the distribution of the last quotients $A_{t-1}, A_{t-2}, \ldots$ is essentially the same as the first.

For example, consider the regular continued fraction expansions for the fractions whose denominator is 29:

$\frac{1}{29} = /29/$	$\frac{8}{29} = /3, 1, 1, 1, 2/$	$\frac{15}{29} = /1, 1, 14/$	$\frac{22}{29} = /1, 3, 7/$
$\frac{2}{29} = /14, 2/$	$\frac{9}{29} = /3, 4, 2/$	$\frac{16}{29} = /1, 1, 4, 3/$	$\frac{23}{29} = /1, 3, 1, 5/$
$\frac{3}{29} = /9, 1, 2/$	$\frac{10}{29} = /2, 1, 9/$	$\frac{17}{29} = /1, 1, 2, 2, 2/$	$\frac{24}{29} = /1, 4, 1, 4/$
$\frac{4}{29} = /7, 4/$	$\frac{11}{29} = /2, 1, 1, 1, 3/$	$\frac{18}{29} = /1, 1, 1, 1, 1, 3/$	$\frac{25}{29} = /1, 6, 4/$
$\frac{5}{29} = /5, 1, 4/$	$\frac{12}{29} = /2, 2, 2, 2/$	$\frac{19}{29} = /1, 1, 1, 9/$	$\frac{26}{29} = /1, 8, 1, 2/$
$\frac{6}{29} = /4, 1, 5/$	$\frac{13}{29} = /2, 4, 3/$	$\frac{20}{29} = /1, 2, 4, 2/$	$\frac{27}{29} = /1, 13, 2/$
$\frac{7}{29} = /4, 7/$	$\frac{14}{29} = /2, 14/$	$\frac{21}{29} = /1, 2, 1, 1, 1, 2/$	$\frac{28}{29} = /1, 28/$

Several things can be observed in this table.

a) As mentioned earlier, the last quotient is always 2 or more. Furthermore, we have the obvious identity

$$/x_1, \ldots, x_{n-1}, x_n + 1/ = /x_1, \ldots, x_{n-1}, x_n, 1/ \tag{38}$$

which shows how partial fractions whose last quotient is unity are related to regular continued fractions.

b) The values in the right-hand columns have a simple relationship to the values in the left-hand columns; can the reader see the correspondence? We have the identity

$$1 - /x_1, x_2, \ldots, x_n/ = /1, x_1 - 1, x_2, \ldots, x_n/; \tag{39}$$

see exercise 9.

c) There is symmetry between left and right in the first two columns: If $/A_1, A_2, \ldots, A_t/$ occurs, so does $/A_t, \ldots, A_2, A_1/$. This will always be the case (see exercise 26).

d) If we examine all of the quotients in the table, we find that there are 96 in all, of which $\frac{39}{96} = 40.6$ percent are equal to 1, $\frac{21}{96} = 21.9$ percent are equal to 2, $\frac{8}{96} = 8.3$ percent are equal to 3; this agrees reasonably well with the probabilities listed above.

The number of division steps. Let us now return to our original problem and investigate T_n, the average number of division steps when $v = n$. (See Eq. (19).) Here are some sample values of T_n:

$$n = \quad 95 \quad 96 \quad 97 \quad 98 \quad 99 \quad 100 \quad 101 \quad 102 \quad 103 \quad 104 \quad 105$$
$$T_n = \quad 5.0 \;\; 4.4 \;\; 5.3 \;\; 4.8 \;\; 4.7 \;\; 4.6 \;\; 5.3 \;\; 4.6 \;\; 5.3 \;\; 4.7 \;\; 4.6$$
$$n = \quad 996 \;\; 997 \;\; 998 \;\; 999 \;\; 1000 \;\; 1001 \cdots 9999 \;\; 10000 \;\; 10001$$
$$T_n = \quad 6.5 \;\; 7.3 \;\; 7.0 \;\; 6.8 \quad 6.4 \quad 6.7 \cdots \quad 8.6 \quad\; 8.3 \quad 9.1$$
$$n = \quad 49999 \;\; 50000 \;\; 50001 \cdots 99999 \;\; 100000 \;\; 100001$$
$$T_n = \quad 10.6 \quad\; 9.7 \quad\; 10.0 \cdots \quad 10.7 \quad\;\; 10.3 \quad\;\; 11.0$$

Note the somewhat erratic behavior; T_n tends to be higher than its neighbors when n is prime, and it is correspondingly lower when n has many divisors. (In this list, 97, 101, 103, 997, and 49999 are primes; $10001 = 73 \cdot 137$, $50001 = 3 \cdot 7 \cdot 2381$, $99999 = 3 \cdot 3 \cdot 41 \cdot 271$, $100001 = 11 \cdot 9091$.) It is not difficult to understand why this happens; if $\gcd(u, v) = d$, Euclid's algorithm applied to u and v behaves essentially the same as if it were applied to u/d and v/d. Therefore, when $v = n$ has several divisors, there are many choices of u for which n behaves as if it were smaller.

Accordingly let us consider *another* quantity, τ_n, which is the average number of division steps when $v = n$ and when u is *relatively prime* to n. Thus

$$\tau_n = \frac{1}{\varphi(n)} \sum_{\substack{0 \le m < n, \\ \gcd(m,n)=1}} T(m, n). \tag{40}$$

It follows that

$$T_n = \frac{1}{n} \sum_{d \backslash n} \varphi(d) \tau_d. \tag{41}$$

Here is a table of τ_n for the same values of n considered above:

$$n = \quad 95 \quad 96 \quad 97 \quad 98 \quad 99 \quad 100 \quad 101 \quad 102 \quad 103 \quad 104 \quad 105$$
$$\tau_n = \quad 5.4 \;\; 5.3 \;\; 5.3 \;\; 5.6 \;\; 5.2 \;\; 5.2 \;\; 5.4 \;\; 5.3 \;\; 5.4 \;\; 5.3 \;\; 5.6$$
$$n = \quad 996 \;\; 997 \;\; 998 \;\; 999 \;\; 1000 \;\; 1001 \cdots 9999 \;\; 10000 \;\; 10001$$
$$\tau_n = \quad 7.2 \;\; 7.3 \;\; 7.3 \;\; 7.3 \quad 7.3 \quad 7.4 \cdots 9.21 \quad 9.21 \quad 9.22$$
$$n = \quad 49999 \;\; 50000 \;\; 50001 \cdots 99999 \;\; 100000 \;\; 100001$$
$$\tau_n = \quad 10.58 \quad 10.57 \quad 10.59 \cdots 11.170 \;\; 11.172 \;\; 11.172$$

Clearly τ_n is much more well-behaved than T_n, and it should be more susceptible to analysis. Inspection of a table of τ_n for small n reveals some curious anomalies; for example, $\tau_{50} = \tau_{100}$ and $\tau_{60} = \tau_{120}$. But as n grows, the values of τ_n behave quite regularly indeed, as the above table indicates, and they show no significant relation to the factorization properties of n. If the reader will plot the values of τ_n versus $\ln n$ on graph paper, for the values of τ_n given above, he will see that the values lie very nearly on a straight line, and that the equation

$$\tau_n \approx 0.843 \ln n + 1.47 \tag{42}$$

gives a very good approximation to τ_n.

We can account for this behavior if we return to the considerations of Theorem L. Note that in Euclid's algorithm as expressed in (15) we have

$$\frac{V_0}{U_0} \frac{V_1}{U_1} \cdots \frac{V_{t-1}}{U_{t-1}} = \frac{V_{t-1}}{U_0},$$

since $U_{k+1} = V_k$; therefore, if $U = U_0$ and $V = V_0$ are relatively prime, and if there are t division steps, we have

$$X_0 X_1 \cdots X_{t-1} = 1/U.$$

Setting $U = N$ and $V = m < N$, we find that

$$\ln X_0 + \ln X_1 + \cdots + \ln X_{t-1} = -\ln N. \tag{43}$$

We know the approximate distribution of $X_0, X_1, X_2, \ldots$, so we can use this equation to estimate

$$t = T(N, m) = T(m, N) - 1.$$

Returning to the formulas preceding Theorem L, we find that the average value of $\ln X_n$, when X_0 is a real number uniformly distributed in $[0, 1)$, is

$$\int_0^1 \ln x \, dF_n(x) = \int_0^1 \ln x \, f_n(x) \, dx, \tag{44}$$

where

$$f_n(x) = F_n'(x) = \int_0^1 \frac{1+y}{(1+yx)^2} \, dG_n(y). \tag{45}$$

[See Eq. (26).] It follows that

$$f_n(x) = \frac{1}{\ln 2 (1 + x)} + O((0.68)^n), \tag{46}$$

using an argument similar to the derivation of (33). Hence the average value of $\ln X_n$ is, to a very good approximation,

$$\frac{1}{\ln 2} \int_0^1 \frac{\ln x}{1+x} \, dx = -\frac{1}{\ln 2} \int_0^\infty \frac{ue^{-u}}{1+e^{-u}} \, du$$

$$= -\frac{1}{\ln 2} \sum_{k \geq 1} (-1)^{k+1} \int_0^\infty ue^{-ku} \, du$$

$$= -\frac{1}{\ln 2} \left(1 - \tfrac{1}{4} + \tfrac{1}{9} - \tfrac{1}{16} + \tfrac{1}{25} - \cdots \right)$$

$$= -\frac{1}{\ln 2} \left(1 + \tfrac{1}{4} + \tfrac{1}{9} + \cdots - 2(\tfrac{1}{4} + \tfrac{1}{16} + \tfrac{1}{36} + \cdots) \right)$$

$$= -\frac{1}{2 \ln 2} \left(1 + \tfrac{1}{4} + \tfrac{1}{9} + \cdots \right)$$

$$= -\pi^2/12 \ln 2. \tag{47}$$

Therefore by (43) we expect to have the approximate formula

$$-t\pi^2/(12 \ln 2) \approx -\ln N;$$

that is, t should be approximately proportional to $(12 \ln 2/\pi^2) \ln N$. This constant $(12 \ln 2/\pi^2) = 0.842765913\ldots$ agrees perfectly with the empirical formula (42) obtained earlier, so we have good reason to believe that the formula

$$\tau_n \approx \frac{12 \ln 2}{\pi^2} \ln n + 1.47 \tag{48}$$

indicates the true asymptotic behavior of τ_n as $n \to \infty$.

If we assume that (48) is valid, we obtain the formula

$$T_n \approx \frac{12 \ln 2}{\pi^2} \left(\ln n - \sum_{d \backslash n} \Lambda(d)/d \right) + 1.47, \tag{49}$$

where $\Lambda(d)$ is *von Mangoldt's function* defined by the rules

$$\Lambda(n) = \begin{cases} \ln p, & \text{if } n = p^r \text{ for } p \text{ prime,} \quad r \geq 1; \\ 0, & \text{otherwise.} \end{cases} \tag{50}$$

For example

$$T_{100} \approx \frac{12 \ln 2}{\pi^2} \left(\ln 100 - \frac{\ln 2}{2} - \frac{\ln 2}{4} - \frac{\ln 5}{5} - \frac{\ln 5}{25} \right) + 1.47$$

$$\approx (0.843)(4.605 - 0.347 - 0.173 - 0.322 - 0.064) + 1.47$$

$$\approx 4.59;$$

the exact value of T_{100} is 4.56.

We can also estimate the average number of division steps when u and v are both uniformly distributed between 1 and N, by calculating

$$\frac{1}{N} \sum_{1 \le n < N} T_n. \tag{51}$$

Assuming formula (49), this sum has the form

$$\frac{12 \ln 2}{\pi^2} \ln N + O(1), \tag{52}$$

(see exercise 27), and empirical calculations with the same numbers used to derive Eq. 4.5.2–45 show good agreement with the formula

$$\frac{12 \ln 2}{\pi^2} \ln N + 0.06. \tag{53}$$

Some refinements of Theorem L have been obtained by Walter Philipp [*Pac. J. Math.* **20** (1967), 109–127], but the results for "almost all real numbers" are not sharp enough to indicate the behavior of rational numbers. By strengthening the theory relating to the essential independence of X_i and X_j when $|i - j|$ is large, John D. Dixon proved that, for all positive numbers ϵ the probability that $|T(m, n) - (12 \ln 2/\pi^2) \ln n| \ge \ln n^{(1/2)+\epsilon}$ is $O(\exp(-c(\log N)^{\epsilon/2}))$ for some c, where $1 \le m \le n \le N$ [*J. Number Theory* **2** (1970), 414–422].

A completely different approach to the problem, which works entirely with integers instead of considering the continuous case, was discovered in 1968 by H. Heilbronn. His idea, which is described in exercises 33 and 34, is based on the fact that T_n can be related to the number of ways to represent n in a certain manner. Heilbronn's paper [*Number Theory and Analysis*, ed. by Paul Turán (New York: Plenum, 1969), 87–96] also shows that the distribution of individual partial quotients 1, 2, ... which we have discussed above actually applies to the entire collection of partial quotients belonging to the fractions having a given denominator; this is a sharper form of Theorem E. As yet no theory accounts for the extremely smooth growth of τ_n.

Summary. We have found that the worst case of Euclid's algorithm occurs when its inputs u and v are consecutive Fibonacci numbers (Theorem F); the number of division steps when $v = n$ will never exceed $\lceil 4.8 \log_{10} N - 0.32 \rceil$. We have determined the frequency of the values of various partial quotients, showing, for example, that the division step finds $\lfloor u/v \rfloor = 1$ about 41 percent of the time (Theorem E). And, finally, the theorems of Heilbronn and Dixon show that the average number T_n of division steps when $v = n$ is approximately

$$(12 \ln 2/\pi^2) \ln n \approx 1.9405 \log_{10} n.$$

Empirical calculations show that T_n is given very accurately by this formula, minus a correction term based on the divisors of n as shown in Eq. (49).

EXERCISES

▶ **1.** [*20*] Since the quotient $\lfloor u/v \rfloor$ is equal to unity over 40 percent of the time, it may be advantageous on some computers to make a test for this case and to avoid the division when the quotient is unity. Is the following MIX program for Euclid's algorithm more efficient than Program 4.5.2A?

	LDX	U	$rX \leftarrow u$.
	JMP	2F	
1H	STX	V	$v \leftarrow rX$.
	SUB	V	$rA \leftarrow u - v$.
	CMPA	V	
	SRAX	5	$rAX \leftarrow rA$.
	JL	2F	Is $u - v < v$?
	DIV	V	$rX \leftarrow rAX \bmod v$.
2H	LDA	V	$rA \leftarrow v$.
	JXNZ	1B	Done if $rX = 0$. ∎

2. [*M21*] Evaluate the matrix product

$$\begin{pmatrix} x_1 & 1 \\ 1 & 0 \end{pmatrix}\begin{pmatrix} x_2 & 1 \\ 1 & 0 \end{pmatrix}\cdots\begin{pmatrix} x_n & 1 \\ 1 & 0 \end{pmatrix}.$$

3. [*M21*] What is the value of

$$\det \begin{matrix} x_1 & 1 & 0 & \cdots & 0 \\ -1 & x_2 & 1 & & 0 \\ 0 & -1 & x_3 & 1 & \\ & & -1 & \ddots & 1 \\ \vdots & & & \ddots & 1 \\ 0 & 0 & \cdots & -1 & x_n \end{matrix} \quad ?$$

4. [*M20*] Prove Eq. (8).

5. [*HM25*] Let $x_1, x_2, \ldots$ be a sequence of real numbers which are each greater than some positive real number ϵ. Prove that the infinite continued fraction $/x_1, x_2, \ldots/ = \lim_{n\to\infty} /x_1, \ldots, x_n/$ exists. Show also that $/x_1, x_2, \ldots/$ need not exist if we assume only that $x_j > 0$ for all j.

6. [*M23*] Prove that the regular continued fraction expansion of a number is *unique* in the following sense: If $B_1, B_2, \ldots$ are positive integers, then the infinite continued fraction $/B_1, B_2, \ldots/$ is an irrational number X between 0 and 1 whose regular continued fraction has $A_n = B_n$ for all $n \geq 1$; and if $B_1, \ldots, B_m$ are positive integers with $B_m > 1$, then the regular continued fraction for $X = /B_1, \ldots, B_m/$ has $A_n = B_n$ for $1 \leq n \leq m$.

7. [*M26*] Find all permutations $p(1)\, p(2) \ldots p(n)$ of $\{1, 2, \ldots, n\}$ such that $Q_n(x_1, x_2, \ldots, x_n) = Q_n(x_{p(1)}, x_{p(2)}, \ldots, x_{p(n)})$ holds for all $x_1, x_2, \ldots, x_n$.

8. [*M20*] If X_n is defined, in the regular continued fraction process, show that $/A_n, \ldots, A_1, -X/ = -1/X_n$.

9. [M21] Show that continued fractions satisfy the following identities:

a) $/x_1, \ldots, x_n/ = /x_1, \ldots, x_k + /x_{k+1}, \ldots, x_n//$, $1 \le k \le n$;

b) $/0, x_1, x_2, \ldots, x_n/ = x_1 + /x_2, \ldots, x_n/$, $n \ge 1$;

c) $/x_1, \ldots, x_{k-1}, x_k, 0, x_{k+1}, x_{k+2}, \ldots, x_n/ =$
$/x_1, \ldots, x_{k-1}, x_k + x_{k+1}, x_{k+2}, \ldots, x_n/$, $1 \le k < n$;

d) $1 - /x_1, x_2, \ldots, x_n/ = /1, x_1 - 1, x_2, \ldots, x_n/$, $n \ge 1$.

10. [M28] By the result of exercise 6, every irrational real number X has a unique regular continued fraction representation of the form

$$X = A_0 + /A_1, A_2, A_3, \ldots/,$$

where A_0 is an integer and $A_1, A_2, A_3, \ldots$ are positive integers. Show that if X has this representation then the regular continued fraction for $1/X$ is

$$1/X = B_0 + /B_1, \ldots, B_m, A_5, A_6, \ldots/$$

for suitable integers $B_0, B_1, \ldots, B_m$. (The case $A_0 < 0$ is of course the most interesting.) Explain how to determine the B's in terms of A_0, A_1, A_2, A_3, and A_4.

11. [M30] (J. Lagrange.) Let $X = A_0 + /A_1, A_2, \ldots/$, $Y = B_0 + /B_1, B_2, \ldots/$ be the regular continued fraction representations of two real numbers X and Y, in the sense of exercise 10. Show that these representations "eventually agree," in the sense that $A_{m+k} = B_{n+k}$ for some m and n and for all $k \ge 0$, if and only if $X = (qY + r)/(sY + t)$ for some integers q, r, s, t with $|qt - rs| = 1$. (This theorem is the analog, for continued fraction representations, of the simple result that the representations of X and Y in the decimal system eventually agree if and only if $X = (10^q Y + r)/10^s$ for some integers q, r, and s.)

▶ 12. [M30] A *quadratic irrationality* is a number of the form $(\sqrt{D} - U)/V$, where D, U, and V are integers, $D > 0$, $V \ne 0$, and D is not a perfect square. We may assume without loss of generality that V is a divisor of $D - U^2$, for otherwise the number may be rewritten as $(\sqrt{DV^2} - U|V|)/V|V|$.

a) Prove that the regular continued fraction expansion (in the sense of exercise 10) of a quadratic irrationality $X = (\sqrt{D} - U)/V$ is obtained by the following formulas:

$$V_0 = V, \qquad A_0 = \lfloor X \rfloor, \qquad U_0 = U + A_0 V;$$
$$V_{n+1} = (D - U_n^2)/V_n, \qquad A_{n+1} = \lfloor (\sqrt{D} + U_n)/V_{n+1} \rfloor,$$
$$U_{n+1} = A_{n+1} V_{n+1} - U_n.$$

[*Note:* An algorithm based on this process has many applications to the solution of quadratic equations in integers; see, for example, H. Davenport, *The higher arithmetic* (London: Hutchinson, 1952); W. J. LeVeque, *Topics in Number Theory* (Reading, Mass.: Addison-Wesley, 1956); and see Section 4.5.4. By exercise 1.2.4–35, $A_{n+1} = \lfloor (\lfloor \sqrt{D} \rfloor + U_n)/V_{n+1} \rfloor$ when $V_{n+1} > 0$, and $A_{n+1} = \lfloor (\lfloor \sqrt{D} \rfloor + 1 + U_n)/V_{n+1} \rfloor$ when $V_{n+1} < 0$; hence such an algorithm need only work with the integer $\lfloor \sqrt{D} \rfloor$.]

b) Prove that $0 < V_n < D$, $0 < U_n < \sqrt{D}$, for all $n > N$, where N is some integer depending on X; hence the regular continued fraction representation of every quadratic irrationality is eventually periodic. [*Hint:* Show that $(-\sqrt{D} - U)/V = A_0 + /A_1, \ldots, A_n, -V_n/(\sqrt{D} + U_n)/$, and use Eq. (5) to prove that $(\sqrt{D} + U_n)/V_n$ is positive when n is large.]

c) Letting $p_n = Q_{n+1}(A_0, A_1, \ldots, A_n)$ and $q_n = Q_n(A_1, \ldots, A_n)$, prove that $V p_n^2 + 2U p_n q_n + ((U^2 - D)/V) q_n^2 = (-1)^{n+1} V_{n+1}$.

d) Prove that the regular continued fraction representation of an irrational number X is eventually periodic if and only if X is a quadratic irrationality. (This is the continued fraction analog of the fact that the decimal expansion of a real number X is eventually periodic if and only if X is rational.)

13. [*M40*] (J. Lagrange, 1797.) Let $f(x) = a_n x^n + \cdots + a_0$, $a_n > 0$, be a polynomial with integer coefficients, having no rational roots, and having exactly one real root $\xi > 1$. Design a computer program to find the first thousand or so partial quotients of ξ, using the following algorithm (which involves only multiprecision addition and subtraction):

L1. Set $A \leftarrow 1$.

L2. Set $k = 0, 1, \ldots, n-1$ (in this order), set $a_k \leftarrow a_n + a_{n-1} + \cdots + a_k$.

L3. If $a_n + a_{n-1} + \cdots + a_0 < 0$, set $A \leftarrow A + 1$ and return to L2.

L4. Output A (which is the value of the next partial quotient). Replace $(a_n, a_{n-1}, \ldots, a_0)$ by $(-a_0, -a_1, \ldots, -a_n)$ and return to L1. ▮

For example, starting with $f(x) = x^3 - 2$, the algorithm will output 1 (changing $f(x)$ to $x^3 - 3x^2 - 3x - 1$); then 3 (changing $f(x)$ to $10x^3 - 6x^2 - 6x - 1$); etc.

14. [*M22*] (A. Hurwitz, 1891.) Show that the following rules make it possible to find the regular continued fraction expansion for $2X$, given the partial quotients of X:

$$2/2a, b, c, \ldots/ = /a, 2b + 2/c, \ldots//;$$
$$2/2a + 1, b, c, \ldots/ = /a, 1, 1 + 2/b - 1, c, \ldots//.$$

Use this idea to find the regular continued fraction expansion of $\frac{1}{2}e$, given the expansion of e in (13).

▶ **15.** [*M26*] Find a rule for obtaining the regular continued fraction representation of $3X$, given the representation of X, analogous to Hurwitz's rule for $2X$ in exercise 14. Use your rule to find the regular continued fraction expansion of $\frac{1}{3}e$.

16. [*HM30*] (L. Euler, 1731.) Let $f_0(z) = (e^z - e^{-z})/(e^z + e^{-z}) = \tanh z$, and let $f_{n+1}(z) = 1/f_n - (2n+1)/z$. Prove that, for all n, $f_n(z)$ is an analytic function of the complex variable z in a neighborhood of the origin, and it satisfies the differential equation $f_n'(z) = 1 - f_n(z)^2 - 2nf_n(z)/z$. Use this fact to prove that

$$\tanh z = \left/ \frac{1}{z}, \frac{3}{z}, \frac{5}{z}, \frac{7}{z}, \cdots \right/ .$$

Then apply Hurwitz's rule (exercise 14) to prove that

$$e^{-1/n} = \left/ 1, (2m+1)n - 1, 1 \right/, \qquad m \geq 0.$$

(This notation denotes the infinite continued fraction

$$/1, n-1, 1, 1, 3n-1, 1, 1, 1, 5n-1, 1, 1, \ldots/.)$$

Also find the regular continued fraction expansion for $e^{-2/n}$ when $n > 0$ is odd.

▶ **17.** [*M23*] (a) Prove that $/x_1, -x_2/ = /x_1 - 1, 1, x_2 - 1/$. (b) Generalize this identity, obtaining a formula for $/x_1, -x_2, x_3, -x_4, \ldots, x_{2n-1}, -x_{2n}/$ in which all partial quotients are positive integers when the x's are large positive integers. (c) The result of exercise 16 implies that $\tan 1 = /1, -3, 5, -7, \ldots/$. Find the regular continued fraction expansion of $\tan 1$.

18. [*M40*] Develop a computer program to find as many partial quotients as possible of a real number given with high precision. Use this program to calculate the first one thousand (or so) partial quotients of Euler's constant γ, based on D. W. Sweeney's 3566-place value [*Math. Comp.* **17** (1963), 170–178]. [*Hint:* Apply Euclid's algorithm to an approximation u/v to γ, where v is a power of 10; if this is done for two values of u, so that $u_1/v < \gamma < u_2/v$, the partial quotients common to both approximations will be partial quotients of γ.]

19. [*M20*] Prove that $F(x) = \log_b(1 + x)$ satisfies Eq. (23).

20. [*HM24*] Let $f(x)$ be a continuous function on the closed interval $[a, b]$, and let $x_1, x_2, \ldots$ be an infinite sequence of values in (a, b) such that, for some function g defined at these points, the series $\sum_{n \geq 1} |g(x_n)|$ converges. Using the definition of Riemann-Stieltjes integration [see, e.g., T. M. Apostol, *Mathematical Analysis* (Reading, Massachusetts: Addison-Wesley, 1957), Chapter 9], prove that

$$\sum_{n \geq 1} f(x_n)g(x_n) = \int_a^b f(t)\,dG(t), \qquad \text{where} \qquad G(x) = \sum_{x_n \leq x} g(x_n).$$

21. [*HM44*] Prepare accurate tables of $F_n(x) - \log_2(1 + x)$ for small values of n, and estimate the true rate of convergence of $F_n(x)$ to $F(x)$.

22. [*HM50*] (K. F. Gauss.) What is the asymptotic behavior of $F_n(x) - \log_2(1 + x)$ as $n \to \infty$?

23. [*HM20*] Derive Eq. (46) from (45).

24. [*M22*] What is the average value of a partial quotient A_n in the regular continued fraction expansion of a random real number?

25. [*HM25*] Find an example of a set $\mathcal{I} = I_1 \cup I_2 \cup I_3 \cup \cdots \subseteq [0, 1]$, where the I's are disjoint intervals, for which (37) does not hold.

26. [*M23*] Show that if the numbers $1/n, 2/n, \ldots, \lfloor n/2 \rfloor/n$ are expressed as regular continued fractions, the result is symmetric between left and right, in the sense that $/A_t, \ldots, A_2, A_1/$ appears whenever $/A_1, A_2, \ldots, A_t/$ does.

27. [*M21*] Derive (49) from (41) and (48).

28. [*M23*] Prove the following identities involving the three number-theoretical functions $\varphi(n)$, $\mu(n)$, $\Lambda(n)$:

a) $\displaystyle\sum_{d \backslash n} \mu(d) = \delta_{n1}.$ b) $\displaystyle\ln n = \sum_{d \backslash n} \Lambda(d), \qquad n = \sum_{d \backslash n} \varphi(d).$

c) $\displaystyle\Lambda(n) = \sum_{d \backslash n} \mu\left(\frac{n}{d}\right) \ln d, \qquad \varphi(n) = \sum_{d \backslash n} \mu\left(\frac{n}{d}\right) d.$

29. [*M23*] Assuming that T_n is given by (49), show that (51) $=$ (52).

▶ **30.** [*HM32*] The following modification of Euclid's algorithm is often suggested: Instead of replacing v by $u \bmod v$ during the division step, replace it by $|(u \bmod v) - v|$ if $u \bmod v > \frac{1}{2}v$. Thus, for example, if $u = 26$ and $v = 7$, we have $\gcd(26, 7) = \gcd(-2, 7) = \gcd(7, 2)$; -2 is the remainder of smallest magnitude when multiples of 7 are subtracted from 26. Compare this procedure with Euclid's algorithm; estimate the number of division steps this method saves, on the average.

▶ **31.** [*M35*] Find the "worst case" of the modification of Euclid's algorithm suggested in exercise 30; what are the smallest inputs $u > v > 0$ which require n division steps?

32. [*20*] (a) A Morse code sequence of length n is a string of r dots and s dashes, where $r + 2s = n$. For example, the Morse code sequences of length 4 are

$$\cdots\cdot \;,\; \cdots - \;,\; \cdot - \cdot \;,\; - \cdots \;,\; - -$$

Noting that the continuant polynomial $Q_4(x_1, x_2, x_3, x_4) = x_1x_2x_3x_4 + x_1x_2 + x_1x_4 + x_3x_4 + 1$, find and prove a simple relation between $Q_n(x_1, \ldots, x_n)$ and Morse code sequences of length n. (b) (J. J. Sylvester, *Philos. Mag.* **6** (1853), 297–299.) Prove that $Q_{m+n}(x_1, \ldots, x_{m+n}) = Q_m(x_1, \ldots, x_m)Q_n(x_{m+1}, \ldots, x_{m+n}) + Q_{m-1}(x_1, \ldots, x_{m-1})Q_{n-1}(x_{m+2}, \ldots, x_{m+n})$.

33. [*M32*] Let $h(n)$ be the number of representations of n in the form

$$n = xx' + yy', \qquad x > y > 0, \quad x' > y' > 0, \quad \gcd(x, y) = 1, \quad \text{integer } x, x', y, y'.$$

(a) Show that if the conditions are relaxed to allow $x' = y'$, the number of representations is $h(n) + \lfloor (n-1)/2 \rfloor$. (b) Show that for fixed $y > 0$ and $0 < t \le y$, where $\gcd(t, y) = 1$, and for each fixed x' such that $x't \equiv n \pmod{y}$ and $0 < x' < n/(y + t)$, there is exactly one representation of n satisfying the restrictions of (a) and the condition $x \equiv t \pmod{y}$. (c) Consequently

$$h(n) = \sum \left[\left(\frac{n}{y + t} - t' \right) \frac{1}{y} \right] - \lfloor (n-1)/2 \rfloor,$$

where the sum is over all positive integers y, t, t' such that $\gcd(t, y) = 1$, $t \le y$, $t' \le y$, $tt' \equiv n \pmod{y}$. (d) Show that each of the $h(n)$ representations can be expressed uniquely in the form

$$x = Q_m(x_1, \ldots, x_m), \qquad y = Q_{m-1}(x_1, \ldots, x_{m-1}),$$
$$x' = Q_k(x_{m+1}, \ldots, x_{m+k})d, \qquad y' = Q_{k-1}(x_{m+2}, \ldots, x_{m+k})d,$$

where m, k, d, and the x_j are positive integers with $x_1 \ge 2$, $x_{m+k} \ge 2$, and d is a divisor of n. The identity of exercise 32 now implies that $n/d = Q_{m+k}(x_1, \ldots, x_{m+k})$. Conversely, every sequence of positive integers $x_1, \ldots, x_{m+k}$ with $x_1 \ge 2$, $x_{m+k} \ge 2$, and $Q_{m+k}(x_1, \ldots, x_{m+k})$ dividing n, corresponds in this way to $m + k - 1$ representations of n. (e) Therefore $nT_n = \lfloor (5n - 3)/2 \rfloor + 2h(n)$.

34. [*HM40*] (H. Heilbronn.) (a) Let $h_a(n)$ be the number of representations of n as in exercise 33 such that $xd < x'$, plus half the number of representations with $xd = x'$. Let $g(n)$ be the number of representations without the requirement $\gcd(x, y) = 1$. Prove that

$$h(n) = \sum_{d \backslash n} \mu(d)g\left(\frac{n}{d}\right), \qquad g(n) = 2\sum_{d \backslash n} h_a\left(\frac{n}{d}\right).$$

(b) Generalizing exercise 33b, show that for $d \ge 1$, $h_a(n) = \sum (n/(y(y + t))) + O(n)$, where the sum is over all integers y, t such that $\gcd(t, y) = 1$, $0 < t \le y < \sqrt{n/d}$. (c) Show that $\sum (y/(y + t)) = \varphi(y) \ln 2 + O(\sigma_{-1}(y))$, where the sum is over the range $0 < t \le y$, $\gcd(t, y) = 1$; and $\sigma_{-1}(y) = \sum_{(d \backslash y)} 1/d$. (d) Show that $\sum_{1 \le y \le n} \varphi(y)/y^2 = \sum_{1 \le d \le n} \mu(d)H_{\lfloor n/d \rfloor}/d^2$. (e) Hence $T_n = (12 \ln 2/\pi^2) \ln n + O(\sigma_{-1}(n)^2)$.

35. [*M35*] (G. H. Bradley.) What is the smallest value of u_n such that the calculation of $\gcd(u_1, \ldots, u_n)$ by steps D1, D2 in Section 4.5.2 requires N divisions, if Euclid's algorithm is used throughout? Assume that $N \ge n$.

4.5.4. Factoring into primes

Several of the computational methods we have encountered in this book rest on the fact that every positive integer n can be expressed in a unique way in the form

$$n = p_1 p_2 \dots p_t, \qquad p_1 \leq p_2 \leq \dots \leq p_t, \tag{1}$$

where each p_k is prime. (When $n = 1$, this equation holds for $t = 0$.) It is unfortunately not a simple matter to find this prime factorization of n, or to determine whether or not n is prime. So far as anyone knows, it is a great deal harder to factor a large number n than to compute the greatest common divisor of two large numbers m and n; therefore we should avoid factoring large numbers whenever possible. But some ingenious ways to simplify the factoring problem have been discovered, and we will now investigate these methods.

Divide and factor. First let us consider the most obvious algorithm for factorization: If $n > 1$, we can divide n by successive primes $p = 2, 3, 5, \dots$ until discovering the smallest p for which $n \bmod p = 0$. Then p is the smallest prime factor of n, and the same process may be applied to $n \leftarrow n/p$ in an attempt to divide this new value of n by p and by higher primes. If at any stage we find that $n \bmod p \neq 0$ but $\lfloor n/p \rfloor \leq p$, we can conclude that n is prime; for if n is not prime then by (1) we must have $n \geq p_1^2$, but condition $p_1 > p$ implies $p_1^2 \geq (p+1)^2 > p(p+1) > p^2 + n \bmod p \geq \lfloor n/p \rfloor p + n \bmod p = n$. This leads us to the following procedure:

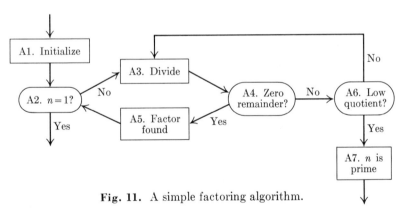

Fig. 11. A simple factoring algorithm.

Algorithm A (*Factoring by division*). Given a positive integer n, this algorithm finds the prime factors $p_1 \leq p_2 \leq \dots \leq p_t$ of n as in Eq. (1). The method makes use of an auxiliary sequence of "trial divisors"

$$2 = d_0 < d_1 < d_2 < d_3 < \dots, \tag{2}$$

which includes all prime numbers $\leq \sqrt{n}$ (and which may also include values which are *not* prime, if it is convenient to do so). The sequence of d's must also include at least one value such that $d_k > \sqrt{n}$.

A1. [Initialize.] Set $t \leftarrow 0$, $k \leftarrow 0$.

A2. [$n = 1$?] If $n = 1$, the algorithm terminates.

A3. [Divide.] Set $q \leftarrow \lfloor n/d_k \rfloor$, $r \leftarrow n \bmod d_k$. (Here q and r are the quotient and remainder obtained when n is divided by d_k.)

A4. [Zero remainder?] If $r \neq 0$, go to step A6.

A5. [Factor found.] Increase t by 1, set $p_t \leftarrow d_k$, set $n \leftarrow q$. Return to step A2.

A6. [Low quotient?] If $q > d_k$, increase k by 1 and return to step A3.

A7. [n is prime.] Increase t by 1, set $p_t \leftarrow n$, and terminate the algorithm. ∎

As an example of Algorithm A, consider the factorization of the number $n = 25852$. We immediately find that $n = 2 \cdot 12926$; hence $p_1 = 2$. Furthermore, $12926 = 2 \cdot 6463$, so $p_2 = 2$. But now 6463 is not divisible by 2, 3, 5, $\ldots$, 19; we find that $n = 23 \cdot 281$, hence $p_3 = 23$. Now $281 = 12 \cdot 23 + 5$ and $12 \leq 23$; hence $p_4 = 281$. This determination of the factors of 25852 requires 12 division operations; on the other hand, if we had tried to factor the slightly smaller number 25849 (which is prime), at least 38 division operations would be necessary. This illustrates the fact that Algorithm A requires a running time roughly proportional to $\max(p_{t-1}, \sqrt{p_t}$ if we set $p_0 = 1$.)

The sequence $d_0, d_1, d_2, \ldots$ of trial divisors used in Algorithm A can be taken to be simply 2, 3, 5, 7, 11, 13, 17, 19, 23, 25, 29, 31, 35, $\ldots$, where we alternately add 2 and 4 after the first three terms. This sequence contains all numbers which are not multiples of 2 or 3; it also includes numbers such as 25, 35, 49, etc., which are not prime, but the algorithm will still give the correct answer. A further savings of 20 percent in computation time can be made by removing the numbers $30m \pm 5$ from the list for $m \geq 1$, thereby eliminating all of the spurious multiples of 5. The exclusion of multiples of 7 shortens the list by 14 percent more, etc. A compact bit table can be used to govern the choice of trial divisors.

If n is known to be small, it is reasonable to have a table of all the necessary primes as part of the program. For example, if n is less than a million, we need only include the 168 primes less than a thousand (followed by the value $d_{168} = 1000$ to terminate the list in case n is a prime larger than 997^2). Such a table can be set up by means of a short auxiliary program, which builds the table just after the factoring program has been loaded into the computer; see Algorithm 1.3.2P, which is reprinted in Appendix A, or see exercise 8.

According to the celebrated "Prime Number Theorem," [see Hardy and Wright, *The Theory of Numbers*, 4th ed. (Oxford, 1960), Chapter 22], the number of primes less than x is asymptotically $x/\ln x$, and tends to be a little higher than this value. For example, we have the following table:

x	primes $< x$	$x/\ln x$
10^3	168	145
10^6	78498	72382
10^9	50847534	48254942

Thus, if we are factoring only six-digit numbers, Algorithm A gives the answer rather quickly; but unfortunately most numbers have more than six digits! If n is a 12-digit prime number, and if the sequence of trial divisors excludes multiples of 2, 3, and 5, Algorithm A requires about 50,000 to 100,000 iterations of steps A3, A4, and A6. When n gets so large that $\sqrt{n}$ cannot fit in one computer word, step A3 becomes a double-precision division, and the amount of computer time rapidly exceeds practical limits.

A modification of Algorithm A has been suggested by G. G. Alway [*Math. Comp.* **6** (1952), 59–60], in which most of the division steps are replaced by a series of five or six additions. This will be an improvement when large numbers are to be factored on computers having a slow division instruction, or on computers which cannot determine $n \bmod d$ without multiplying $\lfloor n/d \rfloor$ by d and subtracting this from n. On some computers the algorithm has reportedly run eight times faster due to this modification. Alway's method is based on the idea that if the trial divisors d_k are in an arithmetic progression, and if they are large enough, then the quotients $q_k = \lfloor n/d_k \rfloor$ will form roughly an arithmetic progression; and hence the remainders $r_k = n - d_k q_k$ will have roughly a quadratic behavior. In order to apply Alway's method, the d_k must be in an arithmetic progression from some point onward; this means we might be doing up to twice as many iterations as necessary, although with appropriate programming it is possible to consider sequences d_k which are obtained by combining several arithmetic progressions having a common difference. For simplicity, Alway's method will be formulated here only for the case $d_{k+1} = d_k + 2$; the more general case $d_{k+1} = d_k + \delta$ is considered in exercise 4.

From the formulas $d_{k+1} = d_k + 2$ and $r_k = n - d_k q_k$, we can easily verify the equation

$$r_{k+1} - 2r_k + r_{k-1} = -4(\nabla q_k) - d_{k+1}(\nabla q_{k+1} - \nabla q_k), \qquad (3)$$

where $\nabla q_k = q_k - q_{k-1}$. Exercise 4 shows that

$$\nabla q_{k+1} - \nabla q_k = -1, \quad 0, \quad 1, \quad \text{or} \quad 2, \quad \text{if} \quad d_k \geq 2\sqrt[3]{n} + 1. \qquad (4)$$

This leads to the following algorithm:

Algorithm B (*Find large odd factor*). Given a positive integer n and an odd number $d \geq 2\sqrt[3]{n} + 1$, this algorithm finds the smallest odd factor f of n such that $d < f \leq \sqrt{n}$, or determines that no such factor exists.

B1. [Initialize.] Set $r \leftarrow n \bmod d$, $r' \leftarrow n \bmod (d-2)$, $q' \leftarrow 4(\lfloor n/(d-2) \rfloor - \lfloor n/d \rfloor)$, $s \leftarrow \lfloor \sqrt{n} \rfloor$. (The quantities d, r, r', q' in this algorithm correspond respectively to $d_k, r_k, r_{k-1}, -4\nabla q_k$ in the formulas of the text.)

B2. [Search complete?] If $d > s$, the algorithm terminates; no factor in the given range exists.

B3. [Step d, r, r'.] Set $d \leftarrow d + 2$, $x \leftarrow r$, $r \leftarrow 2r - r' + q'$, $r' \leftarrow x$.

B4. [$r < 0$?] If $r < 0$, set $r \leftarrow r + d$, increase q' by 4, and go to step B6. (Now r will be nonnegative, by (4).)

B5. $[r \geq d?]$ If $r \geq d$, set $r \leftarrow r - d$, decrease q' by 4, and repeat step B5 again. (This step is repeated at most one more time, because of (4).)

B6. $[r = 0?]$ (At this point $r = n \bmod d$.) If $r = 0$, d is a factor of n; the algorithm terminates with $f = d$. Otherwise, return to step B2. ∎

When n is large, the vast majority of its potential factors are greater than $2\sqrt[3]{n}$; for example, if n is approximately 10^{12}, this algorithm can be used to search for a factor in the range $20{,}000 < f < 1{,}000{,}000$. Upon termination of Algorithm B, the condition $n \bmod d = r$ provides a good check on all of the calculations, since each remainder depends heavily upon the previous ones.

As an example of Algorithm B using small numbers, suppose that $n = 9487$ and $d = 45$. Step B1 sets $r \leftarrow 37$, $r' \leftarrow 27$, $q' \leftarrow 40$, $s \leftarrow 97$. Then in step B3, we set $d \leftarrow 47$, $r \leftarrow 87$, $r' \leftarrow 37$. Step B5 changes r to 40 and q' to 36; hence $9487 \bmod 47 = 40$. On the next iteration, step B3 sets $d \leftarrow 49$, $r \leftarrow 79$, $r' \leftarrow 40$; and then in step B5, $r \leftarrow 30$, $q \leftarrow 32$. The algorithm soon discovers that 9487 is a multiple of 53.

Another modification of Algorithm A, designed to facilitate hand calculation, has been given by N. A. Draim, *Math. Magazine* **25** (1952), 191–194. See also E. W. Dijkstra, *Math. Comp.* **11** (1957), 195–196; G. G. Alway, *Comp. J.* **14** (1971), 166–168.

Fermat's method. Evidently Algorithm A is slow to find large prime factors of n when n is large. Even though a random number n usually has small factors (since n is divisible by 2 with probability $\frac{1}{2}$, by 3 with probability $\frac{1}{3}$, by 5 with probability $\frac{1}{5}$, etc.), it is very unusual for n to have *only* small factors. The number of distinct prime factors of a random integer is asymptotically $\ln \ln n$ [see Hardy and Wright, *The Theory of Numbers*, 4th ed. (Oxford, 1960), Chapter 22]. Algorithm A usually finds a few factors and then begins a long-drawn-out search for the others.

Another approach to the factoring problem, which was used by Pierre de Fermat in 1643, is more suited to finding large factors than small ones. [Fermat's description of his method, translated into English, appears in L. E. Dickson's *History of the Theory of Numbers* **1** (New York: Chelsea, 1952), 357.]

Assume that $n = uv$, where $u \leq v$. For practical purposes we may assume that n is odd; this means that u and v also are odd. We can therefore let

$$x = (u + v)/2, \qquad y = (v - u)/2, \tag{5}$$
$$n = x^2 - y^2, \qquad 0 \leq y < x \leq n. \tag{6}$$

Fermat's method consists of searching for values of x and y that satisfy this equation; the following algorithm shows how factoring can therefore be done *without using any division:*

Algorithm C (*Factoring by addition and subtraction*). Given an odd number n, this algorithm determines the largest factor of n less than or equal to $\sqrt{n}$.

C1. [Initialize.] Set $x' \leftarrow 2\lfloor \sqrt{n} \rfloor + 1$, $y' \leftarrow 1$, $r \leftarrow \lfloor \sqrt{n} \rfloor^2 - n$. (During this algorithm x', y', r correspond respectively to $2x + 1$, $2y + 1$, $x^2 - y^2 - n$ in the formulas above.)

C2. [Test r.] If $r \le 0$, go to step C4.

C3. [Step y.] Set $r \leftarrow r - y'$, $y' \leftarrow y' + 2$, and return to C2.

C4. [Done?] If $r = 0$, the algorithm terminates; we have

$$n = ((x' - y')/2)((x' + y' - 2)/2),$$

and $(x' - y')/2$ is the largest factor of n less than or equal to $\sqrt{n}$.

C5. [Step x.] Set $r \leftarrow r + x'$, $x' \leftarrow x' + 2$, and return to C3. ∎

The number of steps taken by this algorithm to find the factors u, v of $n = uv$ is essentially proportional to $(x' + y' - 2)/2 - \lfloor \sqrt{n} \rfloor = v - \lfloor \sqrt{n} \rfloor$; this can, of course, be a very large number, although each step can be done very rapidly on many computers. The main loop (steps C2 and C3) can often be done with r and y' kept in registers of the machine; for example, if we suppose that MIX has large enough index registers (see exercise 1.3.1–25), the code for the main loop of the computation is

```
C5  INCA  0,2     r ← r + x'.
    INC2  2       x' ← x' + 2.
C3  DECA  0,1     r ← r − y'.
    INC1  2       y' ← y' + 2.
C2  JAP   C3      To C3 if r > 0.
C4  JAN   C5      To C5 if r < 0.  ∎
```

Here r, y', x' are being kept in registers A, I1, and I2, respectively; only three cycles are required to increment x' or y'.

As an example of Algorithm C using small numbers, let us suppose that $n = 9401$. Then in step C1 we begin with $x' \leftarrow 193$, $y' \leftarrow 1$, $r \leftarrow -185$. The computation proceeds as follows:

r	y'	x'	r	y'	x'	r	y'	x'
-185	1	193	167	13	197	7	29	197
8	1	195	154	15	197	-22	31	197
7	3	195	139	17	197	175	31	199
4	5	195	122	19	197	144	33	199
-1	7	195	103	21	197	111	35	199
194	7	197	82	23	197	76	37	199
187	9	197	59	25	197	39	39	199
178	11	197	34	27	197	0	41	199

Thus $9401 = 79 \cdot 119$.

It is not quite correct to call Algorithm C "Fermat's method"; in spite of the fact that the main loop is quite fast on computers, it is not very suitable for hand calculation. Fermat actually did not keep the running value of y; he would look at $x^2 - n$ and tell whether or not this quantity was a perfect square by looking at its least significant digits. (The last two digits of a perfect square must be 00, $a1$, $a4$, 25, $b6$, or $a9$, where a is an even digit and b is an odd digit.)

Therefore he avoided the operations of steps C2 and C3, replacing them by an occasional determination that a certain number is not a perfect square.

Fermat's method of looking at the rightmost digits can, of course, be generalized by using other moduli. Suppose for clarity that $n = 11111$, and consider the following table:

m	if $x \bmod m$ is	then $x^2 \bmod m$ is	and $(x^2 - n) \bmod m$ is
3	$0, 1, 2$	$0, 1, 1$	$1, 2, 2,$
5	$0, 1, 2, 3, 4$	$0, 1, 4, 4, 1$	$4, 0, 3, 3, 0$
7	$0, 1, 2, 3, 4, 5, 6$	$0, 1, 4, 2, 2, 4, 1$	$5, 6, 2, 0, 0, 2, 6$
8	$0, 1, 2, 3, 4, 5, 6, 7$	$0, 1, 4, 1, 0, 1, 4, 1$	$1, 2, 5, 2, 1, 2, 5, 2$
11	$0, 1, 2, 3, 4, 5, 6, 7, 8, 9, 10$	$0, 1, 4, 9, 5, 3, 3, 5, 9, 4, 1$	$10, 0, 3, 8, 4, 2, 2, 4, 8, 3, 0$

If $x^2 - n$ is to be a perfect square y^2, it must have a residue mod m consistent with this fact, for all m. For example, if $x \bmod 3 \neq 0$, then for $n = 11111$, $(x^2 - n) \bmod 3 = 2$, so $x^2 - n$ cannot be a perfect square; therefore x must be a multiple of 3 whenever $11111 = x^2 - y^2$. The table tells us that

$$
\begin{aligned}
x \bmod 3 &= 0; \\
x \bmod 5 &= 0, 1, \text{ or } 4; \\
x \bmod 7 &= 2, 3, 4, \text{ or } 5; \qquad\qquad (7) \\
x \bmod 8 &= 0 \text{ or } 4 \ (\text{hence } x \bmod 4 = 0); \\
x \bmod 11 &= 1, 2, 4, 7, 9, \text{ or } 10.
\end{aligned}
$$

This narrows down the search for x considerably. For example, x must be a multiple of 12. We must have $x \geq \lceil \sqrt{n} \rceil = 106$, and it is easy to verify that the first value of $x \geq 106$ which satisfies all of the conditions in (7) is $x = 144$. Now $144^2 - 11111 = 9625$, and by taking the square root of 9625, we find it is not a square. The first value of $x > 144$ which satisfies (7) is $x = 156$. In this case $156^2 - 11111 = 13225 = 115^2$; so we have the desired solution $x = 156$, $y = 115$. This calculation shows that $11111 = 41 \cdot 271$. The hand calculations involved in this example are comparable to the amount of work required to divide 11111 by 13, 17, 19, 23, 29, 31, 37, and 41, even though the factors 41 and 271 are not very close to each other; thus we can see the advantages of Fermat's method.

In place of the moduli considered in (7) we can use any powers of distinct primes. For example, if we had used 25 in place of 5, we would find that the possible values of $y^2 \bmod 25$ are

$$0, 1, 4, 9, 16, 0, 11, 24, 14, 6, 0, 21, 19, 19, 21, 0, 6, 14, 24, 11, 0, 16, 9, 4, 1$$

while the values of $(x^2 - n) \bmod 25$ are

$$14, 15, 18, 23, 5, 14, 0, 13, 3, 20, 14, 10, 8, 8, 10, 14, 20, 3, 13, 0, 14, 5, 23, 18, 15;$$

so the only permissible values of $x \bmod 25$ are 0, 5, 6, 10, 15, 19, and 20.

This gives more information than (7). In general, we will get more information modulo p^2 than modulo p, for odd primes p, when $x^2 - n \equiv 0$ (modulo p) has a solution x.

The modular method just used is called a *sieve procedure*, since we can imagine passing all integers x through a "sieve" for which only those values with $x \bmod 3 = 0$ come out, then sifting these numbers through another sieve which allows only numbers with $x \bmod 5 = 0$, 1, or 4 to pass, etc. Each sieve by itself will remove about half of the remaining values (see exercise 6); and when we sieve with respect to moduli which are relatively prime in pairs, each sieve is independent of the others because of the Chinese Remainder Theorem (Theorem 4.3.2C). So if we sieve with respect to, say, 30 different primes, only about one value in every 2^{30} will need to be examined to see if $x^2 - n$ is a perfect square y^2.

Algorithm D (*Factoring with sieves*). Given an odd number n, this algorithm determines the largest factor of n less than or equal to $\sqrt{n}$. The procedure uses moduli $m_1, m_2, \ldots, m_r$, which are relatively prime to each other in pairs and relatively prime to n. We assume that r "sieve tables" $S[i, j]$ for $0 \le j < m_i$, $1 \le i \le r$, have been prepared, where

$$S[i, j] = \begin{cases} 1, & \text{if } \quad j^2 - n \equiv y^2 \text{ (modulo } m_i \text{) has a solution } y; \\ 0, & \text{otherwise.} \end{cases} \qquad (8)$$

D1. [Initialize.]   Set $x \leftarrow \lceil \sqrt{n} \rceil$, and $k_i \leftarrow (-x) \bmod m_i$ for $1 \le i \le r$. (Throughout this algorithm the index variables $k_1, k_2, \ldots, k_r$ will be set so that $(-x) \bmod m_i = k_i$.)

D2. [Sieve.]   If $S[i, k_i] = 1$ for $1 \le i \le r$, go to step D4.

D3. [Step x.]   Set $x \leftarrow x + 1$, and $k_i \leftarrow (k_i - 1) \bmod m_i$ for $1 \le i \le r$, and return to step D2.

D4. [Test $x^2 - n$.]   Set $y \leftarrow \lfloor \sqrt{x^2 - n} \rfloor$ or to $\lceil \sqrt{x^2 - n} \rceil$. If $y^2 = x^2 - n$, then $(x - y)$ is the desired factor, and the algorithm terminates. Otherwise return to step D3. ∎

If we compare this algorithm to Algorithm C, which is based on the same underlying idea, we find that Algorithm C requires $x + y - \sqrt{n}$ iterations while Algorithm D requires $x - \sqrt{n}$. Let us compare these running times when $n = uv$, where both u and v are primes. (This is an important case, for even though it is improbable that a random number has this form, such values of n eventually arise during the factorization of most numbers.) If we wish to search for u and v, there are essentially $\sqrt{n}$ divisors to be investigated; half of these are less than $\frac{1}{2}\sqrt{n}$, and half of them are greater than $\frac{1}{2}\sqrt{n}$. If

$$u = \frac{1}{\rho}\sqrt{n} \qquad \text{and} \qquad v = \rho\sqrt{n},$$

then

$$\text{Algorithm C requires } (\rho - 1)\sqrt{n} \text{ iterations;}$$

$$\text{Algorithm D requires } \frac{1}{2}\left(\rho - 2 + \frac{1}{\rho}\right)\sqrt{n} \text{ iterations.}$$

Thus if ρ is 10 or more, Algorithm C takes about twice as many iterations as Algorithm D; if $\rho = 2$, Algorithm C takes about four times as many iterations; and if $\rho = 1 + \epsilon$, where ϵ is small, Algorithm C takes $\epsilon\sqrt{n}$ iterations compared to about $\frac{1}{2}\epsilon^2\sqrt{n}$ for Algorithm D. The division method of Algorithm A takes between $2\sqrt{n}/\rho \ln n$ and $\sqrt{n}/4\rho$ iterations to discover the factorization.

The key factor in the running time is, of course, the speed with which each iteration can be done. We have seen that the main loop of Algorithm C can be made quite fast; on the other hand, the main loop of Algorithm D (which consists of steps D2 and D3) involves r incrementations of k_i, and an average of approximately two inspections of the table entries $S[i, k_i]$. This is clearly more calculation than required by Algorithm C, but there are several ways to speed things up. For example, we have seen that if $n \bmod 3 = 2$, then x must be a multiple of 3; we can set $x = 3x'$, and use a different sieve corresponding to x', increasing the speed by a factor of 3. If $n \bmod 3 = 1$, x must be congruent to ± 1 (modulo 9), so we can run two sieves (for x' and x'', where $x = 9x' + 1$ and $x = 9x'' - 1$) to increase the speed by a factor of $4\frac{1}{2}$. If $n \bmod 4 = 3$, then x must be a multiple of 4 and the speed is increased by an additional factor of 4; in the other case, when $x \bmod 4 = 1$, x must be odd so the speed may be doubled. Another way to double the speed of the algorithm (at the expense of storage space) is to combine pairs of moduli, using $m_{r-k}m_k$ in place of m_k for $1 \le k < \frac{1}{2}r$.

An even more important method of speeding up Algorithm C is to use the "Boolean operations" found on most binary computers. Let us assume for example that MIX is a binary computer with 30 bits per word. The tables $S[i, k_i]$ can be kept in memory with one bit per entry; thus 30 values can be stored in a single word. The operation AND, which replaces the kth bit of the accumulator by zero if the kth bit of a specified word in memory is zero, for $1 \le k \le 30$, can be used to process 30 values of x at once! For convenience, we can make several copies of the tables $S[i, j]$ so that the table entries for m_i involve lcm $(m_i, 30)$ bits; then the sieve tables for each modulus fill an integral number of words. Under these assumptions, 30 executions of the main loop in Algorithm D are equivalent to code of the following form:

D2	LD1	K1	$rI1 \leftarrow k_1$.
	LDA	S1,1	$rA \leftarrow S[1, rI1]$.
	DEC1	1	$rI1 \leftarrow rI1 - 1$.
	J1NN	*+2	
	INC1	M1	If $rI1 < 0$, set $rI1 \leftarrow \text{lcm}(m_1, 30) - 1$.
	ST1	K1	$k_1 \leftarrow rI1$.

```
LD1    K2         rI1 ← k₂.
AND    S2,1       rA ← rA "and" S[2, rI1].
DEC1   1          rI1 ← rI1 − 1.
J1NN   *+2
INC1   M2         If rI1 < 0, set rI1 ← lcm(m₂, 30) − 1.
ST1    K2         k₂ ← rI1.
...
ST1    Kr         kᵣ ← rI1.
INCX   30         x ← x + 30.
JAZ    D2         Repeat if all sieved out. ▮
```

The number of cycles for 30 iterations is essentially $2 + 8r$; if $r = 11$ this means three cycles are being used on each iteration, just as in Algorithm C, and Algorithm C often uses quite a few more iterations.

If the table entries for m_i do not come out to be an integral number of words, further shifting of the table entries will be necessary on each iteration so that the bits are aligned properly. This would add quite a lot of coding to the main loop and it would probably make the program too slow to compete with Algorithm C (see exercise 7).

Sieve procedures can be applied to a variety of other problems, not necessarily having much to do with arithmetic; a survey of these techniques has been prepared by Marvin C. Wunderlich, *JACM* **14** (1967), 10–19.

Special sieve machines (of reasonably low cost) have been constructed by D. H. Lehmer and his associates over a period of many years; see for example *AMM* **40** (1933), 401–406. Lehmer's electronic delay-line sieve which began operating in 1965 processes one million numbers per second; thus, each iteration of the loop in Algorithm D can be performed in one microsecond on this device.

Testing for primality. None of the algorithms we have discussed so far is an efficient way to determine that a large number n is prime. Fortunately, there are other methods available for settling this question; efficient methods for this purpose have been devised by É. Lucas and by D. H. Lehmer [see *Bulletin Amer. Math. Soc.* **33** (1927), 327–340].

According to Fermat's theorem (Theorem 1.2.4F), $x^{p-1} \bmod p = 1$ if p is prime and if x is not a multiple of p. Furthermore, there are efficient ways to calculate $x^{n-1} \bmod n$, requiring only $O(\log n)$ operations of multiplication mod n. Therefore we can often determine that n is *not* prime when this relationship fails.

For example, Fermat verified that the numbers $2^1 + 1$, $2^2 + 1$, $2^4 + 1$, and $2^{16} + 1$ are prime. In a letter to Mersenne written in 1640, Fermat conjectured that $2^{2^n} + 1$ is always prime, but said he was unable to determine definitely whether $4294967297 = 2^{32} + 1$ is a prime or not. We can compute $3^{2^{32}} \bmod (2^{32} + 1)$ by using 32 operations of squaring a number modulo $2^{32} + 1$, and the answer is 3029026160; therefore (by Fermat's own theorem, which he discovered in the same year 1640!) we know that $2^{32} + 1$ is *not* prime. This argument gives us absolutely no idea what the factors are, but it answers Fermat's question.

This technique is a powerful test for showing that n is not a prime. When n is not prime, it is always possible to find a value of $x < n$ such that $x^{n-1} \bmod n \neq 1$; experience shows that in fact such a value can almost always be found very quickly. There are some rare values of n for which $x^{n-1} \bmod n$ is frequently equal to unity, but then n has a factor less than $\sqrt[3]{n}$; see exercise 9. One example is $n = 3 \cdot 11 \cdot 17 = 561$; here $\lambda(n) = \mathrm{lcm}\,(2, 10, 16) = 80$ in the notation of Theorem 3.2.1.2B, so $x^{80} \bmod 561 = 1 = x^{560} \bmod 561$ whenever x is relatively prime to 561.

The same method can be extended to determine rapidly if n *is* a prime, by using the following idea: *If there is a number x for which the order of x modulo n is equal to $n - 1$, then n is prime.* (The order of x modulo n is the smallest positive integer k such that $x^k \bmod n = 1$; see Section 3.2.1.2.) For this condition implies that the numbers $x^k \bmod n$ for $1 \leq k \leq n - 1$ are distinct and relatively prime to n, so they must be the numbers $1, 2, \ldots, n - 1$ in some order; thus n has no proper divisors. If n is prime, such a number x (a "primitive root" of n) will always exist; see exercise 3.2.1.2–16. In fact, the number of such primitive roots is rather numerous; there are $\varphi(n - 1) \approx 6n/\pi^2$ of them.

It is unnecessary to calculate $x^k \bmod n$ for all $k \leq n - 1$ to determine if the order of x is $n - 1$ or not. The order of x will be $n - 1$ if and only if

i) $x^{n-1} \bmod n = 1$
ii) $x^{(n-1)/p} \bmod n \neq 1$ for all primes p which divide $n - 1$.

For $x^s \bmod n = 1$ if and only if s is a multiple of the order of x modulo n. If the two conditions hold, and if k is the order of x modulo n, we therefore know that k is a divisor of $n - 1$, but not a divisor of $(n - 1)/p$ for any prime factor p of $n - 1$; so k must be equal to $n - 1$. This completes the proof that conditions (i) and (ii) suffice to establish the primality of n; see also exercise 10 for an improvement over this testing method.

Conditions (i) and (ii) can be tested efficiently by using the rapid methods for evaluating powers of a number discussed in Section 4.6.3. But it is necessary to know the prime factors of $n - 1$, so we have an interesting situation in which the factorization of n depends on that of $n - 1$!

Another test for primality which uses the factorization of $n + 1$ instead of $n - 1$ is discussed in exercise 15. Such a test can be attempted when the factorization of $n - 1$ is too difficult to complete.

Summary and examples. The remarks made so far in this section now lead us to the following method for factoring a large number n:

1. First see if n has any prime factors less than (say) 1000, by dividing n by all such primes. If so, use the method of Algorithm A to cast out these factors and to reduce n to a number whose prime factors all exceed 1000. (The factorization will now be complete if the new value of n is less than one million $= 1000^2$.)

2. Test if n is prime by computing $3^{n-1} \bmod n$. (Computational experience shows that $3^{n-1} \bmod n \neq 1$ for almost all nonprime n; an exception is $n = \frac{1}{7}(2^{28} - 9) = 2341 \cdot 16381$.) If $3^{n-1} \bmod n \neq 1$, n is not prime, so go to step 4.

3. (Now we want to try to prove that n is prime.) Find the factors of $n - 1$; in other words, start recursively at step 1, with n replaced by $n - 1$, and come back to this point of the procedure when $n - 1$ has been completely factored. Then for each prime factor p of $n - 1$, find a value of $x = 2, 3, 5, 7, 11, \ldots$ such that $x^{n-1} \bmod n = 1$ but $x^{(n-1)/p} \bmod n \neq 1$. (Exercise 10 shows that different values of x may be used for different primes p. We may restrict consideration to primes x, since the order of uv modulo n divides the least common multiple of the orders of u and v.) Continue this process until either finding $x^{n-1} \bmod n \neq 1$ (then go to step 4), or finding a value of x for each specified prime p (then n is prime), or until finding some p dividing $n - 1$ such that $x^{(n-1)/p} \bmod n = 1$ for all primes x less than 1000. (The latter case almost never occurs; we may go to step 4 if it does happen.)

4. Use Algorithm A to cast out any factors of n which are not greater than $\sqrt[3]{n}$. If any factors are found in this way, reduce n and continue again at step 2.

5. (We may now assume that n is the product of two primes.) Prepare for the sieve method, Algorithm D, and test for factors near $\sqrt{n}$ using this method. Algorithm A can also be continued, running concurrently with the sieve method (for example, by putting a division by d_k into the inner loop of Algorithm D).

In step 5 we will discover a large divisor by the sieve method, while a small divisor will be discovered by division; with luck n will be completely factored in a reasonable amount of time.

The easiest way to understand the issues which are involved in factoring large numbers is to consider a typical example; let us try to find the prime factors of $2^{214} + 1$. The factorization of this 65-digit number can be initiated by noticing that

$$2^{214} + 1 = (2^{107} - 2^{54} + 1)(2^{107} + 2^{54} + 1); \qquad (9)$$

this identity is a special case of some factorizations discovered by A. Aurifeuille in 1873 [see Dickson's *History*, **1**, p. 383]. We may now examine each of the 33-digit factors in (9).

A computer program readily discovers that $2^{107} - 2^{54} + 1 = 5 \cdot 857 \cdot n_0$, where

$$n_0 = 3786680906166005726419253397 \qquad (10)$$

is a 29-digit number having no prime factors less than 1000. A multiple-precision calculation using the "binary method" of Section 4.6.3 shows that

$$3^{n_0 - 1} \bmod n_0 = 1,$$

so we suspect that n_0 is prime. It is certainly out of the question to prove that n_0 is prime by trying the 10 million million or so potential divisors, but the method of step 3 gives a feasible test for primality: our next goal is to factor $n_0 - 1$. With little difficulty, a computer program finds that

$$n_0 - 1 = 2 \cdot 2 \cdot 19 \cdot 107 \cdot 353 \cdot n_1, \quad n_1 = 13191270754108226049301. \qquad (11)$$

Here $3^{n_1-1} \bmod n_1 \neq 1$, so n_1 is not prime; continuation of Algorithm A shows that

$$n_1 = 91813 \cdot n_2, \qquad n_2 = 1436754136657196977. \tag{12}$$

This time $3^{n_2-1} \bmod n_2 = 1$, so we will try to prove that n_2 is prime. This requires the factorization

$$n_2 - 1 = 2 \cdot 2 \cdot 2 \cdot 2 \cdot 3 \cdot 3 \cdot 547 \cdot n_3, \qquad n_3 = 1824032775457. \tag{13}$$

Since $3^{n_3-1} \bmod n_3 \neq 1$, we know that n_3 is composite, and Algorithm A finds that

$$n_3 = 1103 \cdot n_4, \qquad n_4 = 1653701519. \tag{14}$$

The number n_4 behaves like a prime (i.e., $3^{n_4-1} \bmod n_4 = 1$), so we calculate

$$n_4 - 1 = 2 \cdot 7 \cdot 19 \cdot 23 \cdot 137 \cdot 1973. \tag{15}$$

This is our first complete factorization; we are ready to backtrack to the previous subproblem, proving that n_4 is prime. The following values are now computed, according to the procedure of exercise 10:

x	p	$x^{(n_4-1)/p} \bmod n_4$	$x^{n_4-1} \bmod n_4$	
2	2	1		
2	7	766408626	1	
2	19	332952683	1	
2	23	1154237810	1	
2	137	373782186	1	(16)
2	1973	490790919	1	
3	2	1		
5	2	1		
7	2	1653701518	1	

Thus n_4 is prime, and $n_2 - 1$ has been completely factored. A similar calculation shows that n_2 is prime, and this complete factorization of $n_0 - 1$ finally shows [after still another calculation like (16)] that n_0 is prime.

The next quantity to be factored is the other half of (9),

$$n_5 = 2^{107} + 2^{54} + 1. \tag{17}$$

Since $3^{n_5-1} \bmod n_5 \neq 1$, we know that n_5 is not prime, and division shows that

$$n_5 = 843589 \cdot n_6,$$
$$n_6 = 1923439931402772993096491917. \tag{18}$$

Unfortunately, $3^{n_6-1} \bmod n_6 \neq 1$, so we are left with a 27-digit nonprime number. Division shows that n_6 has no factors less than its cube root, so it is the product of two primes. It is time to invoke the sieve method; Algorithm D and the modifications discussed above yield

$$n_6 = 8174912477117 \cdot 23528569104401. \tag{19}$$

This result could *not* have been discovered by Algorithm A in a reasonable length of time.

The calculations are complete; the prime factorization of $2^{214} + 1$ is

$$5 \cdot 857 \cdot 843589 \cdot 8174912477117 \cdot 23528569104401 \cdot n_0, \tag{20}$$

where n_0 is the 29-digit prime in (10). A certain amount of good fortune entered into these calculations, because numbers of 27 or more digits often cannot be treated with this factorization method, even with the speeds of modern computers. If we had not started with the known factorization (9), it is quite probable that this factorization would never have been completed because of the order in which the factorization was tried. Algorithm D would have discovered (9), but the procedure we have recommended would first have used Algorithm A to cast out the small factors, reducing n to $n_6 n_0$; this 55-digit number cannot be factored in a reasonable length of time by any of the algorithms we have discussed so far.

Dozens of further numerical examples, and additional comments on this factoring method, can be found in an article by John Brillhart and J. L. Selfridge, *Math. Comp.* **21** (1967), 87–96.

Legendre's method. Numbers which cannot be factored by the algorithms we have seen so far can sometimes be "cracked" by using another technique which is essentially due to A. M. Legendre [*Théorie des nombres* (Paris, 1798), 313–320]. The idea in this case is to search for numbers x and y such that

$$x^2 \equiv y^2 \text{ (modulo } n), \quad 0 < x, y < n, \quad x \neq y, \quad x + y \neq n. \tag{21}$$

Fermat's method imposes the stronger requirement $x^2 - y^2 = n$, but actually the congruence (21) is enough to split n into factors: It implies that n is a divisor of $x^2 - y^2 = (x - y)(x + y)$, yet n divides neither $x - y$ nor $x + y$; hence $\gcd(n, x - y)$ and $\gcd(n, x + y)$ are proper factors of n which can be found by Euclid's algorithm.

One way to discover solutions of (21) is to look for values of x such that $x^2 \equiv a \text{ (modulo } n)$, for small values of $|a|$. As we will see, it is often a simple matter to piece together solutions of this congruence to obtain solutions of (21). Now if $x^2 = a + knd^2$ for some k and d, with small $|a|$, the fraction x/d is a good approximation to $\sqrt{kn}$; conversely, if x/d is an especially good approximation to $\sqrt{kn}$, $|x^2 - knd^2|$ will be small. This observation suggests looking at the continued fraction expansion of $\sqrt{kn}$, since we have seen (Eq. 4.5.3–12) that continued fractions yield good rational approximations.

Continued fractions for quadratic irrationalities have many pleasant properties, which are proved in exercise 4.5.3–12. The algorithm below makes use of these properties to derive solutions to the congruence

$$x^2 \equiv (-1)^{e_0} p_1^{e_1} p_2^{e_2} \cdots p_t^{e_t} \pmod{n}. \tag{22}$$

Here we use the small primes $p_1 = 2$, $p_2 = 3$, $\ldots$, up to p_t, where t is perhaps 20, or perhaps t is as large as 100 or more. (Only primes p such that either $p = 2$ or $(kn)^{(p-1)/2} \bmod p \leq 1$ should appear in this list, since other primes will never be factors of the numbers generated by the algorithm; see exercise 14. It is perhaps a good idea to choose k so that most small primes satisfy this condition.)   If $(x_1, e_{01}, e_{11}, \ldots, e_{t1})$, $\ldots$, $(x_r, e_{0r}, e_{1r}, \ldots, e_{tr})$ are solutions of (22) such that the vector sum

$$(e_{01}, e_{11}, \ldots, e_{t1}) + \cdots + (e_{0r}, e_{1r}, \ldots, e_{tr}) = (2e_0', 2e_1', \ldots, 2e_t') \tag{23}$$

is *even* in each component, then

$$x = (x_1 \ldots x_r) \bmod n, \qquad y = ((-1)^{e_0'} p_1^{e_1'} \ldots p_t^{e_t'}) \bmod n \tag{24}$$

yields a solution to (21), except for the possibility that $x \equiv \pm y$. Condition (23) is equivalent to saying that the vectors are linearly dependent modulo 2, so we must have a solution to (23) if we have found at least $t + 2$ solutions to (22).

Algorithm E (*Factoring via continued fractions*).   Given a positive integer n which is not a perfect square, and a positive integer k which is not divisible by the square of any prime, this algorithm attempts to discover solutions to the congruence (22) for fixed t, by analyzing the convergents to the continued fraction for $\sqrt{kn}$.

E1. [Initialize.]   Set $D \leftarrow kn$, $R \leftarrow \lfloor \sqrt{D} \rfloor$, $U \leftarrow U' \leftarrow R$, $V \leftarrow 1$, $V' \leftarrow D - R^2$, $P \leftarrow R$, $P' \leftarrow 1$, $A \leftarrow 0$, $S \leftarrow 0$. (This algorithm follows the general procedure of exercise 4.5.3–12, finding the continued fraction expansion of $\sqrt{kn}$. The variables U, U', V, V', P, P', A, S represent, respectively, U_j, U_{j-1}, V_j, V_{j-1}, $Q_{j+1}(A_0, A_1, \ldots, A_j) \bmod n$, $Q_j(A_0, A_1, \ldots, A_{j-1}) \bmod n$, A_j, and $j \bmod 2$.)

E2. [Step U, V, S.]   Set $T \leftarrow V$, $V \leftarrow A(U' - U) + V'$, $V' \leftarrow T$, $A \leftarrow \lfloor (R + U)/V \rfloor$, $U' \leftarrow U$, $U \leftarrow AV - U$, $S \leftarrow 1 - S$.

E3. [Factor V.]   Set $(e_0, e_1, \ldots, e_t) \leftarrow (S, 0, \ldots, 0)$, $T \leftarrow V$.   Now do the following, for $1 \leq j \leq t$: If $T \bmod p_j = 0$, set $T \leftarrow T/p_j$ and $e_j \leftarrow e_j + 1$, and repeat this process until $T \bmod p_j \neq 0$.

E4. [Solution?]   If $T = 1$, output the values $(P, e_0, e_1, \ldots, e_t)$, which comprise a solution to (22). (This step may profitably be combined with step E3 when t is large.)

E5. [Step P, P'.] Set $T \leftarrow P$, $P \leftarrow (AP + P') \bmod n$, $P' \leftarrow T$. If $V \neq 1$ or $U \neq R$, return to E2. Otherwise the continued fraction process has started to repeat its cycle, except perhaps for S, so the algorithm terminates. ∎

In practice, Algorithm E might be run for various values of $k = 1, 2, 3, 5,$ 6, 7, 10, 11, etc., in an attempt to factor a large number n. Another algorithm, which uses the outputs $(P, e_0, e_1, \ldots, e_t)$ to complete the factoring process, is the subject of exercise 12.

We can illustrate the application of Algorithm E to relatively small numbers by considering the case $n = 197209$, $k = 1$, $t = 3$, $p_1 = 2$, $p_2 = 3$, $p_3 = 5$. The computation proceeds as follows:

	U	V	A	P	S	T	Output
After E1:	444	1	0	444	0	—	
After E4:	432	73	12	444	1	73	
After E4:	438	145	6	5329	0	29	
After E4:	413	37	23	32418	1	37	
After E4:	307	720	1	159316	0	1	$159316^2 \equiv 2^4 \cdot 3^2 \cdot 5^1$
After E4:	408	143	5	191734	1	143	
After E4:	237	215	3	131941	0	43	
After E4:	419	656	1	193139	1	41	
After E4:	439	33	26	127871	0	11	
After E4:	377	136	6	165232	1	17	
After E4:	433	405	2	133218	0	1	$133218^2 \equiv 2^0 \cdot 3^4 \cdot 5^1$
After E4:	431	24	36	37250	1	1	$37250^2 \equiv -2^3 \cdot 3^1 \cdot 5^0$
After E4:	46	477	1	93755	0	53	

Continuing the computation gives 25 outputs in the first 100 iterations; in other words, the algorithm is finding solutions quite rapidly. Some of the solutions are trivial; for example, if the above computation were continued 13 more times, we would obtain the output $197197^2 \equiv 2^4 \cdot 3^2 \cdot 5^0$, which is of no interest since $197197 \equiv -12$. But the first two solutions above are already enough to complete the factorization: We have found that

$$(159316 \cdot 133218)^2 \equiv (2^2 \cdot 3^3 \cdot 5^1)^2 \qquad (\text{modulo } 197209);$$

hence (21) holds with $x = 159316 \cdot 133218 \bmod 197209 = 126308$, $y = 540$. By Euclid's algorithm, $\gcd(126308 - 540, 197209) = 199$; hence we obtain the pretty factorization

$$197209 = 199 \cdot 991. \qquad (25)$$

This method was used by John Brillhart and Michael E. Morrison in 1971 to discover that $2^{128} + 1 = 59649589127497217 \cdot 5704689200685129054721$.

The theoretical results we have derived in Section 4.5.3 give us a little information about the expected behavior of Algorithm E. The value of variable V will in no case exceed $2\sqrt{kn}$. Assuming that the continued fraction expansion of $\sqrt{kn}$ is essentially random, the probability that $V \leq a$ is a little less than $a/\sqrt{kn}$ when a is small compared to $\sqrt{kn}$.

Even when the continued-fraction method fails to give a satisfactory solution to the congruence $x^2 \equiv y^2$ (modulo n), it can often be used to facilitate factoring by other methods. For example, $n = 2^{101} - 1$ was for many years the smallest number of the form $2^e - 1$ whose least prime factor was not explicitly known, and it seemed to defy attempts at factorization. Several runs of Algorithm E failed to produce any factors, but they did succeed in discovering solutions to the congruences

$$x_1^2 \equiv 2, \quad x_2^2 \equiv -3, \quad x_3^2 \equiv 5, \quad x_4^2 \equiv 7, \quad x_5^2 \equiv -17 \quad \text{(modulo } n\text{)}.$$

These congruences must hold modulo any prime factor p of n. The law of quadratic reciprocity (exercise 1.2.4–47) now implies that

$$p \equiv \pm 1 \text{ (modulo 8)}; \quad p \equiv 1 \text{ (modulo 3)}; \quad p \equiv \pm 1 \text{ (modulo 5)};$$

$$p \equiv \pm 1, \pm 3, \text{ or } \pm 9 \text{ (modulo 28)};$$

$$p \equiv 1, 3, 7, 9, 11, 13, 21, 23, 25, 27, 31, 33, 39, 49, 53, \text{ or } 63 \text{ (modulo 68)};$$

and our other factorization methods can be speeded up with the help of this information. The difficult factorization

$$2^{101} - 1 = 7432339208719 \cdot 341117531003194129$$

was obtained in this way by D. H. Lehmer, John Brillhart, and Gerald D. Johnson.

For further information about factorization via continued fractions, and about many other interesting computational techniques for number-theoretical problems, see the important paper by D. H. Lehmer in *Studies in Number Theory* (1969), 117–151. See also M. Kraitchik, *Recherches sur la Théorie des Nombres* **2** (Paris, 1929).

A completely new method of factorization which requires only $O(N^{(1/4)+\epsilon})$ steps, but with a fairly high constant of proportionality, was discovered about 1968 by Daniel Shanks [*Proc. Symp. Pure Math.* **20** (Amer. Math. Soc., 1971), 415–440. His method is based on composition of binary quadratic forms; the associated theory is unfortunately beyond the scope of this book.

The largest known primes. We have discussed several computational methods elsewhere in this book which require the use of large prime numbers, and the techniques described in this section can be used to discover primes of, say, 24 digits or less, with relative ease. Table 1 shows the ten largest primes which are less than the word size of typical computers.

Actually much larger primes of special forms are known, and it is occasionally important to find primes which are as large as possible. Let us therefore conclude this section by investigating the interesting manner in which the largest explicitly known primes have been discovered. Such primes are of the form $2^n - 1$, for various special values of n, and so they are especially suited to certain applications of binary computers.

Table 1

USEFUL PRIME NUMBERS

N	a_1	a_2	a_3	a_4	a_5	a_6	a_7	a_8	a_9	a_{10}
2^{15}	19	49	51	55	61	75	81	115	121	135
2^{16}	15	17	39	57	87	89	99	113	117	123
2^{17}	1	9	13	31	49	61	63	85	91	99
2^{18}	5	11	17	23	33	35	41	65	75	93
2^{19}	1	19	27	31	45	57	67	69	85	87
2^{20}	3	5	17	27	59	69	129	143	153	185
2^{21}	9	19	21	55	61	69	105	111	121	129
2^{22}	3	17	27	33	57	87	105	113	117	123
2^{23}	15	21	27	37	61	69	135	147	157	159
2^{24}	3	17	33	63	75	77	89	95	117	167
2^{25}	39	49	61	85	91	115	141	159	165	183
2^{26}	5	27	45	87	101	107	111	117	125	135
2^{27}	39	79	111	115	135	187	199	219	231	235
2^{28}	57	89	95	119	125	143	165	183	213	273
2^{29}	3	33	43	63	73	75	93	99	121	133
2^{30}	35	41	83	101	105	107	135	153	161	173
2^{31}	1	19	61	69	85	99	105	151	159	171
2^{32}	5	17	65	99	107	135	153	185	209	267
2^{33}	9	25	49	79	105	285	301	303	321	355
2^{34}	41	77	113	131	143	165	185	207	227	281
2^{35}	31	49	61	69	79	121	141	247	309	325
2^{36}	5	17	23	65	117	137	159	173	189	233
2^{37}	25	31	45	69	123	141	199	201	351	375
2^{38}	45	87	107	131	153	185	191	227	231	257
2^{39}	7	19	67	91	135	165	219	231	241	301
2^{40}	87	167	195	203	213	285	293	299	389	437
2^{41}	21	31	55	63	73	75	91	111	133	139
2^{42}	11	17	33	53	65	143	161	165	215	227
2^{43}	57	67	117	175	255	267	291	309	319	369
2^{44}	17	117	119	129	143	149	287	327	359	377
2^{45}	55	69	81	93	121	133	139	159	193	229
2^{46}	21	57	63	77	167	197	237	287	305	311
2^{47}	115	127	147	279	297	339	435	541	619	649
2^{48}	59	65	89	93	147	165	189	233	243	257
2^{59}	55	99	225	427	517	607	649	687	861	871
2^{60}	93	107	173	179	257	279	369	395	399	453
2^{63}	25	165	259	301	375	387	391	409	457	471
2^{64}	59	83	95	179	189	257	279	323	353	363
10^6	17	21	39	41	47	69	83	93	117	137
10^7	9	27	29	57	63	69	71	93	99	111
10^8	11	29	41	59	69	153	161	173	179	213
10^9	63	71	107	117	203	239	243	249	261	267
10^{10}	33	57	71	119	149	167	183	213	219	231
10^{11}	23	53	57	93	129	149	167	171	179	231
10^{12}	11	39	41	63	101	123	137	143	153	233
10^{16}	63	83	113	149	183	191	329	357	359	369

The ten largest primes less than N are $N - a_1, \ldots, N - a_{10}$.

A number of the form $2^n - 1$ cannot be prime unless n is prime, since $2^{uv} - 1$ is divisible by $2^u - 1$. In 1644, Marin Mersenne astonished his contemporaries by stating, in essence, that the numbers $2^p - 1$ are prime for $p = 2, 3, 5, 7, 13, 17, 19, 31, 67, 127, 257$, and for no other p less than 257. (This statement appeared in connection with a discussion of perfect numbers in the preface to his *Cogitata Physico-Mathematica*. Curiously, he also made the following remark: "To tell if a given number of 15 or 20 digits is prime or not, all time would not suffice for the test, whatever use is made of what is already known.") Mersenne, who had corresponded frequently with Fermat, Descartes, and others about similar topics in previous years, gave no proof of his assertions, and for over 200 years nobody knew whether he was correct or not. Euler showed that $2^{31} - 1$ is prime in 1772, after having tried unsuccessfully to prove this in previous years. About 1875 É. Lucas discovered that $2^{127} - 1$ is prime, but $2^{67} - 1$ is not; therefore Mersenne was not completely accurate. In 1886, I. Pervusin and P. Seelhoff independently proved that $2^{61} - 1$ is prime, and this touched off speculation that Mersenne had only made a copying error, writing 67 for 61. Eventually other errors in Mersenne's statement were discovered; R. E. Powers [*AMM* **18** (1911), 195] found that $2^{89} - 1$ is prime, as had been conjectured by some earlier writers, and three years later he found that $2^{107} - 1$ also is prime. M. Kraitchik showed in 1922 that $2^{257} - 1$ is *not* prime.

At any rate, numbers of the form $2^p - 1$ are now known as *Mersenne numbers*, and it is known [Bryant Tuckerman, *AMS Notices* **18** (1971), 608] that the first 24 Mersenne primes are obtained for p equal to

$$2, 3, 5, 7, 13, 17, 19, 31, 61, 89, 107, 127, 521, 607, 1279,$$
$$2203, 2281, 3217, 4253, 4423, 9689, 9941, 11213, 19937. \qquad (26)$$

(Note that $8191 = 2^{13} - 1$ does not occur in this list; Mersenne had stated that $2^{8191} - 1$ is prime, and others had conjectured that any Mersenne prime could perhaps be used in the exponent.)

Since $2^{19937} - 1$ is a 6002-digit number, it is clear that some special techniques have been used to prove that it is prime. An efficient method for testing whether or not $2^p - 1$ is prime was first given by É. Lucas [*Amer. J. Math.* **1** (1878), 184–239, 289–321, especially p. 316] and improved by D. H. Lehmer ["An extended theory of Lucas' functions," *Annals of Math.* **31** (1930), 419–448, especially p. 443]. The Lucas-Lehmer test is the following:

Theorem L. *Let q be an odd prime, and define the sequence $\langle L_n \rangle$ by the rule*

$$L_0 = 4, \qquad L_{n+1} = (L_n^2 - 2) \bmod (2^q - 1). \qquad (27)$$

Then $2^q - 1$ is prime if and only if $L_{q-2} = 0$.

For example, $2^3 - 1$ is prime since $L_1 = (4^2 - 2) \bmod 7 = 0$. This test is particularly well adapted to binary calculation, using multiple-precision arithmetic when q is large, since calculation mod $(2^q - 1)$ is so convenient; cf. Section 4.3.2.

Proof. We will prove Theorem L using only very simple principles of number theory, by investigating several features of recurring sequences which are of independent interest. Consider the sequences $\langle U_n \rangle$, $\langle V_n \rangle$ defined by the rules

$$U_0 = 0, \qquad U_1 = 1, \qquad U_{n+1} = 4U_n - U_{n-1};$$
$$V_0 = 2, \qquad V_1 = 4, \qquad V_{n+1} = 4V_n - V_{n-1}. \tag{28}$$

We have then

$n =$	0	1	2	3	4	5	6	7	8	$\ldots$,
$U_n =$	0	1	4	15	56	209	780	2911	10864	$\ldots$,
$V_n =$	2	4	14	52	194	724	2702	10084	37634	$\ldots$;

and the following equations are readily proved by induction:

$$V_n = U_{n+1} - U_{n-1}; \tag{29}$$
$$U_n = ((2 + \sqrt{3})^n - (2 - \sqrt{3})^n)/\sqrt{12}; \tag{30}$$
$$V_n = (2 + \sqrt{3})^n + (2 - \sqrt{3})^n; \tag{31}$$
$$U_{m+n} = U_m U_{n+1} - U_{m-1} U_n. \tag{32}$$

Formulas (30) and (31) can be discovered by using generating functions as we did for the Fibonacci sequence in Section 1.2.8. These equations imply immediately that

$$U_{2n} = U_n V_n, \qquad V_{2n} = V_n^2 - 2. \tag{33}$$

First let us prove an auxiliary result, when p is prime and $e \geq 1$:

$$\text{if} \qquad U_n \equiv 0 \ (\text{modulo } p^e) \qquad \text{then} \qquad U_{np} \equiv 0 \ (\text{modulo } p^{e+1}). \tag{34}$$

This follows from the more general considerations of exercise 3.2.2–11, but a simple proof for this case can be given. Assume that $U_n = bp^e$, $U_{n+1} = a$. By (32) and (28), $U_{2n} = bp^e(2a - 4bp^e) \equiv (2a)U_n \ (\text{modulo } p^{e+1})$, while $U_{2n+1} = U_{n+1}^2 - U_n^2 \equiv a^2$. Similarly, $U_{3n} = U_{2n+1}U_n - U_{2n}U_{n-1} \equiv (3a^2)U_n$ and $U_{3n+1} = U_{2n+1}U_{n+1} - U_{2n}U_n \equiv a^3$. In general,

$$U_{kn} \equiv (ka^{k-1})U_n \qquad \text{and} \qquad U_{kn+1} \equiv a^k \ (\text{modulo } p^{e+1}),$$

so (34) follows if we take $k = p$.

From formulas (30) and (31) we can obtain other expressions for U_n and V_n, expanding $(2 \pm \sqrt{3})^n$ by the binomial theorem:

$$U_n = \sum_k \binom{n}{2k+1} 2^{n-2k-1} 3^k, \qquad V_n = \sum_k \binom{n}{2k} 2^{n-2k+1} 3^k. \tag{35}$$

Now if we set $n = p$, where p is an odd prime, and if we use the fact that $\binom{p}{k}$ is a multiple of p except when $k = 0$ or $k = p$, we find that

$$U_p \equiv 3^{(p-1)/2}, \qquad V_p \equiv 4 \quad (\text{modulo } p). \tag{36}$$

If $p \neq 3$, Fermat's theorem tells us that $3^{p-1} \equiv 1$; hence $(3^{(p-1)/2} - 1) \cdot (3^{(p-1)/2} + 1) \equiv 0$, and $3^{(p-1)/2} \equiv \pm 1$. When $U_p \equiv -1$, we have $U_{p+1} = 4U_p - U_{p-1} = 4U_p + V_p - U_{p+1} \equiv -U_{p+1}$; hence $U_{p+1} \bmod p = 0$. When $U_p \equiv +1$, we have $U_{p-1} = 4U_p - U_{p+1} = 4U_p - V_p - U_{p-1} \equiv -U_{p-1}$; hence $U_{p-1} \bmod p = 0$. We have therefore proved that, for all primes p, there is an integer $\epsilon(p)$ such that

$$U_{p+\epsilon(p)} \bmod p = 0, \qquad |\epsilon(p)| \leq 1. \tag{37}$$

Now if N is any positive integer, and if $m = m(N)$ is the smallest positive integer such that $U_{m(N)} \bmod N = 0$, we have

$$U_n \bmod N = 0 \qquad \text{if and only if } n \text{ is a multiple of } m(N). \tag{38}$$

(This number $m(N)$ is called the "rank of apparition" of N in the sequence.) To prove (38), observe that the sequence $U_m, U_{m+1}, U_{m+2}, \ldots$ is congruent (modulo N) to $aU_0, aU_1, aU_2, \ldots$, where $a = U_{m+1} \bmod N$ is relatively prime to N. (The statement in (38) also follows from the more general formula $\gcd(U_m, U_n) = U_{\gcd(m,n)}$, which can be proved for all m and n by means of Euclid's algorithm as in Section 1.2.8. We do not need this general formula in the present proof, except for the special case $\gcd(U_m, U_{m+1}) = 1$ which is easily proved by induction.)

With these preliminaries out of the way, we are ready to prove Theorem L. By (27), (33), and induction,

$$L_n = V_{2^n} \bmod (2^q - 1). \tag{39}$$

Furthermore, it follows from the identity $2U_{n+1} = 4U_n + V_n$ that

$$\gcd(U_n, V_n) \leq 2,$$

since any common factor of U_n and V_n must divide U_n and $2U_{n+1}$, while $\gcd(U_n, U_{n+1}) = 1$. So U_n and V_n have no odd factor in common. Therefore if $L_{q-2} = 0$ we must have

$$U_{2^{q-1}} = U_{2^{q-2}} V_{2^{q-2}} \equiv 0 \quad (\text{modulo } 2^q - 1),$$
$$U_{2^{q-2}} \not\equiv 0 \quad (\text{modulo } 2^q - 1). \tag{40}$$

Now if $m = m(2^q - 1)$ is the rank of apparition of $2^q - 1$, m must be a divisor of 2^{q-1} but not a divisor of 2^{q-2}; thus $m = 2^{q-1}$. We will prove that this implies that $n = 2^q - 1$ is prime: Let the factorization of n be $p_1^{e_1} \ldots p_r^{e_r}$. All primes p_j are greater than 3, since n is odd and congruent to $(-1)^q - 1 =$

-2 (modulo 3). From (34), (37), and (38) we know that $U_t \equiv 0$ (modulo $2^q - 1$), where

$$t = \text{lcm} \; (p_1^{e_1-1}(p_1' + \epsilon_1), \ldots, p_r^{e_r-1}(p_r + \epsilon_r)), \qquad (41)$$

and each ϵ_j is ± 1. Therefore t is a multiple of $m = 2^{q-1}$. Let $n_0 = \prod_{1 \leq j \leq r} p_j^{e_j-1}(p_j + \epsilon_j)$; we have $n_0 \leq \prod_{1 \leq j \leq r} p_j^{e_j-1}(p_j + \frac{1}{5}p_j) = (\frac{6}{5})^r n$. Also, because $p_j + \epsilon_j$ is even, $t \leq n_0/2^{r-1}$, since a factor of 2 is lost each time the least common multiple of two even numbers is taken. Combining these results, we have $m \leq t \leq 2(\frac{3}{5})^r n < 4(\frac{3}{5})^r m < 3m$; hence $r \leq 2$ and $t = m$ or $t = 2m$, a power of 2. Therefore $e_1 = 1$, $e_r = 1$, and if n is not prime we must have $n = 2^q - 1 = (2^k \pm 1)(2^l \mp 1)$ where $2^k \pm 1$ and $2^l \mp 1$ are prime. But the latter is obviously impossible when q is odd, so n is prime.

Conversely, suppose that $n = 2^q - 1$ is prime; we must show that $V_{2^q-2} \equiv 0$ (modulo n). For this purpose it suffices to prove that $V_{2^q-1} \equiv -2$ (modulo n), since $V_{2^q-1} = (V_{2^q-2})^2 - 2$. Now

$$2 \pm \sqrt{3} = ((\sqrt{2} \pm \sqrt{6})/2)^2,$$

so

$$V_{2^q-1} = ((\sqrt{2} + \sqrt{6})/2)^{n+1} + ((\sqrt{2} - \sqrt{6})/2)^{n+1}$$

$$= 2^{-n} \sum_k \binom{n+1}{2k} \sqrt{2}^{n+1-2k} \sqrt{6}^{2k} = 2^{(1-n)/2} \sum_k \binom{n+1}{2k} 3^k.$$

Since n is prime,

$$\binom{n+1}{2k} = \binom{n}{2k} + \binom{n}{2k-1}$$

is divisible by n except when $k = 0$ and $k = (n+1)/2$; hence

$$2^{(n-1)/2} V_{2^q-1} \equiv 1 + 3^{(n+1)/2} \quad \text{(modulo } n\text{)}.$$

Here $2 \equiv (2^{(q+1)/2})^2$, so $2^{(n-1)/2} \equiv (2^{(q+1)/2})^{(n-1)} \equiv 1$ by Fermat's theorem. Finally, by a simple case of the law of quadratic reciprocity (exercise 1.2.4–47), $3^{(n-1)/2} \equiv -1$, since $n \bmod 3 = 1$ and $n \bmod 4 = 3$. This means $V_{2^q-1} \equiv -2$, so $V_{2^q-2} \equiv 0$. ∎

EXERCISES

1. [*10*] If the sequence $d_0, d_1, d_2, \ldots$ of trial divisors in Algorithm A contains a number which is not prime, why will it never appear in the output?

2. [*15*] If it is known that the input n to Algorithm A is equal to 3 or more, could step A2 be eliminated?

3. [*M20*] Show that there is a number N with the following property: If $1000 \leq n \leq 1000000$, then n is prime if and only if $\gcd (n, N) = 1$.

4. [*M30*] Generalize Algorithm B, so that the sequence of trial divisors runs through the arithmetic progression $d_{k+1} = d_k + \delta$, and prove the analogs of (3) and (4) for this general case.

5. [20] Use Fermat's method to find the factors of 10541 by hand.

6. [M24] If p is an odd prime and if n is not a multiple of p, prove that the number of values $0 \leq x < p$, such that $x^2 - n \equiv y^2$ (modulo p) has a solution y, is equal to $(p \pm 1)/2$.

7. [25] Discuss the problems of programming the sieve of Algorithm D on a binary computer when the table entries for modulus m_i do not exactly fill an integral number of memory words.

▶ 8. [23] (The "sieve of Eratosthenes," 3rd century B.C.) The following procedure evidently discovers all odd prime numbers less than a given integer N, since it removes all the nonprime numbers: Start with all the odd numbers less than N; then successively strike out the multiples p_k^2, $p_k(p_k + 2)$, $p_k(p_k + 4)$, ... of the kth prime p_k, for $k = 2, 3, 4, \ldots$, until reaching a prime p_k with $p_k^2 > N$. Show how to adapt the procedure just described into an algorithm which is directly suited to efficient computer calculation, using no multiplication.

9. [M25] Let n be an odd number, $n \geq 3$. Show that if the number $\lambda(n)$ of Theorem 3.2.1.2B is a divisor of $n - 1$ but not equal to $n - 1$, then n must have the form $p_1 p_2 \ldots p_t$ where the p's are distinct primes and $t \geq 3$.

▶ 10. [M26] (John Selfridge.) Prove that if, for each prime divisor p of $n - 1$, there is a number x_p such that $x_p^{(n-1)/p} \bmod n \neq 1$ but $x_p^{n-1} \bmod n = 1$, then n is prime.

11. [M20] What outputs does Algorithm E give when $n = 197209$, $k = 5$, $t = 1$? [Hint: $\sqrt{5 \cdot 197209} = 992 + \overline{/1, 495, 2, 495, 1, 1984/}$.]

▶ 12. [M28] Design an algorithm which uses the outputs of Algorithm E to find a proper factorization of n, if a solution to (21) can be found by combining the outputs of Algorithm E.

13. [M24] (Fermat, 1640.) If $p > 2$ is prime, prove that all divisors d of $2^p - 1$ have the form $2kp + 1$ for some k. [Hint: Consider $2^{d-1} - 1$ when d is prime, in connection with the result of exercise 4.5.2–31.]

14. [M20] Prove that the number T in step E3 of Algorithm E will never be a multiple of an odd prime p for which $(kn)^{(p-1)/2} \bmod p > 1$. [Hint: By exercise 4.5.3.12(c), $P^2 - knQ^2 = (-1)^S V$, where Q is relatively prime to P.]

▶ 15. [M34] (Lucas and Lehmer.) Let P, Q be relatively prime integers, and let $U_0 = 0$, $U_1 = 1$, $U_{n+1} = PU_n - QU_{n-1}$ for $n \geq 1$. Prove that if N is a positive integer relatively prime to $2P^2 - 8Q$, and if $U_{N+1} \bmod N = 0$, while $U_{(N+1)/p} \bmod N \neq 0$ for each prime p dividing $N + 1$, then N is prime. (This gives a test for primality when the factors of $N + 1$ are known instead of the factors of $N - 1$. The value of U_m can be evaluated in $O(\log m)$ steps; cf. exercise 4.6.3–26.) [Hint: See the proof of Theorem L.]

16. [M50] Are there infinitely many Mersenne primes?

4.6. POLYNOMIAL ARITHMETIC

A *polynomial over* S is an expression of the form

$$u(x) = u_n x^n + \cdots + u_1 x + u_0, \qquad (1)$$

where the "coefficients" $u_n, \ldots, u_1, u_0$ are elements of some algebraic system S, and the "variable" x may be regarded as a formal symbol with an indeterminate meaning. We will assume that the algebraic system S is a *commutative ring with identity;* this means that S admits the operations of addition, subtraction, and multiplication, satisfying the customary properties: Addition and multiplication are associative, commutative binary operations defined on S, where multiplication distributes over addition; and subtraction is the inverse of addition. There is an additive identity element 0 such that $a + 0 = a$, and a multiplicative identity element 1 such that $a \cdot 1 = a$, for all a in S. The polynomial $0x^{n+m} + \cdots + 0x^{n+1} + u_n x^n + \cdots + u_1 x + u_0$ is regarded as the same polynomial as (1), although its expression is formally different.

The polynomial (1) is said to have *degree n* and *leading coefficient u_n* if $u_n \neq 0$; and in this case we write

$$\deg (u) = n, \qquad \ell(u) = u_n. \tag{2}$$

By convention, we also set

$$\deg (0) = -\infty, \qquad \ell(0) = 0, \tag{3}$$

where "0" denotes the zero polynomial whose coefficients are all zero. We say $u(x)$ is a *monic polynomial* if $\ell(u) = 1$.

Arithmetic on polynomials consists primarily of addition, subtraction, and multiplication of polynomials; in some cases, further operations such as division, factoring, and taking the greatest common divisor are important. The processes of addition, subtraction, and multiplication are defined in a natural way, as though the variable x were an element of S: Addition and subtraction are done by adding or subtracting the coefficients of like powers of x. Multiplication is done by the rule

$$(u_r x^r + \cdots + u_1 x + u_0)(v_s x^s + \cdots + v_1 x + v_0)$$
$$= (w_{r+s} x^{r+s} + \cdots + w_1 x + w_0),$$

where $\qquad w_k = u_0 v_k + u_1 v_{k-1} + \cdots + u_{k-1} v_1 + u_k v_0. \tag{4}$

In the latter formula u_i or v_j are treated as zero if $i > r$ or $j > s$.

The algebraic system S is usually the set of integers, or the rational numbers; or it may itself be a set of polynomials (in variables other than x); in the latter situation (1) is a multivariate polynomial, a polynomial in several variables. Another important case occurs when the algebraic system S consists of the integers $0, 1, \ldots, m - 1$, with addition, subtraction, and multiplication performed mod m (cf. Eq. 4.3.2–11); this is called *polynomial arithmetic modulo m*. The special case of polynomial arithmetic modulo 2, when each of the coefficients is 0 or 1, is especially important.

The reader should note the similarity between polynomial arithmetic and multiple-precision arithmetic (Section 4.3.1), where the radix b is substituted for x. The chief difference is that the coefficient u_k of x^k in polynomial arithmetic bears little or no essential relation to its neighboring coefficients $u_{k\pm1}$, so the idea of "carrying" from one place to the next is absent. In fact, polynomial arithmetic modulo b is essentially identical to multiple-precision arithmetic with radix b, except that all carries are suppressed. For example, compare the multiplication of $(1101)_2$ by $(1011)_2$ in the binary number system with the analogous multiplication of $x^3 + x^2 + 1$ by $x^3 + x + 1$ modulo 2:

Binary system	Polynomials modulo 2
1101	1101
$\times$ 1011	$\times$ 1011
1101	1101
1101	1101
1101	1101
10001111	1111111

The product of these polynomials modulo 2 is obtained by suppressing all carries, and it is $x^6 + x^5 + x^4 + x^3 + x^2 + x + 1$. If we had multiplied the same polynomials, considered as polynomials over the integers, without taking residues modulo 2, the result would have been $x^6 + x^5 + x^4 + 3x^3 + x^2 + x + 1$; again carries are suppressed, but in this case the coefficients can get arbitrarily large.

In view of this strong analogy with multiple-precision arithmetic, it is unnecessary to discuss polynomial addition, subtraction, and multiplication any further in this section. However, we should point out some factors which often make polynomial arithmetic somewhat different from multiple-precision arithmetic in practice: There is often a tendency to have a large number of zero coefficients, and polynomials of varying degrees, so special forms of representation are desirable; this situation is considered in Section 2.2.4. Furthermore, arithmetic on polynomials in several variables leads to routines which are best understood in a recursive framework; this situation is discussed in Chapter 8.

Although the techniques of polynomial addition, subtraction, and multiplication are comparatively straightforward, there are several other important aspects of polynomial arithmetic which deserve special examination. The following subsections therefore discuss *division* of polynomials, with associated techniques such as finding greatest common divisors; *factoring* of polynomials; and also efficient *evaluation* of polynomials, i.e., finding the value of $u(x)$ when x is a given element of S, using as few operations as possible. The special case of evaluating x^n very rapidly when n is large is quite important, so it is discussed in detail in Section 4.6.3.

The first major set of computer subroutines for doing polynomial arithmetic was the ALPAK system [W. S. Brown, J. P. Hyde, and B. A. Tague, *Bell System Technical Journal* **42** (1963), 2081–2119; **43** (1964), 785–804, 1547–1562]. Another landmark in this field was the PM system of George Collins [*CACM* **9** (1966), 578–589]. See also C. L. Hamblin, *Comp. J.* **10** (1967), 168–171.

EXERCISES

1. [*10*] If we are doing polynomial arithmetic modulo 10, what is $7x + 2$ minus $x^2 + 3$? What is $6x^2 + x + 3$ times $5x^2 + 2$?

2. [*17*] True or false: (a) The product of monic polynomials is monic. (b) The product of polynomials of respective degrees m and n has degree $m + n$. (c) The sum of two polynomials of degree n is of degree n.

3. [*M20*] If each of the coefficients $u_r, \ldots, u_0, v_s, \ldots, v_0$ in (4) is an integer satisfying the conditions $|u_i| \le m_1$, $|v_j| \le m_2$, what is the maximum absolute value of the product coefficients w_k?

▶ **4.** [*21*] Can the multiplication of polynomials modulo 2 be facilitated by using the ordinary arithmetic operations on a binary computer, if coefficients are packed into computer words?

▶ **5.** [*M24*] Show how to multiply two polynomials of degree $\le n$, modulo 2, with an execution time proportional to $O(n^{\log_2 3})$, when n is large, by adapting Karatsuba's method (Section 4.3.3).

4.6.1. Division of polynomials

It is possible to divide one polynomial by another in essentially the same way that we divide one multiple-precision integer by another, when arithmetic is being done on polynomials over a "field." A field S is a commutative ring with identity, in which exact division is possible as well as the operations of addition, subtraction, and multiplication; this means as usual that whenever u and v are elements of S and $v \ne 0$, there is an element w in S such that $u = vw$. The most important fields of coefficients which arise in applications are

a) the rational numbers (represented as fractions, see Section 4.5.1);
b) the real or complex numbers (represented within a computer by means of floating-point approximations; see Section 4.2);
c) the integers modulo p where p is prime (a division algorithm suitable for large values of p appears in exercise 4.5.2–15);
d) "rational functions" over a field (namely, quotients of two polynomials whose coefficients are in that field, the denominator being monic).

Of special importance is the field of integers modulo 2, when the two values 0 and 1 are the only elements of the field. Polynomials over this field (namely polynomials modulo 2) have many analogies to integers expressed in binary notation; and rational functions over this field have striking analogies to rational numbers whose numerator and denominator are represented in binary notation.

Given two polynomials $u(x)$ and $v(x)$ over a field, with $v(x) \ne 0$, we can divide $u(x)$ by $v(x)$ to obtain a quotient polynomial $q(x)$ and a remainder polynomial $r(x)$ satisfying the conditions

$$u(x) = q(x) \cdot v(x) + r(x), \qquad \deg(r) < \deg(v). \qquad (1)$$

[It is easy to see that there is at most one pair of polynomials $(q(x), r(x))$ satisfying these relations; for if (1) holds for both $(q_1(x), r_1(x))$ and $(q_2(x), r_2(x))$ and the same polynomials $u(x)$, $v(x)$, then $q_1(x)v(x) + r_1(x) = q_2(x)v(x) + r_2(x)$, so $(q_1(x) - q_2(x))v(x) = r_2(x) - r_1(x)$. Now if $q_1(x) - q_2(x) \neq 0$, then $\deg((q_1 - q_2) \cdot v) = \deg(q_1 - q_2) + \deg(v) \geq \deg(v) > \deg(r_2 - r_1)$, a contradiction; hence $q_1(x) - q_2(x) = 0$ and $r_1(x) - r_2(x) = 0$.]

The following algorithm, which is essentially the same as Algorithm 4.3.1D for multiple-precision division but without any concerns of carries, may be used to determine $q(x)$ and $r(x)$:

Algorithm D. (*Division of polynomials over a field*). Given polynomials

$$u(x) = u_m x^m + \cdots + u_1 x + u_0, \qquad v(x) = v_n x^n + \cdots + v_1 x + v_0$$

over a field S, where $v_n \neq 0$ and $m \geq n \geq 0$, this algorithm finds the polynomials

$$q(x) = q_{m-n} x^{m-n} + \cdots + q_0, \qquad r(x) = r_{n-1} x^{n-1} + \cdots + r_0$$

over S which satisfy (1).

D1. [Iterate on k.] Do step D2 for $k = m - n, m - n - 1, \ldots, 0$; then the algorithm terminates with $u_{n-1} = r_{n-1}, \ldots, u_0 = r_0$.

D2. [Division loop.]   Set $q_k \leftarrow u_{n+k}/v_n$, and then set $u_j \leftarrow u_j - q_k v_{j-k}$ for $j = n + k - 1, n + k - 2, \ldots, k$.   ∎

An example of Algorithm D appears below in (5). The number of arithmetic operations is essentially proportional to $n(m - n + 1)$. For some reason this procedure has become known as "synthetic division" of polynomials. Note that explicit division of coefficients is only done by v_n; so if $v(x)$ is a monic polynomial (with $v_n = 1$), there is no actual division at all. If multiplication is easier to perform than division it will be preferable to compute $1/v_n$ at the beginning of the algorithm and to multiply by this quantity in step D2.

Unique factorization domains. If we restrict consideration to polynomials over a field, we are not dealing directly with many important cases, such as polynomials over the integers, or polynomials in several variables. Let us therefore now consider the more general situation that the algebraic system S of coefficients is a *unique factorization domain*, not necessarily a field. This means that S is a commutative ring with identity, and that

i) $uv \neq 0$, whenever u and v are nonzero elements of S;
ii) every nonzero element u of S is either a "unit" or has a "unique" representation of the form

$$u = p_1 \ldots p_t, \qquad t \geq 1, \tag{2}$$

where $p_1, \ldots, p_t$ are "primes."

Here a "unit" u is an element such that $uv = 1$ for some v in S; and a "prime" p is a nonunit element such that the equation $p = qr$ can be true only if either q or r is a unit. The representation (2) is to be unique in the sense that if $p_1 \ldots p_t = q_1 \ldots q_s$, where all the p's and q's are primes, then $s = t$ and there is a permutation $\pi_1, \ldots, \pi_t$ such that $p_1 = a_1 q_{\pi_1}, \ldots, p_t = a_t q_{\pi_t}$ for some units $a_1, \ldots, a_t$. In other words, factorization into primes is unique, except for unit multiples and except for the order of the factors.

Any field is a unique factorization domain, in which each nonzero element is a unit and there are no primes. The integers form a unique factorization domain in which the units are $+1$ and -1, and the primes are $\pm 2, \pm 3, \pm 5, \pm 7, \ldots$. The case that S is the set of all integers is of principal importance, because it is often preferable to work with integer coefficients instead of arbitrary rational coefficients.

It is an important fact (see exercise 10) that *the polynomials over a unique factorization domain form a unique factorization domain.* A polynomial which is "prime" in this domain is usually called an *irreducible polynomial.* By using the unique factorization theorem repeatedly, we can prove that multivariate polynomials over the integers, or over any field, in any number of variables, can be uniquely factored into irreducible polynomials. For example, the multivariate polynomial $90x^3 - 120x^2y + 18x^2yz - 24xy^2z$ over the integers is the product of five irreducible polynomials $2 \cdot 3 \cdot x \cdot (3x - 4y) \cdot (5x + yz)$. The same polynomial, as a polynomial over the rationals, is the product of three irreducible polynomials $(6x) \cdot (3x - 4y) \cdot (5x + yz)$; this factorization can also be written $x \cdot (90x - 120y) \cdot (x + \frac{1}{5}yz)$ and in infinitely many other ways, although the factorization is essentially unique.

As usual, we say that $u(x)$ is a multiple of $v(x)$, and $v(x)$ is a divisor of $u(x)$, if $u(x) = v(x)q(x)$ for some polynomial $q(x)$. If we have an algorithm to tell whether or not u is a multiple of v for arbitrary elements u and v of a unique factorization domain S, and to determine w if $u = v \cdot w$, then Algorithm D gives us a method to tell whether or not $u(x)$ is a multiple of $v(x)$ for arbitrary polynomials $u(x)$ and $v(x)$ over S. We have either $u(x) = v(x) = 0$, or else the quotient $u(x)/v(x)$ can be determined by Algorithm D, with u_{n+k} always being a multiple of v_n at the beginning of step D2. (Applying this observation repeatedly, we obtain an algorithm which decides if a given polynomial over S, in any number of variables, is a multiple of another given polynomial over S.)

A set of elements of a unique factorization domain is said to be *relatively prime* if no prime (of the unique factorization domain) divides all of them. A polynomial over a unique factorization domain is called *primitive* if its coefficients are relatively prime. In this connection we have the following important fact:

Lemma G (Gauss's Lemma). *The product of primitive polynomials over a unique factorization domain is primitive.*

Proof. Let $u(x) = u_m x^m + \cdots + u_0$ and $v(x) = v_n x^n + \cdots + v_0$ be primitive polynomials. If p is any prime of the domain, we must show that p does not divide all the coefficients of $u(x)v(x)$. By assumption, there is an index j such that u_j is not divisible by p, and an index k such that v_k is not divisible by p. Let j and k be as small as possible; then the coefficient of x^{j+k} in $u(x)v(x)$ is $u_j v_k + u_{j+1} v_{k-1} + \cdots + u_{j+k} v_0 + u_{j-1} v_{k+1} + \cdots + u_0 v_{k+j}$, and this is not a multiple of p (since its first term isn't, but all of its other terms are). ∎

If a nonzero polynomial $u(x)$ over S is not primitive, we can write $u(x) = p_1 \cdot u_1(x)$, where p_1 is a prime of S dividing all the coefficients of $u(x)$, and where $u_1(x)$ is another nonzero polynomial over S. All of the coefficients of $u_1(x)$ have one less prime factor than the corresponding coefficients of $u(x)$. Now if $u_1(x)$ is not primitive, we can write $u_1(x) = p_2 \cdot u_2(x)$, etc., and this process must ultimately terminate in a representation $u(x) = c \cdot u_k(x)$, where c is an element of S and $u_k(x)$ is primitive. In fact, we have the following lemma:

Lemma H. *Any nonzero polynomial $u(x)$ over a unique factorization domain S can be factored in the form $u(x) = c \cdot v(x)$, where c is in S and $v(x)$ is primitive. Furthermore, this representation is unique, in the sense that if $u = c_1 \cdot v_1(x) = c_2 \cdot v_2(x)$, then $c_1 = ac_2$ and $v_2(x) = av_1(x)$ where a is a unit of S.*

Proof. We have shown that such a representation exists, and so only the uniqueness needs to be proved. Assume that $c_1 \cdot v_1(x) = c_2 \cdot v_2(x)$, where $v_1(x)$ and $v_2(x)$ are primitive and c_1 is not a unit multiple of c_2. By unique factorization there is a prime p of S and an exponent k such that p^k divides one of $\{c_1, c_2\}$ but not the other, say p^k divides c_1 but not c_2. Then p^k divides all of the coefficients of $c_2 \cdot v_2(x)$, so p divides all the coefficients of $v_2(x)$, contradicting the assumption that $v_2(x)$ is primitive. Hence $c_1 = ac_2$, where a is a unit; and $0 = ac_2 \cdot v_1(x) - c_2 \cdot v_2(x) = c_2 \cdot (av_1(x) - v_2(x))$ implies that $av_1(x) - v_2(x) = 0$. ∎

Therefore we may write any nonzero polynomial $u(x)$ as

$$u(x) = \text{cont}\,(u) \cdot \text{pp}\,(u(x)), \tag{3}$$

where cont (u), the "content" of u, is an element of S, and pp $(u(x))$, the "primitive part" of $u(x)$, is a primitive polynomial over S. When $u(x) = 0$, it is convenient to define cont $(u) = \text{pp}\,(u(x)) = 0$. Combining Lemmas G and H gives us the relations

$$\text{cont}\,(u \cdot v) = a \,\text{cont}\,(u)\,\text{cont}\,(v),$$
$$\text{pp}\,(u(x) \cdot v(x)) = b\,\text{pp}\,(u(x))\,\text{pp}\,(v(x)), \tag{4}$$

where a and b are units, depending on u and v, with $ab = 1$. When we are working with polynomials over the integers, the only units are $+1$ and -1, and it is conventional to define $\text{pp}(u(x))$ so that its leading coefficient is positive; then (4) is true with $a = b = 1$, When working with polynomials over a field we may take cont $(u) = \ell(u)$, so that $\text{pp}(u(x))$ is monic; in this case again (4) holds with $a = b = 1$, for all $u(x)$ and $v(x)$.

For example, if we are dealing with polynomials over the integers, let
$u(x) = -26x^2 + 39$, $v(x) = 21x + 14$.   Then

$$\text{cont } (u) = -13, \qquad \text{pp } (u(x)) = 2x^2 - 3,$$
$$\text{cont } (v) = 7, \qquad \text{pp } (v(x)) = 3x + 2,$$
$$\text{cont } (u \cdot v) = -91, \qquad \text{pp}(u(x) \cdot v(x)) = 6x^3 + 4x^2 - 9x - 6.$$

Greatest common divisors. When there is unique factorization, it makes sense
to speak of a "greatest common divisor" of two elements; this is a common
divisor which is divisible by as many primes as possible. (Cf. Eq. 4.5.2–6.)
Since a unique factorization domain may have many units, there is a certain
amount of ambiguity in this definition of greatest common divisor; if w is a
greatest common divisor of u and v, so is $a \cdot w$, when a is a unit. Conversely,
the assumption of unique factorization implies that if w_1 and w_2 are both
greatest common divisors of u and v, then $w_1 = a \cdot w_2$ for some unit a. Therefore it does not make sense, in general, to speak of "the" greatest common
divisor of u and v; there is a set of greatest common divisors, each one being a
unit multiple of the others.

Let us now consider the problem of finding a greatest common divisor of two
given polynomials over an algebraic system S. If S is a field, the problem is
relatively simple; our division algorithm, Algorithm D, can be extended to an
algorithm which computes greatest common divisors, just as Euclid's algorithm
(Algorithm 4.5.2A) yields the greatest common divisor of two given integers
based on a division algorithm for integers: If $v(x) = 0$, then gcd $(u(x), v(x)) =$
$u(x)$; otherwise gcd $(u(x), v(x)) =$ gcd $(v(x), r(x))$, where $r(x)$ is given by (1).
This procedure is called Euclid's algorithm for polynomials over a field; it was
first used by Simon Stevin in 1585 [*Les œuvres mathématiques de Simon Stevin*,
ed. by A. Girard, **1** (Leyden, 1634), 56].

For example, let us determine the gcd of $x^8 + x^6 + 10x^4 + 10x^3 + 8x^2 +$
$2x + 8$ and $3x^6 + 5x^4 + 9x^2 + 4x + 8$, mod 13, by using Euclid's algorithm
for polynomials over the integers modulo 13. First, writing only the coefficients
to show the steps of Algorithm D, we have

$$
\begin{array}{r}
9\ 0\ 7 \\
3\ 0\ 5\ 0\ 9\ 4\ 8\ \overline{)\ 1\ 0\ 1\ 0\ 10\ 10\quad 8\ 2\ 8} \\
1\ 0\ 6\ 0\quad 3\ 10\quad 7 \\
\hline
0\ 8\ 0\quad 7\quad 0\quad 1\ 2\ 8 \\
8\ 0\quad 9\quad 0\ 11\ 2\ 4 \\
\hline
0\ 11\quad 0\quad 3\ 0\ 4
\end{array}
\tag{5}
$$

and hence

$$x^8 + x^6 + 10x^4 + 10x^3 + 8x^2 + 2x + 8$$
$$= (9x^2 + 7)(3x^6 + 5x^4 + 9x^2 + 4x + 8) + 11x^4 + 3x^2 + 4.$$

Similarly,

$$3x^6 + 5x^4 + 9x^2 + 4x + 8 = (5x^2 + 5)(11x^4 + 3x^2 + 4) + 4x + 1;$$
$$11x^4 + 3x^2 + 4 \quad = (6x^3 + 5x^2 + 6x + 5)(4x + 1) + 12;$$
$$4x + 1 = (9x + 12) \cdot 12 + 0. \tag{6}$$

(The equality sign here means congruence modulo 13, since all arithmetic on the coefficients has been done mod 13.) This computation shows that 12 is a greatest common divisor of the two original polynomials. Now any nonzero element of a field is a unit of the domain of polynomials over that field, so any nonzero multiple of a greatest common divisor is also a greatest common divisor (over a field). It is therefore conventional in this case to divide the result of the algorithm by its leading coefficient, producing a *monic* polynomial which is called *the* greatest common divisor of the two given polynomials. The gcd computed in (6) is accordingly taken to be 1, not 12. The last step in (6) could have been omitted, for if deg $(v) = 0$, then gcd $(u(x), v(x)) = 1$, no matter what polynomial is chosen for $u(x)$.

Let us now turn to the more general situation in which our polynomials are given over a unique factorization domain which is not a field. From Eqs. (4) we can deduce the important relations

$$\text{cont } (\text{gcd } (u, v)) = a \cdot \text{gcd } (\text{cont } (u), \text{cont } (v)),$$
$$\text{pp } (\text{gcd } (u(x), v(x))) = b \cdot \text{gcd } (\text{pp } (u(x)), \text{pp } (v(x))), \tag{7}$$

where a and b are units. Here gcd $(u(x), v(x))$ denotes any particular polynomial in x which is a greatest common divisor of $u(x)$ and $v(x)$. Equations (7) reduce the problem of finding greatest common divisors of arbitrary polynomials to the problem of finding greatest common divisors of *primitive* polynomials. Clearly cont (u) is a greatest common divisor of the coefficients of u, and pp $(u(x)) = u(x)/\text{cont } (u)$.

Algorithm D for division of polynomials over a field can be generalized to a pseudo-division of polynomials over any algebraic system which is a commutative ring with identity. We can observe that Algorithm D requires explicit division only by $\ell(v)$, the leading coefficient of $v(x)$, and that step D2 is carried out exactly $m - n + 1$ times; thus if $u(x)$ and $v(x)$ start with integer coefficients, and if we are working over the rational numbers, then the only denominators which appear in the coefficients of $q(x)$ and $r(x)$ are divisors of $\ell(v)^{m-n+1}$. This suggests that we can always find polynomials $q(x)$ and $r(x)$ such that

$$\ell(v)^{m-n+1} u(x) = q(x)v(x) + r(x), \quad \deg (r) < n, \tag{8}$$

where $m = \deg (u)$, $n = \deg (v)$, and $m \geq n$, for any polynomials $u(x)$ and $v(x) \neq 0$.

Algorithm R (*Pseudo-division of polynomials*). Given polynomials

$$u(x) = u_m x^m + \cdots + u_1 x + u_0, \qquad v(x) = v_n x^n + \cdots + v_1(x) + v_0,$$

where $v_n \neq 0$ and $m \geq n \geq 0$, this algorithm finds polynomials $q(x) = q_{m-n} x^{m-n} + \cdots + q_0$ and $r(x) = r_{n-1} x^{n-1} + \cdots + r_0$ satisfying (8).

R1. [Iterate on k.] Do step R2 for $k = m - n, m - n - 1, \ldots, 0$; then the algorithm terminates with $u_{n-1} = r_{n-1}, \ldots, u_0 = r_0$.

R2. [Multiplication loop.] Set $q_k \leftarrow u_{n+k} v_n^k$, and then set $u_j \leftarrow v_n u_j - u_{n+k} v_{j-k}$ for $j = n + k - 1, n + k - 2, \ldots, 0$. (When $j < k$ this means that $u_j \leftarrow v_n u_j$, since we treat $v_{-1}, v_{-2}, \ldots$, as zero.) ∎

An example calculation appears below in (10); it is easy to prove the validity of Algorithm R by induction on $m - n$. Note that no division whatever is used in this algorithm; the coefficients of $q(x)$ and $r(x)$ are themselves certain polynomial functions of the coefficients of $u(x)$ and $v(x)$. If $v_n = 1$, the algorithm is identical to Algorithm D. If $v_n \neq 0$ and if $u(x)$ and $v(x)$ are polynomials over a unique factorization domain, we can prove as before that the polynomials $q(x)$ and $r(x)$ of (8) are unique; therefore another way to do the pseudo-division over a unique factorization domain, which may sometimes be preferable, is to multiply $u(x)$ by v_n^{m-n+1} and apply Algorithm D, knowing that all the quotients in step D2 will exist.

Algorithm R can be extended to a "generalized Euclidean algorithm" for primitive polynomials over a unique factorization domain, in the following way: Let $u(x)$ and $v(x)$ be primitive polynomials with deg $(u) \geq$ deg (v), and determine $r(x)$ satisfying (8) by means of Algorithm R. Now gcd $(u(x), v(x)) =$ gcd $(v(x), r(x))$: For any common divisor of $u(x)$ and $v(x)$ divides $v(x)$ and $r(x)$; conversely, any common divisor of $v(x)$ and $r(x)$ divides $\ell(v)^{m-n+1} u(x)$, and it must be primitive (since $v(x)$ is primitive) so it divides $u(x)$. If $r(x) = 0$, we therefore have gcd $(u(x), v(x)) = v(x)$; if $r(x) \neq 0$, we have gcd $(v(x), r(x)) =$ gcd $(v(x), \text{pp} (r(x)))$ since $v(x)$ is primitive, so the process can be iterated.

Algorithm E (*Generalized Euclidean algorithm*). Given nonzero polynomials $u(x)$ and $v(x)$ over a unique factorization domain S, this algorithm calculates a greatest common divisor of $u(x)$ and $v(x)$. We assume that auxiliary algorithms exist to calculate greatest common divisors of elements of S, and to divide a by b in S, when $b \neq 0$ and a is a multiple of b.

E1. [Reduce to primitive.] Set $d \leftarrow$ gcd $(\text{cont} (u), \text{cont} (v))$, using the assumed algorithm for calculating greatest common divisors in S. (Note that cont (u) is the greatest common divisor of the coefficients of $u(x)$.) Replace $u(x)$ by the polynomial $u(x)/\text{cont} (u) = \text{pp} (u(x))$; similarly, replace $v(x)$ by pp $(v(x))$.

E2. [Pseudo-division.]   Calculate $r(x)$ using Algorithm R.   (It is unnecessary to calculate the quotient polynomial $q(x)$.)   If $r(x) = 0$, go to E4.   If $\deg(r) = 0$, replace $v(x)$ by the constant polynomial "1" and go to E4.

E3. [Make remainder primitive.]   Replace $u(x)$ by $v(x)$, and replace $v(x)$ by pp $(r(x))$. Go back to step E2. (This is the "Euclidean step," analogous to the other instances of Euclid's algorithm that we have seen.)

E4. [Attach the content.]   The algorithm terminates, with $d \cdot v(x)$ as the answer. ∎

As an example of Algorithm E, let us calculate the greatest common divisor of

$$u(x) = x^8 + x^6 - 3x^4 - 3x^3 + 8x^2 + 2x - 5,$$
$$v(x) = 3x^6 + 5x^4 - 4x^2 - 9x + 21, \tag{9}$$

over the integers. These polynomials are primitive, so step E1 sets $d \leftarrow 1$. In step E2 we have the pseudo-division

$$
\begin{array}{r}
1\ \ 0\quad -6 \\
3\ 0\ 5\ 0\ -4\ -9\ 21\ )\ \overline{1\ 0\quad\ \ 1\ 0\ -3\ -3\ \ 8\ \ 2\quad -5} \\
3\ 0\qquad\ 3\ 0\ -9\ -9\ 24\ \ 6\quad -15 \\
3\ 0\qquad\ 5\ 0\ -4\ -9\ 21 \\
\hline
0\ -2\ 0\ \ -5\quad 0\ \ 3\ \ 6\quad -15 \\
0\ -6\ 0\ -15\quad 0\ \ 9\ 18\quad -45 \\
0\quad 0\ 0\quad\ \ 0\quad 0\ \ 0\ \ 0\qquad 0 \\
\hline
-6\ 0\ -15\quad 0\ \ 9\ 18\quad -45 \\
-18\ 0\ -45\quad 0\ 27\ 54\ -135 \\
-18\ 0\ -30\quad 0\ 24\ 54\ -126 \\
\hline
-15\quad 0\ \ 3\ \ 0\qquad -9
\end{array}
\tag{10}
$$

Here the quotient $q(x)$ is $1 \cdot 3^2 x^2 + 0 \cdot 3^1 x + -6 \cdot 3^0$; we have

$$27u(x) = v(x)(9x^2 - 6) + (-15x^4 + 3x^2 - 9). \tag{11}$$

Now step E3 replaces $u(x)$ by $v(x)$ and $v(x)$ by pp $(r(x)) = 5x^4 - x^2 + 3$. The subsequent calculation may be summarized as follows, writing only the coefficients:

$u(x)$	$v(x)$	$r(x)$
$1, 0, 1, 0, -3, -3, 8, 2, -5$	$3, 0, 5, 0, -4, -9, 21$	$-15, 0, 3, 0, -9$
$3, 0, 5, 0, -4, -9, 21$	$5, 0, -1, 0, 3$	$-585, -1125, 2205$
$5, 0, -1, 0, 3$	$13, 25, -49$	$-233150, 307500$
$13, 25, -49$	$4663, -6150$	143193869

(12)

It is instructive to compare this calculation with the computation of the same greatest common divisor over the *rational* numbers, instead of over the integers, by using Euclid's algorithm for polynomials over a field as described earlier in this section. The following sequence of results occurs:

$u(x)$ $v(x)$

1, 0, 1, 0, -3, -3, 8, 2, -5 3, 0, 5, 0, -4, -9, 21

3, 0, 5, 0, -4, -9, 21 $-\frac{5}{9}$, 0, $\frac{1}{9}$, 0, $-\frac{1}{3}$

$-\frac{5}{9}$, 0, $\frac{1}{9}$, 0, $-\frac{1}{3}$ $-\frac{117}{25}$, -9, $\frac{441}{25}$

$-\frac{117}{25}$, -9, $\frac{441}{25}$ $\frac{233150}{6591}$, $-\frac{102500}{2197}$

$\frac{233150}{6591}$, $-\frac{102500}{2197}$ $\frac{1288744821}{543589225}$ (13)

To improve that algorithm, we can reduce $u(x)$ and $v(x)$ to monic polynomials at each step, since this removes "unit" factors which make the coefficients more complicated than necessary; this is actually Algorithm E over the rationals:

$u(x)$ $v(x)$

1, 0, 1, 0, -3, -3, 8, 2, -5 1, 0, $\frac{5}{3}$, 0, $-\frac{4}{3}$, -3, 7

1, 0, $\frac{5}{3}$, 0, $-\frac{4}{3}$, -3, 7   1, 0, $-\frac{1}{5}$, 0, $\frac{3}{5}$

1, 0, $-\frac{1}{5}$, 0, $\frac{3}{5}$ 1, $\frac{25}{13}$, $-\frac{49}{13}$

1, $\frac{25}{13}$, $-\frac{49}{13}$ 1, $-\frac{6150}{4663}$

1, $-\frac{6150}{4663}$ 1   (14)

In both (13) and (14) the sequence of polynomials is essentially the same as (12), which was obtained by Algorithm E over the integers; the only difference is that the polynomials have been multiplied by certain rational numbers. Whether we have $5x^4 - x^2 + 3$ or $-\frac{5}{9}x^4 + \frac{1}{9}x^2 - \frac{1}{3}$ or $x^4 - \frac{1}{5}x^2 + \frac{3}{5}$, the computations are essentially the same. But either algorithm using rational arithmetic will run noticeably slower than the all-integer Algorithm E, since rational arithmetic requires many more evaluations of gcd's of integers within each step. Therefore it is definitely preferable to use the all-integer algorithm instead of rational arithmetic, when the gcd of polynomials with integer or rational coefficients is desired.

It is also instructive to compare the above calculations with (6) above, where we determined the gcd of the same polynomials $u(x)$ and $v(x)$ modulo 13 with considerably less labor. Since $\ell(u)$ and $\ell(v)$ are not multiples of 13, the fact that $\gcd\big(u(x), v(x)\big) = 1$ modulo 13 is sufficient to prove that $u(x)$ and $v(x)$ are relatively prime over the integers (and therefore over the rational numbers); we will return to this time-saving observation at the close of Section 4.6.2.

Collins's Algorithm. An ingenious algorithm which is generally superior to Algorithm E, and which gives us further information about Algorithm E's behavior, has been discovered by George E. Collins [*JACM* **14** (1967), 128–142]. His algorithm avoids the calculation of primitive part in step E3, dividing instead by an element of S which is known to be a factor of $r(x)$:

Algorithm C (*Greatest common divisor over a unique factorization domain*). This algorithm has the same input and output assumptions as Algorithm E, and has the advantage that fewer calculations of greatest common divisors of coefficients are needed.

C1. [Reduce to primitive.] As in step E1 of Algorithm E, set $d \leftarrow$ gcd $(\text{cont } (u),$ $\text{cont}(v))$, and replace $(u(x), v(x))$ by $(\text{pp}(u(x)), \text{pp}(v(x)))$. Also set $a \leftarrow 1$.

C2. [Pseudo-division.] Set $b \leftarrow \ell(v)^{\deg(u)-\deg(v)+1}$. Calculate $r(x)$ using Algorithm R. If $r(x) = 0$, go to C4. If deg $(r) = 0$, replace $v(x)$ by the constant polynomial "1" and go to C4.

C3. [Adjust remainder.] Replace the polynomial $u(x)$ by $v(x)$, and replace $v(x)$ by $r(x)/a$. (At this point all coefficients of $r(x)$ are multiples of a.) Also set $a \leftarrow b$; then return to C2.

C4. [Attach the content.] The algorithm terminates, with $d \cdot \text{pp}(v(x))$ as the answer. ∎

If we apply this algorithm to the polynomials (9) considered earlier, the following sequence of results is obtained:

$u(x)$	$v(x)$	a
1, 0, 1, 0, -3, -3, 8, 2, -5	3, 0, 5, 0, -4, -9, 21	1
3, 0, 5, 0, -4, -9, 21	-15, 0, 3, 0, -9	27
-15, 0, 3, 0, -9	585, 1125, -2205	$(-15)^3$
585, 1125, -2205	-18885150, 24907500	$(585)^3$ (15)

At the conclusion of the algorithm, $r(x)/a = 527933700$.

The sequence of polynomials consists of integral multiples of the polynomials in the sequence produced by Algorithm E. The numbers in (15) actually constitute an unusually bad example of the efficiency of Algorithm C, for reasons explained later; but our example shows that, in spite of the fact that the polynomials are not reduced to primitive form, the coefficients are kept to a reasonable size because the reduction factor a is large.

In order to analyze Algorithm C and to prove that it is valid, let us call the sequence of polynomials it produces $u_1(x), u_2(x), u_3(x), \ldots$, where $u_1(x) = u(x)$ and $u_2(x) = v(x)$. Let $t_1 = 0$ and let $t_{j+1} = n_j - n_{j+1} + 1$, where $n_j = \deg(u_j), j \geq 1$. Then if $\ell_j = \ell(u_j)$, we have

$$\ell_2^{t_2} u_1(x) = u_2(x)q_1(x) + \ell_1^{t_1} u_3(x), \qquad n_3 < n_2;$$
$$\ell_3^{t_3} u_2(x) = u_3(x)q_2(x) + \ell_2^{t_2} u_4(x), \qquad n_4 < n_3; \qquad (16)$$
$$\ell_4^{t_4} u_3(x) = u_4(x)q_3(x) + \ell_3^{t_3} u_5(x), \qquad n_5 < n_4;$$

and so on. The process terminates when $n_{k+1} = \deg(u_{k+1}) \leq 0$. We must show that $u_4(x), u_5(x), \ldots$, have coefficients in A, i.e., that the factor $\ell_{k+1}^{t_{k+1}}$ introduced into the dividend in the kth step can be divided from the remainder in the $(k+1)$st step. The proof is rather involved, and it can be most easily understood by considering an example.

Table 1

COEFFICIENTS IN COLLINS'S ALGORITHM

Row name							Row								Multiply by	Replace by row
A_5	a_8	a_7	a_6	a_5	a_4	a_3	a_2	a_1	a_0	0	0	0	0	0	b_6^3	C_5
A_4	0	a_8	a_7	a_6	a_5	a_4	a_3	a_2	a_1	a_0	0	0	0	0	b_6^3	C_4
A_3	0	0	a_8	a_7	a_6	a_5	a_4	a_3	a_2	a_1	a_0	0	0	0	b_6^3	C_3
A_2	0	0	0	a_8	a_7	a_6	a_5	a_4	a_3	a_2	a_1	a_0	0	0	b_6^3	C_2
A_1	0	0	0	0	a_8	a_7	a_6	a_5	a_4	a_3	a_2	a_1	a_0	0	b_6^3	C_1
A_0	0	0	0	0	0	a_8	a_7	a_6	a_5	a_4	a_3	a_2	a_1	a_0	b_6^3	C_0
B_7	b_6	b_5	b_4	b_3	b_2	b_1	b_0	0	0	0	0	0	0	0		
B_6	0	b_6	b_5	b_4	b_3	b_2	b_1	b_0	0	0	0	0	0	0		
B_5	0	0	b_6	b_5	b_4	b_3	b_2	b_1	b_0	0	0	0	0	0		
B_4	0	0	0	b_6	b_5	b_4	b_3	b_2	b_1	b_0	0	0	0	0		
B_3	0	0	0	0	b_6	b_5	b_4	b_3	b_2	b_1	b_0	0	0	0	c_4^3/b_6^3	D_3
B_2	0	0	0	0	0	b_6	b_5	b_4	b_3	b_2	b_1	b_0	0	0	c_4^3/b_6^3	D_2
B_1	0	0	0	0	0	0	b_6	b_5	b_4	b_3	b_2	b_1	b_0	0	c_4^3/b_6^3	D_1
B_0	0	0	0	0	0	0	0	b_6	b_5	b_4	b_3	b_2	b_1	b_0	c_4^3/b_6^3	D_0
C_5	0	0	0	0	c_4	c_3	c_2	c_1	c_0	0	0	0	0	0		
C_4	0	0	0	0	0	c_4	c_3	c_2	c_1	c_0	0	0	0	0		
C_3	0	0	0	0	0	0	c_4	c_3	c_2	c_1	c_0	0	0	0		
C_2	0	0	0	0	0	0	0	c_4	c_3	c_2	c_1	c_0	0	0		
C_1	0	0	0	0	0	0	0	0	c_4	c_3	c_2	c_1	c_0	0	d_2^2/c_4^3	E_1
C_0	0	0	0	0	0	0	0	0	0	c_4	c_3	c_2	c_1	c_0	d_2^2/c_4^3	E_0
D_3	0	0	0	0	0	0	0	0	d_2	d_1	d_0	0	0	0		
D_2	0	0	0	0	0	0	0	0	0	d_2	d_1	d_0	0	0		
D_1	0	0	0	0	0	0	0	0	0	0	d_2	d_1	d_0	0		
D_0	0	0	0	0	0	0	0	0	0	0	0	d_2	d_1	d_0	e_2^2/d_2^2	F_0
E_1	0	0	0	0	0	0	0	0	0	0	0	e_1	e_0	0		
E_0	0	0	0	0	0	0	0	0	0	0	0	0	e_1	e_0		
F_0	0	0	0	0	0	0	0	0	0	0	0	0	0	f_0		

Suppose, as in our example (15), that $n_1 = 8$, $n_2 = 6$, $n_3 = 4$, $n_4 = 2$, $n_5 = 1$, $n_6 = 0$, so that $t_1 = 0$, $t_2 = t_3 = t_4 = 3$, $t_5 = t_6 = 2$. Let us write $u_1(x) = a_8x^8 + a_7x^7 + \cdots + a_0$, $u_2(x) = b_6x^6 + b_5x^5 + \cdots + b_0$, $\ldots$, $u_5(x) = e_1x + e_0$, $u_6(x) = f_0$, and consider the array shown in Table 1. For simplicity, let us assume that the coefficients of the polynomials are integers. We have $b_6^3u_1(x) = u_2(x)q_1(x) + u_3(x)$; so if we multiply row A_5 by b_6^3 and subtract appropriate multiples of rows B_7, B_6, and B_5 [corresponding to the coefficients of $q_1(x)$] we will get row C_5. Similarly, if we multiply row A_4 by b_6^3 and subtract multiples of rows B_6, B_5, and B_4, we get row C_4. In a similar way, we have $c_4^3u_2(x) = u_3(x)q_2(x) + b_6^3u_4(x)$; so we can multiply row B_3 by c_4^3, subtract integer multiples of rows C_5, C_4, and C_3, then divide by b_6^3 to obtain row D_3.

In order to prove that $u_4(x)$ has integer coefficients, let us consider the matrix

$$
\begin{matrix}
A_2 \\
A_1 \\
A_0 \\
B_4 \\
B_3 \\
B_2 \\
B_1 \\
B_0
\end{matrix}
\begin{pmatrix}
a_8 & a_7 & a_6 & a_5 & a_4 & a_3 & a_2 & a_1 & a_0 & 0 & 0 \\
0 & a_8 & a_7 & a_6 & a_5 & a_4 & a_3 & a_2 & a_1 & a_0 & 0 \\
0 & 0 & a_8 & a_7 & a_6 & a_5 & a_4 & a_3 & a_2 & a_1 & a_0 \\
b_6 & b_5 & b_4 & b_3 & b_2 & b_1 & b_0 & 0 & 0 & 0 & 0 \\
0 & b_6 & b_5 & b_4 & b_3 & b_2 & b_1 & b_0 & 0 & 0 & 0 \\
0 & 0 & b_6 & b_5 & b_4 & b_3 & b_2 & b_1 & b_0 & 0 & 0 \\
0 & 0 & 0 & b_6 & b_5 & b_4 & b_3 & b_2 & b_1 & b_0 & 0 \\
0 & 0 & 0 & 0 & b_6 & b_5 & b_4 & b_3 & b_2 & b_1 & b_0
\end{pmatrix} = M. \qquad (17)
$$

The indicated row operations and a permutation of rows will transform M into

$$
\begin{matrix}
B_4 \\
B_3 \\
B_2 \\
B_1 \\
C_2 \\
C_1 \\
C_0 \\
D_0
\end{matrix}
\begin{pmatrix}
b_6 & b_5 & b_4 & b_3 & b_2 & b_1 & b_0 & 0 & 0 & 0 & 0 \\
0 & b_6 & b_5 & b_4 & b_3 & b_2 & b_1 & b_0 & 0 & 0 & 0 \\
0 & 0 & b_6 & b_5 & b_4 & b_3 & b_2 & b_1 & b_0 & 0 & 0 \\
0 & 0 & 0 & b_6 & b_5 & b_4 & b_3 & b_2 & b_1 & b_0 & 0 \\
0 & 0 & 0 & 0 & c_4 & c_3 & c_2 & c_1 & c_0 & 0 & 0 \\
0 & 0 & 0 & 0 & 0 & c_4 & c_3 & c_2 & c_1 & c_0 & 0 \\
0 & 0 & 0 & 0 & 0 & 0 & c_4 & c_3 & c_2 & c_1 & c_0 \\
0 & 0 & 0 & 0 & 0 & 0 & 0 & 0 & d_2 & d_1 & d_0
\end{pmatrix} = M'. \qquad (18)
$$

Because of the way M' has been derived from M, we have

$$
b_6^3 \cdot b_6^3 \cdot b_6^3 \cdot (c_4^3/b_6^3) \cdot \det M_0 = \pm \det M_0', \qquad (19)
$$

if M_0 and M_0' represent any square matrices obtained by selecting eight corresponding columns from M and M'. For example, let us select the first seven columns and the column containing d_1; then

$$
b_6^3 \cdot b_6^3 \cdot b_6^3 \cdot (c_4^3/b_6^3) \cdot \det
\begin{pmatrix}
a_8 & a_7 & a_6 & a_5 & a_4 & a_3 & a_2 & 0 \\
0 & a_8 & a_7 & a_6 & a_5 & a_4 & a_3 & a_0 \\
0 & 0 & a_8 & a_7 & a_6 & a_5 & a_4 & a_1 \\
b_6 & b_5 & b_4 & b_3 & b_2 & b_1 & b_0 & 0 \\
0 & b_6 & b_5 & b_4 & b_3 & b_2 & b_1 & 0 \\
0 & 0 & b_6 & b_5 & b_4 & b_3 & b_2 & 0 \\
0 & 0 & 0 & b_6 & b_5 & b_4 & b_3 & b_0 \\
0 & 0 & 0 & 0 & b_6 & b_5 & b_4 & b_1
\end{pmatrix} = \pm b_6^4 \cdot c_4^3 \cdot d_1.
$$
$$(20)$$

Since $b_6 c_4 \neq 0$, this proves that d_1 is an integer; in fact, d_1 is a multiple of b_6^2. Similarly, d_2 and d_0 are integer multiples of b_6^2. This accounts for the fact that the numbers 585, 1125, -2205 in (15) are multiples of 9.

In general, we can show that $u_{j+1}(x)$ has integer coefficients in a similar manner. If we start with the matrix M, consisting of rows $A_{n_2-n_j}$ through A_0 and $B_{n_1-n_j}$ through B_0, and if we perform the row operations indicated in Table 1, we will obtain a matrix M' consisting in some order of rows $B_{n_1-n_j}$ through $B_{n_3-n_j+1}$, $C_{n_2-n_j}$ through $C_{n_4-n_j+1}$, . . . , $F_{n_{j-2}-n_j}$ through F_1, $G_{n_{j-1}-n_j}$

through G_0, and finally H_0 [a row containing the coefficients of $u_{j+1}(x)$].
Extracting appropriate columns shows that

$$(\ell_2^{t_2})^{n_2-n_j+1}(\ell_3^{t_3}/\ell_2^{t_2})^{n_3-n_j+1}\cdots(\ell_j^{t_j}/\ell_{j-1}^{t_{j-1}})^{n_j-n_j+1}\det M_0$$
$$= \pm\ell_2^{n_1-n_3}\ell_3^{n_2-n_4}\cdots\ell_{j-1}^{n_{j-2}-n_j}\ell_j^{n_{j-1}-n_j+1}h_i, \qquad (21)$$

where h_i is a given coefficient of $u_{j+1}(x)$ and M_0 is a submatrix of M. Thus
every coefficient of $u_{j+1}(x)$ is a multiple of

$$\ell_2^{(t_2-1)(t_3-2)}\ell_3^{(t_3-1)(t_4-2)}\cdots\ell_{j-1}^{(t_{j-1}-1)(t_j-2)}. \qquad (22)$$

(This proof, although stated for the domain of integers, obviously applies to
any unique factorization domain.)

The quantity (22) appears to be an extremely large number which makes
the coefficients of $u_{j+1}(x)$ much larger than they need to be. It would be
possible to rewrite Algorithm C in an appropriate manner so that this factor
would be eliminated. But actually *the quantity* (22) *is almost always equal to
unity!* This follows from the fact that $t_3, t_4, \ldots, t_k$ will all be equal to 2, for
almost every given pair of polynomials $u_1(x), u_2(x)$; the particular polynomials
in example (15) are very exceptional indeed, since both t_3 and t_4 are equal to 3.
In order to prove this fact, note for example that we could have chosen the first
eight columns of M and M' in (17) and (18), and then we would have found in
place of Eq. (19) that $u_4(x)$ has degree less than 3 if and only if $d_3 = 0$, that
is, if and only if

$$\det\begin{pmatrix} a_8 & a_7 & a_6 & a_5 & a_4 & a_3 & a_2 & a_1 \\ 0 & a_8 & a_7 & a_6 & a_5 & a_4 & a_3 & a_2 \\ 0 & 0 & a_8 & a_7 & a_6 & a_5 & a_4 & a_3 \\ b_6 & b_5 & b_4 & b_3 & b_2 & b_1 & b_0 & 0 \\ 0 & b_6 & b_5 & b_4 & b_3 & b_2 & b_1 & b_0 \\ 0 & 0 & b_6 & b_5 & b_4 & b_3 & b_2 & b_1 \\ 0 & 0 & 0 & b_6 & b_5 & b_4 & b_3 & b_2 \\ 0 & 0 & 0 & 0 & b_6 & b_5 & b_4 & b_3 \end{pmatrix} = 0. \qquad (23)$$

In general, $\deg(u_{j+1}) < \deg(u_j) - 1$ if and only if a similar determinant in
the coefficients of $u_1(x)$ and $u_2(x)$ is zero. Since such a determinant is a multivariate polynomial in the coefficients, it will be nonzero "almost always," or
"with probability one." (See exercise 16 for a more precise formulation of this
statement.)

Now if we assume that $t_3 = t_4 = \cdots = t_k = 2$, since this almost always
happens, we find from the above argument that each coefficient of $u_{j+1}(x)$ is
given by the value of a certain determinant in the a's and b's, i.e., a determinant
whose entries are the coefficients of the original polynomials. A well-known
theorem of Hadamard (see exercise 15) states that

$$|\det(a_{ij})| \le \prod_{1\le i\le n}\left(\sum_{1\le j\le n} a_{ij}^2\right)^{1/2}; \qquad (24)$$

therefore an upper bound for the maximum coefficient appearing during Algorithm C when $t_3 = t_4 = \cdots = t_k = 2$ is

$$N^{m+n}(m + 1)^{n/2}(n + 1)^{m/2}, \qquad (25)$$

if all coefficients of the given polynomials $u(x)$ and $v(x)$ are bounded by N in absolute value. This same upper bound applies to the coefficients of all polynomials $u(x)$, $v(x)$ computed during the execution of Algorithm E, regardless of the value of $t_3, t_4, \ldots$, since the polynomials obtained in Algorithm E are always divisors of the polynomials obtained in Algorithm C, with the factor (22) removed.

This upper bound on the coefficients is extremely gratifying, because it is much better than we would ordinarily have a right to expect. For example, suppose we performed Algorithm E and Algorithm C with *no* correction in step E3 or C3 [so that we just replace $v(x)$ by $r(x)$]. This is the simplest gcd algorithm, and it is one which traditionally appears in textbooks on algebra (for theoretical purposes, not intended for practical calculations). If we suppose that $t_2 = t_3 = \cdots = 2$, we find that the coefficients of $u_3(x)$ are bounded by N^3, the coefficients of $u_4(x)$ are bounded by N^7, those of $u_5(x)$ by $N^{17}, \ldots$, where the size of the coefficients is N^{a_k}, for $a_k = 2a_{k-1} + a_{k-2}$. The upper bound in place of (25) for $m = n + 1$ would be approximately

$$N^{0.5(2.414)^n}, \qquad (26)$$

and experiments show that the simple algorithm does in fact have this behavior; the number of digits in the coefficients grows exponentially at each step! In Algorithm E the growth in number of digits is only linear.

The paper of Collins cited above reports on several computational experiments using Algorithm E and Algorithm C; for example, the gcd of two polynomials of degree 35, whose coefficients were integers selected at random in the range $-100 < $ coefficient < 100, was calculated in 1.48 minutes by Algorithm C, compared to 3.25 minutes by Algorithm E. The main difference is the additional time to calculate the primitive part in step E3 (which requires a calculation of the gcd of all coefficients).

When the algorithms were tried on polynomials in two variables, Algorithm C was of course far superior; it found the gcd of two polynomials of total degree 5 in x and y in 0.23 minutes, compared to 4.29 minutes for Algorithm E. (The calculation of $\mathrm{pp}(r(x))$ in step E3 now is the dominant factor, since it requires calculating a gcd of polynomials, not merely of integers.) For polynomials of total degree 6, with one-digit coefficients, Algorithm E was found to be too slow to be practical, while Algorithm C did the calculation in about 40 seconds.

The considerations above can be used to derive the well-known fact that two polynomials are relatively prime if and only if their "resultant" is nonzero; the resultant is a determinant having the form of rows A_5 through A_0 and B_7 through B_0 in Table 1. (This is "Sylvester's determinant"; see B. L. van der Waerden, *Modern Algebra*, tr. by Fred Blum (New York: Ungar, 1949), Sections

27–28.) From the standpoint discussed above, we would say that the gcd is "almost always" of degree zero, since Sylvester's determinant is almost never zero. But many calculations of practical interest would never be undertaken if there weren't some reasonable chance that the gcd would be a polynomial of positive degree.

We can see exactly what happens during Algorithms E and C when the gcd is not 1 by considering $u(x) = w(x)u_1(x)$, $v(x) = w(x)u_2(x)$, where $u_1(x)$ and $u_2(x)$ are relatively prime and $w(x)$ is primitive. Then if the polynomials $u_1(x)$, $u_2(x)$, $u_3(x)$, $\ldots$ are obtained when Algorithm E works on $u(x) = u_1(x)$ and $v(x) = u_2(x)$, it is easy to show that the sequence obtained for $u(x) = w(x)u_1(x)$, $v(x) = w(x)u_2(x)$ is simply $w(x)u_1(x)$, $w(x)u_2(x)$, $w(x)u_3(x)$, $w(x)u_4(x)$, $\ldots$ With Algorithm C the behavior is different; if the polynomials $u_1(x)$, $u_2(x)$, $u_3(x)$, $\ldots$ are obtained when Algorithm C is applied to $u(x) = u_1(x)$ and $v(x) = u_2(x)$, and if we assume that $\deg(u_{j+1}) = \deg(u_j) - 1$ (which is almost always true when $j > 1$), then the sequence

$$w(x)u_1(x), \ w(x)u_2(x), \ \ell^2 w(x)u_3(x), \ \ell^4 w(x)u_4(x), \ \ldots,$$
$$\ell^{2(j-2)} w(x)u_j(x), \ \ldots \qquad (27)$$

is obtained when Algorithm C is applied to $u(x) = w(x)u_1(x)$ and $v(x) = w(x)u_2(x)$, where $\ell = \ell(w)$. (See exercise 13.) So Algorithm E may be superior to Algorithm C when the primitive part of the greatest common divisor has a large enough leading coefficient.

To find the gcd of polynomials over the integers, a somewhat more complicated algorithm which has advantages over both Algorithms C and E is mentioned at the close of Section 4.6.2. For further study of polynomial algebra over finite fields, see the book *Algebraic Coding Theory* (New York: McGraw-Hill, 1968) by E. R. Berlekamp.

For a general determinant-evaluation algorithm, which essentially includes Algorithm C as a special case, see E. H. Bareiss, *Math. Comp.* **22** (1968), 565–578.

EXERCISES

1. [*10*] Compute the pseudo-quotient $q(x)$ and pseudo-remainder $r(x)$, namely, the polynomials satisfying (8), when $u(x) = x^6 + x^5 - x^4 + 2x^3 + 3x^2 - x + 2$ and $v(x) = 2x^3 + 2x^2 - x + 3$, over the integers.

2. [*15*] What is the greatest common divisor of $3x^6 + x^5 + 4x^4 + 4x^3 + 3x^2 + 4x + 2$ and its "reverse" $2x^6 + 4x^5 + 3x^4 + 4x^3 + 4x^2 + x + 3$, modulo 7?

▶ **3.** [*M25*] Show that Euclid's algorithm for polynomials over a field S can be extended to find polynomials $U(x)$, $V(x)$ over S such that

$$u(x)V(x) + U(x)v(x) = \gcd\big(u(x), v(x)\big).$$

(Cf. Algorithm 4.5.2X.) What are the degrees of the polynomials $U(x)$ and $V(x)$ which are computed by this extended algorithm? Prove that if S is the field of rational numbers, and if $u(x) = x^m - 1$ and $v(x) = x^n - 1$, then the extended algorithm yields polynomials $U(x)$ and $V(x)$ having *integer* coefficients. Find $U(x)$ and $V(x)$ when $u(x) = x^{21} - 1$, $v(x) = x^{13} - 1$.

▶ **4.** [*M30*] Let p be prime, and suppose that Euclid's algorithm applied to the polynomials $u(x)$ and $v(x)$ modulo p yields a sequence of polynomials having respective degrees $m, n, n_1, \ldots, n_t, -\infty$, where $m = \deg(u)$, $n = \deg(v)$, and $n_t \geq 0$. Assume that $m \geq n$. If $u(x)$ and $v(x)$ are monic polynomials, independently and uniformly distributed over all the p^{m+n} pairs of monic polynomials having respective degrees m and n, what is the average value of the three quantities t, $n_1 + \cdots + n_t$, $(n - n_1)n_1 + \cdots + (n_{t-1} - n_t)n_t$, as a function of m, n, and p? (These three quantities are the fundamental factors in the running time of Euclid's algorithm applied to polynomials modulo p, assuming that division is done by Algorithm D.) [*Hint:* Show that $u(x) \bmod v(x)$ is uniformly distributed and independent of $v(x)$.]

5. [*M22*] What is the probability that $u(x)$ and $v(x)$ are relatively prime modulo p, if $u(x)$ and $v(x)$ are independently and uniformly distributed monic polynomials of degree n?

6. [*M23*] We have seen that Euclid's Algorithm 4.5.2A for integers can be directly adapted to an algorithm for the greatest common divisor of polynomials. Can the "binary gcd algorithm," Algorithm 4.5.2B, be adapted in an analogous way to an algorithm which applies to polynomials?

7. [*M10*] What are the units in the domain of all polynomials over a unique factorization domain S?

▶ **8.** [*M22*] Show that if a polynomial with integer coefficients is irreducible over the domain of integers, it is irreducible when considered as a polynomial over the field of rational numbers.

9. [*M25*] Let $u(x)$ and $v(x)$ be primitive polynomials over a unique factorization domain S. Prove that $u(x)$ and $v(x)$ are relatively prime if and only if there are polynomials $U(x)$ and $V(x)$ such that $u(x)V(x) + U(x)v(x)$ is a polynomial of degree zero. [*Hint:* Extend Algorithm E, as Algorithm 4.5.2E is extended in exercise 3.]

10. [*M28*] Prove that the polynomials over a unique factorization domain form a unique factorization domain. [*Hint:* Use the result of exercise 9 to help show that there is at most one kind of factorization possible.]

11. [*M22*] What row names would have appeared in Table 1 if the sequence of degrees had been 9, 6, 5, 2, $-\infty$ instead of 8, 6, 4, 2, 1, 0?

▶ **12.** [*M24*] Let $u_1(x), u_2(x), u_3(x), \ldots$ be a sequence of polynomials obtained by Algorithm C. "Sylvester's matrix" is the square matrix formed from rows A_{n_2-1} through A_0, B_{n_1-1} through B_0 (in a notation analogous to that of Table 1). Show that if $u_1(x)$ and $u_2(x)$ have a common factor of positive degree, then the determinant of Sylvester's matrix is zero; conversely, given that $\deg(u_k) = 0$ for some k, show that the determinant of Sylvester's matrix is nonzero by deriving a formula for its absolute value in terms of $\ell(u_j)$ and $\deg(u_j)$, $1 \leq j \leq k$.

13. [*M20*] Show that the leading coefficient ℓ of the primitive part of $\gcd(u(x), v(x))$ enters into Algorithm C's polynomial sequence as shown in (27), when $t_2 = t_3 = \cdots = t_k = 2$.

14. [*M28*] Let p be prime. Prove that if $\deg(u) = m > n = \deg(v) > \deg(w) + 1$, $\ell(v) = p$, and $p^{m-n+1}u(x) = v(x)q(x) + w(x)$, then $w(x)$ is a multiple of p.

15. [*M26*] Prove Hadamard's inequality (24). [*Hint:* Consider the matrix AA^T.]

16. [*HM22*] Let $f(x_1, \ldots, x_n)$ be a multivariate polynomial with real coefficients, not all zero, and let a_N be the number of solutions to the equation $f(x_1, \ldots, x_n) = 0$ such that $|x_1| \leq N, \ldots, |x_n| \leq N$, and each x_j is an integer. Prove that

$$\lim_{N \to \infty} a_N / (2N + 1)^n = 0.$$

17. [*M32*] (P. M. Cohn's algorithm for division of string polynomials.) Let A be an "alphabet," i.e., a set of symbols. A *string* α on A is a sequence of $n \geq 0$ symbols, $\alpha = a_1 \cdots a_n$, where each a_j is in A. The length of α, denoted by $|\alpha|$, is the number n of symbols. A *string polynomial* on A is a finite sum $U = \sum_k r_k \alpha_k$, where each r_k is a nonzero rational number and each α_k is a string on A. The *degree* of U, deg (U), is defined to be $-\infty$ if $U = 0$ (i.e., if the sum is empty), and max $|\alpha_k|$ if $U \neq 0$. The sum and product of string polynomials are defined in an obvious manner, e.g., $(\sum_j r_j \alpha_j)(\sum_k s_k \beta_k) = \sum_{j,k} r_j s_k \alpha_j \beta_k$, where the product of two strings is obtained by simply juxtaposing them. For example, if $A = \{a, b\}$, $U = ab + ba - 2a - 2b$, $V = a + b - 1$, then deg $(U) = 2$, deg $(V) = 1$, $V^2 = aa + ab + ba + bb - 2a - 2b + 1$, and $V^2 - U = aa + bb + 1$. Clearly, deg $(UV) = $ deg $(U) + $ deg (V), deg $(U + V) \leq$ max $($deg $(U),$ deg $(V))$, with equality in the latter formula if deg $(U) \neq$ deg (V). (String polynomials may be regarded as ordinary multivariate polynomials over the field of rational numbers, except that the variables are *not commutative* under multiplication. In the conventional language of pure mathematics, the set of string polynomials with the operations defined here is the free associative algebra generated by A over the rationals.)

a) Let Q_1, Q_2, U, V be string polynomials with deg $(U) \geq$ deg (V) and deg $(Q_1 U - Q_2 V) <$ deg $(Q_1 U)$. Give an algorithm to find a string polynomial Q such that deg $(U - QV) <$ deg (U). (Thus if we are given U and V such that $Q_1 U = Q_2 V + R$ and deg $(R) <$ deg $(Q_1 U)$, for some Q_1 and Q_2, there is a solution to these conditions with $Q_1 = 1$.)

b) Given that U and V are string polynomials with deg $(V) >$ deg $(Q_1 U - Q_2 V)$ for some Q_1 and Q_2, show that the result of (a) can be improved to find a quotient Q such that $U = QV + R$, deg $(R) <$ deg (V). (This is the analog of (1) for string polynomials; part (a) showed that we can make deg $(R) <$ deg (U), under weaker hypotheses.)

c) A "homogeneous" polynomial is one in which all terms have the same degree (length). If U_1, U_2, V_1, V_2 are homogeneous string polynomials with $U_1 V_1 = U_2 V_2$ and deg $(V_1) \geq$ deg (V_2), show that there is a homogeneous string polynomial U such that $U_2 = U_1 U$, $V_1 = U V_2$.

d) Given that U and V are homogeneous string polynomials with $UV = VU$, prove that there is a homogeneous string polynomial W such that $U = rW^m$, $V = sW^n$ for some integers m, n, and rational numbers r, s. Give an algorithm to compute such a W having the largest possible degree. (This algorithm is of interest, for example, when $U = \alpha$ and $V = \beta$ are strings satisfying $\alpha\beta = \beta\alpha$; then W is simply a string γ. When $U = x^m$ and $V = x^n$, the solution of largest degree is $W = x^{\gcd(m,n)}$, so this algorithm includes a gcd algorithm for integers as a special case.)

▶ **18.** [*M24*] (Euclid's algorithm for string polynomials.) Given string polynomials V_1, V_2 not both zero, which have a "common left multiple," i.e., for which there exist U_1, U_2, not both zero, such that $U_1 V_1 = U_2 V_2$, the purpose of this exercise is to find an algorithm to compute their "greatest common right divisor" gcrd (V_1, V_2)

as well as their "least common left multiple" lclm (V_1, V_2). The latter quantities are defined as follows: gcrd (V_1, V_2) is a common right divisor of V_1 and V_2 (that is, $V_1 = W_1 \, \mathrm{gcrd} \, (V_1, V_2)$ and $V_2 = W_2 \, \mathrm{gcrd} \, (V_1, V_2)$ for some W_1, W_2), and any common right divisor of V_1 and V_2 is a right divisor of gcrd (V_1, V_2); lclm (V_1, V_2) is a common left multiple of V_1 and V_2 (that is, lclm $(V_1, V_2) = Z_1 V_1 = Z_2 V_2$ for some Z_1, Z_2), and any common left multiple of V_1 and V_2 is a left multiple of lclm (V_1, V_2).

For example, let $U_1 = abbbab + abbab - bbab + ab - 1$, $V_1 = babab + abab + ab - b$; $U_2 = abb + ab - b$, $V_2 = babbabab + bababab + babab + abab - babb - 1$. Then we have $U_1 V_1 = U_2 V_2 = abbbabbabab + abbabbabab + abbbababab + abbababab - bbabbabab + abbbabab - bbababab + 2abbabab - abbbabb + ababab - abbabb - bbabab - babab + bbabb - abb - ab + b$. For these string polynomials it can be shown that gcrd $(V_1, V_2) = ab + 1$, and lclm $(V_1, V_2) = U_1 V_1$.

The division algorithm of exercise 17 may be restated thus: If V_1 and V_2 are string polynomials, with $V_2 \neq 0$, and if $U_1 \neq 0$ and U_2 satisfy the equation $U_1 V_1 = U_2 V_2$, then there exist string polynomials Q and R such that $V_1 = QV_2 + R$, deg $(R) <$ deg (V_2). *Note:* It follows readily that Q and R are uniquely determined, they do not depend on U_1 and U_2; furthermore the result is right-left symmetric in the sense that

$$U_2 = U_1 Q + R' \quad \text{where} \quad \deg (R') = \deg (U_1) - \deg (V_2) + \deg (R) < \deg (U_1).$$

Show that this division algorithm can be extended to an algorithm which computes lclm (V_1, V_2) and gcrd (V_1, V_2), and, in fact, it finds string polynomials Z_1, Z_2 such that $Z_1 V_1 + Z_2 V_2 = \mathrm{gcrd} \, (V_1, V_2)$. [*Hint:* Use auxiliary variables u_1, u_2, v_1, v_2, w_1, w_2, w_1', w_2', z_1, z_2, z_1', z_2', whose values are string polynomials; start by setting $u_1 \leftarrow U_1$, $u_2 \leftarrow U_2$, $v_1 \leftarrow V_1$, $v_2 \leftarrow V_2$, and throughout the algorithm maintain the conditions

$$U_1 w_1 + U_2 w_2 = u_1, \qquad\qquad z_1 V_1 + z_2 V_2 = v_1,$$
$$U_1 w_1' + U_2 w_2' = u_2, \qquad\qquad z_1' V_1 + z_2' V_2 = v_2,$$
$$u_1 z_1 - u_2 z_1' = (-1)^n U_1, \qquad w_1 v_1 - w_1' v_2 = (-1)^n V_1,$$
$$-u_1 z_2 + u_2 z_2' = (-1)^n U_2, \qquad -w_2 v_1 + w_2' v_2 = (-1)^n V_2$$

at the nth iteration.]

19. [*M39*] (Euclid's algorithm for matrices.) Exercise 18 shows that the concept of greatest common right divisor can be meaningful when multiplication is not commutative. Prove that any two $n \times n$ matrices A and B of integers have a greatest common right matrix divisor D. (*Suggestion:* Design an algorithm whose inputs are A and B, and whose outputs are integer matrices P, Q, X, Y, where $A = PD$, $B = QD$, $D = XA + YB$.) Find a greatest common right divisor of $\left(\begin{smallmatrix} 1 & 2 \\ 3 & 4 \end{smallmatrix}\right)$ and $\left(\begin{smallmatrix} 4 & 3 \\ 2 & 1 \end{smallmatrix}\right)$.

20. [*M40*] What can be said about calculation of the greatest common divisor of polynomials whose coefficients are floating-point numbers? (Investigate the accuracy of Euclid's algorithm.)

21. [*M24*] Assuming that $t_3 = t_4 = \cdots = t_k = 2$ (see (25)), prove that the computation time required by Algorithm C to compute the gcd of two nth degree polynomials over the integers is $O(n^4 (\log Nn)^2)$, if the coefficients of the given polynomials are bounded by N in absolute value.

*4.6.2. Factorization of polynomials

Factoring modulo p. Let us now consider the problem of *factoring* polynomials, not merely finding the greatest common divisor of two or more of them. As in the case of integer numbers (Sections 4.5.2, 4.5.4), the problem of factoring is definitely more difficult than finding the greatest common divisor; but factorization of polynomials modulo a prime integer p is not as hard to do as we might expect. It is much easier to find the factors of an arbitrary polynomial of degree n, modulo 2, than to use any known method to find the factors of an arbitrary n-bit binary number. This surprising situation is a consequence of an important factorization algorithm discovered in 1967 by Elwyn R. Berlekamp [*Bell System Technical J.* **46** (1967), 1853–1859].

Let p be a prime number; all arithmetic on polynomials in the following discussion will be done modulo p. Suppose that someone has given us a polynomial $u(x)$, whose coefficients are chosen from the set $\{0, 1, \ldots, p-1\}$; we may assume that $u(x)$ is monic. Our goal is to express $u(x)$ in the form

$$u(x) = p_1(x)^{e_1} \ldots p_r(x)^{e_r}, \tag{1}$$

where $p_1(x), \ldots, p_r(x)$ are distinct, monic irreducible polynomials.

As a first step, we can use a standard technique to determine whether any of $e_1, \ldots, e_r$ are greater than unity. If

$$u(x) = u_n x^n + \cdots + u_0 = v(x)^2 w(x), \tag{2}$$

then its "derivative" formed in the usual way (but modulo p) is

$$u'(x) = nu_n x^{n-1} + \cdots + u_1 = 2v(x)v'(x)w(x) + v(x)^2 w'(x), \tag{3}$$

and this is a multiple of the squared factor $v(x)$. Therefore our first step in factoring $u(x)$ is to form

$$\gcd\big(u(x), u'(x)\big) = d(x). \tag{4}$$

If $d(x)$ is equal to 1, we know that $u(x)$ is "squarefree," the product of distinct primes $p_1(x) \ldots p_r(x)$. If $d(x)$ is not equal to 1 and $d(x) \neq u(x)$, then $d(x)$ is a proper factor of $u(x)$; so the process can be continued by factoring $d(x)$ and $u(x)/d(x)$ separately. Finally, if $d(x) = u(x)$, we must have $u'(x) = 0$; hence the coefficient u_k of x^k is nonzero only when k is a multiple of p. This means that $u(x)$ can be written as a polynomial of the form $v(x^p)$, and in such a case we have

$$u(x) = v(x^p) = \big(v(x)\big)^p; \tag{5}$$

the factorization process can be continued by finding the irreducible factors of $v(x)$.

Identity (5) may appear somewhat strange to the reader, and it is an important fact which is basic to Berlekamp's algorithm and which is easily proved: If $v_1(x)$ and $v_2(x)$ are any polynomials modulo p, then

$$\left(v_1(x)v_2(x)\right)^p = v_1(x)^p v_2(x)^p,$$

$$\left(v_1(x) + v_2(x)\right)^p = v_1(x)^p + \binom{p}{1} v_1(x)^{p-1} v_2(x) + \cdots$$

$$+ \binom{p}{p-1} v_1(x) v_2(x)^{p-1} + v_2(x)^p$$

$$= v_1(x)^p + v_2(x)^p;$$

since the binomial coefficients

$$\binom{p}{1}, \ldots, \binom{p}{p-1}$$

are all multiples of p. Furthermore if a is any integer, $a^p \equiv a$ (modulo p). Therefore when $v(x) = v_m x^m + v_{m-1} x^{m-1} + \cdots + v_0$, we find that

$$v(x)^p = (v_m x^m)^p + (v_{m-1} x^{m-1})^p + \cdots + (v_0)^p$$

$$= v_m x^{mp} + v_{m-1} x^{(m-1)p} + \cdots + v_0 = v(x^p).$$

This proves (5).

The above remarks show that the problem of factoring a polynomial reduces to the problem of factoring a squarefree polynomial. Let us therefore assume that

$$u(x) = p_1(x)p_2(x) \cdots p_r(x) \tag{6}$$

is the product of distinct primes. How can we be clever enough to discover the $p_j(x)$'s when only $u(x)$ is given? Berlekamp's idea is to make use of the Chinese remainder theorem, which is valid for polynomials just as it is valid for integers (see exercise 3). If $(s_1, s_2, \ldots, s_r)$ is any r-tuple of integers mod p, the Chinese remainder theorem implies that *there is a unique polynomial $v(x)$ such that*

$$v(x) \equiv s_1 \;(\text{modulo } p_1(x)), \quad \ldots, \quad v(x) \equiv s_r \;(\text{modulo } p_r(x)),$$

$$\deg(v) < \deg(p_1) + \deg(p_2) + \cdots + \deg(p_r) = \deg(u). \tag{7}$$

[The statement $g(x) \equiv h(x)$ (modulo $f(x)$) here is the same as $g(x) \equiv h(x)$ (modulo $f(x)$ and p) in the notation of exercise 3.2.2–11, since we are considering polynomial arithmetic modulo p.] The polynomial $v(x)$ in (7) gives us a way to get at the factors of $u(x)$, for if $r \geq 2$ and $s_1 \neq s_2$, we will have $\gcd(u(x), v(x) - s_1)$ divisible by $p_1(x)$ but not by $p_2(x)$.

Since this observation shows that we can get information about the factors of $u(x)$ from appropriate solutions $v(x)$ of (7), let us analyze (7) more closely. In the first place we can observe that the polynomial $v(x)$ satisfies the condition $v(x)^p \equiv s_j^p = s_j \equiv v(x)$ (modulo $p_j(x)$) for $1 \leq j \leq r$, therefore

$$v(x)^p \equiv v(x) \quad (\text{modulo } u(x)), \qquad \deg(v) < \deg(u). \tag{8}$$

In the second place we have the basic identity

$$x^p - x \equiv (x - 0)(x - 1) \cdots (x - (p-1)) \quad (\text{modulo } p) \qquad (9)$$

(see exercise 6); hence

$$v(x)^p - v(x) = (v(x) - 0)(v(x) - 1) \cdots (v(x) - (p-1)) \qquad (10)$$

is an identity for any polynomial $v(x)$, when we are working modulo p. If $v(x)$ satisfies (8), it follows that $u(x)$ divides the left-hand side of (10), so every irreducible factor of $u(x)$ must divide one of the p relatively prime factors of the right-hand side of (10). *All* solutions of (8) must therefore have the form (7), for some $s_1, s_2, \ldots, s_r$; *there are exactly p^r solutions of* (8).

Therefore the solutions $v(x)$ to congruence (8) provide a key to the factorization of $u(x)$. It may seem harder to find all solutions to (8) than to factor $u(x)$ in the first place, but in fact this is not true, since the set of solutions to (8) is closed under addition. Let $\deg(u) = n$; we can construct the $n \times n$ matrix

$$Q = \begin{pmatrix} q_{0,0} & q_{0,1} & \cdots & q_{0,n-1} \\ \vdots & \vdots & & \vdots \\ q_{n-1,0} & q_{n-1,1} & \cdots & q_{n-1,n-1} \end{pmatrix} \qquad (11)$$

where

$$x^{pk} \equiv q_{k,n-1}x^{n-1} + \cdots + q_{k,1}x + q_{k,0} \quad (\text{modulo } u(x)). \qquad (12)$$

Then $v(x) = v_{n-1}x^{n-1} + \cdots + v_1 x + v_0$ is a solution to (8) if and only if

$$(v_0, v_1, \ldots, v_{n-1})Q = (v_0, v_1, \ldots, v_{n-1}). \qquad (13)$$

For the latter equation holds if and only if

$$v(x) = \sum_j v_j x^j = \sum_j \sum_k v_k q_{k,j} x^j \equiv \sum_k v_k x^{pk} = v(x^p) = v(x)^p \quad (\text{modulo } u(x)).$$

Berlekamp's factoring algorithm therefore proceeds as follows:

B1. Ensure that $u(x)$ is squarefree; i.e., if $\gcd(u(x), u'(x)) \neq 1$, reduce the problem of factoring $u(x)$, as stated earlier in this section.

B2. Form the matrix Q defined by (12). This can be done in one of two ways, depending on whether or not p is very large, as explained below.

B3. "Triangularize" the matrix $Q - I$, where $I = (\delta_{ij})$ is the $n \times n$ identity matrix, finding its rank $n - r$ and finding linearly independent vectors $v^{[1]}, \ldots, v^{[r]}$ such that $v^{[j]}(Q - I) = (0, 0, \ldots, 0)$ for $1 \leq j \leq r$. (The first vector $v^{[1]}$ may always be taken as $(1, 0, \ldots, 0)$, representing the trivial solution $v^{[1]}(x) = 1$ to (8). The "triangularization" needed in this step can be done using appropriate column operations, as explained in Algorithm N below.) *At this point, r is the number of irreducible factors of*

$u(x)$, because the solutions to (8) are the p^r polynomials corresponding to the vectors $t_1 v^{[1]} + \cdots + t_r v^{[r]}$ for all choices of integers $0 \leq t_1, \ldots, t_r < p$. Therefore if $r = 1$ we know that $u(x)$ is irreducible, and the procedure terminates.

B4. Calculate $\gcd\left(u(x), v^{[2]}(x) - s\right)$ for $0 \leq s < p$, where $v^{[2]}(x)$ is the polynomial represented by vector $v^{[2]}$. The result will be a nontrivial factorization of $u(x)$, because $v^{[2]}(x) - s$ is nonzero and has degree less than $\deg(u)$, and by exercise 7 we have

$$u(x) = \prod_{0 \leq s < p} \gcd\left(v(x) - s, u(x)\right) \tag{14}$$

whenever $v(x)$ satisfies (8).

If the use of $v^{[2]}(x)$ does not succeed in splitting $u(x)$ into r factors, further factors can be obtained by calculating $\gcd\left(v^{[k]}(x) - s, w(x)\right)$ for $0 \leq s < p$ and all factors $w(x)$ found so far, for $k = 3, 4, \ldots$, until r factors are obtained. (If we choose $s_i \neq s_j$ in (7), we obtain a solution $v(x)$ to (8) which distinguishes $p_i(x)$ from $p_j(x)$; some $v^{[k]}(x) - s$ will be divisible by $p_i(x)$ and not by $p_j(x)$, so this procedure will eventually find all of the factors.) ∎

As an example of this procedure, let us now determine the factorization of

$$u(x) = x^8 + x^6 + 10x^4 + 10x^3 + 8x^2 + 2x + 8 \tag{15}$$

modulo 13. (This polynomial appears in several of the examples in Section 4.6.1.) A quick calculation using Algorithm 4.6.1E shows that $\gcd\left(u(x), u'(x)\right) = 1$; therefore $u(x)$ is squarefree, and we turn to step B2. Step B2 involves calculating the Q matrix, which in this case is an 8×8 array. The first row of Q is always $(1, 0, 0, \ldots, 0)$, representing the polynomial $x^0 \bmod u(x) = 1$. The second row represents $x^{13} \bmod u(x)$, and, in general, $x^k \bmod u(x)$ may readily be determined as follows (for relatively small values of k): If

$$u(x) = x^n + u_{n-1}x^{n-1} + \cdots + u_1 x + u_0$$

and if

$$x^k \equiv a_{k,n-1}x^{n-1} + \cdots + a_{k,1}x + a_{k,0} \quad (\text{modulo } (u(x))),$$

then

$$
\begin{aligned}
x^{k+1} &\equiv a_{k,n-1}x^n + \cdots + a_{k,1}x^2 + a_{k,0}x \\
&\equiv a_{k,n-1}(-u_{n-1}x^{n-1} - \cdots - u_1 x - u_0) + a_{k,n-2}x^{n-1} + \cdots + a_{k,0}x \\
&= a_{k+1,n-1}x^{n-1} + \cdots + a_{k+1,1}x + a_{k+1,0},
\end{aligned}
$$

where

$$a_{k+1,j} = a_{k,j-1} - a_{k,n-1}u_j. \tag{16}$$

In this formula $a_{k,-1}$ is treated as zero, so that $a_{k+1,0} = -a_{k,n-1}u_0$. The

simple "shift register" recurrence (16) makes it easy to calculate x^1, x^2, x^3, ...
mod $u(x)$. Inside a computer, this calculation would of course be done by
keeping a one-dimensional array $(a_{n-1}, \ldots, a_1, a_0)$ and repeatedly setting
$t \leftarrow a_{n-1}$, $a_{n-1} \leftarrow (a_{n-2} - tu_{n-1})$ mod p, ..., $a_1 \leftarrow (a_0 - tu_1)$ mod p, $a_0 \leftarrow$
$(-tu_0)$ mod p. (We have seen similar procedures in connection with random-
number generation; cf. Section 3.2.2.) For our example polynomial $u(x)$ in (15),
we obtain the following sequence of coefficients of x^k mod $u(x)$, using arithmetic
modulo 13:

k	$a_{k,7}$	$a_{k,6}$	$a_{k,5}$	$a_{k,4}$	$a_{k,3}$	$a_{k,2}$	$a_{k,1}$	$a_{k,0}$
0	0	0	0	0	0	0	0	1
1	0	0	0	0	0	0	1	0
2	0	0	0	0	0	1	0	0
3	0	0	0	0	1	0	0	0
4	0	0	0	1	0	0	0	0
5	0	0	1	0	0	0	0	0
6	0	1	0	0	0	0	0	0
7	1	0	0	0	0	0	0	0
8	0	12	0	3	3	5	11	5
9	12	0	3	3	5	11	5	0
10	0	4	3	2	8	0	2	8
11	4	3	2	8	0	2	8	0
12	3	11	8	12	1	2	5	7
13	11	5	12	10	11	7	1	2

Therefore the second row of Q is (2, 1, 7, 11, 10, 12, 5, 11). Similarly we may
determine x^{26} mod $u(x)$, ..., x^{91} mod $u(x)$, and we find that

$$Q = \begin{pmatrix} 1 & 0 & 0 & 0 & 0 & 0 & 0 & 0 \\ 2 & 1 & 7 & 11 & 10 & 12 & 5 & 11 \\ 3 & 6 & 4 & 3 & 0 & 4 & 7 & 2 \\ 4 & 3 & 6 & 5 & 1 & 6 & 2 & 3 \\ 2 & 11 & 8 & 8 & 3 & 1 & 3 & 11 \\ 6 & 11 & 8 & 6 & 2 & 7 & 10 & 9 \\ 5 & 11 & 7 & 10 & 0 & 11 & 7 & 12 \\ 3 & 3 & 12 & 5 & 0 & 11 & 9 & 12 \end{pmatrix},$$

$$Q - I = \begin{pmatrix} 0 & 0 & 0 & 0 & 0 & 0 & 0 & 0 \\ 2 & 0 & 7 & 11 & 10 & 12 & 5 & 11 \\ 3 & 6 & 3 & 3 & 0 & 4 & 7 & 2 \\ 4 & 3 & 6 & 4 & 1 & 6 & 2 & 3 \\ 2 & 11 & 8 & 8 & 2 & 1 & 3 & 11 \\ 6 & 11 & 8 & 6 & 2 & 6 & 10 & 9 \\ 5 & 11 & 7 & 10 & 0 & 11 & 6 & 12 \\ 3 & 3 & 12 & 5 & 0 & 11 & 9 & 11 \end{pmatrix}.$$

(17)

The next step of Berlekamp's procedure requires finding the "null space" of $Q - I$. In general, suppose that A is an $n \times n$ matrix over a field, whose rank $n - r$ is to be determined; suppose further that we wish to determine linearly independent vectors $v^{[1]}, v^{[2]}, \ldots, v^{[r]}$ such that $v^{[1]}A = v^{[2]}A = \cdots = v^{[r]}A = (0, \ldots, 0)$. An algorithm for this calculation can be based on the observation that any column of A may be multiplied by a nonzero quantity, and any multiple of one of its columns may be added to a different column, without changing the rank or the vectors $v^{[1]}, \ldots, v^{[r]}$. (These transformations amount to replacing A by AB, where B is a nonsingular matrix.) The following well-known "triangularization" procedure may therefore be used.

Algorithm N (*Null space algorithm*). Let $A = (a_{ij})$ be an $n \times n$ matrix, whose elements a_{ij} belong to a field and have subscripts in the range $0 \le i, j < n$. This algorithm outputs r vectors $v^{[1]}, \ldots, v^{[r]}$, which are linearly independent over the field and satisfy $v^{[j]}A = (0, \ldots, 0)$, where $n - r$ is the rank of A.

N1. [Initialize.] Set $c_0 \leftarrow c_1 \leftarrow \cdots \leftarrow c_{n-1} \leftarrow -1$, $r \leftarrow 0$. (During the calculation we will have $c_j \ge 0$ only if $a_{c_j j} = -1$ and all other entries of row c_j are zero.)

N2. [Loop on k.] Do step N3 for $k = 0, 1, \ldots, n - 1$, and then terminate the algorithm.

N3. [Scan row for dependence.] If there is some j in the range $0 \le j < n$ such that $a_{kj} \ne 0$ and $c_j < 0$, then do the following: Multiply column j of A by $-1/a_{kj}$ (so that a_{kj} becomes equal to -1); then add a_{ki} times column j to column i for all $i \ne j$; finally set $c_j \leftarrow k$. (Since it is not difficult to show that $a_{sj} = 0$ for all $s < k$, these operations have no effect on rows $0, 1, \ldots, k - 1$ of A.)

On the other hand, if there is no j in the range $0 \le j < n$ such that $a_{kj} \ne 0$ and $c_j < 0$, then set $r \leftarrow r + 1$ and output the vector

$$v^{[r]} = (v_0, v_1, \ldots, v_{n-1})$$

defined by the rule

$$v_j = \begin{cases} a_{ks}, & \text{if} \quad c_s = j \ge 0; \\ 1, & \text{if} \quad j = k; \\ 0, & \text{otherwise.} \end{cases} \qquad (18) \quad \blacksquare$$

An example will reveal the mechanism of this algorithm; let A be the matrix $Q - I$ of (17) over the field of integers modulo 13. When $k = 0$, we output the vector $v^{[1]} = (1, 0, 0, 0, 0, 0, 0, 0)$. When $k = 1$, we may take j in step N3 to be either 0, 2, 3, 4, 5, 6, or 7; the choice here is completely arbitrary, although it affects the particular vectors which are chosen to be output by the algorithm. For hand calculation, it is most convenient to pick $j = 5$, since $a_{15} = 12 = -1$ already; the column operations of step N3 then change

A to the matrix

$$
\begin{pmatrix}
0 & 0 & 0 & 0 & 0 & 0 & 0 & 0 \\
0 & 0 & 0 & 0 & 0 & \textcircled{12} & 0 & 0 \\
11 & 6 & 5 & 8 & 1 & 4 & 1 & 7 \\
3 & 3 & 9 & 5 & 9 & 6 & 6 & 4 \\
4 & 11 & 2 & 6 & 12 & 1 & 8 & 9 \\
5 & 11 & 11 & 7 & 10 & 6 & 1 & 10 \\
1 & 11 & 6 & 1 & 6 & 11 & 9 & 3 \\
12 & 3 & 11 & 9 & 6 & 11 & 12 & 2
\end{pmatrix}.
$$

(The circled element in column "5," row "1," may be used to indicate that $c_5 = 1$. Remember that Algorithm N numbers the rows and columns of the matrix starting with 0, not 1.) When $k = 2$, we may choose $j = 4$ and proceed in a similar way, obtaining the following matrices, which all have the same null space as $Q - I$:

$$k = 2$$

$$
\begin{pmatrix}
0 & 0 & 0 & 0 & 0 & 0 & 0 & 0 \\
0 & 0 & 0 & 0 & 0 & \textcircled{12} & 0 & 0 \\
0 & 0 & 0 & 0 & \textcircled{12} & 0 & 0 & 0 \\
8 & 1 & 3 & 11 & 4 & 9 & 10 & 6 \\
2 & 4 & 7 & 1 & 1 & 5 & 9 & 3 \\
12 & 3 & 0 & 5 & 3 & 5 & 4 & 5 \\
0 & 1 & 2 & 5 & 7 & 0 & 3 & 0 \\
11 & 6 & 7 & 0 & 7 & 0 & 6 & 12
\end{pmatrix}
$$

$$k = 3$$

$$
\begin{pmatrix}
0 & 0 & 0 & 0 & 0 & 0 & 0 & 0 \\
0 & 0 & 0 & 0 & 0 & \textcircled{12} & 0 & 0 \\
0 & 0 & 0 & 0 & \textcircled{12} & 0 & 0 & 0 \\
0 & \textcircled{12} & 0 & 0 & 0 & 0 & 0 & 0 \\
9 & 9 & 8 & 9 & 11 & 8 & 8 & 5 \\
1 & 10 & 4 & 11 & 4 & 4 & 0 & 0 \\
5 & 12 & 12 & 7 & 3 & 4 & 6 & 7 \\
2 & 7 & 2 & 12 & 9 & 11 & 11 & 2
\end{pmatrix}
$$

$$k = 4$$

$$
\begin{pmatrix}
0 & 0 & 0 & 0 & 0 & 0 & 0 & 0 \\
0 & 0 & 0 & 0 & 0 & \textcircled{12} & 0 & 0 \\
0 & 0 & 0 & 0 & \textcircled{12} & 0 & 0 & 0 \\
0 & \textcircled{12} & 0 & 0 & 0 & 0 & 0 & 0 \\
0 & 0 & 0 & 0 & 0 & 0 & 0 & \textcircled{12} \\
1 & 10 & 4 & 11 & 4 & 4 & 0 & 0 \\
8 & 2 & 6 & 10 & 11 & 11 & 0 & 9 \\
1 & 6 & 4 & 11 & 2 & 0 & 0 & 10
\end{pmatrix}
$$

$$k = 5$$

$$
\begin{pmatrix}
0 & 0 & 0 & 0 & 0 & 0 & 0 & 0 \\
0 & 0 & 0 & 0 & 0 & \textcircled{12} & 0 & 0 \\
0 & 0 & 0 & 0 & \textcircled{12} & 0 & 0 & 0 \\
0 & \textcircled{12} & 0 & 0 & 0 & 0 & 0 & 0 \\
0 & 0 & 0 & 0 & 0 & 0 & 0 & \textcircled{12} \\
\textcircled{12} & 0 & 0 & 0 & 0 & 0 & 0 & 0 \\
5 & 0 & 0 & 0 & 5 & 5 & 0 & 9 \\
12 & 9 & 0 & 0 & 11 & 9 & 0 & 10
\end{pmatrix}
$$

Now every column which has no circled entry is completely zero; so when $k = 6$ and $k = 7$ the algorithm outputs two more vectors, namely

$$v^{[2]} = (0, 5, 5, 0, 9, 5, 1, 0), \qquad v^{[3]} = (0, 9, 11, 9, 10, 12, 0, 1). \qquad (19)$$

From the form of matrix *A* after $k = 5$, it is evident that these vectors satisfy the equation $vA = (0, \ldots, 0)$. Since the computation has produced three linearly independent vectors, $u(x)$ must have exactly three irreducible factors.

We now go to step B4 of the factoring procedure. Calculation of gcd $(u(x), v^{[2]}(x) - s)$ for $s = 0, 1, \ldots, 12$, where $v^{[2]}(x) = x^6 + 5x^5 + 9x^4 + 5x^2 + 5x$, gives the answer $x^5 + 5x^4 + 9x^3 + 5x + 5$ when $s = 0$, $x^3 + 8x^2 + 4x + 12$ when $s = 2$, and unity for other values of s. Therefore $v^{[2]}(x)$ gives us only two of the three factors. Turning to gcd $(v^{[3]}(x) - s, x^5 + 5x^4 + 9x^3 + 5x + 5)$, where $v_3(x) = x^7 + 12x^5 + 10x^4 + 9x^3 + 11x^2 + 9x$, we obtain the value $x^4 + 2x^3 + 3x^2 + 4x + 6$ when $s = 6$, $x + 3$ when $s = 8$, and unity otherwise. Thus the complete factorization is

$$u(x) = (x^4 + 2x^3 + 3x^2 + 4x + 6)(x^3 + 8x^2 + 4x + 12)(x + 3). \quad (20)$$

Let us now estimate the running time of Berlekamp's method when an nth degree polynomial is factored modulo p. First assume that p is relatively small, so that the four arithmetic operations can be done modulo p in essentially a fixed length of time. (Division modulo p can be converted to multiplication, by storing a table of reciprocals; modulo 13, for example, we have $\frac{1}{2} = 7$, $\frac{1}{3} = 9$, etc.) The computation in step B1 takes $O(n^2)$ units of time; step B2 takes $O(pn^2)$. For step B3 we use Algorithm N, which requires $O(n^3)$ units of time at most. Finally, in step B4 we can observe that the calculation of gcd $(f(x), g(x))$ by Euclid's algorithm takes $O(\deg (f) \deg (g))$ units of time; hence the calculation of gcd $(v^{[j]}(x) - s, w(x))$ for fixed j and s and for all factors $w(x)$ of $u(x)$ found so far takes $O(n^2)$ units. Step B4 therefore requires $O(prn^2) = O(pn^3)$ units of time at most. *Berlekamp's procedure factors an arbitrary polynomial of degree n, modulo p, in $O(pn^3)$ steps, when p is a small prime.* This is much faster than any known methods of factoring n-digit numbers in the p-ary number system.

Of course, when n is small, a trial-and-error factorization procedure analogous to Algorithm 4.5.4A will be even faster than Berlekamp's method. Exercise 1 implies that it is worth while to cast out factors of small degree first, before going to any more complicated procedure, even when n is large.

When p is large, a different implementation of Berlekamp's procedure would be used for the calculations. Division modulo p would not be done with an auxiliary table of reciprocals; instead the method of exercise 4.5.2–15, which takes $O((\log p)^2)$ steps, would probably be used. Then step B1 would take $O(n^2(\log p)^2)$ units of time; similarly, step B3 takes $O(n^3(\log p)^2)$. In step B2, we can form $x^p \bmod u(x)$ in a more efficient way than (16) when p is large: Section 4.6.3 shows that this value can essentially be obtained by using $O(\log p)$ operations of "squaring mod $u(x)$," i.e., going from $x^k \bmod u(x)$ to $x^{2k} \bmod u(x)$. This squaring operation is relatively easy to perform if we first make an auxiliary table of $x^m \bmod u(x)$ for $m = n, n + 1, \ldots, 2n - 2$; if

$$x^k \bmod u(x) = c_{n-1}x^{n-1} + \cdots + c_1 x + c_0,$$

then

$$x^{2k} \bmod u(x) = c_{n-1}^2 x^{2n-2} + \cdots + (c_1 c_0 + c_1 c_0)x + c_0^2,$$

where $x^{2n-2}, \ldots, x^n$ can be replaced by polynomials in the auxiliary table. The net time to compute $x^p \bmod u(x)$ comes to $O(n^2(\log p)^3)$ units, and we obtain the second row of Q. To get further rows of Q, we form $x^{2p} \bmod u(x)$, $x^{3p} \bmod u(x)$, $\ldots$ simply by multiplying repeatedly by $x^p \bmod u(x)$, in a fashion analogous to squaring mod $u(x)$; step B2 is completed in $O(n^2(\log p)^3 + n^3(\log p)^2)$ units of time. The same upper bound applies to steps B1, B2, and B3 taken as a whole; these three steps tell us the number of factors of $u(x)$.

But when p is large and we get to step B4, we are asked to calculate a greatest common divisor for p different values of s, and this is out of the question if p is very large. It is possible to do some of the calculations for all s at once, as shown in exercise 14, but only until getting to polynomials of degree about $n/2$; in fortunate circumstances this will yield factors of $u(x)$, but it cannot be guaranteed to work.

Another technique. A partial method of factoring modulo p, which is especially useful for large p when Berlekamp's method is too slow, can be based on the fact that an irreducible polynomial $q(x)$ of positive degree d is a divisor of $x^{p^d} - x$, and it is not a divisor of $x^{p^t} - x$ for $t < d$. [A proof of this relation follows from the theory of finite fields; see exercise 16.] This factorization procedure was fairly well known in 1960, but there seem to be few references to it in the "open literature." For example, see Solomon W. Golomb, Lloyd R. Welch, Alfred Hales, "On the factorization of trinomials over GF(2)," Jet Propulsion Laboratory memo 20–189 (July 14, 1959).

S1. Rule out squared factors and find the matrix Q, as in Berlekamp's method. Also set $v(x) \leftarrow u(x)$, $w(x) \leftarrow$ "x", and $d \leftarrow 0$. (Here $v(x)$ and $w(x)$ are variables which have polynomials as values.)

S2. Increase d by 1 and replace $w(x)$ by $w(x)^p \bmod u(x)$. (In other words, the coefficients $(w_0, \ldots, w_{n-1})$ are replaced by $(w_0, \ldots, w_{n-1})Q$. At this point $w(x) = x^{p^d} \bmod u(x)$.)

S3. Find $g(x) = \gcd\left(w(x) - x, v(x)\right)$. (This is the product of all the irreducible factors of $u(x)$ whose degree is d.) Replace $v(x)$ by $v(x)/g(x)$.

S4. If $d < \deg(v)/2$, return to S2. Otherwise $v(x)$ is irreducible, and the procedure terminates. ∎

The total running time, analyzed as we have done for Berlekamp's method, is $O(n^3(\log p)^2 + n^2(\log p)^3)$; this procedure determines the product of all irreducible factors of degree d, and it also tells us how many factors there are of each degree. Since the three factors of our example polynomial (20) have different degrees, they would all have been discovered.

When the degree of $g(x)$ in step S3 turns out to be dk for some $k \geq 2$, this method has not given us the complete factorization. Suppose that $g(x) = p_1(x) \ldots p_k(x)$ where each $p_j(x)$ is irreducible of degree d. Of course, we may assume that the constant term of $g(x)$ is nonzero, so that $p_j(x) \neq x$. Let $0 \leq s < p$; in a field with p^d elements, the roots of $p_j(x + s)$ have order f_{js}, for some integer f_{js} which divides $p^d - 1$, but f_{js} does not divide $p^t - 1$ for $t < d$. If f_{js} is not a multiple of f_{is}, it is not difficult to prove that $p_j(x)$ divides $(x - s)^{f_{js}} - 1$, but $p_i(x)$ does not. Therefore H. Zassenhaus has suggested that we calculate gcd $(g(x), (x - s)^f - 1)$ for suitable f and for small s. (The same procedure was used for $p = 2$, $s = 1$ by Golomb, Welch, and Hales in the reference cited earlier.) Exercise 16 shows that the multiplicative structure of a field with p^d elements is cyclic of order $p^d - 1$, so we can see that this method is quite likely to give factorizations quickly, even if we do not know the complete factorization of $p^d - 1$. See exercises 15 and 17; factors will usually be found when $f = (p^d - 1)/q$, where q is a relatively small divisor of $p^d - 1$. For further information, see E. R. Berlekamp, *Math. Comp.* **24** (1970), 713–735.

Factoring over the integers. It is somewhat more difficult to find the complete factorization of polynomials with integer coefficients when we are *not* working modulo p, but some reasonably efficient methods are available for this purpose.

Isaac Newton gave a method for finding linear and quadratic factors of polynomials with integer coefficients in his *Arithmetica Universalis* (1707). This method was extended by an astronomer named Friedrich von Schubert in 1793, who showed how to find all factors of degree n in a finite number of steps; see M. Cantor, *Geschichte der Mathematik* **4** (Leipzig: Teubner, 1908), 136–137. L. Kronecker rediscovered von Schubert's method independently about 90 years later; but unfortunately the method is very inefficient when n is five or more. Much better results can be obtained with the help of the "mod p" factorization methods presented above.

Suppose that we want to find the irreducible factors of a given polynomial

$$u(x) = u_n x^n + u_{n-1} x^{n-1} + \cdots + u_0, \qquad u_n \neq 0,$$

over the integers. As a first step, we can divide by the greatest common divisor of the coefficients, and this leaves us with a *primitive* polynomial. We may assume, in fact, that $u(x)$ is *monic;* for it is not difficult to relate the factors of the monic polynomial $u_n^{n-1} \cdot u(x/u_n)$ to the factors of $u(x)$; see exercise 18.

Now if $u(x) = v(x)w(x)$, where all of these polynomials are monic and have integer coefficients, we obviously have $u(x) \equiv v(x)w(x)$ (modulo p), for all primes p, so there is a nontrivial factorization modulo p. Berlekamp's efficient algorithm for factoring $u(x)$ modulo p can therefore be used in an attempt to reconstruct possible factorizations of $u(x)$ over the integers.

For example, let

$$u(x) = x^8 + x^6 - 3x^4 - 3x^3 + 8x^2 + 2x - 5. \tag{21}$$

We have seen above in (20) that

$$u(x) \equiv (x^4 + 2x^3 + 3x^2 + 4x + 6)(x^3 + 8x^2 + 4x + 12)(x + 3)$$

$$\text{(modulo 13);} \tag{22}$$

and the complete factorization of $u(x)$ modulo 2 shows one factor of degree 6 and another of degree 2 (see exercise 10). From (22) we can see that $u(x)$ has no factor of degree 6, so it must be irreducible over the integers.

In practice, we would probably not have factored $u(x)$ modulo 13 first; smaller primes 2, 3, 5, 7, 11 would have been tried. The case $p = 13$ has been considered here only because it was already treated above.

This particular example was perhaps too simple; it is not always so easy to show that a polynomial is irreducible! For example, there are polynomials which are irreducible over the integers, but which can be properly factored modulo p for all primes p, with consistent degrees of the factors (see exercise 12).

Suppose that we have been able to find the factors of $u(x)$ modulo p for several primes p, and suppose that a consistent pattern has been found suggesting that $u(x) = v(x)w(x)$, where the degrees of $v(x)$ and $w(x)$ are known. We may assume that u, v, and w are monic. An important help in ruling out possible alternatives for v and w is to note that the product of their constant terms (the coefficients of x^0) is the constant term of u. This gives a finite number of possibilities for the constant coefficients, and each can be tried to see if it is compatible with the factorization modulo p for all p. For each compatible factorization of this form, we can determine the values of the coefficients of v and w modulo m, where m is the product of the primes p, by means of the Chinese remainder theorem (see Section 4.3.2).

Furthermore we can often make use of the fact that the coefficients of x^0, x^1, $\ldots$ in v and w are multiples of some rather large numbers which are prime to m, as shown in exercise 18(b); this observation effectively increases m, and it considerably improves our chances of success in discovering a factorization rapidly.

All we need to complete the factorization is to obtain bounds on the possible sizes of the coefficients of v and/or w; then if m is large enough, each coefficient will be determined.

Two methods for finding bounds on the sizes of the coefficients have been suggested by G. E. Collins. For simplicity, let us suppose that $v(x) = x^3 + v_2 x^2 + v_1 x + v_0$ is a factor of $u(x)$ having degree 3; other degrees can be handled similarly. In matrix notation we have

$$\begin{pmatrix} 1 & -1 & 1 \\ 0 & 0 & 1 \\ 1 & 1 & 1 \end{pmatrix} \begin{pmatrix} v_2 \\ v_1 \\ v_0 \end{pmatrix} = \begin{pmatrix} v(-1) + 1 \\ v(0) \\ v(1) - 1 \end{pmatrix}; \qquad (23)$$

therefore, by taking the inverse of the Vandermonde matrix on the left (see exercise 1.2.3–40), we have

$$\begin{pmatrix} v_2 \\ v_1 \\ v_0 \end{pmatrix} = \begin{pmatrix} \frac{1}{2} & -1 & \frac{1}{2} \\ -\frac{1}{2} & 0 & \frac{1}{2} \\ 0 & 1 & 0 \end{pmatrix} \begin{pmatrix} v(-1) + 1 \\ v(0) \\ v(1) - 1 \end{pmatrix}. \qquad (24)$$

Now $v(-1)$, $v(0)$, and $v(1)$ are divisors of $u(-1)$, $u(0)$, and $u(1)$; and since we may assume that none of the latter are zero (lest $u(x)$ have the factor $x + 1$, x, or $x - 1$), we have $|v(-1)| \leq |u(-1)|$, $|v(0)| \leq |u(0)|$, etc. We may now derive upper bounds on the coefficients of v, using (24); for example, $v_2 = \frac{1}{2}v(-1) - v(0) + \frac{1}{2}v(1)$; hence

$$|v_2| \leq \tfrac{1}{2}|u(-1)| + |u(0)| + \tfrac{1}{2}|u(1)|. \tag{25}$$

This method gives a finite bound on the size of each coefficient; hence the factorization can be done in finitely many steps. Moreover, if we know (from factorizations mod p) that there is only one possible value of v_2 mod m, where m is suitably large, then there will be at most one possible value for v_2.

Collins's second suggestion for bounding the coefficients is based on the fact that a monic polynomial of degree 3 can be factored

$$v(z) = (z - \zeta_1) \cdot (z - \zeta_2) \cdot (z - \zeta_3) \tag{26}$$

over the complex numbers. Here ζ_1, ζ_2, and ζ_3 are roots of the equation $v(z) = 0$, so they are roots also of $u(z) = 0$. We have

$$\begin{aligned} v_2 &= -(\zeta_1 + \zeta_2 + \zeta_3), \\ v_1 &= (\zeta_1\zeta_2 + \zeta_1\zeta_3 + \zeta_2\zeta_3), \\ v_0 &= -(\zeta_1\zeta_2\zeta_3). \end{aligned} \tag{27}$$

Therefore if we can determine by some numerical method that all complex numbers ζ which are roots of the equation $u(z) = 0$ satisfy $|\zeta| \leq \alpha$, we have

$$|v_2| \leq 3\alpha, \qquad |v_1| \leq 3\alpha^2, \qquad |v_0| \leq \alpha^3. \tag{28}$$

Two useful estimates for α appear in exercises 19 and 20.

As one example of these remarks, consider again the polynomial $u(x)$ in (21); let us try to prove that $u(x)$ is irreducible over the integers just by knowing (22), its complete factorization modulo 13. It is easy to see, by the divisibility constraint on constant coefficients, that the only possibility for a factorization consistent with (22) is

$$u(x) = (x^3 + v_2x^2 + v_1x - 1)(x^5 + w_4x^4 + w_3x^3 + w_2x^2 + w_1x + 5), \tag{29}$$

where v_2 mod $13 = 8$. Now the estimate for v_2 in (25) can be improved to

$$|v_2| \leq \tfrac{1}{2}|u(-1)| + |v_0| + \tfrac{1}{2}|u(1)|,$$

since v_0 is known, and therefore we must have $|v_2| \leq 3$. This is inconsistent with the condition v_2 mod $13 = 8$, so $u(x)$ is irreducible.

(From (24) we also get the bound $|v_1| \leq 3$; the bounds in (28) are not as useful in this case, because $u(-1)$ and $u(1)$ are so small, but (28) is often superior to the Vandermonde bound.)

Of course, it may happen that neither of the above methods gives a very small upper bound on the magnitude of the coefficients, so we might have a long wait if we try to factor $u(x)$ modulo different primes $p_1, p_2, p_3, \ldots, p_k$ until $p_1 p_2 \ldots p_k$ is large enough. The use of different primes is often helpful for proving that $u(x)$ is irreducible, but it is not so efficient for actually finding factors when they exist. An important time-saving device has been suggested by H. Zassenhaus: Find a prime p such that $u(x)$ has no multiple factors modulo p. Then if $u(x) \equiv v_0(x) w_0(x)$ (modulo p), where $v_0(x)$ and $w_0(x)$ are relatively prime polynomials modulo p, use the algorithm in exercise 4.6.1–3 followed by the method of exercise 22 to find sequences of monic polynomials $\langle v_n(x) \rangle$, $\langle w_n(x) \rangle$ with integer coefficients such that

$$v_n(x) \equiv v_0(x), \qquad w_n(x) \equiv w_0(x) \qquad (\text{modulo } p);$$
$$u(x) \equiv v_n(x) w_n(x) \qquad (\text{modulo } p^{2^n}). \qquad (30)$$

The results of exercise 22 show that $v_n(x)$ and $w_n(x)$ are uniquely determined by the conditions in (30); therefore if $u(x)$ can be factored over the integers into monic polynomials $v(x) w(x)$, where $v(x) \equiv v_0(x)$ and $w(x) \equiv w_0(x)$ (modulo p), the coefficients of $v(x)$ and $w(x)$ will be found rapidly by taking n large enough in (30).

Greatest common divisors. Finally, let us observe that similar techniques can be used to facilitate calculating greatest common divisors of polynomials over the integers: If $\gcd\big(u(x), v(x)\big) = d(x)$ over the integers, and if $\gcd\big(u(x), v(x)\big) = q(x)$ modulo p, then $d(x)$ is a common divisor of $u(x)$ and $v(x)$ modulo p; hence

$$d(x) \quad \text{divides} \quad q(x) \quad (\text{modulo } p). \qquad (31)$$

If p does not divide the leading coefficients of both u and v, it does not divide the leading coefficient of d; in such a case $\deg(d) \le \deg(q)$. When $q(x) = 1$ for such a prime p, we must therefore have $\deg(d) = 0$, and $d(x) = \gcd\big(\text{cont}(u), \text{cont}(v)\big)$. This justifies the remark made in Section 4.6.1 that the simple calculation (4.6.1–6) of $\gcd\big(u(x), v(x)\big)$ modulo 13 is enough to prove that $u(x)$ and $v(x)$ are relatively prime over the integers; the comparatively laborious calculations of Algorithm 4.6.1E or Algorithm 4.6.1C are unnecessary. Since two random primitive polynomials are almost always relatively prime over the integers, and since they are relatively prime modulo p with probability $1 - 1/p$, it is usually a good idea to do the computations modulo p.

Of course the polynomials which arise in practice are *not* "random," since there is probably some reason to expect a nontrivial greatest common divisor or else the calculation would not have been attempted in the first place. Therefore we wish to sharpen our techniques and discover how to find $\gcd\big(u(x), v(x)\big)$ in general, over the integers, based entirely on information that we obtain working modulo primes p. We may assume that $u(x)$ and $v(x)$ are primitive.

Instead of calculating gcd $(u(x), v(x))$ directly, it will be convenient to search instead for the polynomial

$$d(x) = c \cdot \text{gcd}\ (u(x), v(x)), \qquad (32)$$

where the constant c is chosen so that

$$\ell(d) = \text{gcd}\ (\ell(u), \ell(v)). \qquad (33)$$

This condition will always hold for suitable c, since the leading coefficient of any common divisor of $u(x)$ and $v(x)$ must be a divisor of gcd $(\ell(u), \ell(v))$. Once $d(x)$ has been found satisfying these conditions, we can readily compute pp $(d(x))$, which is the true greatest common divisor of $u(x)$ and $v(x)$. Condition (32) is convenient since it avoids the uncertainty of unit multiples of the gcd, which would otherwise be present when we are working modulo p.

To begin the calculation of gcd $(u(x), v(x))$, let p be a prime number which is not a divisor of $\ell(u)$ or $\ell(v)$. After applying Euclid's algorithm for polynomials modulo p to $u(x)$ and $v(x)$, we obtain a polynomial $q(x)$ which satisfies (31), and by multiplying it by a suitable constant we may assume that $\ell(q) = $ gcd $(\ell(u), \ell(v))$. Now $q(x)$ has the leading coefficient we seek for $d(x)$, but we have been working modulo p instead of doing exact integer arithmetic; under what circumstances can we be sure that the congruence

$$q(x) \equiv d(x) \qquad (\text{modulo } p) \qquad (34)$$

is valid? A study of Algorithm 4.6.1E reveals that this will be the case if p does not divide any of the leading coefficients of the nonzero remainders which would appear in step E2 if exact integer arithmetic were used; for in such a case Euclid's algorithm modulo p deals with precisely the same sequence of polynomials as Algorithm 4.6.1E, except for nonzero constant multiples (modulo p). So all but a finite number of primes will satisfy (34), and if we choose a large prime p (near the word size of our computer) we will almost always find that (34) is true.

Once $q(x)$ has been computed, we can attempt to find $d(x)$ by assuming that (34) holds; we may choose the coefficients of $d(x)$ to be the smallest possible numbers in absolute value such that (34) is satisfied. Now if the polynomial pp $(d(x))$ is a bona fide divisor of both $u(x)$ and $v(x)$, condition (31) implies that our assumption (34) must indeed have been true, and pp $(d(x))$ is the desired gcd. On the other hand if pp $(d(x))$ does not divide both $u(x)$ and $v(x)$, we can repeat the calculation using further primes p. Eventually $d(x)$ will be determined modulo so many primes, it will be determined completely (using the Chinese remainder algorithm).

When we apply this method there is always a chance that we will have hit one of the rare primes p which violates assumption (34); but this will be recognized by the fact that any polynomial $q(x)$ obtained using such primes will have a higher degree than the degree which is obtained modulo all of the primes which do satisfy (34). Thus if the algorithm obtains $q_1(x)$ of degree k_1, working modulo p_1, and $q_2(x)$ of degree $k_2 < k_1$, working modulo p_2, the polynomial $q_1(x)$ should be discarded. The Chinese remainder algorithm should be applied only to those polynomials $q(x)$ of smallest degree that have been obtained so far. It is clear that the algorithm sketched here will always find the gcd, and on the average it will be much faster than the algorithms of Section 4.6.1. [This algorithm is based in part on ideas suggested to the author by W. S. Brown and G. E. Collins.]

We may find gcd $(u(x, y), v(x, y))$ for bivariate polynomials over the integers in essentially the same way, considering u and v as polynomials in x whose coefficients are polynomials in y, if we replace "p" in the preceding discussion by the irreducible polynomial "$y - y_0$" for any integer y_0. The computation of gcd $(u(x, y), v(x, y))$ modulo $(y - y_0)$ is exactly the same as the computation of gcd $(u(x, y_0), v(x, y_0))$, so it involves polynomials in just one variable over the integers. The Chinese remainder algorithm for this case is called "interpolation"; see Sections 4.3.3 and 4.6.4. An example appears in exercise 28. In this way we may reduce the problem of calculating the gcd of polynomials in n variables to the $(n - 1)$-variable problem, so we have an algorithm which finds the gcd of polynomials in any number of variables over the integers.

For a complete discussion of the modular method of gcd calculation, see the article by W. S. Brown, "On Euclid's algorithm and the computation of polynomial greatest common divisors," *JACM* (to appear).

EXERCISES

▶ **1.** [*M24*] Let p be prime. What is the probability that a random polynomial of degree n has a linear factor (a factor of degree 1), when $n \geq p$? (Assume that each of the p^n monic polynomials is equally probable.)

▶ **2.** [*M25*] (a) Show that any monic polynomial $u(x)$, over a unique factorization domain, may be expressed uniquely in the form

$$u(x) = v(x)^2 w(x),$$

where $w(x)$ is squarefree (has no factor of positive degree of the form $d(x)^2$) and both $v(x)$ and $w(x)$ are monic.

b) (E. R. Berlekamp.) How many monic polynomials of degree n are squarefree modulo p, when p is prime?

3. [*M25*] Let $u_1(x), \ldots, u_r(x)$ be polynomials over a field S, with $u_j(x)$ relatively prime to $u_k(x)$ for all $j \neq k$. For any given polynomials $w_1(x), \ldots, w_r(x)$ over S, prove that there is a unique polynomial $v(x)$ over S such that

$$\deg (v) < \deg (u_1) + \cdots + \deg (u_r)$$

and

$$v(x) \equiv w_j(x) \pmod{u_j(x)}$$

for $1 \leq j \leq r$. (Compare with Theorem 4.3.2C.)

4. [*HM28*] Let a_n be the number of monic irreducible polynomials of degree n, modulo a prime p. Find a formula for the generating function $G(z) = \sum_n a_n z^n$. [*Hint:* Prove the following identity connecting power series: $f(z) = \sum_{j \geq 1} g(z^j)/j^i$ if and only if $g(z) = \sum_{n \geq 1} \mu(n)f(z^n)/n^t$.]

5. [*HM25*] Let A_{np} be the average number of factors of a randomly selected polynomial of degree n, modulo a prime p. Show that

$$\lim_{p \to \infty} A_{np} = H_n.$$

6. [*M21*] (J. L. Lagrange, 1771.) Prove the congruence (9). [*Hint:* Factor $x^p - x$ in the field of p elements.]

7. [*M22*] Prove Eq. (14).

8. [*HM20*] How can we be sure that the vectors output by Algorithm N are linearly independent?

9. [*20*] Explain how to construct a table of reciprocals mod 101 in a simple way, given that 2 is a primitive root of 101.

▶ **10.** [*21*] Find the complete factorization of the polynomial $u(x)$ in (21), modulo 2, using Berlekamp's procedure.

11. [*22*] Find the complete factorization of the polynomial $u(x)$ in (21), modulo 5.

▶ **12.** [*M22*] Use Berlekamp's algorithm to determine the number of factors of $u(x) = x^4 + 1$ modulo p, for all primes p. [*Hint:* Consider the cases $p = 2$, $p = 8k + 1$, $p = 8k + 3$, $p = 8k + 5$, $p = 8k + 7$ separately; what is the matrix Q? You need not discover the factors; just determine how many there are.]

13. [*M25*] Give an explicit formula for the factors of $x^4 + 1$, modulo p, for all primes p, in terms of the quantities $\sqrt{-1}$, $\sqrt{2}$, $\sqrt{-2}$ (if such quantities exist modulo p).

14. [*M30*] Berlekamp has observed that some of the calculations of

$$\gcd \left(u(x), v^{[2]}(x) - s \right)$$

in step B4 can be done for all s at once, by retaining s as a parameter. For example, consider applying Euclid's algorithm to the polynomials $u(x)$ and $v^{[3]}(x) - s$ in the example of the text; writing only the coefficients, we have the following sequence of polynomials:

1	0	1	0	10	10	8	2	8
	1	0	12	10	9	11	9	$12s$
		2	3	1	12	12	$2+s$	8
			4	8	12	$6+6s$	$4s$	$6+12s$
				2	$5+2s$	$4+9s$	$5+7s$	$11+11s$

Now when attempting to carry out the next step we run into difficulties, since it becomes necessary to go to quadratic polynomials in s. If we terminate the division before this happens, the partial division is

	4	8	12	$6+6s$	$4s$	$6+12s$
$-$	4	$10+4s$	$8+5s$	$10+\ s$	$9+9s$	
		$11+9s$	$4+8s$	$9+5s$	$4+8s$	$6+12s.$

From this calculation we know that

$$\gcd\left(u(x), v^{[3]}(x) - s\right)$$
$$= \gcd\left(2x^4 + (5 + 2s)x^3 + (4 + 9s)x^2 + (5 + 7s)x + (11 + 11s),\right.$$
$$\left.(11 + 9s)x^4 + (4 + 8s)x^3 + (9 + 5s)x^2 + (4 + 8s)x + (6 + 12s)\right).$$

Here we may observe that if $s = 6$, all coefficients of the latter polynomial are zero! This determines the greatest common divisor when $s = 6$, and so it determines a factor of $u(x)$. In general, we can keep s as a parameter during Euclid's algorithm until an s appears in the leading coefficient of one of the polynomials at some stage; in fortunate circumstances, the value of s which makes this leading coefficient zero will make *all* coefficients zero, and the greatest common divisor will be determined for this value of s.

Show that this procedure will determine a factor of $u(x)$ whenever there is exactly one value of s such that $\gcd\left(u(x), v(x) - s\right)$ has degree greater than or equal to $\deg(u)/2$.

▶ **15.** [*M35*] Design an algorithm to calculate the "square root" of a given integer u modulo a given prime p, i.e., to find an integer U such that $U^2 \equiv u \pmod{p}$ whenever such a U exists. Your algorithm should be efficient even for very large primes p. (A solution to this problem leads to a procedure for solving any quadratic equation modulo p, using the quadratic formula in the usual way.)

16. [*M30*] (a) Given that $f(x)$ is an irreducible polynomial modulo a prime p, of degree n, prove that the p^n polynomials of degree less than n form a field under arithmetic modulo $f(x)$ and p. (*Note:* The existence of irreducible polynomials of each degree is proved in exercise 4; therefore fields with p^n elements exist for all primes p and all $n \geq 1$.)

b) Show that any field with p^n elements has a "primitive root" element ξ such that the elements of the field are $\{0, 1, \xi, \xi^2, \ldots, \xi^{p^n-2}\}$.

c) If $f(x)$ is an irreducible polynomial modulo p, of degree n, prove that $x^{p^m} - x$ is divisible by $f(x)$ if and only if m is a multiple of n.

17. [*M23*] Let F be a field with 13^2 elements. How many elements of F have order f, for each integer f with $1 \leq f < 13^2$? (The "order" of an element a is the least positive integer m such that $a^m = 1$.)

▶ **18.** [*M25*] Let $u(x) = u_n x^n + \cdots + u_0$, $u_n \neq 0$, be a primitive polynomial with integer coefficients, and let

$$v(x) = u_n^{n-1} \cdot u(x/u_n) = x^n + u_{n-1}x^{n-1} + u_{n-2}u_n x^{n-2} + \cdots + u_0 u_n^{n-1}.$$

(a) Given that $v(x)$ has the complete factorization $p_1(x) \cdots p_r(x)$ over the integers, where each $p_j(x)$ is monic, what is the complete factorization of $u(x)$ over the integers?

(b) If $w(x) = x^m + w_{m-1}x^{m-1} + \cdots + w_0$ is a factor of $v(x)$, prove that w_k is a multiple of u_n^{m-1-k} for $0 \leq k < m$.

19. [*HM23*] (H. Zassenhaus.) Let $u(z) = z^n + u_{n-1}z^{n-1} + \cdots + u_0$ for a monic polynomial with complex coefficients, and let

$$\alpha = \max_{1 \leq k \leq n} |u_{n-k}/\binom{n}{k}|^{1/k}/(\sqrt[n]{2} - 1).$$

Given that $u(z_0) = 0$, prove that $|z_0| \leq \alpha$.

20. [*HM21*] Under the hypotheses of exercise 19, prove also that

$$|z_0| < 2 \max_{1 \leq k \leq n} |u_{n-k}|^{1/k}.$$

(This often gives a better bound than the estimate in exercise 19, when the constant term u_0 is the largest coefficient.)

21. [*HM21*] Prove that if $u(z_0) = 0$, where $u(x)$ is the polynomial (21) and z_0 is a complex number, then $|z_0| < 2$.

▶ **22.** [*M30*] (Factoring modulo p^e.) Let $u(x)$, $v(x)$, $w(x)$, $a(x)$, $b(x)$, $c(x)$, $d(x)$ be polynomials with integer coefficients, satisfying the relations

$$v(x)w(x) = u(x) + p^m c(x), \qquad a(x)v(x) + b(x)w(x) = 1 + p^m d(x),$$

where p is a prime, m is a positive integer; $u(x)$, $v(x)$, and $w(x)$ are monic; deg $(a) <$ deg (w), deg $(b) <$ deg (v), deg $(c) <$ deg (u), and deg $(d) <$ deg (u). Show how to find polynomials $V(x)$, $W(x)$, $A(x)$, $B(x)$, $C(x)$, $D(x)$ which satisfy the same conditions but with m replaced by $2m$. Use your method for $p = 2$ to prove that (21) is irreducible over the integers, starting with its factorization mod 2 found in exercise 10.

23. [*HM23*] Let $u(x)$ be a primitive polynomial with integer coefficients, which is "squarefree." (Cf. exercise 2; the latter condition is equivalent to saying that gcd $(u(x), u'(x)) = 1$.) Prove that there are only finitely many primes p such that $u(x)$ is not squarefree modulo p.

24. [*M20*] The text speaks only of factorization over the integers, not over the field of rational numbers. Explain how to find the complete factorization of a polynomial with rational coefficients, over the field of rational numbers.

25. [*M25*] What is the complete factorization of $x^5 + x^4 + x^2 + x + 2$ over the field of rational numbers?

26. [*HM46*] Starting with the algorithm of exercise 4.6.1–18, it can be proved that string polynomials can be "uniquely" factored into irreducible string polynomials, in an appropriate sense. (See P. M. Cohn, *Transactions Amer. Math. Soc.* **109** (1963), 313–356.) Design an algorithm which does this factorization, and experiment with its efficiency.

27. [*HM30*] Prove that a random primitive polynomial over the integers is "almost always" irreducible, in some appropriate sense.

28. [*M24*] Explain how the gcd algorithm sketched in the text would determine gcd $(x^2 - y^2, x^2 + 2xy + y^2)$ over the integers, using the irreducible polynomials $y - 0$, $y - 1$, $y - 2$ as moduli.

4.6.3. Evaluation of powers

In this section we shall study the interesting problem of computing x^n efficiently, given x and n, where n is a positive integer. Suppose, for example, that we are asked to compute x^{16}; we could simply start with x and multiply by x fifteen times. But it is possible to obtain the same answer with only four multiplications, if we repeatedly take the square of each partial result, successively forming x^2, x^4, x^8, x^{16}.

The same idea applies, in general, to any value of n, in the following way: Write n in the binary number system (suppressing zeros at the left). Then replace each "1" by the pair of letters SX, replace each "0" by S, and cross off the "SX" which now appears at the left. The result is a rule for computing x^n, if "S" is interpreted as the operation of *squaring*, and if "X" is interpreted as the operation of *multiplying by x*. For example, if $n = 23$, its binary representation is 10111; so we form the sequence SX S SX SX SX and remove the leading SX to obtain the rule SSXSXSX. This rule states that we should "square, square, multiply by x, square, multiply by x, square, and multiply by x"; in other words, we should successively compute x^2, x^4, x^5, x^{10}, x^{11}, x^{22}, x^{23}.

This "binary method" is easily justified by a consideration of the sequence of exponents in the calculation: If we reinterpret "S" as the operation of multiplying by 2 and "X" as the operation of adding 1, and if we start with 1 instead of x, the rule will lead to a computation of n because of the properties of the binary number system. The method is quite ancient; it appeared before 200 B.C. in Pingala's Hindu classic Chandah-sûtra [see B. Datta and A. N. Singh, *History of Hindu Mathematics* 1 (Bombay, 1935), 76]; however, there seem to be no other references to this method outside of India during the next 2000 years!

The S-and-X binary method for obtaining x^n requires no temporary storage except for x and the current partial result, so it is well suited for incorporation in the hardware of a binary computer. The method can also be readily programmed for either binary or decimal computers; but it requires that the binary representation of n be scanned from left to right, while it is usually more convenient to do this from right to left. For example, with a binary computer we can shift the binary value of n to the right one bit at a time until zero is reached; with a decimal computer we can divide by 2 (or, equivalently, multiply by 5 or $\frac{1}{2}$) to deduce the binary representation from right to left. Therefore the following algorithm, based on a right-to-left scan of the number, is often more convenient:

Algorithm A (*Right-to-left binary method for exponentiation*). This algorithm evaluates x^n, where n is a positive integer. (Here x belongs to any algebraic system in which an associative multiplication, with identity element 1, has been defined.)

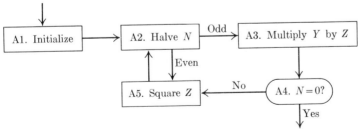

Fig. 12. Evaluation of x^n, based on a right-to-left scan of the binary notation for n.

A1. [Initialize.] Set $N \leftarrow n$, $Y \leftarrow 1$, $Z \leftarrow x$.

A2. [Halve N.]   (At this point, we have the relation $x^n = Y \cdot Z^N$.)   Set $N \leftarrow \lfloor N/2 \rfloor$, and at the same time determine whether N was even or odd. If N was even, skip to step A5.

A3. [Multiply Y by Z.] Set $Y \leftarrow Z$ times Y.

A4. [$N = 0$?] If $N = 0$, the algorithm terminates, with Y as the answer.

A5. [Square Z.] Set $Z \leftarrow Z$ times Z, and return to step A2. ∎

As an example of Algorithm A, consider the steps in the evaluation of x^{23}:

	N	Y	Z
After step A1	23	1	x
After step A4	11	x	x
After step A4	5	x^3	x^2
After step A4	2	x^7	x^4
After step A4	0	x^{23}	x^{16}

A MIX program corresponding to Algorithm A appears in exercise 2.

The great calculator al-Kashî stated Algorithm A about 1414 A.D. [*Istoriko-Mat. Issledovaniya* **7** (1954), 256–257]. It is closely related to a procedure for multiplication which was used by Egyptian mathematicians as early as 1800 B.C. If we change step A3 to "$Y \leftarrow Y + Z$" and step A5 to "$Z \leftarrow Z + Z$", and if we set Y to zero instead of unity in step A1, the algorithm terminates with $Y = nx$. This is a practical method for multiplication by hand, since it involves only the simple operations of doubling, halving, and adding. It is often called the "Russian peasant method" of multiplication, since Western visitors to Russia in the nineteenth century found the method in wide use there.

The number of multiplications required by Algorithm A is $\lfloor \log_2 n \rfloor + \nu(n)$, where $\nu(n)$ is the number of ones in the binary representation of n. This is one more multiplication than the left-to-right binary method mentioned at the beginning of this section would require, due to the fact that the first execution of step A3 is simply a multiplication by unity.

Because of the bookkeeping time required by this algorithm, the binary method is usually not of importance for small values of n, say $n \leq 10$, unless the time for a multiplication is comparatively large. If the value of n is known in advance, the left-to-right binary method is preferable. In some situations, such as the calculation of $x^n \bmod u(x)$ discussed in Section 4.6.2, it is much easier to multiply by x than to perform a general multiplication or to square a value, so binary methods for exponentiation are primarily suited for quite large n in such cases. There are, of course, many applications in which multiplication requires a significant amount of time (for example, multiple-precision multiplication, the multiplication of polynomials or matrices, etc.) when it is important to minimize the number of multiplications; and in all applications with large exponents n the serial method is far inferior to the binary method even though multiplication is fairly rapid, since the execution time can be reduced from order n to order $\log n$.

Fewer multiplications. Several authors have published statements (without proof) that the binary method actually gives the *minimum* possible number of multiplications. But this is not true. The smallest counterexample is $n = 15$, when the binary method needs 6 multiplications, yet we can calculate $y = x^3$ in two multiplications and $x^{15} = y^5$ in three more, achieving the desired result with only 5 multiplications. Let us now discuss some other procedures for evaluating x^n, which are appropriate in situations (e.g., within an optimizing compiler) when n is known in advance.

The *factor method* is based on a factorization of n. If $n = pq$, where p is the smallest prime factor of n, and $q > 1$, we may calculate x^n by first calculating x^p and then raising this quantity to the qth power. If n is prime, we may calculate x^{n-1} and multiply by x. And, of course, if $n = 1$, we have x^n with no calculation at all. Repeated application of these rules gives a procedure for evaluating x^n, for any given n. For example, if we want to calculate x^{55}, we first evaluate $y = x^5 = x^4x = (x^2)^2x$; then we form $y^{11} = y^{10}y = (y^2)^5y$. The whole process takes eight multiplications, while the binary method would have required nine. The factor method is better than the binary method on the average, but there are cases ($n = 33$ is the smallest example) where the binary method excels.

The binary method can be generalized to an *m-ary method* as follows: Let $n = d_0 m^t + d_1 m^{t-1} + \cdots + d_t$, where $0 \le d_j < m$ for $0 \le j \le t$. The computation begins by forming $x, x^2, x^3, \ldots, x^{m-1}$. (Actually, only those powers x^{d_j}, for d_j in the representation of n, are needed, and this observation often saves some of the work.) Then raise x^{d_0} to the mth power and multiply by x^{d_1}; we have computed $y_1 = x^{d_0 m + d_1}$. Next, raise y_1 to the mth power and multiply by x^{d_2}, obtaining $y_2 = x^{d_0 m^2 + d_1 m + d_2}$. The process continues in this way until $y_t = x^n$ has been computed. Whenever $d_j = 0$, it is, of course, unnecessary to multiply by x^{d_j}. Note that this method reduces to the left-to-right binary method discussed earlier, when $m = 2$; but no right-to-left m-ary method will give as few multiplications when $m > 2$. When m is a small prime, the m-ary method will be particularly efficient for calculating powers of polynomials whose coefficients are treated modulo m (see Eq. 4.6.2–5).

A systematic method which gives the minimum number of multiplications for all of the relatively small values of n (in particular, for most n which occur in practical applications) is indicated in Fig. 13. To calculate x^n, find n in this tree, and the path from the root to n indicates the sequence of exponents which occur in an efficient evaluation of x^n. The rule for generating this "power tree" appears in exercise 5. Computer tests have shown that the power tree gives optimal results for all of the n listed in the figure. But for large enough values

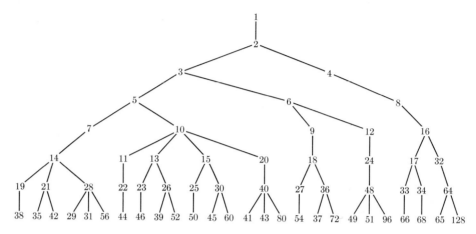

Fig. 13. "The power tree."

of n the power tree method is not always an optimal procedure; the smallest examples are $n = 77, 154, 233$. The first case for which the power tree is superior to both the binary method and the factor method is $n = 23$.

Addition chains. The most economical way to compute x^n by multiplication is a mathematical problem with an interesting history. We shall now examine it in detail, not only because it is interesting in its own right, but because it is an excellent example of the theoretical questions which arise in a study of "optimal methods of computation."

Although we are concerned with multiplication of powers of x, the problem can easily be reduced to addition, since the exponents are additive. This leads us to the following abstract formulation: An *addition chain for* n is a sequence of integers

$$1 = a_0, \quad a_1, \quad a_2, \quad \ldots, \quad a_r = n \tag{1}$$

with the property that

$$a_i = a_j + a_k, \quad \text{for some} \quad k \leq j < i, \tag{2}$$

for all $i = 1, 2, \ldots, r$. One way of looking at this definition is to consider a simple computer which has an accumulator and which is capable of the three operations LDA, STA, and ADD; the machine begins with the number 1 in its accumulator, and it proceeds to compute the number n by adding together previous results. Note that a_1 must equal 2, and a_2 is either 2, 3, or 4.

The smallest length, r, for which an addition chain for n exists is denoted by $l(n)$. Our goal in the remainder of this section is to discover as much as we can about this function $l(n)$. The values of $l(n)$ for small n are displayed in tree form in Fig. 14, which shows how to calculate x^n with the fewest possible multiplications for all $n \leq 100$.

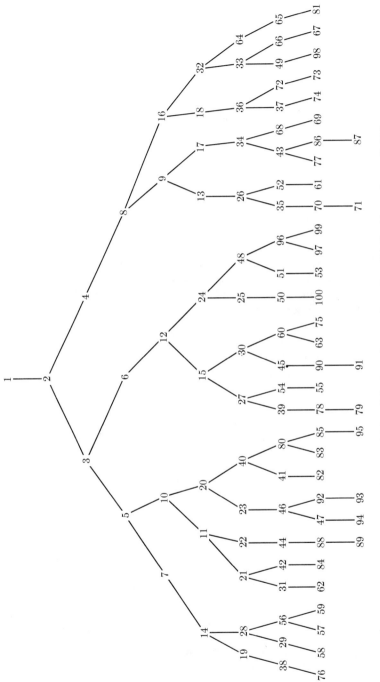

Fig. 14. A tree which minimizes the number of multiplications, for $n \leq 100$.

The methods discussed earlier in this section give us upper bounds for $l(n)$. From the factor method, we see that

$$l(mn) \leq l(m) + l(n), \tag{3}$$

since we can take the chains $1, a_1, \ldots, a_r = m$ and $1, b_1, \ldots, b_s = n$ and form the chain $1, a_1, \ldots, a_r, a_r b_1, \ldots, a_r b_s = mn$.

We can also adapt the m-ary method into the form of an addition chain. Consider the case $m = 2^k$, and write $n = d_0 m^t + d_1 m^{t-1} + \cdots + d_t$ in the m-ary number system; the corresponding addition chain takes the form

$$
\begin{aligned}
&1, \ 2, \ 3, \ \ldots, \ m-2, \ m-1, \\
&\quad 2d_0, \ 4d_0, \ \ldots, \ md_0, \ md_0 + d_1, \\
&\qquad 2(md_0 + d_1), \ 4(md_0 + d_1), \ \ldots, \ m(md_0 + d_1), \ m^2 d_0 + md_1 + d_2, \\
&\qquad\qquad \ldots, \qquad\qquad m^t d_0 + m^{t-1} d_1 + \cdots + d_t.
\end{aligned}
\tag{4}
$$

The length of this chain is $m - 2 + t(k + 1)$; this number can often be reduced by deleting certain elements of the first row which do not occur among the coefficients d_j, plus elements among $2d_0, 4d_0, \ldots$ which already appear in the first row; and whenever a d_j is zero, the step at the right end of the corresponding line may be dropped.

The simplest case of the m-ary method is the binary method ($m = 2$), when the general scheme (4) simplifies to the "S" and "X" rule mentioned at the beginning of this section: The binary addition chain for $2n$ is the binary chain for n followed by $2n$; for $2n + 1$ it is the binary chain for $2n$ followed by $2n + 1$. From the binary method we conclude that

$$l(2^{e_0} + 2^{e_1} + \cdots + 2^{e_t}) \leq e_0 + t, \quad \text{if} \quad e_0 > e_1 > \cdots > e_t \geq 0. \tag{5}$$

Let us now define two auxiliary functions for convenience in our subsequent discussion:

$$\lambda(n) = \lfloor \log_2 n \rfloor; \tag{6}$$

$$\nu(n) = \text{number of 1's in the binary representation of } n. \tag{7}$$

Thus $\lambda(17) = 4$, $\nu(17) = 2$; these functions may be defined by the recurrence relations

$$\lambda(1) = 0, \quad \lambda(2n) = \lambda(2n + 1) = \lambda(n) + 1; \tag{8}$$

$$\nu(1) = 1, \quad \nu(2n) = \nu(n), \quad \nu(2n + 1) = \nu(n) + 1. \tag{9}$$

In terms of these functions, the binary addition chain for n requires $\lambda(n) + \nu(n) - 1$ steps, and (5) becomes

$$l(n) \leq \lambda(n) + \nu(n) - 1. \tag{10}$$

Special classes of chains. We may assume without any loss of generality that an addition chain is "ascending,"

$$1 = a_0 < a_1 < a_2 < \cdots < a_r = n; \tag{11}$$

for if any two a's are equal, one of them may be dropped; and we can also rearrange the sequence (1) into ascending order and remove terms $> n$ without destroying the addition chain property (2). *From now on we shall consider only ascending chains,* without explicitly mentioning this assumption.

It is convenient at this point to define a few special terms relating to addition chains. By definition we have, for $1 \leq i \leq r$,

$$a_i = a_j + a_k \tag{12}$$

for some j and k, $0 \leq k \leq j < i$. Let us say that step i of (11) is a *doubling*, if $j = k = i - 1$; then a_i has the maximum possible value $2a_{i-1}$ which can follow the ascending chain $1, a_1, \ldots, a_{i-1}$. If j (but not necessarily k) equals $i - 1$, let us say that step i is a *star step*. The importance of star steps is explained below. Finally let us say that step i is a *small step* if $\lambda(a_i) = \lambda(a_{i-1})$. Since $a_{i-1} < a_i \leq 2a_{i-1}$, $\lambda(a_i)$ is always equal to either $\lambda(a_{i-1})$ or $\lambda(a_{i-1}) + 1$; it follows that, in any chain (11), *the length r is equal to $\lambda(n)$ plus the number of small steps.*

Several elementary relations hold between these types of steps: Step 1 is always a doubling. A doubling obviously is a star step, but never a small step. A doubling must be followed by a star step. Furthermore if step i is *not* a small step, then step $i + 1$ is either a small step or a star step, or both; putting this another way, if step $i + 1$ is neither small nor star, step i must be small.

A *star chain* is an addition chain which involves only star steps. This means that each term a_i is the sum of a_{i-1} and a previous a_k; the simple "computer" discussed above after Eq. (2) makes use only of the two operations STA and ADD (no LDA) in a star chain, since each new term of the sequence utilizes the preceding result in the accumulator. Most of the addition chains we have discussed so far are star chains. The minimum length of a star chain for n is denoted by $l^*(n)$; clearly

$$l(n) \leq l^*(n). \tag{13}$$

We are now ready to derive some nontrivial facts about addition chains. First we can show that there must be fairly many doublings if r is not far from $\lambda(n)$:

Theorem A. *If the addition chain* (11) *includes d doublings and $t = r - d$ nondoublings, then*

$$n \leq 2^{d-1}F_{t+3}. \tag{14}$$

Proof. By induction on $r = d + t$, we see that (14) is certainly true when $r = 1$. When $r > 1$, there are three cases: If step r is a doubling, then $\frac{1}{2}n =$

$a_{r-1} \leq 2^{d-2}F_{t+3}$; hence (14) follows. If steps r and $r-1$ are both non-doublings, then $a_{r-1} \leq 2^{d-1}F_{t+2}$ and $a_{r-2} \leq 2^{d-1}F_{t+1}$; hence $n = a_r \leq a_{r-1} + a_{r-2} \leq 2^{d-1}(F_{t+2} + F_{t+1}) = 2^{d-1}F_{t+3}$ by the definition of the Fibonacci sequence. Finally, if step r is a nondoubling but step $r-1$ is a doubling, then $a_{r-2} \leq 2^{d-2}F_{t+2}$ and $n = a_r \leq a_{r-1} + a_{r-2} = 3a_{r-2}$. Now $2F_{t+3} - 3F_{t+2} = F_{t+1} - F_t \geq 0$; hence $n \leq 2^{d-1}F_{t+3}$ in all cases.   ∎

The method of proof we have used shows that inequality (14) is "best possible" under the stated assumptions; for the addition chain

$$1, 2, \ldots, 2^{d-1}, 2^{d-1}F_3, 2^{d-1}F_4, \ldots, 2^{d-1}F_{t+3} \tag{15}$$

has d doublings and t nondoublings.

Corollary. *If the addition chain (11) includes t nondoublings and s small steps, then*

$$s \leq t \leq 3.271s. \tag{16}$$

Proof. Obviously $s \leq t$. We have $2^{\lambda(n)} \leq n \leq 2^{d-1}F_{t+3} \leq 2^d\phi^t = 2^{\lambda(n)+s}(\phi/2)^t$, since $d + t = \lambda(n) + s$, and since $F_{t+3} \leq 2\phi^t$ when $t \geq 0$. Hence $0 \leq s \ln 2 + t \ln (\phi/2)$, and (16) follows from the fact that

$$\ln 2/\ln (2/\phi) \approx 3.2706.   ∎$$

Values of $l(n)$ for special n. It is easy to show by induction that $a_i \leq 2^i$, and therefore $\log_2 n \leq r$ in any addition chain (11). Hence

$$l(n) \geq \lceil \log_2 n \rceil. \tag{17}$$

This lower bound, together with the upper bound (10) given by the binary method, gives us the values

$$l(2^A) = A; \tag{18}$$

$$l(2^A + 2^B) = A + 1, \quad \text{if} \quad A > B. \tag{19}$$

Thus when $\nu(n) \leq 2$, the binary method is optimal. With some further calculation we can extend these formulas to the case $\nu(n) = 3$:

Theorem B (A. A. Gioia, M. V. Subbarao, M. Sugunumma, *Duke Math. J.* **29** (1962), 481–487.)

$$l(2^A + 2^B + 2^C) = A + 2, \quad \text{if} \quad A > B > C. \tag{20}$$

Proof. We can, in fact, prove a stronger result which will be of use to us later in this section: *All addition chains with exactly one small step have one of the following six types* (where all steps indicated by "$\cdots$" represent doublings):

Type 1. $1, \cdots, 2^A, 2^A + 2^B, \cdots, 2^{A+C} + 2^{B+C}$; $A > B \geq 0, C \geq 0$.

Type 2. $1, \cdots, 2^A, 2^A + 2^B, 2^{A+1} + 2^B, \cdots, 2^{A+C+1} + 2^{B+C}$; $A > B \geq 0$, $C \geq 0$.

Type 3. $1, \cdots, 2^A, 2^A + 2^{A-1}, 2^{A+1} + 2^{A-1}, 2^{A+2}, \cdots, 2^{A+C}$; $A > 0$, $C \geq 2$.

Type 4. $1, \cdots, 2^A, 2^A + 2^{A-1}, 2^{A+1} + 2^A, 2^{A+2}, \cdots, 2^{A+C}$; $A > 0$, $C \geq 2$.

Type 5. $1, \cdots, 2^A, 2^A + 2^{A-1}, \cdots, 2^{A+C} + 2^{A+C-1}, 2^{A+C+1} + 2^{A+C-2}, \cdots, 2^{A+C+D+1} + 2^{A+C+D-2}$; $A > 0$, $C > 0$, $D \geq 0$.

Type 6. $1, \cdots, 2^A, 2^A + 2^B, 2^{A+1}, \cdots, 2^{A+C}$; $A > B \geq 0$, $C \geq 1$.

A straightforward hand calculation shows that these six types exhaust all possibilities. (Note that, by the corollary to Theorem A, there are at most three nondoublings when there is one small step; this maximum of three is attained only in sequences of type 3. All of the above are star chains, except type 6 when $B < A - 1$.)

The theorem now follows from the observation that $l(2^A + 2^B + 2^C) \leq A + 2$; and it must be greater than $A + 1$ since none of the six possible types have $\nu(n) > 2$. ∎

The calculation of $l(2^A + 2^B + 2^C + 2^D)$, when $A > B > C > D$, is more involved; by the binary method it is at most $A + 3$, and by the proof of Theorem B it is at least $A + 2$. The value $A + 2$ is possible, since we know that the binary method is not optimal when $n = 15$ or $n = 23$. The complete behavior when $\nu(n) = 4$ can be determined, as we shall now see.

Theorem C. *If $\nu(n) \geq 4$ then $l(n) \geq \lambda(n) + 3$, except in the following circumstances when $A > B > C > D$ and $l(2^A + 2^B + 2^C + 2^D)$ equals $A + 2$:*

    Case 1. $A - B = C - D$. (Example: $n = 15$.)
    Case 2. $A - B = C - D + 1$. (Example: $n = 23$.)
    Case 3. $A - B = 3$, $C - D = 1$. (Example: $n = 39$.)
    Case 4. $A - B = 5$, $B - C = C - D = 1$. (Example: $n = 135$.)

Proof. When $l(n) = \lambda(n) + 2$, there is an addition chain for n having just two small steps; such an addition chain starts out as one of the six types in the proof of Theorem B, followed by a small step, followed by a sequence of nonsmall steps. Let us say that n is "special" if $n = 2^A + 2^B + 2^C + 2^D$ for one of the four cases listed in the theorem. We can obtain addition chains of the required form for each special n, as shown in exercise 13; therefore it remains for us to prove that no chain with exactly two small steps contains any elements with $\nu(a_i) \geq 4$ except when a_i is special.

Let a "counterexample chain" be an addition chain with two small steps such that $\nu(a_r) \geq 4$, but a_r is not special. If counterexample chains exist, let $1 = a_0 < a_1 < \cdots < a_r = n$ be a counterexample chain of shortest possible length. Then step r is not a small step, since none of the six types in the proof of

Theorem B can be followed by a small step with $\nu(n) \geq 4$ except when n is special. Furthermore, step r is not a doubling, otherwise $a_0, \ldots, a_{r-1}$ would be a shorter counterexample chain; and step r is a star step, otherwise $a_0, \ldots, a_{r-2}, a_r$ would be a shorter counterexample chain. Thus

$$a_r = a_{r-1} + a_{r-k}, \qquad k > 1. \tag{21}$$

Let $m = \lambda(a_{r-1})$. Let c be the number of carries which occur when a_{r-1} is added to a_{r-k} in the binary number system. Then we have the fundamental relation

$$\nu(a_r) = \nu(a_{r-1}) + \nu(a_{r-k}) - c. \tag{22}$$

We can now prove that *step* $r - 1$ *is not a small step* (see exercise 14).

Since neither r nor $r - 1$ is a small step, $c \geq 2$; and $c = 2$ can hold only when $a_{r-1} \geq 2^m + 2^{m-1}$.

Now let us suppose that $r - 1$ is not a star step; then $r - 2$ is a small step, and $a_0, \ldots, a_{r-3}, a_{r-1}$ is a chain with only one small step; hence $\nu(a_{r-1}) \leq 2$ and $\nu(a_{r-2}) \leq 4$. The relation (22) can now hold only if $\nu(a_r) = 4$, $\nu(a_{r-1}) = 2$, $k = 2$, $c = 2$, $\nu(a_{r-2}) = 4$. From $c = 2$ we conclude that $a_{r-1} = 2^m + 2^{m-1}$; hence $a_0, a_1, \ldots, a_{r-3} = 2^{m-1} + 2^{m-2}$ is an addition chain with only one small step, and it must be of Type 1 so a_r belongs to Case 3. Thus $r - 1$ *is a star step.*

Now assume that $a_{r-1} = 2^t a_{r-k}$ for some t. If $\nu(a_{r-1}) \leq 3$, then by (22), $c = 2$, $k = 2$, and we see that a_r must belong to Case 3. If $\nu(a_{r-1}) = 4$ then a_{r-1} is special, and it is easy to see by considering each case that a_r also belongs to one of the four cases. (Case 4 arises, for example, when $a_{r-1} = 90$, $a_{r-k} = 45$; or $a_{r-1} = 120$, $a_{r-k} = 15$.) Therefore we may conclude that $a_{r-1} \neq 2^t a_{r-k}$ for any t.

We have proved that $a_{r-1} = a_{r-2} + a_{r-q}$ for some $q \geq 2$. If $k = 2$, then $q > 2$ and $a_0, a_1, \ldots, a_{r-2}, 2a_{r-2}\ 2a_{r-2} + a_{r-q} = a_r$ is a counterexample sequence in which $k > 2$; therefore we may assume that $k > 2$.

Let us now suppose that $\lambda(a_{r-k}) = m - 1$; the case $\lambda(a_{r-k}) < m - 1$ may be ruled out by similar arguments, as shown in exercise 14. If $k = 4$, both $r - 2$ and $r - 3$ are small steps; hence $a_{r-4} = 2^{m-1}$, and (22) is impossible. Therefore $k = 3$; so step $r - 2$ is small, $\nu(a_{r-3}) = 2$, $c = 2$, $a_{r-1} \geq 2^m + 2^{m-1}$, and $\nu(a_{r-1}) = 4$. There must be at least two carries when a_{r-2} is added to $a_{r-1} - a_{r-2}$; hence $\nu(a_{r-2}) = 4$, and a_{r-2} (being special and $\geq \frac{1}{2} a_{r-1}$) has the form $2^{m-1} + 2^{m-2} + 2^{d+1} + 2^d$ for some d. Now a_{r-1} is either $2^m + 2^{m-1} + 2^{d+1} + 2^d$ or $2^m + 2^{m-1} + 2^{d+2} + 2^{d+1}$, and in both cases a_{r-3} must be $2^{m-1} + 2^{m-2}$, so a_r belongs to Case 3. $\blacksquare$

Asymptotic values. Theorem C shows that it is apparently quite difficult to get exact values of $l(n)$ for large n, when $\nu(n) > 4$; however, we can determine the approximate behavior in the limit as $n \to \infty$.

Theorem D (A. Brauer, *Bull. AMS* **45** (1939), 736–739).

$$\lim_{n\to\infty} l^*(n)/\lambda(n) = \lim_{n\to\infty} l(n)/\lambda(n) = 1. \tag{23}$$

Proof. The addition chain (4) for the 2^k-ary method is a star chain if we delete the second occurrence of any element which appears twice in the chain; for if a_i is the first element among $2d_0, 4d_0, \ldots$ of the second line which is not present in the first line, we have $a_i \leq 2(m-1)$; hence $a_i = (m-1) + a_j$ for some a_j in the first line. By totaling up the length of the chain, we therefore have

$$\lambda(n) \leq l(n) \leq l^*(n) < \left(1 + \frac{1}{k}\right) \log_2 n + 2^k \tag{24}$$

for all $k \geq 1$. The theorem follows if we choose, say, $k = \lfloor \frac{1}{2} \log_2 \lambda(n) \rfloor$. ∎

If we let $k = \lfloor \log_2 \lambda(n) - 2 \log_2 \left(\lambda(\lambda(n))\right) \rfloor$ in (24), for large n, we obtain the stronger asymptotic formulas

$$l(n) \leq l^*(n) \leq \lambda(n) + \lambda(n)/\lambda\left(\lambda(n)\right) + O\left(\lambda(n)\lambda(\lambda(\lambda(n)))\right)/\left(\lambda(\lambda(n))\right)^2). \tag{25}$$

The second term $\lambda(n)/\lambda\left(\lambda(n)\right)$ is essentially the best that can be obtained from (24). A much deeper analysis of lower bounds can be carried out, to show that this term $\lambda(n)/\lambda\left(\lambda(n)\right)$ is, in fact, essential in (25). In order to see why this is so, let us consider the following fact:

Theorem E (Paul Erdös, *Acta Arithmetica* **6** (1960), 77–81). *Let ϵ be a positive real number. The number of addition chains (11) such that*

$$\lambda(n) = m, \qquad r \leq m + (1 - \epsilon)m/\lambda(m) \tag{26}$$

is less than $2^m/\alpha^m$, for some $\alpha > 1$, for all suitably large m. (In other words, the number of addition chains which are so short that (26) is satisfied is substantially less than the number of values of n such that $\lambda(n) = m$, when m is large.)

Proof. We want to estimate the number of possible addition chains, and for this purpose our first goal is to get an improvement of Theorem A which enables us to deal more satisfactorily with nondoublings:

Lemma P. *Let $\delta < \sqrt{2} - 1$ be a fixed positive real number. Call step i of an addition chain a "ministep" if $a_i < a_j(1 + \delta)^{i-j}$ for some j, $0 \leq j < i$. If the addition chain contains s small steps and t ministeps, then*

$$t \leq s/(1 - \theta), \qquad \text{where} \qquad (1 + \delta)^2 = 2^\theta. \tag{27}$$

Proof. For each ministep i_k, $1 \leq k \leq t$, we have $a_{i_k} < a_{j_k}(1 + \delta)^{i_k - j_k}$ for some $j_k < i_k$. Let $I_1, \ldots, I_t$ be the intervals $(j_1, i_1], \ldots, (j_t, i_t]$, where the notation $(j, i]$ stands for the set of all integers k such that $j < k \leq i$. It is possible (see exercise 17) to find nonoverlapping intervals $J_1, \ldots, J_h = (j'_1, i'_1], \ldots, (j'_h, i'_h]$ such that

$$I_1 \cup \cdots \cup I_t = J_1 \cup \cdots \cup J_h,$$
$$a_{i'_k} < a_{j'_k}(1 + \delta)^{2(i'_k - j'_k)}, \qquad \text{for} \qquad 1 \le k \le h. \qquad (28)$$

Now for all steps i outside of the intervals $J_1, \ldots, J_h$ we have $a_i \le 2a_{i-1}$; hence if

$$q = (i'_1 - j'_1) + \cdots + (i'_h - j'_h)$$

we have

$$2^{\lambda(n)} \le n \le 2^{r-q}(1 + \delta)^{2q} = 2^{\lambda(n)+s-(1-\theta)q} \le 2^{\lambda(n)+s-(1-\theta)t}. \quad \blacksquare$$

Returning to the proof of Theorem E, let us choose $\delta = 2^{\epsilon/4} - 1$, and let us divide the r steps of each addition chain into three classes:

$$t \text{ ministeps}, \qquad u \text{ doublings}, \qquad \text{and } v \text{ other steps}, \qquad t + u + v = r. \quad (29)$$

Counting another way, we have s small steps, where $s + m = r$. By the hypotheses, Theorem A, and Lemma P, we obtain the relations

$$t \le s/(1 - \epsilon/2), \qquad t + v \le 3.271s, \qquad s \le (1 - \epsilon)m/\lambda(m). \quad (30)$$

Given s, t, u, v satisfying these conditions, there are at most

$$\binom{r}{t+v}\binom{t+v}{v} \qquad (31)$$

ways to decide which steps belong to which class. Given such a distribution of the steps, let us consider how the non-ministeps can be selected: If step i is one of the "other" steps in (29), $a_i \ge (1 + \delta)a_{i-1}$, so $a_i = a_j + a_k$, where $\delta a_{i-1} \le a_k \le a_j \le a_{i-1}$. Also $a_j \le a_i/(1 + \delta)^{i-j} \le 2a_{i-1}/(1 + \delta)^{i-j}$, so $\delta \le 2/(1 + \delta)^{i-j}$. This gives at most β choices for j, where β is a constant that depends only on δ. There are also at most β choices for k, so the number of ways to assign j and k for each of the non-ministeps is at most

$$\beta^{2v}. \qquad (32)$$

Finally, once the "j" and "k" have been selected for each of the non-ministeps, there are less than

$$\binom{r^2}{t} \qquad (33)$$

ways to choose the j and the k for the ministeps: We select t distinct pairs $(j_1, k_1), \ldots, (j_t, k_t)$ of indices in the range $0 \le k_h \le j_h < r$, in less than (33)

ways. Then for each ministep i, in turn, we use a pair of indices (j_h, k_h) such
that

   a) $j_h < i$;
   b) $a_{j_h} + a_{k_h}$ is as small as possible of the pairs not already used for smaller
      ministeps i;
   c) $a_i = a_{j_h} + a_{k_h}$ satisfies the definition of ministep.

If no such pair (j_h, k_h) exists, we get no addition chain; on the other hand, any
addition chain with ministeps in the designated places must be selected in one
of these ways, so (33) is an upper bound on the possibilities.

Thus the total number of possible addition chains satisfying (26) is bounded
by (31) times (32) times (33), summed over all s, t, u, and v satisfying (30).
The proof of Theorem E can now be completed by means of a rather standard
estimation of these functions (exercise 18). ∎

Corollary. $l(n)$ *is asymptotically* $\lambda(n) + \lambda(n)/\lambda(\lambda(n))$, *for almost all* n. *More
precisely, there is a function* $f(n)$ *such that* $f(n) \to 0$ *as* $n \to \infty$, *and*

$$\Pr\left(|l(n) - \lambda(n) - \lambda(n)/\lambda(\lambda(n))| \geq f(n)\lambda(n)/\lambda(\lambda(n))\right) = 0. \qquad (34)$$

(See Section 3.5 for this definition of the probability "Pr".)

Proof. The upper bound (25) shows that (34) holds without the absolute value
signs. The lower bound comes from Theorem E, if we let $f(n)$ decrease to zero
slowly enough so that, when $f(N) \leq 1/M$, the value N is so large that at most
N/M values $n \leq N$ have $l(n) \leq \lambda(n) + (1 - 1/M)\lambda(n)/\lambda(\lambda(n))$. ∎

Star chains. Optimistic people find it reasonable to suppose that $l(n) = l^*(n)$;
given an addition chain of minimal length $l(n)$, it appears hard to believe that
we cannot find one of the same length which satisfies the (apparently mild) star
condition. In support of this hypothesis, computer tests have verified that
$l(n) = l^*(n)$ for all $n \leq 2500$. But in 1958 Walter Hansen proved the remarkable theorem that, for certain large values of n, $l(n)$ is definitely less than $l^*(n)$,
and he also proved several related theorems which we will now investigate.

Hansen's theorems begin with an investigation of the detailed structure of
a star chain. This structure is given in terms of other sequences and sets constructed from the given chain. Let $n = 2^{e_0} + 2^{e_1} + \cdots + 2^{e_t}, e_0 > e_1 > \cdots >
e_t \geq 0$, and let $1 = a_0 < a_1 < \cdots < a_r = n$ be a star chain for n. If there
are d doublings in this chain, we define the auxiliary sequence

$$d = b_0 \geq b_1 \geq b_2 \geq \cdots \geq b_r = 0 \qquad (35)$$

where b_i is the number of doublings among steps $i+1$, $i+2$, $\ldots$, r. We
also define a sequence of "multisets" $S_0, S_1, \ldots, S_r$, which keep track of the
powers of 2 present in the chain, (A *multiset* is a mathematical entity which is
like a set but it is allowed to contain repeated elements; an object may be an

element of a multiset several times, and its multiplicity of occurrences is relevant. See exercise 19 for familiar examples of multisets.) S_i is defined by the rules

a) $S_0 = 0$;
b) If $a_{i+1} = 2a_i$, then $S_{i+1} = \{x \mid x - 1 \in S_i\}$;
c) If $a_{i+1} = a_i + a_k$, $k < i$, then $S_{i+1} = S_i \uplus S_k$.

(The symbol $\uplus$ means that the multisets are combined, adding the multiplicities.) From this definition it follows that

$$a_i = \sum_{x \in S_i} 2^x, \tag{36}$$

where the terms in this sum are not necessarily distinct, and in particular

$$n = 2^{e_0} + 2^{e_1} + \cdots + 2^{e_t} = \sum_{x \in S_r} 2^x. \tag{37}$$

The number of elements in the latter sum is at most 2^f, where $f = r - d$ is the number of nondoublings.

Since n has two different binary representations in (37), we can partition the multiset S_r into multisets $M_0, M_1, \ldots, M_t$ such that

$$2^{e_j} = \sum_{x \in M_j} 2^x, \qquad 0 \le j \le t. \tag{38}$$

This can be done by arranging the elements of S_r into nondecreasing order $x_1 \le x_2 \le \cdots$, taking $M_t = \{x_1, x_2, \ldots, x_k\}$ where $2^{x_1} + \cdots + 2^{x_k} = 2^{e_t}$. This must be possible since e_t is the smallest of the e's. Similarly, $M_{t-1} = \{x_{k+1}, x_{k+2}, \ldots, x_{k'}\}$, and so on. This process is easily visualized in the binary number system.

Let M_j contain m_j elements (counting multiplicities); then $m_j \le 2^f - t$, since S_r has at most 2^f elements and it has been partitioned into $(t+1)$ non-empty multisets. By Eq. (38), we can see that

$$e_j \ge x > e_j - m_j, \qquad \text{for all} \qquad x \in M_j. \tag{39}$$

Our examination of the star chain's structure is completed by forming the multisets M_{ij} which record the ancestral history of M_j. The multiset S_i is partitioned into $(t + 1)$ multisets M_{ij} as follows:

a) $M_{rj} = M_j$;
b) If $a_{i+1} = 2a_i$, then $M_{ij} = \{x \mid x + 1 \in M_{(i+1)j}\}$;
c) If $a_{i+1} = a_i + a_k$, $k < i$, then (since $S_{i+1} = S_i + S_k$) we let $M_{ij} = M_{(i+1)j}$ minus S_k, that is, we remove the elements of S_k from $M_{(i+1)j}$. If some element of S_k appears in two different multisets $M_{(i+1)j}$, we remove it from the set with the largest possible value of j; this rule uniquely defines M_{ij} for each j, when i is fixed.

From this definition it follows that

$$e_j - b_i \ge x > e_j - b_i - m_j, \qquad \text{for all} \qquad x \in M_{ij}. \tag{40}$$

As an example of this detailed construction, here is the pertinent information for the addition chain 1, 2, 3, 5, 10, 20, 23:

$$t = 3, \ r = 6, \ d = 3, \ f = 3$$

	S_0	S_1	S_2	S_3	S_4	S_5	S_6		
$(b_0, b_1, \ldots, b_6)$:	3	2	2	2	1	0	0		
$(a_0, a_1, \ldots, a_6)$:	1	2	3	5	10	20	23		
$(M_{03}, M_{13}, \ldots, M_{63})$:							0	M_3	$e_3 = 0, \ m_3 = 1$
$(M_{02}, M_{12}, \ldots, M_{62})$:							1	M_2	$e_2 = 1, \ m_2 = 1$
$(M_{01}, M_{11}, \ldots, M_{61})$:				0	0	1	2 2	M_1	$e_1 = 2, \ m_1 = 1$
$(M_{00}, M_{10}, \ldots, M_{60})$:	0	1	1	1	2	3	3	M_0	$e_0 = 4, \ m_0 = 2$
					1	2	3 3		

Thus $M_{40} = \{2, 2\}$, etc. From the construction we can see that $r - f - b_i$ is the largest element of S_i; hence

$$r - f - b_i \in M_{i0}. \tag{41}$$

The most important part of this structure comes from Eq. (40); one of the useful properties is

Lemma K. *If M_{ij} and M_{uv} both contain a common integer x, then*

$$-m_v < (e_j - e_v) - (b_i - b_u) < m_j. \tag{42}$$

This is proved immediately by considering the two equations $e_j - b_i \geq x > e_j - b_i - m_j$, $e_v - b_u \geq x > e_v - b_u - m_v$. ∎

Although Lemma K may not look extremely powerful, it says (when M_{ij} contains an element in common with M_{uv} and when m_j, m_v are reasonably small) that the number of doublings between steps i and u is approximately equal to the difference between the exponents e_j and e_v. This imposes a certain amount of regularity on the addition chain; and it suggests that we might be able to prove a result analogous to Theorem B above, that $l^*(n) = e_0 + t$, provided that the e_j are far enough apart. The next theorem shows how this can be done.

Theorem H (W. Hansen, *J. reine angew. Math.* **202** (1959), 129–136.) *Let* $n = 2^{e_0} + 2^{e_1} + \cdots + 2^{e_t}$, $e_0 > e_1 > \cdots > e_t \geq 0$. *If*

$$e_0 > 2e_1 + 3.271(t - 1) \quad \text{and} \quad e_{i-1} \geq e_i + 2m$$
$$\text{for} \quad 1 \leq i \leq t, \ \text{where} \ m = 2^{\lfloor 3.271(t-1) \rfloor} - t, \tag{43}$$

then $l^*(n) = e_0 + t$.

Proof. We may assume that $t > 2$, since the result of the theorem is true without restriction on the e's when $t \leq 2$. Suppose that we have a star chain $1 = a_0 < a_1 < \cdots < a_r = n$ for n with $r \leq e_0 + t - 1$. Let the integers d, f, $b_0, \ldots, b_r$, and the multisets M_{ij} and S_i, reflect the structure of this chain, as defined earlier. By the corollary to Theorem A, we know that $f \leq \lfloor 3.271(t - 1) \rfloor$; therefore the m in (43) is a bona fide upper bound for the number of elements m_j in each multiset M_j.

In the summation

$$a_i = \left(\sum_{x \in M_{i0}} 2^x \right) + \left(\sum_{x \in M_{i1}} 2^x \right) + \cdots + \left(\sum_{x \in M_{it}} 2^x \right),$$

no carries propagate from the term corresponding to M_{ij} to the term corresponding to $M_{i(j-1)}$, if we think of this sum as being carried out in the binary number system, since the e's are so far apart. [Cf. (40).] In particular, the sum of all the terms for $j \neq 0$ will not carry up to affect the terms for $j = 0$, so we must have

$$a_i \geq \sum_{x \in M_{i0}} 2^x \geq 2^{\lambda(a_i)}, \qquad 0 \leq i \leq r. \tag{44}$$

In order to prove Theorem H, we would like to show that in some sense the t extra powers of 2 occurring in the binary representation of n must be put in "one at a time," so we want to find a way to tell at which step each of these terms essentially enters the addition chain.

Let j be a number between 1 and t. Since M_{0j} is empty and $M_{rj} = M_j$ is nonempty, we can find the *first* step i for which M_{ij} is not empty.

From the way in which the M_{ij} are defined, we know that step i is a nondoubling; $a_i = a_{i-1} + a_u$ for some $u < i - 1$. We also know that all of the elements of M_{ij} are elements of S_u. We will prove that a_u must be relatively small compared to a_i.

Let x_j be an element of M_{ij}. Then since $x_j \in S_u$, there is some v for which $x_j \in M_{uv}$. It follows that

$$b_u - b_i > m, \tag{45}$$

i.e., at least $m + 1$ doublings occur between steps u and i. For if $b_u - b_i \leq m$, Lemma K tells us that $|e_j - e_v| < 2m$; hence $v = j$. But this is impossible because M_{uj} is empty by our choice of step i.

All elements of S_u are less than or equal to $e_1 - b_i$. For if $x \in S_u \subseteq S_i$ and $x > e_1 - b_i$, then $x \in M_{u0}$ and $x \in M_{i0}$ by (40); so Lemma K implies that $|b_u - b_i| < m$, contradicting (45). In fact, this argument proves that M_{i0} has no elements in common with S_u, so $M_{(i-1)0} = M_{i0}$. From (44) we have $a_{i-1} \geq 2^{\lambda(a_i)}$, and therefore *step i is a small step*.

We can now prove that *all elements of S_u are in M_{u0}*. Let x be any element of S_u and let $y = r - f - b_u$. We know [cf. (41)] that y is the largest element of S_u. If $x \notin M_{u0}$, then by (40) $e_1 \geq b_u$; but this implies that

$$e_1 \geq e_1 - b_i \geq y > e_0 - f - b_u \geq e_0 - 3.271(t-1) - e_1,$$

contradicting the hypothesis (43) about e_0. (This is probably the key step of the entire proof.)

Going back to our element x_j in M_{ij}, we have $x_j \in M_{uv}$; and we have proved that $v = 0$. Therefore, by equation (40) again,

$$e_0 - b_u \geq x_j > e_0 - b_u - m. \tag{44}$$

For all $j = 1, 2, \ldots, t$ we have determined a number x_j satisfying (44), and a small step i at which (in some sense) the term 2^{e_j} in n has entered into the addition chain. If $j \neq j'$, the step i at which this occurs cannot be the same for both j and j'; for (44) would tell us that $|x_j - x_{j'}| < m$, while elements of M_{ij} and $M_{ij'}$ must differ by more than m, since e_j and $e_{j'}$ are so far apart. Therefore the chain contains at least t small steps, but this is a contradiction. ∎

Theorem F (W. Hansen).

$$l(2^A + xy) \leq A + \nu(x) + \nu(y) - 1, \qquad \text{if} \qquad \lambda(x) + \lambda(y) \leq A. \tag{45}$$

Proof. An addition chain (which is *not* a star chain in general) may be constructed by combining the binary method and the factor method. Let $x = 2^{x_1} + \cdots + 2^{x_u}$, $y = 2^{y_1} + \cdots + 2^{y_v}$, where $x_1 > \cdots > x_u \geq 0$, $y_1 > \cdots > y_v \geq 0$.

The first steps of the chain form successive powers of 2, until 2^{A-y_1} is reached; in between these steps, the values $2^{x_{u-1}} + 2^{x_u}, 2^{x_{u-2}} + 2^{x_{u-1}} + 2^{x_u}, \ldots$, and x are inserted in the appropriate places. After a chain up to $2^{A-y_i} + x(2^{y_1-y_i} + \cdots + 2^{y_{i-1}-y_i})$ has been formed, continue by adding x and doubling the latter result $y_i - y_{i+1}$ times; this yields

$$2^{A-y_{i+1}} + x(2^{y_1-y_{i+1}} + \cdots + 2^{y_i-y_{i+1}}).$$

If this construction is done for $i = 1, 2, \ldots, v$, assuming for convenience that $y_{v+1} = 0$, we have an addition chain for $2^A + xy$ as desired. ∎

Theorem F enables us to find values of n for which $l(n) < l^*(n)$, since Theorem H gives an explicit value of $l^*(n)$ in certain cases. For example, let $x = 2^{1016} + 1$, $y = 2^{2032} + 1$, and let $n = 2^{6106} + xy = 2^{6106} + 2^{3048} + 2^{2032} + 2^{1016} + 1$. According to Theorem F, $l(n) \leq 6109$. But Theorem H also applies, with $m = 508$, and this proves that $l^*(n) = 6110$.

It would be possible to extend Theorem C, determining all values of n such that $l(n) = \lambda(n) + 3$; but the number of cases to be examined is so great that the calculations must be done with a sophisticated computer program that has not yet been written. The result of such a calculation would, in particular, characterize all n with $\nu(n) = 5$ such that $l(n) \neq l^*(n)$, and it is likely that such n do not have to be extremely large.

Extensive computer calculations have shown that $n = 12509$ is the smallest value with $l(n) < l^*(n)$. No star chain for this value of n is as short as the sequence 1, 2, 4, 8, 16, 17, 32, 64, 128, 256, 512, 1024, 1041, 2082, 4164, 8328, 8345, 12509.

Some conjectures. Although it is reasonable to guess at first glance that $l(n) = l^*(n)$, we have now seen that this is false. Another plausible conjecture [W. R. Utz, *Proc. Amer. Math. Soc.* **4** (1953), 462–463] is that $l(2n) = l(n) + 1$; a doubling step is so efficient, it seems unlikely that there could be any shorter chain for $2n$ than to add a doubling step to the shortest chain for n. But computer calculations have shown that this conjecture also is false; it is a remarkable fact that $l(191) = l(382) = 11$. (A star chain of length 11 for 382 is not difficult to find; for example, 1, 2, 4, 5, 9, 14, 23, 46, 92, 184, 198, 382 is one of them. The number 191 is the smallest n such that $l(n) = 11$, and it is very difficult to prove by hand that $l(191) > 10$; the computer's proof of this fact required a detailed examination of 948 cases, so a proof cannot be given here.) The smallest four values of n such that $l(2n) = l(n)$ are $n = 191, 701, 743, 1111$. It seems reasonable to conjecture that $l(2n) \geq l(n)$, and more generally that $l(mn) \geq l(n)$, but even this may be false.

Let $c(r)$ be the smallest value of n such that $l(n) = r$. We have the following table:

r	$c(r)$	r	$c(r)$	r	$c(r)$
1	2	6	19	11	191
2	3	7	29	12	379
3	5	8	47	13	607
4	7	9	71	14	1087
5	11	10	127	15	1903

For $r \leq 11$, $c(r)$ is approximately equal to $c(r - 1) + c(r - 2)$, and this fact led to speculation by several people that $c(r)$ grows like the function ϕ^r; but the result of Theorem D (with $n = c(r)$) implies that $r/\log_2 c(r) \to 1$ as $n \to \infty$. Several people also have conjectured that $c(r)$ is always a prime number; but $c(15) = 1903 = 11 \cdot 173$. Perhaps no conjecture about addition chains is safe!

An algorithm for the computation of $l(n)$ is discussed in Chapter 7. The basic idea is to obtain a fairly low value r such that $l(n) \leq r$, then to show by trial and error that no chain of length $r - 1$ exists. Such an algorithm can be made reasonably efficient, using a table of values of $l(m)$ for all $m < n$; but for values of n near $c(l(n))$, where c is the function tabulated above, the calculation slows down appreciably, due to the large number of cases which must be examined. As remarked above, the algorithm's proof that $l(191) = 11$ involves 948 cases; the proof that $l(1903) = 15$ involves 57571 cases!

Tabulated values of $l(n)$ show that this function is surprisingly smooth; for example, $l(n) = 13$ for all n in the range $1125 \leq n \leq 1148$. The computer

Table 1

VALUES OF n FOR SPECIAL ADDITION CHAINS

23	165	281	349	382	437	551	611	667	713	787	841	901
43	179	283	355	395	451	553	619	669	715	803	845	903
59	203	293	359	403	453	557	623	677	717	809	849	905
77	211	311	367	413	455	561	631	683	739	813	863	923
83	213	317	371	419	457	569	637	691	741	825	869	941
107	227	319	373	421	479	571	643	707	749	835	887	947
149	229	323	377	423	503	573	645	709	759	837	893	955
163	233	347	381	429	509	581	659	711	779	839	899	983
						599						

calculations show that a table of $l(n)$ may be prepared for all $n \leq 1000$ by using the formula

$$l(n) = \min \left(l(n-1) + 1, l \right) - \delta, \qquad (46)$$

where $l = \infty$ if n is prime, otherwise $l = l(p) + l(n/p)$ if p is the smallest prime dividing n; and $\delta = 1$ for n in Table 1, $\delta = 0$ otherwise.

Let $d(r)$ be the number of solutions n to the equation $l(n) = r$. We have the following table:

r	$d(r)$	r	$d(r)$	r	$d(r)$
1	1	5	9	9	78
2	2	6	15	10	136
3	3	7	26	11	246
4	5	8	44	12	432

Surely $d(r)$ must be an increasing function of r, but there is no evident way to prove this seemingly simple assertion, much less to determine the asymptotic growth of $d(r)$ for large r.

The most famous problem about addition chains which is still outstanding is the "Scholz-Brauer conjecture," which states that

$$l(2^n - 1) \leq n - 1 + l(n). \qquad (47)$$

Computer calculations show, in fact, that equality holds in (47) for $1 \leq n \leq 11$; no other n have yet been tested. Most of the research on addition chains has been devoted to attempts to prove this conjecture; addition chains for the number $2^n - 1$, which has so many ones in its binary representation, are of special interest, since this is the worst case for the binary method. Arnold Scholz coined the name "addition chain" and posed (47) as a problem in 1937 [*Jahresbericht der deutschen Mathematiker-Vereinigung*, class II, **47** (1937),

41–42]; Alfred Brauer proved in 1939 that

$$l^*(2^n - 1) \leq n - 1 + l^*(n). \tag{48}$$

Hansen's theorems show that $l(n)$ can be less than $l^*(n)$, so more work is definitely necessary in order to prove or disprove (47). As a step in this direction, Hansen has defined the concept of an l^0-*chain*, which lies "between" l-chains and l^*-chains. In an l^0-chain, certain of the elements are underlined; the condition is that $a_i = a_j + a_k$, where a_j is the largest underlined element less than a_i.

As an example of an l^0-chain (certainly not a minimal one), consider

$$\underline{1},\ \underline{2},\ \underline{4},\ 5,\ \underline{8},\ 10,\ 12,\ \underline{18}; \tag{49}$$

it is easy to verify that the difference between each element and the previous underlined element is in the chain. We let $l^0(n)$ denote the minimum length of an l^0-chain for n. Clearly $l(n) \leq l^0(n) \leq l^*(n)$.

The chain constructed in Theorem F is an l^0-chain (see exercise 22); hence we have $l^0(n) < l^*(n)$ for certain n. It is not known whether or not $l(n) = l^0(n)$ in all cases; if this equation were true, the Scholz-Brauer conjecture would be settled, because of another theorem due to Hansen:

Theorem G. $l^0(2^n - 1) \leq n - 1 + l^0(n)$.

Proof. Let $1 = a_0, a_1, \ldots, a_r = n$ be an l^0-chain of minimum length for n, and let $1 = b_0, b_1, \ldots, b_t = n$ be the subsequence of underlined elements. (We may assume that n is underlined.) Then we can get an l^0-chain for $2^n - 1$ as follows:

a) Include the $l^0(n)$ numbers $2^{a_i} - 1$, for $1 \leq i \leq r$, underlined if and only if a_i is underlined.

b) Include the numbers $2^i(2^{b_j} - 1)$, for $0 \leq j < t$ and for $0 < i \leq b_{j+1} - b_j$, all underlined. (This is a total of $b_1 - b_0 + \cdots + b_t - b_{t-1} = n - 1$ numbers.)

c) Sort the numbers from (a) and (b) into ascending sequence.

We may easily verify that this gives an l^0-chain: The numbers of (b) are all equal to twice some other element of (a) or (b); furthermore, this element is the preceding underlined element. If $a_i = b_j + a_k$, where b_j is the largest underlined element less than a_i, then $a_k = a_i - b_j \leq b_{j+1} - b_j$, so $2^{a_k}(2^{b_j} - 1) = 2^{a_i} - 2^{a_k}$ appears underlined in the chain, just preceding $2^{a_i} - 1$. Since $2^{a_i} - 1 = (2^{a_i} - 2^{a_k}) + (2^{a_k} - 1)$ and both of these appear in the chain, we have an addition chain with the l^0 property. ∎

The chain corresponding to (49), constructed in the proof of Theorem G, is

$\underline{1}, \underline{2}, \underline{3}, \underline{6}, \underline{12}, \underline{15}, \underline{30}, 31, \underline{60}, \underline{120}, \underline{240}, \underline{255}, \underline{510}, \underline{1020}, 1023, \underline{2040},$

$\underline{4080}, 4095, \underline{8160}, \underline{16320}, \underline{32640}, \underline{65280}, \underline{130560}, \underline{261120}, \underline{262143}.$

EXERCISES

1. [*15*] What is the value of Z when Algorithm A terminates?

2. [*24*] Write a MIX program for Algorithm A, to calculate x^n mod w, given integers n and x, where w is the word size. Assume that MIX has the binary operations SLB, JAE, etc., which are described in Section 4.5.2. Write another program which computes x^n mod w in a serial manner (multiplying repeatedly by x), and compare the running times of these programs.

▶ **3.** [*22*] How is x^{975} calculated by (a) the binary method? (b) the ternary method? (c) the quaternary method? (d) the factor method?

4. [*M20*] Find a number n for which the octal (2^3-ary) method gives ten less multiplications than the binary method.

▶ **5.** [*24*] Fig. 13 shows the first eight levels of the "power tree." The $(k+1)$-st level of this tree is defined as follows, assuming that the first k levels have been constructed: Take each node n of the kth level, from left to right in turn, and attach below it the nodes

$$n+1, n+a_1, n+a_2, \ldots, n+a_{k-1} = 2n$$

(in this order), where $1, a_1, a_2, \ldots, a_{k-1}$ is the path from the root of the tree to n; but discard any node which duplicates a number that has already appeared in the tree.

Design an efficient algorithm which constructs the first $r+1$ levels of the power tree. [*Hint:* Make use of two sets of variables LINKU[j], LINKR[j] for $0 \le j \le 2^r$; these point upwards and to the right, respectively, if j is a number in the tree.]

6. [*M26*] If a slight change is made to the definition of the power tree which is given in exercise 5, so that the nodes below n are attached in *decreasing* order

$$n+a_{k-1}, \ldots, n+a_2, n+a_1, n+1$$

instead of increasing order, we get a tree whose first five levels are

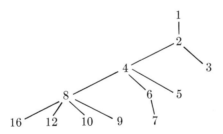

Show that this tree gives a method of computing x^n which requires exactly as many multiplications as the binary method; therefore it is not as good as the power tree, although it has been constructed in almost the same way.

7. [*M21*] Prove that there are infinitely many values of n

a) for which the factor method is better than the binary method;
b) for which the binary method is better than the factor method;
c) for which the power tree method is better than both the binary and factor methods.
(Here the "better" method shows how to compute x^n using fewer multiplications.)

8. [*M21*] Prove that the power tree (exercise 5) never gives more multiplications for the computation of x^n than the binary method.

9. [*M46*] Is the power tree method ever worse than the factor method? (Cf. exercises 7 and 8.)

10. [*10*] Fig. 14 shows a tree that indicates one way to compute x^n with the fewest possible number of multiplications, for all $n \leq 100$. How can this tree be conveniently represented within a computer, in just 100 memory locations?

▶ **11.** [*M26*] The tree of Fig. 14 depicts addition chains $a_0, a_1, \ldots, a_r$ such that $l(a_i) = i$ for all i in the chain. Find all addition chains for n which have this property, when $n = 43$ and when $n = 77$. Show that any tree such as Fig. 14 must either include the path 1, 2, 4, 8, 9, 17, 34, 43, 77 or the path 1, 2, 4, 8, 9, 17, 34, 68, 77.

12. [*M10*] Is it possible to extend the tree shown in Fig. 14 to an infinite tree that yields a minimum-multiplication rule for computing x^n, for all positive integers n?

13. [*M21*] Find a star chain of length $A + 2$ for each of the four cases listed in Theorem C. (Consequently Theorem C holds also with l replaced by l^*.)

14. [*M35*] Complete the proof of Theorem C, by demonstrating that (a) step $r - 1$ is not a small step; and (b) $\lambda(a_{r-k})$ cannot be less than $m - 1$.

15. [*M48*] Write a computer program to extend Theorem C, characterizing all n such that $l(n) = \lambda(n) + 3$ and characterizing all n such that $l^*(n) = \lambda(n) + 3$. What is the smallest n of this type such that $l(n) \neq l^*(n)$?

16. [*HM15*] Show that Theorem D is not trivially true just because of the binary method; if $l^B(n)$ denotes the length of the addition chain for n produced by the binary S-and-X method, $l^B(n)/\lambda(n)$ does not approach a limit as $n \to \infty$.

17. [*M25*] Explain how to find the intervals $J_1, \ldots, J_h$ which are required in the proof of Lemma P.

18. [*HM29*] Let β be a positive constant. Show that

$$\binom{m+s}{t+v}\binom{t+v}{v}\beta^{2v}\binom{(m+s)^2}{t},$$

summed over all s, t, u, v satisfying (26), (29), and (30), is less than $2^m/\alpha^m$ for some $\alpha > 1$, for all large m.

19. [*M23*] A "multiset" is like a set, but it may contain identical elements repeated a finite number of times. If A and B are multisets, we define new multisets $A \uplus B$, $A \cup B$, $A \cap B$ in the following way: An element x occurring exactly a times in A and b times in B occurs exactly $a + b$ times in $A \uplus B$, max (a, b) times in $A \cup B$, min (a, b) times in $A \cap B$. (A "set" is a multiset which contains no elements more than once; if A and B are sets, so are $A \cup B$ and $A \cap B$, and the definition given in this exercise agrees with the customary definition of set union and intersection.)

   a) The prime factorization of an integer $n > 0$ is a multiset N whose elements are primes, where $\prod_{p \in N} = n$. The fact that every positive integer can be uniquely factored into primes gives us a one-to-one correspondence between the positive integers and the finite multisets of prime numbers; for example if $n = 2^2 \cdot 3^3 \cdot 17$, the corresponding multiset is $N = \{2, 2, 3, 3, 3, 17\}$. If M and N are the multisets corresponding to m and n, what multisets correspond to gcd (m, n), lcm (m, n), and mn?

b) Every monic polynomial $f(z)$ over the complex numbers corresponds in a natural way to the multiset F of its "roots"; we have $f(z) = \prod_{\zeta \in F} (z - \zeta)$. If $f(z)$ and $g(z)$ are the polynomials corresponding to the finite multisets F and G of complex numbers, what polynomials correspond to $F \uplus G$, $F \cup G$, and $F \cap G$?

c) Find as many interesting identities as you can which hold between multisets, with respect to the three operations $\uplus$, $\cup$, and $\cap$.

20. [*M20*] What are the sequences b_i, S_i, M_{ij} ($0 \leq i \leq r$, $0 \leq j \leq t$) arising in Hansen's structural decomposition of star chains (a) of Type 3? (b) of Type 5? (The six "types" are defined in the proof of Theorem B.)

▶ **21.** [*M25*] (W. Hansen.) Let q be any positive integer. Find a value of n such that $l(n) \leq l^*(n) - q$.

22. [*M20*] Prove that the addition chain constructed in the proof of Theorem F is an l^0-chain.

23. [*M20*] Show that $l^*(2^n - 1) \leq n - 1 + l^*(n)$.

▶ **24.** [*M22*] Generalize the proof of Theorem G to show that

$$l^0 \left(\frac{B^n - 1}{B - 1} \right) \leq (n - 1)l^0(B) + l^0(n),$$

for any integer B greater than unity; and prove that $l(2^{mn} - 1) \leq l(2^m - 1) + mn - m + l^0(n)$.

25. [*20*] Let y be a fraction, $0 < y < 1$, expressed in the binary number system as $y = (.d_1 \ldots d_k)$. Design an algorithm to compute x^y using the operations of multiplication and square-root extraction.

▶ **26.** [*M24*] Design an efficient algorithm to compute the nth Fibonacci number F_n exactly, when n is very large.

▶ **27.** [*24*] (E. G. Straus.) Find a way to compute a general *monomial* $x_1^{n_1} x_2^{n_2} \ldots x_m^{n_m}$ in at most $2\lambda(\max (n_1, n_2, \ldots, n_m)) + 2^m - m - 1$ multiplications.

28. [*M39*] (Kenneth B. Stolarsky.) (a) Assume that $1 = a_0 < a_1 < \cdots < a_q < \cdots < a_r = n$ is an addition chain in which step q is the last small step, and $q + 1 < r$. (According to an observation made in the text, steps $q + 2, \ldots, r$ must therefore be star steps.) Prove that steps $q + 2, \ldots, r$ contain at most two nondoublings; and if two nondoublings occur, the second must be a "Fibonacci step," namely a step of the form $a_i = a_{i-1} + a_{i-2}$. (Therefore we have only a limited number of possibilities between consecutive small steps of an addition chain.) (b) Prove that $\nu(n) \leq 8^{l(n) - \lambda(n)}$, for all $n > 0$.

29. [*M50*] Is $\nu(n) \leq 2^{l(n) - \lambda(n)}$ for all positive integers n? (If so, we have the lower bound $l(2^n - 1) \geq n - 1 + \lceil \log_2 n \rceil$; cf. (17) and (47).)

30. [*20*] (Peter Ungar.) An *addition-subtraction chain* has the rule $a_i = a_j \pm a_k$ in place of (2); the imaginary computer described in the text has a new operation code, SUB. (This corresponds in practice to evaluating x^n using both multiplications and divisions.) Show that there can be addition-subtraction chains for n whose length is less than $l(n)$.

31. [*M46*] (D. H. Lehmer.) Explore the problem of minimizing $\epsilon q + (r - q)$ in an addition chain (1), where q is the number of "simple" steps in which $a_i = a_{i-1} + 1$, given a small positive "weight" ϵ. (This problem comes closer to reality for many calculations of x^n, if multiplication by x is simpler than a general multiplication.)

32. [*M50*] Explore the problem of computing a *set* of values $x^{n_1}, \ldots, x^{n_m}$, instead of just an isolated value x^n, with the fewest number of multiplications, for various sets of exponents $\{n_1, \ldots, n_m\}$. For example, consider the exponents $\{1, 4, 9, \ldots, m^2\}$; or consider the general case when $m = 2$.

33. [*M50*] What is the asymptotic value of the function $d(r)$ defined in the text?

34. [*M50*] Is $l(2^n - 1) \leq n - 1 + l(n)$ for all positive integers n? Does equality hold? Does $l(n) = l^0(n)$?

4.6.4. Evaluation of polynomials

Now that we have considered efficient means of evaluating the special polynomial x^n, let us consider the general problem of computing an nth degree polynomial

$$u(x) = u_n x^n + u_{n-1} x^{n-1} + \cdots + u_1 x + u_0, \qquad u_n \neq 0, \qquad (1)$$

for given values of x. This problem arises frequently in practice.

In the following discussion we shall concentrate on minimizing the number of operations required to evaluate polynomials by computer, blithely assuming that all arithmetic operations are exact. Polynomials are most commonly evaluated using floating-point arithmetic, which is not exact, and different schemes for the evaluation will, in general, give different answers. A numerical analysis of the accuracy achieved depends on the coefficients of the particular polynomial being considered, and is beyond the scope of this book; the reader should be careful to investigate the accuracy of any calculations undertaken with floating-point arithmetic. [See J. H. Wilkinson, *Rounding errors in algebraic processes* (Englewood Cliffs: Prentice-Hall, 1963), Chapter 2; John R. Rice, *Numer. Math.* **7** (1965), 426–435; W. Kahan and I. Farkas, *CACM* **6** (1963), 164–165; C. Mesztenyi and C. Witzgall, *J. Res. Nat. Bur. Stand.* **71B** (1967), 11–17.]

A beginning programmer will often evaluate the polynomial (1) in a manner corresponding directly to its conventional textbook form: First $u_n x^n$ is calculated, then $u_{n-1} x^{n-1}, \ldots, u_1 x$, and finally all of the terms of (1) are added together. But even if the efficient methods of Section 4.6.3 for calculating powers are used in this approach, the resulting calculation is needlessly slow unless nearly all of the coefficients u_k are zero. If the coefficients are all nonzero, an obvious alternative would be to form a table containing $x^2, x^3, \ldots, x^n$ first, and then to multiply each power of x by an appropriate coefficient and add up the results. Such a process involves $2n - 1$ multiplications and n additions, plus further instructions to store intermediate results into temp storage and to retrieve them again.

Horner's rule. One of the first things a novice programmer is usually taught is an elegant way to rearrange this computation, by evaluating $u(x)$ as follows:

$$u(x) = ((\cdots (u_n x + u_{n-1})x + \cdots)x + u_0. \tag{2}$$

Start with u_n, multiply by x, add u_{n-1}, multiply by x, ..., multiply by x, add u_0. This form of the computation is usually called "Horner's rule"; we have already seen it used in connection with radix conversion in Section 4.4. The entire process requires n multiplications and n additions, minus one addition for each coefficient that is zero. Furthermore, there is no need to store partial results, since each quantity arising during the calculation is used immediately after it has been computed.

W. G. Horner gave this rule early in the nineteenth century [*Philosophical Transactions, Royal Society of London* **109** (1819), 308–335] in connection with an efficient method for finding the coefficients of $u(x + c)$. The fame of the latter method accounts for the fact that Horner's name has been attached to (2); but actually (2) was given by Isaac Newton over 100 years earlier [*Analysis per Quantitatem Series* (London, 1711), 10]. Newton wrote, for example,

$$\overline{\overline{y - 4} \times y + 5} \times y - 12 \times y + 17$$

for the polynomial $y^4 - 4y^3 + 5y^2 - 12y + 17$; at that time it was customary to denote grouping by using horizontal lines instead of parentheses.

Several generalizations of Horner's rule have been suggested. Let us first consider evaluating $u(z)$ when z is a complex number, while the coefficients u_k are real. Complex addition and multiplication can obviously be reduced to a sequence of ordinary operations on real numbers:

real + complex	requires	1 addition
complex + complex	requires	2 additions
real × complex	requires	2 multiplications
complex × complex	requires	4 multiplications, 2 additions
	or	3 multiplications, 5 additions

(See exercise 41. Subtraction is here considered as if it were equivalent to addition.) Therefore Horner's rule (2) uses either $4n - 2$ multiplications and $3n - 2$ additions or $3n - 1$ multiplications and $6n - 5$ additions to evaluate $u(z)$ when $z = x + iy$ is complex. An alternative procedure for evaluating $u(x + iy)$ is to let

$$a_1 = u_n, \quad b_1 = u_{n-1}, \quad r = x + x, \quad s = x^2 + y^2;$$
$$a_j = b_{j-1} + r a_{j-1}, \quad b_j = u_{n-j} - s a_{j-1}, \quad 1 < j \le n. \tag{3}$$

Then it is easy to prove by induction that $u(z) = z a_n + b_n$. This scheme [*BIT* **5** (1965), 142] requires $2n + 2$ multiplications and $2n + 1$ additions, so it is an improvement over Horner's rule when $n \geq 3$. (A good programmer should therefore be careful not to make indiscriminate use of the built-in "complex arithmetic" features of high-level programming languages!)

Consider the process of dividing the polynomial $u(x)$ by the polynomial $x - x_0$, obtaining $u(x) = (x - x_0)q(x) + r(x)$; here deg $(r) < 1$, so $r(x)$ is a constant independent of x, and $u(x_0) = 0 \cdot q(x_0) + r = r$. An examination of this division process reveals that the computation is essentially the same as Horner's rule for evaluating $u(x_0)$. Similarly, if we divide $u(z)$ by the polynomial $(z - z_0)(z - \bar{z}_0) = z^2 - 2x_0z + x_0^2 + y_0^2$, the resulting computation is equivalent to (3); we obtain

$$u(z) = (z - z_0)(z - \bar{z}_0)q(z) + a_nz + b_n;$$

hence

$$u(z_0) = a_nz_0 + b_n.$$

In general, if we divide $u(x)$ by $f(x)$, to obtain $u(x) = f(x)q(x) + r(x)$, then if $f(x_0) = 0$, we have $u(x_0) = r(x_0)$; this observation leads to further generalizations of Horner's rule. For example, we may let $f(x) = x^2 - x_0^2$; this yields the "second-order" Horner's rule

$$u(x) = (\cdots (u_{2\lfloor n/2 \rfloor}x^2 + u_{2\lfloor n/2 \rfloor - 2})x^2 + \cdots)x^2 + u_0$$
$$+ ((\cdots (u_{2\lceil n/2 \rceil - 1}x^2 + u_{2\lceil n/2 \rceil - 3})x^2 + \cdots)x^2 + u_1)x. \qquad (4)$$

The second-order rule uses $n + 1$ multiplications and n additions; so it is no improvement over Horner's rule from this standpoint. But there are at least two circumstances in which (4) is useful: If we want to evaluate both $u(x)$ and $u(-x)$, this means of computation yields $u(-x)$ with just one more addition operation; two values can be obtained almost as cheaply as one. Moreover, if we have a computer which allows parallel computations, the two lines of (4) may be evaluated independently, so the entire polynomial will be evaluated about twice as fast.

When the computer allows parallel computation on k arithmetic units at once, a "kth-order" Horner's rule (obtained in a similar manner from $f(x) = x^k - x_0^k$) may be used. Another attractive method for parallel computation has been suggested by G. Estrin [Proc. Western Joint Computing Conf. **17** (May, 1960), 33–40]; for $n = 7$, Estrin's method is:

Processor 1	Processor 2	Processor 3	Processor 4	Processor 5
$a_1 = u_7x + u_6$	$b_1 = u_5x + u_4$	$c_1 = u_3x + u_2$	$d_1 = u_1x + u_0$	x^2
$a_2 = a_1x^2 + b_1$		$c_2 = c_1x^2 + d_1$		x^4
$a_3 = a_2x^4 + c_2$				

Here $a_3 = u(x)$. However, an interesting analysis by W. S. Dorn [*IBM J. Res. and Devel.* **6** (1962), 239–245] shows that these methods often are no improvement over the second-order rule, if each arithmetic unit must access the same core memory (where the latter can communicate with only one processor at a time).

Still another generalization of Horner's rule applies when we want to evaluate both $u(x)$ and its derivative

$$u'(x) = nu_nx^{n-1} + (n-1)u_{n-1}x^{n-2} + \cdots + u_1. \tag{5}$$

This can be done in a convenient manner by letting

$$a_0 = u_n, \qquad b_0 = 0,$$
$$a_j = a_{j-1}x + u_{n-j}, \qquad b_j = b_{j-1}x + a_{j-1}, \qquad 1 \le j \le n. \tag{6}$$

Here $a_n = u(x)$ and $b_n = u'(x)$; the calculation takes $2n - 1$ multiplications and $2n - 1$ additions, and it does not require that the new coefficients ju_j be stored in memory.

Special polynomials. When $u(x)$ has a special form, special techniques may be employed for its calculation. For example, if $u(x) = x^n + \cdots + x + 1$, we know that $u(x) = (x^{n+1} - 1)/(x - 1)$, so the calculation can be done in $l(n + 1)$ multiplications (see Section 4.6.3), two subtractions, and one division. Note that division is an essential part of the technique for evaluating this special polynomial, although polynomials are defined only in terms of addition and multiplication.

The *determinant* of an $n \times n$ matrix may be considered to be a polynomial in n^2 variables x_{ij}, $1 \le i, j \le n$. If $x_{11} \ne 0$, we have

$$\det \begin{pmatrix} x_{11} & x_{12} & \cdots & x_{1n} \\ x_{21} & x_{22} & \cdots & x_{2n} \\ \vdots & & & \vdots \\ x_{n1} & x_{n2} & \cdots & x_{nn} \end{pmatrix}$$

$$= x_{11} \det \begin{pmatrix} x_{22} - (x_{21}/x_{11})x_{12} & \cdots & x_{2n} - (x_{21}/x_{11})x_{1n} \\ \vdots & & \vdots \\ x_{n2} - (x_{n1}/x_{11})x_{12} & \cdots & x_{nn} - (x_{n1}/x_{11})x_{1n} \end{pmatrix}. \tag{7}$$

The determinant of an $n \times n$ matrix may therefore be evaluated by evaluating the determinant of an $(n-1) \times (n-1)$ matrix and using an additional $(n-1)^2 + 1$ multiplications, $(n-1)^2$ additions, and $n - 1$ divisions. Since a 2×2 determinant can be evaluated with two multiplications and one addition, we see that the determinant of almost all matrices (namely those for which no division by zero is required) can be computed with at most $(2n^3 - 3n^2 + 7n - 6)/6$ multiplications, $(2n^3 - 3n^2 + n)/6$ additions, and $(n^2 - n - 2)/2$ divisions.

When zero occurs, the determinant is even easier to compute; for example if $x_{11} = 0$ but $x_{21} \neq 0$, we have

$$\det \begin{pmatrix} 0 & x_{12} & \ldots & x_{1n} \\ x_{21} & x_{22} & \ldots & x_{2n} \\ x_{31} & x_{32} & \ldots & x_{3n} \\ \vdots & & & \vdots \\ x_{n1} & x_{n2} & \ldots & x_{nn} \end{pmatrix} = -x_{21} \det \begin{pmatrix} x_{12} & \ldots & x_{1n} \\ x_{32} - (x_{31}/x_{21})x_{22} & \ldots & x_{3n} - (x_{31}/x_{21})x_{2n} \\ \vdots & & \vdots \\ x_{n2} - (x_{n1}/x_{21})x_{22} & \ldots & x_{nn} - (x_{n1}/x_{21})x_{2n} \end{pmatrix}. \quad (8)$$

Here the reduction to an $(n-1) \times (n-1)$ determinant saves $n-1$ of the multiplications and $n-1$ of the additions used in (7), and this certainly compensates for the additional bookkeeping required to recognize this case. Therefore any determinant can be evaluated with roughly $\frac{2}{3}n^3$ arithmetic operations (including division); this is remarkable, since it is a polynomial with $n!$ terms and n variables in each term.

If we want to evaluate the determinant of a matrix with *integer* elements, the above process appears to be unattractive since it requires rational arithmetic. However, we can use the method to evaluate the determinant mod p, for any prime p, since division mod p is possible; if this is done for several primes p, the exact value of the determinant can be found as explained in Section 4.3.2, since Hadamard's inequality (Eq. 4.6.1–24) gives an upper bound for the absolute value of the determinant. In fact, if p is large enough (see Section 4.5.4), computation modulo this single prime p suffices to determine the value of the determinant.

The *permanent* of a matrix is a polynomial which is very similar to the determinant; the only difference is that all of its coefficients are $+1$:

$$\text{per} \begin{pmatrix} x_{11} & \ldots & x_{1n} \\ \vdots & & \vdots \\ x_{n1} & \ldots & x_{nn} \end{pmatrix} = \sum x_{1j_1} x_{2j_2} \ldots x_{nj_n} \quad (9)$$

summed over all permutations $(j_1, j_2, \ldots, j_n)$ of $\{1, 2, \ldots, n\}$. No way to evaluate the permanent as efficiently as the determinant is known; exercises 9 and 10 show that substantially less than $n!$ operations will suffice, for large n, but the execution time still grows exponentially with the size of the matrix.

Another important operation involving matrices is, of course, *matrix multiplication:* If $X = (x_{ij})$ is an $m \times n$ matrix, $Y = (y_{jk})$ is an $n \times r$ matrix, and $Z = (z_{ik})$ is an $m \times r$ matrix, then $Z = XY$ means that

$$z_{ik} = \sum_{1 \leq j \leq n} x_{ij} y_{jk}, \qquad 1 \leq i \leq m, \qquad 1 \leq k \leq r. \quad (10)$$

This equation may be regarded as the computation of mr simultaneous polynomials in $mn + nr$ variables; each polynomial is the "inner product" of two n-place vectors. A brute-force calculation would involve mnr multiplications and $mr(n-1)$ additions; but S. Winograd has discovered an ingenious way to save about half of the multiplications:

$$z_{ik} = \sum_{1 \le j \le n/2} (x_{i,2j} + y_{2j-1,k})(x_{i,2j-1} + y_{2j,k}) - a_i - b_k + c_{ik};$$

$$a_i = \sum_{1 \le j \le n/2} x_{i,2j} x_{i,2j-1}; \qquad b_k = \sum_{1 \le j \le n/2} y_{2j-1,k} y_{2j,k};$$

$$c_{ik} = \begin{cases} 0, & n \text{ even;} \\ x_{in} y_{nk}, & n \text{ odd.} \end{cases}$$

This uses $\lceil n/2 \rceil mr + \lfloor n/2 \rfloor (m + r)$ multiplications, $(n + 2)mr + (\lfloor n/2 \rfloor - 1)$ $\times (mr + m + r)$ additions; the total number of operations has increased slightly, while the number of multiplications is roughly halved.

An even better scheme, discovered by V. Strassen in 1968, is based on the identity

$$\begin{pmatrix} a & b \\ c & d \end{pmatrix} \begin{pmatrix} A & -C \\ -B & -D \end{pmatrix}$$

$$= \begin{pmatrix} (a+d)(A+D) - (b+d)(B+D) - d(A-B) - (a-b)D & (a-b)D - a(D-C) \\ (d-c)A - d(A-B) & (a+d)(A+D) - (a+c)(A+C) - a(D-C) - (d-c)A \end{pmatrix}.$$

This formula, unlike Winograd's, does not depend on commutativity of multiplication, so $(2n \times 2n)$ matrices can be partitioned into four $(n \times n)$ matrices and the identity can be used recursively. Hence we can multiply $(2^n \times 2^n)$ matrices using only 7^n multiplications and $6(7^n - 4^n)$ additions; the total number of operations required to multiply $(n \times n)$ matrices is thereby reduced to $O(n^{\log_2 7}) = O(n^{2.81})$, a substantial saving for large n. (Yet Winograd's method is superior on 4×4 matrices.)

The *finite Fourier transform* of a complex-valued function of n variables, from a set of m elements, has been defined in Section 3.3.4. [See Eqs. (1) and (2) in that section.] In the special case $m = 2$ we have

$$f(s_1, \ldots, s_n) = \sum_{0 \le t_1, \ldots, t_n \le 1} F(t_1, \ldots, t_n)(-1)^{s_1 t_1 + \cdots + s_n t_n}, \qquad (11)$$

$$0 \le s_1, \ldots, s_n \le 1,$$

and this may be regarded as a simultaneous evaluation of 2^n linear polynomials in 2^n variables $F(t_1, \ldots, t_n)$. A well-known technique due to F. Yates [*The Design and Analysis of Factorial Experiments* (Harpenden: Imperial Bureau of Soil Sciences, 1937)] can be used to reduce the number of additions implied in (11) from $2^n(2^n - 1)$ to $n2^n$. Yates's method can be understood by considering the case $n = 3$: Let $x_{t_1 t_2 t_3} = F(t_1, t_2, t_3)$.

Given	First step	Second step	Third step
x_{000}	$x_{000} + x_{001}$	$x_{000} + x_{001} + x_{010} + x_{011}$	$x_{000} + x_{001} + x_{010} + x_{011} + x_{100} + x_{101} + x_{110} + x_{111}$
x_{001}	$x_{010} + x_{011}$	$x_{100} + x_{101} + x_{110} + x_{111}$	$x_{000} - x_{001} + x_{010} - x_{011} + x_{100} - x_{101} + x_{110} - x_{111}$
x_{010}	$x_{100} + x_{101}$	$x_{000} - x_{001} + x_{010} - x_{011}$	$x_{000} + x_{001} - x_{010} - x_{011} + x_{100} + x_{101} - x_{110} - x_{111}$
x_{011}	$x_{110} + x_{111}$	$x_{100} - x_{101} + x_{110} - x_{111}$	$x_{000} - x_{001} - x_{010} + x_{011} + x_{100} - x_{101} - x_{110} + x_{111}$
x_{100}	$x_{000} - x_{001}$	$x_{000} + x_{001} - x_{010} - x_{011}$	$x_{000} + x_{001} + x_{010} + x_{011} - x_{100} - x_{101} - x_{110} - x_{111}$
x_{101}	$x_{010} - x_{011}$	$x_{100} + x_{101} - x_{110} - x_{111}$	$x_{000} - x_{001} + x_{010} - x_{011} - x_{100} + x_{101} - x_{110} + x_{111}$
x_{110}	$x_{100} - x_{101}$	$x_{000} - x_{001} - x_{010} + x_{011}$	$x_{000} + x_{001} - x_{010} - x_{011} - x_{100} - x_{101} + x_{110} + x_{111}$
x_{111}	$x_{110} - x_{111}$	$x_{100} - x_{101} - x_{110} + x_{111}$	$x_{000} - x_{001} - x_{010} + x_{011} - x_{100} + x_{101} + x_{110} - x_{111}$

To get from the "Given" to the "First step" requires four additions and four subtractions; and the interesting feature of Yates's method is that exactly the same transformation which takes us from "Given" to "First step" will take us from "First step" to "Second step" and from "Second step" to "Third step." In each case we do four additions, then four subtractions; and after three steps we have the desired Fourier transform $f(s_1, s_2, s_3)$ in the place originally occupied by $F(s_1, s_2, s_3)$!

Yates's method can be generalized to the evaluation of any finite Fourier transform, and, in fact, to the evaluation of any sum which can be written

$$f(s_1, s_2, \ldots, s_n) = \sum_{0 \le t_1, \ldots, t_n < m} g_1(s_1, s_2, \ldots, s_n, t_1) g_2(s_2, \ldots, s_n, t_2) \cdots$$

$$g_n(s_n, t_n) F(t_1, t_2, \ldots, t_n), \qquad 0 \le s_1, s_2, \ldots, s_n < m, \qquad (12)$$

for given functions $g_j(s_j, \ldots, s_n, t_j)$. Proceed as follows:

$$f^{[0]}(t_1, t_2, t_3, \ldots, t_n) = F(t_1, t_2, t_3, \ldots, t_n);$$

$$f^{[1]}(s_n, t_1, t_2, \ldots, t_{n-1}) = \sum_{0 \le t_n < m} g_n(s_n, t_n) f^{[0]}(t_1, t_2, \ldots, t_n);$$

$$f^{[2]}(s_{n-1}, s_n, t_1, \ldots, t_{n-2}) = \sum_{0 \le t_{n-1} < m} g_{n-1}(s_{n-1}, s_n, t_{n-1}) f^{[1]}(s_n, t_1, \ldots, t_{n-1});$$

$$\cdots$$

$$f^{[n]}(s_1, s_2, s_3, \ldots, s_n) = \sum_{0 \le t_1 < m} g_1(s_1, \ldots, s_n, t_1) f^{[n-1]}(s_2, s_3, \ldots, s_n, t_1);$$

$$f(s_1, s_2, s_3, \ldots, s_n) = f^{[n]}(s_1, s_2, s_3, \ldots, s_n). \qquad (13)$$

The computation takes place in nm^n steps, instead of approximately $(m^n)^2$. For Yates's method as shown above, $g_j(s_j, \ldots, s_n, t_j) = (-1)^{s_j t_j}$; $f^{[0]}(t_1, t_2, t_3)$ represents the "Given"; $f^{[1]}(s_3, t_1, t_2)$ represents the "First step"; etc. Whenever a sum can be put into the form of (12), for reasonably simple functions $g_j(s_j, \ldots, s_n, t_j)$, the scheme (13) will reduce the amount of computation from order N^2 to order $N \log N$, where $N = m^n$. Note that approximately $2N$ storage locations are required; hence N (and especially m and n) must be of a reasonable size.

Let us consider one more special case of polynomial evaluation. *Lagrange's interpolation polynomial* of order n,

$$u^{[n]}(x) = y_0 \frac{(x - x_1)(x - x_2) \cdots (x - x_n)}{(x_0 - x_1)(x_0 - x_2) \cdots (x_0 - x_n)}$$

$$+ y_1 \frac{(x - x_0)(x - x_2) \cdots (x - x_n)}{(x_1 - x_0)(x_1 - x_2) \cdots (x_1 - x_n)} + \cdots$$

$$+ y_n \frac{(x - x_0)(x - x_1) \cdots (x - x_{n-1})}{(x_n - x_0)(x_n - x_1) \cdots (x_n - x_{n-1})}, \qquad (14)$$

is the only polynomial of degree $\le n$ in x which takes on the respective values

$y_0, y_1, \ldots, y_n$ at the $n + 1$ distinct points $x = x_0, x_1, \ldots, x_n$. (For it is evident from (14) that $u^{[n]}(x_k) = y_k$ for $0 \leq k \leq n$. If $f(x)$ is any such polynomial of degree $\leq n$, then $g(x) = f(x) - u^{[n]}(x)$ is of degree $\leq n$, and $g(x)$ is zero for $x = x_0, x_1, \ldots, x_n$; therefore $g(x)$ is a multiple of the polynomial $(x - x_0)(x - x_1) \cdots (x - x_n)$. The degree of the latter polynomial is greater than n, so $g(x) = 0$.) If we assume that the values of a function in some table are well-approximated by a polynomial, Lagrange's formula (14) may therefore be used to "interpolate" for values of the function at points x not appearing in the table. Unfortunately, there seem to be quite a few additions, subtractions, multiplications, and divisions in Lagrange's formula; in fact, there are n additions, $2n^2 + 2$ subtractions, $2n^2 + n - 1$ multiplications, and $n + 1$ divisions. Fortunately (as we might suspect), there are some methods available which substantially reduce the amount of calculation.

The basic idea for simplifying (14) is to note that $u^{[n]}(x) - u^{[n-1]}(x)$ is zero for $x = x_0, \ldots, x_{n-1}$; thus $u^{[n]}(x) - u^{[n-1]}(x)$ is a polynomial of degree $\leq n$ and a multiple of $(x - x_0) \cdots (x - x_{n-1})$. We conclude that $u^{[n]}(x) = \alpha_n(x - x_0) \cdots (x - x_{n-1}) + u^{[n-1]}(x)$, where α_n is a constant. This leads us to *Newton's interpolation formula*

$$u^{[n]}(x) = \alpha_n(x - x_0)(x - x_1) \cdots (x - x_{n-1}) + \cdots$$
$$+ \alpha_2(x - x_0)(x - x_1) + \alpha_1(x - x_0) + \alpha_0, \qquad (15)$$

where the α's are some constants we should like to determine from $x_0, x_1, \ldots,$ $x_n, y_0, y_1, \ldots, y_n$. Note that this formula holds for all n; α_k does not depend on $x_{k+1}, \ldots, x_n, y_{k+1}, \ldots, y_n$. Once the α's are known, Newton's interpolation formula is convenient for calculation, since we may generalize Horner's rule once again and write

$$u^{[n]}(x) = ((\cdots (\alpha_n(x - x_{n-1}) + \alpha_{n-1})(x - x_{n-2}) + \cdots)(x - x_1) + \alpha_1)$$
$$\times (x - x_0) + \alpha_0. \qquad (16)$$

This requires n multiplications and $2n$ additions. Alternatively, we may evaluate each of the individual terms of (15); thereby with $2n - 1$ multiplications and $2n$ additions we can calculate all of the values $u^{[0]}(x), u^{[1]}(x), \ldots, u^{[n]}(x)$, and this will indicate whether or not the interpolation process is "converging."

Let us therefore investigate how to find the α's in Newton's formula. This may be done by finding the "divided differences" in the following tableau (shown for $n = 3$):

y_0

$\qquad (y_1 - y_0)/(x_1 - x_0) = y_1'$

$y_1 \qquad\qquad\qquad\qquad (y_2' - y_1')/(x_2 - x_0) = y_2''$

$\qquad (y_2 - y_1)/(x_2 - x_1) = y_2' \qquad\qquad\qquad (y_3'' - y_2'')/(x_3 - x_0) = y_3'''$

$y_2 \qquad\qquad\qquad\qquad (y_3' - y_2')/(x_3 - x_1) = y_3''$

$\qquad (y_3 - y_2)/(x_3 - x_2) = y_3'$

$y_3 \qquad\qquad\qquad\qquad\qquad\qquad\qquad\qquad\qquad\qquad\qquad\qquad (17)$

It is possible, although not extremely easy, to prove that $\alpha_0 = y_0$, $\alpha_1 = y_1'$, $\alpha_2 = y_2''$, etc.; see exercise 15. Therefore the following calculation (corresponding to (17)) may be used to obtain the α's:

> Start with $(\alpha_0, \alpha_1, \ldots, \alpha_n) \leftarrow (y_0, y_1, \ldots, y_n)$; then, for $k = 1, 2, \ldots, n$ (in this order), set $y_j \leftarrow (y_j - y_{j-1})/(x_j - x_{j-k})$ for $j = n, n-1, \ldots, k$ (in this order).

This process requires $\frac{1}{2}(n^2 + n)$ divisions and $n^2 + n$ subtractions, so about three-fourths of the work implied in (14) has been saved.

For example, suppose that we want to estimate $\frac{3}{2}!$ from the values of $0!$, $1!$, $2!$, and $3!$, using a cubic polynomial. The divided differences are

x	y	y'	y''	y'''
0	1			
		0		
1	1		$\frac{1}{2}$	
		1		$\frac{1}{3}$
2	2		$\frac{3}{2}$	
		4		
3	6			

so $u^{[0]}(x) = u^{[1]}(x) = 1$, $u^{[2]}(x) = \frac{1}{2}x(x-1) + 1$, $u^{[3]}(x) = \frac{1}{3}x(x-1)(x-2) + \frac{1}{2}x(x-1) + 1$. Setting $x = \frac{3}{2}$ gives $-\frac{1}{8} + \frac{3}{8} + 1 = 1.25$; presumably the "correct" value is

$$\Gamma(\tfrac{5}{2}) = \tfrac{3}{4}\sqrt{\pi} \approx 1.33.$$

When $y_k = u(x_k)$, $0 \le k \le n$, for some polynomial $u(x)$ of degree n, we must have $u(x) = u^{[n]}(x)$, and so the nth divided difference $y_n^{(n)}$ must equal the leading coefficient of $u(x)$. Therefore in any table of values of a polynomial of degree n, the nth divided differences are all equal; every $(n+1)$st divided difference must be zero, regardless of what distinct points $x_0, x_1, \ldots, x_{n+1}$ are used for the calculation.

It is instructive to note that evaluation of the interpolation polynomial is just a special case of the Chinese remainder algorithm of Section 4.3.2, since we know the value of $u^{[n]}(x)$ mod the relatively prime polynomials $x - x_0, \ldots,$ $x - x_n$. (As we have seen above, $f(x)$ mod $(x - x_0) = f(x_0)$.) Under this interpretation, Newton's formula (15) is precisely the "mixed radix representation" of Eq. 4.3.2–24; and 4.3.2–23 yields another way to compute $\alpha_0, \ldots, \alpha_n$ using the same number of operations as (17), but with less opportunities for parallelism.

A remarkable modification of the method of divided differences, which applies to rational functions instead of polynomials, was introduced by T. N. Thiele in 1909. For a discussion of Thiele's method of "reciprocal differences," see L. M. Milne-Thomson, *Calculus of Finite Differences* (London: MacMillan, 1933), Chapter 5; R. W. Floyd, *CACM* **3** (1960), 508.

Tabulating polynomial values. If we wish to evaluate the nth degree polynomial $u(x)$ for many values of x in an arithmetic progression (i.e., if we want to calculate $u(x_0)$, $u(x_0 + h)$, $u(x_0 + 2h)$, ...), the process can be reduced to addition only, after the first few steps, because of the fact that the "nth difference" of the polynomial is constant.

P1. Find the coefficients $\beta_n, \ldots, \beta_1, \beta_0$ when Newton's interpolation polynomial is written in the form

$$u(x) = \frac{\beta_n}{n!h^n}(x - x_0 - (n-1)h) \cdots (x - x_0 - h)(x - x_0) + \cdots$$
$$+ \frac{\beta_2}{2!h^2}(x - x_0 - h)(x - x_0) + \frac{\beta_1}{h}(x - x_0) + \beta_0. \qquad (18)$$

(This can be done by taking repeated differences, just as the α's of (15) were found above, with $x_j = x_0 + jh$, except that all divisions by $x_j - x_{j-k}$ are suppressed from the calculation.)

P2. Set $x \leftarrow x_0$.

P3. β_0 is now the value of $u(x)$.

P4. Set $\beta_j \leftarrow \beta_j + \beta_{j+1}$ for $j = 0, 1, \ldots, n - 1$ (in this order). Increase x by h and return to step P3. ∎

(The quantity x need not actually be computed; it has been included in this procedure just to illustrate what is taking place.)

Suppose, for example, that we wish to evaluate the simple polynomial $u(x) = x^3 + x$ for $x = 0, \frac{1}{2}, 1, 1\frac{1}{2}$, etc. The preliminary calculation of step P1 proceeds as follows, by taking differences without division:

$$
\begin{array}{llll}
u(0) = 0.000 & & & \\
& 0.625 & & \\
u(\tfrac{1}{2}) = 0.625 & & 0.750 & \\
& 1.375 & & 0.750 \\
u(1) = 2.000 & & 1.500 & \\
& 2.875 & & \\
u(\tfrac{3}{2}) = 4.875 & & &
\end{array}
$$

Therefore $\beta_0 = 0.000$, $\beta_1 = 0.625$, $\beta_2 = 0.750$, $\beta_3 = 0.750$. We now can tabulate the values desired:

x	$u(x) = \beta_0$	β_1	β_2	β_3
0.0	0.000	0.625	0.750	0.750
0.5	0.625	1.375	1.500	0.750
1.0	2.000	2.875	2.250	0.750
1.5	4.875	5.125	3.000	0.750
2.0	10.000	8.125	3.750	0.750

Caution: When carrying out an evaluation by this method using floating-point arithmetic, rounding errors can build up rather rapidly. An error in β_k reflects an error in the coefficient of x^k in the polynomial being computed. Therefore the accuracy of this method should be carefully investigated before it is used rashly.

Adaptation of coefficients. Let us now return to our original problem of evaluating a given polynomial $u(x)$ as rapidly as possible, for "random" values of x. The importance of this problem is due partly to the fact that standard functions such as $\sin x$, $\cos x$, e^x, etc., are usually computed by subroutines which rely on the evaluation of certain polynomials; these polynomials are evaluated so often, it is desirable to find the fastest possible way to do the computation.

Arbitrary polynomials of degree five and higher can be evaluated with less operations than Horner's rule requires, if we first "adapt" the coefficients $u_0, u_1, \ldots, u_n$. Such an adaptation process might involve a fairly complex calculation, as explained below; but the preliminary calculation is not wasted, since it must be done only once while the polynomial will be evaluated many times. For examples of "adapted" polynomials for standard functions, see V. Ya. Pan, *USSR Computational Math. and Math. Physics* **2** (1963), 137–146.

The simplest case for which adaptation of coefficients is helpful occurs for a fourth degree polynomial:

$$u(x) = u_4 x^4 + u_3 x^3 + u_2 x^2 + u_1 x + u_0, \qquad u_4 \neq 0. \qquad (19)$$

This equation can be rewritten in a form suggested by Motzkin,

$$y = (x + \alpha_0)x + \alpha_1, \qquad u(x) = ((y + x + \alpha_2)y + \alpha_3)\alpha_4, \qquad (20)$$

for suitably "adapted" coefficients $\alpha_0, \alpha_1, \alpha_2, \alpha_3, \alpha_4$. The computation in this form obviously involves three multiplications, five additions, and one instruction to store the partial result y into temp storage; by comparison to Horner's rule, we have traded a multiplication for an addition and a storing. Even this comparatively small savings is worth while if the polynomial is to be evaluated often. (Of course, if the time for multiplication is comparable to the time for addition, (20) gives no improvement; we will see that a general fourth-degree polynomial always requires at least eight arithmetic operations in its evaluation.)

By comparing coefficients in (19) and (20), we obtain formulas for computing the α_j's in terms of the u_k's:

$$\alpha_0 = \tfrac{1}{2}(u_3/u_4 - 1), \qquad \beta = u_2/u_4 - \alpha_0(\alpha_0 + 1), \qquad \alpha_1 = u_1/u_4 - \alpha_0\beta,$$
$$\alpha_2 = \beta - 2\alpha_1, \qquad \alpha_3 = u_0/u_4 - \alpha_1(\alpha_1 + \alpha_2), \qquad \alpha_4 = u_4. \qquad (21)$$

A similar scheme, which evaluates a fourth-degree polynomial in the same number of steps as (20), appears in exercise 18; this alternative method will give greater numerical accuracy than (20) in certain cases, although it yields poorer accuracy in other cases.

Polynomials which arise in practice often have a rather small leading coefficient, so that the division by u_4 in (21) leads to instability. In such a case it is usually preferable to replace x by $|u_4|^{1/4}x$ as the first step, reducing (19) to $\pm$ a monic polynomial. A similar transformation applies to polynomials of higher degrees. This idea is due to C. T. Fike [*CACM* **10** (1967), 175–178], who has presented several interesting examples.

Any polynomial of the fifth degree may be evaluated using four multiplications, six additions, and one storing, by using the rule $u(x) = U(x)x + u_0$, where $U(x) = u_5x^4 + u_4x^3 + u_3x^2 + u_2x + u_1$ is evaluated as in (20). Alternatively, we can do the evaluation with four multiplications, five additions, and three storings, if the calculations take the form

$$y = (x + \alpha_0)^2, \qquad u(x) = (((y + \alpha_1)y + \alpha_2)(x + \alpha_3) + \alpha_4)\alpha_5. \qquad (22)$$

The determination of the α's this time requires the solution of a cubic equation (see exercise 19).

On many computers the number of "storing" operations required by (22) is less than 3; for example, we may be able to compute $(x + \alpha_0)^2$ without storing $x + \alpha_0$. Many computers have more than one arithmetic register for the calculations. Because of the wide variety of features available for arithmetic on different computers, we shall henceforth in this section count only the arithmetic operations, not the operations of storing and loading the accumulator. The computation schemes can usually be adapted in a straightforward manner to any particular computer, so that very few of these auxiliary operations are necessary.

A polynomial $u(x) = u_6x^6 + \cdots + u_1x + u_0$ of degree six can always be evaluated using four multiplications and seven additions, with the scheme

$$z = (x + \alpha_0)x + \alpha_1, \qquad w = (x + \alpha_2)z + \alpha_3,$$
$$u(x) = ((w + z + \alpha_4)w + \alpha_5)\alpha_6. \qquad (23)$$

This saves two of the six multiplications required by Horner's rule. Here again we must solve a cubic equation: Since $\alpha_6 = u_6$, we may assume that $u_6 = 1$. Under this assumption, let $\beta_1 = \frac{1}{2}(u_5 - 1)$, $\beta_2 = u_4 - \beta_1(\beta_1 + 1)$, $\beta_3 = u_3 - \beta_1\beta_2$, $\beta_4 = \beta_1 - \beta_2$, $\beta_5 = u_2 - \beta_1\beta_3$. Let β_6 be a real root of the cubic equation

$$2y^3 + (2\beta_4 - \beta_2 + 1)y^2 + (2\beta_5 - \beta_2\beta_4 - \beta_3)y + (u_1 - \beta_2\beta_5) = 0. \qquad (24)$$

(This equation always has a real root, since the polynomial on the left approaches $+\infty$ for large positive y, and it approaches $-\infty$ for large negative y; it must assume the value zero somewhere in between.) Now if we define

$$\beta_7 = \beta_6^2 + \beta_4\beta_6 + \beta_5, \qquad \beta_8 = \beta_3 - \beta_6 - \beta_7,$$

we have finally

$$\alpha_0 = \beta_2 - 2\beta_6, \qquad \alpha_2 = \beta_1 - \alpha_0, \qquad \alpha_1 = \beta_6 - \alpha_0\alpha_2,$$
$$\alpha_3 = \beta_7 - \alpha_1\alpha_2, \qquad \alpha_4 = \beta_8 - \beta_7 - \alpha_1, \qquad \alpha_5 = u_0 - \beta_7\beta_8. \qquad (25)$$

We can illustrate this procedure with a contrived example: Suppose that we want to evaluate $x^6 + 13x^5 + 49x^4 + 33x^3 - 61x^2 - 37x + 3$. We obtain $\alpha_6 = 1$, $\beta_1 = 6$, $\beta_2 = 7$, $\beta_3 = -9$, $\beta_4 = -1$, $\beta_5 = -7$, and so we meet with the cubic equation

$$2y^3 - 8y^2 + 2y + 12 = 0. \tag{26}$$

This equation has $\beta_6 = 2$ as a root, and we continue to find $\beta_7 = -5$, $\beta_8 = -6$, $\alpha_0 = 3$, $\alpha_2 = 3$, $\alpha_1 = -7$, $\alpha_3 = 16$, $\alpha_4 = 6$, $\alpha_5 = -27$. The resulting scheme is

$$z = (x+3)x - 7, \qquad w = (x+3)z + 16, \qquad u(x) = (w+z+6)w - 27.$$

By sheer coincidence the quantity $x + 3$ appears twice here, so we have found a method which uses three multiplications and six additions.

Another method for doing sixth-degree equations has been suggested by V. Ya. Pan [see L. A. Lyusternik, O. A. Chervonenkis, A. R. Yanpol'skii, *Handbook for Computing Elementary Functions*, tr. by G. J. Tee (Oxford: Pergamon, 1965), 10–16]. Pan's method requires one more addition operation, but it does not involve first solving a cubic equation in the preliminary steps. We may proceed as follows:

$$z = (x + \alpha_0)x + \alpha_1, \qquad w = z + x + \alpha_2,$$
$$u(x) = (((z - x + \alpha_3)w + \alpha_4)z + \alpha_5)\alpha_6. \tag{27}$$

To determine the α's, once again divide the polynomial by $u_6 = \alpha_6$ so that $u(x)$ becomes monic. It can then be verified that $\alpha_0 = u_5/3$ and that

$$\alpha_1 = (u_1 - \alpha_0 u_2 + \alpha_0^2 u_3 - \alpha_0^3 u_4 + 2\alpha_0^5)/(u_3 - 2\alpha_0 u_4 + 5\alpha_0^3). \tag{28}$$

Note that Pan's method requires that the denominator does not vanish; i.e.,

$$27u_3 u_6^2 - 18u_6 u_5 u_4 + 5u_5^3 \neq 0; \tag{29}$$

in fact, this quantity should not be so small that α_1 becomes too large. Once α_1 has been determined, the remaining α's may be determined from the equations

$$\beta_1 = 2\alpha_0, \qquad \beta_2 = u_4 - \alpha_0\beta_1 - \alpha_1, \qquad \beta_3 = u_3 - \alpha_0\beta_2 - \alpha_1\beta_1,$$
$$\beta_4 = u_2 - \alpha_0\beta_3 - \alpha_1\beta_2,$$
$$\alpha_3 = \tfrac{1}{2}(\beta_3 - (\alpha_0 - 1)\beta_2 + (\alpha_0 - 1)(\alpha_0^2 - 1)) - \alpha_1,$$
$$\alpha_2 = \beta_2 - (\alpha_0^2 - 1) - \alpha_3 - 2\alpha_1, \qquad \alpha_4 = \beta_4 - (\alpha_2 + \alpha_1)(\alpha_3 + \alpha_1),$$
$$\alpha_5 = u_0 - \alpha_1\beta_4. \tag{30}$$

We have discussed the cases of degree $n = 4$, 5, 6 in detail because the smaller values of n arise most frequently in applications. Let us now consider a general evaluation scheme for nth degree polynomials, which always takes $\lfloor n/2 \rfloor + 2$ multiplications and n additions; the idea is to generalize the "second-order" rule (4).

Theorem E. *Every nth degree polynomial* (1) *with real coefficients, $n \geq 3$, can be evaluated by the scheme*

$$y = x + c, \qquad w = y^2;$$
$$z = (u_n y + \alpha_0)y + \beta_0 \quad (n \text{ even}), \qquad z = u_n y + \beta_0 \quad (n \text{ odd});$$
$$u(x) = (\cdots ((z(w - \alpha_1) + \beta_1)(w - \alpha_2) + \beta_2) \cdots)(w - \alpha_m) + \beta_m; \quad (31)$$

for suitable real parameters c, α_k and β_k, where $m = \lceil n/2 \rceil - 1$. In fact it is possible to select these parameters so that $\beta_m = 0$.

Proof. Let us first examine the circumstances under which the α's and β's can be chosen in (31), if c is fixed: let

$$p(x) = u(x - c) = a_n x^n + a_{n-1}x^{n-1} + \cdots + a_1 x + a_0. \quad (32)$$

We want to show that $p(x)$ has the form $p_1(x)(x^2 - \alpha_m) + \beta_m$ for some polynomial $p_1(x)$ and some constants α_m, β_m. If we divide $p(x)$ by $x^2 - \alpha_m$, we can see that the remainder β_m is a constant only if the auxiliary polynomial

$$q(x) = a_{2m+1}x^m + a_{2m-1}x^{m-1} + \cdots + a_1, \quad (33)$$

formed from every odd-numbered coefficient of $p(x)$, is a multiple of $x - \alpha_m$. Conversely, if $q(x)$ has $x - \alpha_m$ as a factor, then $p(x) = p_1(x)(x^2 - \alpha_m) + \beta_m$, for some constant β_m, which may be determined by division.
    Similarly, we want $p_1(x)$ to have the form $p_2(x)(x^2 - \alpha_{m-1}) + \beta_{m-1}$, and this is the same as saying that $q(x)/(x - \alpha_m)$ is a multiple of $x - \alpha_{m-1}$; for if $q_1(x)$ is the polynomial corresponding to $p_1(x)$ as $q(x)$ corresponds to $p(x)$, we have $q_1(x) = q(x)/(x - \alpha_m)$. Continuing in the same way, we find that the parameters $\alpha_1, \beta_1, \ldots, \alpha_m, \beta_m$ will exist if and only if

$$q(x) = a_{2m+1}(x - \alpha_1) \cdots (x - \alpha_m). \quad (34)$$

In other words, either $q(x)$ is identically zero (and this can happen only when n is even), or else $q(x)$ is an mth degree polynomial having all real roots.
    Now we have a surprising fact discovered by J. Eve [*Numerische Mathematik* **6** (1964), 17–21]: *If $p(x)$ has at least $n - 1$ complex roots whose real parts are all nonnegative, or all nonpositive, then the corresponding polynomial $q(x)$ is identically zero or has all real roots.* (See exercise 23.) Now $u(x) = 0$ if and only if $p(x + c) = 0$; so we need merely choose the parameter c large enough that at least $n - 1$ of the roots of $u(x) = 0$ have a real part $\geq -c$, and (31) will be satisfied whenever $a_{n-1} = u_{n-1} - ncu_n \neq 0$.
    Now we can determine c so that these conditions are fulfilled *and* that $\beta_m = 0$: First determine the n roots of $u(x) = 0$. If $a + bi$ is a root having the largest or the smallest real part, and $b \neq 0$, let $c = -a$ and $\alpha_m = -b^2$; then $x^2 - \alpha_m$ is a factor of $u(x - c)$. If the root with smallest or largest real part is real, but the root with *second* smallest (or second largest) real part is nonreal,

the same transformation applies. If the two roots with smallest (or largest) real parts are both real, they can be expressed in the form $a - b$ and $a + b$, respectively; let $c = -a$ and $\alpha_m = b^2$. Again $x^2 - \alpha_m$ is a factor of $u(x - c)$. (Still other values of c are often possible; see exercise 24.) The coefficient a_{n-1} will be nonzero for at least one of these alternatives, unless $q(x)$ is identically zero. ▮

Note that this method of proof usually gives at least two values of c, and the chance to permute $\alpha_1, \ldots, \alpha_{m-1}$ in $(m - 1)!$ ways. Some of these alternatives may give more desirable numerical accuracy than others.

Polynomial chains. Now let us consider questions of optimality. What are the *best possible* schemes for evaluating polynomials of various degrees, in terms of the minimum possible number of arithmetic operations? This question was first analyzed by A. M. Ostrowski [*Studies in Mathematics and Mechanics presented to R. von Mises* (New York: Academic Press, 1954), 40–48] for the case when no preliminary adaptation of coefficients is made, and by T. S. Motzkin [*Bull. Amer. Math. Soc.* **61** (1955), 163] for the case of adapted coefficients.

In order to investigate this question, we can extend the concept of addition chains, defined in Section 4.6.3, to that of *polynomial chains*. A polynomial chain is a sequence of the form

$$x = \lambda_0, \qquad \lambda_1, \qquad \ldots, \qquad \lambda_r = u(x), \tag{35}$$

where $u(x)$ is some polynomial in x, and for $1 \le i \le r$

$$\text{either} \quad \lambda_i = (\pm\lambda_j) \circ \lambda_k, \qquad 0 \le j, k < i$$
$$\text{or} \quad \lambda_i = \alpha_j \circ \lambda_k, \qquad 0 \le k < i. \tag{36}$$

Here "$\circ$" denotes any of the three operations "$+$", "$-$", or "$\times$", and α_j denotes a so-called "parameter." Steps of the first kind are called *chain steps*, and steps of the second kind are called *parameter steps*. We shall assume that a different parameter α_j is used in each parameter step; and that if there are s parameter steps, they involve $\alpha_1, \alpha_2, \ldots, \alpha_s$ in this order. It follows that the polynomial $u(x)$ at the end of the chain has the form

$$u(x) = q_n x^n + \cdots + q_1 x + q_0, \tag{37}$$

where $q_n, \ldots, q_1, q_0$ are polynomials in $\alpha_1, \alpha_2, \ldots, \alpha_s$ with integer coefficients. We shall interpret the parameters $\alpha_1, \alpha_2, \ldots, \alpha_s$ as real numbers, and we shall therefore restrict ourselves to considering the evaluation of polynomials with real coefficients. The *result set* R of a polynomial chain is defined to be the set of all vectors $(q_n, \ldots, q_1, q_0)$ of real numbers which occur as $\alpha_1, \alpha_2, \ldots, \alpha_s$ independently assume all possible real values.

If for every choice of $t + 1$ distinct integers $j_0, \ldots, j_t \in \{0, 1, \ldots, n\}$ there is a nonzero multivariate polynomial f with integer coefficients such that $f(q_{j_0}, \ldots, q_{j_t}) = 0$ for all $(q_n, \ldots, q_1, q_0)$ in R, let us say that the result set R has at most t *degrees of freedom*, and that the chain (35) has at most t degrees of freedom. We also say that the chain (35) *computes* a given polynomial $u(x)$ with real coefficients if $u(x) = u_n x^n + \cdots + u_1 x + u_0$, where $(u_n, \ldots, u_1, u_0)$ is in R. It follows that a polynomial chain with at most n degrees of freedom cannot compute all nth degree polynomials (see exercise 27).

As an example of a polynomial chain, consider the following chain corresponding to Theorem E, when n is odd:

$$
\left.
\begin{aligned}
\lambda_0 &= x \\
\lambda_1 &= \alpha_1 + \lambda_0 \\
\lambda_2 &= \lambda_1 \times \lambda_1 \\
\lambda_3 &= \alpha_2 \times \lambda_1 \\
\lambda_{1+3i} &= \alpha_{1+2i} + \lambda_{3i} \\
\lambda_{2+3i} &= \alpha_{2+2i} + \lambda_2 \\
\lambda_{3+3i} &= \lambda_{1+3i} \times \lambda_{2+3i}
\end{aligned}
\right\} \quad 1 \leq i < n/2
\tag{38}
$$

There are $\lfloor n/2 \rfloor + 2$ multiplications and n additions; $\lfloor n/2 \rfloor + 1$ chain steps and $n + 1$ parameter steps. By Theorem E, the result set R includes the set of *all* $(u_n, \ldots, u_1, u_0)$ with $u_n \neq 0$; so (38) computes all polynomials of degree n. We cannot prove that R has at most n degrees of freedom, since the result set has $n + 1$ independent components.

In a sense it is obvious that a polynomial chain with s parameter steps has at most s degrees of freedom; we can't compute a function with t degrees of freedom with less than t arbitrary parameters. This fact is not easy to prove formally; for example, there are continuous functions ("space-filling curves") which map the real line onto a plane, and which therefore map a single parameter into two independent parameters. No polynomial functions with integer coefficients can have such a property, however; a proof appears in exercise 28.

Theorem M (T. S. Motzkin, 1954). *A polynomial chain with $m > 0$ multiplications has at most $2m$ degrees of freedom.*

Proof. Let $\mu_1, \mu_2, \ldots, \mu_m$ be the λ_i's of the chain which correspond to multiplication operations. Then

$$
\begin{aligned}
\mu_i &= S_{2i-1} \times S_{2i}, \quad 1 \leq i \leq m, \\
u(x) &= S_{2m+1},
\end{aligned}
\tag{39}
$$

where each S_j is a certain sum of μ's, x's, and α's. Write $S_j = T_j + \beta_j$, where T_j is a sum of μ's and x's while β_j is a sum of α's.

Now $u(x)$ is expressible as a polynomial in $x, \beta_1, \ldots, \beta_{2m+1}$ with integer coefficients. Since the β's are expressible as linear functions of $\alpha_1, \ldots, \alpha_s$, the set of values represented by all real values of $\beta_1, \ldots, \beta_{2m+1}$ contains the result set of the chain. Therefore there are at most $2m + 1$ degrees of freedom; this can be improved to $2m$ when $m > 0$, as shown in exercise 30. ∎

An example of the construction in the proof of Theorem M appears in exercise 25. A similar result can be proved for additions:

Theorem A (É. G. Belaga, 1958). *A polynomial chain containing q additions and subtractions has at most $q + 1$ degrees of freedom.*

Proof. [*Problemi Kibernetiki* **5** (1961), 7–15.] Let $\kappa_1, \ldots, \kappa_q$ be the λ_i's of the chain which correspond to addition or subtraction operations. Then

$$\kappa_i = \pm T_{2i-1} \pm T_{2i}, \qquad 1 \leq i \leq q,$$
$$u(x) = T_{2q+1}, \tag{40}$$

where each T_j is a product of κ's, x's, and α's. We may write $T_j = A_j B_j$, where A_j is a product of α's and B_j is a product of κ's and x's. The following transformation may now be made to the chain, successively for $i = 1, 2, \ldots, q$: Let $\beta_i = A_{2i}/A_{2i-1}$, so that $\kappa_i = A_{2i-1}(\pm B_{2i-1} \pm \beta_i B_{2i})$. Then change κ_i to $\pm B_{2i-1} \pm \beta_i B_{2i}$, and replace each occurrence of κ_i in future formulas $T_{2i+1}, T_{2i+2}, \ldots, T_{2q+1}$ by $A_{2i-1}\kappa_i$. (This replacement may change the values of $A_{2i+1}, A_{2i+2}, \ldots, A_{2q+1}$.)

After the above transformation has been done for all i, let $\beta_{q+1} = A_{2q+1}$; then $u(x)$ can be expressed as a polynomial in $\beta_1, \ldots, \beta_{q+1}$, and x, with integer coefficients. The proof is not yet complete, for the values representable by this polynomial in $\beta_1, \ldots, \beta_{q+1}$ may not include all values obtainable by the original chain (see exercise 26); it is possible to have $A_{2i-1} = 0$ in the above discussion, for some values of the α's.

To complete the proof, let us observe that the result set R of the original chain can be written $R = R_1 \cup R_2 \cup \cdots \cup R_q \cup R'$, where R_i is the set of result vectors possible when $A_{2i-1} = 0$, and R' is the set of result vectors possible when all α's are nonzero. The above argument proves that R' has at most $q + 1$ degrees of freedom. When $A_{2i-1} = 0$, then $T_{2i-1} = 0$, so addition step κ_i may be dropped to obtain another chain computing the result set R_i; by induction we see that each R_i has at most q degrees of freedom. Hence by exercise 29, R has at most $q + 1$ degrees of freedom. ∎

Theorem C. *If a polynomial chain computes all nth degree polynomials $u(x) = u_n x^n + \cdots + u_0$, $u_n \neq 0$, for some $n \geq 2$, then it includes at least $\lfloor n/2 \rfloor + 1$ multiplications and at least n addition-subtractions.*

Proof. Let there be m multiplication steps. By Theorem M, the chain has at most $2m$ degrees of freedom, so $2m \geq n + 1$. Similarly, by Theorem A there are $\geq n$ addition-subtractions. ∎

This theorem states that no *single* method having less than $\lfloor n/2 \rfloor + 1$ multiplications or less than n additions can evaluate *all* nth degree polynomials. On the other hand, some special polynomials can be evaluated with fewer operations. We can prove that "almost all" polynomials of degree n require the number of operations mentioned in the theorem, using simple principles of measure theory; for the proof of Theorem C shows that each chain having fewer multiplications or fewer additions will have a result set with at most n degrees of freedom. Such a set has Lebesgue measure zero in $(n + 1)$-dimensional space. There are only countably many polynomial chains having less than $\lfloor n/2 \rfloor + 1$ multiplications or less than n additions, so the set of coefficients of polynomials that can be computed more efficiently than stated in Theorem C has measure zero.

Unfortunately, a gap still remains between the lower bounds of Theorem C and the actual operation counts known to be achievable, except in the trivial case $n = 2$. Theorem E gives $\lfloor n/2 \rfloor + 2$ multiplications, not $\lfloor n/2 \rfloor + 1$, although it does achieve the minimum number of additions. Our special methods for $n = 4$ and $n = 6$ have the minimum number of multiplications, but one extra addition. When n is odd, it is not difficult to prove that the lower bounds of Theorem C cannot be achieved simultaneously for both multiplications and additions; see exercise 33. For $n = 3$, 5, and 7, it is possible to show that at least $\lfloor n/2 \rfloor + 2$ multiplications are necessary. Exercises 35 and 36 show that the lower bounds of Theorem C cannot both be achieved when $n = 4$, $n = 6$; thus the methods of this section are best possible, for $n < 8$. For $n \geq 8$ the existence of general methods with $\lfloor n/2 \rfloor + 1$ multiplications is unknown; when n is odd it is unlikely that such schemes exist. Pan has extended (27) to a method for $n = 8$, having five multiplications and 10 additions, but a condition analogous to (29) keeps this scheme from being completely general.

It is clear that the results we have obtained about chains for polynomials in a single variable can be extended without difficulty to multivariate polynomials. With minor variations, the methods can also be extended to rational functions as well as polynomials; curiously, the rational function analog of Horner's rule turns out to be optimal from an operation-count standpoint, if multiplication and division speeds are comparable (see exercise 37).

EXERCISES

1. [*15*] What is a good way to evaluate an "odd" polynomial

$$u(x) = u_{2n+1}x^{2n+1} + u_{2n-1}x^{2n-1} + \cdots + u_1 x?$$

2. [*M21*] (W. G. Horner.) Design a computational method to find the coefficients of $u(x + c)$, given a constant c and the coefficients of an nth degree polynomial $u(x)$, using approximately n^2 arithmetic operations. (For example if $u(x) = 3x^2 + 2x - 1$, then $u(x - 2) = 3x^2 - 10x + 7$.)

3. [*20*] Give a method analogous to Horner's rule, for evaluating a polynomial in two variables $\sum_{i+j\leq n} u_{ij}x^iy^j$. (This polynomial has $(n+1)(n+2)/2$ coefficients, and "total degree" n.) Count the number of additions and multiplications in your method.

4. [*M20*] The text shows that scheme (3) is superior to Horner's rule when we are evaluating a polynomial with real coefficients at a complex point z. Compare (3) to Horner's rule when *both* the coefficients and the variable z are complex numbers; how many (real) multiplications and addition-subtractions are required by each method?

5. [*M20*] Count the number of multiplications and additions required by the second-order rule (4).

6. [*HM21*] The scheme in (6) shows how a polynomial $u(x)$ and its derivative $u'(x)$ can be evaluated at the same time, by generalizing Horner's rule. Generalize this scheme still further, by showing how to evaluate $u(x)$, $u'(x)$, and $u''(x)/2$ all at the same time.

▶ **7.** [*HM24*] Show that (3) can be generalized to the evaluation of both $u(z)$ and $u'(z)$, when z is complex and $u(z)$ has real coefficients, by using a method analogous to (6).

8. [*M20*] The factorial power $x^{\underline{k}}$ is defined to be $k!\binom{x}{k} = x(x-1)\cdots(x-k+1)$. Explain how to evaluate $u_n x^{\underline{n}} + \cdots + u_1 x^{\underline{1}} + u_0$ with at most n multiplications and $2n-1$ additions, starting with x and the $n+3$ constants $u_n, \ldots, u_0, 1, n-1$.

9. [*M24*] (H. J. Ryser.) Show that if $X = (x_{ij})$ is an $n \times n$ matrix, then

$$\text{per}\,(X) = \sum (-1)^{n-\epsilon_1-\cdots-\epsilon_n} \prod_{1\leq i\leq n} \sum_{1\leq j\leq n} \epsilon_j x_{ij}$$

summed over all 2^n choices of $\epsilon_1, \ldots, \epsilon_n$ equal to 0 or 1 independently. Count the number of addition and multiplication operations required to evaluate per (X) by this formula.

10. [*M21*] The permanent of an $n \times n$ matrix $X = (x_{ij})$ may be calculated as follows: Start with the n quantities $x_{11}, x_{12}, \ldots, x_{1n}$. For $1 \leq k < n$, assume that the $\binom{n}{k}$ quantities A_{kS} have been computed, for all k-element subsets S of $\{1, 2, \ldots, n\}$, where $A_{kS} = \sum x_{1j_1} \ldots x_{kj_k}$ summed over all $k!$ permutations $j_1 \ldots j_k$ of the elements of S; then form all of the sums

$$A_{(k+1)S} = \sum_{j \in S} A_{k(S-\{j\})} x_{(k+1)j}.$$

We have per $(X) = A_{n\{1,\ldots,n\}}$.

How many additions and multiplications does this method require? How much temporary storage is needed?

11. [*M50*] Is there any way to evaluate the permanent of a general $n \times n$ matrix using a number of operations which does not grow exponentially with n?

12. [*M50*] What is the minimum number of multiplications required to multiply two $n \times n$ matrices? Can the total number of operations be reduced to less than $O(n^{\log_2 7})$?

13. [*M22*] (I. J. Good.) Yates's method, as described in the text, performs a certain transformation n times, on a column of 2^n numbers; the result is the n-dimensional finite Fourier transform of the original 2^n numbers. What happens if the same transformation is applied n more times to the result?

▶ **14.** [*HM28*] (a) (*"Fast Fourier transforms."*) Show that the scheme (13) can be used to evaluate the one-dimensional Fourier transform

$$f(s) = \sum_{0 \le t < 2^n} F(t) \omega^{st}, \qquad \omega = e^{2\pi i/2^n}, \qquad 0 \le s < 2^n,$$

using arithmetic on complex numbers. (b) Apply this result to prove that we can obtain the coefficients of the product of two given polynomials, $u(z)v(z)$, in $O(n \log n)$ operations of (exact) addition and multiplication of complex numbers, if $u(z)$ and $v(z)$ are nth degree polynomials with complex coefficients. [*Hint:* Consider the product of Fourier transforms of the coefficients.]

15. [*M25*] Prove that, after the computations indicated in (17), the coefficients in Newton's interpolation polynomial are given by $\alpha_0 = y_0$, $\alpha_1 = y_1'$, $\alpha_2 = y_2''$, etc.

16. [*M22*] How can we readily compute the coefficients of $u^{[n]}(x) = u_n x^n + \cdots + u_0$, if we are given the values of $x_0, x_1, \ldots, x_{n-1}, \alpha_0, \alpha_1, \ldots, \alpha_n$ in Newton's interpolation polynomial (15)?

17. [*M46*] Is there a way to evaluate the polynomial

$$\sum_{1 \le i < j \le n} x_i x_j = x_1 x_2 + \cdots + x_{n-1} x_n$$

with fewer than $n - 1$ multiplications and $2n - 4$ additions? (There are $\binom{n}{2}$ terms.)

18. [*M20*] If the fourth-degree scheme (20) were changed to

$$y = (x + \alpha_0)x + \alpha_1, \qquad u(x) = ((y - x + \alpha_2)y + \alpha_3)\alpha_4,$$

what formulas for computing the α_j's in terms of the u_k's would take the place of (21)?

▶ **19.** [*M24*] Explain how to determine the adapted coefficients $\alpha_0, \alpha_1, \ldots, \alpha_5$ in (22) from the coefficients $u_5, \ldots, u_1, u_0$ of $u(x)$, and find the α's for the particular polynomial $u(x) = x^5 + 5x^4 - 10x^3 - 50x^2 + 13x + 60$.

▶ **20.** [*21*] Write a MIX program which evaluates a fifth-degree polynomial according to scheme (22); try to make the program as efficient as possible, by making slight modifications to (22). Use MIX's floating-point arithmetic operators.

21. [*20*] Find two more ways to evaluate the polynomial $x^6 + 13x^5 + 49x^4 + 33x^3 - 61x^2 - 37x + 3$ by scheme (23), using the two roots of (26) which were not considered in the text.

22. [*18*] What is the scheme for evaluating $x^6 - 3x^5 + x^4 - 2x^3 + x^2 - 3x - 1$, using Pan's method (27)?

23. [*HM30*] (J. Eve.) Let $f(z) = a_n z^n + a_{n-1} z^{n-1} + \cdots + a_0$ be a polynomial of degree n with real coefficients, having at least $n - 1$ roots with a nonnegative real part. Let

$$g(z) = a_n z^n + a_{n-2} z^{n-2} + \cdots + a_{n \bmod 2} z^{n \bmod 2},$$
$$h(z) = a_{n-1} z^{n-1} + a_{n-3} z^{n-3} + \cdots + a_{(n-1) \bmod 2} z^{(n-1) \bmod 2}.$$

Assume that $h(z)$ is not identically zero.

a) Show that $g(z)$ has at least $n - 2$ imaginary roots (i.e., roots whose real part is zero), and $h(z)$ has at least $n - 3$ imaginary roots. [*Hint:* Consider the number of times the path $f(z)$ circles the origin as z goes around the path shown in Fig. 15, for a sufficiently large radius R.]

b) Prove that the squares of the roots of $g(z) = 0$, $h(z) = 0$ are all real.

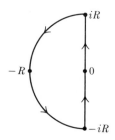

Fig. 15. Proof of Eve's theorem.

▶ **24.** [*M24*] Find values of c and α_k, β_k satisfying the conditions of Theorem E, for the polynomial $u(x) = (x + 7)(x^2 + 6x + 10)(x^2 + 4x + 5)(x + 1)$. Choose these values so that $\beta_2 = 0$. Give two different solutions to this problem!

25. [*M20*] When the construction in the proof of Theorem M is applied to the (inefficient) polynomial chain

$$\lambda_1 = \alpha_1 + \lambda_0, \qquad \lambda_2 = -\lambda_0 - \lambda_0, \qquad \lambda_3 = \lambda_1 + \lambda_1, \qquad \lambda_4 = \alpha_2 \times \lambda_3,$$

$$\lambda_5 = \lambda_0 - \lambda_0, \qquad \lambda_6 = \alpha_6 - \lambda_5, \qquad \lambda_7 = \alpha_7 + \lambda_6, \qquad \lambda_8 = \lambda_7 \times \lambda_7,$$

$$\lambda_9 = \lambda_1 \times \lambda_4, \qquad \lambda_{10} = \alpha_8 - \lambda_9, \qquad \lambda_{11} = \lambda_3 - \lambda_{10},$$

how can $\beta_1, \beta_2, \ldots, \beta_9$ be expressed in terms of $\alpha_1, \ldots, \alpha_8$?

▶ **26.** [*M21*] (a) Give the polynomial chain corresponding to Horner's rule for evaluating polynomials of degree $n = 3$. (b) Using the construction in the proof of Theorem A, express κ_1, κ_2, κ_3, and the result polynomial $u(x)$ in terms of β_1, β_2, β_3, β_4, and x. (c) Show that the result set obtained in (b), as $\beta_1, \beta_2, \beta_3, \beta_4$ independently assume all real values, omits certain vectors in the result set of (a).

27. [*M22*] Let R be a set which includes all $(n + 1)$-tuples $(q_n, \ldots, q_1, q_0)$ of real numbers such that $q_n \neq 0$; prove that R does not have at most n degrees of freedom.

28. [*HM20*] Show that if $f_0(\alpha_1, \ldots, \alpha_s), \ldots, f_s(\alpha_1, \ldots, \alpha_s)$ are multivariate polynomials with integer coefficients, then there is a nonzero polynomial $g(x_0, \ldots, x_s)$ with integer coefficients such that $g\big(f_0(\alpha_1, \ldots, \alpha_s), \ldots, f_s(\alpha_1, \ldots, \alpha_s)\big) = 0$ for all real $\alpha_1, \ldots, \alpha_s$. (Hence any polynomial chain with s parameters has at most s degrees of freedom.) [*Hint:* Use the theorems about "algebraic dependence" which are found, for example, in B. L. van der Waerden's *Modern Algebra*, tr. by Fred Blum (New York: Ungar, 1949), section 64.]

▶ **29.** [*M20*] Let $R_1, R_2, \ldots, R_m$ all be sets of $(n + 1)$-tuples of real numbers having at most t degrees of freedom. Show that the union $R_1 \cup R_2 \cup \cdots \cup R_m$ also has at most t degrees of freedom.

▶ **30.** [*M28*] Prove that a polynomial chain with m_1 chain multiplications and m_2 parameter multiplications has at most $2m_1 + m_2 + \delta_{0m_1}$ degrees of freedom. [*Hint:* Generalize Theorem M, showing that the first chain multiplication and each parameter multiplication can essentially introduce only one new parameter into the result set.]

31. [*M23*] Prove that a polynomial chain capable of computing all *monic* polynomials of degree n has at least $\lfloor n/2 \rfloor$ multiplications and at least n addition-subtractions.

32. [*M24*] Find a polynomial chain of minimum possible length which can compute all polynomials of the form $u_4 x^4 + u_2 x^2 + u_0$; prove that its length is minimal.

▶ **33.** [*M25*] Let $n \geq 3$ be odd. Prove that a polynomial chain with $\lfloor n/2 \rfloor + 1$ multiplication steps cannot compute all polynomials of degree n unless it has at least $n + 2$ addition-subtraction steps. [*Hint:* See exercise 30.]

34. [*M26*] Let $\lambda_0, \lambda_1, \ldots, \lambda_r$ be a polynomial chain in which all addition and subtraction steps are parameter steps, and which contains at least one parameter multiplication. Assume that this scheme has m multiplications and $k = r - m$ addition-subtractions, and that the polynomial computed by the chain has maximum degree n. Prove that all polynomials computable by this chain, for which the coefficient of x^n is not zero, can be computed by another chain which has at most m multiplications and at most k additions, and no subtractions; and whose last step is the only parameter multiplication.

▶ **35.** [*M25*] Show that any polynomial chain which computes a general fourth degree polynomial using only three multiplications must have at least five addition-subtractions. [*Hint:* Assume that there are only four addition-subtractions, and show that exercise 34 applies; this means the scheme must have a particular form which is incapable of representing all fourth degree polynomials.]

36. [*M27*] Show that any polynomial chain which computes a general sixth-degree polynomial using only four multiplications must have at least seven addition-subtractions. (Cf. exercise 35.)

37. [*M21*] (T. S. Motzkin.) Show that "almost all" rational functions of the form

$$(u_n x^n + u_{n-1} x^{n-1} + \cdots + u_1 x + u_0)/(x^n + v_{n-1} x^{n-1} + \cdots + v_1 x + v_0),$$

with coefficients in a field S, can be evaluated using the scheme

$$\alpha_1 + \beta_1/(x + \alpha_2 + \beta_2/(x + \ldots \beta_n/(x + \alpha_{n+1}) \cdots)),$$

for suitable α_j, β_j in S. (This continued fraction scheme has n divisions and $2n$ additions; by "almost all" rational functions we mean all except those whose coefficients satisfy some nontrivial polynomial equation.) Determine the α's and β's for the rational function $(x^2 + 10x + 29)/(x^2 + 8x + 19)$.

▶ **38.** [*HM32*] (A. M. Garsia, 1962.) The purpose of this exercise is to prove that Horner's rule is really optimal if no preliminary adaptation of coefficients is made; we need n multiplications and n additions to compute $u_n x^n + \cdots + u_1 x + u_0$, if the variables $u_n, \ldots, u_1, u_0, x$, and arbitrary constants are given. Consider chains which are as before except that $u_n, \ldots, u_1, u_0, x$ are each considered to be variables; we may say, for example, that $\lambda_{-j-1} = u_j$, $\lambda_0 = x$. In order to show that Horner's rule is best, it is convenient to prove a somewhat more general theorem: Let $A =$

(a_{ij}), $0 \le i \le m$, $0 \le j \le n$, be an $(m+1) \times (n+1)$ matrix of real numbers, of rank $n+1$; and let $B = (b_0, \ldots, b_m)$ be a vector of real numbers. Prove that *any polynomial chain which computes*

$$P(x; u_0, \ldots, u_n) = \sum_{0 \le i \le m} (a_{i0}u_0 + \cdots + a_{in}u_n + b_i)x^i$$

involves at least n chain multiplications. (Note that this does not mean only that we are considering some fixed chain in which the parameters α_j are assigned values depending on A and B; it means that both the chain *and* the values of the α's may depend on the given matrix A and vector B. No matter how A, B, and the values of α_j are chosen, it is impossible to compute $P(x; u_0, \ldots, u_n)$ without doing n "chain-step" multiplications.) The assumption that A has rank $n+1$ implies that $m \ge n$. [*Hint:* Show that from any such scheme we can derive another which has one less chain multiplication and which has n decreased by one.]

39. [*HM40*] Discuss efficient ways to evaluate the coefficients $a_1, \ldots, a_n$ of the *characteristic polynomial* $\det(xI - A) = x^n + a_1 x^{n-1} + \cdots + a_n$ of a given $n \times n$ matrix A. Can it be done in less than $O(n^4)$ arithmetic operations?

40. [*M45*] Can the lower bound in the number of multiplications in Theorem C be raised from $\lfloor n/2 \rfloor + 1$ to $\lceil n/2 \rceil + 1$? (Cf. exercise 33. An unfounded remark that Belaga [*Dokladi Akad. Nauk SSSR* **123** (1958), 775–777] gave such an improvement has appeared several times in the literature.)

41. [*22*] (Oscar Buneman.) Show that the real and imaginary parts of $(a + bi)(c + di)$ can be obtained using 3 multiplications and 5 additions of real numbers.

*4.7. MANIPULATION OF POWER SERIES

If we are given two power series

$$U(z) = U_0 + U_1 z + U_2 z^2 + \cdots, \qquad V(z) = V_0 + V_1 z + V_2 z^2 + \cdots, \quad (1)$$

whose coefficients belong to a field, we can form their sum, their product, their quotient, etc., to obtain new power series. A polynomial is obviously a special case of a power series, in which there are only finitely many terms. Of course, only a finite number of terms can be represented and stored within a computer, so it makes sense to ask whether power series arithmetic is even possible on computers; and if it is possible, what makes it different from polynomial arithmetic? The answer is that we work only with the first N coefficients of the power series, where N is a parameter which may in principle be arbitrarily large; instead of ordinary polynomial arithmetic, we are essentially doing polynomial arithmetic modulo z^N, and this often leads to a somewhat different point of view. Furthermore, special operations, like "reversion," can be performed on power series but not on polynomials, since polynomials are not closed under these operations.

Manipulation of power series has several applications to numerical analysis, but perhaps its greatest use is the determination of asymptotic expansions (as we have seen in Section 1.2.11.3), or the calculation of quantities defined by certain generating functions. The latter applications make it desirable to calculate the coefficients exactly, instead of with floating-point arithmetic. All of the algorithms in this section, with obvious exceptions, can be done using rational operations only, so the techniques of Section 4.5.1 can be used to obtain exact results when desired.

The calculation of $W(z) = U(z) \pm V(z)$ is, of course, trivial, since we have $W_n = U_n \pm V_n$ for $n = 0, 1, 2, \ldots$. It is also easy to calculate $W(z) = U(z)V(z)$, using the familiar "Cauchy product rule":

$$W_n = \sum_{0 \leq k \leq n} U_k V_{n-k} = U_0 V_n + U_1 V_{n-1} + \cdots + U_n V_0. \tag{2}$$

The quotient $W(z) = U(z)/V(z)$, when $V_0 \neq 0$, can be obtained by interchanging U and W in (2); we obtain the rule

$$W_n = \left(U_n - \sum_{0 \leq k < n} W_k V_{n-k} \right) \Big/ V_0$$

$$= (U_n - W_0 V_n - W_1 V_{n-1} - \cdots - W_{n-1} V_1)/V_0. \tag{3}$$

This recurrence relation for the W's makes it easy to determine $W_0, W_1, W_2, \ldots$ successively, without inputting U_n and V_n until after W_{n-1} has been computed. Let us say that a power-series manipulation algorithm with the latter property is "sequential"; a sequential algorithm can be used to determine N coefficients $W_0, W_1, \ldots, W_{N-1}$ of the result without knowing N in advance, so it is possible in theory to run the algorithm indefinitely and compute the entire power series; or to run it until a certain condition is met. (The term "on-line" is sometimes used instead of "sequential.")

Let us now consider the operation of computing $W(z) = V(z)^\alpha$, where α is an "arbitrary" power. For example, we could calculate the square root of $V(z)$ by taking $\alpha = \frac{1}{2}$, or we could find $V(z)^{-10}$ or even $V(z)^\pi$. If V_m is the first nonzero coefficient of $V(z)$, we have

$$\begin{aligned} V(z) &= V_m z^m (1 + (V_{m+1}/V_m)z + (V_{m+2}/V_m)z^2 + \cdots), \\ V(z)^\alpha &= V_m^\alpha z^{\alpha m} (1 + (V_{m+1}/V_m)z + (V_{m+2}/V_m)z^2 + \cdots)^\alpha. \end{aligned} \tag{4}$$

This will be a power series if and only if αm is a nonnegative integer. From (4) we can see that the problem of computing general powers can be reduced to the case that $V_0 = 1$; then the problem is to find the coefficients of

$$W(z) = (1 + V_1 z + V_2 z^2 + V_3 z^3 + \cdots)^\alpha. \tag{5}$$

Clearly $W_0 = 1^\alpha = 1$.

The obvious way to find the coefficients of (5) is to use the binomial theorem (Eq. 1.2.9–19), but a much simpler and more efficient device for calculating powers has been suggested by J. C. P. Miller. [See P. Henrici, *JACM* **3** (1956), 10–15.] If $W(z) = V(z)^\alpha$, we have by differentiation

$$W_1 + 2W_2 z + 3W_3 z^2 + \cdots = W'(z) = \alpha V(z)^{\alpha-1} V'(z); \tag{6}$$

therefore

$$W'(z)V(z) = \alpha W(z)V'(z). \tag{7}$$

If we now equate the coefficients of z^{n-1} in (7), we find that

$$\sum_{0 \le k \le n} kW_k V_{n-k} = \alpha \sum_{0 \le k \le n} (n - k)W_k V_{n-k}, \tag{8}$$

and this gives us a useful computational rule,

$$W_n = \sum_{1 \le k \le n} \left(\left(\frac{\alpha + 1}{n}\right)k - 1 \right) V_k W_{n-k}$$

$$= \left(\left(\frac{\alpha + 1}{n}\right) - 1 \right) V_1 W_{n-1} + \left(2\left(\frac{\alpha + 1}{n}\right) - 1 \right) V_2 W_{n-2} + \cdots + \alpha V_n,$$

$$n \ge 1. \tag{9}$$

This equation leads to a sequential algorithm by which we can successively determine $W_1, W_2, \ldots$, using approximately $2n$ multiplications to compute the nth coefficient. Note the special case $\alpha = -1$, in which (9) becomes the special case $U(z) = V_0 = 1$ of (3).

A similar technique can be used to form $f(V(z))$ when f is any function which satisfies a simple differential equation. (For example, see exercise 4.) A comparatively straightforward "power-series method" is often used to obtain the solution of differential equations; this technique is explained in nearly all textbooks about differential equations.

The transformation of power series which is perhaps of greatest interest is called "reversion of series." This problem is to solve the equation

$$z = t + V_2 t^2 + V_3 t^3 + V_4 t^4 + \cdots \tag{10}$$

for t, obtaining the coefficients of the power series

$$t = z + W_2 z^2 + W_3 z^3 + W_4 z^4 + \cdots. \tag{11}$$

Several interesting ways to achieve such a reversion are known; consider the following algorithm, for example:

Algorithm R (*Reversion of series*). Given the first N coefficients $V_1 = 1$, $V_2, \ldots, V_N$ of (10), this algorithm finds the first N coefficients $W_1 = 1$, $W_2, \ldots, W_N$ of (11).

R1. [Initialize.] Set $W_0 \leftarrow 0$, $W_1 \leftarrow 1$, and set $W_k \leftarrow -V_k$ for $2 \le k \le N$.

R2. [Loop on k.] Perform step R3 for $k = 1, 2, 3, \ldots, N - 1$ (in this order). Then the algorithm terminates, having found $W_0, W_1, \ldots, W_N$ of the solution (11).

R3. [Loop on j.] Perform step R4 for $j = k, k + 1, \ldots, N - 1$ (in this order); but omit the value $j = 1$ when $k = 1$.

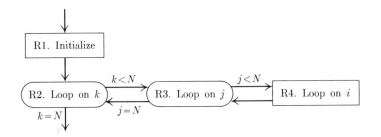

Fig. 16. Power-series reversion by Algorithm R.

R4. [Loop on i.] Set $W_i \leftarrow W_i - W_j V_{i-j+1}$, for $j + 1 \le i \le N$. ∎

Algorithm R is due to J. N. Bramhall, with simplifications by M. A. Chapple [*CACM* **4** (1961), 503]. It is based on the identity

$$t = z - V_2 t^2 - V_3 t^3 - V_4 t^4 - \cdots ; \tag{12}$$

the basic idea is to continue replacing the t's on the right-hand side of this equation by the entire right-hand side, ignoring terms of degree more than N, until all of the t's have disappeared. We can easily prove by induction that the equation

$$t = z + W_2 z^2 + \cdots + W_k z^k + W_{k+1} t z^k + \cdots + W_N t^{N-k} z^k + O(z^{N+1}) \tag{13}$$

holds at the conclusion of step R3.

As an illustration of Algorithm R, let us compute the first five coefficients of the reverted series for $z = t - t^2$:

k	j	Condition at the end of step R4	
(1)	(1)	$t = z + t^2 + 0 \cdot t^3 + 0 \cdot t^4 + 0 \cdot t^5 + O(z^6)$	
1	2	$t = z + tz + t^3 + 0 \cdot t^4 + 0 \cdot t^5 + O(z^6)$	
1	3	$t = z + tz + t^2 z + t^4 + 0 \cdot t^5 + O(z^6)$	
1	4	$t = z + tz + t^2 z + t^3 z + t^5 + O(z^6)$	
(1)	(5)	$t = z + tz + t^2 z + t^3 z + t^4 z + O(z^6)$	
2	2	$t = z + z^2 + 2t^2 z + t^3 z + t^4 z + O(z^6)$	
2	3	$t = z + z^2 + 2tz^2 + 3t^3 z + t^4 z + O(z^6)$	
2	4	$t = z + z^2 + 2tz^2 + 3t^2 z^2 + 4t^4 z + O(z^6)$	
(2)	(5)	$t = z + z^2 + 2tz^2 + 3t^2 z^2 + 4t^3 z^2 + O(z^6)$	
3	3	$t = z + z^2 + 2z^3 + 5t^2 z^2 + 4t^3 z^2 + O(z^6)$	
3	4	$t = z + z^2 + 2z^3 + 5tz^3 + 9t^3 z^2 + O(z^6)$	
(3)	(5)	$t = z + z^2 + 2z^3 + 5tz^3 + 9t^2 z^3 + O(z^6)$	
4	4	$t = z + z^2 + 2z^3 + 5z^4 + 14t^2 z^3 + O(z^6)$	
(4)	(5)	$t = z + z^2 + 2z^3 + 5z^4 + 14tz^4 + O(z^6)$	
(5)	(5)	$t = z + z^2 + 2z^3 + 5z^4 + 14z^5 + O(z^6)$	(14)

(Parenthesized values of k and j have been suppressed from the algorithm, in order to make it slightly more efficient, but it is convenient to imagine that the computations are done as shown here, in order to understand the procedure.) Note that identity (12) is applied by Algorithm R in a clever manner. A more obvious procedure would have been to replace the t in "tz", not the t in "t^3", by $z + t^2$, after the second line of (14); but this would give $t = z + z^2 + zt^2 + t^3$, and it would be inconvenient to keep track of two different terms of degree 3. The procedure in Algorithm R ensures that there is at most one term of each degree, throughout the computations. Algorithm R is not "sequential" in the sense defined above, since N must be known in advance. The running time is of order N^3; in fact, the basic step $W_i \leftarrow W_i - W_j V_{i-j+1}$ is executed exactly

$$\binom{N+1}{3} - N + 1$$

times. (See exercise 6.)

Another method for power series reversion can be based on Lagrange's inversion formula [*Mémoires Acad. royale des Sciences et Belles-Lettres de Berlin* **24** (1768), 251–326], which states that

$$W_n = U_{n-1}/n,$$

if

$$U_0 + U_1 t + U_2 t^2 + \cdots = (1 + V_2 t + V_3 t^2 + \cdots)^{-n}. \tag{15}$$

For example, we have $(1 - t)^{-5} = \binom{4}{4} + \binom{5}{4}t + \binom{6}{4}t^2 + \cdots$; hence W_5 in the reversion of $z = t - t^2$ is equal to $\binom{8}{4}/5 = 14$; this checks with the result found by Algorithm R in (14).

Relation (15) shows that we can revert the series (10) if we compute $(1 + V_2 t + V_3 t^2 + \cdots)^{-n}$ for $n = 1, 2, 3, \ldots$. A straightforward application of this idea would lead to a nonsequential power-series algorithm which uses approximately $N^3/2$ multiplications to find N coefficients, so it would be inferior to Algorithm R; but Eq. (9) makes it possible to work with only the first n coefficients of $(1 + V_2 t + V_3 t^2 + \cdots)^{-n}$, obtaining a *sequential* algorithm which has roughly the same number of steps as Algorithm R.

Algorithm S (*Sequential power-series reversion*). This algorithm inputs the value of V_n in (10) and outputs the value of W_n in (11), for $n = 2, 3, 4, \ldots, N$. (The number N need not be specified in advance; some other termination condition may be substituted.)

S1. [Initialize.] Set $n \leftarrow 1$, $T_0 \leftarrow 1$, $U_0 \leftarrow 1$. (During the course of this algorithm, we will have $T_0 = V_1, \ldots, T_{n-1} = V_n$, and

$$(1 + V_2 t + V_3 t^2 + \cdots)^{-n} = U_0 + U_1 t + \cdots + U_{n-1} t^{n-1} + O(t^n).) \tag{16}$$

S2. [Input V_n.] Increase n by 1. If $n > N$, the algorithm terminates; otherwise set T_{n-1} to the next element of the input (namely, V_n).

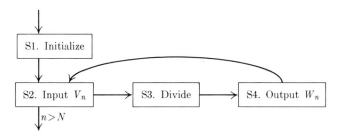

Fig. 17. Power-series reversion by Algorithm S.

S3. [Divide.]   Set $U_k \leftarrow U_k - U_{k-1}T_1 - \cdots - U_1 T_{k-1} - T_k$,   for $k = 1, 2, \ldots, n-2$ (in this order); then set

$$U_{n-1} \leftarrow -2U_{n-2}T_1 - 3U_{n-3}T_2 - \cdots - (n-1)U_1 T_{n-2} - nT_{n-1}.$$

(We have thereby divided $U(z)$ by $T(z)$; cf. (3) and (9).)

S4. [Output W_n.] Output U_{n-1}/n (which is W_n) and return to S2. ∎

A comparison of Algorithm R with Algorithm S makes it hard to believe that they calculate the same function, since the methods are so different. When applied to the example $z = t - t^2$, the computation in Algorithm S is

n	$V_n = T_{n-1}$	U_0	U_1	U_2	U_3	U_4	W_n	
1	1	1					1	
2	−1	1	2				1	
3	0	1	3	6			2	
4	0	1	4	10	20		5	
5	0	1	5	15	35	70	14	(17)

Still another power-series reversion algorithm should be mentioned here, since it provides a solution to Equation (11) given the more general conditions

$$U_1 z + U_2 z^2 + U_3 z^3 + \cdots = t + V_2 t^2 + V_3 t^3 + \cdots . \tag{18}$$

Eq. (10) is the special case $U_1 = 1$, $U_2 = U_3 = \cdots = 0$. If $U_1 \neq 0$, we may assume that $U_1 = 1$, if we replace z by $(U_1 z)$; but it is worth while to consider the general equation (18), since U_1 might equal zero.

Algorithm T (*General power-series reversion*). This algorithm inputs the values of U_n and V_n in (18) and outputs the value of W_n in (11), for $n = 1, 2, 3, \ldots, N$. An auxiliary matrix T_{mn}, $1 \leq m \leq n \leq N$, is used in the calculations.

T1. [Initialize.] Set $n \leftarrow 1$. Let the first two inputs (namely, U_1 and V_1) be stored in T_{11} and V_1, respectively. (We must have $V_1 = 1$).

T2. [Output W_n.] Output the value of T_{1n} (which is W_n).

T3. [Input U_n, V_n.] Increase n by 1. If $n > N$, the algorithm terminates; otherwise store the next two inputs (namely U_n and V_n) in T_{1n} and V_n, respectively.

T4. [Multiply.] Set

$$T_{mn} \leftarrow T_{11}T_{m-1,n-1} + T_{12}T_{m-1,n-2} + \cdots + T_{1,n-m+1}T_{m-1,m-1}$$

and $T_{1n} \leftarrow T_{1n} - V_m T_{mn}$, for $2 \leq m \leq n$. (After this step we have the conditions

$$t^m = T_{mm}z^m + T_{m,m+1}z^{m+1} + \cdots + T_{mn}z^n + O(z^{n+1}), \quad 1 \leq m \leq n. \tag{19}$$

It is easy to verify (19) by induction for $m \geq 2$, and when $m = 1$, we have $U_n = T_{1n} + V_2 T_{2n} + \cdots + V_n T_{nn}$ by (18) and (19).) Return to step T2. ∎

Equation (19) explains the mechanism of this algorithm, which is due to Henry C. Thacher, Jr. [*CACM* **9** (1966), 10–11]. The running time is essentially the same as Algorithms R and S, but considerably more storage space is required. An example of this algorithm is worked in exercise 9. See exercise 11 for another sequential way to revert (18), using less storage space.

EXERCISES

1. [*M10*] The text explains how to divide $U(z)$ by $V(z)$ when $V_0 \neq 0$; how should the division be done when $V_0 = 0$?

2. [*M15*] Does formula (9) give the right results when $\alpha = 0$? When $\alpha = 1$?

▸ **3.** [*M24*] Compare the efficiency of (9) versus the methods of Section 4.6.3, when $V(z)$ is a *polynomial* which is being raised to the mth power.

▸ **4.** [*HM23*] Show that simple modifications of (9) can be used to calculate $e^{V(z)}$ and $\ln(1 + V(z))$, when $V(z) = V_1 z + V_2 z^2 + \cdots$.

5. [*M00*] What happens when a power series is reverted twice; i.e., if the output of Algorithm R or S is reverted again?

6. [*M21*] Determine the number of times the operation $W_i \leftarrow W_i - W_j V_{i-j+1}$ is performed in Algorithm R.

7. [*M22*] Use Lagrange's inversion formula (15) to find a simple expression for the coefficient W_n in the reversion of $z = t - t^2$.

▸ **8.** [*M25*] Lagrange's inversion formula can be generalized as follows: If $W(z) = W_1 z + W_2 z^2 + \cdots = G_1 t + G_2 t^2 + G_3 t^3 + \cdots = G(t)$, where z and t are related by Eq. (10), then $W_n = U_{n-1}/n$ where

$$U_0 + U_1 z + U_2 z^2 + \cdots = (G_1 + 2G_2 z + 3G_3 z^2 + \cdots)(1 + V_2 z + V_3 z^2 + \cdots)^{-n}.$$

(Equation (15) is the special case $G_1 = 1$, $G_2 = G_3 = \cdots = 0$. This equation can be proved, for example, by using tree-enumeration formulas as in exercise 2.3.4.4–33.)

Extend Algorithm S so that it obtains the coefficients $W_1, W_2, \ldots$, in this more general situation, without substantially increasing its running time.

9. [*20*] Find the values of T_{mn} computed by Algorithm T as it determines the first five coefficients in the reversion of $z = t - t^2$.

10. [*M20*] Given that $y = x^\alpha + a_1 x^{\alpha+1} + a_2 x^{\alpha+2} + \cdots$, $\alpha \neq 0$, show how to compute the coefficients in the expansion $x = y^{1/\alpha} + b_2 y^{2/\alpha} + b_3 y^{3/\alpha} + \cdots$.

▶ **11.** [*M25*] Let

$$U(z) = U_0 + U_1 z + U_2 z^2 + \cdots \quad \text{and} \quad V(z) = V_1 z + V_2 z^2 + V_3 z^3 + \cdots\,;$$

design an algorithm which computes the first N coefficients of $U(V(z))$.

ANSWERS TO EXERCISES

NOTES ON THE EXERCISES

1. An average problem for a mathematically inclined reader.

3. See W. J. LeVeque, *Topics in Number Theory* **2** (Reading, Mass.: Addison-Wesley, 1956), Chapter 3. (*Note:* One of the men who read a preliminary draft of the manuscript for this book reported that he had discovered a truly remarkable proof, which the margin of his copy was too small to contain.)

SECTION 3.1

1. (a) This will usually fail, since "round" telephone numbers are often selected by the telephone user when possible. In some areas, telephone numbers are perhaps assigned randomly. But it is a mistake in any case to try to get several successive random numbers from the same page, since many telephone numbers are listed several times in sequence.

(b) But do you use the left-hand page or the right-hand page? Say, use the left-hand page number, divide by 2, and take the units digit.

(c) The markings on the faces will slightly bias the die, but for practical purposes this method is quite satisfactory (and it has been used by the author in the preparation of several examples in this book). See *Math. Comp.* **15** (1961), 94–95 for further discussion of these dice.

(d) (This is a hard question thrown in purposely as a surprise.) The number is not random; if the average number of emissions per minute is m, the probability that the counter registers k is $e^{-m}m^k/k!$ (the Poisson distribution). So the digit 0 is selected with probability $e^{-m} \sum_{k \geq 0} m^{10k}/(10k)!$, etc. Even digits are selected with probability $e^{-m} \cosh m = \frac{1}{2} + \frac{1}{2}e^{-2m}$, and this is not equal to $\frac{1}{2}$, regardless of the value of m (although the error may be so small as to be considered negligible when m is very large).

(e) Okay, provided that the time since the last digit selected in this way is random. However, there is possible bias in borderline cases.

(f, g) No, people usually think of certain digits (like 7) with higher probability.

(h) Okay; your assignment of numbers to the horses had probability $\frac{1}{10}$ of assigning a given digit to the winning horse.

452

2. The number of such sequences is the multinomial coefficient $1000000!/(100000!)^{10}$; the probability is this number divided by the total number of sequences of a million digits, namely $10^{1000000}$. By Stirling's approximation we get the estimate

$$1/(16\pi^4 10^{22}\sqrt{2\pi}) \approx 2.55 \times 10^{-26},$$

about one chance in 4×10^{25}.

3. 3040504030.

4. Step K11 can be entered only from step K10 or step K2, and in either case we find it impossible for X to be zero by a simple argument. If X could be zero at that point, the algorithm would not terminate.

5. Since only 10^{10} ten-digit numbers are possible, a value of X must be repeated during the first $10^{10} + 1$ steps; and as soon as a value is repeated, the sequence continues periodically.

6. (a) Arguing as in the previous exercise, the sequence must eventually repeat a value; let this repetition occur for the first time at step $\mu + \lambda$, where $X_{\mu+\lambda} = X_\mu$. (This condition defines μ and λ.) We have $0 \le \mu < m$, $0 < \lambda \le m$, $\mu + \lambda \le m$. The values $\mu = 0$, $\lambda = m$ are attained if f is a permutation; $\mu = m - 1$, $\lambda = 1$ occurs if $X_0 = 0$, $f(x) = x + 1$ for $x < m - 1$, and $f(m - 1) = m - 1$.

(b) We have, for $r \ge n$, $X_r = X_n$ iff $r - n$ is a multiple of λ and $n \ge \mu$. Hence $X_{2n} = X_n$ iff n is a multiple of λ and $n \ge \mu$. The desired results now follow immediately. (*Note:* This is essentially a proof of the familiar mathematical result that the powers of an element in a finite semigroup include a unique idempotent element: take $X_0 = a$, $f(x) = ax$.)

7. F1. Set $X \leftarrow Y \leftarrow X_0$.

F2. Output X.

F3. Set $X \leftarrow f(X)$, $Y \leftarrow f(f(Y))$.

F4. If $X \ne Y$, return to F2. ▮

The number of elements output is the smallest multiple of λ which is $\ge \mu$.

8. (a, b) $00, 00, \ldots$ [62 starting values]; $10, 10, \ldots$ [19]; $60, 60, \ldots$ [15]; $50, 50, \ldots$ [1]; $24, 57, 24, 57, \ldots$ [3]. (c) 42 or 69; these both lead to a set of fifteen distinct values, namely (42 or 69), 76, 77, 92, 46, 11, 12, 14, 19, 36, 29, 84, 05, 02, 00.

9. Since $X < b^n$, $X^2 < b^{2n}$, so the middle square is $\lfloor X^2/b^n \rfloor \le X^2/b^n$. If $X > 0$, $X^2/b^n < Xb^n/b^n = X$.

10. If $X = ab^n$, the next number of the sequence has the same form; it is $(a^2 \bmod b^n)b^n$. If a is a multiple of b, the sequence will soon degenerate to zero; if a is not a multiple of b, the sequence will degenerate into a cycle of numbers having the same general form as X.

Further facts about the middle-square method have been found by B. Jansson, *Random Number Generators* (Stockholm: Almqvist and Wiksell, 1966), Section 3A. Numerologists will be interested to learn that the number 3792 is self-reproducing in the four-digit middle-square method, since $3792^2 = 14379264$; furthermore (as Jansson has observed), it is "self-reproducing" in another sense, since its prime factorization is $3 \cdot 79 \cdot 2^4$!

11. The probability that $\lambda = 1, \mu = 0$, is the probability that $X_1 = X_0$, namely $1/m$. The probability that $\lambda = 1, \mu = 1$, or $\lambda = 2, \mu = 0$, is the probability that $X_1 \neq X_0$ and that X_2 has a certain value, so it is $(1 - 1/m)(1/m)$. Similarly, the probability that the sequence has any given μ and λ is

$$P(\mu, \lambda) = \frac{1}{m} \prod_{1 \leq k < \lambda + \mu} \left(1 - \frac{k}{m}\right).$$

For the probability that $\lambda = 1$, we have

$$\sum_{\mu \geq 0} \frac{1}{m} \prod_{1 \leq k \leq \mu} \left(1 - \frac{k}{m}\right) = \frac{1}{m} Q(m),$$

where $Q(m)$ is defined in Section 1.2.11.3, Eq. (2). By Eq. (25) in that section, the probability is approximately $\sqrt{\pi/2m} \approx 1.25/\sqrt{m}$. The chance of Algorithm K converging as it did is only about one in 80000; the author was decidedly unlucky. But see exercise 15 for comments on the "colossalness."

12.
$$\sum_{\substack{1 \leq \lambda \leq m \\ 0 \leq \mu < m}} \lambda P(\mu, \lambda) = \frac{1}{m}\left(1 + 3\left(1 - \frac{1}{m}\right) + 6\left(1 - \frac{1}{m}\right)\left(1 - \frac{2}{m}\right) + \cdots\right)$$
$$= \frac{1 + Q(m)}{2}.$$

(See the previous answer and formulas related to "open addressing" in Chapter 6.) Therefore the average value of λ (and, by symmetry of $P(\mu, \lambda)$, also of $\mu + 1$) is approximately $\sqrt{\pi m/8} + \frac{1}{3}$. The average value of $\mu + \lambda$ is exactly $Q(m)$, approximately $\sqrt{\pi m/2} - \frac{1}{3}$. For alternative derivations and further results, including asymptotic values for the moments, see B. Harris, *Annals of Math. Statistics* **31** (1960), 1045–1062; see also I. M. Sobol, *Theory of Probability and Its Applications*, **9** (1964), 333–338. Sobol discusses the asymptotic period length for the more general sequence $X_{n+1} = f(X_n)$ if $n \not\equiv 0$ (modulo m); $X_{n+1} = g(X_n)$ if $n \equiv 0$ (modulo m); with both f and g random.

13. (Solution by Paul Purdom and John Williams.) Let T_{mn} be the number of functions which have n one-cycles and no cycles of length greater than one. Then

$$T_{mn} = \binom{m-1}{n-1} m^{m-n}.$$

(See Section 2.3.4.4.) *Any* function is such a function followed by a permutation of the n elements that were the one-cycles. Hence $\sum_{n \geq 1} T_{mn} n! = m^m$.

Let P_{nk} be the number of permutations of n elements in which the longest cycle is of length k. Then the number of functions with a maximum cycle of length k is $\sum_{n \geq 1} T_{mn} P_{nk}$. To get the average value of k, we compute $\sum_{k \geq 1} \sum_{n \geq 1} k T_{mn} P_{nk}$, which by the result of exercise 1.3.3–23 is $\sum_{n \geq 1} T_{mn} n! \, c(n + \frac{1}{2} + \epsilon_n)$ where $\epsilon_n \to 0$ as $n \to \infty$. Summing, we get the average value $cQ(m) + \frac{1}{2}c + \delta_m$, where $\delta_m \to 0$ as $m \to \infty$. (This is not substantially larger than the average value when X_0 is selected at random. The average value of $\max(\mu + \lambda)$ is still unknown.)

14. Let $c_r(m)$ be the number of functions with exactly r different final cycles. From the recurrence $c_1(m) = (m-1)! - \sum_{k>0} \binom{m}{k}(-1)^k(m-k)^k c_1(m-k)$, which comes by counting the number of functions whose image contains at most $m - k$ elements, we find the solution $c_1(m) = m^{m-1}Q(m)$. (Cf. exercise 1.2.11.3–16.) Another way to obtain the value of $c_1(m)$, which is perhaps more elegant and revealing, is given in exercise 2.3.4.4–17. The value of $c_r(m)$ may be determined by solving the recurrence

$$c_r(m) = \sum_{0<k<m} \binom{m-1}{k-1} c_1(k) c_{r-1}(m-k),$$

which has the solution

$$c_r(m) = m^{m-1}\left(\frac{1}{0!}\begin{bmatrix}1\\r\end{bmatrix} + \frac{1}{1!}\begin{bmatrix}2\\r\end{bmatrix}\frac{m-1}{m} + \frac{1}{2!}\begin{bmatrix}3\\r\end{bmatrix}\frac{m-1}{m}\frac{m-2}{m} + \cdots\right).$$

The desired average value can now be computed; it is

$$E_m = \frac{1}{m}\left(H_1 + 2H_2\frac{m-1}{m} + 3H_3\frac{m-1}{m}\frac{m-2}{m} + \cdots\right)$$

$$= 1 + \frac{1}{2}\frac{m-1}{m} + \frac{1}{3}\frac{m-1}{m}\frac{m-2}{m} + \cdots.$$

This latter formula was obtained by quite different means by Martin D. Kruskal, *AMM* **61** (1954), 392–397. Using the integral representation

$$E_m = \int_0^\infty \left(\left(1+\frac{x}{m}\right)^m - 1\right) e^{-x}\frac{dx}{x},$$

he proved the asymptotic relation

$$\lim_{m\to\infty} (E_m - \tfrac{1}{2}\ln m) = \tfrac{1}{2}(\gamma + \ln 2).$$

For further results and references, see John Riordan, *Annals Math. Stat.* **33** (1962), 178–185.

15. The probability that $f(x) \neq x$ for all x is $(m-1)^m/m^m$, which is approximately $1/e$. The existence of a self-repeating value in an algorithm like Algorithm K is therefore not "colossal" at all—it occurs with probability $1 - 1/e \approx .63212$; the only "colossal" thing was that the author happened to hit such a value when X_0 was chosen at random (see exercise 11).

16. The sequence will repeat when a pair of successive elements occurs for the second time. The maximum period is m^2. (Cf. next exercise.)

17. Select $X_0, \ldots, X_{k-1}$ arbitrarily; let $X_{n+1} = f(X_n, \ldots, X_{n-k+1})$, where $0 \leq x_1, \ldots, x_k < m$ implies that $0 \leq f(x_1, \ldots, x_k) < m$. The maximum period is m^k. This is an obvious upper bound, but it is not obvious that it can be attained; for a proof that it can always be attained for suitable f, see exercise 3.2.2–17.

18. Let μ, λ be the smallest positive integers for which $X_{\mu+\lambda-j} = X_{\mu-j}$ for $0 \leq j < k$. We have $X_{2n-j} = X_{n-j}$, $0 \leq j < k$, for those values of n which are multiples of λ, and $n \geq \mu$.

SECTION 3.2.1

1. Take X_0 even, a even, c odd. Then X_n is odd for $n > 0$.

2. Let X_r be the first repeated value in the sequence. If $X_r = X_k$ for $0 < k < r$, we could prove that $X_{r-1} = X_{k-1}$, since X_n uniquely determines X_{n-1} when a is prime to m. Hence, $k = 0$.

3. If d is the greatest common divisor of a and m, the quantity aX_n can take on at most m/d values, modulo m, and therefore $(aX_n + c) \bmod m$ can take on at most m/d values. The situation can be even worse; e.g., if $m = 2^e$ and if a is even, Eq. (6) shows the sequence is eventually constant.

4. Induction on k.

5. If a is relatively prime to m, there is a number a' for which $aa' \equiv 1 \pmod{m}$. Then $X_{n-1} = (a'X_n - a'c) \bmod m$, and in general,

$$X_{n-k} = \left((a')^k X_n - c(a' + \cdots + (a')^k)\right) \bmod m$$

$$= \left((a')^k X_n - c\frac{(a')^{k+1} - a'}{a' - 1}\right) \bmod m$$

when $k > 0$, $n - k \geq 0$. If a is not relatively prime to m, it is not possible to determine X_{n-1} when X_n is given; multiples of $m/(\gcd(a, m))$ may be added to X_{n-1} without changing the value of X_n. (See also exercise 3.2.1.3–7.)

SECTION 3.2.1.1

1. Let c' be a solution to the congruence $ac' \equiv c \pmod{m}$. (Thus, $c' = a'c \bmod m$, if a' is the number in the answer to exercise 3.2.1–5.) Then we have

```
        LDA    X
        ADD    CPRIME
        MUL    A        ▌
2. RANDM STJ    1F
        LDA    XRAND
        MUL    A
        SLAX   5
        ADD    C           (or, INCA c, if c is small)
        STA    XRAND
   1H   JNOV   *
        JMP    *-1
   XRAND CON    X
   A     CON    a
   C     CON    c        ▌
```

Note: Locations A and C should probably be named 2H and 3H to avoid conflict with other symbols, if this subroutine is to be used by other programmers.

3. For example,

```
BSIZE   EQU   1(4:4)
1H      EQU   BSIZE-1
ANSWER  CON   1B,1B(4:4),1B(3:3),1B(2:2),1B(1:1)  ▮
```

A shorter solution appears at the end of Program 4.2.3D.

4. Line 00 shuts off overflow in case it was on. Line 08 jumps out unless the result of $(aX) \bmod (w+1)$ was equal to w. Lines 09–10 cause the next value of the sequence,

$$(aw) \bmod (w+1) = (-a) \bmod (w+1)$$

to be used; i.e., the value w is bypassed.

5. (a) Subtraction: LDA X, SUB Y, JANN *+2, ADD M. (b) Addition: LDA X, SUB M, ADD Y, JANN *+2, ADD M. (Note that if m is near the word size, the instruction "SUB M" must precede the instruction "ADD Y".)

6. The sequences are not essentially different, since adding the constant $(m - c)$ has the same effect as subtracting the constant c. Since the operation must be combined with multiplication, a subtractive process has little merit over the additive one (at least in MIX's case), except when it is necessary to avoid affecting the overflow toggle.

7. The prime factors of $z^k - 1$ appear in the factorization of $z^{kr} - 1$. If r is odd, the prime factors of $z^k + 1$ appear in the factorization of $z^{kr} + 1$. And $z^{2k} - 1$ equals $(z^k - 1)(z^k + 1)$.

8.
```
   JOV   *+1       (Ensure that overflow is off.)
   LDA   X
   MUL   A
   STX   TEMP
   ADD   TEMP      Add lower half to upper half.
   JNOV  *+2       If ≥w, subtract w − 1.
   INCA  1         (Overflow is impossible in this step.)  ▮
```

9. The pairs $(0, w - 2)$, $(1, w - 1)$ are treated as equivalent in input and output:

```
   JOV   *+1
   LDA   X
   MUL   A          aX = qw + r
   SLC   5          rA ← r, rX ← q
   STX   TEMP
   ADD   TEMP
   JNOV  *+2        Get (r + q) mod (w − 2).
   INCA  2          Overflow is possible in one case.
   ADD   TEMP
   JNOV  *+3        Get (r + 2q) mod (w − 2).
   INCA  2          Overflow is possible.
   JOV   *-1                                  ▮
```

SECTION 3.2.1.2

1. Period length m, by Theorem A. (Cf. exercise 3.)

2. Yes, these conditions imply the conditions in Theorem A, since the only prime divisor of 2^e is 2, and any odd number is relatively prime to 2^e. (In fact, the conditions of the exercise are *necessary* and sufficient, if $e \neq 2$.)

3. By Theorem A, we need $a \equiv 1$ (modulo 4) and $a \equiv 1$ (modulo 5). By Law D of Section 1.2.4, this is equivalent to $a \equiv 1$ (modulo 20).

4. We know $X_{2^{e-1}} \equiv 0$ (modulo 2^{e-1}) by using Theorem A in the case $m = 2^{e-1}$. Also, using Theorem A for $m = 2^e$, we know that $X_{2^{e-1}} \not\equiv 0$ (modulo 2^e). It follows that $X_{2^{e-1}} = 2^{e-1}$. More generally, we can use this argument to prove that the second half of the period is essentially like the first half, since $X_{n+2^{e-1}} = (X_n + 2^{e-1}) \bmod 2^e$.

5. We need $a \equiv 1$ (modulo p) for $p = 3, 11, 43, 281, 86171$. By Law D of Section 1.2.4, this is equivalent to $a \equiv 1$ (modulo $3 \cdot 11 \cdot 43 \cdot 281 \cdot 86171$) so the *only* solution is the terrible multiplier $a = 1$!

6. (Cf. previous exercise.) $a \equiv 1$ (modulo $3 \cdot 7 \cdot 11 \cdot 13 \cdot 37$) implies that the solutions are $a = 1 + 111111k$, $0 \leq k \leq 8$.

7. Using the notation of the proof of Lemma Q, μ is the smallest value such that $X_{\mu+\lambda} = X_\mu$; so it is the smallest value such that $Y_{\mu+\lambda} = Y_\mu$ and $Z_{\mu+\lambda} = Z_\mu$. This shows that $\mu = \max(\mu_1, \ldots, \mu_t)$. The highest achievable μ is $\max(e_1, \ldots, e_t)$, but nobody really wants to achieve it.

8. $a^2 \equiv 1$ (modulo 8); so $a^4 \equiv 1$ (modulo 16), $a^8 \equiv 1$ (modulo 32), etc. If $a \bmod 4 = 3$, $a - 1$ is twice an odd number; thus, we have $(a^{2^{e-1}} - 1)/2 \equiv 0$ (modulo $2^{e+1}/2$), and this yields the desired result.

9. Substitute for X_n in terms of Y_n and simplify. If $X_0 \bmod 4 = 3$, the formulas of the exercise do not apply, but they do apply to the sequence $Z_n = (-X_n) \bmod 2^e$, which has essentially the same behavior.

10. Only $m = 1, 2, 4, p^e$, and $2p^e$, for odd primes p. In all other cases, the result of Theorem B is an improvement over Euler's theorem (exercise 1.2.4–28).

11. (a) Either $x + 1$ or $x - 1$ (not both) will be a multiple of 4, so $x \mp 1 = q2^f$, where q is odd and f is greater than 1. (b) In the given circumstances, $f < e$ and so $e \geq 3$. We have $\pm x \equiv 1$ (modulo 2^f) and $\pm x \not\equiv 1$ (modulo 2^{f+1}) and $f > 1$. Hence by applying Lemma P, we find that $(\pm x)^{2^{e-f-1}} \not\equiv 1$ (modulo 2^e), while $x^{2^{e-f}} = (\pm x)^{2^{e-f}} \equiv 1$ (modulo 2^e). So the order is a divisor of 2^{e-f}, but not a divisor of 2^{e-f-1}. (c) 1 has order 1; $2^e - 1$ has order 2; the maximum period when $e \geq 3$ is therefore 2^{e-2}, and this requires that $f = 2$, that is, $x \equiv 4 \pm 1$ (modulo 8).

12. If k is a proper divisor of $p - 1$ and if $a^k \equiv 1$ (modulo p), then by Lemma P $a^{kp^{e-1}} \equiv 1$ (modulo p^e). Similarly, if $a^{p-1} \equiv 1$ (modulo p^2), we find that $a^{(p-1)p^{e-2}} \equiv 1$ (modulo p^e). So in these cases a is *not* primitive. Conversely, if $a^{p-1} \not\equiv 1$ (modulo p^2), Theorem 1.2.4F and Lemma P tell us that $a^{(p-1)p^{e-2}} \not\equiv 1$ (modulo p^e), but $a^{(p-1)p^{e-1}} \equiv 1$ (modulo p^e). So the order is a divisor of $(p - 1)p^{e-1}$ but not of $(p - 1)p^{e-2}$; it therefore has the form kp^{e-1}, where k divides $p - 1$. But if a is primitive modulo p, $a^{kp^{e-1}} \equiv a^k \equiv 1$ (modulo p) implies that $k = p - 1$.

13. Let λ be the order of a modulo p. By Theorem 1.2.4F, λ is a divisor of $p - 1$. If $\lambda < p - 1$, $(p - 1)/\lambda$ has a prime factor, q.

14. Let $0 < k < p$. If $a^{p-1} \equiv 1$ (modulo p^2), then $(a + kp)^{p-1} \equiv a^{p-1} + (p - 1)a^{p-2}kp$ (modulo p^2), and this is $\not\equiv 1$, since $(p - 1)a^{p-2}k$ is not a multiple of p. By exercise 12, $a + kp$ is primitive modulo p^e.

15. (a) If $\lambda_1 = p_1^{e_1} \ldots p_t^{e_t}$, $\lambda_2 = p_1^{f_1} \ldots p_t^{f_t}$, let $\kappa_1 = p_1^{g_1} \ldots p_t^{g_t}$, $\kappa_2 = p_1^{h_1} \ldots p_t^{h_t}$, where

$$g_j = \begin{cases} e_j, & \text{if} \quad e_j < f_j, \\ 0, & \text{if} \quad e_j \geq f_j; \end{cases}$$

$$h_j = \begin{cases} 0, & \text{if} \quad e_j < f_j, \\ f_j, & \text{if} \quad e_j \geq f_j. \end{cases}$$

Now $a_1^{\kappa_1}$, $a_2^{\kappa_2}$ have periods λ_1/κ_1, λ_2/κ_2, and the latter are relatively prime. Also, $(\lambda_1/\kappa_1)(\lambda_2/\kappa_2) = \lambda$, so it suffices to consider the case when λ_1 is relatively prime to λ_2, that is, when $\lambda = \lambda_1\lambda_2$. Now since $(a_1a_2)^\lambda \equiv 1$, we have $1 \equiv (a_1a_2)^{\lambda\lambda_1} \equiv a_2^{\lambda\lambda_1}$; hence $\lambda\lambda_1$ is a multiple of λ_2. This implies that λ is a multiple of λ_2, since λ_1 is relatively prime to λ_2. Similarly, λ is a multiple of λ_1; hence λ is a multiple of $\lambda_1\lambda_2$. But obviously $(a_1a_2)^{\lambda_1\lambda_2} \equiv 1$, so $\lambda = \lambda_1\lambda_2$.

(b) If a_1 has order $\lambda(m)$ and if a_2 has order λ, by part (a) $\lambda(m)$ must be a multiple of λ, otherwise we could find an element of higher order, namely, l.c.m. $(\lambda, \lambda(m))$.

16. (a) $f(x) = (x - a)(x^{n-1} + (a + c_1)x^{n-2} + \cdots + (a^{n-1} + \cdots + c_{n-1})) + f(a)$.
(b) The statement is clear when $n = 0$. If a is one root,

$$f(x) \equiv (x - a)q(x);$$

if a' is any other root,

$$0 \equiv f(a') \equiv (a' - a)q(a'),$$

and since $a' - a$ is not a multiple of p, a' must be a root of $q(x)$. So if $f(x)$ has more than n distinct roots, $q(x)$ has more than $n - 1$ distinct roots. (c) $\lambda(p) \geq p - 1$, since $f(x)$ must have degree $\geq p - 1$ in order to possess so many roots. But by Theorem 1.2.4F, $\lambda(p) \leq p - 1$.

17. By Lemma P, $11^5 \equiv 1$ (modulo 25), $11^5 \not\equiv 1$ (modulo 125), etc.; so the order of 11 is 5^{e-1} modulo 5^e, not the maximum value $\lambda(5^e) = 4 \cdot 5^e$. But by Lemma Q the total period length is the least common multiple of the period modulo 2^e (namely 2^{e-2}) and the period modulo 5^e (namely 5^{e-1}), and this is $2^{e-2}5^{e-1} = \lambda(10^e)$. The period modulo 5^e may be 5^{e-1}, $2 \cdot 5^{e-1}$, or $4 \cdot 5^{e-1}$, without affecting the length of period modulo 10^e, since the least common multiple is taken. The values which are primitive modulo 5^e are those congruent to 2, 3, 8, 12, 13, 17, 22, 23 modulo 25 (cf. exercise 12), namely, 3, 13, 27, 37, 53, 67, 77, 83, 117, 123, 133, 147, 163, 173, 187, 197.

18. (Cf. previous exercise.) ($a \bmod 8$) must be 3 or 5. Knowing the period of a modulo 5 and modulo 25 allows us to apply Lemma P to determine admissible values of $a \bmod 25$. Period $= 4 \cdot 5^{e-1}$: 2, 3, 8, 12, 13, 17, 22, 23; period $= 2 \cdot 5^{e-1}$: 4, 9, 14, 19; period $= 5^{e-1}$: 6, 11, 16, 21. Each of these 16 values yields one value of a, $0 \leq a < 200$, with $a \bmod 8 = 3$, and another value of a with $a \bmod 8 = 5$.

20. $X_n = ((a^n - 1)/(a - 1))c \bmod m$. The period length is the least common multiple of λ' and $\lambda(m_1)$, where m_1 is odd, $m = 2^e m_1$, $\lambda' = 1$ if $e \leq 1$, $\lambda' = 2$ if $e = 2$, and if $e \geq 3$ $\lambda' = 2^e$ for $a \bmod 8 = 5$, $\lambda' = 2^{e-1}$ for $a \bmod 8 = 3$.

SECTION 3.2.1.3

1. $c = 1$ is always relatively prime to B^5; and every prime dividing $m = B^5$ is a divisor of B, so it divides $b = B^2$ to at least the second power.

2. 3, so the generator is not recommended in spite of its long period.

3. The potency is 18 in both cases (see next exercise).

4. Since $a \bmod 4 = 1$, we must have $a \bmod 8 = 1$ or 5, so $b \bmod 8 = 0$ or 4. If b is an odd multiple of 4, and if b_1 is a multiple of 8, clearly $b^s \equiv 0$ (modulo 2^e) implies that $b_1^s \equiv 0$ (modulo 2^e), so b_1 cannot have higher potency.

5. The potency is the smallest value of s such that $f_j s \geq e_j$ for all j.

6. m must be divisible by 2^7 or by p^4 (for odd prime p) in order to have a potency as high as 4. The only values are $m = 2^{27} + 1$ and $10^9 - 1$.

7. $a' = (1 - b + b^2 - \cdots) \bmod m$, where the terms in b^s, b^{s+1}, etc., are dropped (if s is the potency).

8. Since X_n is always odd,

$$X_{n+2} = (2^{34} + 3 \cdot 2^{18} + 9)X_n \bmod 2^{35} = (2^{34} + 6X_{n+1} - 9X_n) \bmod 2^{35}.$$

Given Y_n and Y_{n+1}, the possibilities for

$$Y_{n+2} \approx (5 + 6(Y_{n+1} + \epsilon_1) - 9(Y_n + \epsilon_2)) \bmod 10,$$

with $0 \leq \epsilon_1 < 1$, $0 \leq \epsilon_2 < 1$, are limited and nonrandom.

Note: If the multiplier suggested in the text were, say, $2^{23} + 2^{18} + 2^2 + 1$, instead of $2^{23} + 2^{14} + 2^2 + 1$, we would similarly find $X_{n+2} - 10X_{n+1} + 25X_n \equiv$ constant (modulo 2^{35}). In general, we do not want $a \pm \delta$ to be divisible by high powers of 2 when δ is small, else we get "second order impotency." See Section 3.3.4 for a more detailed discussion.

The generator which appears in this exercise may be found in the article by MacLaren and Marsaglia, *JACM* **12** (1965), 83–89.

SECTION 3.2.2

1. The method is not to be recommended. In the first place, aU_n is likely to be so large that the addition of c/m which follows will lose almost all significance, and the "mod 1" operation will nearly destroy any vestiges of significance which might remain. We conclude that double-precision floating point would be necessary. And even in this case, one must be sure that no rounding, etc., occurs to affect the numbers of the sequence in any way, since that would destroy the theoretical grounds for the good behavior of the sequence.

2. X_{n+1} equals either $X_{n-1} + X_n$ or $X_{n-1} + X_n - m$; if $X_{n+1} < X_n$ we must have $X_{n+1} = X_{n-1} + X_n - m$; hence $X_{n+1} < X_{n-1}$.

3. (Underlined numbers are $V[j]$ in step M3.)

Output: initial		0 4 5 6 2 0 3 (2 7 4 1 6 3 0 5) and repeats.
$V[0]$:	0	4̲ 7 7 7 7 7 7 7 4̲ 7̲ 7 7 7 7 7 7 4̲ 7̲ ...
$V[1]$:	3	3 3 3 3 3 3 2̲ 5̲ 5 5 5 5 5 5 2̲ 5̲ 5 5 ...
$V[2]$:	2	2 2 2 2 0̲ 3̲ 3 3 3 3 3 3 0̲ 3̲ 3 3 3 3 ...
$V[3]$:	5	5 5 6̲ 1̲ 1 1 1 1 1 1 6̲ 1̲ 1 1 1 1 1 1 ...
X:		4 7 6 1 0 3 2 5 4 7 6 1 0 3 2 5 4 7 ...
Y:		0 1 6 7 4 5 2 3 0 1 6 7 4 5 2 3 0 1 ...

4. The low-order byte of many random sequences (e.g., linear congruential sequences with m = word size) is much less random than the high-order byte. See Section 3.2.1.1.

5. The randomizing effect would be quite minimized, because $V[j]$ would always contain a number in a certain range, essentially $j/k \le V[j]/m < (j+1)/k$. However, some similar approaches could be used: we could take $Y_n = X_{n-1}$, or we could choose j from X_n by extracting some digits from about the middle instead of at the extreme left. This is an inexpensive way to improve the randomness of a sequence which is already reasonably good.

6. For example, if $X_n/m < \frac{1}{2}$, then $X_{n+1} = 2X_n$.

7.

	$00 \cdots 01$		$00 \cdots 01$
The subsequence of	$00 \cdots 10$		$00 \cdots 10$
X values:	$\cdots$	becomes:	$\cdots$
	$10 \cdots 00$		$10 \cdots 00$
	CONTENTS (C)		$00 \cdots 00$
			CONTENTS (C)

8. We may assume that $X_0 = 0$, $m = p^e$, as in the proof of Theorem 3.2.1.2A. First suppose that the sequence has period length p^e; it follows that the period of the sequence mod p^f has length p^f, for $1 \le f \le e$, otherwise some residues mod p^f would never occur. Clearly, c is not a multiple of p, for otherwise each X_n would be a multiple of p. If $p \le 3$, it is easy to establish the necessity of conditions (iii) and (iv) by trial and error, so we may assume that $p \ge 5$. If $d \not\equiv 0$ modulo p then $dx^2 + ax + c \equiv d(x + a_1)^2 + c_1$ (modulo p^e) for some integers a_1 and c_1 and for all x; this quadratic takes the same value at the points x and $-x - 2a_1$, so it cannot assume all values modulo p^e. Hence $d \equiv 0$ (modulo p); and if $a \not\equiv 1$, we would have $dx^2 + ax + c \equiv x$ (modulo p) for some x, contradicting the fact that the sequence mod p has period length p.

To show the sufficiency of the conditions, we may assume by Theorem 3.2.1.2A and consideration of some trivial cases that $m = p^e$ where $e \ge 2$. If $p = 2$, we have $X_{n+p} \equiv X_n + pc$ (modulo p^2), by trial; and if $p = 3$, we have either $X_{n+p} \equiv X_n + pc$ (modulo p^2), for all n, or $X_{n+p} \equiv X_n - pc$ (modulo p^2), for all n. For $p \ge 5$, we can prove that $X_{n+p} \equiv X_n + pc$ (modulo p^2): Let $d = pr$, $a = 1 + ps$.

Then if $X_n \equiv cn + pY_n$ (modulo p^2), we must have $Y_{n+1} \equiv n^2c^2r + ncs + Y_n$ (modulo p); hence $Y_n \equiv \binom{n}{2}2c^2r + \binom{n}{2}(c^2r + cs)$ (modulo p). Thus $Y_p \bmod p = 0$, and the desired relation has been proved.

Now we can prove that the sequence $\langle X_n \rangle$ of integers defined in the "hint" satisfies the relation

$$X_{n+p^f} \equiv X_n + tp^f \quad (\text{modulo } p^{f+1}), \qquad n \geq 0,$$

for some t with $t \bmod p \neq 0$, and for all $f \geq 1$. This suffices to prove that the sequence $\langle X_n \bmod p^e \rangle$ has period length p^e, for the length of the period is a divisor of p^e but not a divisor of p^{e-1}. The above relation has already been established for $f = 1$, and for $f > 1$ it can be proved by induction in the following manner: Let

$$X_{n+p^f} \equiv X_n + tp^f + Z_n p^{f+1} \quad (\text{modulo } p^{f+2});$$

then the quadratic law for generating the sequence, with $d = pr$, $a = 1 + ps$, yields

$$Z_{n+1} \equiv 2rtnc + st + Z_n \quad (\text{modulo } p).$$

It follows that $Z_{n+p} \equiv Z_n \pmod{p}$; hence

$$X_{n+kp^f} \equiv X_n + k(tp^f + Z_n p^{f+1}) \quad (\text{modulo } p^{f+2})$$

for $k = 1, 2, 3, \ldots$; setting $k = p$ completes the proof.

Notes: If $f(x)$ is a polynomial of degree higher than 2 and $X_{n+1} = f(X_n)$, the analysis is more complicated, although we can use the fact that $f(m + p^k) = f(m) + p^k f'(m) + p^{2k} f''(m)/2! + \cdots$ to prove that many polynomial recurrences give the maximum period. For example, Coveyou has proved that the period is $m = 2^e$ if $f(0)$ is odd, $f'(j) \equiv 1$, $f''(j) \equiv 0$, and $f(j+1) \equiv f(j) + 1$ (modulo 4) for $j = 0, 1, 2, 3$.

9. Let $X_n = 4Y_n + 2$; then the sequence Y_n satisfies the quadratic recurrence $Y_{n+1} = (4Y_n^2 + 5Y_n + 1) \bmod 2^{e-2}$.

10. *Case 1:* $X_0 = 0$, $X_1 = 1$; hence $X_n = F_n$. We seek the smallest n for which $F_n \equiv 0$ and $F_{n+1} \equiv 1$ (modulo 2^e). Since $F_{2n} = F_n(F_{n-1} + F_{n+1})$, $F_{2n+1} = F_n^2 + F_{n+1}^2$, we find by induction on e that for $e > 1$

$$F_{3 \cdot 2^{e-1}} \equiv 0 \quad (\text{modulo } 2^{e+1}),$$

$$F_{3 \cdot 2^{e-1}+1} \equiv 2^e + 1 \quad (\text{modulo } 2^{e+1}).$$

This implies the period is a divisor of $3 \cdot 2^{e-1}$ but not a divisor of $3 \cdot 2^{e-2}$, so it is either $3 \cdot 2^{e-1}$ or 2^{e-1}. But $F_{2^{e-1}}$ is always odd (since only F_{3n} is even).

Case 2: $X_0 = a$, $X_1 = b$. Then $X_n \equiv aF_{n-1} + bF_n$; we need to find the smallest positive n with $a(F_{n+1} - F_n) + bF_n \equiv a$, $aF_n + bF_{n+1} \equiv b$. This implies that $(b^2 - ab - a^2)F_n \equiv 0$, $(b^2 - ab - a^2)(F_{n+1} - 1) \equiv 0$; and $b^2 - ab - a^2$ is odd (i.e., prime to m) so the condition is equivalent to $F_n \equiv 0$, $F_{n+1} \equiv 1$.

Methods to determine the period of F_n for any modulus appear in the article by D. D. Wall, *AMM* **67** (1960), 525–532. Further facts about the Fibonacci sequence mod 2^e have been derived by B. Jansson [*Random Number Generators* (Stockholm: Almqvist and Wiksell, 1966), Section 3C1].

11. (a) We have $z^\lambda = 1 + f(z)u(z) + p^e v(z)$ for some $u(z)$, $v(z)$, where $v(z) \not\equiv 0$ (modulo $f(z)$ and p). By the binomial theorem

$$z^{\lambda p} = 1 + p^{e+1}v(z) + p^{2e+1}v(z)^2(p-1)/2$$

plus further terms congruent to zero (modulo $f(z)$ and p^{e+2}). Since $p^e > 2$, we have $z^{\lambda p} \equiv 1 + p^{e+1}v(z)$ (modulo $f(z)$ and p^{e+2}). If $p^{e+1}v(z) \equiv 0$ (modulo $f(z)$ and p^{e+2}), there exist polynomials $a(z)$, $b(z)$ such that $p^{e+1}(v(z) + pa(z)) = f(z)b(z)$, since $f(0) = 1$, this implies that $b(z)$ is a multiple of p^{e+1} (by Gauss's Lemma 4.6.1G); hence $v(z) \equiv 0$ (modulo $f(z)$ and p), a contradiction.

(b) If $z^\lambda - 1 = f(z)u(z) + p^e v(z)$, then

$$G(z) = u(z)/(z^\lambda - 1) + p^e v(z)/f(z)(z^\lambda - 1);$$

hence $A_{n+\lambda} \equiv A_n$ (modulo p^e) for large n. Conversely, if $\langle A_n \rangle$ has the latter property then $G(z) = u(z) + v(z)/(1 - z^\lambda) + p^e H(z)$, for some polynomials $u(z)$ and $v(z)$, and some power series $H(z)$, all with integer coefficients. This implies that $1 - z^\lambda = u(z)f(z)(1 - z^\lambda) + v(z)f(z) + p^e H(z)f(z)(1 - z^\lambda)$; and $H(z)f(z)(1 - z^\lambda)$ is a polynomial since the other terms of the equation are polynomials.

(c) It suffices to prove that $\lambda(p^e) \neq \lambda(p^{e+1})$ implies that $\lambda(p^{e+1}) = p\lambda(p^e) \neq \lambda(p^{e+2})$. Applying (a) and (b), we know that $\lambda(p^{e+2}) \neq p\lambda(p^e)$, and that $\lambda(p^{e+1})$ is a divisor of $p\lambda(p^e)$ but not of $\lambda(p^e)$. Hence if $\lambda(p^e) = p^f q$, where $q \bmod p \neq 0$, then $\lambda(p^{e+1})$ must be $p^{f+1}d$, where d is a divisor of q. But now $X_{n+p^{f+1}d} \equiv X_n$ (modulo p^e); hence $p^{f+1}d$ is a multiple of $p^f q$, hence $d = q$. (*Note:* The hypothesis $p^e > 2$ is necessary; for example, let $a_1 = 4$, $a_2 = -1$, $k = 2$; then $\langle A_n \rangle = 1, 4, 15, 56, 209, 780, \ldots$; $\lambda(2) = 2$, $\lambda(4) = 4$, $\lambda(8) = 4$.)

(d) $g(z) = X_0 + (X_1 - a_1 X_0)z + \cdots$
$$+ (X_{k-1} - a_1 X_{k-2} - a_2 X_{k-3} - \cdots - a_{k-1}X_0)z^{k-1}.$$

(e) The derivation in (b) can be generalized to the case $G(z) = g(z)/f(z)$; then the assumption of period length λ implies that $g(z)(1 - z^\lambda) \equiv 0$ (modulo $f(z)$ and p^e); we treated only the special case $g(z) = 1$ above. But both sides of this congruence can be multiplied by Hensel's $b(z)$, and we obtain $1 - z^\lambda \equiv 0$ (modulo $f(z)$ and p^e).

Note: A more "elementary" proof of the result in (c) can be given without using generating functions, using methods analogous to those in the answer to exercise 8: If $A_{\lambda+n} = A_n + p^e B_n$, for $n = r, r+1, \ldots, r+k-1$ and some integers B_n, then this same relation holds for *all* $n \geq r$ if we define $B_{r+k}, B_{r+k+1}, \ldots$ by the given recurrence relation. Since the resulting sequence of B's is some linear combination of shifts of the sequence of A's, we will have $B_{\lambda+n} \equiv B_n$ (modulo p^e) for all large enough values of n. Now $\lambda(p^{e+1})$ must be some multiple of $\lambda = \lambda(p^e)$; for all large enough n we have $A_{n+j\lambda} = A_n + p^e(B_n + B_{n+\lambda} + B_{n+2\lambda} + \cdots + B_{n+(j-1)\lambda}) \equiv A_n + jp^e B_n$ (modulo p^{2e}) for $j = 1, 2, 3, \ldots$. No k consecutive B's are multiples of p; hence $\lambda(p^{e+1}) = p\lambda(p^e) \neq \lambda(p^{e+2})$ follows immediately when $e \geq 2$. We still must prove that $\lambda(p^{e+2}) \neq p\lambda(p^e)$ when p is odd and $e = 1$; here we let $B_{\lambda+n} = B_n + pC_n$, and observe that $C_{n+\lambda} \equiv C_n$ (modulo p) when n is large enough. Then $A_{n+p} \equiv A_n + p^2(B_n + \binom{p}{2}C_n)$ (modulo p^3), and the proof is readily completed.

For the history of this problem, see Morgan Ward, *Trans. Amer. Math. Soc.* **35** (1933), 600–628; see also D. W. Robinson, *AMM* **73** (1966), 619–621.

The results of this exercise and exercise 16 suggest the following way to obtain an efficiently computable sequence with a very long period, generalizing Algorithm A: If $x^k + x^j + 1$ is a primitive polynomial modulo 2, the sequence defined by $X_n = (X_{n-j} + X_{n-k}) \bmod 2^e$ has a period length between $2^k - 1$ and $2^{e-1}(2^k - 1)$, if $X_0, \ldots, X_{k-1}$ are not all even. For example, the sequence defined by $X_n = (X_{n-24} + X_{n-55}) \bmod 2^e$ will have a period length of at least $2^{55} - 1$, which is the period of its units digits, and the actual period length will probably be much longer. This idea, due to G. J. Mitchell and D. P. Moore in 1958 [unpublished], has been used with apparent success for many years to generate full-word random numbers; the sequence clearly has some locally nonrandom subsequences which hardly ever occur. When $k = 127$ we may take $j = 1, 7, 15, 30, 63, 64, 97, 112, 120$, or 126.

12. The period length mod 2 can be at most 4; and the maximum period length mod p^{e+1} is at most twice the maximum length mod p^e, by the considerations of the previous exercise. So the maximum conceivable period length is 2^{e+1}; this is achievable, for example, in the trivial case $a = 0$, $b = c = 1$.

13, 14. Clearly $Z_{n+\lambda} = Z_n$, so λ' certainly is a divisor of λ. Let the least common multiple of λ' and λ_1 be λ_1', and define λ_2' similarly. $X_n + Y_n \equiv Z_n \equiv Z_{n+\lambda_1'} \equiv X_n + Y_{n+\lambda_1'}$ so λ_1' is a multiple of λ_2. Similarly, λ_2' is a multiple of λ_1. This yields the desired result. (The result is "best possible" in the sense that sequences for which $\lambda' = \lambda_0$ can be constructed, as well as sequences for which $\lambda' = \lambda$.)

16. There are several methods of proof.

(1) Using the theory of finite fields. In the field with 2^k elements let ξ satisfy $\xi^k = a_1\xi^{k-1} + \cdots + a_k$. Let $f(b_1\xi^{k-1} + \cdots + b_k) = b_k$, where each b_j is either zero or one; this is a linear function. If word X in the generation algorithm is $b_1b_2 \ldots b_k$ before (10) is executed, and if $b_1\xi^{k-1} + \cdots + b_k\xi = \xi^n$, then word X represents ξ^{n+1} after (10) is executed. Hence the sequence is $f(\xi^n), f(\xi^{n+1}), f(\xi^{n+2}), \ldots$ And $f(\xi^{n+k}) = f(\xi^n\xi^k) = f(a_1\xi^{n+k-1} + \cdots + a_k\xi^n) = a_1f(\xi^{n+k-1}) + \cdots + a_kf(\xi^n)$.

(2) Using brute force, or elementary ingenuity. We are given a sequence X_{nj}, $n \geq 0$, $1 \leq j \leq k$, satisfying

$$X_{(n+1)j} \equiv X_{n(j+1)} + a_jX_{n1}, \quad 1 \leq j < k; \qquad X_{(n+1)k} \equiv a_kX_{n1} \quad \text{(modulo 2)}.$$

We must show this implies $X_{nk} \equiv a_1X_{(n-1)k} + \cdots + a_kX_{(n-k)k}$, for $n \geq k$. Indeed, it implies $X_{nj} \equiv a_1X_{(n-1)j} + \cdots + a_kX_{(n-k)j}$ when $1 \leq j \leq k$. This is clear for $j = 1$, since $X_{n1} \equiv a_1X_{(n-1)1} + X_{(n-1)2} \equiv a_1X_{(n-1)1} + a_2X_{(n-2)2} + X_{(n-2)3}$, etc. For $j > 1$, we have by induction

$$X_{nj} \equiv X_{(n+1)(j-1)} - a_jX_{n1}$$
$$\equiv \sum_{1 \leq i \leq k} a_iX_{(n+1-i)(j-1)} - a_j \sum_{1 \leq i \leq k} a_iX_{(n-i)1}$$
$$\equiv \sum_{1 \leq i \leq k} a_i(X_{(n+1-i)(j-1)} - a_jX_{(n-i)1})$$
$$\equiv a_1X_{(n-1)j} + \cdots + a_kX_{(n-k)j}.$$

This proof does *not* depend on the fact that operations were done modulo 2, or modulo any prime number.

17. (a) When the sequence terminates, the $(k-1)$-tuple $(X_{n+1}, \ldots, X_{n+k-1})$ occurs for the $(m+1)$st time. A given $(k-1)$-tuple $(X_{r+1}, \ldots, X_{r+k-1})$ can have only m distinct predecessors X_r, so one of these occurrences must be for $r = 0$. (b) Since the $(k-1)$-tuple $(0, \ldots, 0)$ occurs $(m+1)$ times, each possible predecessor appears, so the k-tuple $(a_1, 0, \ldots, 0)$ appears for all a_1, $0 \le a_1 < m$. Let $1 \le s < k$ and suppose we have proved that all k-tuples $(a_1, \ldots, a_s, 0, \ldots, 0)$ appear in the sequence when $a_s \ne 0$. By the construction, this k-tuple would not be in the sequence unless $(a_1, \ldots, a_s, 0, \ldots, 0, y)$ had appeared earlier for $1 \le y < m$. Hence the $(k-1)$-tuple $(a_1, \ldots, a_s, 0, \ldots, 0)$ has appeared m times, and all m possible predecessors appear; this means $(a, a_1, \ldots, a_s, 0, \ldots, 0)$ appears for $0 \le a < m$. The proof is now complete by induction.

The result also follows from Theorem 2.3.4.2D, using the directed graph of exercise 2.3.4.2–23; the set of arcs from $(x_1, \ldots, x_j, 0, \ldots, 0)$ to $(x_2, \ldots, x_j, 0, \ldots, 0, 0)$, where $x_j \ne 0$ and $1 \le j \le k$, forms an oriented subtree related neatly to Dewey decimal notation.

18. The third most significant bit of U_{n+1} is completely determined by the first and third bits of U_n, so only 32 of the 64 possible pairs $(\lfloor 8U_n \rfloor, \lfloor 8U_{n+1} \rfloor)$ occur. (*Notes:* If we had used, say, 11-bit numbers $U_n = (.Y_{11n}Y_{11n+1} \cdots Y_{11n+10})_2$, the sequence *would* be satisfactory for many applications. If another constant appears in C, having more "one" bits, the spectral test gives some indication of its suitability. See exercise 3.3.4–26; we would calculate $f(s_1, \ldots, s_n)$ for $n = 36, 37, 38, \ldots$ The Fourier transform 3.3.4–4 can also be examined with X_k/m replaced by U_k, to get an indication of the randomness of such sequences compared to linear congruential sequences.)

21. [*J. London Math. Soc.* **21** (1946), 169–172.] Any sequence of period length $m^k - 1$ with no k consecutive zeroes leads to a sequence of period length m^k by inserting a zero in the appropriate place, as in exercise 7; conversely, we can start with a sequence of period length m^k and delete an appropriate zero from the period, to form a sequence of the other type. Let us call these "(m, k) sequences" of types A and B; the hypothesis assures us of the existence of (p, k) sequences of type A, for all primes p and all $k \ge 1$; hence we have (p, k) sequences of type B for all such p and k.

To get a (p^e, k) sequence of type B, let $e = qr$, where q is a power of p and r is not a multiple of p. Start with a (p, qrk) sequence of type A, namely $X_0, X_1, X_2, \ldots$; then (using the p-ary number system) the grouped digits $(X_0 \ldots X_{q-1})_p$, $(X_q \ldots X_{2q-1})_p$, $\ldots$ form a (p^q, rk) sequence of type A, since q is relatively prime to $p^{qrk} - 1$ and the sequence therefore has a period length of $p^{qrk} - 1$. This leads to a (p^q, rk) sequence $\langle Y_n \rangle$ of type B; and $(Y_0 Y_1 \ldots Y_{r-1})_{p^q}$, $(Y_r Y_{r+1} \ldots Y_{2r-1})_{p^q}$, $\ldots$ is a (p^{qr}, k) sequence of type B by a similar argument, since r is relatively prime to p^{qk}.

To get an (m, k) sequence of type B for arbitrary m, we can combine (p^e, k) sequences for each of the prime power factors of m using the Chinese remainder theorem; but a simpler method is available. Let $\langle X_n \rangle$ be an (r, k) sequence of type B, and let $\langle Y_n \rangle$ be an (s, k) sequence of type B, where r and s are relatively prime; then $\langle sX_n + Y_n \rangle$ is an (rs, k) sequence of type B.

22. By the Chinese remainder theorem, we can find $a_1, \ldots, a_k$, having desired residues mod each prime divisor of m. If $m = p_1 p_2 \ldots p_t$, then the period length can be as high as $\mathrm{lcm}\,(p_1^k - 1, \ldots, p_t^k - 1)$. In fact, we can achieve reasonably long periods for arbitrary m (not necessarily squarefree), as shown in exercise 11.

SECTION 3.3.1

1. There are $k = 11$ categories, so the line $\nu = 10$ should be used.

2. $\frac{2}{49}$, $\frac{3}{49}$, $\frac{4}{49}$, $\frac{5}{49}$, $\frac{6}{49}$, $\frac{9}{49}$, $\frac{6}{49}$, $\frac{5}{49}$, $\frac{4}{49}$, $\frac{3}{49}$, $\frac{2}{49}$.

3. $V = 7\frac{173}{240}$, only very slightly higher than that obtained from the good dice! There are two reasons why we do not detect the weighting: (a) The new probabilities (cf. exercise 2) are not really very far from the old ones in Eq. (1). The sum of the two dice tends to smooth out the probabilities; if we considered instead each of the 36 possible pairs of values, and counted these, we would probably detect the difference quite rapidly (assuming the two dice are distinguishable). (b) A far more important reason is that n is too small for a significant difference to be detected. If the same experiment is done for large enough n, the faulty dice will be discovered (see exercise 12).

4. $p_s = \frac{1}{12}$ for $2 \le s \le 12$ and $s \ne 7$; $p_7 = \frac{1}{6}$. The value of V is $15\frac{1}{2}$, which falls between the 25-percent level and the 10-percent level; so it is reasonable, in spite of the fact that not too many sevens actually turned up.

5. $K_{20}^+ = 1.15$; $K_{20}^- = 0.215$; these do not differ significantly from random behavior (being at about the 6-percent and 14-percent levels), but they are mighty close. (The data values in this exercise come from Appendix B, Table 1.)

6. The probability that $X_j \le x$ is $F(x)$, so we have essentially the coin-tossing problem discussed in Section 1.2.10. $F_n(x) = s/n$ with probability

$$\binom{n}{s}F(x)^s\big(1 - F(x)\big)^{n-s};$$

the mean is $F(x)$; the standard deviation is $\sqrt{F(x)\big(1 - F(x)\big)/n}$. (Cf. Eq. 1.2.10–19. This suggests that a slightly better statistic would be to define

$$K_n^+ = \sqrt{n} \max_{-\infty < x < \infty} \big((F_n(x) - F(x))/\sqrt{F(x)\big(1 - F(x)\big)},$$

etc.) *Notes:* Similarly, we can calculate the mean and standard deviation of $F_n(x) - F_n(y)$, for $x < y$, and obtain the covariance of $F_n(x)$, $F_n(y)$. Using these facts, it can be shown that for large values of n the function $F_n(x)$ behaves as a "Brownian motion," and techniques from this branch of probability theory may be used to study it. The situation is exploited in articles by J. L. Doob and M. D. Donsker, *Annals of Mathematical Statistics* **20** (1949), 393–403 and **23** (1952), 277–281; this type of examination of the KS tests is generally regarded as the most enlightening one.

7. (Cf. Eq. (13).) Take $j = n$ to see that K_n^+ is never negative and that it can get as high as $\sqrt{n}$. Similarly, take $j = 1$ to make the same observations about K_n^-.

8. The new KS statistic was computed for 20 observations. The distribution of K_{10}^+ was used as $F(x)$ when the KS statistic was computed.

9. This idea is erroneous, because all of the observations must be *independent*. There is a relation between the statistics K_n^+ and K_n^- on the same data, so each test should be judged separately. (A high value of one tends to give a low value of the other.) Similarly, the data in Figs. 2 and 5 (which show 15 tests for each generator) does not show 15 independent observations, because the "maximum of 5" test is not independent of the "maximum of 4" test. The three tests of each horizontal row are independent (because they were done on different parts of the sequence) but the five tests in a column are somewhat correlated. The net effect of this is that the 95-percent probability levels, etc., which apply to one test, cannot be legitimately applied to a whole group of tests on the same data. Moral: When testing a random-number generator, we may expect it to "pass" each of several tests, e.g., the frequency test, maximum test, run test, etc.; but an array of data from several different tests should not be considered as a unit since the tests themselves may not be independent. The K_n^+ and K_n^- statistics should be considered as two separate tests; a good source of random numbers will pass both tests.

10. Each Y_s is doubled, and np_s is doubled, so the numerators of (6) are quadrupled while the denominators only double. Hence the new value of V is twice as high as the old one.

11. The empirical distribution function stays the same; the values of K_n^+ and K_n^- are multiplied by $\sqrt{2}$.

12. Let $Z_s = (Y_s - nq_s)/\sqrt{nq_s}$. The value of V is n times

$$\sum_{1 \le s \le k} (q_s - p_s + \sqrt{nq_s}Z_s)^2/p_s,$$

and the latter quantity stays bounded away from zero as n increases (since Z_s is bounded with probability 1). Hence the value of V will increase to a value which is extremely improbable under the p_s assumption.

For the KS test, let $F(x)$ be the assumed distribution, $G(x)$ the actual distribution, and let $h = \max |G(x) - F(x)|$. Take n large enough so that the probability that $|F_n(x) - G(x)| < h/2$ is very small; hence $F_n(x) - F(x)$ will be improbably high under the assumed distribution $F(x)$.

13. (The "max" notation should really be replaced by "sup" since a least upper bound is meant; however, "max" was used in the text to avoid confusing too many readers by the less familiar "sup" notation.) For convenience, let $X_0 = -\infty$, $X_{n+1} = +\infty$. When $X_j \le x < X_{j+1}$, we have $F_n(x) = j/n$, and in this interval $\max (F_n(x) - F(x)) = j/n - F(X_j)$; $\max (F(x) - F_n(x)) = F(X_{j+1}) - j/n$. For $0 \le j \le n$, all real values of x are considered; this proves that

$$K_n^+ = \sqrt{n} \max_{0 \le j \le n} \left(\frac{j}{n} - F(X_j)\right),$$

$$K_n^- = \sqrt{n} \max_{1 \le j \le n+1} \left(F(X_j) - \frac{j-1}{n}\right).$$

These are equivalent to (13), since the extra term under the maximum signs is non-positive and it must be redundant by exercise 7.

14. The logarithm of the left-hand side simplifies to

$$-\sum_{1 \le s \le k} Y_s \ln\left(1 + \frac{Z_s}{\sqrt{np_s}}\right) + \frac{1-k}{2} \ln(2\pi n)$$

$$-\frac{1}{2}\sum_{1 \le s \le k} \ln p_s - \frac{1}{2}\sum_{1 \le s \le k} \ln\left(1 + \frac{Z_s}{\sqrt{np_s}}\right) + O\left(\frac{1}{n}\right)$$

and this further simplifies (upon expanding $\ln (1 + Z_s/\sqrt{np_s})$ and realizing that $\sum_{1 \le s \le k} Z_s \sqrt{np_s} = 0$) to

$$-\frac{1}{2} \sum_{1 \le s \le k} Z_s^2 + \frac{1-k}{2} \ln (2\pi n) - \frac{1}{2} \ln (p_1 \ldots p_k) + O\left(\frac{1}{\sqrt{n}}\right).$$

15. The corresponding Jacobian determinant is easily evaluated by (a) removing the factor r^{n-1} from the determinant, (b) expanding the resulting determinant by the cofactors of the row containing "$\cos \theta_1 - \sin \theta_1 \, 0 \ldots 0$" [each of the cofactor determinants may be evaluated by induction], and (c) using the identity $\sin^2 \theta_1 + \cos^2 \theta_1 = 1$.

16.
$$\int_0^{z\sqrt{2x}+y} \exp\left(-\frac{u^2}{2x} + \cdots\right) du = ye^{-z^2} + O\left(\frac{1}{\sqrt{x}}\right)$$

$$+ \int_0^{z\sqrt{2x}} \exp\left(-\frac{u^2}{2x} + \cdots\right) du.$$

The latter integral is

$$\int_0^{z\sqrt{2x}} e^{-u^2/2x} \, du + \frac{1}{3x^2} \int_0^{z\sqrt{2x}} e^{-u^2/2x} u^3 \, du + O\left(\frac{1}{\sqrt{x}}\right).$$

When all is put together, the final result is

$$\frac{\gamma(x+1, x+z\sqrt{2x}+y)}{\Gamma(x+1)} = \frac{1}{\sqrt{2\pi}} \int_{-\infty}^{z\sqrt{2}} e^{-u^2/2} \, du + \frac{e^{-z^2}}{\sqrt{2\pi x}} (y - \tfrac{2}{3} - \tfrac{2}{3}z^2) + O\left(\frac{1}{x}\right).$$

If we let

$$\frac{1}{\sqrt{2\pi}} \int_{-\infty}^{z\sqrt{2}} e^{-u^2/2} \, du = p, \qquad x+1 = \frac{\nu}{2}, \qquad \gamma\left(\frac{\nu}{2}, \frac{t}{2}\right) \Big/ \Gamma\left(\frac{\nu}{2}\right) = p,$$

and

$$\frac{t}{2} = x + z\sqrt{2x} + y,$$

we can solve for y to obtain $y = \frac{2}{3}(1 + z^2) + O(1/\sqrt{x})$, which is consistent with the above analysis. The solution is therefore $t = \nu + 2\sqrt{\nu z} + \frac{4}{3}z^2 - \frac{2}{3} + O(1/\sqrt{\nu})$.

17. (a) Change of variable, $x_j \leftarrow x_j + t$. (b) Induction on n; by definition,

$$P_{n0}(x - t) = \int_n^x P_{(n-1)0}(x_n - t) \, dx_n.$$

(c) The left-hand side is

$$\int_n^{x+s} dx_n \cdots \int_{k+1}^{x_{k+2}} dx_{k+1} \quad \text{times} \quad \int_t^k dx_k \int_t^{x_k} dx_{k-1} \cdots \int_t^{x_2} dx_1.$$

(d) From (b), (c) we have

$$P_{nk}(x) = \sum_{0 \le r \le k} \frac{(r-t)^r}{r!} \frac{(x+t-r)^{n-r-1}}{(n-r)!} (x+t-n).$$

The numerator in (24) is $P_{n\lfloor t\rfloor}(n)$.

18. If $0 \le X_1 \le \cdots \le X_n \le 1$, let $Z_j = 1 - X_{n+1-j}$. We have $0 \le Z_1 \le \cdots \le Z_n \le 1$; and K_n^+ evaluated for $X_1, \ldots, X_n$ equals K_n^- evaluated for $Z_1, \ldots, Z_n$. This symmetrical relation gives a one-to-one correspondence between sets of equal volume for which K_n^+ and K_n^- fall in a given range.

SECTION 3.3.2

1. The observations for a chi-square test must be independent, and in the second sequence successive observations are manifestly dependent, since the second component of one equals the first component of the next.

2. Form n t-tuples, $(Y_{j_t}, \ldots, Y_{j_t+t-1})$, $0 \le j < n$, and count how many of these equal each possible value. Apply the chi-square test with $k = d^t$ and with probability $1/d^t$ in each case. The number of observations, n, should be at least $5d^t$.

3. The probability that j values are examined, i.e., the probability that U_{j-1} is the nth element of the sequence lying in the range $\alpha \le U_{j-1} < \beta$, is easily seen to be

$$\binom{j-1}{n-1} p^n (1-p)^{j-n},$$

by enumeration of the possible places in which the other $n-1$ occurrences can appear and by evaluating the probability of such a pattern. The generating function is $G(z) = (pz/(1-(1-p)z))^n$, which makes sense since the given distribution is the n-fold convolution of the same thing for $n = 1$. Hence the mean and variance are proportional to n; the number of U's to be examined is now easily found to have the characteristics $(\min n, \text{ave } n/p, \max \infty, \text{dev } \sqrt{n(1-p)}/p)$. A more detailed discussion of this probability distribution when $n = 1$ may be found in the answer to exercise 3.4.1–17; see also the considerably more general results of exercise 2.3.4.2–26.

4. The probability of a gap of length $\ge r$ is the probability that r consecutive U's lie outside the given range, i.e., $(1-p)^r$. The probability of a gap of length exactly r is the above value for length $\ge r$ minus the value for length $\ge (r+1)$.

5. As N goes to infinity, so does n (with probability 1), and so this test is just the same as the gap test described in the text except for the length of the very last gap. And the text's gap test certainly is asymptotic to the chi-square distribution stated, since the length of each gap is clearly independent of the length of the others. [*Notes:* A quite complicated proof of this result is by E. Bofinger and V. J. Bofinger in *Annals of Mathematical Statistics* **32** (1961), 524–534. Their paper is of interest because it discusses several interesting variations of the gap test; they show, for example, that the quantity

$$\sum_{0 \le r \le t} \frac{(Y_r - (Np)p_r)^2}{(Np)p_r}$$

does *not* approach a chi-square distribution, although others had suggested this statistic as a "stronger" test because Np is the expected value of n.

7. 5, 3, 5, 6, 5, 5, 4.

8. See exercise 10, with $w = d$.

9. (Step C4 is changed so that the test is "$q < w$" rather than "$q < d$".) We have

$$p_r = \frac{d(d-1)\cdots(d-w+1)}{d^r} \begin{Bmatrix} r-1 \\ w-1 \end{Bmatrix} \qquad \text{for} \qquad w \le r < t;$$

$$p_t = 1 - \frac{d!}{d^{t-1}} \left(\frac{1}{0!} \begin{Bmatrix} t-1 \\ d \end{Bmatrix} + \cdots + \frac{1}{(d-w)!} \begin{Bmatrix} t-1 \\ w \end{Bmatrix} \right).$$

10. As in exercise 3, we really need consider only the case $n = 1$. The generating function for the probability that a coupon set has length r is, by the previous exercise,

$$G(z) = \frac{d!}{(d-w)!} \sum_{r>0} \begin{Bmatrix} r-1 \\ w-1 \end{Bmatrix} \left(\frac{z}{d} \right)^r = z^w \left(\frac{d-1}{d-z} \right) \cdots \left(\frac{d-w+1}{d-(w-1)z} \right)$$

by Eq. 1.2.9(28). The mean and variance are readily computed using Theorem 1.2.10A and exercise 3.4.1–17. We find that

$$\text{mean}\,(G) = w + \left(\frac{d}{d-1} - 1 \right) + \cdots + \left(\frac{d}{d-w+1} - 1 \right) = d(H_d - H_{d-w}) = \mu;$$

$$\text{var}\,(G) = d^2(H_d^{(2)} - H_{d-w}^{(2)}) - d(H_d - H_{d-w}) = \sigma^2.$$

The number of U's examined, as the search for a coupon set is repeated n times, therefore has the characteristics (min wn, ave μn, max ∞, dev $\sigma\sqrt{n}$).

11. $|1|2|9\ 8\ 5\ 3|6|7\ 0|4|$.

12. Algorithm R (*Data for run test*).

R1. [Initialize.] Set $j \leftarrow -1$, and set COUNT[1] $\leftarrow$ COUNT[2] $\leftarrow \cdots \leftarrow$ COUNT[6] $\leftarrow$ 0. Also set U_n to some value which is *less than* U_{n-1}, for convenience in terminating the algorithm.

R2. [Set r zero.] Set $r \leftarrow 0$.

R3. [Is $U_j > U_{j+1}$?] Increase r and j by 1. If $U_j < U_{j+1}$, repeat this step.

R4. [Record the length.] If $r \ge 6$, increase COUNT[6] by one, otherwise increase COUNT[r] by one.

R5. [Done?] If $j < n - 1$, return to step R2. ∎

13. There are $(p+q+1)\binom{p+q}{p}$ ways to have $U_{i-1} \gtrless U_i < \cdots < U_{i+p-1} \gtrless U_{i+p} < \cdots < U_{i+p+q-1}$; subtract $\binom{p+q+1}{p+1}$ of these in which $U_{i-1} < U_i$, and subtract $\binom{p+q+1}{q+1}$ for those in which $U_{i+p-1} < U_{i+p}$; then add in 1 for the case that both $U_{i-1} < U_i$ and $U_{i+p-1} < U_{i+p}$, since this case has been subtracted out twice. (This is a special case of the "inclusion-exclusion" principle, which is explained further in Section 1.3.3.)

14. A run of length r occurs with probability $1/r! - 1/(r+1)!$.

15. This is always true of $F(X)$ when F is continuous and X has distribution F; see Section 3.3.1.C.

16. (a) $Z_{jt} = \max\,(Z_{j(t-1)}, Z_{(j+1)(t-1)})$. If the $Z_{j(t-1)}$ are stored in memory, it is therefore a simple matter to transform this array into the set of Z_{jt} with no auxiliary storage required. (b) With his "improvement," each of the V's should indeed have the stated distribution, but the observations are no longer *independent*. In fact, when U_j is a relatively large value, all of $Z_{jt}, Z_{(j-1)t}, \ldots, Z_{(j-t+1)t}$ will be equal to U_j; so we almost have the effect of repeating the same data t times (and that would multiply V by t, cf. exercise 3.3.1–10).

17. (b) By Lagrange's identity, the difference is $\sum_{1 \le k < j \le n} (X'_k Y'_j - X'_j Y'_k)^2$ and this is certainly positive. (c) Therefore if $D^2 = N^2$, we must have $X'_k Y'_j - X'_j Y'_k = 0$, for all pairs j, k. This means the matrix

$$\begin{pmatrix} X'_1\ X'_2 \cdots\ X'_n \\ Y'_1\ Y'_2 \cdots\ Y'_n \end{pmatrix}$$

has rank <2, so its rows are linearly dependent. (A more elementary proof can be given, using the fact that $X'_1 Y'_j - X'_j Y'_1 = 0$, $2 \le j \le n$, implies the existence of constants a, b such that $aX'_j + bY'_j = 0$ for all j, provided that X'_1 and Y'_1 are not both zero; the latter case can be avoided by a suitable renumbering.)

18. (a) The numerator is $-(U_0 - U_1)^2$, the denominator is $(U_0 - U_1)^2$. (b) The numerator is $-(U_0^2 + U_1^2 + U_2^2 - U_0U_1 - U_1U_2 - U_2U_0)$; the denominator is $2(U_0^2 + \cdots - U_2U_0)$. (c) The denominator always equals $\sum_{0 \le j < k < n} (U_j - U_k)^2$, by exercise 1.2.3–30 or 1.2.3–31.

21. The C-table contains 0, 0, 1, 2, 2, 4, 6, 7, 3, 2; $f = 992822$.

22. The C-table contains 0, 0, 0, 0, 0, 0, 1, 3, 3, 4; the permutation is

$$(5,\ 6,\ 1,\ 8,\ 9,\ 4,\ 0,\ 2,\ 7,\ 3).$$

SECTION 3.3.3

1. If m and n are integers, the number of integers x with $m \le x < n$ is $n - m$; so (7) is just the sum of the number of solutions appearing in each of the disjoint cases in (6).

2. See exercise 1.2.4–38, and 1.2.4–39 (a), (b), (g).

3. $f(x) = \sum_{n \ge 1} (-\sin 2\pi nx)/n$, which converges for all x. (The representation in Eq. (24) may be considered a "finite" Fourier series, for the case when x is rational. Any periodic function has such a "finite" Fourier series, as discussed in Section 3.3.4.)

4. $d = 2^9 \cdot 5$. But the probability that $X_{n+1} < X_n$ is $\frac{1}{2} + \epsilon$, where

$$|\epsilon| < d/2 \cdot 10^{10} = 1/(4 \cdot 5^9),$$

so it is still a respectable generator from the standpoint of Theorem P.

5. An intermediate result:

$$\sum_{0 \le x < m} \frac{x}{m}\,\frac{s(x)}{m} = \frac{1}{12}\,\sigma(a, m, c) + \frac{m}{4} - \frac{c}{2m} - \frac{x'}{2m}.$$

6. (a) Use induction and the formula

$$\left(\!\!\left(\frac{hj+c}{k}\right)\!\!\right) - \left(\!\!\left(\frac{hj+c-1}{k}\right)\!\!\right) = \frac{1}{k} - \frac{1}{2}\delta\left(\frac{hj+c}{k}\right) - \frac{1}{2}\delta\left(\frac{hj+c-1}{k}\right).$$

(b)
$$\frac{k'c}{h} = -\left(\frac{h'c}{k} - \frac{c}{hk}\right).$$

Now $h'c/k$ is not an integer, and $|c/hk| < 1/k$, so

$$\left(\!\!\left(\frac{h'c}{k} - \frac{c}{hk}\right)\!\!\right) = \left(\!\!\left(\frac{h'c}{k}\right)\!\!\right) - \frac{c}{hk}.$$

More generally, if $0 \le c < h$,

$$\left(\!\!\left(\frac{h'c}{k}\right)\!\!\right) + \left(\!\!\left(\frac{k'c}{h}\right)\!\!\right) = \frac{c}{hk} - \frac{1}{2}\delta\left(\frac{c}{k}\right).$$

(c) $12 \displaystyle\sum_{0 \le j < c} \frac{j}{hk} + 6\frac{c}{hk} = 6\frac{c^2}{hk}.$

7. Take $m = h$, $n = k$, $k = 2$ in the second formula of exercise 1.2.4–45:

$$\sum_{0 < j < k}\left(\frac{hj}{k} - \left(\!\!\left(\frac{hj}{k}\right)\!\!\right) + \frac{1}{2}\right)\left(\frac{hj}{k} - \left(\!\!\left(\frac{hj}{k}\right)\!\!\right) - \frac{1}{2}\right) + 2\sum_{0 < j < h}\left(\frac{kj}{h} - \left(\!\!\left(\frac{kj}{h}\right)\!\!\right) + \frac{1}{2}\right)j$$
$$= kh(h-1).$$

The sums on the left simplify, and by standard manipulations we get

$$h^2 k - hk - \frac{h}{2} + \frac{h^2}{6k} + \frac{k}{12} + \frac{1}{4} - \frac{h}{6}\sigma(h, k, 0) - \frac{h}{6}\sigma(k, h, 0) + \frac{1}{12}\sigma(1, k, 0) = h^2 k - hk.$$

Since $\sigma(1, k, 0) = (k-1)(k-2)/k$, this reduces to the reciprocity law.

8. See *Duke Mathematical Journal* **21** (1954), 391–397.

10. Obviously [cf. (10)] $\sigma(-h, k, c) = -\sigma(h, k, -c)$. Replace j by $k - j$ in definition (16), to deduce $\sigma(h, k, c) = \sigma(h, k, -c)$.

11. Let $c_1 = \lfloor c/d \rfloor$, $c_0 = c \bmod d$. Then

$$\sigma(h, k, c) = 12 \sum_{\substack{0 \le i < d \\ 0 \le j < k/d}} \left(\!\!\left(\frac{ik/d + j}{k}\right)\!\!\right)\left(\!\!\left(\frac{h(ik/d + j) + c}{k}\right)\!\!\right)$$

$$= 12 \sum_{0 \le j < k/d} \left(\!\!\left(\frac{hj + c}{k}\right)\!\!\right) \sum_{0 \le i < d} \left(\!\!\left(\frac{i}{d} + \frac{j}{k}\right)\!\!\right)$$

$$= 12 \sum_{0 \le j < k/d} \left(\!\!\left(\!\!\left(\frac{(h/d)j + c_1}{k/d}\right)\!\!\right) + \frac{c_0}{k} - f(j)\right)\left(\!\!\left(\frac{j}{k/d}\right)\!\!\right)$$

where $f(j) = \frac{1}{2}$ if $c_0 \neq 0$ and $hj + c_1 d \equiv 0$ (modulo k); $f(j) = 0$ otherwise. (We have used Eq. (12) to do the sum on i.) If $c_0 \neq 0$, $f(j) = \frac{1}{2}$ only when

$$j = (-c_1 h') \bmod (k/d).$$

12. Since $\left(\left(\frac{hj+c}{k}\right)\right)$ runs through the same values as $\left(\left(\frac{j}{k}\right)\right)$ in some order, Schwarz's inequality (exercise 1.2.3–30) implies $\sigma(h, k, c)^2 \leq \sigma(1, k, 0)^2$; and $\sigma(1, k, 0)$ may be summed directly, it is $(k-1)(k-2)/k$.

13. $(2^{68} + 5)/(2^{70} - 1)$.

14. $(5 \cdot 2^{36} - 9 \cdot 2^{18} - 1 - 3/(2^{18} + 1))/(2^{71} - 2) \approx 5/2^{35}$. An extremely satisfactory value, in spite of the local nonrandomness!

15. $1 - 6\dfrac{cm}{m^2 - 1} + \dfrac{6c^2}{m^2 - 1} = \dfrac{m^2}{m^2 - 1}\left(1 - 6\dfrac{c}{m} + 6\left(\dfrac{c}{m}\right)^2\right) - \dfrac{1}{m^2 - 1}$.

16. When $a = 1$, the probability in Theorem P is $c/2m$, so we can take $c/m \approx \frac{1}{2}$ to satisfy that test. Similarly, we can take c/m as in Eq. (41), to satisfy the serial correlation test. But we can't satisfy both at once; the best choice of c (which minimizes the maximum of the deviation of the $X_{n+1} < X_n$ probability from $\frac{1}{2}$, and the deviation of C from 0) is $(7 - \sqrt{13})/12$, which results in a probability of about 0.38 and a C of about -0.22.

17. Choose a so that $ac \equiv 1$ or 3 (modulo m). This is always possible at least when $m = 2^e$, with a little finagling.

18. (a) $S(h, k, c, z) = S(h \bmod k, k, c \bmod k, z \bmod k)$.   (We assume that $z \geq 0$.) Also

$$S(-h, k, -c, z) = -S(h, k, c, z) = S(h, -k, c, z).$$

(b) Let $c \equiv c'h$ modulo k), where $0 \leq c' < k$. Then

$$S(h, k, c, z) = \sum_{0 \leq j < z}\left(\left(\frac{h(j+c')}{k}\right)\right) = S(h, k, 0, z + c') - S(h, k, 0, c').$$

It therefore suffices to consider the case $c = 0$.   (c) Let $hh' + kk' = 1$.   Then

$$S(k', h, 0, z) + S(h', k, 0, z) = \frac{z(z-1)}{2hk} - \frac{1}{2}\left\lfloor\frac{z-1}{k}\right\rfloor,$$

if $0 \leq z < h$, by exercise 6. This "reciprocity law" has a different form than the one considered in the text. It may also be written as

$$S(h, k, 0, z) + S(k', h', 0, z) = \frac{z(z-1)}{2h'k} - \frac{1}{2}\left\lfloor\frac{z-1}{h'}\right\rfloor, \qquad 0 \leq z < k;$$

and we can ensure that $-k/2 < h' \leq k/2$, so the number of reduction steps is at worst $\log_2 k$.

19. Proceeding as in the proof of Theorem P, we find $\gamma \leq s(x) < \delta$ if and only if $\lceil(km + \gamma - c)/a\rceil \leq x < \lceil(km + \delta - c)/a\rceil$ for some k. We also require that

$\alpha \leq x < \beta$; and the total number of such x's, divided by m, is

$$\frac{1}{m}\left(\sum_{k_0 \leq k < k_1}\left(\left\lceil\frac{km+\delta-c}{a}\right\rceil - \left\lceil\frac{km+\gamma-c}{a}\right\rceil\right) + \max\left(0, \beta - \left\lceil\frac{k_1 m+\gamma-c}{a}\right\rceil\right)\right.$$

$$\left. - \max\left(0, \alpha - \left\lceil\frac{k_0 m+\gamma-c}{a}\right\rceil\right)\right),$$

where $k_0 = \lceil(a\alpha + c - \delta)/m\rceil$, $k_1 = \lceil(a\beta + c - \delta)/m\rceil$. The two "max" terms are each bounded by $\lfloor(\delta-\gamma)/a\rfloor$, and each term in the sum is $(\delta-\gamma)/a + \epsilon$, $|\epsilon| \leq 1$. In applications of the serial test, we have $\beta - \alpha = \delta - \gamma = m/d$, and the formula has the value $1/d^2 + \epsilon$, $|\epsilon| < 1/ad + 1/m + 1/ad + a/md$. Thus if $a \approx \sqrt{m}$, the serial test has a reasonable outcome taken over the entire period. (Even the serial test for the locally nonrandom generator in exercise 14 will be satisfactory over the full period.) The first sum in the above formula for the exact value of the probability can be rewritten as $S(m, a, \delta + k_0 m - c, k_1 - k_0) - S(m, a, \gamma + k_0 m - c, k_1 - k_0) + d$, where the S-function is the subject of exercise 18, and

$$d = \frac{1}{2}\sum_{k_0 \leq k < k_1}\left(\delta\left(\frac{km+\gamma-c}{a}\right) - \delta\left(\frac{km+\delta-c}{a}\right)\right); \qquad \text{hence} \qquad |d| \leq \frac{1}{2}.$$

20. $$\sigma(h, k, c) + \frac{3(k-1)}{k} = \frac{12}{k}\sum_{0 < j < k}\frac{\omega^{-cj}}{(\omega^{-hj}-1)(\omega^j-1)} + \frac{6}{k}(c \bmod k)$$

$$- 6\left(\left(\frac{h'c}{k}\right)\right), \qquad \text{if} \qquad hh' \equiv 1 \pmod{k}.$$

21. $$\int_0^1 x\,dx = \frac{1}{2}; \qquad \int_0^1 x^2\,dx = \frac{1}{3};$$

$$s_n = \int_{x_n}^{x_{n+1}} x\{ax+\theta\}\,dx = \frac{1}{a^2}\left(\frac{1}{3} - \frac{\theta}{2} + \frac{n}{2}\right), \qquad \text{if} \qquad x_n = \frac{n-\theta}{a};$$

$$s = \int_0^1 x\{ax+\theta\}\,dx = s_0 + s_1 + \cdots + s_{a-1} + \int_{-\theta/a}^0 (ax+\theta)\,dx$$

$$= \frac{1}{3a} - \frac{\theta}{2a} + \frac{a-1}{4a} + \frac{\theta^2}{2a}.$$

Therefore

$$C = \frac{s - \frac{1}{4}}{\frac{1}{3} - \frac{1}{4}} = \frac{1}{a}(1 - 6\theta + 6\theta^2).$$

22. Let $[u, v)$ denote the set $\{x \mid u \leq x < v\}$. We have $s(x) < x$ in the disjoint intervals

$$\left[\frac{1-\theta}{a}, \frac{1-\theta}{a-1}\right), \left[\frac{2-\theta}{a}, \frac{2-\theta}{a-1}\right), \cdots, \left[\frac{a-\theta}{a}, 1\right),$$

which have total length

$$1 + \sum_{0 < j \le (a-1)} \left(\frac{j-\theta}{a-1}\right) - \sum_{0 < j \le a} \left(\frac{j-\theta}{a}\right) = 1 + \frac{a}{2} - \theta - \frac{a+1}{2} + \theta = \frac{1}{2}.$$

23. We have $s(s(x)) < s(x) < x$ when x is in

$$\left[\frac{k-\theta}{a}, \frac{k-\theta}{a-1}\right)$$

and $ax + \theta - k$ is in

$$\left[\frac{j-\theta}{a}, \frac{j-\theta}{a-1}\right),$$

for $0 < j \le k < a$, or when x is in

$$\left[\frac{a-\theta}{a}, 1\right)$$

and $ax + \theta - a$ is in either

$$\left[\frac{j-\theta}{a}, \frac{j-\theta}{a-1}\right),$$

for $0 < j \le \lfloor a\theta \rfloor$, or

$$\left[\frac{\lfloor a\theta \rfloor + 1 - \theta}{a}, \theta\right).$$

The desired probability is

$$\sum_{0 < j \le k < a} \frac{j-\theta}{a^2(a-1)} + \sum_{0 < j \le \lfloor a\theta \rfloor} \frac{j-\theta}{a^2(a-1)} + \frac{1}{a^2} \max\left(0, \{a\theta\} + \theta - 1\right)$$

$$= \frac{1}{6} + \frac{1}{6a} - \frac{\theta}{2a} + \frac{1}{a^2}\left(\frac{\lfloor a\theta \rfloor(\lfloor a\theta \rfloor + 1 - 2\theta)}{2(a-1)} + \max\left(0, \{a\theta\} + \theta - 1\right)\right)$$

$$= \frac{1}{6} + \frac{1}{6a}(1 - 3\theta + 3\theta^2) + O\left(\frac{1}{a^2}\right),$$

for large a. Note that $1 - 3\theta + 3\theta^2 \ge \frac{1}{4}$, so θ can't be chosen to make the probability come out right.

24. (a) Proceed as in the previous exercise; the sum of the interval lengths is

$$\sum_{0 < j_1 \le \cdots \le j_{t-1} < a} \frac{j_1}{a^{t-1}(a-1)} = \frac{1}{a^{t-1}(a-1)} \binom{a+t-2}{t}.$$

(b) Let p_k be the probability of a run of length $\ge k$. The average is

$$\sum_{k \ge 1} p_k = \sum_{k \ge 1} \binom{a+k-2}{k} \frac{1}{a^{k-1}(a-1)} = \left(\frac{a}{a-1}\right)^a - \frac{a}{a-1}.$$

The value for a truly random sequence is $e - 1$; our value is $e - 1 + (e/2 - 1)/a + O(1/a^2)$. *Note:* The same result holds for an ascending run, since $U_n > U_{n+1}$ if and only if $1 - U_n < 1 - U_{n+1}$. This would lead us to suspect that run lengths in linear congruential sequences tend to be slightly longer than normal, so the run test should be applied to such generators.

25. Let $k_0 = \lceil a\alpha + \theta - \delta \rceil$, $k_1 = \lceil a\beta + \theta - \delta \rceil$. x must be in the intervals $[(k + \gamma - \theta)/a, (k + \delta - \theta)/a)$ for some k, and also in the interval $[\alpha, \beta)$. With due regard to boundary conditions, we get the probability

$$(k_1 - k_0)(\delta - \gamma)/a + \max\left(0, \beta - (k_1 + \gamma - \theta)/a\right) - \max\left(0, \alpha - (k_0 + \gamma - \theta)/a\right).$$

This is $(\beta - \alpha)(\delta - \gamma) + \epsilon$, where $|\epsilon| < 2(\delta - \gamma)/a$.

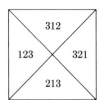

Fig. A–1. Permutation regions for Fibonacci generator.

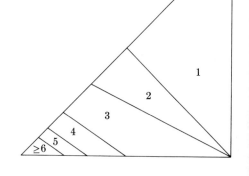

Fig. A–2. Run-length regions for Fibonacci generator.

26. See Fig. A-1; the orderings $U_1 < U_3 < U_2$ and $U_2 < U_3 < U_1$ are impossible; the other four each have probability $\frac{1}{4}$.

27. $U_n = (F_{n-1}U_0 + F_n U_1) \bmod 1$. We need $F_{k-1}U_0 + F_k U_1 < 1$ and $F_k U_0 + F_{k+1}U_1 > 1$. The half-unit-square in which $U_0 > U_1$ is broken up as shown in Fig. A-2, with various values of k indicated. The probability for a run of length k is $\frac{1}{2}$ if $k = 1$; $1/F_{k-1}F_{k+1} - 1/F_k F_{k+2}$, if $k > 1$. The corresponding probabilities for a random sequence are $2k/(k + 1)! - 2(k + 1)/(k + 2)!$; the following table compares the first few values.

k:	1	2	3	4	5
Probability in Fibonacci case:	$\frac{1}{2}$	$\frac{1}{3}$	$\frac{1}{10}$	$\frac{1}{24}$	$\frac{1}{65}$
Probability in random case:	$\frac{1}{3}$	$\frac{5}{12}$	$\frac{11}{60}$	$\frac{19}{360}$	$\frac{29}{2520}$

28. Fig. A-3 shows the various regions in the general case. The "213" region means $U_2 < U_1 < U_3$, if U_1 and U_2 are chosen at random; the "321" region means $U_3 < U_2 < U_1$, etc. The probabilities for 123 and 321 are $\frac{1}{4} - \alpha/2 + \alpha^2/2$; the probabilities for all other cases are $\frac{1}{8} + \alpha/4 - \alpha^2/4$. To have all equal to $\frac{1}{6}$, we must have $\frac{1}{4} - \alpha/2 + \alpha^2/2 = \frac{1}{8} + \alpha/4 - \alpha^2/4$, that is, $1 - 6\alpha + 6\alpha^2 = 0$. It is interesting to note that this is exactly the equation for c/m which minimizes the serial

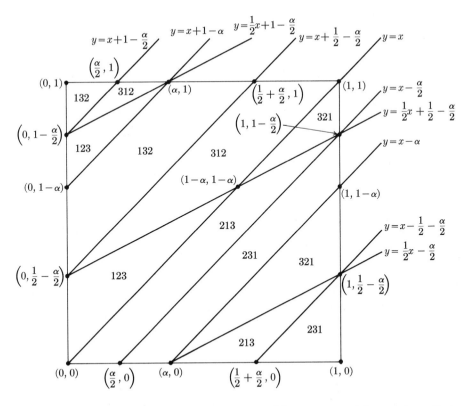

Fig. A–3. Permutation regions for a generator with a potency of 2; $\alpha = (a - 1)c/m$.

correlation [cf. Eq. (40), (41)]! [The connection is not very strong, since this application requires $(a - 1)c/m$ to have this value, not c/m.] *Notes:* The result of this exercise gives a simple proof of a theorem due to J. Franklin (*Math. Comp.* **17** (1963), 43–47, Theorem 13); other results of that paper are related to some of the exercises above: Theorem 2 (pp. 32–33) is exercise 22, Theorem 5 (pp. 34–35) is a special case of exercise 23.

SECTION 3.3.4

1. $\displaystyle\sum_{0 \le s_1, \ldots, s_n < m} \; \sum_{0 \le u_1, \ldots, u_n < m} \exp\left(\frac{2\pi i}{m}\bigl(s_1(t_1 - u_1) + \cdots + s_n(t_n - u_n)\bigr)\right) F(u_1, \ldots, u_n)$

$$= \sum_{0 \le u_1, \ldots, u_n < m} F(u_1, \ldots, u_n)\left(\sum_{0 \le s_1 < m} \omega^{s_1(t_1 - u_1)}\right) \cdots \left(\sum_{0 \le s_n < m} \omega^{s_n(t_n - u_n)}\right)$$

$$= m^n F(t_1, \ldots, t_n),$$

where $\omega = \exp(2\pi i/m)$.

2. Like exercise 1.

3. (a) $\omega^s f(s)$, $\omega = \exp(2\pi i/m)$.   (b) $f(s)g(s)$.

4.
$$\frac{m-1}{2m} f(0,0) + \frac{1}{m} \sum_{0<s<m} \frac{f(s,0) - f(0,s) + f(m-s,s)}{\omega^s - 1},$$

where $\omega = \exp(2\pi i/m)$. (This is the sum of $F(t_1, t_2)$ over the range $0 \le t_2 < t_1 < m$.)

5. $m \sum_{0 \le t_1, t_2 < m} t_1 t_2 F(t_1, t_2) = (m-1)\binom{m}{2} f(0,0) + \binom{m}{2} \sum_{0<s<m} \frac{f(0,s) + f(s,0)}{\omega^s - 1}$

$$+ m \sum_{0<s_1,s_2<m} \frac{f(s_1, s_2)}{(\omega^{s_1} - 1)(\omega^{s_2} - 1)}.$$

6. The formula in exercise 4 becomes

$$\frac{m-1}{2m} + \frac{1}{m} \sum_{0<s<m} \delta\left(\frac{(m-s) + as}{m}\right) \omega^{-sc} \bigg/ (\omega^s - 1).$$

The latter sum requires that $(a-1)s \equiv 0 \pmod{m}$, so if $d = \gcd(a-1, m)$ and $\zeta = \exp(2\pi i/d)$, it becomes

$$\sum_{0 \le j < d} \frac{\zeta^{jc}}{\zeta^j - 1};$$

by Eq. 3.3.3(23) this is $c \bmod d - \frac{1}{2}(d-1)$.

8. $m(m-1)^2/2 + m \sum_{0<s<m} \omega^{-cs}/(\omega^{-as} - 1)(\omega^s - 1)$.   This formula and the result of exercise 3.3.3–20 leads directly into Dedekind sums and the reciprocity law of the previous section.

9. Following the hint, let the first column of U be a nonzero integer vector y for which $Ay \cdot Ay$ comes within ϵ of its greatest lower bound. We clearly may assume the entries of y are relatively prime. Any such column vector can be completed to an integer matrix of determinant 1 (prove this, for example, by induction on the magnitude of the smallest nonzero entry in the column). Thus U may be constructed as stated, and we may assume without loss of generality that A has been replaced by AU for some such matrix U. Let $Q = A^T A$, and write $x^T Q x$ as

$$Q_{11}\left(x_1 + \frac{1}{Q_{11}}(Q_{12}x_2 + \cdots + Q_{1n}x_n)\right)^2 + g(x_2, \ldots, x_n).$$

Here g is a positive definite quadratic form corresponding to an $(n-1) \times (n-1)$ matrix Q' where $\det Q' = (\det Q)/Q_{11}$. Hence by induction (the result being clear for $n = 1$) there are integers $x_2, \ldots, x_n$ which make

$$g(x_2, \ldots, x_n) \le (\tfrac{4}{3})^{(n-2)/2}((\det Q)/Q_{11})^{1/(n-1)}.$$

Now choose x_1 so that

$$\left(x_1 + \frac{1}{Q_{11}}(Q_{12}x_2 + \cdots + Q_{1n}x_n)\right)^2 \le \tfrac{1}{4}.$$

It follows that for this choice of x we have

$$Q_{11} - \epsilon \leq x^T Q x \leq Q_{11}/4 + (\tfrac{4}{3})^{(n-2)/2}((\det Q)/Q_{11})^{1/(n-1)}.$$

The desired inequality now follows readily. *Notes:* When $n = 2$, the inequality is best possible, since we may have $A_{11} = 1$, $A_{12} = \tfrac{1}{2}$, $A_{21} = 0$, $A_{22} = \sqrt{\tfrac{3}{4}}$. A fundamental theorem due to Minkowski ("An n-dimensional convex set symmetric about the origin with volume $\geq 2^n$ contains a nonzero integer point") says that $C_n \leq 2^n$; this is stronger than Hermite's theorem when n is large, but still it is far from best possible. For further details, see the book by Cassels quoted in the text; Hermite's theorem is discussed on pages 30ff.

10. $\lim \dfrac{1}{N} \displaystyle\sum_{0 \leq k < N} \exp\left(2\pi i \left(s_1 \dfrac{X_k}{m} + \cdots + s_n \dfrac{X_{k+n-1}}{m}\right)\right) \cdot$

$$\exp\left(2\pi i(s_1 \lambda V_k + \cdots + s_n \lambda V_{k+n-1})\right)$$

$$= f(s_1, \ldots, s_n) \prod_{1 \leq j \leq n} \lim \frac{1}{N} \sum_{0 \leq k < N} \exp\left(2\pi i s_j \lambda V_{k+j-1}\right),$$

since by independence of the V's from the X's, the average value is multiplicative. (Cf. exercise 3(b), which is a special case of this multiplicative property.) The sum remaining is $\int_0^1 \exp(2\pi i s_j \lambda v)\, dv$ (cf. Theorem 3.5B); and this is $(\exp(2\pi i s_j \lambda) - 1)/2\pi i s_j \lambda$. The absolute value of this quantity is $|\sin(\pi \lambda s_j)/\pi \lambda s_j|$. For $\lambda \to 0$ the new Fourier coefficients are in absolute value

$$|f(s_1, \ldots, s_n)|(1 - \tfrac{1}{6}(s_1^2 + \cdots + s_n^2)\lambda^2 + O(\lambda^4));$$

for λ increasing so that $\lambda\sqrt{s_1^2 + \cdots + s_n^2}$ is $\gg 1$, the new Fourier coefficients approach zero.

11. Since $F(t_1, \ldots, t_n)$ is uniquely determined by $f(s_1, \ldots, s_n)$ and conversely, there can be no genuine cancelling between waves with different wavelength components. But certain statistical quantities calculated from $f(s_1, \ldots, s_n)$, such as the serial correlation coefficient, can be significantly affected by the choice of c. (It would appear that changing c will affect some statistics favorably at the expense of others.)

12. $B_k' \cdot B_k' = R_{kk} + 2\sum_{j \neq k} c_j R_{kj} + \sum_{i \neq k}\sum_{j \neq k} c_i c_j R_{ij}$. The partial derivative with respect to c_j is twice the left-hand side of (28). If the minimum can be achieved, these partial derivatives must all vanish.

13. (a) The inequality (21) shows that only finitely many vectors x give a value smaller than $\min_{1 \leq j \leq n} Q_{jj}$. (b) Let $A_{11} = \phi$, $A_{12} = -1$, $A_{21} = A_{22} = 0$. The vector $(x_1, x_2) = (F_r, F_{r+1})$ gives the value ϕ^{2r}, by exercise 1.2.8–28, and this gets arbitrarily close to zero which cannot be attained.

14.

							Search limits: (step S3)	Transform numbers: (step S4)

Matrix Q:			Matrix R:			Search limits: (step S3)	Transform numbers: (step S4)
10000	−4100	1900	2043	4100	−1900		
−4100	1682	−779	4100	10000	0		
1900	−779	362	−1900	0	10000	(8 19 19)	(5 −2 *)

(continued)

		Search limits: (step S3)	Transform numbers: (step S4)

Matrix Q: Matrix R:

$$\begin{pmatrix} 50 & -25 & 90 \\ -25 & 14 & -55 \\ 90 & -55 & 362 \end{pmatrix} \quad \begin{pmatrix} 2043 & 4100 & 115 \\ 4100 & 10000 & 500 \\ 115 & 500 & 75 \end{pmatrix} \quad (1 \quad 3 \quad 0) \quad (-2 \quad * \quad -4)$$

$$\begin{pmatrix} 6 & 3 & -8 \\ 3 & 14 & 1 \\ -8 & 1 & 146 \end{pmatrix} \quad \begin{pmatrix} 2043 & -446 & 115 \\ -446 & 812 & -30 \\ 115 & -30 & 75 \end{pmatrix} \quad (1 \quad 0 \quad 0). \quad q = 6.$$

15. The first step of the iteration would produce no change, then k would be set to n and the process would proceed exactly as it does now. (*Note:* This follows since step S1 sets X_j to a^{j-1} reduced (modulo m) so that $|X_j| \leq m/2$. If we had not done this and if k is initially 1, the effect would be that the first iteration brings the matrices to the form now used in step S1.)

16. Every time c_j is nonzero in step S4, Q_{jj} decreases. (See the equation in the text just preceding (27).)

17. In step S1, also set $A[1, 1] \leftarrow m$; $A[1, j] \leftarrow -X[j]$, $A[j, j] \leftarrow 1$ for $1 < j \leq n$; set other entries of A zero. In step S2, when q changes, also set $s_i \leftarrow A[i, j]$, $1 \leq i \leq n$. In step S4, also set $A[r, j] \leftarrow A[r, j] - c[j]A[r, k]$, for $1 \leq r \leq n$. And in step S8 if q changes, also set $s_i \leftarrow \sum_{1 \leq j \leq n} A[i, j]X[j]$. For example, in the solution to exercise 14, the A-matrix would have the successive values

$$\begin{pmatrix} 100 & -41 & 19 \\ 0 & 1 & 0 \\ 0 & 0 & 1 \end{pmatrix}, \quad \begin{pmatrix} 5 & -3 & 19 \\ 0 & 1 & 0 \\ -5 & 2 & 1 \end{pmatrix}, \quad \begin{pmatrix} -1 & -3 & 7 \\ 2 & 1 & 4 \\ -1 & 2 & 9 \end{pmatrix}.$$

18. Let

$$A = \begin{pmatrix} 1 + 2y & 1 - y & 1 - y \\ 1 - y & 1 + 2y & 1 - y \\ 1 - y & 1 - y & 1 + 2y \end{pmatrix}, \qquad \text{for arbitrarily large } y.$$

19. $f(1, 1, -1) = 1.$

20.
$$\frac{4}{m} \sum_{0 \leq k < m/4} \exp\left(\frac{2\pi i}{m}(s_1 X_k + \cdots + s_n X_{k+n-1})\right)$$

$$= \frac{4}{m} \sum_{0 \leq k < m/4} \exp\left(\frac{2\pi i}{m} X_k s(a)\right)$$

$$= \frac{4}{m} \sum_{0 \leq k < m/4} \exp\left(\frac{2\pi i}{m}(4k + 1)s(a)\right)$$

$$= \delta\left(\frac{4s(a)}{m}\right) \exp\left(\frac{2\pi i s(a)}{m}\right).$$

Change m to $\frac{1}{4}m$ in both Eqs. (11) and (15) to get a spectral test for this case analogous to the spectral test for the case of maximum period.

21. After a calculation similar to that in exercise 20, using the fact that the values which occur in the period are precisely the values of the forms $8k + 1$ and $8k + 3$, we find that

$$f(s_1, \ldots, s_n) = \delta\left(\frac{8s(a)}{m}\right) \exp\left(\frac{4\pi i s(a)}{m}\right) \cos\frac{2\pi s(a)}{m}.$$

So in this case $f(s_1, \ldots, s_n)$ can be nonzero when $s(a) \equiv 0$ (modulo $m/8$) while $s(a) \not\equiv 0$ (modulo $m/4$); but in this case, the magnitude of $f(s_1, \ldots, s_n)$ is $1/\sqrt{2}$, not 1. Since the sequence $((-1)^k X_k) \bmod m$ is a sequence of the type in exercise 20, we would expect the randomness of the sequences considered in exercises 20 and 21 to be essentially the same. If we were to say the value of $|f(s_1, \ldots, s_n)|^2$ is to be divided by $\sqrt{s_1^2 + \cdots + s_n^2}$ when we are assessing the "badness" of a nonzero $f(s_1, \ldots, s_n)$, we find the definitions of ν_n and C_n become identical for exercises 20 and 21. The "right" way to define ν_n in this case is not clear, since $f(s_1, \ldots, s_n)$ has different intensities in different bands of the spectrum.

24. If $\omega = e^{2\pi i/m}$, $f(s_1, \ldots, s_n) = g(\omega^{s_1}, \ldots, \omega^{s_n})$. Thus, finite Fourier transforms may be regarded as a special case of generating functions.

25. Let $P = 2^{e-2}$; then $f(s_1, 0)$ is zero unless s_1 is a multiple of P. Assume that $s_1 s_2 \neq 0$; we have

$$f(s_1, s_2) = \frac{1}{P} \sum_{0 \le k < P} \exp\left(\frac{-2\pi i}{4P}(s_1 X_k + s_2 X_{k+1})\right)$$

$$= \frac{1}{P} \sum_{0 \le k < P} \exp\left(\frac{-2\pi i}{4P}(A + 4Bk + 4Ck^2)\right),$$

where $A = 2s_1 + 6s_2$, $B = s_1 + 5s_2$, $C = 4s_2$, since the numbers X_k which occur in the period are $2, 6, \ldots, 4P - 2$ in some order. It follows that

$$|f(s_1, s_2)|^2 = \frac{1}{P^2} \sum_{0 \le j,k < P} \exp\left(\frac{-2\pi i}{P}(B(j - k) + C(j^2 - k^2))\right)$$

$$= \frac{1}{P^2} \sum_{0 \le j,k < P} \exp\left(\frac{-2\pi i}{P}(Bj + C(j^2 + 2jk))\right),$$

replacing j by $j + k$. Summing on k gives zero unless $2Cj$ is a multiple of P. Let $s_1 = 2^{r_1} \cdot q_1$, $s_2 = 2^{r_2} \cdot q_2$, where q_1 and q_2 are odd; then the sum is taken over the 2^{r_2+3} values of j, which are multiples of 2^{e-r_2-5}. If $2r_2 \le e - 8$, the j^2 terms may be removed since j^2 is a multiple of P, and we have $f(s_1, s_2) = 0$ in this case except when $r_1 = r_2$, $(q_1 + q_2) \bmod 8 = 4$; in the latter case $|f(s_1, s_2)|^2 = 2^{r_2+5-e}$. Thus when e is reasonably large, the nonvanishing waves of highest wavelength occur for $(s_1, s_2) = (3, 1), (1, 3), (-1, -3), (-3, -1), (1, -5), (5, -1), (-1, 5), (-5, 1)$, having amplitude $\sqrt{32/2^e}$. (The amplitude for $(s_1, s_2) = (2, 6)$ is $\sqrt{64/2^e}$, when $e \ge 10$.) In general, if we replace the sum on j by the upper bound 2^{r_2+3}, we find that $|f(s_1, s_2)|^2 < 32\sqrt{s_1^2 + s_2^2}/2^e$. Compare with the analysis in exercise 21; this sequence seems to pass the spectral test nicely, when both the amplitudes and wave lengths are considered. (But the results of this exercise give no indication that the quadratic sequence is any better a source of random numbers than a good linear sequence. Furthermore, the calculation of $f(s_1, s_2, s_3)$ seems to be very difficult.)

26. $f(s_1, s_2, s_3) = (1/(p^2 - 1)) \sum_{0 \leq l < p^2 - 1} \exp\left(-2\pi i(s_1 X_l + s_2 X_{l+1} + s_3 X_{l+2})/p\right)$.
Now each pair (X_l, X_{l+1}) occurs exactly once in the period, except $(0, 0)$, so this
sum is

$$\frac{1}{p^2 - 1}\left(-1 + \sum_{0 \leq j,l < p} \exp\left(\frac{-2\pi i}{p}\left(s_1 j + s_2 l + s_3(a_1 l + a_2 j)\right)\right)\right).$$

The resulting sum is zero unless both $s_1 + a_2 s_3$ and $s_2 + a_1 s_3$ are multiples of p;
in the latter case, the sum is p^2. When the sum is zero, $f(s_1, s_2, s_3)$ has the negligible
value $-1/(p^2 - 1)$; hence the nonrandomness is due to the other values (s_1, s_2, s_3),
when $f(s_1, s_2, s_3) = 1$. As in the simpler case $k = 1$ considered in the text, a good
indication of the randomness would seem to be the minimum value of

$$(px_1 - a_2 x_3)^2 + (px_2 - a_1 x_3)^2 + x_3^2$$

over all sets of integers (x_1, x_2, x_3) not all zero. (Cf. Eq. (17).)
A computational method for determining this minimum value can clearly be
contrived, analogous to Algorithm S; the determinant of Q in this case is p^2, not p.
We would expect ν_3 to be on the order of $p^{2/3}$, not $p^{1/3}$ as in the linear case.
It is clear that this procedure can be extended to $f(s_1, \ldots, s_n)$ in a similar manner.

27. $s_1 X_k + \cdots + s_n X_{k+n-1} \equiv s(a) X_k + (s(a) - s(1))c/(a - 1)$ (modulo m), as in
(6). The distance between the planes is the minimum value of $\sqrt{x_1^2 + \cdots + x_n^2}$ when
$s_1 x_1 + \cdots + s_n x_n = 1$, and this is $1/\sqrt{s_1^2 + \cdots + s_n^2}$ by Schwarz's inequality.
Finally to count the number of planes, consider the difference between the smallest
and largest achievable values of N [*Proc. Nat. Acad. Sci.* **61** (1968), 25–28].
Note: In 1959, while deriving upper bounds for the error in the evaluation of
k-dimensional integrals by the Monte Carlo method, N. M. Korobov devised a way to
rate the multiplier of a linear congruential sequence. His formula (which is rather
complicated) is related to the spectral test since it is strongly influenced by "small"
solutions to (11); but the results of this exercise suggest that his formula is not as
appropriate for the problem as ν_k, since it lacks spherical symmetry. A table of multipliers which give the best rating in Korobov's test, for small m, appears in *Zhurnal
Vych. Mat. i Mat. Fiziki* **3** (1963), 181–186.

SECTION 3.4.1

1. $\alpha + (\beta - \alpha)U$.

2. Let $U = X/m$; $\lfloor kU \rfloor = r$ if and only if $r \leq kX/m < r + 1$, that is $mr/k \leq X < m(r + 1)/k$, that is $\lceil mr/k \rceil \leq X < \lceil m(r + 1)/k \rceil$. The exact probability is
$(1/m)(\lceil m(r + 1)/k \rceil - \lceil mr/k \rceil) = 1/k + \epsilon$, $|\epsilon| < 1/m$.

3. If full-word random numbers are given, the result will be sufficiently random as
in exercise 2. But if a linear congruential sequence is used, k must be relatively prime
to the modulus m, lest the numbers have a very short period (e.g., if $k = 2$ and m is
even, the numbers will at best be alternately 0 and 1), by the results of Section 3.2.1.1.
The method is slower than (1) in nearly every case, so it is not recommended.

4. $\max(X_1, X_2) \leq x$ if and only if $X_1 \leq x$ and $X_2 \leq x$; $\min(X_1, X_2) \geq x$ if and
only if $X_1 \geq x$ and $X_2 \geq x$. The probability that two independent events both
happen is the product of the individual probabilities.

5. Obtain independent uniform deviates U_1, U_2. Set $X \leftarrow U_2$. If $U_1 \geq p$, set $X \leftarrow \max(X, U_3)$, where U_3 is a third uniform deviate. If $U_1 \geq p + q$, also set $X \leftarrow \max(X, U_4)$, where U_4 is a fourth uniform deviate. This method can obviously be generalized to any polynomial, and indeed even to infinite power series (as shown for example in Algorithm E, which uses minimization instead of maximization).

Alternatively, we could proceed as follows (suggested by M. D. MacLaren): If $U_1 < p$, set $X \leftarrow U_1/p$; otherwise if $U_1 < p + q$, set $X \leftarrow \max((U_1 - p)/q, U_2)$; otherwise set $X \leftarrow \max((U_1 - p - q)/r, U_2, U_3)$. This method requires less time than the other to obtain the uniform deviates, although it involves further arithmetical operations and it is less stable numerically.

6. $F(x) = A_1/(A_1 + A_2)$, where A_1 and A_2 are the areas in Fig. A–4; so

$$F(x) = \frac{\int_0^x \sqrt{1 - y^2}\, dy}{\int_0^1 \sqrt{1 - y^2}\, dy} = \frac{2}{\pi} \arcsin x + \frac{2}{\pi} x\sqrt{1 - x^2}.$$

The probability of termination at step 2 is $p = \pi/4$, each time step 2 is encountered, so the number of executions of step 2 has the geometric distribution. The characteristics of this number are (min 1, ave $4/\pi$, max ∞, dev $(4/\pi)\sqrt{1 - \pi/4}$), by exercise 17.

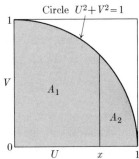

Fig. A–4. Region of "acceptance" for the algorithm of exercise 6.

7. The altitude of $p_j f_j(x)$ must not exceed $f(x)$, since each component $f_k(x)$ of Eq. (14) must be ≥ 0. Now p_j is the area under the curve $p_j f_j(x)$, that is, the altitude of $p_j f_j(x)$ times $\frac{1}{4}$. So the given value is the largest multiple of $\frac{1}{256}$ which satisfies the conditions.

8. The idea is to make the selection of the large rectangle distributions as efficient as possible. (An analogous situation is given in exercise 21.)

9. Consider the sign of $f''(x) = \sqrt{2/\pi}(x^2 - 1)e^{-x^2/2}$.

10. Figure 10 shows how a_j and b_j must be chosen for the cases that $f(x)$ is concave up or down. The rest is simple manipulation.

11. More generally, replace 3 by q and 9 by q^2 in that algorithm. We proceed as in the proof of the polar method, finding that $R' \leq r$ with probability $1 - e^{-(r^2 - q^2)/2}$, if $r \geq q$; and Θ' is uniformly distributed between 0 and $\pi/2$. The probability density in the plane is now $f(r, \theta) = (2/\pi)re^{-(r^2 - q^2)/2}$ if $r \geq q$ and $0 \leq \theta \leq \pi/2$; $f(r, \theta) = 0$ elsewhere. If the value of $X = U \times T = R' \cos \Theta'$ is $\geq q$, it has the distribution

$$F(x) = \frac{\int_{A_x} f(r, \theta)\, dr\, d\theta}{\int_{A_\infty} f(r, \theta)\, dr\, d\theta},$$

where A_x is the set $\{(r, \theta) \mid q \leq r \cos \theta \leq x\}$; this is

$$\frac{\int_q^x e^{-x^2/2}\, dx}{\int_q^\infty e^{-x^2/2}\, dx},$$

since the integral for $0 \leq y < \infty$ cancels from numerator and denominator. If $U \times T$ is $<q$ but $V \times T$ is $\geq q$, the value of X has the distribution

$$F(x) = \frac{\int_{B_x} f(r, \theta) \, dr \, d\theta}{\int_{B_\infty} f(r, \theta) \, dr \, d\theta},$$

where $B_x = \{(r, \theta) \mid r \cos \theta \leq q, \ q \leq r \sin \theta \leq x\}$; and again this is the correct probability if we transform from polar to cartesian coordinates, since the integral over $0 \leq x \leq q$ cancels from numerator and denominator. (*Note:* The probability that *both* $U \times T$ and $V \times T$ are less than three is approximately 51 percent; the probability that both are greater than three is only 0.0006 percent. If both were greater than three, they would be independent, but this occurs so infrequently it is not worth putting into the algorithm.)

13. Take $b_j = \mu_j$; consider now the problem with $\mu_j = 0$ for each j. In matrix notation, if $Y = AX$, where $A = (a_{ij})$, we need $AA^T = C = (c_{ij})$. [In other notation, if $Y_j = \sum a_{jk}X_k$, then the average value of Y_iY_j is $\sum a_{ik}a_{jk}$.] If this matrix equation can be solved for A, it can be solved when A is triangular, since $A = BU$ for some orthogonal matrix U and some triangular B, and $BB^T = C$. The desired triangular solution can be obtained by solving the equations $a_{11}^2 = c_{11}, a_{11}a_{21} = c_{12}, a_{21}^2 + a_{22}^2 = c_{22}, a_{11}a_{31} = c_{13}, a_{21}a_{31} + a_{22}a_{32} = c_{23}, \ldots$, successively for $a_{11}, a_{21}, a_{22}, a_{31}, a_{32}$, etc. [*Note:* The covariance matrix must be positive semidefinite, since the average value of $(\sum y_jY_j)^2$ is $\sum c_{ij}y_iy_j$, which must be nonnegative. And there is always a solution when C is positive semidefinite, since $C = U^{-1} \text{diag}\, (\lambda_1, \ldots, \lambda_n)U$, where the eigenvalues λ_j are nonnegative, and $U^{-1} \text{diag}\, (\sqrt{\lambda_1}, \ldots, \sqrt{\lambda_n})U$ is a solution.]

14. $F(x/c)$ if $c > 0$, a step function if $c = 0$, and $1 - F(x/c)$ if $c < 0$.

15. Distribution $\int_{-\infty}^{\infty} F_1(x - t) \, dF_2(t)$. Density $\int_{-\infty}^{\infty} f_1(x - t)f_2(t) \, dt$. This is called the *convolution* of the given distributions.

16. (a) X^2 has the distribution

$$F(x) = \frac{1}{\sqrt{2\pi}} \int_{-\sqrt{x}}^{\sqrt{x}} e^{-t^2/2} \, dt.$$

Substitute $y = t^2$ in this integral. (b) The density function of $X_1 + X_2$ is, by the previous exercise,

$$\int_0^x \left(\frac{1}{\sqrt{2\pi}} e^{-t/2}t^{-1/2}\right)\left(\frac{1}{\sqrt{2\pi}} e^{-(x-t)/2}(x+t)^{-1/2}\right) dt = \frac{1}{\pi} e^{-x/2} \arctan \frac{t}{x-t}\bigg|_0^x = \tfrac{1}{2}e^{-x/2}.$$

Hence $F(x) = 1 - e^{-x/2}$. (c) Similar argument, induction on $\nu_1 + \nu_2$.

17. (a) $F(x) = 1 - (1 - p)^{\lfloor x \rfloor}$, for $x \geq 0$. (b) $G(z) = pz/(1 - (1 - p)z)$. (c) Mean $1/p$, standard deviation $\sqrt{1 - p}/p$. To do the latter calculation, observe that if $H(z) = q + (1 - q)z$, then $H'(1) = 1 - q$ and $H''(1) + H'(1) - (H'(1))^2 = q(1 - q)$, so the mean and variance of $1/H(z)$ are $q - 1$ and $q(q - 1)$, respectively. (See Section 1.2.10.) In this case, $q = 1/p$; the extra factor z in the numerator of $G(z)$ increases the mean by one.

18. $N = N_1 + N_2$, where N_1 and N_2 independently have the geometric distribution. Similarly, $p^{r+1}\binom{n}{r}(1 - p)^{n-r}$ can be treated by adding $r + 1$ geometric variables, and subtracting 1. (Consider the generating function.)

19. (a) The generating function is

$$G(z) = \frac{1}{e-1} \sum_{n\geq 1} \frac{z^n}{n!} = \frac{e^z - 1}{e-1}.$$

(min 1, ave $e/(e-1)$, max ∞, dev $\sqrt{e(e-2)}/(e-1)$). (b) The generating function is $(e-1)/(e-z)$. [This is essentially a geometric distribution.]

$$(\text{min } 0, \text{ ave } 1/(e-1), \text{ max } \infty, \text{ dev } \sqrt{e}/(e-1)).$$

20. If $0 \leq b_1 b_2 b_3 < 6$, set $X \leftarrow A[b_1 b_2 b_3]$; if $48 \leq b_1 b_2 b_3 b_4 b_5 b_6 < 63$, set $X \leftarrow B[b_1 b_2 b_3 b_4 b_5 b_6]$; otherwise set $X \leftarrow C[b_1 b_2 b_3 b_4 b_5 b_6 b_7 b_8 b_9]$. Here $(A[0], \ldots, A[5]) = (x_1, x_2, x_3, x_3, x_6, x_6)$; $(B[48], \ldots, B[62]) = (x_1, x_1, x_1, x_2, x_2, x_4, x_5, x_5, x_5, x_5, x_6, x_6, x_6, x_6, x_6)$; $(C[504], \ldots, C[511]) = (x_1, x_1, x_2, x_3, x_3, x_3, x_4, x_4)$.

This method uses 29 auxiliary table entries, which can be overlapped so that only 21 are actually required: Let $(D[0], \ldots, D[20]) = (x_1, x_1, x_2, x_4, x_5, x_5, x_5, x_5, x_6, x_6, x_6, x_6, x_6, x_2, x_1, x_3, x_3, x_1, x_3, x_4, x_4)$; $A[j] = D[j+11]$, $B[j] = D[j-48]$, $C[j] = D[j-491]$.

21. If $0 \leq b_1 b_2 b_3 < 6$, set $X \leftarrow A[b_1 b_2 b_3]$; if $48 \leq b_1 b_2 b_3 b_4 b_5 b_6 < 63$, set $X \leftarrow B[b_1 b_2 b_3 b_4 b_5 b_6]$; if $504 \leq b_1 b_2 b_3 b_4 b_5 b_6 b_7 b_8 b_9 < 510$, set $X \leftarrow C[b_1 b_2 b_3 b_4 b_5 b_6 b_7 b_8 b_9]$; otherwise find the smallest j such that $U < P[j]$, and set $X \leftarrow x_j$. Here

$$(A[0], \ldots, A[5]) = (x_3, x_6, x_6, x_1, x_2, x_3);$$
$$(B[48], \ldots, B[62]) = (x_4, x_5, x_6, x_1, x_1, x_1, x_2, x_2, x_5, x_5, x_5, x_6, x_6, x_6, x_6);$$
$$(C[504], \ldots, C[509]) = (x_1, x_3, x_3, x_3, x_4, x_4).$$

And $P[1] = \frac{510}{512} + \epsilon_1$, $P[2] = \frac{510}{512} + \epsilon_1 + \epsilon_2, \ldots, P[6] = \frac{510}{512} + \epsilon_1 + \cdots + \epsilon_6 = 1$. This makes 33 table entries; strictly speaking, however, the elements x_j should also be considered auxiliary table entries, since we say "set $X \leftarrow x_j$" in the algorithm. The x_j-table appears as the last three entries of A and the first three entries of B, so we arrange to have these tables adjacent in the computer memory. Further overlap is also possible, as in exercise 20. Note that the technique considered in this exercise is essentially the "rectangle" part of the rectangle-wedge-tail method.

23. Yes. The second method calculates $|\cos 2\theta|$, where θ is uniformly distributed between 0 and $\pi/2$. (Let $U = r \cos \theta$, $V = r \sin \theta$.) Therefore it is unnecessary to calculate the sine or cosine of a uniform random variable; other methods in the text show that it is unnecessary to calculate the square root or logarithm of a uniform random variable. For a generalization, see J. M. Cook, *Math. Comp.* **11** (1957), 81–82.

25. $\frac{21}{32} = (.10101)_2$. In general, the binary representation is formed by using 1 for $\vee$, 0 for $\wedge$, from left to right, then suffixing 1. This technique [cf. K. D. Tocher, *J. Roy. Stat. Soc.* **B-16** (1954), 49] can lead to efficient generation of independent bits having a given probability p, as well as the geometric and binomial distributions.

SECTION 3.4.2

1. There are $\binom{N-t}{n-m}$ ways to pick $n-m$ records from the last $N-t$; $\binom{N-t-1}{n-m-1}$ ways to pick $n-m-1$ from $N-t-1$ after selecting the $(t+1)$st item.

2. (a) At most n records are selected, since the algorithm terminates when $m = n$. (b) At least n records are selected, since step S3 will never go to step S5 when the number of records left to be examined is equal to $n-m$.

3. We should not confuse "a priori" and "a posteriori" probabilities. The quantity m depends randomly on the selections which took place among the first t elements; if we take the average over all possible choices which could have occurred among these elements, we will find $(n - m)/(N - t)$ is exactly n/N on the average. For example, consider the second element; if the first element was selected in the sample (this happens with probability n/N), the second element is selected with probability $(n - 1)/(N - 1)$; if the first element was not selected, the second is selected with probability $n/(N - 1)$. The overall probability of selecting the second element is $(n/N)(n - 1)/(N - 1) + (1 - n/N)(n)/(N - 1) = n/N$.

4. From the algorithm,

$$p(m, t+1) = \left(1 - \frac{n - m}{N - t}\right) p(m, t) + \frac{n - (m - 1)}{N - t} p(m - 1, t).$$

The desired formula can be proved by induction on t. In particular, $p(n, N) = 1$.

5. In the notation of exercise 4, the probability that $t = k$ at the conclusion is

$$q_k = p(n, k) - p(n, k - 1) = \binom{k - 1}{n - 1} \bigg/ \binom{N}{n}.$$

The average is

$$\sum_{0 \le k \le N} k q_k = \left(n \sum_{0 \le k \le N} \binom{k}{n}\right) \bigg/ \binom{N}{n} = \frac{n}{n + 1}(N + 1).$$

6. Similarly,

$$\sum_{0 \le k \le N} k(k + 1)q_k = \frac{n}{n + 2}(N + 2)(N + 1);$$

the standard deviation comes to

$$\sqrt{\frac{n}{n + 2} \frac{N + 1}{n + 1} \frac{N - n}{n + 1}}.$$

7. Suppose the choice is $0 < x_1 < x_2 < \cdots < x_n \le N$. Let $x_0 = 0$, $x_{n+1} = N + 1$. The choice is obtained with probability $p = \prod_{1 \le t \le N} p_t$, where

$$p_t = \frac{N - (t - 1) - n + m}{N - (t - 1)} \qquad \text{for} \qquad x_m < t < x_{m+1},$$

$$p_t = \frac{n - m}{N - (t - 1)} \qquad \text{for} \qquad t = x_{m+1}.$$

The denominator of the product p is $N!$; the numerator contains the terms $N - n$, $N - n - 1, \ldots, 1$ for those t's which are not x's, and the terms $n, n - 1, \ldots, 1$ for those t's which *are* x's. Hence $p = (N - n)!n!/N!$. *Example:* $n = 3$, $N = 8$, $(x_1, x_2, x_3) = (2, 3, 7)$; $p = \frac{5}{8}\frac{3}{7}\frac{2}{6}\frac{4}{5}\frac{3}{4}\frac{2}{3}\frac{1}{2}\frac{1}{1}$.

8. Zero entries will be present in the internal table, and all records present will be in the reservoir. The condition is easily recognized by the fact that I_1 will be zero after the sort in step R8.

9. The reservoir gets seven records: 1, 2, 3, 5, 9, 13, 16. The second, fourth, and seventh of these are selected, namely 2, 5, 16.

11. The $N!$ permutations of the U's are assumed equally probable. The value of m is the number of elements of the permutation which have less than n inversions (see Chapter 5). Arguing as in Section 1.2.10, which considers the special case $n = 1$, we see that the tth item goes into the reservoir with probability $n/\max{(t, n)}$. The generating function is then

$$G(z) = z^n \left(\frac{1}{n+1} + \frac{n}{n+1}z\right)\left(\frac{2}{n+2} + \frac{n}{n+2}z\right) \cdots \left(\frac{N-n}{N} + \frac{n}{N}z\right).$$

The mean is

$$n + \frac{n}{n+1} + \cdots + \frac{n}{N} = n(1 + H_N - H_n);$$

the variance is $n(H_N - H_n) - n^2(H_N^{(2)} - H_n^{(2)})$.

12. (a) $\pi_1^n = (12)$ if n is any odd number which is a multiple of the lengths of all the other cycles. (b) If $\pi_2 = (2\, x_2 \ldots x_{t-1})$, we find $\pi_2^{t-1-j}(1\,2)\pi_2^j = (1\, x_{j+1})$. (c) $(j\, k) = (1\, j)(1\, k)(1\, j)$. (d) Any permutation is a product of exchange operations $(j\, k)$.

13. Renumbering the deck $0, 1, \ldots, 2n - 2$, we find s takes x into $(2x) \bmod (2n - 1)$, c takes x into $(x + 1) \bmod (2n - 1)$. We have $(c$ followed by $s) = cs = sc^2$. Therefore any product of c's and s's can be transformed into the form $s^j c^k$. Also $2^{\varphi(2n-1)} \equiv 1$ (modulo $(2n - 1)$); since $s^{\varphi(2n-1)}$ and c^{2n-1} are the identity permutation, at most $(2n - 1)\varphi(2n - 1)$ arrangements are possible. (The *exact* number of different arrangements is $(2n - 1)k$, where k is the order of 2 modulo $(2n - 1)$. For if $s^k = c^j$, then c^j fixes the card 0, so $s^k = c^j = $ identity.) For futher details, see *SIAM Review* **3** (1961), 293–297.

SECTION 3.5

1. A b-ary sequence, yes (cf. exercise 2); a $[0, 1)$ sequence, no (since only finitely many values are assumed by the elements).

2. It is 1-distributed and 2-distributed, but not 3-distributed (the binary number 111 never appears).

3. Cf. exercise 3.2.2–17; repeat the sequence there with a period of length 27.

4. The sequence begins $\frac{1}{3}, \frac{2}{3}, \frac{2}{3}, \frac{1}{3}, \frac{1}{3}, \frac{1}{3}, \frac{1}{3}, \frac{2}{3}, \frac{2}{3}, \frac{2}{3}, \frac{2}{3}, \frac{2}{3}, \frac{2}{3}, \frac{2}{3}, \frac{2}{3}$, etc. When $n = 1, 3, 7, 15, \ldots$ we have $\nu(n) = 1, 1, 5, 5, \ldots$ so that $\nu(2^{2k-1} - 1) = \nu(2^{2k} - 1) = (2^{2k} - 1)/3$; hence $\nu(n)/n$ oscillates between $\frac{1}{3}$ and approximately $\frac{2}{3}$, and no limit exists. So the probability is undefined.

The methods of Section 4.2.4 show that a numerical value *can* be meaningfully assigned to $\Pr(U_n < \frac{1}{2}) = \Pr$ (leading digit of the radix-4 representation of n is 1) $= \log_4 2 = \frac{1}{2}$.

5. If $\nu_1(n), \nu_2(n), \nu_3(n), \nu_4(n)$ are the counts corresponding to the four probabilities, we have $\nu_1(n) + \nu_2(n) = \nu_3(n) + \nu_4(n)$ for all n. So the desired result follows by addition of limits.

6. By exercise 5 and induction,

$$\Pr\left(S_j(n)\ \text{for some}\ j,\ 1 \le j \le k\right) = \sum_{1 \le j \le k} \Pr\left(S_j(n)\right).$$

As $k \to \infty$, the latter is a monotone sequence bounded by 1, so it converges; and

$$\underline{\Pr}\left(S_j(n)\ \text{for some}\ j \ge 1\right) \ge \sum_{1 \le j \le k} \Pr\left(S_j(n)\right)$$

for all k. For a counterexample to equality, it is not hard to arrange things so that $S_j(n)$ is always true for *some* j, yet $\Pr\left(S_j(n)\right) = 0$ for *all* j. More strongly, let $\langle U_n \rangle$ be the sequence of exercise 4, and for $n > 3$ let $V_n = U_n + 1/n$. Let $S_j(n)$ be the statement

$$\frac{1}{3} + \frac{1}{j} \le V_n < \frac{1}{3} + \frac{1}{j-1};$$

then $\Pr\left(S_j(n)\right) = 0$, and $S_j(n)$ holds for some $j \ge 1$ only when $n \ge 6$ and $U_n = \frac{1}{3}$, so $\Pr\left(S_j(n)\ \text{for some}\ j \ge 1\right)$ does not exist.

7. Let $p_i = \sum_{j \ge 1} \Pr\left(S_{ij}(n)\right)$. The result of the preceding exercise can be generalized to $\underline{\Pr}\left(S_j(n)\ \text{for some}\ j \ge 1\right) \ge \sum_{j \ge 1} \underline{\Pr}\left(S_j(n)\right)$, for *any* disjoint statements $S_j(n)$. So we have $1 = \Pr\left(S_{ij}(n)\ \text{for some}\ i, j \ge 1\right) \ge \sum_{i \ge 1} \underline{\Pr}\left(S_{ij}(n)\ \text{for some}\ j \ge 1\right) \ge \sum_{i \ge 1} p_i = 1$, and hence $\underline{\Pr}\left(S_{ij}(n)\ \text{for some}\ j \ge 1\right) = p_i$. Given $\epsilon > 0$, let I be large enough so that $\sum_{1 \le i \le I} p_i \ge 1 - \epsilon$. Let $\phi_i(N) = $ (number of n with $S_{ij}(n)$ true, for some $j \ge 1$, $1 \le n \le N$)$/N$. We have $\sum_{1 \le i \le I} \phi_i(N) \le 1$, and for all large enough N, $\sum_{2 \le i \le I} \phi_i(N) \ge \sum_{2 \le i \le I} p_i - \epsilon$; hence $\phi_1(N) \le 1 - \phi_2(N) - \cdots - \phi_I(N) \le 1 - p_2 - \cdots - p_I + \epsilon \le 1 - (1 - \epsilon - p_1) + \epsilon = p_1 + 2\epsilon$. This proves $\overline{\Pr}\left(S_{1j}(n),\ \text{some}\ j \ge 1\right) \le p_1 + 2\epsilon$; hence $\Pr\left(S_{1j}(n),\ \text{some}\ j \ge 1\right) = p_1$, and the desired result holds for $i = 1$. By symmetry of the hypotheses, it holds for any value of i.

8. Add together the probabilities for $j,\ j + d,\ j + 2d,\ \ldots$ in Definition E.

9.
$$\limsup_{n \to \infty} (a_n + b_n) \le \limsup_{n \to \infty} a_n + \limsup_{n \to \infty} b_n;$$

hence we get

$$\limsup_{n \to \infty} \left((y_{1n} - \alpha)^2 + \cdots + (y_{mn} - \alpha)^2\right) \le m\alpha^2 - 2m\alpha^2 + m\alpha^2 = 0,$$

and this can happen only if each $(y_{jn} - \alpha)$ tends to zero.

10. In the evaluation of the sum in Eq. (22).

11. $\langle U_{2n} \rangle$ is k-distributed if $\langle U_n \rangle$ is $(2, 2k)$-distributed.

12. Let $f(x_1, \ldots, x_k) = 1$ if $u \le \max(x_1, \ldots, x_k) < v$, $f(x_1, \ldots, x_k) = 0$ otherwise. Then apply Theorem B.

13. Let

$$\begin{aligned} p_k &= \Pr\left(U_n\ \text{begins a gap of length}\ k - 1\right) \\ &= \Pr\left(U_{n-1} \in [\alpha, \beta),\ U_n \notin [\alpha, \beta),\ \ldots,\ U_{n+k-2} \notin [\alpha, \beta),\ U_{n+k-1} \in [\alpha, \beta)\right) \\ &= p^2(1 - p)^{k-1}. \end{aligned}$$

It remains to translate this into the probability that $f(n) - f(n-1) = k$. Let $\nu_k(n) =$ (number of $j \leq n$ with $f(j) - f(j-1) = k$); let $\mu_k(n) =$ (number of $j \leq n$ with U_j the beginning of a gap of length $k-1$); and let $\mu(n)$ similarly count the number of $1 \leq j \leq n$ with $U_j \in [\alpha, \beta)$. We have

$$\nu_k(n) = \mu_k\big(f(n)\big), \qquad \mu\big(f(n)\big) = n.$$

As $n \to \infty$, $f(n) \to \infty$, and we find that

$$\nu_k(n)/n = \big(\mu_k(f(n))/f(n)\big) \cdot \big(f(n)/\mu(f(n))\big) \to p_k/p = p(1-p)^{k-1}.$$

[We have only made use of the fact that the sequence is $(k+1)$-distributed.]

14. Let

$$\begin{aligned}
p_k &= \Pr\,(U_n \text{ begins a run of length } k) \\
&= \Pr\,(U_{n-1} > U_n < \cdots < U_{n+k-1} > U_{n+k}) \\
&= \frac{1}{(k+2)!}\left(\binom{k+2}{1}\binom{k+1}{1} - \binom{k+2}{1} - \binom{k+2}{1} + 1\right) \\
&= \frac{k}{(k+1)!} - \frac{k+1}{(k+2)!}
\end{aligned}$$

(cf. exercise 3.3.2–13). Now proceed as in the previous exercise to transfer this to $\Pr\,(f(n) - f(n-1) = k)$. [We have assumed only that the sequence is $(k+2)$-distributed.]

15. For $s, t \geq 0$ let

$$\begin{aligned}
p_{st} &= \Pr\,(X_{n-2t-3} = X_{n-2t-2} \neq X_{n-2t-1} \neq \cdots \neq X_{n-1} \\
&\qquad \text{and} \qquad X_n = \cdots = X_{n+s} \neq X_{n+s+1}) \\
&= (\tfrac{1}{2})^{s+2t+3};
\end{aligned}$$

for $t \geq 0$ let $q_t = \Pr\,(X_{n-2t-2} = X_{n-2t-1} \neq \cdots \neq X_{n-1}) = (\tfrac{1}{2})^{2t+1}$. By exercise 7, $\Pr\,(X_n \text{ is not beginning of a coupon set}) = \sum_{t \geq 0} q_t = \tfrac{2}{3}$, $\Pr\,(X_n \text{ is beginning of coupon set of length } s+2) = \sum_{t \geq 0} p_{st} = \tfrac{1}{3}(\tfrac{1}{2})^{s+1}$. Now proceed as in exercise 13.

16. (Solution by R. P. Stanley.) Whenever the subsequence $S = (b-1), (b-2), \ldots, 1, 0, 0, 1, \ldots, (b-2), (b-1)$ appears, a coupon set must end at the right of S, since some coupon set is completed in the first half of S. We now proceed to calculate the probability that a coupon set begins at position n by manipulating the probabilities that the last prior appearance of S ends at position $n-1$, $n-2$, etc., as in exercise 15.

18. Proceed as in the proof of Theorem A to calculate $\underline{\Pr}$ and $\overline{\Pr}$.

21. $\Pr\,(Z_n \in M_1, \ldots, Z_{n+k-1} \in M_k) = p(M_1) \ldots p(M_k)$, for all $M_1, \ldots, M_k \in \mathfrak{M}$.

22. If the sequence is k-distributed, the limit is zero by integration and Theorem B. Conversely, note that if $f(x_1, \ldots, x_k)$ has an absolutely convergent Fourier series

$$f(x_1, \ldots, x_k) = \sum_{-\infty < c_1, \ldots, c_k < \infty} a(c_1, \ldots, c_k) \exp\big(2\pi i (c_1 x_1 + \cdots + c_k x_k)\big),$$

we have

$$\lim_{N \to \infty} \frac{1}{N} \sum_{0 \le n < N} f(U_n, \ldots, U_{n+k-1}) = a(0, \ldots, 0) + \epsilon_r,$$

where

$$|\epsilon_r| \le \sum_{|c_1|, \ldots, |c_k| > r} |a(c_1, \ldots, c_k)|,$$

so ϵ_r can be made arbitrarily small; hence this limit is equal to $a(0, \ldots, 0) = \int_0^1 \cdots \int_0^1 f(x_1, \ldots, x_k) \, dx_1 \ldots dx_k$. So Eq. (8) holds for all sufficiently smooth functions f. The remainder of the proof shows that the function in (9) can be approximated by smooth functions to any desired accuracy.

23. Use finite Fourier series: Eq. (5) in Section 3.3.4 is the m-ary analog of Weyl's criterion of exercise 22. See also *AMM* **75** (1968), 260–264.

24. This follows immediately from exercise 22. No direct proof of this interesting result (which does not use advanced calculus via Fourier series) is known to the author.

25. If the sequence is equidistributed, the denominator in Corollary S approaches $\frac{1}{12}$, and the numerator approaches the quantity in this exercise.

26. See *Math. Comp.* **17** (1963), 50–54.

28. Let $\langle U_n \rangle$ be ∞-distributed, and use the sequence $\langle \frac{1}{2}(X_n + U_n) \rangle$. This is 3-distributed, using the fact that $\langle U_n \rangle$ is (16, 3)-distributed.

29. If $x = x_1 x_2 \ldots x_l$ is any binary number, we can consider the number $\nu_x^E(n)$ of times $X_p \ldots X_{p+l-1} = x$, where $1 \le p \le n$ and p is even. Similary, let $\nu_x^O(n)$ count the number of times when p is odd. Let $\nu_x^E(n) + \nu_x^O(n) = \nu_x(n)$. Now

$$\nu_0^E(n) = \sum \nu_{0**\ldots*}^E(n) \approx \sum \nu_{*0*\ldots*}^O(n) \approx \sum \nu_{**0\ldots*}^E(n) \approx \cdots \approx \sum \nu_{***\ldots 0}^O(n)$$

where the ν's in these summations have $2k$ subscripts, where asterisks denote summation over all 2^{2k-1} combinations of zeros and ones, and where "$\approx$" denotes approximate equality (except for an error of at most $2k$ due to end conditions). Therefore we find that

$$\frac{1}{n} 2k \nu_0^E(n) = \frac{1}{n} \left(\sum \nu_{*0*\ldots*}(n) + \cdots + \sum \nu_{***\ldots 0}(n) \right)$$
$$+ \frac{1}{n} \sum_{\text{all } x} (r(x) - s(x)) \nu_x^E(n) + O\left(\frac{1}{n}\right),$$

where $x = x_1 \ldots x_{2k}$ contains $r(x)$ zeros in odd positions and $s(x)$ zeros in even positions. By $(2k)$-distribution, the parenthesized quantity tends to $k(2^{2k-1})/2^{2k} = k/2$. The remaining sum is clearly a maximum if $\nu_x^E(n) = \nu_x(n)$, when $r(x) > s(x)$, and if $\nu_x^E(n) = 0$, when $r(x) < s(x)$. So the maximum of the right-hand side becomes

$$\frac{k}{2} + \sum_{0 \le s < r \le k} (r - s) \binom{k}{r} \binom{k}{s} \bigg/ 2^{2k} = \frac{k}{2} + k \binom{2k-1}{k} \bigg/ 2^{2k}.$$

Now $\overline{\mathrm{Pr}}\,(X_{2n} = 0) \le \lim \sup_{n\to\infty} \nu_0^E(2n)/n$, so the proof is complete. *Note:*

$$\sum_{r,s} \binom{n}{r}\binom{n}{s} \max\,(r, s) = 2n2^{2n-2} + n\binom{2n-1}{n};$$

$$\sum_{r,s} \binom{n}{r}\binom{n}{s} \min\,(r, s) = 2n2^{2n-2} - n\binom{2n-1}{n}.$$

30. Let $f(x_1, x_2, \ldots, x_{2k}) = \mathrm{sign}\,(x_1 - x_2 + x_3 - x_4 + \cdots - x_{2k})$. Construct a directed graph with 2^{2k} nodes labeled $(E; x_1, \ldots, x_{2k-1})$ and $(O; x_1, \ldots, x_{2k-1})$, where each x is either 0 or 1. Let there be $1 + f(x_1, x_2, \ldots, x_{2k})$ directed arcs from $(E; x_1, \ldots, x_{2k-1})$ to $(O; x_2, \ldots, x_{2k})$, and $1 - f(x_1, x_2, \ldots, x_{2k})$ directed arcs leading from $(O; x_1, \ldots, x_{2k-1})$ to $(E; x_2, \ldots, x_{2k})$. We find each node has the same number of arcs leading into it as those leading out; for example, $(E; x_1, \ldots, x_{2k-1})$ has $1 - f(0, x_1, \ldots, x_{2k-1}) + 1 - f(1, x_1, \ldots, x_{2k-1})$ leading in and $1 + f(x_1, \ldots, x_{2k-1}, 0) + 1 + f(x_1, \ldots, x_{2k-1}, 1)$ arcs leading out, and $f(x, x_1, \ldots, x_{2k-1}) = -f(x_1, \ldots, x_{2k-1}, x)$. Drop all nodes which have no paths leading either in or out, i.e., $(E; x_1, \ldots, x_{2k-1})$ if $f(0, x_1, \ldots, x_{2k-1}) = +1$, or $(O; x_1, \ldots, x_{2k-1})$ if $f(1, x_1, \ldots, x_{2k-1}) = -1$. The resulting directed graph is seen to be connected, since we can get from any node to $(E; 1, 0, 1, 0, \ldots, 1)$ and from this to any desired

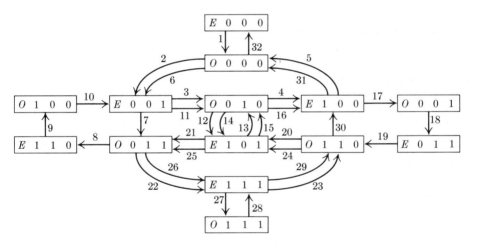

Fig. A–5. Directed graph for the construction in exercise 30.

node. By Theorem 2.3.4.2G, there is a cyclic path traversing each arc; this path has length 2^{2k+1}, and we may assume it starts at node $(E; 0, \ldots, 0)$. Construct the cyclic sequence with $X_1 = \cdots = X_{2k-1} = 0$, and $X_{n+2k-1} = x_{2k}$ if the nth arc of the path is from $(E; x_1, \ldots, x_{2k-1})$ to $(O; x_2, \ldots, x_{2k})$ or from $(O; x_1, \ldots, x_{2k-1})$ to $(E; x_2, \ldots, x_{2k})$. For example, the graph for $k = 2$ is shown in Fig. A–5; the arcs of the cyclic path are numbered from 1 to 32, and the cyclic sequence is $(00001000110010101001101110111110)(000\ldots)$. Note that $\mathrm{Pr}\,(X_{2n} = 0) = \frac{11}{16}$ in this sequence. The sequence is clearly $(2k)$-distributed, since each $(2k)$-tuple

$x_1 x_2 \ldots x_{2k}$ occurs $1 + f(x_1, \ldots, x_{2k}) + 1 - f(x_1, \ldots, x_{2k}) = 2$ times in the cycle. The fact that $\Pr(X_{2n} = 0)$ has the desired value comes from the fact that the maximum value on the right-hand side in the proof of the preceding exercise has been achieved by this construction.

31. Use Algorithm W with rule $\mathcal{R}_1$ selecting the entire sequence.

Note: For a generalization of this type of nonrandom behavior in $R5$-sequences, see Jean Ville, *Étude Critique de la notion de Collectif* (Paris, 1939), 55–62. Perhaps $R6$ is also too weak, from this standpoint.

32. If $\mathcal{R}, \mathcal{R}'$ are computable subsequence rules, so is $\mathcal{R}'' = \mathcal{R}\mathcal{R}'$ defined by the following functions: $f_n''(x_0, \ldots, x_{n-1}) = 1$ if and only if $\mathcal{R}$ defines the subsequence $x_{r_1}, \ldots, x_{r_k}$ of $x_0, \ldots, x_{n-1}$, where $k \geq 0$ and $0 \leq r_1 < \cdots < r_k < n$ and $f_k'(x_{r_1}, \ldots, x_{r_k}) = 1$.

Now $\langle X_n \rangle \mathcal{R}\mathcal{R}'$ is $(\langle X_n \rangle \mathcal{R})\mathcal{R}'$. The result follows immediately.

33. Given $\epsilon > 0$, find N_0 such that $N > N_0$ implies that both $|\nu_r(N)/N - p| < \epsilon$ and $|\nu_s(N)/N - p| < \epsilon$. Then find N_1 such that $N > N_1$ implies that t_N is r_M or s_M for some $M > N_0$. Now $N > N_1$ implies that

$$\left| \frac{\nu_t(N)}{N} - p \right| = \left| \frac{\nu_r(N) + \nu_s(N)}{N} - p \right| = \left| \frac{\nu_r(N) - pN_r + \nu_s(N) - pN_s}{N_r + N_s} \right| < 2\epsilon.$$

34. For example, if the binary representation of t is $1\,0^{b-2}\,1\,0^{a_1}\,1\,0^{a_2}\,1 \ldots 1\,0^{a_k}$, where "$0^a$" stands for a sequence of a consecutive zeros, let the rule $\mathcal{R}_t$ accept U_n if and only if $\lfloor bU_{n-k} \rfloor = a_1, \ldots, \lfloor bU_{n-1} \rfloor = a_k$.

36. Let b and k be arbitrary but fixed integers greater than 1. Let $Y_n = \lfloor bU_n \rfloor$. An arbitrary infinite subsequence $\langle Z_n \rangle = \langle Y_{s_n} \rangle \mathcal{R}$ determined by algorithms S and $\mathcal{R}$ (as in the proof of Theorem M) corresponds in a straightforward but notationally hopeless manner to algorithms $S', \mathcal{R}'$ which inspect $X_t, X_{t+1}, \ldots, X_{t+s}$ and/or select $X_t, X_{t+1}, \ldots, X_{t+\min(k-1,s)}$ of $\langle X_n \rangle$ if and only if S and $\mathcal{R}$ inspect and/or select Y_s, where $U_s = 0.X_t X_{t+1} \cdots X_{t+s}$. Algorithms S' and $\mathcal{R}'$ determine an infinite 1-distributed subsequence of $\langle X_n \rangle$ and in fact (as in exercise 32) this subsequence is ∞-distributed so it is $(k, 1)$-distributed. Hence we find $\underline{\Pr}(Z_n = a)$ and $\overline{\Pr}(Z_n = a)$ differ from $1/b$ by less than $1/2^k$.

Note: The result of this exercise is true if "$R6$" is replaced consistently by "$R4$" or "$R5$"; but it is false if "$R1$" is used, since $X\binom{n}{2}$ might be identically zero.

39. See *Mathematika* **1** (1954), 73–79. It is unknown whether $\sqrt{\log N}$ is "best possible"; an improvement of Lemma T shows that the lower bound must be $O(\log N)$.

SECTION 3.6

1.

RANDI	STJ	9F	Store exit location.
	STA	8F	Store value of k.
	LDA	XRAND	$rA \leftarrow X$.
	MUL	7F	$rAX \leftarrow aX$.
	SLAX	5	$rA \leftarrow (aX) \bmod m$.
	INCA	1	$rA \leftarrow (aX + c) \bmod m$.
	STA	XRAND	Store X.

```
        MUL    8F          rA ← ⌊kX/m⌋.
        INCA   1           Add 1, so 1 ≤ Y ≤ k.
9H      JNOV   *           Return.
        JMP    *-1
XRAND   CON    1           Value of X; X₀ = 1.
8H      CON    0           Temp storage of k.
7H      CON    3141596521  Multiplier, a. ∎
```

2. Putting a random-number generator into a program makes the results essentially unpredictable to the programmer. If the behavior of the machine on each problem were known in advance, few programs would ever be written. As Turing has said, the actions of a computer quite often *do* surprise the programmer, especially when he is debugging his program.

SECTION 4.1

1. 1010, 1011, 1000, ..., 11000, 11001, 11110.

2. (a) -110001, $-11.001001001001\ldots$, $11.0010010000111111\ldots$.
   (b) 11010011, 1101.001011001011..., 111.011001000100000....
   (c) $\bar{1}11\bar{1}\bar{1}$, $\bar{1}0.0\bar{1}\bar{1}011011\bar{1}011\ldots$, $10.011\bar{1}1111\bar{1}\ldots$.
   (d) -9.4, $-\ldots7582417582413$, $\ldots562951413$.

3. 1010113.2.

4. (a) Between rA and rX; (b) the remainder in rX has radix point between bytes 3 and 4; the quotient in rA has radix point one byte to the right of the least significant portion of the register.

5. It has been subtracted from $999\ldots9 = 10^p - 1$, instead of from $1000\ldots0 = 10^p$.

6. (a) $2^p - 1$, $-(2^p - 1)$; (b) $2^{p-1} - 1$, -2^{p-1}; (c) $2^{p-1} - 1$, $-(2^{p-1} - 1)$.

7. A ten's complement representation for a negative number x can be obtained by considering $10^n + x$ (where n is large enough for this to be positive) and extending it on the left with infinitely many nines. The nines' complement representation can be obtained in the usual manner. (These two representations are equal for nonterminating decimals, otherwise the nines' complement representation has the form $..a99999...$ while the ten's complement representation has the form $..(a+1)0000....$) These representations may be considered sensible if we regard the value of the infinite sum $N = 9 + 90 + 900 + 9000 + \cdots$ as -1, since $N - 10N = 9$.

See also exercise 31, which considers p-adic number systems. The latter agree with the p's complement notations considered here, for numbers whose radix-p representation is terminating, but there is no simple relation between the field of p-adic numbers and the field of real numbers.

8. $\sum_j a_j b^i = \sum_j (a_{kj+k-1}b^{k-1} + \cdots + a_{kj})b^{kj}$.

9. A BAD ADOBE FACADE FADED. (*Note:* Other possible "number sentences" would be, DO A DEED A DECADE; A CAD FED A BABE BEEF, COCOA, COFFEE; BOB FACED A DEAD DODO.)

10. $\begin{pmatrix} \ldots, a_3, a_2, a_1, a_0; \ a_{-1}, a_{-2}, \ldots \\ \ldots, b_3, b_2, b_1, b_0; \ b_{-1}, b_{-2}, \ldots \end{pmatrix} = \begin{pmatrix} \ldots, A_3, A_2, A_1, A_0; \ A_{-1}, A_{-2}, \ldots \\ \ldots, B_3, B_2, B_1, B_0; \ B_{-1}, B_{-2}, \ldots \end{pmatrix}$

if $A_j = \begin{pmatrix} a_{k_{j+1}-1}, \ a_{k_{j+1}-2}, \ \ldots, \ a_{k_j} \\ b_{k_{j+1}-2}, \ \ldots, \ b_{k_j} \end{pmatrix}$, $B_j = b_{k_{j+1}-1} \ldots b_{k_j}$,

where $\langle k_n \rangle$ is any infinite sequence of integers with $k_{j+1} > k_j$.

11. (The following algorithm works both for addition or subtraction, depending on whether the plus or minus sign is chosen.)

Start by setting $k \leftarrow a_{n+1} \leftarrow a_{n+2} \leftarrow b_{n+1} \leftarrow b_{n+2} \leftarrow 0$; then for $m = 0, 1, \ldots,$ $n+2$ do the following: Set $c_m \leftarrow a_m \pm b_m + k$; then if $c_m \geq 2$, set $k \leftarrow -1$ and $c_m \leftarrow c_m - 2$; if $c_m < 0$, set $k \leftarrow 1$ and $c_m \leftarrow c_m + 2$; and if $0 \leq c_m \leq 1$, set $k \leftarrow 0$.

12. (a) Subtract $(\ldots a_3 0 a_1 0)$ from $(\ldots a_4 0 a_2 0 a_0)$ in the negative binary system. (b) Subtract $(\ldots b_3 0 b_1 0)$ from $(\ldots b_4 0 b_2 0 b_0)$ in the binary system.

13. $(1.90909090 \cdots) = (0.090909 \cdots) = \frac{1}{11}$.

14.

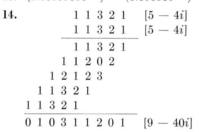

$$\begin{array}{r} 1\ 1\ 3\ 2\ 1 \quad [5-4i] \\ 1\ 1\ 3\ 2\ 1 \quad [5-4i] \\ \hline 1\ 1\ 3\ 2\ 1 \\ 1\ 1\ 2\ 0\ 2 \\ 1\ 2\ 1\ 2\ 3 \\ 1\ 1\ 3\ 2\ 1 \\ 1\ 1\ 3\ 2\ 1 \\ \hline 0\ 1\ 0\ 3\ 1\ 1\ 2\ 0\ 1 \quad [9-40i] \end{array}$$

15. $[-\frac{10}{11}, \frac{1}{11}]$, and the rectangle shown in Fig. A–6.

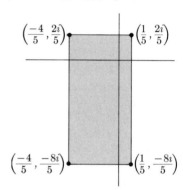

$\left(\frac{-4}{5}, \frac{2i}{5}\right)$ $\left(\frac{1}{5}, \frac{2i}{5}\right)$

$\left(\frac{-4}{5}, \frac{-8i}{5}\right)$ $\left(\frac{1}{5}, \frac{-8i}{5}\right)$

Fig. A–6. Fundamental region for quater-imaginary numbers.

16. It is tempting to try to do this in a very simple way, by using the rule $2 = (1100)_{i-1}$ to take care of carries; but that leads to a nonterminating method if we try to add one to $(11101)_{i-1} = -1$. We need to use the further identity $(11)_{i-1} + (111)_{i-1} = 0$. The following solution does this, by providing essentially three algorithms (not only for adding one, but for adding or subtracting i) and the relationships between them.

It is convenient to describe the algorithm by using the following notation: If α is a string of zeros and ones, then α^P is the string of zeros and ones such that $(\alpha^P)_{i-1} = (\alpha)_{i-1} + 1$; the operations Q and R are defined similarly with $(\alpha^Q)_{i-1} = (\alpha)_{i-1} + i$, $(\alpha^R)_{i-1} = (\alpha)_{i-1} - i$. Then

$(\alpha 0)^P = \alpha 1$; $(\alpha x 1)^P = \alpha^Q x 0$.
$(\alpha 0)^Q = \alpha^P 1$; $(\alpha 1)^Q = \alpha^R 0$.
$(\alpha 0)^R = \alpha^Q 1$; $(\alpha x 01)^R = \alpha^R x 10$; $(\alpha 11)^R = \alpha 00$.

Here x stands for either 0 or 1, and the strings are extended on the left with zeros if necessary. The processes will clearly always terminate.

As a result of this exercise, we can also subtract one (by adding $i - 1$ and subtracting i); so every number of the form $a + bi$ with a, b integers is representable in the $i - 1$ system.

17. No; the number -1 cannot be so represented. (This can be proved by constructing a set S as in Fig. 1. We have the representations $-i = (0.1111111 \ldots)_{1+i}$, $i =$

$(100.1111111\dots)_{1+i}$, but S contains no integer points besides 0 and $-i$.) See exercise 28, however.

18. Let S_0 be the set of points $(a_7a_6a_5a_4a_3a_2a_1a_0)_{i-1}$, where each a_k is 0 or 1. (S_0 is given by the 256 interior dots shown in Fig. 1, if that picture is multiplied by 16.) We first show that S is closed: If $y_1, y_2, \dots$ is an infinite subset of S, we have $y_n = \sum_{k \geq 1} a_{nk} 16^{-k}$, where each a_{nk} is in S_0. Construct a tree whose nodes are $(a_{n1}, \dots, a_{nr})$, and let a node of this tree be an ancestor of another node if it is an initial subsequence of that node. By the infinity lemma this tree has an infinite path $(a_1, a_2, a_3, \dots)$, and so $\sum_{k \geq 1} a_k 16^{-k}$ is a limit point of $\{y_1, y_2, \dots\}$ in S.

Now since each point of S_0 is contained in the square $\{x + yi \mid |x|, |y| < 1\}$, we can see by the distribution of S_0 that every point which is representable and which lies within the square $\{x + yi \mid |x|, |y| < \frac{1}{16}\}$ must be in S. (Note, for example, that the point $-1 - i$ is in S_0; a representable point not in S must lie in $m + ni + S$, for some integers m, n, but $m + ni + S$ cannot have any points inside the small square stated.) Now we can show that every point within that small square is in S, since it is a limit of representable points (by exercise 16 every number of the form $(m + ni)/16^k$ is representable).

20, 21. We can convert to such a representation by using a method like that suggested in the text for converting to balanced ternary.

For the system of exercise 21, zero can be represented in infinitely many ways, which are all obtained from $\frac{1}{2} + \sum_{k \geq 1} -4\frac{1}{2} \cdot 10^{-k}$ or the negative of this representation, by multiplying it by a power of ten. The representations of unity are $1\frac{1}{2} - \frac{1}{2}$, and also infinitely many representations of the form $\frac{1}{2} + \frac{1}{2}$, $5 - 4\frac{1}{2} + \frac{1}{2}$, $50 - 45 - 4\frac{1}{2} + \frac{1}{2}$, etc., where $\frac{1}{2} = (4\frac{1}{2})(10^{-1} + 10^{-2} + \cdots)$.

22. Assuming that we have some approximation $b_n \cdots b_1 b_0$ with error $\sum_{0 \leq k \leq n} b_k 10^k - x > 10^{-t}$ for $t > 0$, we will show how to reduce the error by approximately 10^{-t}. (The process can be started by finding a suitable $\sum_{0 \leq k \leq n} b_k 10^k > x$; then a finite number of reductions of this type will make the error less than ϵ.) Simply choose $m > n$ so large that the decimal representation of $-10^m \alpha$ has a one in position 10^{-t} and no ones in positions $10^{-t+1}, 10^{-t+2}, \dots, 10^n$. Then $10^m \alpha +$ (a suitable sum of powers of 10 between 10^m and 10^n) $+ \sum_{0 \leq k \leq n} b_k 10^k \approx \sum_{0 \leq k \leq n} b_k 10^k - 10^{-t}$.

23. Let the elements of D be $0 = \alpha_0 < \alpha_1 < \cdots < \alpha_r$, where $r \leq 9$. We may divide each of these elements by $\alpha_r/9$, so that we obtain a set with $\alpha_r = 9$. Let $S = \{\sum_{k \geq 0} a_k 10^{-k} \mid a_k \in D\}$; it follows that $S \subseteq [0, 10]$. If $[0, 10] \backslash S$ is not empty, let $x = \inf ([0, 10] \backslash S)$. If $x = 0$, then D cannot represent all numbers less than α_1, so we may assume that $x > 0$. Now suppose there is a $k \geq 0$ such that $\alpha_1 - \alpha_0 \leq x/10$, $\alpha_2 - \alpha_1 \leq x/10, \dots, \alpha_k - \alpha_{k-1} \leq x/10$, but $\alpha_{k+1} - \alpha_k > x/10$. Then $\alpha_k \leq kx/10$, and $\alpha_k + x/10 \geq x$, since all numbers $\leq \alpha_k + x/10$ are in S. This implies that $9x/10 \leq kx/10$, that is, $k \geq 9$, and this is a contradiction. It follows that $r = 9$. The only such sets D are multiples of $\{0, 1, 2, 3, 4, 5, 6, 7, 8, 9\}$.

24. It is not difficult to show that the sets $\{x, x + 1, x + 2, \dots, x + 9\}$ and multiples of these sets will work, for $-9 < x \leq +1$. Are there other sets? When less than 10 numbers are used, S can be shown to have measure zero.

25. A positive number whose base b representation has m consecutive $(b - 1)$'s to the right of the decimal point must have the form $c/b^n + (b^m - \theta)/b^{n+m}$, where $0 < \theta \leq 1$ and c, n are nonnegative integers. So if u/v has this form, we find that $b^{m+n} u = b^m cv + b^m v - \theta v$. Therefore θv is an integer which is a multiple of b^m. But

$0 < \theta v \leq v < b^m$. (There can be arbitrarily long runs of other digits $dddddddd$, if $0 \leq d < b - 1$, for example in the representation of $d/(b - 1)$.)

26. The proof of "sufficiency" is a straightforward generalization of the usual proof for base b, by successively constructing the desired representation. The proof of "necessity" breaks into two parts: If β_{n+1} is greater than $\sum_{k \leq n} c_k \beta_k$ for some n, then $\beta_{n+1} - \epsilon$ has no representation for small ϵ. If $\beta_{n+1} \leq \sum_{k \leq n} c_k \beta_k$ for all n, but equality does not always hold, we can show there are two representations for certain x. [See *Transactions of the Royal Society of Canada*, series III, **46** (1952), 45–55.]

27. Proof by induction on $|n|$: If n is even we must take $e_0 > 0$, and the result then follows by induction, since $n/2$ has a unique such representation. If n is odd, we must take $e_0 = 0$, and the problem reduces to representing $-(n - 1)/2$; the latter quantity is either zero or one, when there is obviously only one way to proceed, or it has a unique reversing representation by induction.

28. A proof like that of exercise 27 may be given. Note that $a + bi$ is a multiple of $1 + i$ by a complex integer if and only if $a + b$ is even.

29. It suffices to prove that any collection $T_0, T_1, T_2, \ldots$ satisfying Property B may be obtained by collapsing some collection $S_0, S_1, S_2, \ldots$, where $S_0 = \{0, 1, \ldots, b - 1\}$ and all elements of $S_1, S_2, \ldots$ are multiples of b.

To prove the latter statement, we may assume that $1 \in T_0$ and that there is a least element $b > 1$ such that $b \notin T_0$. We will prove, by induction on n, that if $nb \notin T_0$, then $nb + 1$, $nb + 2$, ..., $nb + b - 1$ are not in any of the T_j's; but if $nb \in T_0$, then so are $nb + 1, \ldots, nb + b - 1$. The result then follows with $S_1 = \{nb \mid nb \in T_0\}$, $S_2 = T_1, S_3 = T_2$, etc.

If $nb \notin T_0$, then $nb = t_0 + t_1 + \cdots$, where $t_1, t_2, \ldots$ are multiples of b; hence $t_0 < nb$ is a multiple of b. By induction, $(t_0 + k) + t_1 + t_2 + \cdots$ is the representation of $nb + k$, for $0 < k < b$; hence $nb + k \notin T_j$ for any j.

If $nb \in T_0$ and $0 < k < b$, let the representation of $nb + k$ be $t_0 + t_1 + \cdots$. We cannot have $t_j = nb + k$ for $j \geq 1$, lest $nb + b$ have two representations $(b - k) + \cdots + (nb + k) + \cdots = (nb) + \cdots + b + \cdots$. Hence by induction, $t_0 \bmod b = k$; and the representation $nb = (t_0 - k) + t_1 + \cdots$ implies that $t_0 = nb + k$.

[Reference: *Nieuw Archief voor Wiskunde* (3) **4** (1956), 15–17. A finite analog of this result was derived by P. A. McMahon, *Combinatory Analysis* **1** (Cambridge, 1915), 217–223.]

30. a) Let A_j be the set of numbers n whose representation does not involve b_j; then by the uniqueness property, $n \in A_j$ iff $n + b_j \notin A_j$. Consequently $n \in A_j$ iff $n + 2b_j \in A_j$. It follows that, for $j \neq k$, $n \in A_j \cap A_k$ iff $n + 2b_j b_k \in A_j \cap A_k$. Let m be the number of integers $n \in A_j \cap A_k$ such that $0 \leq n < 2b_j b_k$. Then there are exactly m integers in the same range which are in A_j but not A_k, exactly m in A_k but not A_j, and exactly m in neither A_j nor A_k; hence $4m = 2b_j b_k$. Therefore b_j and b_k cannot both be odd. But at least one b_j is odd, of course, since odd numbers can be represented.

b) According to (a) we can renumber the b's so that b_0 is odd and $b_1, b_2, \ldots$ are even; then $\frac{1}{2} b_1, \frac{1}{2} b_2, \ldots$ must also be a binary basis, and the process can be iterated.

c) If it is a binary basis, we must have positive and negative d_k's for arbitrarily large k, in order to represent $\pm 2^n$ when n is large. Conversely, the following algorithm may be used:

S1. Set $k \leftarrow 0$.

S2. If $n = 0$, terminate.

S3. If n is odd, include $2^k d_k$ in the representation, and replace n by $(n - d_k)/2$. Otherwise replace n by $n/2$.

S4. Increase k by 1 and return to S2. ∎

Here step S3 decreases $|n|$ unless $n = -d_k$; hence the algorithm must terminate.

d) Two iterations of steps S2–S4 in the preceding algorithm will transform $4t \to t$, $4t + 1 \to t + 5$, $4t + 2 \to t + 7$, $4t + 3 \to t - 1$. If n is thereby transformed into n', note that $n < n'$ implies that $n' < \max(0, \frac{19}{3}, \frac{26}{3}, -\frac{7}{3}) = \frac{26}{3}$; $n = n'$ implies that $n' = 0$; and $n > n'$ implies that $n' > \min(0, \frac{19}{3}, \frac{26}{3}, -\frac{7}{3}) = -\frac{7}{3}$. Hence we need only show that the algorithm terminates for $-2 \le n \le 8$; all other values of n are moved toward this interval. In this range we have the following tree structure:

$$3 \to -1 \to -2$$
$$\searrow$$
$$6 \to 8 \to 2 \to 7 \to 0.$$
$$\nearrow$$
$$4 \to 1 \to 5$$

Thus

$$1 = -19 + 4(-19 + 4(-26 + 4(0 + 4(-26 + 4(7)))))$$
$$= 7 \cdot 2^0 - 13 \cdot 2^1 + 7 \cdot 2^2 - 13 \cdot 2^3 - 13 \cdot 2^5 - 13 \cdot 2^9 + 7 \cdot 2^{10}.$$

Note: The choice $d_0, d_1, d_2, \ldots = 5, -3, 3, 5, -3, 3, \ldots$ also yields a binary basis. In general a similar test applies, for any periodic sequence $\langle d_n \rangle$; the corresponding sequence $\langle 2^n d_n \rangle$ is a binary basis if and only if a certain set of numbers forms an oriented tree. For further details see *Math. Comp.* **18** (1964), 537–546; A. D. Sands, *Acta Mathematica*, Acad. Sci. Hung., **8** (1957), 65–86.

31. (See also the related exercises 3.2.2–11, 4.3.2–13, 4.6.2–22.)

a) By multiplying numerator and denominator by suitable powers of 2, we may assume that $u = (\ldots u_2 u_1 u_0)_2$ and $v = (\ldots v_2 v_1 v_0)_2$, where $v_0 = 1$. The following computational method now determines w, using the notation $u^{(n)}$ to stand for the integer $(u_{n-1} \ldots u_0)_2 = u \bmod 2^n$ when $n > 0$:

Let $w_0 = u_0$. For $n = 1, 2, \ldots$, assume that we have found an integer $w^{(n)} = (w_{n-1} \ldots w_0)_2$ such that $u^{(n)} \equiv v^{(n)} w^{(n)}$ (modulo 2^n). Then $u^{(n+1)} \equiv v^{(n+1)} w^{(n)}$ (modulo 2^n), hence we may let $w_n = 0$ or 1 according as $(u^{(n+1)} - v^{(n+1)} w^{(n)}) \bmod 2^{n+1}$ is 0 or 2^n.

b) Find the smallest integer k such that $2^k \equiv 1$ (modulo $2n + 1$). Then $1/(2n + 1) = m/(2^k - 1)$ for some integer m, $1 \le m < 2^{k-1}$. Let α be the k-bit binary representation of m; then $(0.\alpha\alpha\alpha \ldots)_2$ times $2n + 1$ is $(0.111 \ldots)_2 = 1$ in the binary system, and $(\ldots \alpha\alpha\alpha)_2$ times $2n + 1$ is $(\ldots 111)_2 = -1$ in the 2-adic system.

c) If u is rational, say $u = m/2^t n$ where n is odd and positive, the 2-adic representation of u is periodic, because the set of numbers with periodic expansions includes $-1/n$ and is closed under the operations of negation, division by 2, and addition.

d) The square of any number of the form $(\ldots u_2 u_1 1)_2$ has the form $(\ldots 001)_2$, hence the condition is necessary. To show the sufficiency, use the following procedure to compure $v = \sqrt{n}$ when $n \bmod 8 = 1$:

H1. Set $n \leftarrow (n - 1)/8$, $k \leftarrow 2$, $v_0 \leftarrow 1$, $v_1 \leftarrow 0$, $v \leftarrow 1$.

H2. If n is even, set $v_k \leftarrow 0$, $n \leftarrow n/2$. Otherwise set $v_k \leftarrow 1$, $n \leftarrow (n - v - 2^{k-1})/2$, $v \leftarrow v + 2^k$.

H3. Increase k by 1 and return to H2. ∎

SECTION 4.2.1

1. $N = (62, +.60\ 22\ 50\ 00)$; $\hbar = (37, +.10\ 54\ 50\ 00)$. Note that $10\hbar$ would be $(38, +.01\ 05\ 45\ 00)$.

2. $b^{E-q}(1 - b^{-p})$, b^{-q-p}; $b^{E-q}(1 - b^{-p})$, b^{-q-1}.

3. When e does not have its smallest value, the most significant "one" bit (which appears in all such normalized numbers) need not appear in the computer word. [I. Bennett Goldberg, *CACM* **10** (1967), 105–106.]

4. $(51, +.10209877)$; $(50, +.12346000)$; $(53, +.99999999)$. The third answer would be $(54, +.10000000)$ if the first operand had been $(45, -.50000000)$.

5. This transformation has no effect unless $f_v \neq 0$ and $e_u \geq e_v + 3$, since otherwise $b^{p+2}f_v$ would be an integer; and since u is normalized these conditions imply $u \neq 0$. Now $|f_u + f_v| > 1/b - 1/b^3 > 1/b^2$; so the leading nonzero digit of $f_u + f_v$ must not be more than two positions to the right of the radix point, and the digit which governs rounding must not be more than $p + 2$ positions to the right of the radix point. The proof is completed by showing that the effect of this transformation is precisely equivalent to zeroing out the digits of $f_u + f_v$ which are more than $p + 2$ digits to the right of the radix point.

Notes: The third example in exercise 4 shows that, in general, $2p + 1$ digits must be retained in the addition, if a transformation of this type is not made. It is easy to find examples to show that a $(p + 1)$-digit accumulator is not sufficient, even when transformations of this type are considered, although it is true that the $(p + 2)$nd digit is never used except to control rounding. When b is even, a single digit suffices to control rounding; this is not the case when b is odd, except for "balanced" notations, where rounding simply reduces to truncation.

6. No, since it would cause register A to have precisely the wrong sign, and this will make the answer have the wrong sign.

7. One solution is merely to delete step A4; an accurate answer will be given, but this is very impractical since it requires a large accumulator (with about as many digits as the range of exponents). Alternatively, we could give another trivial solution to the problem by normalizing both u and v between steps A1 and A2. Another solution which seems more aesthetically satisfying is to change step A4 to:

> **A4′.** If $f_u = 0$, set $e_w \leftarrow e_v$ and $f_w \leftarrow f_v$ and go to A7. Otherwise if $e_u - e_v \geq 2p + 1$, set $f_w \leftarrow f_u$ and go to A7.

This means that the accumulator needed for the addition needs $3p$ digits instead of $2p + 1$ digits as presently; this many digits are needed in general, as shown by the example in the answer to the next exercise, unless some prenormalization has been done.

8. For example, let the byte size be 100. We have $(60, 0) \oplus (50, x) = (0, 0)$ regardless of x. Also consider the following example related to the third example of exercise 4: $(50, -.50\ 00\ 00\ 01) \oplus (58, +.00\ 00\ 00\ 01)$ gives $(55, +.01\ 00\ 00\ 00)$, which is improperly rounded; the correct answer is $(54, +.99\ 99\ 99\ 99)$.

9. $a = (-50, +.10000000)$, $b = (-41, +.20000000)$, $c = a$, $d = (-41, +.80000000)$, $y = (11, +.10000000)$.

10. $(50, +.99999000) \oplus (55, +.99999000)$.

11. $(50, +.10000001) \otimes (50, +.99999990)$.

12. If $0 < |f_u| < |f_v|$, then $|f_u| \le |f_v| - b^{-p}$; hence $1/b < |f_u/f_v| \le 1 - b^{-p}/|f_v| < 1 - b^{-p}$. If $0 < |f_v| \le |f_u|$, we have $1/b < |f_u/f_v|/b \le (1 - b^{-p})/(1/b)/b = 1 - b^{-p}$.

13. If u is very small, the instruction "FADD FUDGE" results in $\mathtt{CONTENTS(FUDGE)} - \frac{1}{2}$, since no rounding is required. The Quick trick would work if the 2:2 byte of $\mathtt{CONTENTS(FUDGE)}$ were 2 instead of 1, for $0 < u < (b - 2)b^3$.

14.

FIX	STJ	9F	Float-to-fix subroutine:
	STA	TEMP	
	LD1	TEMP(EXP)	$\text{rI1} \leftarrow e$.
	SLA	1	$\text{rA} \leftarrow \pm f f f f 0$.
	JAZ	9F	Is input zero?
	DEC1	1	
	CMPA	=0=(1:1)	If leading byte is zero,
	JE	*-4	shift left again.
	ENN1	-Q-4,1	
	J1N	FIXOVFLO	Is magnitude too large?
	ENTX	0	
	SRAX	0,1	
	CMPX	=1//2=	
	JL	9F	
	STA	*+1(0:0)	Round, if necessary.
	INCA	1	Add ± 1 (overflow is impossible).
9H	JMP	*	Exit from subroutine. ∎

15.

FP	STJ	EXITF	Fractional part subroutine:
	JOV	OFLO	Ensure overflow is off.
	STA	TEMP	$\text{TEMP} \leftarrow u$.
	SLA	1	$\text{rA} \leftarrow f_u$.
	LD2	TEMP(EXP)	$\text{rI2} \leftarrow e_u$.
	DEC2	Q	
	J2NP	*+3	
	SLA	0,2	Remove integer part of u.
	ENT2	0	
	JANN	1F	
	ENN2	0,2	Fraction is negative: find
	SRA	0,2	its complement.
	ENT2	0	*Note:* rX is not needed for the
	JAZ	*+2	precision required.
	INCA	1	
	ADD	WM1	Add word size minus one.
1H	INC2	Q	Prepare to normalize the answer.
	ENTX	0	
	JMP	NORM	Normalize, round, and exit.
8H	EQU	1(1:1)	
WM1	CON	8B-1,8B-1(1:4)	Word size minus one ∎

16. If $|c| \geq |d|$, then set $r \leftarrow d \oslash c, s \leftarrow c \oplus (r \otimes d); x \leftarrow (a \oplus (b \otimes r)) \oslash s;$ $y \leftarrow (b \ominus (a \otimes r)) \oslash s.$ Otherwise set $r \leftarrow c \oslash d, s \leftarrow d \oplus (r \otimes c); x \leftarrow ((a \otimes r) \oplus b) \oslash s, y \leftarrow ((b \otimes r) \ominus a) \oslash s.$ Then $x + iy$ is the desired approximation to $(a + bi)/(c + di)$. [*CACM* **5** (1963), 435.] Other algorithms for complex arithmetic and function evaluation are given by P. Wynn, *BIT* **2** (1962), 232–255; see also Paul Friedland, *CACM* **10** (1967), 665.

SECTION 4.2.2

1. $u \ominus v = u \oplus -v = -v \oplus u = -(v \oplus -u) = -(v \ominus u).$

2. $u \oplus x \geq u \oplus 0 = u$, by (8), (2), (6); hence by (8) again, $(u \oplus x) \oplus v \geq u \oplus v$. Similarly, (8) and (6) together with (2) imply that $(u \oplus x) \oplus (v \oplus y) \geq (u \oplus x) \oplus v.$

3. $u = 8.0000001$, $v = 1.2500008$, $w = 8.0000008$; $(u \otimes v) \otimes w = 80.000064$, $u \otimes (v \otimes w) = 80.000057.$

4. Yes, assuming that a typical range of exponents is allowed, e.g., if w and v are very large but u is very small in magnitude.

5. Not always; in decimal arithmetic take $u = v = 9.$

6. (a) Yes, by (2) $-$ (6) or (11). (b) Try $u = .99999999$; $1 \oslash u = 1.0000000.$

7. Begin with the inequality $b^{e-p}(b^{p-e}x - \frac{1}{2}) < $ round $(x, p) \leq b^{e-p}(b^{p-e}x + \frac{1}{2})$, when $b^{e-1} \leq x < b^e$, which follows immediately from the definition (9).

8. (a) $\sim$, $\approx$; (b) $\sim$, $\approx$; (c) $\sim$, $\approx$; (d) $\sim$; (e) $\sim$.

9. $|u - w| \leq |u - v| + |v - w| \leq \epsilon_1 \min (b^{e_u-q}, b^{e_v-q}) + \epsilon_2 \min (b^{e_v-q}, b^{e_w-q}) \leq \epsilon_1 b^{e_u-q} + \epsilon_2 b^{e_w-q} \leq (\epsilon_1 + \epsilon_2) \max (b^{e_u-q}, b^{e_w-q})$. The result cannot be strengthened in general, since for example we might have e_u very small compared to both e_v and e_w, and this means $u - w$ might be fairly large under the hypotheses.

10. If $b > 2$, we have

$$(0, +.a_1 \ldots a_{p-1}a_p) \times (0, +.9 \ldots 99) = (0, +.a_1 \ldots a_{p-1}(a_p - 1))$$

if we take $a_p \geq 2$; here "9" stands for $b - 1$. Furthermore, $(0, +.a_1 \ldots a_{p-1}a_p) \times (1, +.10 \ldots 0) = (0, +.a_1 \ldots a_{p-1}0)$, so the multiplication is not monotone. But when $b = 2$, this argument can be extended to show that multiplication *is* monotone; obviously the "certain computer" had $b > 2$.

11. This floating-point operation may be defined as $u \ominus v = $ round$'(u - v, p)$ for a suitable function round$'(x, p)$. The idea of this problem is to find a value of x such that round$'(-x, p) \neq -$round$'(x, p)$. One answer is to take $u = (2, 0.1010 \ldots 0)$, $v = (0, 0.10 \ldots 01)$. Then $u \ominus v = (2, 0.1000 \ldots 0) = 2$, but $v \ominus u = (1, 1.00 \ldots 01) = -2 + 2^{1-p}.$

12. Since round (round $(x, p), p$) $=$ round (x, p), we may round each of the three components of the equation.

13. Note that round $(x, p) = 0$ if and only if $x = 0$; hence $u \bmod 1 = 0$ if and only if u is an integer. Therefore the problem becomes one of determining which integers m, n will have the property that $m \oslash n$ is an integer if and only if m/n is an integer.

Assume first that $|m|, |n| < b^p$. Now if m/n is an integer, so is $m \oslash n = m/n$, since $|m/n| < b^p$. The only problem arises therefore if m/n is not an integer, but $m \oslash n$ is. Thus $|m \oslash n - m/n| \geq 1/|n|$, that is,

$$|m \oslash n - m/n| \geq \left|\frac{m}{n}\right| \left|\frac{1}{m}\right|.$$

But we know that

$$|m \oslash n - m/n| \leq \left|\frac{m}{n}\right| \cdot \frac{1}{2} b^{1-p};$$

hence the test will fail only if $|1/m| \leq \frac{1}{2}b^{1-p}$.

Our answer is therefore to require $|m| < 2b^{p-1}$ and $0 < |n| < b^p$. Slightly more liberal conditions on n could be given, namely that n can be represented exactly as a floating-point number and that $1 \oslash n$ does not cause exponent underflow.

14. $|(u \otimes v) \otimes w - uvw| \leq |(u \otimes v) \otimes w - (u \otimes v)w| + |w| \, |u \otimes v - uv| \leq \delta_{(u \otimes v) \otimes w} + b^{e_w - q - l_w} \delta_{u \otimes v} \leq (1 + b)\delta_{(u \otimes v) \otimes w}$. Now $|e_{(u \otimes v) \otimes w} - e_{u \otimes (v \otimes w)}| \leq 2$, so we may take

$$\epsilon = \frac{1 + b}{2} b^{2-p}.$$

15. $u \leq v$ implies that $(u \oplus u) \oslash 2 \leq (u \oplus v) \oslash 2 \leq (v \oplus v) \oslash 2$, so the condition holds for all u and v iff it holds whenever $u = v$. For base $b = 2$, the condition is therefore always satisfied (barring overflow); but for $b > 2$ there are numbers $v \neq w$ such that $v \oplus v = w \oplus w$, hence the condition fails. On the other hand, the formula $u \oplus ((v \ominus u) \oslash 2)$ does give a midpoint in the correct range.

16. (a) $u_0 v_0$.

(b) $$\frac{1}{4\delta\epsilon} \int_{u_0 - \delta}^{u_0 + \delta} \int_{v_0 - \epsilon}^{v_0 + \epsilon} \left(\frac{x}{y}\right) dx\, dy = \frac{u_0}{2\epsilon} \ln\left(\frac{v_0 + \epsilon}{v_0 - \epsilon}\right)$$
$$= \frac{u_0}{v_0}\left(1 + \frac{1}{3}\left(\frac{\epsilon}{v_0}\right)^2 + \frac{1}{5}\left(\frac{\epsilon}{v_0}\right)^4 + O\left(\left(\frac{\epsilon}{v_0}\right)^6\right)\right).$$

This suggests that it is perhaps better to round up than down after a floating-point division, but actually the assumption of uniform distribution is usually not justified. The *median* value is u_0/v_0, i.e., half the time the answer is less than u_0/v_0, and half the time the answer is greater.

17.

```
FCMP  STJ   9F              Floating point comparison subroutine:
      JOV   OFLO            Ensure overflow is off.
      STA   TEMP
      LDAN  TEMP
```

(Copy here lines 07–27 of Program 4.2.1A, except replace line 23 by the following four lines:

```
      DEC1  5
      J1N   *+2
      ENT1  0               Replace large difference in exponents
      SRAX  5,1                by a smaller one.
```

After this portion of the program, the difference $u - v$ has been calculated.)

```
          JOV    7F              Fraction overflow: not approx. equal.
          CMPA   EPSILON(1:5)
          JG     8F              Jump if not approximately equal.
          JL     6F              Jump if approximately equal.
          JXP    1F              If rA = ε, check if rX has same sign.
          JXZ    9F              Jump if approximately equal.
          JAP    9F
          JMP    5F
    1H    JANN   5F
    6H    CMPA   *(0:0)          Set equal condition.
          JMP    9F
    5H    JANZ   8F
          LDA    FV              Special case, ε = 0 = rA ≠ rX: Use sign of v.
    7H    ENTX   1
          SRC    1               Make rA nonzero with same sign.
    8H    CMPA   =0=             Set greater or less condition.
    9H    JMP    *               Exit from subroutine. ∎
```

SECTION 4.2.3

1. First, $(w_m, w_l) = (.573, .248)$; then $w_m v_l / v_m = .290$; so the answer is $(.572, .958)$. This in fact is the correct result to six decimals.

2. In Program A there is no effect, since the $(1:1)$ byte of rX is zero at this point. In Program M, the answer is not affected, since the normalization routine truncates to eight places and can never look at this particular byte position. (Scaling to the left occurs at most once during normalization, since we are assuming the inputs are normalized.)

3. Overflow obviously cannot occur at line 10, since we are adding two-byte quantities, or at line 24, since we are adding four-byte quantities. In line 32 we are computing the sum of three four-byte quantities, so this cannot overflow. Finally, in line 34, overflow is impossible because the product of f_u times f_v must be less than unity.

4. Insert "JOV OFLO; ENT1 0" between lines 04 and 05. Delete lines 19 and 23. Insert "JNOV *+2; INC1 1" between lines 24 and 25. Change lines 30–33 to "SLAX 5; ADD TEMP; JNOV *+2; INC1 1; SRAX 5; ENTA 0,1". This adds a net total of four lines; the running time is virtually unchanged (since it may be decreased by one unit, or increased by one unit, if there are zero or two carries from the least significant halves into the most significant half of the answer). So this seems a worth-while improvement.

5. First solution: We can add the lower halves as in Program D, but instead of using "1" and "WM1" it is slightly more convenient to assume the byte size is even and to use a constant "HALF" as in Program 4.2.1A. Insert "JOV OFLO" between lines 04 and 05. Change line 19 to "INC1 -9,2". Change lines 22–24 to "SRAX 10,1".

Change lines 32–33 to "SLAX 2". Then change lines 36–41 to the following:

```
          STA   HALF(0:0)
          STA   1F(0:0)
          SLAX  5
          ADD   ARGX
          JOV   2F
          ADD   HALF         Add word size, with
          ADD   HALF           sign of u.
          JNOV  4F           (Now rX = 0)
          ENTX  1            Let rX record the
          JMP   4F             number of carries.
     2H   ENTX  2
     4H   SLC   5                            ▮
```

Now the status is just as it would be after line 41 of Program A, except that we have done the computations with one further byte of precision. This modification requires a net change of four more lines of code (HALF takes the place of TRICK), and the execution time is either decreased by $5u$, decreased by $1u$, or increased by $1u$, depending on whether the setting of rX at step 4H is 2, 0, or 1. So this seems worth while. (If WM1 were used instead of HALF, add $1u$ to the execution time.)

Second solution: We can continue to add the lower halves as in Program A, but can extend the precision of the upper-half calculation by one byte. Insert "JOV OFLO" between lines 04 and 05. Change line 19 to "INC1 -9,2". Change lines 32–36 to "STX TRICK(2:4); SLAX 2; STA ACC; ENTX 0". Between lines 44 and 45, insert "DEC2 1; JNOV DNORM; INC2 1; INCX 1; SRC 1". This modification requires a net change of five more lines of code, and the execution time is either unchanged or it increases by $4u$ when there is fraction overflow. The first solution is therefore better. (*Note:* In the second solution, it is tempting to replace lines 28–29 by "STZ EXPO" instead of adding "DEC2 1" after line 44; but this would mean that index register 2 would not have enough capacity to hold the exponent when there is exponent overflow.)

```
6.  DOUBLE  STJ   EXITDF        Convert to double precision:
            ENTX  0             Clear rX.
            STA   TEMP
            LD2   TEMP(EXP)     rI2 ← e.
            INC2  QQ-Q          Correct for difference in excess.
            STZ   EXPO          EXPO ← 0.
            SLAX  1             Remove exponent.
            JMP   DNORM         Normalize and exit.
    SINGLE  STJ   EXITF         Convert to single precision:
            JOV   OFLO          Ensure overflow is off.
            STA   TEMP
            LD2   TEMP(EXPD)    rI2 ← e.
            DEC2  QQ-Q          Correct for difference in excess.
            SLAX  2             Remove exponent.
            JMP   NORM          Normalize, round, and exit.  ▮
```

7. All three routines give zero as the answer if and only if the exact result would be zero, so we need not worry about zero denominators in the expressions for relative error. The worst case of the addition routine is pretty bad: Visualized in decimal notation, if the inputs are 1.0000000 and .99999999, the answer is 10^{-7} instead of 10^{-8}; thus the maximum relative error δ_1 is $b - 1$, where b is the byte size.

For multiplication and division, we may assume that the exponents of both operands are QQ, and that both operands are positive. The maximum error in multiplication is readily bounded by considering (2): When $uv \geq 1/b$, we have $0 \leq uv - u \bigotimes v < 3b^{-9} + (b - 1)b^{-9}$, so the relative error is bounded by $(b + 2)b^{-8}$. When $1/b^2 \leq uv < 1/b$, we have $0 \leq uv - u \bigotimes v < 3b^{-9}$, so the relative error in this case is bounded by $3b^{-9}/uv \leq 3b^{-7}$. We take δ_2 to be the larger of the two estimates, namely $3b^{-7}$.

Division requires a more careful analysis of Program D. The quantity actually computed by the subroutine is $\alpha - \delta - b\epsilon((\alpha - \delta'')(\beta - \delta') - \delta''') - \delta_n$ where $\alpha = (u_m + \epsilon u_l)/bv_m$, $\beta = v_l/bv_m$, and δ, δ', δ'', δ''', δ_n are nonnegative "rounding" errors with $\delta < b^{-10}$, $\delta' < b^{-5}$, $\delta'' < b^{-5}$, $\delta''' < b^{-6}$, and δ_n (which occurs during normalization) is either less than b^{-9} or b^{-8} depending on whether scaling occurs or not. The actual value of the quotient is $\alpha/(1 + b\epsilon\beta) = \alpha - b\epsilon\alpha\beta + b^2\alpha\beta^2\delta''''$, where δ'''' is the nonnegative error due to truncation of the infinite series (3); $\delta'''' < \epsilon^2$, since it is an alternating series. The relative error is therefore the absolute value of $(b\epsilon\delta' + b\epsilon\delta''\beta/\alpha + b\epsilon\delta'''/\alpha) - (\delta/\alpha + b\epsilon\delta'\delta''/\alpha + b^2\beta^2\delta'''' + \delta_n/\alpha)$, times $(1 + b\epsilon\beta)$. The positive terms in this expression are bounded by $b^{-9} + b^{-8} + b^{-8}$, and the negative terms are bounded by $b^{-8} + b^{-12} + b^{-8}$ plus the contribution by the normalizing phase, which can be about b^{-7} in magnitude. It is therefore clear that the potentially greatest part of the relative error comes during the normalization phase, and that we may let $\delta_3 = (b + 2)b^{-8}$ be a safe upper bound for the relative error.

8. Addition: If $e_u \leq e_v + 1$, the entire relative error occurs during the normalization phase, so it is bounded above by b^{-7}. If $e_u \geq e_v + 2$, and if the signs are the same, again the entire error may be ascribed to normalization; if the signs are opposite, the error due to shifting digits out of the register is in the opposite direction from the subsequent error introduced during normalization. The first of these errors is at its worst when we are subtracting $1 - b^{-8}$ from b^2, and this is bounded by b^{-7}. The second error occurs only when the value of $|f_u + f_v/b^{e_u - e_v}|$ is $1/b - 1/b^9$, so it cannot contribute more than b^{-7} to the relative error. Hence $\delta_1 = b^{-7}$. (This is substantially better than the result in exercise 7!)

Multiplication: An analysis as in exercise 7 gives $\delta_2 = (b + 2)b^{-8}$.

SECTION 4.2.4

1. Since fraction overflow can occur only when the operands have the same sign, this is the probability that fraction overflow occurs divided by the probability that the operands have the same sign, namely, $7\%/\frac{1}{2}92\% \approx 15\%$.

3. $\log_{10} 2.3 - \log_{10} 2.2 \approx 1.930516\%$.

4. The pages would be uniformly gray (same as "random point on a slide rule").

5. The probability that $10f_U \le r$ is

$$\left(\frac{r}{10} - \frac{1}{10}\right) + \left(\frac{r}{100} - \frac{1}{100}\right) + \cdots = \frac{r-1}{9}.$$

So in this case the leading digits are *uniformly* distributed, e.g., leading digit 1 occurs with probability $\frac{1}{9}$.

6. The probability that there are three leading zero bits is $\log_{16} 2 = \frac{1}{4}$; the probability that there are two leading zero bits is $\log_{16} 4 - \log_{16} 2 = \frac{1}{4}$; and similarly for the other two cases. The "average" number of leading zero bits is $1\frac{1}{2}$, so the "average" number of "significant bits" is $p + \frac{1}{2}$. The worst case, $p - 1$ bits, occurs however with rather high probability. In practice, it is usually necessary to base error estimates on the worst case; in the error analysis of Section 4.2.2, the upper bounds for the relative error of rounding are 2^{1-p} in the hexadecimal case and 2^{-p} in the binary case, so binary arithmetic permits better bounds on the accuracy. See also the trick for increasing precision when $b = 2$, in exercise 4.2.1–3.

Sweeney's tables show that hexadecimal arithmetic can be done a little faster, since fewer normalization cycles are needed. (Furthermore, it turns out that computer circuitry for the hexadecimal case is less expensive than for the binary case, to achieve the same speed of operation.) Thus the trade-off is between accuracy and the speed or cost of operation.

7. Suppose that

$$\sum_m \left(F(10^{km} \cdot 5^k) - F(10^{km})\right) = \frac{\log 5^k}{\log 10^k},$$

and

$$\sum_m \left(F(10^{km} \cdot 4^k) - F(10^{km})\right) = \frac{\log 4^k}{\log 10^k};$$

then

$$\sum_m \left(F(10^{km} \cdot 5^k) - F(10^{km} \cdot 4^k)\right) = \log_{10} \tfrac{5}{4}$$

for all k. But now let ϵ be a small positive number, and choose $\delta > 0$ so that $F(x) < \epsilon$ for $0 < x < \delta$, and choose $M > 0$ so that $F(x) > 1 - \epsilon$ for $x > M$. We can take k so large that $10^{-k} \cdot 5^k < \delta$ and $4^k > M$; hence by the monotonicity of F,

$$\sum_m \left(F(10^{km} \cdot 5^k) - F(10^{km} \cdot 4^k)\right) \le \sum_{m<0} \left(F(10^{km} \cdot 5^k) - F(10^{k(m-1)} \cdot 5^k)\right)$$
$$+ \sum_{m \ge 0} \left(F(10^{k(m+1)} \cdot 4^k) - F(10^{km} \cdot 4^k)\right)$$
$$= F(10^{-k} 5^k) + 1 - F(10^k 4^k) < 2\epsilon.$$

8. When $s > r$, $P_0(10^n s)$ is 1 for small n, and 0 when $\lfloor 10^n s \rfloor > \lfloor 10^n r \rfloor$. The least n for which this happens may be arbitrarily large, so no uniform bound can be given for $N_0(\epsilon)$ independent of s. (In general, calculus textbooks prove that such a uniform bound would imply that the limit function $S_0(s)$ would be continuous, and it isn't.)

When $m \ge 1$, the uniform bound which is stated in Lemma Q is necessary for the inductive argument to be valid, as shown by the next exercise.

9. No; let $x_n = a$ for $n = 2, 12, 112, 1112, 11112,$ etc., and $x_n = b$ otherwise. Then for any s, $x_{\lfloor 10^n s\rfloor}$ equals b for all large n, while $\liminf_{n\to\infty} x_n = a$ (when $a < b$).

10. The generating function $C(z)$ has simple poles at the points $1 + w_n$, $w_n = 2\pi n i/\ln 10$. We can expand it in partial fractions as follows:

$$C(z) = \frac{(\log_{10} r - 1)}{1 - z} - \frac{e^{-w_1 \ln r}}{(\ln 10)w_1((1 - z)/(1 + w_1))}$$
$$- \frac{e^{-w_{-1} \ln r}}{(\ln 10)w_{-1}((1 - z)/(1 + w_{-1}))} - \cdots.$$

So

$$c_m = (\log_{10} r - 1) + \frac{2}{\ln 10} \operatorname{Re}\left(e^{-w_1 \ln r}(1 + w_1)^{-m}\right)$$
$$+ \frac{2}{\ln 10} \operatorname{Re}\left(e^{-w_2 \ln r}(1 + w_2)^{-m}\right) + \cdots.$$

The second term in the expansion is

$$\frac{2 \cos\left(-2\pi \log_{10} r - m \arctan(2\pi/\ln 10)\right)}{((\ln 10)^2 + 4\pi^2)^{m/2}} + O\left(\frac{1}{((\ln 10)^2 + 16\pi^2)^{m/2}}\right).$$

11. When $(\log_b U) \bmod 1$ is uniformly distributed in $[0, 1)$, so is $(\log_b 1/U) \bmod 1 = 1 - (\log_b U) \bmod 1$. (The latter formula holds except when U is a power of b, but that occurs with probability zero.)

12. It satisfies the logarithmic law too. (Show that if X, Y are independently and uniformly distributed in $[0, 1]$, so is $(X + Y) \bmod 1$.)

13. Let $X = (\log_b U) \bmod 1$, $Y = (\log_b V) \bmod 1$, so that X and Y are independently and uniformly distributed in $[0, 1)$. No left shift is needed if and only if $X + Y \geq 1$, and this occurs with probability $\frac{1}{2}$.

(Similarly, the result of floating-point division by Algorithm M needs no normalization shifts, with probability $\frac{1}{2}$; this needs only the weaker assumption that both operands have the *same* distribution.)

14. For convenience, the calculations are shown here for $b = 10$. If $k = 0$, the probability of a carry is

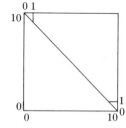

$$\left(\frac{1}{\ln 10}\right)^2 \int_{\substack{1\leq x, y\leq 10 \\ x+y\geq 10}} \frac{dx\,dy}{x\,y}.$$

(See Fig. A–7.) The value of the integral is

$$\int_0^{10} \frac{dy}{y} \int_{10-y}^{10} \frac{dx}{x} - 2\int_0^1 \frac{dy}{y} \int_{10-y}^{10} \frac{dx}{x},$$

and

$$\int_0^t \frac{dy}{y} \ln\left(\frac{1}{1 - y/10}\right) = \int_0^t \left(\frac{1}{10} + \frac{y}{200} + \frac{y^2}{3000} + \cdots\right) dy = \frac{t}{10} + \frac{t^2}{400} + \frac{t^3}{9000} + \cdots.$$

Figure A–7

(The latter integral is essentially a "dilogarithm" integral.) Hence the probability of a carry when $k = 0$ is $(1/\ln 10)^2(\pi^2/6 - 2\sum_{n\geq 1} 1/n^2 10^n) = .25332$. [*Note:* When $b = 2$ and $k = 0$, fraction overflow *always* occurs, so this derivation proves that $\sum_{n\geq 1} 1/n^2 2^n = \pi^2/12 - \frac{1}{2}(\ln 2)^2$.]

When $k > 0$, the probability is

$$\left(\frac{1}{\ln 10}\right)^2 \int_{10^{-k}}^{10^{1-k}} \frac{dy}{y} \int_{10-y}^{10} \frac{dx}{x} = \left(\frac{1}{\ln 10}\right)^2 \left(\sum_{n\geq 1} \frac{1}{n^2 10^{nk}} - \sum_{n\geq 1} \frac{1}{n^2 10^{n(k+1)}}\right);$$

thus when $b = 10$, the fraction overflow should occur with probability $.253p_0 + .055p_1 + .002p_2 + \cdots$. When $b = 2$ the corresponding figures are $p_0 + .684p_1 + .261p_2 + .137p_3 + .067p_4 + \cdots$.

Now if we use the probabilities from Sweeney's first table, we predict a probability of a little over 13 percent when $b = 10$, instead of the 15 percent in exercise 1. For $b = 2$, we predict about 45 percent, while the second table yields 44 percent. These results are certainly in agreement within the limits of experimental error.

15. (Presumably, this somewhat paradoxical situation is ameliorated by the fact that subsequent operations such as subtraction will tend to correct this imbalance.)

When $k = 0$, the leading digit is 1 if and only if there is a carry. (It is possible for fraction overflow and subsequent rounding to yield a leading digit of 2, when $b \geq 4$, but we are ignoring rounding in this exercise.) The probability of fraction overflow is $.253 < \log_{10} 2$, as shown in the previous exercise.

When $k > 0$, the leading digit is 1 with probability

$$\left(\frac{1}{\ln 10}\right)^2 \left(\int_{10^{-k}}^{10^{1-k}} \frac{dy}{y} \int_{\substack{1\leq x<2-y \\ \text{or } 10-y\leq x<10}} \frac{dx}{x}\right) < \left(\frac{1}{\ln 10}\right)^2 \left(\int_{10^{-k}}^{10^{1-k}} \frac{dy}{y} \int_{1\leq x\leq 2} \frac{dx}{x}\right)$$

$$= \log_{10} 2.$$

(*Note:* As a consequence of this exercise, there are unfortunately no grounds for the suspicion that the logarithmic law might help to determine p_0, p_1, p_2, etc.)

SECTION 4.3.1

2. If the ith number to be added is $u_i = (u_{i1}u_{i2}\ldots u_{in})_b$, use Algorithm A with step A2 changed to the following:

A2′. [Add digits.] Set

$$w_j \leftarrow (u_{1j} + \cdots + u_{mj} + k) \bmod b, \quad \text{and} \quad k \leftarrow \lfloor(u_{1j} + \cdots + u_{mj} + k)/b\rfloor.$$

(The maximum value of k is $m - 1$, so step A3 will have to be altered if $m \geq b$.)

3.

ENT1	N	1	
JOV	OFLO	1	Ensure overflow is off.
ENTA	0	1	$k \leftarrow 0$.

2H	ENT2	0	N	(rI2 $\equiv$ next value of k)
	ENT3	M*N-N,1	N	$(\text{LOC}(u_{ij}) \equiv \text{U} + n(i-1) + j)$
3H	ADD	U,3	MN	rA $\leftarrow$ rA $+ u_{ij}$.
	JNOV	*+2	MN	
	INC2	1	K	Carry one.
	DEC3	N	MN	Repeat for $m \geq i \geq 0$.
	J3P	3B	MN	$(\text{rI3} \equiv n(i-1) + j)$
	STA	W,1	N	$w_j \leftarrow$ rA.
	ENTA	0,2	N	$k \leftarrow$ rI2.
	DEC1	1	N	
	J1P	2B	N	Repeat for $n \geq j \geq 0$.
	STA	W	1	Store final carry in w_0. ∎

Running time, assuming that $K = \frac{1}{2}MN$, is $5.5MN + 7N + 4$ cycles.

4. We may make the following assertion before A1: "$n \geq 1$; and $0 \leq u_i, v_i < b$ for $1 \leq i \leq n$." Before A2, we assert: "$1 \leq j \leq n$; $0 \leq u_i, v_i < b$ for $1 \leq i \leq n$; $0 \leq w_i < b$ for $j < i \leq n$; $0 \leq k \leq 1$; and

$$(u_{j+1} \cdots u_n)_b + (v_{j+1} \cdots v_n)_b = (k\, w_{j+1} \cdots w_n)_b.\text{"}$$

The latter statement means more precisely that

$$\sum_{j < t \leq n} u_t b^{n-t} + \sum_{j < t \leq n} v_t b^{n-t} = kb^{n-j} + \sum_{j < t \leq n} w_t b^{n-t}.$$

Before A3, we assert: "$1 \leq j \leq n$; $0 \leq u_i, v_i < b$ for $1 \leq i \leq n$; $0 \leq w_i < b$ for $j \leq i \leq n$; $0 \leq k \leq 1$; and $(u_j \cdots u_n)_b + (v_j \cdots v_n)_b = (k\, w_j \cdots w_n)_b$." After A3, we assert that $0 \leq w_i < b$ for $1 \leq i \leq n$; $0 \leq w_0 \leq 1$; and $(u_1 \cdots u_n)_b + (v_1 \cdots v_n)_b = (w_0 \cdots w_n)_b$.

It is a simple matter to complete the proof by verifying the necessary implications between the assertions and by showing that the algorithm always terminates.

5. B1. Set $j \leftarrow 1$, $w_0 \leftarrow 0$.

B2. Set $t \leftarrow u_j + v_j$, $w_j \leftarrow t \bmod b$, $i \leftarrow j$.

B3. If $t \geq b$, set $i \leftarrow i - 1$, $t \leftarrow w_i + 1$, $w_i \leftarrow t \bmod b$, and repeat this step until $t < b$.

B4. Increase j by one, and if $j \leq n$ go back to B2. ∎

6. C1. Set $j \leftarrow 1$, $i \leftarrow 0$, $r \leftarrow 0$.

C2. Set $t \leftarrow u_j + v_j$. If $t \geq b$, set $w_i \leftarrow r + 1$, $w_k \leftarrow 0$ for $i < k < j$, set $i \leftarrow j$, and $r \leftarrow t \bmod b$. Otherwise if $t < b - 1$, set $w_i \leftarrow r$, $w_k \leftarrow b - 1$ for $i < k < j$, set $i \leftarrow j$, and $r \leftarrow t$.

C3. Increase j by one. If $j \leq n$, go back to C2; otherwise set $w_i \leftarrow r$, and $w_k \leftarrow b - 1$ for $i < k \leq n$. ∎

7. When $j = 3$, for example, we have $k = 0$ with probability $(b+1)/2b$, $k = 1$ with probability

$$\frac{b-1}{2b}\left(1 - \frac{1}{b}\right)$$

[this is the probability that a carry occurs and that the preceding digit wasn't $b - 1$]; $k = 2$ with probability

$$\frac{b-1}{2b}\left(\frac{1}{b}\right)\left(1 - \frac{1}{b}\right);$$

and $k = 3$ with probability

$$\frac{b-1}{2b}\left(\frac{1}{b}\right)\left(\frac{1}{b}\right)(1).$$

For fixed k we may add the probabilities as j varies from 1 to n; this gives the mean number of times the carry propagates back k places,

$$m_k = \frac{b-1}{2b^k}\left((n+1-k)\left(1-\frac{1}{b}\right)+\frac{1}{b}\right).$$

As a check, we find that the average number of carries is

$$m_1 + 2m_2 + \cdots + nm_n = \frac{1}{2}\left(n - \frac{1}{b-1}\left(1 - \left(\frac{1}{b}\right)^n\right)\right),$$

in agreement with (6).

8.

	ENN1	N	1	2H	LDA	W+N+1,2	K
	JOV	OFLO	1		INCA	1	K
	STZ	W	1		STA	W+N+1,2	K
1H	LDA	U+N+1,1	N		DEC2	1	K
	ADD	V+N+1,1	N		JOV	2B	K
	STA	W+N+1,1	N	3H	INC2	1	N
	JNOV	3F	N		J2N	1B	N
	ENT2	-1,1	L				∎

The running time depends on L, the number of positions in which $u_j + v_j \geq b$; and on K, the total number of carries. It is not difficult to see that K is the same quantity which appears in Program A. The analysis in the text shows that L has the average value $N((b-1)/2b)$, and K has the average value $\frac{1}{2}(N - b^{-1} - b^{-2} - \cdots - b^{-n})$. So if we ignore terms of order $1/b$, the running time is $9N + L + 7K + 3 \approx 13N + 3$ cycles.

Note: Since a carry occurs almost half of the time, it would be more efficient to delay storing the result by one step. This leads to a somewhat longer program whose running time is approximately $12N + 5$ cycles, based on the somewhat more detailed information calculated in exercise 7.

9. Replace "b" by "b_{n-j}" everywhere in step A2.

10. If lines 06 and 07 were interchanged, we would almost always have overflow, but register A might have a negative value at line 08, so this would not work. If the instructions on lines 05 and 06 were interchanged, the sequence of overflows occurring in the program would be slightly different in some cases, but the program would still be right.

11. (a) Set $j \leftarrow 1$; (b) if $u_j < v_j$, terminate [$u < v$]; if $u_j = v_j$ and $j = n$, terminate [$u = v$]; if $u_j = v_j$ and $j < n$, set $j \leftarrow j+1$ and repeat (b); if $u_j > v_j$, terminate

$[u > v]$. This algorithm tends to be quite fast, since there is usually low probability that j will have to get very high before we encounter a case with $u_j \neq v_j$.

12. Use Algorithm S with $u_j = 0$ and $v_j = w_j$. Another "borrow" will occur at the end of the algorithm, which this time should be ignored.

13.

	ENT1	N	1
	JOV	OFLO	1
	ENTX	0	1
2H	STX	CARRY	N
	LDA	U,1	N
	MUL	V	N
	SLC	5	N
	ADD	CARRY	N
	JNOV	*+2	N
	INCX	1	K
	STA	W,1	N
	DEC1	1	N
	J1P	2B	N
	STX	W	1 ∎

The running time is $23N + K + 5$ cycles, and K is roughly $\frac{1}{2}N$.

14. The key inductive assertion is the one which should be valid at the beginning of step M4; all others are readily filled in from this one, which is as follows: "$1 \leq i \leq n$; $1 \leq j \leq m$; $0 \leq u_r < b$ for $1 \leq r \leq n$; $0 \leq v_r < b$ for $1 \leq r \leq m$; $0 \leq w_r < b$ for $j < r \leq m + n$; $0 \leq k < b$; and

$$(w_{j+1} \cdots w_{m+n})_b + kb^{m+n-i-j} = u \times (v_{j+1} \cdots v_m)_b + (u_{i+1} \cdots u_n)_b \times v_j b^{m-j}."$$

(For the precise meaning of this notation, see the answer to exercise 4.)

15. The error is nonnegative and less than $(n-1)b^{-n-1}$. (Similarly, if we ignore the products with $i + j > n + 3$, the error is bounded by $(n-2)b^{-n-2}$, etc.; but, in general, we must compute all of the products if we want to get the true rounded result.)

16. **S1.** Set $r \leftarrow 0$, $j \leftarrow 1$.

    S2. Set $w_j \leftarrow \lfloor (rb + u_j)/v \rfloor$, $r \leftarrow (rb + u_j) \bmod v$.

    S3. Increase j by 1, and return to S2 if $j \leq n$. ∎

17. $u/v > u_0 b^n/(v_1 + 1)b^{n-1} = b(1 - 1/(v_1 + 1)) > b(1 - 1/(b/2)) = b - 2$.

18. $(u_0 b + u_1)/(v_1 + 1) \leq u/(v_1 + 1)b^{n-1} < u/v$.

19. $u - \hat{q}v \leq u - \hat{q}v_1 b^{n-1} - \hat{q}v_2 b^{n-2} = u_2 b^{n-2} + \cdots + u_n + \hat{r}b^{n-1} - \hat{q}v_2 b^{n-2} < b^{n-2}(u_2 + 1 + \hat{r}b - \hat{q}v_2) \leq 0$. Since $u - \hat{q}v < 0$, $q < \hat{q}$.

20. If $q \leq \hat{q} - 2$, then $u < (\hat{q} - 1)v < \hat{q}(v_1 b^{n-1} + (v_2 + 1)b^{n-2}) - v < \hat{q}v_1 b^{n-1} + \hat{q}v_2 b^{n-2} + b^{n-1} - v \leq \hat{q}v_1 b^{n-1} + (b\hat{r} + u_2)b^{n-2} + b^{n-1} - v = u_0 b^n + u_1 b^{n-1} + u_2 b^{n-2} + b^{n-1} - v \leq u_0 b^n + u_1 b^{n-1} + u_2 b^{n-2} \leq u$. In other words, $u < u$, and this is a contradiction.

21. Assume that $u - qv < (1 - 3/b)v$; then $\hat{q}(v - b^{n-2}) < \hat{q}(v_1 b^{n-1} + v_2 b^{n-2}) \leq u_0 b^n + u_1 b^{n-1} + u_2 b^{n-2} \leq u$; hence $1 = \hat{q} - q < u/(v - b^{n-2}) - u/v + 1 + 3/b$;

that is,

$$\frac{3}{b} < \frac{u}{v}\left(\frac{b^{n-2}}{v - b^{n-2}}\right) < \left(\frac{b^{n-1}}{v - b^{n-2}}\right).$$

Finally, therefore, $b^n + 3b^{n-2} > 3v$; but this contradicts the size of v_1, unless $b = 3$, and the exercise is obviously true when $b = 3$.

22. Let $u = 4100, v = 588$. We first try $\hat{q} = \frac{41}{5} = 8$, but $8 \cdot 8 > 10(41 - 40) + 0 = 10$. Then we set $\hat{q} = 7$, and now we find $7 \cdot 8 < 10(41 - 35) + 0 = 60$. But 7 times 588 equals 4116, so the true quotient is $q = 6$. (Incidentally, this example shows that Theorem B cannot be improved under the given hypotheses, when $b = 10$.)

23. Obviously $v\lfloor b/(v+1)\rfloor < (v+1)\lfloor b/(v+1)\rfloor \le (v+1)b/(v+1) = b$; also if $v \ge \lfloor b/2\rfloor$ we obviously have $v\lfloor b/(v+1)\rfloor \ge v \ge \lfloor b/2\rfloor$. Finally, assume that $1 \le v < \lfloor b/2\rfloor$. Then $v\lfloor b/(v+1)\rfloor > v(b/(v+1) - 1) \ge b/2 - 1 \ge \lfloor b/2\rfloor - 1$, because $v(b/(v+1) - 1) - b/2 + 1 = (b/2 - v - 1)(v - 1)/(v+1) \ge 0$. Finally since $v\lfloor b/(v+1)\rfloor > \lfloor b/2\rfloor - 1$, we must have $v\lfloor b/(v+1)\rfloor \ge \lfloor b/2\rfloor$.

24. The probability is only $\log_b 2$, *not* $\frac{1}{2}$! (For example, if $b = 2^{35}$, the probability is approximately $\frac{1}{35}$; this is still high enough to warrant the special test for $d = 1$ in steps D1 and D8.)

25.

002		ENTA	1	1	
003		ADD	V+1	1	
004		STA	TEMP	1	
005		ENTA	1	1	
006		JOV	1F	1	Jump if $v_1 = b - 1$.
007		ENTX	0	1	
008		DIV	TEMP	1	Otherwise compute $b/(v_1 + 1)$.
009		JOV	DIVBYZERO	1	
010	1H	STA	D	1	
011		DECA	1	1	
012		JANZ	*+3	1	Jump if $d \ne 1$.
013		STZ	U	$1 - A$	
014		JMP	D2	$1 - A$	
015		ENT1	N	A	Multiply v by d.
016		ENTX	0	A	
017	2H	STX	CARRY	AN	
018		LDA	V,1	AN	
019		MUL	D	AN	
...					(as in exercise 13)
026		J1P	2B	AN	
027		ENT1	M+N	A	(Now rX = 0.)
028	2H	STX	CARRY	$A(M + N)$	Multiply u by d.
029		LDA	U,1	$A(M + N)$	
...					(as in exercise 13)
037		J1P	2B	$A(M + N)$	
038		STX	U	A	∎

26. (See the algorithm of exercise 16.)

101	D8	LDA	D	1	(Remainder will be left in loca-
102		DECA	1	1	tions U+M+1 through U+M+N)
103		JAZ	DONE	1	Terminate if $d = 1$.
104		ENN1	N	A	$rI1 \equiv j - n - 1$; $j \leftarrow 1$.
105		ENTA	0	A	$r \leftarrow 0$.
106	1H	LDX	U+M+N+1,1	AN	$rAX \leftarrow rb + u_{m+j}$.
107		DIV	D	AN	
108		STA	U+M+N+1,1	AN	
109		SLAX	5	AN	$r \leftarrow (rb + u_{m+j}) \bmod d$.
110		INC2	1	AN	$j \leftarrow j + 1$.
111		J2N	1B	AN	Repeat for $1 \leq j \leq n$. ∎

At this point, the division routine is complete; and by the next exercise, register AX is zero.

27. It is $du \bmod dv = d(u \bmod v)$.

28. For convenience, let us assume v has a decimal point at the left, that is, $v = (v_0.v_1v_2\cdots)$. After step N1 we have $\frac{1}{2} \leq v < 1 + 1/b$: for

$$v\left\lfloor \frac{b+1}{v_1+1} \right\rfloor \leq \frac{v(b+1)}{v_1+1} = \frac{v(1+1/b)}{(1/b)(v_1+1)} < 1 + \frac{1}{b},$$

and

$$v\left\lfloor \frac{b+1}{v_1+1} \right\rfloor \geq \frac{v(b+1-v_1)}{v_1+1} \geq \frac{1}{b}\frac{v_1(b+1-v_1)}{v_1+1}.$$

The latter quantity takes its smallest value when $v_1 = 1$, since it is a convex function and the other extreme value is greater.

The formula in step N2 may be rewritten

$$v \leftarrow \left\lfloor \frac{b(b+1)}{v_1+1} \right\rfloor \frac{v}{b},$$

so we see as above that v will never become $\geq 1 + 1/b$.

The minimum value of v after one iteration of step N2 is $\geq$

$$\left(\frac{b(b+1)-v_1}{v_1+1}\right)\frac{v}{b} \geq \left(\frac{b(b+1)-v_1}{v_1+1}\right)\frac{v_1}{b^2} = \left(\frac{b(b+1)+1-t}{t}\right)\left(\frac{t-1}{b^2}\right)$$

$$= 1 + \frac{1}{b} + \frac{2}{b^2} - \frac{1}{b^2}\left(t + \frac{b(b+1)+1}{t}\right),$$

if $t = v_1 + 1$. The minimum of this quantity occurs for $t = b/2 + 1$; a lower bound is $1 - 3/2b$. Hence $v_1 \geq b - 2$, after one iteration of step N2. Finally, we have $(1 - 3/2b)(1 + 1/b)^2 > 1$, when $b \geq 5$, so at most two more iterations are needed. The assertion is easily verified when $b < 5$.

29. True, since $(u_j \cdots u_{j+n})_b < v$. (But in Program D, u_j is left as $b - 1$ if step D6 is necessary, since there was no need to reset u_j when it was never being used in the subsequent calculation.)

30. In Algorithms A and S, such overlap is possible if the algorithms are rewritten slightly; e.g., in Algorithm A, we could rewrite step A2 thus: "Set $t \leftarrow u_j + v_j + k$, $w_j \leftarrow t \bmod b$, $k \leftarrow \lfloor t/b \rfloor$."

In Algorithm M, v_j may be in the same location as w_j. In Algorithm D, it is most convenient (as in Program D, exercise 26) to let $r_1 \cdots r_n$ be the same as $u_{m+1} \cdots u_{m+n}$; and we can also have $q_0 \cdots q_m$ the same as $u_0 \cdots u_m$, provided that no alteration of u_j is made in step D6. (Such an alteration is unnecessary, as mentioned in exercise 29.)

31. One possibility is this, if we consider the situation of Fig. 6 with $u = u_j u_{j+1} \cdots u_{j+n}$ as in Algorithm D: If the leading nonzero digits of u and v have the same sign, set $r \leftarrow u - v$, $q \leftarrow 1$; otherwise set $r \leftarrow u + v$, $q \leftarrow -1$. Now if $|r| > |u|$, or if $|r| = |u|$ and the first nonzero digit of $u_{j+n+1} \cdots u_{m+n}$ has the same sign as the first nonzero digit of r, set $q \leftarrow 0$; otherwise set $u_j \cdots u_{j+n} \leftarrow r$.

SECTION 4.3.2

1. The solution is unique since $7 \cdot 11 \cdot 13 = 1001$. The "constructive" proof of Theorem C tells us that the answer is $\left((11 \cdot 13)^6 + 6 \cdot (7 \cdot 13)^{10} + 5 \cdot (7 \cdot 11)^{12}\right)$ $\bmod 1001$. This answer is perhaps not explicit enough! By (23) we have $v_1 = 1$, $v_2 = (6 - 1) \cdot 8 \bmod 11 = 7$, $v_3 = \left((5 - 1) \cdot 2 - 7\right) \cdot 6 \bmod 13 = 6$, so $u = 6 \cdot 7 \cdot 11 + 7 \cdot 7 + 1 = 512$.

2. Yes (by the "constructive" proof given, but not by the "nonconstructive" proof).

3. $u \equiv u_i$ (modulo m_i) implies that $u \equiv u_i$ (modulo gcd (m_i, m_j)), so the condition $u_i \equiv u_j$ (modulo gcd (m_i, m_j)) must surely hold if there is a solution. Furthermore if $u \equiv v$ (modulo m_j) for all j, then $u - v$ is a multiple of lcm $(m_1, \ldots, m_r) = m$; hence there is at most one solution.

The proof can now be completed in a nonconstructive manner by counting the number of different sets $(u_1, \ldots, u_r)$ satisfying the conditions $0 \le u_j < m_j$ and $u_i \equiv u_j$ (modulo gcd (m_i, m_j)). If this number is m, there must be a solution since $(u \bmod m_1, \ldots, u \bmod m_r)$ takes on m distinct values as u goes from a to $a + m$. Assume that $u_1, \ldots, u_{r-1}$ have been chosen satisfying the given conditions; we must now pick $u_r \equiv u_j$ (modulo gcd (m_j, m_r)) for $1 \le j < r$, and by the generalized Chinese Remainder Theorem for $r - 1$ elements there are

$$m_r/\text{lcm} \left(\gcd (m_1, m_r), \ldots, \gcd (m_{r-1}, m_r)\right) = m_r/\gcd \left(\text{lcm} (m_1, \ldots, m_{r-1}), m_r\right)$$
$$= \text{lcm} (m_1, \ldots, m_r)/\text{lcm} (m_1, \ldots, m_{r-1})$$

ways to do this. [This proof is based on identities (10), (11), (12), and (14) of Section 4.5.2.]

A constructive proof [A. S. Fraenkel, *Proc. AMS* **14** (1963), 790–791] generalizing (24) can be given as follows. Let $M_j = \text{lcm} (m_1, \ldots, m_j)$; we wish to find $u = v_r M_{r-1} + \cdots + v_2 M_1 + v_1$, where $0 \le v_j < M_j/M_{j-1}$. Assume that $v_1, \ldots, v_{j-1}$ have already been determined; then we must solve the congruence

$$v_j M_{j-1} + v_{j-1} M_{j-2} + \cdots + v_1 \equiv u_j \pmod{m_j}.$$

Here $v_{j-1} M_{j-2} + \cdots + v_1 \equiv u_i \equiv u_j$ (modulo gcd (m_i, m_j)) for $i < j$ by hypothesis, so $c = u_j - (v_{j-1}M_{j-2} + \cdots + v_1)$ is a multiple of

$$\text{lcm}\big(\gcd(m_1, m_j), \ldots, \gcd(m_{j-1}, m_j)\big) = \gcd(M_{j-1}, m_j) = d_j.$$

We therefore must solve $v_j M_{j-1} \equiv c \pmod{m_j}$. By Euclid's algorithm there is a number c_j such that $c_j M_{j-1} \equiv d_j \pmod{m_j}$; hence we may take

$$v_j = (c_j c)/d_j \bmod (m_j/d_j).$$

Note that, as in the nonconstructive proof, we have $m_j/d_j = M_j/M_{j-1}$.

4. (After $m_4 = 91 = 7 \cdot 13$, we have used up all products of two or more odd primes that can be less than 100, so $m_5, \ldots$ must all be prime.)

$m_7 = 79,$	$m_8 = 73,$	$m_9 = 71,$	$m_{10} = 67,$	$m_{11} = 61,$
$m_{12} = 59,$	$m_{13} = 53,$	$m_{14} = 47,$	$m_{15} = 43,$	$m_{16} = 41,$
$m_{17} = 37,$	$m_{18} = 31,$	$m_{19} = 29,$	$m_{20} = 23,$	$m_{21} = 17,$

and then we are stuck ($m_{22} = 1$ does no good).

5. No; an obvious upper bound is

$$3^4 5^2 7^2 11^1 \cdots = \prod_{\substack{p \text{ odd} \\ p \text{ prime}}} p^{\lfloor \log_p 100 \rfloor}.$$

This upper bound is attained if we choose $m_1 = 3^4$, $m_2 = 5^2$, etc. (It is more difficult, however, to maximize $m_1 \cdots m_r$ when r is fixed, or to maximize $m_1 + \cdots + m_r$ as we would using moduli $2^{m_j} - 1$.)

6. (a) If $e = f + kg$, then $2^e = 2^f(2^g)^k \equiv 2^f \cdot 1^k \pmod{2^g - 1}$. So if $2^e \equiv 2^f \pmod{2^g - 1}$, then $2^{e \bmod g} \equiv 2^{f \bmod g} \pmod{2^g - 1}$, and since the latter quantities lie between zero and $2^g - 1$ we must have $e \bmod g = f \bmod g$.

(b) By part (a),

$$(1 + 2^d + \cdots + 2^{(c-1)d}) \cdot (2^e - 1) \equiv (1 + 2^d + \cdots + 2^{(c-1)d}) \cdot (2^d - 1)$$
$$= 2^{cd} - 1 \equiv 2^{ce} - 1 \equiv 2^1 - 1 = 1 \pmod{2^f - 1}.$$

7. $(u_j - (v_1 + m_1(v_2 + m_2(v_3 + \cdots + m_{j-2}v_{j-1}) \ldots)))c_{1j} \cdots c_{(j-1)j}$
$$= (u_j - v_1)c_{1j} \cdots c_{(j-1)j} - m_1 v_2 c_{1j} \cdots c_{(j-1)j} - \cdots$$
$$\qquad\qquad - m_1 \cdots m_{j-2} v_{j-1} c_{1j} \cdots c_{(j-1)j}$$
$$\equiv (u_j - v_1)c_{1j} \cdots c_{(j-1)j} - v_2 c_{2j} \cdots c_{(j-1)j} - \cdots - v_{j-1}c_{(j-1)j}$$
$$= (\cdots ((u_j - v_1)c_{1j} - v_2)c_{2j} - \cdots - v_{j-1})c_{(j-1)j} \pmod{m_j}.$$

This method of rewriting the formulas uses the same number of arithmetic operations and fewer constants; but the number of constants is fewer only if we order the moduli so that $m_1 < m_2 < \cdots < m_r$, otherwise we would need a table of $m_i \bmod m_j$. This ordering of the moduli might seem to require more computation than if we made m_1 the largest, m_2 the next largest, etc., since there are many more operations to be done modulo m_r than modulo m_1; but since v_j can be as large as $m_j - 1$, we are better off with $m_1 < m_2 < \cdots < m_r$ in (23) also. So this idea appears to be preferable to the formulas in the text, although the formulas in the text are advantageous when the moduli have the form (14), as shown in Section 4.3.3.

8. $m_{j-1} \cdots m_1 v_j \equiv m_{j-1} \cdots m_1 (\cdots ((u_j - v_1)c_{1j} - v_2)c_{2j} - \cdots - v_{j-1})c_{(j-1)j} \equiv$
$m_{j-2} \cdots m_1 (\cdots (u_j - v_1)c_{1j} - \cdots - v_{j-2})c_{(j-2)j} - v_{j-1}m_{j-2} \cdots m_1 \equiv \cdots \equiv$
$u_j - v_1 - v_2 m_1 - \cdots - v_{j-1}m_{j-2} \cdots m_1$ (modulo m_j).

9. $u_r \leftarrow ((\cdots (v_r m_{r-1} + v_{r-1})m_{r-2} + \cdots)m_1 + v_1) \bmod m_r, \ldots,$
$$u_2 \leftarrow (v_2 m_1 + v_1) \bmod m_2, \quad u_1 \leftarrow v_1 \bmod m_1.$$

(The computation should be done in this order, if we want to let u_j and v_j share the same memory locations as is possible in (23).)

10. If we redefine the "mod" operator so that it produces residues in the symmetrical range, the basic formulas (2), (3), (4) for arithmetic and (23), (24) for conversion remain the same, and the number u in (24) lies in the desired range (10). (Here (24) is a *balanced mixed radix* notation, generalizing "balanced-ternary" notation.) The comparison of two numbers may still be done from left to right, in the simple manner described in the text. Furthermore, it is possible to retain the value u_j in a single computer word, if we have signed magnitude representation within the computer, even if m_j is almost twice the word size. But the arithmetic operations analogous to (11) and (12) are more difficult, so it appears that on most computers this idea would result in slightly slower operation.

11. Multiply by

$$\frac{m+1}{2} = \left(\frac{m_1+1}{2}, \ldots, \frac{m_r+1}{2} \right).$$

Note that

$$2t \cdot \frac{m+1}{2} \equiv t \quad (\text{modulo } m).$$

In general if v is relatively prime to m, then we can find (by Euclid's algorithm) a number $v' = (v'_1, \ldots, v'_r)$ such that $vv' \equiv 1$ (modulo m); and then if u is known to be a multiple of v we have $u/v = uv'$, where the latter is computed with modular multiplication. When v is not relatively prime to m, division is much more difficult.

12. Obvious from (11), if we replace m_j by m.

13. (a) $x^2 - x = (x-1)x \equiv 0$ (modulo 10^n)   is equivalent to $(x-1)x \equiv 0$ (modulo p^n) for $p = 2$ and 5. Either x or $x - 1$ must be a multiple of p, and then the other is relatively prime to p^n; so either x or $x - 1$ must be a multiple of p^n. If $x \bmod 2^n = x \bmod 5^n = 0$ or 1, we must have $x = 0$ or 1; hence $x \bmod 2^n \not\equiv x \bmod 5^n$. (b) If $x = qp^n + r$, where $r = 0$ or 1, then $r \equiv r^2 \equiv r^3$, so $3x^2 - 2x^3 \equiv (6qp^n r + 3r) - (6qp^n r + 2r) \equiv r(\text{modulo } p^n)$.   (c) Let $c' = (3(cx^n)^2 - 2(cx^n)^3)/x^{2n} = 3c^2 - 2c^3 x^n$.

Note: Since the last k digits of an n-digit automorph form a k-digit automorph, it makes sense to speak of the two ∞-digit automorphs, x and $1 - x$, which are 10-adic numbers (cf. exercise 4.1–31). The set of 10-adic numbers is equivalent under modular arithmetic to the set of ordered pairs (u_1, u_2), where u_1 is a 2-adic number and u_2 is a 5-adic number.

SECTION 4.3.3

1.

27 × 47:	18 × 42:	09 × 05:	2718 × 4742:
08	04	00	1269
08	04	00	1269
−15	14	−45	−0045
49	16	45	0756
49	16	45	0756
1269	0756	0045	12888756

2. $\sqrt{Q + \lfloor\sqrt{Q}\rfloor} \leq \sqrt{Q + \sqrt{Q}} < \sqrt{Q + 2\sqrt{Q} + 1} = \sqrt{Q} + 1$, so $\lfloor\sqrt{Q + R}\rfloor \leq \lfloor\sqrt{Q}\rfloor + 1$.

3. When $k \leq 2$, the result is true, so assume that $k > 2$. Let $q_k = 2^{Q_k}$, $r_k = 2^{R_k}$, so that $R_k = \lfloor\sqrt{Q_k}\rfloor$ and $Q_k = Q_{k-1} + R_{k-1}$. We must show that $1 + (R_k + 1)2^{R_k} \leq 2^{Q_{k-1}}$; that is, $2^{R_{k-1}} + (R_k + 1)2^{R_k + R_{k-1}} \leq 2^{Q_k}$. But $1 + (R_k + 1)2^{R_k} \leq 1 + 2^{2R_k}$ and $2R_k < Q_{k-1}$ when $k > 2$. (The fact that $2R_k < Q_{k-1}$ is readily proved by induction since $R_{k+1} - R_k \leq 1$ and $Q_k - Q_{k-1} \geq 2$.)

This inequality is not close at all, so there are many other ways to prove it.

4. For $j = 1, \ldots, r$, calculate $U_e(j^2)$, $jU_o(j^2)$, $V_e(j^2)$, $jV_o(j^2)$; and by recursively calling the multiplication algorithm, calculate

$$W(j) = (U_e(j^2) + jU_o(j^2))(V_e(j^2) + jV_o(j^2)),$$

$$W(-j) = (U_e(j^2) - jU_o(j^2))(V_e(j^2) - jV_o(j^2));$$

and then we have $W_e(j^2) = \frac{1}{2}(W(j) + W(-j))$, $W_o(j^2) = \frac{1}{2}(W(j) - W(-j))$. Also calculate $W_e(0) = U(0)V(0)$. Now construct difference tables for W_e and W_o, which are polynomials whose respective degrees are r and $r - 1$.

This method reduces the size of the numbers being handled, and reduces the number of additions and multiplications. Its only disadvantage is a longer program (since the control is somewhat more complex, and some of the calculations must be done with signed numbers).

Another possibility would perhaps be to evaluate W_e and W_o at $1^2, 2^2, 4^2, \ldots,$ $(2^r)^2$; although the numbers involved are larger, the calculations are faster, since all multiplications are replaced by shifting and all divisions are by binary numbers of the form $2^j(2^k - 1)$. (Simple procedures are available for dividing by such numbers.)

5. Start the q, r sequences out with q_0, q_1 large enough so that the inequality in exercise 3 is valid. Then we will find in the formulas analogous to those preceding Theorem C that $\eta_1 \to 0$ and $\eta_2 = (1 + 1/2r_k)2^{1+\sqrt{2Q_k}-\sqrt{2Q_{k+1}}}(Q_k/Q_{k+1})$. The factor $Q_k/Q_{k+1} \to 1$ as $k \to \infty$, so we can ignore it if we want to show $\eta_2 < 1 - \epsilon$ for all large k. Now $\sqrt{2Q_{k+1}} = \sqrt{2Q_k + 2\lceil\sqrt{2Q_k}\rceil + 2} \geq \sqrt{(2Q_k + 2\sqrt{2Q_k} + 1) + 1} \geq \sqrt{2Q_k} + 1 + 1/3R_k$. Hence $\eta_2 \leq (1 + 1/2r_k)2^{-1/3R_k}$, and $\log_2 \eta_2 \leq 1/2(\ln 2)r_k - 1/3R_k < 0$ for large enough k.

Note: Algorithm C can also be modified to define a sequence $q_0, q_1, \ldots$ of a similar type which is based on n, so that $n \approx q_k + q_{k+1}$ after step C1. (Algorithm S essentially does this.) This would establish (39).

6. Any common divisor of $6q + d_1$ and $6q + d_2$ must also divide their difference $d_2 - d_1$. The $\binom{6}{2}$ differences are 2, 3, 4, 6, 8, 1, 2, 4, 6, 1, 3, 5, 2, 4, 2, so we must only show that at most one of the given numbers is divisible by each of the primes 2, 3, 5. Clearly only $6q + 2$ is even, and only $6q + 3$ is a multiple of 3; and there is at most one multiple of 5, since $q_k \not\equiv 3$ (modulo 5).

7. $t_k \le 6t_{k-1} + ck3^k$ for some constant c; so $t_k/6^k \le t_{k-1}/6^{k-1} + ck/2^k \le t_0 + c \sum_{j \ge 1} (j/2^j) = M$. Thus $t_k \le M \cdot 6^k$.

8. The analog of Eq. (39) would break down. In fact it is impossible to "untransform" the finite Fourier transform of (a_0, a_1, a_2, a_3) mod 15 when $\omega = 2$, since information is lost during the transformation process.

9. W_r''' is the unique number in the range $0 \le W_r''' < 2^k(2^{2L} + 1)$ which is $\equiv W_r'$ (modulo 2^k) and $\equiv W_r''$ (modulo $2^{2L} + 1$).

10. The most essential use is in the definition of ψ. [Schönhage observes that we could actually allow $k = \ell + 2$, since $(2^{3 \cdot 2^{n-2}} - 2^{2^{n-2}})^2 \equiv 2$ (modulo $2^{2^n} + 1$).]

11. An automaton cannot have $z_2 = 1$ until it has $c \ge 2$, and this occurs first for M_j at time $3j - 1$. It follows that M_j cannot have $z_2 z_1 z_0 \ne 000$ until time $3(j - 1)$. Furthermore, if M_j has $z_0 \ne 0$ at time t, we cannot change this to $z_0 = 0$ without affecting the output; but the output cannot be affected by this value of z_2 until at least time $t + j - 1$, so we must have $t + j - 1 \le 2n$. Since the first argument we gave proves that $3(j - 1) \le t$, we must have $4(j - 1) \le 2n$, that is, $j - 1 \le n/2$, i.e., $j \le \lfloor n/2 \rfloor + 1$.

Furthermore, this is the best possible bound, since the inputs $u = v = q = 2^n - 1$ require the use of M_j for all $j \le \lfloor n/2 \rfloor + 1$. (For example, note from Table 1 that M_2 is needed to multiply two-bit numbers, at time 3.)

13. If it takes $T(n)$ cycles to multiply n-bit numbers, we can accomplish the multiplication of m-bit by n-bit by breaking the n-bit number into $\lceil n/m \rceil$ m-bit groups, using $\lceil n/m \rceil T(m) + O(n + m)$ operations. The results of this section therefore give an estimated running time of $O(n \log m \log \log m)$.

SECTION 4.4

1. We compute $(\cdots (a_m b_{m-1} + a_{m-2})b_{m-2} + \cdots + a_1)b_1 + a_0$ by adding and multiplying in the B_j-system.

	T.	= 20(cwt.	= 8(st.	= 14(lb.	= 16 oz.)))
Start with zero	0	0	0	0	0
Add 3	0	0	0	0	3
Multiply by 24	0	0	0	4	8
Add 9	0	0	0	5	1
Multiply by 60	0	2	5	9	12
Add 12	0	2	5	10	8
Multiply by 60	8	3	1	0	0
Add 37	8	3	1	2	5

(Addition and multiplication by a constant in a mixed radix system are readily done using a simple generalization of the usual carry rule; cf. exercise 4.3.1–9.)

2. We compute $\lfloor u/B_0 \rfloor$, $\lfloor \lfloor u/B_0 \rfloor /B_1 \rfloor$, etc., and the remainders are A_0, A_1, etc. The division is done in the b_j-system.

	d.	= 24(h.	= 60(m.	= 60s.))	
Start with u	3	9	12	37	
Divide by 16	0	5	4	32	Remainder = 5
Divide by 14	0	0	21	45	Remainder = 2
Divide by 8	0	0	2	43	Remainder = 1
Divide by 20	0	0	0	8	Remainder = 3
Divide by ∞	0	0	0	0	Remainder = 8

Answer: 8 T. 3 cwt. 1 st. 2 lb. 5 oz.

3. [*CACM* **2** (July, 1959), 27.]

A1. Set $k \leftarrow 0$, $U_0 \leftarrow 0$.

A2. If $u < \epsilon$, terminate the algorithm with $U = (U_0.U_{-1} \cdots U_{-k})_B$. If $1 - u < \epsilon$, go to A4.

A3. Set $k \leftarrow k+1$, $U_{-k} \leftarrow \lfloor Bu \rfloor$, $u \leftarrow Bu \bmod 1$, $\epsilon \leftarrow B\epsilon$, and return to A2.

A4. Set $j \leftarrow k$, and then set $U_{-j} \leftarrow 0$, $j \leftarrow j - 1$ zero or more times until $U_{-j} \neq B - 1$. Finally, set $U_{-j} \leftarrow U_{-j} + 1$, and terminate the algorithm with $U = (U_0.U_{-1} \cdots U_{-k})_B$. ∎

(*Note:* It is possible for this algorithm to yield $U = 1$; if this is undesirable, we could modify the algorithm as follows: In step A1, set a new variable $t \leftarrow 1$. In step A2, do not go to A4 if $t = 1$. In step A3, set $t \leftarrow 0$ if $U_{-k} \neq B - 1$. The resulting algorithm now converts the fewest number of places, subject to the condition that the answer must be less than unity.)

4. (a) $1/2^k = 5^k/10^k$. (b) Every prime divisor of b divides B.

5. If any number rounds to unity, so does the largest number, $1 - b^{-m}$. This number rounds to unity if and only if it is $\geq 1 - \frac{1}{2}B^{-M}$.

7. $\alpha u \leq ux \leq \alpha u + u/w \leq \alpha u + 1$, hence $\lfloor \alpha u \rfloor \leq \lfloor ux \rfloor \leq \lfloor \alpha u + 1 \rfloor$.

8.
```
       ENT1  0
       LDA   U
  1H   MUL   =1//10=
       STA   TEMP
       MUL   =-10=
       SLAX  5
       ADD   U
       ENT2  0
       JANN  2F
       INCA  10
       ENT2  1
  2H   STA   ANSWER,1        (May be minus zero.)
       LDA   TEMP
       DECA  0,2
       INC1  1
       JAP   1B        ∎
```

9. If x' is an integer, $x - \epsilon \leq x' \leq x$, then $(1 + 1/n)x - ((1 + 1/n)\epsilon + 1 - 1/n) \leq x' + \lfloor x'/n \rfloor \leq (1 + 1/n)x$. Hence if α is the binary fraction

$$(.00011001100110011001100110011)_2,$$

we find that at the end of the computation, $\alpha u - \epsilon \leq v \leq \alpha u$, where

$$\epsilon = \tfrac{7}{8} + (.10001000101010001100\cdots)_2 < \tfrac{3}{2}.$$

Hence $u/10 - 2 < u/10 - (\epsilon + (1/10 - \alpha)u) \leq v \leq \alpha u < u/10$. Since v is an integer, the proof is complete.

10. (a) Shift right one; (b) Extract left bit of each group; (c) Shift result of (b) right two; (d) Shift result of (c) right one, and add to result of (c); (e) Subtract result of (c) from result of (a).

11.

```
    5.7 7 2 1
  - 1 0
    ─────────
  4 7.7 2 1
  -   9 4
    ─────────
  3 8 3.2 1
  -     7 6 6
    ─────────
  3 0 6 6.1
  -     6 1 3 2
    ─────────
  2 4 5 2 9     Answer: (24529)₁₀.
```

$Answer: (24529)_{10}.$

12. First convert the ternary number to nonary (radix 9) notation, then proceed as in octal-to-decimal conversion but without doubling. In the given example, we have

```
      1.7 6 4 7 2 3
  -   1
    ───────────────
    1 6.6 4 7 2 3
  -   1 6
    ───────────────
  1 5 0.4 7 2 3
  -   1 5 0
    ───────────────
  1 3 5 4.7 2 3
  -   1 3 5 4
    ───────────────
1 2 1 9 3.2 3
  - 1 2 1 9 3
    ───────────────
1 0 9 7 3 9.3
- 1 0 9 7 3 9
    ───────────────
  9 8 7 6 5 4     Answer: (987654)₁₀.
```

$Answer: (987654)_{10}.$

13.

START	JOV	OFLO	Ensure overflow is off.
	ENT2	-24	Set buffer pointer.
8H	ENT3	10	Set loop counter.
1H	ENT1	m	Begin multiplication routine.
	ENTX	0	
2H	STX	CARRY	

```
               . . .                    (See exercise 4.3.1–13, with
        J1P    2B                           v = 10⁹ and W = U.)
        SLAX   5                        rA ← next nine digits.
        CHAR
        STA    BUF2,2(2:5)              Store next nine digits.
        STX    BUF2+1,2
        INC2   2                        Increase buffer pointer.
        DEC3   1
        J3P    1B                       Repeat ten times.
        OUT    -20,2(PRINTER)
        J2P    DONE                     Have we printed both lines?
        ENT2   0                        Set buffer pointer for 2nd buffer.
        JMP    8B
BUF1    ALF    .                        (Radix point on first line)
        ORIG   *+23
BUF2    ORIG   *+24             ▮
```

14. Let $K(n)$ be the number of steps required to convert an n-digit decimal number to binary and at the same time to compute the binary representation of 10^n. Then we have $K(2n) \leq 2K(n) + O(M(n))$. *Proof:* Given the number $U = (u_{2n-1} \cdots u_0)_{10}$, compute $U_1 = (u_{2n-1} \cdots u_n)_{10}$ and $U_0 = (u_{n-1} \cdots u_0)_{10}$ and 10^n, in $2K(n)$ cycles, then compute $U = 10^n U_1 + U_0$ and $10^{2n} = 10^n \cdot 10^n$ in $O(M(n))$ cycles. It follows that $K(2^n) = O(M(2^n) + 2M(2^{n-1}) + 4M(2^{n-2}) + \ldots) = O(M(2^n) + M(2^n) + M(2^n) + \ldots, n$ terms.

(Similarly, Schönhage has observed that we can convert a $(2^n \log_2 10)$-bit number U from binary to decimal, in $O(nM(2^n))$ steps. First form $V = 10^{2^{n-1}}$ in $O(M(2^{n-1}) + M(2^{n-2}) + \ldots) = O(M(2^n))$ steps, then compute $U_0 = (U \bmod V)$ and $U_1 = \lfloor U/V \rfloor$ in $O(M(2^n))$ further steps, then convert U_0 and U_1.)

18. Let $U = \text{round}_B(u, P)$ and $v = \text{round}_b(U, p)$. We may assume that $u \neq 0$, so that $U \neq 0$ and $v \neq 0$. *Case 1, $v < u$:* Determine e and E such that $b^{e-1} < u \leq b^e$, $B^{E-1} \leq U < B^E$. Then $u < U + \frac{1}{2}B^{E-P}$ and $U < u - \frac{1}{2}b^{e-p}$; hence $B^{P-1} \leq B^{P-E}U < B^{P-E}u < b^{p-e}u \leq b^p$. *Case 2, $v > u$:* Determine e and E such that $b^{e-1} \leq u < b^e$, $B^{E-1} < U \leq B^E$. Then $u \geq U - \frac{1}{2}B^{E-P}$ and $U \geq u + \frac{1}{2}b^{e-p}$; hence $B^{P-1} \leq B^{P-E}(U - B^{E-P}) < B^{P-E}u \leq b^{p-e}u < b^p$. Thus we have proved that $B^{P-1} < b^p$ whenever $v \neq u$.

Conversely, if $B^{P-1} < b^p$, the above proof suggests that the most likely example for which $u \neq v$ will occur when u is a power of b and at the same time u is close to a power of B. We have $B^{P-1}b^p < B^{P-1}b^p + \frac{1}{2}b^p - \frac{1}{2}B^{P-1} - \frac{1}{4} = (B^{P-1} + \frac{1}{2}) \times (b^p - \frac{1}{2})$; hence $1 < \alpha = 1/(1 - \frac{1}{2}b^{-p}) < 1 + \frac{1}{2}B^{1-P} = \beta$. There is a solution to the equation $\log_B \alpha < e \log_B b - E < \log_B \beta$, since Weyl's theorem (exercise 3.5–22) implies that there is an integer e with $0 < \log_B \alpha < (e \log_B b) \bmod 1 < \log_B \beta < 1$ when $\log_B b$ is irrational. Hence $\alpha < b^e/B^E < \beta$, for some e and E. (Such e and E may also be found by applying the theory of continued fractions, see Section 4.5.3.) Now we have $\text{round}_B(b^e, P) = B^E$, and $\text{round}_b(B^E, p) < b^e$. [*CACM* **11** (1968), 47–50; *Proc. Amer. Math. Soc.* **19** (1968), 716–723.]

SECTION 4.5.1

1. Test whether or not $uv' < u'v$.

2. If $c > 1$ divides both u/d and v/d, then cd divides both u and v.

3. Let p be prime. If p^e is a divisor of uv and $u'v'$ for $e \geq 1$, then either $p^e \backslash u$ and $p^e \backslash v'$ or $p^e \backslash u'$ and $p^e \backslash v$; hence $p^e \backslash \gcd (u, v') \gcd (u', v)$. The converse follows by reversing the argument.

4. Let $d_1 = \gcd (u, v)$, $d_2 = \gcd (u', v')$; the answer is $w = (u/d_1)(v'/d_2) \operatorname{sign} (v)$, $w' = |(u'/d_2)(v/d_1)|$, with a "divide by zero" error message if $v = 0$.

5. $d_1 = 10$, $t = 17 \cdot 7 - 27 \cdot 12 = -205$, $d_2 = 5$, $w = -41$, $w' = 168$.

6. Let $u'' = u'/d_1$, $v'' = v'/d_1$; we want to show that $\gcd (uv'' + u''v, d_1) = \gcd (uv'' + u''v, d_1 u''v'')$. If p is a prime which divides u'', then p does not divide u or v'', so p does not divide $uv'' + u''v$. A similar argument holds for prime divisors of v'', so no prime divisors of $u''v''$ affect the given gcd.

7. $(N - 1)^2 + (N - 2)^2 = 2N^2 - (6N - 5)$. If the inputs are n-bit binary numbers, $2n + 1$ bits may be necessary to represent t.

8. For multiplication and division these quantities will obey the rules $x/0 = \operatorname{sign} (x) \infty$, $(\pm \infty) \times x = x \times (\pm \infty) = (\pm \infty)/x = \pm \operatorname{sign} (x) \infty$, $x/(\pm \infty) = 0$, provided that x is finite and nonzero, without change to the algorithms described. Furthermore, the algorithms can readily be modified so that $0/0 = 0 \times (\pm \infty) = (\pm \infty) \times 0 = $ "$(0/0)$", where the latter is a representation of "undefined"; and so that if either operand is "undefined" the result will be "undefined" also. Since the multiplication and division subroutines can yield these fairly natural rules of "extended arithmetic," it is sometimes worth while to modify the addition and subtraction operations so that they satisfy the rules $x \pm \infty = \pm \infty$, $x \pm (-\infty) = \mp \infty$, for x finite; $(\pm \infty) + (\pm \infty) = \pm \infty - (\mp \infty) = \pm \infty$, $(\pm \infty) + (\mp \infty) = (\pm \infty) - (\pm \infty) = (0/0)$; and if either or both operands is $(0/0)$, so is the result. Equality tests and comparisons may be treated in a similar manner.

    The above remarks are independent of "overflow" indications. If ∞ is being used to suggest overflow, it is incorrect to let $1/\infty$ be equal to zero, or to let $\infty - \infty$ be equal to zero, lest inaccurate results be regarded as true answers! It is far better to represent overflow by $(0/0)$, and to use the convention that the result of any operation is undefined if at least one of the inputs is undefined. This type of overflow indication has the advantage that final results of an extended calculation reveal exactly which answers are defined and which are not.

9. If $u/u' \neq v/v'$, then

$$1 \leq |uv' - u'v| = u'v'|(u/u') - (v/v')| < |2^{2n}(u/u') - 2^{2n}(v/v')|;$$

two quantities differing by more than unity cannot have the same "floor." (In other words, the first $2n$ bits to the right of the binary point is enough to characterize the value of the fraction, when there are n-bit denominators. We cannot improve this to $2n - 1$ bits, for if $n = 4$ we have $\frac{1}{13} = (.00010011 \cdots)_2$, $\frac{1}{14} = (.00010010 \cdots)_2$.)

11. To divide by $(v + v'\sqrt{5})/v''$, when v and v' are not both zero, multiply by $(v - v'\sqrt{5})v''/(v^2 - 5v'^2)$ and reduce to lowest terms.

SECTION 4.5.2

1. Substitute min, max, $+$ consistently for gcd, lcm, $\times$.

2. For prime p, let $u_p, v_{1p}, \ldots, v_{np}$ be the exponents of p in the canonical factorization of $u, v_1, \ldots, v_n$. By hypothesis, $u_p \leq v_{1p} + \cdots + v_{np}$. We must show that $u_p \leq \min(u_p, v_{1p}) + \cdots + \min(u_p, v_{np})$, and this is certainly true if u_p is greater than or equal to each v_{jp}, or if u_p is less than some v_{jp}.

3. *Solution 1:* A one-to-one correspondence is obtained if we set $u = \gcd(d, n)$, $v = n^2/\text{lcm}(d, n)$ for each divisor d of n^2. *Solution 2:* If $n = p_1^{e_1} \cdots p_r^{e_r}$, the number in each case is $(2e_1 + 1) \cdots (2e_r + 1)$.

4. See exercise 3.2.1.2–15(a).

5. Shift u, v right until neither is a multiple of 3; each iteration sets $t \leftarrow u + v$ or $t \leftarrow u - v$ (whichever is a multiple of 3), and shifts t right until it is not a multiple of 3, then replaces max (u, v) by the result.

u	v	t
13634	24140	10506, 3502;
13634	3502	17136, 5712, 1904;
1904	3502	5406, 1802;
1904	1802	102, 34;
34	1802	1836, 612, 204, 68;
34	68	102, 34;
34	34	0.

6. The probability that both u and v are even is $\frac{1}{4}$; the probability that both are multiples of four is $\frac{1}{16}$; etc. Thus A has the distribution given by the generating function

$$\tfrac{3}{4} + \tfrac{3}{16}z + \tfrac{3}{64}z^2 + \cdots = \frac{\frac{3}{4}}{1 - z/4}.$$

The mean is $\frac{1}{3}$, and the standard deviation is $\sqrt{\frac{2}{9} + \frac{1}{3} - \frac{1}{9}} = \frac{2}{3}$. If u, v are independently and uniformly distributed with $1 \leq u, v < 2^N$, then some small correction terms are needed; the mean is then actually

$$(2^N - 1)^{-2} \sum_{1 \leq k \leq N} (2^{N-k} - 1)^2 = \tfrac{1}{3} - \tfrac{4}{3}(2^N - 1)^{-1} + N(2^N - 1)^{-2}.$$

7. When u, v are not both even, each of the cases (even, odd), (odd, even), (odd, odd) is equally probable, and $B = 1, 0, 0$ in these cases. Hence $B = \frac{1}{3}$ on the average. [Actually, as in exercise 6, a small correction could be given to be strictly accurate when $1 \leq u, v < 2^N$; the probability that $B = 1$ is actually

$$(2^N - 1)^{-2} \sum_{1 \leq k \leq N} (2^{N-k} - 1)2^{N-k} = \tfrac{1}{3} - \tfrac{1}{3}(2^N - 1)^{-1}.$$

8. E is the number of subtraction cycles in which $u > v$, plus one if u is odd after step B1. If we change the inputs from (u, v) to (v, u), the value of C stays unchanged, while E becomes $C - E$ or $C - E - 1$; the latter case occurs iff u and v are both odd after step B1, and this has probability $\frac{1}{3} + \frac{2}{3}/(2^N - 1)$. Hence

$$E_{\text{ave}} = C_{\text{ave}} - E_{\text{ave}} - \tfrac{1}{3} - \tfrac{2}{3}/(2^N - 1).$$

9. The binary algorithm first gets to B6 with $u = 1963$, $v = 1359$; then $t \leftarrow 604$, $302, 151$, etc. The gcd is 302. Using Algorithm X we find that $2 \cdot 31408 - 23 \cdot 2718 = 302$.

10. (a) Two integers are relatively prime iff they are not both divisible by any prime number. (b) Rearrangement of the sum in (a), in terms of the denominators $k = p_1 \cdots p_r$. (Note that each of the sums in (a), (b) is actually finite.) (c) $(n/k)^2 - \lfloor n/k \rfloor^2 = O(n/k)$, so $q_n - \sum_{1 \le k \le n} \mu(k)(n/k)^2 = \sum_{1 \le k \le n} O(n/k) = O(nH_n)$. (d) $\sum_{d \backslash n} \mu(d) = \delta_{1n}$. [In fact we have the more general result

$$\sum_{d \backslash n} \mu(d) \left(\frac{n}{d} \right)^s = n^s - \sum \left(\frac{n}{p} \right)^s + \cdots,$$

as in part (b), where the sums are over the prime divisors of n, and this equals $n^s(1 - 1/p_1^s) \cdots (1 - 1/p_r^s)$ if $n = p_1^{e_1} \cdots p_r^{e_r}$.]

Note: Similarly, a set of k integers is relatively prime with probability $1/(\sum_{n \ge 1} 1/n^k)$.

11. (a) $6/\pi^2$ times $1 + \frac{1}{4} + \frac{1}{9}$, namely $49/6\pi^2 \approx .828$. (b) $6/\pi^2$ times $1/1 + 2/4 + 3/9 + \cdots$, namely ∞! (This is true in spite of the result of exercise 12, and in spite of the fact that the average value of $\ln \gcd(u, v)$ is a small, finite number.)

12. Let $\sigma(n)$ be the number of positive divisors of n. The answer is

$$\sum_{k \ge 1} \sigma(k) \cdot \frac{6}{\pi^2 k^2} = \frac{6}{\pi^2} \left(\sum_{k \ge 1} \frac{1}{k^2} \right)^2 = \frac{\pi^2}{6} \cdot$$

[Thus, the average is *less* than 2, although there are always at least two common divisors when u, v are not relatively prime.]

13. $1 + \frac{1}{9} + \frac{1}{25} + \cdots = 1 + \frac{1}{4} + \frac{1}{9} + \cdots - \frac{1}{4}(1 + \frac{1}{4} + \frac{1}{9} + \cdots)$.

14. $v_1 = \pm v/u_3$, $v_2 = \mp u/u_3$ (the sign depends on whether the number of iterations is even or odd). This follows from the fact that v_1 and v_2 are relatively prime to each other (throughout the algorithm), and that $v_1 u = -v_2 v$. [Hence $v_1 u = \text{lcm}(u, v)$ at the close of the algorithm, but this is not an especially efficient way to compute the least common multiple. For a generalization, see exercise 4.6.1–18.]

G. E. Collins has observed that $|u_1| \le \frac{1}{2} v/u_3$, $|u_2| \le \frac{1}{2} u/u_3$, at the termination of Algorithm X, except in certain trivial cases, since the final value of q is usually ≥ 2. This bounds the size of $|u_1|$, $|u_2|$ throughout the execution of the algorithm.

15. Apply Algorithm X to v and m, thus obtaining a value x such that $xv \equiv 1$ (modulo m). (This can be done by simplifying Algorithm X so that u_2 and v_2 are not computed, since they are never used in the answer.) Then set $w \leftarrow ux \bmod m$. [It follows, as in exercise 30, that this process requires $O(n^2)$ units of time, when it is applied to large n-bit numbers.]

16. (a) Set $t_1 = x + 2y + 3z$; then $3t_1 + y + 2z = 1$, $5t_1 - 3y - 20z = 3$. Eliminate y, then $14t_1 - 14z = 6$: No solution. (b) This time $14t_1 - 14z = 0$. Divide by 14, eliminate t_1; the general solution is $x = 8z - 2$, $y = 1 - 5z$, z arbitrary.

17. Let u_1, u_2, u_3, v_1, v_2, v_3 be multiprecision variables, not just u and v. The extended algorithm will act the same on u_3 and v_3 as Algorithm L does on u and v. New multiprecision operations are to set $t \leftarrow Au_j$, $t \leftarrow t + Bv_j$, $w \leftarrow Cu_j$, $w \leftarrow w + Dv_j$, $u_j \leftarrow t$, $v_j \leftarrow w$ for *all* j, in step L4; also if $B = 0$ in that step to set $t \leftarrow u_j - qv_j$, $u_j \leftarrow v_j$, $v_j \leftarrow t$ for all j and for $q = \lfloor u_3/v_3 \rfloor$. A similar modification is made to step L1 if v_3 is small. The inner loop (steps L2 and L3) is unchanged.

18. If $mn = 0$, the probabilities of the lattice-point model in the text are exact, so we may assume that $m \geq n > 0$. *Valida vi*, the following values have been obtained:

Case 1, $m = n$. From (n, n) we go to $(n - t, n)$ with probability $t/2^t - 5/2^{t+1} + 3/2^{2t}$, for $2 \leq t < n$. (These values are $\frac{1}{16}$, $\frac{7}{64}$, $\frac{27}{256}$, ...) To $(0, n)$ the probability is $n/2^{n-1} - 1/2^n + 1/2^{2n-2}$. To (n, k) the probability is the same as to (k, n). The algorithm terminates with probability $1/2^{n-1}$.

Case 2, $m = n + 1$. From $(n + 1, n)$ we get to (n, n) with probability $\frac{1}{8}$ when $n > 1$, or 0 when $n = 1$; to $(n - t, n)$ with probability $11/2^{t+3} - 3/2^{2t+1}$, for $1 \leq t < n - 1$. (These values are $\frac{5}{16}$, $\frac{1}{4}$, ...) We get to $(1, n)$ with probability $5/2^{n+1} - 3/2^{2n-1}$, for $n > 1$; to $(0, n)$ with probability $3/2^n - 1/2^{2n-1}$.

Case 3, $m \geq n + 2$. The probabilities are given by the following table:

$$(m - 1, n): \quad 1/2 - 3/2^{m-n+2} - \delta_{n1}/2^{m+1};$$
$$(m - t, n): \quad 1/2^t + 3/2^{m-n+t+1}, \ 1 < t < n;$$
$$(m - n, n): \quad 1/2^n + 1/2^m, \ n > 1;$$
$$(m - n - 1, n): 1/2^{n+1} + 1/2^{m-1};$$
$$(m - n - t, n): 1/2^{n+t}, \ 1 < t < m - n;$$
$$(0, n): \quad 1/2^{m-1}.$$

[*Note:* Although these exact probabilities will certainly improve on the lattice-point model considered in the text, they lead to recurrence relations of much greater complexity; and they will not provide the true behavior of Algorithm B, since for example the probability that gcd $(u, v) = 5$ is different from the probability that gcd $(u, v) = 7$.]

19.
$$A_{n+1} = a + \sum_{1 \leq k \leq n} 2^{-k} A_{(n+1)(n-k)} + 2^{-n}b$$
$$= a + \sum_{1 \leq k \leq n} 2^{-k} A_{n(n-k)} + \frac{c}{2}(1 - 2^{-n}) + 2^{-n}b$$
$$= a + \tfrac{1}{2}A_{n(n-1)} + \tfrac{1}{2}(A_n - a) + \frac{c}{2}(1 - 2^{-n});$$

now substitute for $A_{n(n-1)}$ from (36).

20. The paths described in the hint have the same probability, only the subsequent termination of the algorithm has a different probability; thus $\lambda = k + 1$ with probability 2^{-k} times the probability that $\lambda = 1$. Let the latter probability be p. We know from the text that $\lambda = 0$ with approximate probability $\frac{3}{5}$; hence $\frac{2}{5} = p(1 + \frac{1}{2} + \frac{1}{4} + \frac{1}{8} + \cdots) = 2p$. The average is $p(1 + \frac{2}{2} + \frac{3}{4} + \frac{4}{8} + \cdots) = p(1 + \frac{1}{2} + \frac{1}{4} + \frac{1}{8} + \cdots)^2 = 4p$. [The exact probability that $\lambda = 1$ is $\frac{1}{5} - \frac{6}{5}(-\frac{1}{4})^n$ if $m > n \geq 1$, $\frac{1}{5} - \frac{16}{5}(-\frac{1}{4})^n$ if $m = n \geq 2$.]

21. Show that for fixed v and for $2^m < u < 2^{m+1}$, when m is large, each subtraction-shift cycle of the algorithm reduces $\lfloor \log_2 u \rfloor$ by two on the average.

22. Exactly $(N - m)2^{m-1+\delta_{m0}}$ integers u in the range $1 \leq u < 2^N$ have $\lfloor \log_2 u \rfloor = m$ after u has been shifted right until it is odd.

23. The first sum is

$$2^{2N-2} \sum_{0 \leq m < n < N} mn2^{-m-n}((\alpha + \beta)N + \gamma - \alpha m - \beta n).$$

Since

$$\sum_{0 \leq m < n} m2^{-m} = 2 - (n+1)2^{1-n}$$

and

$$\sum_{0 \leq m < n} m(m-1)2^{-m} = 4 - (n^2 + n + 2)2^{1-n},$$

the sum on m is

$$2^{2N-2} \sum_{0 \leq n < N} n2^{-n}((\gamma - \alpha + (\alpha + \beta)N)(2 - (n+1)2^{1-n})$$
$$- \alpha(4 - (n^2 + n + 2)2^{1-n}) - \beta n)$$
$$= 2^{2N-2}\left((\alpha + \beta)N \sum_{0 \leq n < N} n2^{-n}(2 - (n+1)2^{1-n}) + O(1)\right).$$

Thus the coefficient of $(\alpha + \beta)N$ in the answer is found to be $2^{-2}(4 - (\frac{4}{3})^3) = \frac{11}{27}$. A similar argument applies to the other sum.

[*Note:* The *exact* value of the sums may be obtained after some tedious calculation by means of the general formula

$$\sum_{0 \leq k < n} k^m z^k = \frac{m! z^m}{(1-z)^{m+1}} - \sum_{0 \leq k \leq m} \frac{\overline{m-n}^{k} \underline{m-k} z^{n+k}}{(1-z)^{k+1}}$$

which follows from summation by parts.]

24. Solving a recurrence similar to (34), we find that the number of times is A_{mn}, where $A_{00} = 1$, $A_{0n} = (n+3)/2$, $A_{nn} = \frac{8}{5} - (3n + 13)/(9 \cdot 2^n) + \frac{128}{45}(-\frac{1}{4})^n$ if $n \geq 1$, $A_{mn} = \frac{8}{5} - 2/(3 \cdot 2^n) + \frac{16}{15}(-\frac{1}{4})^n$ if $m > n \geq 1$. Since the condition $u = 1$ or $v = 1$ is therefore satisfied only about 1.6 times in an average run, it is not worth making the suggested test each time step B5 is performed. (Of course the lattice model is not completely accurate, but it seems reasonable to believe that it is not too inaccurate for this application.)

26. By induction, the length is $m + \lfloor n/2 \rfloor$ when $m \geq n$, except that when $m = n = 1$ there is *no* path to $(0, 0)$.

27. Let $a_n = (2^n - (-1)^n)/3$; then $a_0, a_1, a_2, \ldots = 0, 1, 1, 3, 5, 11, 21, \ldots$ (This sequence of numbers has an interesting pattern of zeros and ones in its binary representation. Note that $a_n = a_{n-1} + 2a_{n-2}$, and $a_n + a_{n+1} = 2^n$.) For $m > n$, let $u = 2^{m+1} - a_{n+2}$, $v = a_{n+2}$. For $m = n > 0$, let $u = a_{n+2}$, $v = 2a_{n+1}$, or $u = 2a_{n+1}$, $v = a_{n+2}$ (depending on which is larger). Another example for $m = n > 0$ is $u = 2^{n+1} - 1$, $v = 2^{n+1} - 2$; this takes more shifts, and gives $C = n + 1$, $D = 2n$, $E = 1$.

28. This is a problem where it appears to be necessary to prove *more* than was asked just to prove what was asked. Let us prove the following: *If u, v are positive integers, Algorithm B takes $\leq 1 + \lfloor \log_2 \max(u, v) \rfloor$ subtraction cycles; and if equality holds, then $\lfloor \log_2 (u + v) \rfloor > \lfloor \log_2 \max(u, v) \rfloor$.*

For convenience, let us assume that $u \geq v$; let $m = \lfloor \log_2 u \rfloor$, $n = \lfloor \log_2 v \rfloor$; and let us use the "lattice-point" terminology, saying that we are "at point (m, n)." The proof is by induction on $m + n$.

Case 1, $m = n$. Clearly, $\lfloor \log_2 (u + v) \rfloor > \lfloor \log_2 u \rfloor$ in this case. If $u = v$ the result is trivial; otherwise the next subtraction-shift cycle takes us to a point $(m - k, m)$. By induction, at most $m + 1$ further subtraction steps will be required; but if $m + 1$ more *are* needed, we have $\lfloor \log_2((u - v)2^{-r} + v) \rfloor > \lfloor \log_2 v \rfloor$, where $r \geq 1$ is the number of right shifts that were made. This is impossible, since $(u - v)2^{-r} + v < (u - v) + v = u$, so at most m further steps are needed.

Case 2, $m > n$. The next subtraction step takes us to $(m - k, n)$, and at most $1 + \max(m - k, n) \leq m$ further steps will be required. Now if m further steps *are* required, then u has been replaced by $u' = (u - v)2^{-r}$ for some $r \geq 1$. By induction, $\lfloor \log_2 (u' + v) \rfloor \geq m$; hence

$$\lfloor \log_2 (u + v) \rfloor = \lfloor \log_2 2((u - v)/2 + v) \rfloor \geq \lfloor \log_2 2(u' + v) \rfloor \geq m + 1 > \lfloor \log_2 u \rfloor.$$

29. Subtract the kth column from the $2k$th, $3k$th, $4k$th, etc., for $k = 1, 2, 3, \cdots$. The result is a triangular matrix with x_k on the diagonal in column k, where $m = \sum_{d \backslash m} x_d$. It follows that $x_m = \varphi(m)$, so the determinant is $\varphi(1)\varphi(2) \cdots \varphi(n)$. [In general, "Smith's determinant," in which the (i, j) element is $f(\gcd(i, j))$ for an arbitrary function f, is equal to $\prod_{1 \leq m \leq n} \sum_{d \backslash m} \mu(m/d)f(d)$, by the same argument. See L. E. Dickson, *History of the Theory of Numbers* **1** (New York: Chelsea, 1952), 122–123.]

30. To determine A and r such that $u = Av + r$, $0 \leq r < v$, takes

$$O\big((1 + \log A)(\log u)\big)$$

units of time, using ordinary long division. If the quotients during the algorithm are $A_1, A_2, \ldots, A_m$, then $A_1 A_2 \ldots A_m \leq u$, so $\log A_1 + \cdots + \log A_m \leq \log u$. Also $m = O(\log u)$.

31. Since $(a^u - 1) \bmod (a^v - 1) = a^{u \bmod v} - 1$ (cf. Eq. 4.3.2–19), we find that $\gcd(a^m - 1, a^n - 1) = a^{\gcd(m,n)} - 1$ for all positive integers a.

32. Yes, to $O\big(n(\log n)^2(\log \log n)\big)$; see A. Schönhage, *Acta Informatica* **1** (1971), (to appear).

SECTION 4.5.3

1. The running time is about $19.02T + 6$, just a trifle slower than Program 4.5.2A.

2. $\begin{pmatrix} Q_n(x_1, \ldots, x_n) & Q_{n-1}(x_1, \ldots, x_{n-1}) \\ Q_{n-1}(x_2, \ldots, x_n) & Q_{n-2}(x_2, \ldots, x_{n-1}) \end{pmatrix}.$

3. $Q_n(x_1, \ldots, x_n)$.

4. By induction, or by taking the determinant of the matrix product in exercise 2.

5. When the x's are positive, the q's of (9) are positive, and $q_{n+1} > q_{n-1}$; hence (9) is an alternating series of decreasing terms, and it converges if and only if $q_n \to \infty$. By induction, if the x's are greater than ϵ, $q_n \geq c(1 + \epsilon/2)^n$, where c is chosen such that $1 \geq c$, $\epsilon \geq c(1 + \epsilon/2)$. But if $x_n = 1/2^n$ then $q_n \leq 2 - 1/2^n$.

6. It suffices to prove that $A_1 = B_1$; and from the fact that $0 \leq /x_1, \ldots, x_n/ < 1$ whenever $x_1, \ldots, x_n$ are positive integers, $A_1 = \lfloor 1/X_0 \rfloor$ is the only possibility.

7. Only $1\,2 \cdots n$ and $n \cdots 2\,1$. (The variable x_k appears in exactly $F_k F_{n-k}$ terms; hence x_1 and x_n can only be permuted into x_1 and x_n. If x_1 and x_n are fixed by the permutation, it follows by induction that $x_2, \ldots, x_{n-1}$ are also fixed.)

8. This is equivalent to

$$\frac{Q_{n-2}(A_{n-1}, \ldots, A_2) - XQ_{n-1}(A_{n-1}, \ldots, A_1)}{Q_{n-1}(A_n, \ldots, A_2) - XQ_n(A_n, \ldots, A_1)} = -\frac{1}{X_n},$$

and by (6) this is equivalent to

$$X = \frac{Q_{n-1}(A_2, \ldots, A_n) + X_n Q_{n-2}(A_2, \ldots, A_{n-1})}{Q_n(A_1, \ldots, A_n) + X_n Q_{n-1}(A_1, \ldots, A_{n-1})}.$$

9. (a) By definition. (b), (d) Prove this when $n = 1$, then apply (a) to get the result for general n. (c) Prove when $n = k + 1$, then apply (a).

10. If $A_0 > 0$, then $B_0 = 0$, $B_1 = A_0$, $B_2 = A_1$, $B_3 = A_2$, $B_4 = A_3$, $B_5 = A_4$, $m = 5$. If $A_0 = 0$, then $B_0 = A_1$, $B_1 = A_2$, $B_2 = A_3$, $B_3 = A_4$, $m = 3$. If $A_0 = -1$ and $A_1 = 1$, then $B_0 = -(A_2 + 2)$, $B_1 = 1$, $B_2 = A_3 - 1$, $B_3 = A_4$, $m = 3$. If $A_0 = -1$ and $A_1 > 1$, then $B_0 = -2$, $B_1 = 1$, $B_2 = A_1 - 2$, $B_3 = A_2$, $B_4 = A_3$, $B_5 = A_4$, $m = 5$. If $A_0 < -1$, then $B_0 = -1$, $B_1 = 1$, $B_2 = -A_0 - 2$, $B_3 = 1$, $B_4 = A_1 - 1$, $B_5 = A_2$, $B_6 = A_3$, $B_7 = A_4$. [Actually, the last three cases involve eight subcases; if any of the B's is set to zero, the values should be "collapsed together" by using the rule of exercise 9(c). For example, if $A_0 = -1$, $A_1 = A_3 = 1$, we actually have $B_0 = -(A_2 + 2)$, $B_1 = A_4 + 1$, $m = 1$. Double collapsing occurs when $A_0 = -2$, $A_1 = 1$.]

11. Let $q_n = Q_n(A_1, \ldots, A_n)$, $q'_n = Q_n(B_1, \ldots, B_n)$, $p_n = Q_{n+1}(A_0, \ldots, A_n)$, $p'_n = Q_{n+1}(B_0, \ldots, B_n)$. We have $X = (p_m + p_{m-1}X_m)/(q_m + q_{m-1}X_m)$, $Y = (p'_n + p'_{n-1}Y_n)/(q'_n + q'_{n-1}Y_n)$; therefore if $X_m = Y_n$, the stated relation between X and Y holds by (8). Conversely, if $X = (qY + r)/(sY + t)$, $|qt - rs| = 1$, we may assume that $s \geq 0$, and we can show that the partial quotients of X and Y eventually agree by induction on s. The result is clear when $s = 0$, by exercise 9(d).

If $s > 0$, let $q = as + s'$ $(0 \le s' < s)$. Then $X = a + 1/((sY + t)/(s'Y + r - at))$; since $s(r - at) - ts' = sr - tq$, and $s' < s$, we know by induction and exercise 10 that the partial quotients of X and Y eventually agree. [*Note:* The fact that m is always odd in exercise 10 shows, by a close inspection of this proof, that $X_m = Y_n$ if and only if $X = (qY + r)/(sY + t)$, where $qt - rs = (-1)^{m-n}$.]

12. (a) Since $V_n V_{n+1} = D - U_n^2$, we know that $D - U_{n+1}^2$ is a multiple of V_{n+1}; hence by induction $X_n = (\sqrt{D} - U_n)/V_n$, where U_n, V_n are integers. [Note that the identity $V_{n+1} = A_n(U_{n-1} - U_n) + V_{n-1}$ makes it unnecessary to divide when V_{n+1} is being determined.]

(b) Let $Y = (-\sqrt{D} - U)/V$, $Y_n = (-\sqrt{D} - U_n)/V_n$. The stated identity obviously holds by replacing $\sqrt{D}$ by $-\sqrt{D}$ in the proof of (a). We have

$$Y = (p_n/Y_n + p_{n-1})/(q_n/Y_n + q_{n-1}),$$

where

$$p_n = Q_{n+1}(A_0, \ldots, A_n), \qquad q_n = Q_n(A_1, \ldots, A_n);$$

hence

$$Y_n = (-q_n/q_{n-1})(Y - p_n/q_n)/(Y - p_{n-1}/q_{n-1}).$$

But by (12), p_{n-1}/q_{n-1} and p_n/q_n are extremely close to X; since $X \ne Y$, $Y - p_n/q_n$ and $Y - p_{n-1}/q_{n-1}$ will have the same sign as $Y - X$ for all large n. This proves that $Y_n < 0$ for all large n; hence $0 < X_n < X_n - Y_n = 2\sqrt{D}/V_n$; V_n must be positive. Also $V_n < D$, since V_n divides $D - U_n^2$; and $U_n < \sqrt{D}$, since $X_n > 0$.

Finally, we want to show that $U_n > 0$. Since $X_n < 1$, we have $U_n > \sqrt{D} - V_n$, so we need only consider the case $V_n > \sqrt{D}$; then $U_n = A_n V_n - U_{n-1} \ge V_n - U_{n-1} > \sqrt{D} - U_{n-1}$, and this is positive as we have already observed.

Note: In the repeating cycle we have $\sqrt{D} + U_n = A_n V_n + (\sqrt{D} - U_{n-1}) > V_n$; hence $\lfloor(\sqrt{D} + U_{n+1})/V_{n+1}\rfloor = \lfloor A_{n+1} + V_n/(\sqrt{D} + U_n)\rfloor = A_{n+1} = \lfloor(\sqrt{D} + U_n)/V_{n+1}\rfloor$. Thus, A_{n+1} is determined by U_{n+1} and V_{n+1}; we can determine (U_n, V_n) from (U_{n+1}, V_{n+1}) in the period. In fact, when $0 < V_n < \sqrt{D} + U_n$, $0 < U_n < \sqrt{D}$, the arguments above prove that $0 < V_{n+1} < \sqrt{D} + U_{n+1}$ and $0 < U_{n+1} < \sqrt{D}$; moreover, if the pair (U_{n+1}, V_{n+1}) follows (U', V') with $0 < V' < \sqrt{D} + U'$ and $0 < U' < \sqrt{D}$, then $U' = U_n$ and $V' = V_n$. Hence (U_n, V_n) *is part of the cycle if and only if* $0 < V_n < \sqrt{D} + U_n$ and $0 < U_n < \sqrt{D}$.

(c) $$\frac{-V_{n+1}}{V_n} = X_n Y_n = \frac{(q_n X - p_n)(q_n Y - p_n)}{(q_{n-1}X - p_{n-1})(q_{n-1}Y - p_{n-1})}.$$

[A companion identity is

$$V p_n p_{n-1} + U(p_n q_{n-1} + p_{n-1}q_n) + ((U^2 - D)/V)q_n q_{n-1} = (-1)^n U_n.]$$

(d) If $X_n = X_m$ for $n \ne m$, then X is an irrational number which satisfies the quadratic equation $(q_n X - p_n)/(q_{n-1}X - p_{n-1}) = (q_m X - p_m)/(q_{m-1}X - p_{m-1})$.

14. As in exercise 9, we need only verify the stated identities when c is the last partial quotient, and this verification is trivial. Now Hurwitz's rule gives $2/e = /1, 2, 1, 2, 0, 1, 1, 1, 1, 1, 0, 2, 3, 2, 0, 1, 1, 3, 1, 1, 0, 2, 5, \ldots/$. Taking the reciprocal, collapsing out the zeros as in exercise 9, and taking note of the pattern that appears, we find that (cf. exercise 16) $e/2 = 1 + /2, \overline{2m + 1, 3, 1, 2m + 1, 1, 3}/$, $m \geq 0$. [*Schriften der phys.-ökon. Gesellschaft zu Königsberg* **32** (1891), 59–62.]

15.
$$3/3a, b, c, \ldots/ = /a, 3b + 3/c, \ldots//;$$
$$3/3a + 1, b, c, \ldots/ = /a, 2, 1 + 3/b - 2, c, \ldots //, \text{ if } b \neq 1;$$
$$3/3a + 1, 1, c, \ldots/ = /a, 1, 1, 1 + 3/c - 1, \ldots//;$$
$$3/3a + 2, b, c, \ldots/ = /a, 1, 2 + 3/b - 1, c, \ldots//.$$

This yields
$$e/3 = /1, 9, \overline{1, 1, 2m + 1, 5, 1, 2m + 1, 1, 1, 18m + 26}/, \quad m \geq 0.$$

16. It is not difficult to prove by induction that $f_n(z) = z/(2n + 1) + O(z^3)$ is an odd function with a convergent power series in a neighborhood of the origin, and that it satisfies the given differential equation. Hence

$$f_0(z) = \left/ \frac{1}{z} + f_1(z) \right/ = \left/ \frac{1}{z}, \frac{3}{z} + f_2(z) \right/ = \cdots$$

$$= \left/ \frac{1}{z}, \frac{3}{z}, \ldots, \frac{2n + 1}{z} + f_{n+1}(z) \right/ .$$

It remains to prove that

$$\lim_{n \to \infty} \left/ \frac{1}{z}, \frac{3}{z}, \ldots, \frac{2n + 1}{z} \right/ = f_0(z).$$

[Actually Euler, age 24, obtained continued fraction expansions for the considerably more general differential equation $f_n'(z) = az^m + bf_n(z)z^{m-1} + cf_n(z)^2$; but he did not bother to prove convergence, since formal manipulation and intuition were good enough in the eighteenth century.]

There are several ways to prove the desired limiting equation. First, letting $f_n(z) = \sum_k a_{nk}z^k$, we can argue from the equation

$$(2n + 1)a_{n1} + (2n + 3)a_{n3}z^2 + (2n + 5)a_{n5}z^4 + \cdots$$
$$= 1 - (a_1z + a_3z^3 + a_5z^5 + \cdots)^2$$

that $(-1)^k a_{n(2k+1)}$ is a sum of terms of the form $c_k/(2n + 1)^{k+1}(2n + b_{k1}) \cdots$ $(2n + b_{kk})$, where the c_k and b_{km} are positive integers independent of n. For example, $-a_{n7} = 4/(2n + 1)^4(2n + 3)(2n + 5)(2n + 7) + 1/(2n + 1)^4(2n + 3)^2(2n + 7)$. Thus $|a_{(n+1)k}| \leq |a_{nk}|$, and $|f_n(z)| \leq \tan|z|$ for $|z| < \pi/2$. This uniform bound on $f_n(z)$ makes the convergence proof very simple. Careful study of this argument reveals that the power series for $f_n(z)$ actually converges for $|z| < \pi\sqrt{2n + 1}/2$; this is interesting, since it shows that the singularities of $f_n(z)$ get farther and farther away from the origin as n grows, so the continued fraction actually represents $\tanh z$ *throughout* the complex plane.

Another proof gives further information of a different kind: If we let

$$A_n(z) = n! \sum_{0 \leq k \leq n} \binom{2n - k}{n} z^k \left/ k! \right. = \sum_{k \geq 0} \frac{(n + k)!z^{n-k}}{k!(n - k)!},$$

then

$$A_{n+1}(z) = \sum_{k \geq 0} \frac{(n+k-1)!((4n+2)k+(n+1-k)(n-k))}{k!(n+1-k)!} z^{n+1-k}$$

$$= (4n+2)A_n(z) + z^2 A_{n-1}(z).$$

It follows, by induction, that

$$Q_n\left(\frac{1}{z}, \frac{3}{z}, \ldots, \frac{2n-1}{z}\right) = \frac{A_n(2z) + A_n(-2z)}{2^{n+1}z^n},$$

$$Q_{n-1}\left(\frac{3}{z}, \ldots, \frac{2n-1}{z}\right) = \frac{A_n(2z) - A_n(-2z)}{2^{n+1}z^n}.$$

Hence

$$\Big/ \frac{1}{z}, \frac{3}{z}, \ldots, \frac{2n-1}{z} \Big/ = \frac{A_n(2z) - A_n(-2z)}{A_n(2z) + A_n(-2z)},$$

and we want to show that this ratio approaches $\tanh z$. By Eqs. 1.2.9–11 and 1.2.6–24,

$$e^z A_n(-z) = n! \sum_{m \geq 0} z^m \left(\sum_{0 \leq k \leq n} \binom{m}{k}\binom{2n-k}{n}(-1)^k\right) = \sum_{m \geq 0} \binom{2n-m}{n} z^m \frac{n!}{m!}.$$

Hence

$$e^z A_n(-z) - A_n(z) = R_n(z) = (-1)^n x^{2n+1} \sum_{k \geq 0} \frac{(n+k)!x^k}{(2n+k+1)!k!}.$$

We now have $(e^{2z}-1)(A_n(2z) + A_n(-2z)) - (e^{2z}+1)(A_n(2z) - A_n(-2z)) = 2R_n(2z)$; hence

$$\tanh z - \Big/ \frac{1}{z}, \frac{3}{z}, \ldots, \frac{2n-1}{z} \Big/ = \frac{2R_n(2z)}{(A_n(2z) + A_n(-2z))(e^{2z}+1)}.$$

Thus we have an exact formula for the difference. When $|z| \leq 1$, $e^{2z}+1$ is bounded away from zero, $|R_n(z)| \leq en!/(2n+1)!$, and

$$\tfrac{1}{2}|A_n(2z) + A_n(-2z)| \geq n!\left(\binom{2n}{n} - 2\binom{2n-2}{n} - \binom{2n-4}{n} - \cdots\right)$$

$$\geq \frac{(2n)!}{n!}\left(1 - \tfrac{1}{2} - \tfrac{1}{16}(1 + \tfrac{1}{4} + \tfrac{1}{16} + \cdots)\right) = \frac{5}{12}\frac{(2n)!}{n!}.$$

Thus convergence is very rapid, even for complex values of z.

To go from this continued fraction to the continued fraction for e^z, we have $\tanh z = 1 - 2/(e^{2z}+1)$; hence we get the continued fraction representation for $(e^{2z}+1)/2$ by simple manipulations. Hurwitz's rule gives the expansion of $e^{2z}+1$, from which we may subtract unity. For n odd,

$$e^{-2/n} = \overline{/1, 3mn + \lfloor n/2 \rfloor, (12m+6)n, (3m+2)n + \lfloor n/2 \rfloor, 1/}, \qquad m \geq 0.$$

Another derivation has been given by C. S. Davis, *J. London Math. Soc.* **20** (1945), 194–198.

17. (b) $/x_1 - 1, 1, x_2 - 2, 1, x_3 - 2, 1, \ldots, 1, x_{2n-1} - 2, 1, x_{2n} - 1/$.

(c) $1 + /1, 1, 3, 1, 5, 1, \ldots/$.

19. The sum for $1 \le k \le N$ is $\log_b (2 \cdot 3 \cdots (N+1)) - \log_b (1 \cdot 2 \cdots N) - \log_b ((2+x)(3+x) \cdots (N+1+x)) + \log_b ((1+x)(2+x) \cdots (N+x)) = \log_b ((1+x)(N+1)/(N+1+x))$.

20. The function $G(x)$ is well-defined, since any rearrangement of a subseries of an absolutely convergent series has the same value. Given $\epsilon > 0$, we must show that there is a partition $P = \{T_0, T_1, \ldots, T_m\} \subseteq [a, b]$ such that for any finer partition P' we have $|\sum' - \int| < \epsilon$, where $\sum'$ is any appropriate "approximating sum" and $\int$ is the stated value of the integral. Now since f is uniformly continuous there is a partition P such that if $T_k \le t' \le t'' \le T_{k+1}$, then $|f(t') - f(t'')| < \epsilon$. Also f is bounded, so $\sum_{n \ge 1} |f(x_n) g(x_n)|$ exists. For any partition $a = t_0 < t_1 < \cdots < t_N = b$ finer than P, we must examine the approximating sum $\sum_{1 \le k \le N} f(u_k)(G(t_k) - G(t_{k-1}))$, where u_k is a point between t_{k-1} and t_k. Now

$$f(u_k)(G(t_k) - G(t_{k-1})) = f(u_k) \sum_{t_{k-1} < x_n \le t_k} g(x_n) = \sum_{t_{k-1} \le x_n < t_k} (f(x_n) g(x_n) + \xi_n g(x_n)),$$

where $|\xi_n| < \epsilon$ for all n. Thus

$$\left| \sum_{1 \le k \le N} f(u_k)(G(t_k) - G(t_{k-1})) - \sum_{n \ge 1} f(x_n) g(x_n) \right| = \left| \sum_{n \ge 1} \xi_n g(x_n) \right| \le \epsilon \sum_{n \ge 1} |g(x_n)|.$$

23. $f_n(x) - \dfrac{1}{(1+x) \ln 2} = \displaystyle\int_0^1 \frac{1+y}{(1+yx)^2} \, d(G_n(y) - G(y))$

$$= \int_0^1 (G_n(y) - G(y)) \, d\left(\frac{1+y}{(1+yx)^2} \right).$$

24. ∞.

25. Any union of intervals can be written as a union of disjoint intervals, since $\bigcup_{k \ge 1} I_k = \bigcup_{k \ge 1} (I_k - \bigcup_{1 \le j < k} I_j)$, and this is a disjoint union in which $I_k - \bigcup_{1 \le j < k} I_j$ can be expressed as a finite union of disjoint intervals. Therefore we may take $\mathcal{I} = \bigcup I_k$, where I_k is an interval of length $\epsilon/2^k$ containing the kth rational number in $[0, 1]$, using some enumeration of the rationals. In this case $\mu(\mathcal{I}) \le \epsilon$, but $\|\mathcal{I} \cap P_n\| = n$ for all n.

26. The continued fractions $/A_1, \ldots, A_t/$ which appear are precisely those for which $A_1 > 1$, $A_t > 1$, and $Q_t(A_1, A_2, \ldots, A_t)$ is a divisor of n. Therefore (6) completes the proof. (*Note:* If $m_1/n = /A_1, \ldots, A_t/$ and $m_2/n = /A_t, \ldots, A_1/$, where m_1 and m_2 are relatively prime to n, then $m_1 m_2 \equiv \pm 1$ (modulo n); this rule defines the correspondence. When $A_1 = 1$ an analogous symmetry is valid, according to (38).)

27. First prove the result for $n = p^e$, then for $n = rs$, where r and s are relatively prime. Alternatively, use the formula in the next exercise.

28. (a) The left-hand side is multiplicative (see exercise 1.2.4–31), and it is easily evaluated when n is a power of a prime. (c) From (a), we have *Möbius' inversion formula:* If $f(n) = \sum_{d\backslash n} g(d)$, then $g(n) = \sum_{d\backslash n} \mu(n/d)f(d)$.

29. The sum is approximately $(12 \ln 2/\pi^2) \ln N!/N - \sum_{d\geq 1} \Lambda(d)/d^2 + 1.47$; here $\sum_{d\geq 1} \Lambda(d)/d^2$ converges to the constant value $-\zeta'(2)/\zeta(2)$, while $\ln N! = N \ln N - N + O(\log N)$ by Stirling's approximation.

30. The modified algorithm affects the calculation if and only if the following division step in the unmodified algorithm would have the quotient 1, and in this case it avoids the following division step. The probability that a given division step is avoided is the probability that $A_k = 1$ and that this quotient is preceded by an even number of quotients equal to 1. By the symmetry condition, this is the probability that $A_k = 1$ and is *followed* by an even number of quotients equal to 1. The latter happens if and only if $X_{k-1} > \phi - 1 = 0.618\ldots$, where ϕ is the golden ratio: For $A_k = 1$, $A_{k+1} > 1$ iff $\frac{2}{3} \leq X_{k-1} < 1$; $A_k = A_{k+1} = A_{k+2} = 1$, $A_{k+3} > 1$ iff $\frac{5}{8} \leq X_{k-1} < \frac{2}{3}$; etc. Thus we save approximately $F_{k-1}(1) - F_{k-1}(\phi - 1) \approx 1 - \log_2 \phi \approx 0.306$ of the division steps. The average number of steps is approximately $(12 \ln \phi/\pi^2) \ln n$, when $v = n$ and u is relatively prime to n. Kronecker [*Vorlesungen über Zahlentheorie* **1** (Leipzig: Teubner, 1901), 118] observed that this choice of least remainder in absolute value always gives the shortest possible number of iterations, over all algorithms which replace u by $(\pm u) \bmod v$ at each iteration. For further results see N. G. de Bruijn and W. M. Zaring, *Nieuw Archief voor Wiskunde* (3) **1** (1953), 105–112.

On many computers, the modified algorithm makes each division step longer; the idea of exercise 1, which saves *all* division steps when the quotient is unity, would be preferable in such cases. Thus, the modified algorithm is not especially attractive on many computers.

31. Let $a_0 = 0$, $a_1 = 1$, $a_{n+1} = 2a_n + a_{n-1}$; then $a_n = \big((1 + \sqrt{2})^n - (1 - \sqrt{2})^n\big)/2\sqrt{2}$, and the worst case (in the sense of Theorem F) occurs when $u = a_n + a_{n-1}$, $v = a_n$, $n \geq 2$.

This result is due to A. Dupré [*J. de Math.* **11** (1846), 41–64], who also investigated more general "look-ahead" procedures suggested by J. Binet. See P. Bachmann, *Niedere Zahlentheorie* **1** (Leipzig: Teubner, 1902), 99–118, for a discussion of early analyses of Euclid's algorithm.

32. $Q_{m-1}(x_1, \ldots, x_{m-1})Q_{n-1}(x_{m+2}, \ldots, x_{m+n})$ corresponds to Morse code sequences of length $(m + n)$ in which a dash occupies positions m and $(m + 1)$; the other term corresponds to the opposite case. (Alternatively, use exercise 2.)

33. (a) The new representations are $x = m/d$, $y = (n - m)/d$, $x' = y' = d = \gcd(m, n - m)$, for $\frac{1}{2}n < m < n$. (b) The relation $(n/x') - y \leq x < n/x'$ defines x. (c) Count the x' satisfying (b). (d) A pair of integers $x > y > 0$, $\gcd(x, y) = 1$, can be uniquely written in the form $x = Q_m(x_1, \ldots, x_m)$, $y = Q_{m-1}(x_1, \ldots, x_{m-1})$, where $x_1 \geq 2$, $m \geq 1$; here $y/x = /x_m, \ldots, x_1/$. (e) It suffices to show that $\sum_{1\leq k\leq n/2} T(k, n) = 2\lceil n/2 \rceil + h(n)$. For $1 \leq k \leq n/2$ the continued fractions $k/n = /x_1, \ldots, x_m/$ run through all sequences $(x_1, \ldots, x_m)$ such that $m \geq 1$, $x_1 \geq 2$, $x_m \geq 2$, $Q_m(x_1, \ldots, x_m)\backslash n$.

34. (a) Dividing x, y by $\gcd(x, y)$ yields $g(n) = \sum_{d\backslash n} h(n/d)$; apply exercise 28(c), and use the symmetry between primed and unprimed variables. (b) For fixed y and t the representations with $xd \geq x'$ have $x' < \sqrt{nd}$; hence there are $O(\sqrt{nd}/y)$ such representations. Now sum for $0 < t \le y < \sqrt{n/d}$. (c) If $s(y)$ is the given sum, then $\sum_{d\backslash y} s(d) = y(H_{2y} - H_y) = k(y)$, say; hence $s(y) = \sum_{d\backslash y} k(y/d)$. Now $k(y) = y \ln 2 - \frac{1}{4} + O(1/y)$. (d) $\sum_{1 \le y \le n} \varphi(y)/y^2 = \sum_{1 \le y \le n,\ d\backslash y} \mu(d)/yd = \sum_{cd \le n} \mu(d)/cd^2$. (Similarly, $\sum_{1 \le y \le n} \sigma_{-1}(y)/y^2 = O(1)$.) (e) $\sum_{1 \le k \le n} \mu(k)/k^2 = 6/\pi^2 + O(1/n)$ (see exercise 4.5.2–10d); and $\sum_{1 \le k \le n} \mu(k) \log k/k^2 = O(1)$. Hence $h_d(n) = n(3 \ln 2/\pi^2)$ $\ln(n/d) + O(n)$ for $d \ge 1$. So $h(n) = 2\sum_{cd \backslash n} \mu(d) h_d(n/cd) = (6 \ln 2/\pi^2)(n \ln n + \sum) + O(n\sigma_{-1}(n)^2)$, where $\sum = \sum_{cd\backslash n} \mu(d) \ln(cd)/cd = 0$. [It is well-known that $\sigma_{-1}(n) = O(\log \log n)$.]

35. Working the algorithm backwards, we want to choose $k_1, \ldots, k_{n-1}$ so that $u_k \equiv F_{k_1} \ldots F_{k_{i-1}} F_{k_i - 1}$ (modulo gcd $(u_{i+1}, \ldots, u_n)$) for $1 \le i < n$, with $u_n = F_{k_1} \ldots F_{k_{n-1}}$ a minimum, where the k's are positive, $k_1 \ge 3$, and $k_1 + \ldots + k_{n-1} = N + n - 1$. The solution is $k_2 = \cdots = k_{n-1} = 2$, $u_n = F_{N-n+3}$. [See *CACM* **13** (1970), 433–436, 447–448.]

SECTION 4.5.4

1. If d_k isn't prime, its prime divisors will already have been factored out before d_k is tried.

2. No, the algorithm would fail if $p_{t-1} = p_t$, giving "1" as a spurious prime factor.

3. Let N be the product of the first 168 primes. (*Note:* Although N is a large number, it is considerably faster on many computers to calculate this gcd than to do the 168 divisions, if we just want to test whether or not n is prime.)

4. Let $\nabla r_k = r_k - r_{k-1}$, $\nabla d_k = \delta$. Since $r_k = n - q_k d_k$, we have $\nabla r_{k+1} = -d_{k+1}\nabla q_{k+1} - \delta q_k$; thus $\nabla^2 r_{k+1} = -d_{k+1}\nabla^2 q_{k+1} - 2\delta\nabla q_k$ is the equation corresponding to (3). Now observe that $\nabla^2(n/d_{k+1}) = 2\delta^2 n/d_{k+1}d_k d_{k-1}$. Since $n/d_k - 1 < q_k \le n/d_k$, we have $\nabla^2(n/d_{k+1}) - 2 < \nabla^2 q_{k+1} < \nabla^2(n/d_{k+1}) + 2$. Now $\nabla^2(n/d_{k+1})$ is nonnegative, and it is ≤ 1 if and only if $d_{k+1}d_k d_{k-1} \ge 2\delta^2 n$, that is, $d_k^3 - \delta^2 d_k \ge 2\delta^2 n$. For small δ, and $n \ge 1$, this will be true for $d_k \ge \sqrt[3]{2\delta^2 n} + 1$; for all δ, $d_k \ge \sqrt[3]{2\delta^2 n} + \delta$ will be sufficient. These formulas make the generalization of Algorithm B straightforward. (Note that our argument proves that $\nabla^2 q_{k+1} \ge 0$, when d_k is *less* than $\sqrt[3]{2\delta^2 n}$, since $\nabla^2(n/d_{k+1}) > 1$ in such a case.)

5. $x \bmod 3 = 0$; $x \bmod 5 = 0, 1, 4$; $x \bmod 7 = 0, 1, 6$; $x \bmod 8 = 1, 3, 5, 7$; $x > 103$. The first try is $x = 105$: $(105)^2 - 10541 = 484 = 22^2$. This would also have been found by Algorithm C in a relatively short time. Thus $10541 = 83 \cdot 127$.

6. Let us count the number of solutions (x, y) of the congruence $n \equiv (x - y)(x + y)$ (modulo p), where $0 \le x, y < p$. Since $n \not\equiv 0$ and p is prime, $x + y \not\equiv 0$. For each $v \not\equiv 0$ there is a unique u (modulo p) such that $n \equiv uv$. The congruences $x - y \equiv u$, $x + y \equiv v$ now uniquely determine $x \bmod p$ and $y \bmod p$, since p is odd. Thus the stated congruence has exactly $p - 1$ solutions (x, y). If (x, y) is a solution, so is $(x, p - y)$ if $y \ne 0$, since $(p - y)^2 \equiv y^2$; and if (x, y_1) and (x, y_2) are solutions with $y_1 \ne y_2$, we have $y_1^2 \equiv y_2^2$; hence $y_1 = p - y_2$. Thus the number of different x values among the solutions (x, y) is $(p - 1)/2$ if $n \equiv x^2$ has no solutions, or $(p + 1)/2$ if $n \equiv x^2$ has solutions.

7. One procedure is to keep two indices for each modulus, one for the current word position and one for the current bit position; loading two words of the table and doing an indexed shift command will bring the table entries into proper alignment. A trickier (and probably a little slower) method is to shift the table entries in memory: For example, assume a 30-bit word and a 67-bit table; then a typical iteration could change the memory configuration from

$$\boxed{s_0 \quad s_1 \quad \cdots \quad s_{29}} \qquad \boxed{s_{30} \; s_{31} \; \cdots \; s_{59}} \leftarrow k \; \boxed{s_{37} \; s_{38} \; \cdots \; s_{66}}$$

to $\boxed{s_0 \quad s_1 \quad \cdots \quad s_{29}} \leftarrow k \; \boxed{s_7 \quad s_8 \quad \cdots \quad s_{36}} \qquad \boxed{s_{37} \; s_{38} \; \cdots \; s_{66}}$.

In either case the shifting operation requires the use of two registers, and requires further memory accesses, so it probably slows down the main loop too much.

8. (We may assume that $N = 2M$ is even.) The following algorithm uses an auxiliary table $X[1], X[2], \ldots, X[M]$, where $X[k]$ represents the primality of $2k + 1$.

S1. Set $X[k] \leftarrow 1$ for $1 \le k \le M$. Also set $j \leftarrow 1$, $p \leftarrow 3$, $q \leftarrow 4$. (During this algorithm $p = 2j + 1$, $q = 2j + 2j^2$; the integer variables j, k, p, q may readily be manipulated in index registers.)

S2. If $X[j] = 0$, go to S4. Otherwise output p, which is prime, and set $k \leftarrow q$.

S3. If $k \le M$, then set $X[k] \leftarrow 0$, $k \leftarrow k + p$, and repeat this step.

S4. Set $j \leftarrow j + 1$, $p \leftarrow p + 2$, $q \leftarrow q + 2p - 2$. If $j \le M$, return to S2. ∎

A major part of this calculation could be made noticeably faster if q (instead of j) were tested against M in step S4, and if a new loop were appended which outputs $2j + 1$ for all remaining $X[j]$ that equal 1, suppressing the manipulation of p and q.

9. If p^2 is a divisor of n for some prime p, then p is a divisor of $\lambda(n)$, but not of $n - 1$. If $n = p_1 p_2$, where $p_1 < p_2$ are primes, then $p_2 - 1$ is a divisor of $\lambda(n)$ and therefore $p_1 p_2 - 1 \equiv 0 \pmod{(p_2 - 1)}$. Since $p_2 \equiv 1$, $p_1 - 1$ is a multiple of $p_2 - 1$; but this is impossible. (Values of n for which $\lambda(n)$ properly divides $n - 1$ are called "Carmichael numbers." See D. Shanks, *Solved and Unsolved Problems in Number Theory* **1** (Washington, D.C.: Spartan, 1962), Sections 39–40.)

10. Let k_p be the order of x_p modulo n, and let λ be the least common multiple of all the k_p's. Then λ is a divisor of $n - 1$ but not of any $(n - 1)/p$, so $\lambda = n - 1$. Since $x_p^{\varphi(n)} \bmod n = 1$, $\varphi(n)$ is a multiple of k_p for all p, so $\varphi(n) \ge \lambda$. But $\varphi(n) < n - 1$ when n is not prime. (Another way to carry out the proof is to construct an element x of order $n - 1$ from the x_p's, by the method of exercise 3.2.1.2–15.)

11.

	U	V	A	P	S	T	Output
	992	1	0	992	0	—	
	989	1981	1	992	1	1981	
	991	4	495	993	0	1	$993^2 \equiv 2^2$
	991	991	2	98109	1	991	
	989	4	495	2	0	1	$2^2 \equiv 2^2$
	992	1981	1	99099	1	1984	
	992	1	1984	99101	0	1	$99101^2 \equiv 2^0$

The factorization $199 \cdot 991$ is evident from the first or last outputs. The shortness of the cycle, and the appearance of the well-known number 1984, are probably just coincidences.

12. The following algorithm makes use of an auxiliary $(t+1) \times (t+1)$ matrix of single-precision integers E_{jk}, $0 \le j, k \le t$; a single-precision vector $(b_0, b_1, \ldots, b_t)$; and a multiple precision vector $(x_0, x_1, \ldots, x_t)$ whose entries lie in the range $0 \le x_k < n$.

F1. [Initialize.] Set $b_i \leftarrow -1$ for $0 \le i \le t$; then set $j \leftarrow 0$.

F2. [Next solution.] Get the next solution $(x, e_0, e_1, \ldots, e_t)$ produced by Algorithm E. (It is convenient to regard Algorithms E and F as coroutines.) Set $k \leftarrow 0$.

F3. [Search for odd.] If $k > t$, go to step F5. Otherwise if e_k is even, set $k \leftarrow k + 1$ and repeat this step.

F4. [Linear dependence?] If $b_k \ge 0$, then set $i \leftarrow b_k$, $x \leftarrow (x_i x) \bmod n$, $e_r \leftarrow e_r + E_{ir}$ for $0 \le r \le t$; set $k \leftarrow k + 1$ and return to F3. Otherwise set $b_k \leftarrow j$, $x_j \leftarrow x$, $E_{jr} \leftarrow e_r$ for $0 \le r \le t$; set $j \leftarrow j + 1$ and return to F2. (In the latter case we have a new linearly independent solution, modulo 2, whose first odd component is e_k.)

F5. [Try to factor.] (Now $e_0, e_1, \ldots, e_t$ are even.) Set

$$y \leftarrow \left((-1)^{e_0/2} p_1^{e_1/2} \cdots p_t^{e_t/2} \right) \bmod n.$$

If $x = y$ or if $x + y = n$, return to F2. Otherwise compute $\gcd(x - y, n)$, which is a proper factor of n, and terminate the algorithm. ∎

13. It suffices to consider the case d prime. Then d divides $\gcd(2^p - 1, 2^{d-1} - 1)$ so p and $d - 1$ are not relatively prime. Thus $d - 1$ is a multiple of p, so d has the form $pk + 1$; finally k is even, since d is odd. [*Note:* Moreover we can prove that $d \equiv \pm 1 \pmod 8$, since $(2^{(p+1)/2})^2 \equiv 2 \pmod d$. These results simplify searching for divisors of Mersenne numbers, in Algorithms A, B, C, or D. The complete factorization of $2^n - 1$ has several applications, e.g., to random-number generation (Sections 3.2.1.1, 3.2.2) or to the construction of finite fields with 2^n elements (exercise 4.6.2–16). The argument we have used proves in general that any divisor d of $2^{2^r p} - 1$ is either a divisor of $2^{2^r} - 1$ or has the form $2kp + 1$. Hence the proof that n_6 in (18) has no primes less than its cube root requires testing less than 3 million divisors, in spite of the fact that $\sqrt[3]{n_6} \approx 576 \times 10^6$.]

14. Since $P^2 \equiv knQ^2 \pmod p$ for any prime divisor p of V, we have $1 \equiv P^{2(p-1)/2} \equiv (knQ^2)^{(p-1)/2} \equiv (kn)^{(p-1)/2} \pmod p$, if $P \not\equiv 0$.

15. $U_n = (a^n - b^n)/\sqrt{D}$, where $a = \frac{1}{2}(P + \sqrt{D})$, $b = \frac{1}{2}(P - \sqrt{D})$, $D = P^2 - 4Q$. Then $2^{n-1}U_n = \sum_k \binom{n}{2k+1} P^{n-2k-1} D^k$; so if p is an odd prime, $U_p \equiv D^{(p-1)/2} \pmod p$. Similarly, if $V_n = a^n + b^n = U_{n+1} - QU_{n-1}$, then $2^{n-1}V_n = \sum_k \binom{n}{2k} P^{n-2k} D^k$, and $V_p \equiv P^n \equiv P$. Thus if $U^p \equiv -1$, we find that $U_{p+1} \bmod p = 0$. If $U_p \equiv 1$, we find that $(QU_{p-1}) \bmod p = 0$; here if Q is a multiple of p, $U_n \equiv P^{n-1} \pmod p$ for $n > 0$, so U_n is never a multiple of p; if Q is not a multiple of p, $U_{p-1} \bmod p = 0$. Therefore as in Theorem L, $U_t \bmod N = 0$ if $N = p_1^{e_1} \cdots p_r^{e_r}$, $\gcd(N, Q) = 1$, and $t = \operatorname{lcm}_{1 \le j \le r} (p_j^{e_j - 1}(p_j + \epsilon_j))$. Under the assumptions of this exercise, the rank of apparition of N is $N + 1$; hence N is prime to Q and t is a multiple of $N + 1$. Also $t \leq \prod p_j^{e_j-1}(p_j + 1) < 2N$, so $t = \prod_{1 \leq j \leq r} p_j^{e_j-1}(p_j + \epsilon_j) = N + 1$. The assumptions of this exercise imply that each p_j is odd and each ϵ_j is ± 1; hence $r = 1$ and $t = p_1^{e_1} + \epsilon_1 p_1^{e_1-1}$. Finally, therefore $e_1 = 1$ and $\epsilon_1 = 1$.

Note: Obviously if this test for primality is to be any good, we must choose P and Q in some way which makes it likely that the test will work. Lehmer suggests taking $P = 1$ so that $D = 1 - 4Q$, and choosing Q so that $\gcd(N, QD) = 1$. (If the latter condition fails, we know already that N is not prime, unless $|D| \geq N$.) Furthermore, the derivation above shows that we will want $\epsilon_1 = 1$, that is, $D^{(N-1)/2} \equiv -1$ (modulo N). This is another condition which determines the choice of Q. Furthermore, if D satisfies this condition, and if $U_{N+1} \bmod N \neq 0$, we know that N is *not* prime.

Example: If $P = 1$, $Q = -1$, we have the Fibonacci sequence, with $D = 5$. Since $5^{11} \equiv -1$ (modulo 23), we might attempt to prove that 23 is prime by using the Fibonacci sequence:

$$\langle F_n \bmod 23 \rangle = 0, 1, 1, 2, 3, 5, 8, 13, 21, 11, 9, 20, 6, 3, 9, 12, 21, 10, 8, 18, 3, 21, 1, 22, 0, \ldots$$

so 24 is the rank of apparition of 23 and the test works. However, the Fibonacci sequence cannot be used in this way to prove the primality of 13 or 17, since $F_7 \bmod 13 = 0$ and $F_9 \bmod 17 = 0$. When $p \equiv \pm 1$ (modulo 10), we have $5^{(p-1)/2} \bmod p = 1$, so F_{p-1} (not F_{p+1}) is divisible by p.

SECTION 4.6

1. $9x^2 + 7x + 9$; $5x^3 + 7x^2 + 2x + 6$.

2. (a) True. (b) False if the algebraic system S contains "zero divisors", nonzero numbers whose product is zero, as in exercise 1; otherwise true. (c) False; $(x + 1) + ((-1)x + 0) = 1$.

3. Assume that $r \leq s$. For $0 \leq k \leq r$ the maximum is $m_1 m_2 (k + 1)$; for $r \leq k \leq s$ it is $m_1 m_2 (r + 1)$; for $s \leq k \leq r + s$ it is $m_1 m_2 (r + s + 1 - k)$. The least upper bound valid for all k is $m_1 m_2 (r + 1)$. (The solver of this exercise will know how to factor the polynomial $x^7 + 2x^6 + 3x^5 + 3x^4 + 3x^3 + 3x^2 + 2x + 1$.)

4. If the polynomials each have less than 2^t nonzero coefficients, the product can be formed by putting exactly $t - 1$ zeros between each of the coefficients, then multiplying in the binary number system, and finally using a logical AND operation (present on most binary computers, cf. Section 4.5.4) to zero out the extra bits. For example, if $t = 3$, the multiplication in the text would become $(1001000001)_2 \times (1000001001)_2 = (1001001011001001001)_2$; if we AND the result with the constant $(1001001 \cdots 1001)_2$, the desired answer is obtained. A similar technique can be used to multiply polynomials with nonnegative coefficients when it is known that the coefficients will not be too large.

5. Polynomials of degree $\leq 2n$ can be represented as $U_1(x)x^n + U_0(x)$ where $\deg(U_1)$, $\deg(U_0) \leq n$; and

$$(U_1(x)x^n + U_0(x))(V_1(x)x^n + V_0(x))$$
$$= U_1(x)V_1(x)(x^{2n} + x^n) + (U_1(x) + U_0(x))(V_1(x) + V_0(x))x^n$$
$$+ U_0(x)V_0(x)(x^n + 1).$$

(This equation assumes that arithmetic is being done modulo 2.) Thus Eqs. 4.3.3–3, 4, 5 hold.

Note: S. A. Cook has shown that Algorithm 4.3.3C can be extended in a similar way. In fact, a modular method of computation may also be used, in a manner analogous to Algorithm 4.3.3S, using relatively prime polynomials as moduli.

SECTION 4.6.1

1.

						1	0	−2	8
2	2	−1	3)	1	1	−1	2	3	−1	2
				2	2	−2	4	6	−2	4
				2	2	−1	3			
				0	−1	1	6	−2	4	
				0	−2	2	12	−4	8	
				0	0	0	0			
					−2	2	12	−4	8	
					−4	4	24	−8	16	
					−4	−4	2	−6		
						8	22	−2	16	
						16	44	−4	32	
						16	16	−8	24	
							28	4	8	

Thus

$$q(x) = 1 \cdot 2^3 x^3 + 0 \cdot 2^2 x^2 - 2 \cdot 2x + 8 = 8x^3 - 4x + 8;$$

$$r(x) = 28x^2 + 4x + 8.$$

2. The monic sequence of polynomials produced during Euclid's algorithm has the coefficients $(1, 5, 6, 6, 1, 6, 3)$, $(1, 2, 5, 2, 2, 4, 5)$, $(1, 5, 6, 2, 3, 4)$, $(1, 3, 4, 6)$, 0. Hence the greatest common divisor is $x^3 + 3x^2 + 4x + 6$. (The greatest common divisor of a polynomial and its reverse is always symmetric, in the sense that it is a unit multiple of its own reverse.)

3. The procedure of Algorithm 4.5.2X is valid, with polynomials over S substituted for integers. When the algorithm terminates, we have $U(x) = u_2(x)$, $V(x) = u_1(x)$. Let $m = \deg(u)$, $n = \deg(v)$. It is easy to prove by induction that $\deg(u_3) + \deg(t_1) = n$, $\deg(u_3) + \deg(v_2) = m$, after step X3, throughout the execution of the algorithm, provided that $m \geq n$. Hence if m and n are positive we have $\deg(U) < m - d$, $\deg(V) < n - d$, where $d = \deg(\gcd(u, v))$; the exact degrees are $m - d_1$ and $n - d_1$, where d_1 is the degree of the second-last nonzero remainder.

When $u(x) = x^m - 1$ and $v(x) = x^n - 1$, the identity $(x^m - 1) \bmod (x^n - 1) = x^{m \bmod n} - 1$ shows that all polynomials occurring during the calculation are monic with integer coefficients. When $u(x) = x^{21} - 1$, $v(x) = x^{13} - 1$, we have $V(x) = (x^{11} + x^8 + x^6 + x^3 + 1)$, $U(x) = -(x^{19} + x^{16} + x^{14} + x^{11} + x^8 + x^6 + x^3 + x)$. [See also Eq. 3.3.3–29, which gives an alternative formula for $U(x)$, $V(x)$.]

4. Since the quotient $q(x)$ depends only on $v(x)$ and the first $m - n$ coefficients of $u(x)$, the remainder $r(x) = u(x) - q(x)v(x)$ is uniformly distributed and independent of $v(x)$. Hence each step of the algorithm may be regarded as independent of the others; this algorithm is much more well-behaved than Euclid's algorithm over the integers.

The probability that $n_1 = n - k$ is $p^{1-k}(1 - 1/p)$, and $t = 0$ with probability p^{-n}. Each succeeding step has essentially the same behavior; hence we can see that any given sequence of degrees $n, n_1, \ldots, n_t, -\infty$ occurs with probability $(p - 1)^t/p^n$. To find the average value of $f(n_1, \ldots, n_t)$, let S_t be the sum of $f(n_1, \ldots, n_t)$ over all sequences $n > n_1 > \cdots > n_t \geq 0$ having a given value of t; then the average is $\sum_t S_t(p - 1)^t/p^n$.

Let $f(n_1, \ldots, n_t) = t$; then $S_t = \binom{n}{t}(t + 1)$, so the average is $n(1 - 1/p)$. Similarly, if $f(n_1, \ldots, n_t) = n_1 + \cdots + n_t$, then $S_t = \binom{n}{2}\binom{n-1}{t-1}$, and the average is $\binom{n}{2}(1 - 1/p)$. Finally, if $f(n_1, \ldots, n_t) = (n - n_1)n_1 + \cdots + (n_{t-1} - n_t)n_t$ then $S_t = \binom{n+2}{t+2} - (n+1)\binom{n+1}{t+1} + \binom{n+1}{2}\binom{n}{t}$, and the average is $\binom{n+1}{2} - (n+1)p/(p-1) + (p/(p-1))^2(1 - 1/p^{n+1})$.

As a consequence we can see that if p is large there is very high probability that $n_{j+1} = n_j - 1$ for all j. (If this condition fails over the rational numbers, it fails for all p, so we have further evidence for the text's claim that Algorithm C almost always finds $t_3 = \cdots = 2$.)

5. Using the formulas developed in exercise 4, with $f(n_1, \ldots, n_t) = \delta_{n_t 0}$, we find the probability is $1 - 1/p$ if $n > 0$, 1 if $n = 0$.

Alternative proof: Let there be x_n pairs of monic relatively prime polynomials of degree n; counting *all* pairs by the degree of their gcd we have $p^{2n} = x_n + px_{n-1} + \cdots + p^n x_0$. Hence $x_n = p^{2n} - p^{2n-1}$ for $n > 0$.

6. Assuming that the constant terms $u(0)$ and $v(0)$ are nonzero, imagine a "right-to-left" division algorithm, $u(x) = v(x)q(x) + x^{m-n}r(x)$, $\deg(r) < \deg(v)$. We obtain a gcd algorithm analogous to Algorithm 4.5.2B, which is essentially Euclid's algorithm applied to the "reverse" of the original inputs (cf. exercise 2), afterwards reversing the answer and multiplying by an appropriate power of x.

7. The units of S (as polynomials of degree zero).

8. If $u(x) = v(x)w(x)$, where $u(x)$ has integer coefficients while $v(x)$ and $w(x)$ have rational coefficients, there are integers m and n such that $m \cdot v(x)$ and $n \cdot w(x)$ have integer coefficients. Now $u(x)$ is primitive, so we have

$$u(x) = \pm \text{ pp } (m \cdot v(x)) \text{ pp } (n \cdot w(x)).$$

9. We can extend Algorithm E as follows: Let $(u_1(x), u_2(x), u_3, u_4(x))$ and $(v_1(x), v_2(x), v_3, t_4(x))$ be quadruples satisfying the relations $u_1(x)u(x) + u_2(x)v(x) = u_3 u_4(x)$, $v_1(x)u(x) + v_2(x)v(x) = v_3 v_4(x)$. The extended algorithm starts with $(1, 0, \text{cont}(u), \text{pp}(u(x)))$ and $(0, 1, \text{cont}(v), \text{pp}(v(x)))$ and manipulates these quadruples in such a way as to preserve the above conditions, where $u_4(x), v_4(x)$ run through the same sequence as $u(x), v(x)$ do in Algorithm E. If $au_4(x) = q(x)v_4(x) + br(x)$, we have $av_3(u_1(x), u_2(x)) - q(x)u_3(v_1(x), v_2(x)) = (r_1(x), r_2(x))$, where $r_1(x)u(x) + r_2(x)v(x) = bu_3 v_3 r(x)$, so the extended algorithm can preserve the desired relations. If $u(x)$ and $v(x)$ are relatively prime, the extended algorithm eventually finds $r(x)$ of degree zero, and we obtain $U(x) = r_2(x)$, $V(x) = r_1(x)$ as desired.

Conversely, if such $U(x)$ and $V(x)$ exist, then $u(x)$ and $v(x)$ have no common prime divisors, since they are primitive and have no common divisors of positive degree.

10. By successively factoring polynomials which are reducible into polynomials of smaller degree, we must obtain a finite factorization of any polynomial into irreducibles. The factorization of the *content* is unique. To show that there is at most one factorization of the primitive part, the key result is to prove that if $u(x)$ is an irreducible factor of $v(x)w(x)$, but not a unit multiple of the irreducible polynomial $v(x)$, then $u(x)$ is a factor of $w(x)$. This can be proved by observing that $u(x)$ is a factor of $v(x)w(x)U(x) = rw(x) - w(x)u(x)V(x)$ by the result of exercise 9, where r is a nonzero constant.

11. The only row names needed to show that $u_4(x)$ has integer coefficients are A_1, A_0, B_4, B_3, B_2, B_1, B_0, C_1, C_0, D_0. In general, let $u_{j+2}(x) = 0$; then the rows needed for the proof are $A_{n_2-n_j}$ through A_0, $B_{n_1-n_j}$ through B_0, $C_{n_2-n_j}$ through C_0, $D_{n_3-n_j}$ through D_0, etc.

12. If $n_k = 0$, the argument in the text can be modified to show that the absolute value of the determinant is $\ell_k^{n_{k-1}}/\prod_{1<j<k} \ell_j^{(t_j-1)(t_{j+1}-2)}$. If the polynomials have a factor of positive degree, we can artificially assume that the polynomial zero has degree zero and use the above formula for $\ell_k = 0$. *Note:* The value $R(u, v)$ of Sylvester's determinant is called the *resultant* of u and v, and the quantity $(-1)^{\deg(u)(\deg(u)-1)/2} \ell(u)R(u, u')$ is called the *discriminant* of u.

If we replace each row A_i in Sylvester's matrix by

$$(b_0 A_i + b_1 A_{i+1} + \cdots + b_{n_2-1-i}A_{n_2-1}) - (a_0 B_i + a_1 B_{i+1} + \cdots + a_{n_2-1-i}B_{n_2-1}),$$

and then delete rows B_{n_2-1} through B_0 and the last n_2 columns, we obtain an $n_1 \times n_1$ determinant for the resultant instead of the original $(n_1 + n_2) \times (n_1 + n_2)$ determinant. In some cases the resultant can be evaluated efficiently by means of this determinant; see *CACM* **12** (1969), 23–30, 302–303.

13. The induction step: If $\ell_k^2 u_{k-1}(x) = u_k(x)q_{k-1}(x) + \ell_{k-1}^2 u_{k+1}(x)$, then

$$(\ell^{2(k-2)+1}\ell_k)^2\big(\ell^{2(k-3)}u_{k-1}(x)w(x)\big) = \big(\ell^{2(k-2)}u_k(x)w(x)\big)\big(\ell^{4k-8}q_{k-1}(x)\big)$$

$$+ \big(\ell^{2(k-3)+1}\ell_{k-1}\big)^2 u_{k+1}(x)w(x).$$

14. In general, $r(x)$ in (8) is a multiple of $l(v) = v_n$ if u_{n+k} is a multiple of v_n at the beginning of step R2 for some value of k. In this case we can prove by induction that $u_{n+k} \equiv (-1)^{m-n-k}u_m v_{n-1}^{m-n-k}$ (modulo p) at the beginning of step R2; thus $\deg(w) < \deg(v) - 1$ implies that $w_{n-1} = 0$; hence $u_m v_{n-1}^{m-n+1} \equiv 0$ (modulo p). So either u_m or v_{n-1} is a multiple of p; and u_{n+k} is a multiple of p on the second time step R2 is performed. [To see that this result need not hold if p is not prime, consider $u(x) = x^3 + 1$, $v(x) = 4x^2 + 2x + 1$, $w(x) = 18$.]

15. Let $c_{ij} = a_{i1}a_{j1} + \cdots + a_{in}a_{jn}$; we may assume that $c_{ii} > 0$ for all i. If $c_{ij} \neq 0$ for some $i \neq j$, we can replace row i by $(a_{i1} - ca_{j1}, \ldots, a_{in} - ca_{jn})$, where $c = c_{ij}/c_{jj}$; this does not change the value of the determinant, and it decreases the value of the upper bound we wish to prove, since c_{ii} is replaced by $c_{ii} - c_{ij}^2/c_{jj}$. These replacements can be done in a systematic way for increasing i and for $j < i$, until $c_{ij} = 0$ for all $i \neq j$. (The latter algorithm is called "Schmidt's orthogonalization process"; see *Math. Annalen* **63** (1907), 442.) Then $\det(A)^2 = \det(AA^T) = c_{11} \cdots c_{nn}$.

16. Let $f(x_1, \ldots, x_n) = g_m(x_2, \ldots, x_n)x^m + \cdots + g_0(x_2, \ldots, x_n)$, and let $g(x_2, \ldots, x_n) = g_m(x_2, \ldots, x_n)^2 + \cdots + g_0(x_2, \ldots, x_n)^2$; the latter is not identically zero. We have $a_N \leq m(2N+1)^{n-1} + (2N+1)b_N$, where b_N is the number of integer solutions of $g(x_2, \ldots, x_n) = 0$ with variables bounded by N. Hence $\lim_{N \to \infty} a_N/(2N+1)^n = \lim_{N \to \infty} b_N/(2N+1)^{n-1}$, and this is zero by induction.

17. (a) For convenience, let us describe the algorithm only for $A = \{a, b\}$. The hypotheses imply that $\deg(Q_1 U) = \deg(Q_2 V) \geq 0$, and $\deg(Q_1) \leq \deg(Q_2)$. If $\deg(Q_1) = 0$, then Q_1 is just a nonzero rational number, so we set $Q = Q_2/Q_1$. Otherwise let $Q_1 = aQ_{11} + bQ_{12} + r_1$, $Q_2 = aQ_{21} + bQ_{22} + r_2$, where r_1 and r_2 are rational numbers; it follows that

$$Q_1 U - Q_2 V = a(Q_{11} U - Q_{21} V) + b(Q_{12} U - Q_{22} V) + r_1 U - r_2 V.$$

We must have either $\deg(Q_{11}) = \deg(Q_1) - 1$ or $\deg(Q_{12}) = \deg(Q_1) - 1$. In the former case, $\deg(Q_{11} U - Q_{21} V) < \deg(Q_{11} U)$, by considering the terms of highest degree which start with a; so we may replace Q_1 by Q_{11}, Q_2 by Q_{21}, and repeat the process. Similarly in the latter case, we may replace (Q_1, Q_2) by (Q_{12}, Q_{22}) and repeat the process.

(b) We may assume that $\deg(U) \geq \deg(V)$. If $\deg(R) \geq \deg(V)$, note that $Q_1 U - Q_2 V = Q_1 R - (Q_2 - Q_1 Q)V$ has degree less than $\deg(V) \leq \deg(Q_1 R)$, so we can repeat the process with U replaced by R; we obtain $R = Q'V + R'$, $U = (Q + Q')V + R'$, where $\deg(R') < \deg(R)$, so eventually a solution will be obtained.

(c) The algorithm of (b) gives $V_1 = UV_2 + R$, $\deg(R) < \deg(V_2)$; by homogeneity $R = 0$ and U is homogeneous.

(d) We may assume that $\deg(V) \leq \deg(U)$. If $\deg(V) = 0$, set $W \leftarrow U$; otherwise use (c) to find $U = QV$, so that $QVV = VQV$, $(QV - VQ)V = 0$. This implies that $QV = VQ$, so we can set $U \leftarrow V$, $V \leftarrow Q$ and repeat the process.

For further details about the subject of this exercise, see P. M. Cohn, *Proc. Cambridge Phil. Soc.* **57** (1961), 18–30. The considerably more difficult problem of characterizing *all* string polynomials such that $UV = VU$ has been solved by G. M. Bergman [Ph.D. thesis, Harvard University, 1967].

18. [P. M. Cohn, *Transactions Amer. Math. Soc.* **109** (1963), 332–356.]

C1. Set $u_1 \leftarrow U_1$, $u_2 \leftarrow U_2$, $v_1 \leftarrow V_1$, $v_2 \leftarrow V_2$, $z_1 \leftarrow z_2' \leftarrow w_1 \leftarrow w_2' \leftarrow 1$, $z_1' \leftarrow z_2 \leftarrow w_1' \leftarrow w_2 \leftarrow 0$, $n \leftarrow 0$.

C2. (At this point the identities given in the exercise hold, and also $u_1 v_1 = u_2 v_2$; $v_2 = 0$ if and only if $u_1 = 0$.) If $v_2 = 0$, the algorithm terminates with $\gcrd(V_1, V_2) = v_1$, $\operatorname{lclm}(V_1, V_2) = z_1' V_1 = -z_2' V_2$. (Also by symmetry $\gcld(U_1, U_2) = u_2$, $\operatorname{lcrm}(U_1, U_2) = U_1 w_1 = -U_2 w_2$.)

C3. Find Q and R such that $v_1 = Qv_2 + R$, where $\deg(R) < d(v_2)$. (We have $u_1(Qv_2 + R) = u_2 v_2$, so $u_1 R = (u_2 - u_1 Q)v_2 = R'v_2$.)

C4. Set

$$(w_1, w_2, w_1', w_2', z_1, z_2, z_1', z_2', u_1, u_2, v_1, v_2)$$
$$\leftarrow (w_1' - w_1 Q, w_2' - w_2 Q, w_1, w_2, z_1', z_2', z_1 - Qz_1', z_2 - Qz_2',$$
$$u_2 - u_1 Q, u_1, v_2, v_1 - Qv_2)$$

and $n \leftarrow n + 1$. Go back to C2. ∎

This extension of Euclid's algorithm includes most of the features we have seen in previous extensions, all at the same time, so it provides new insight into the special cases already considered. To prove that it is valid, note first that deg (v_2) decreases in step C4, so the algorithm certainly terminates. At the conclusion of the algorithm, v_1 is a common right divisor of V_1 and V_2, since $w_1 v_1 = (-1)^n V_1$ and $-w_2 v_1 = (-1)^n V_2$; also if d is any common right divisor of V_1 and V_2, it is a right divisor of $z_1 V_1 + z_2 V_2 = v_1$. Hence $v_1 = \text{gcrd}\,(V_1, V_2)$. Also if m is any common left multiple of V_1 and V_2, we may assume without loss of generality that $m = U_1 V_1 = U_2 V_2$, since the sequence of values of Q does not depend on U_1 and U_2. Hence $m = (-1)^n(-u_2 z_1')V_1 = (-1)^n(u_2 z_2')V_2$ is a multiple of $z_1' V_1$.

In practice, if we just want to calculate gcrd (V_1, V_2), we may suppress the computation of n, w_1, w_2, w_1', w_2', z_1, z_2, z_1', z_2'; these additional quantities were added to the algorithm primarily to make its validity more readily established.

Note: Nontrivial factorizations of string polynomials, such as the example given with this exercise, can be found from matrix identities such as

$$\begin{pmatrix} a & 1 \\ 1 & 0 \end{pmatrix}\begin{pmatrix} b & 1 \\ 1 & 0 \end{pmatrix}\begin{pmatrix} c & 1 \\ 1 & 0 \end{pmatrix}\begin{pmatrix} 0 & 1 \\ 1 & -c \end{pmatrix}\begin{pmatrix} 0 & 1 \\ 1 & -b \end{pmatrix}\begin{pmatrix} 0 & 1 \\ 1 & -a \end{pmatrix} = \begin{pmatrix} 1 & 0 \\ 0 & 1 \end{pmatrix},$$

which hold even when multiplication is not commutative. For example,

$$(abc + a + c)(1 + ba) = (ab + 1)(cba + a + c).$$

(Compare this with the "continuant polynomials" of Section 4.5.3.)

19. (Solution by Michael Fredman.) If such an algorithm exists, D is a gcrd by the argument in exercise 18. Let us regard A and B as a single $2n \times n$ matrix C whose first n rows are those of A, and whose second n rows are those of B. Similarly, P and Q can be combined into a $2n \times n$ matrix R; X and Y can be combined into an $n \times 2n$ matrix Z; the desired conditions now reduce to two equations $C = RD$, $D = ZC$. If we can find a $2n \times 2n$ integer matrix U of determinant ± 1 such that the last n rows of $U^{-1}C$ are all zero, then $R = $ (first n columns of U), $D = $ (first n rows of $U^{-1}C$), $Z = $ (first n rows of U^{-1}) solves the desired conditions. Hence, for example, the following triangularization algorithm may be used ($m = 2n$):

Algorithm T. Let C be an $m \times n$ matrix of integers. This algorithm finds $m \times m$ integer matrices U and V such that $UV = I$ and VC is "upper triangular." (The entry in row i and column j of VC is zero if $i > j$.)

T1. Set $U \leftarrow V \leftarrow I$, the $m \times m$ identity matrix; and set $T \leftarrow C$. (Throughout the algorithm we will have $T = VC$, $UV = I$.)

T2. Do step T3 for $j = 1, 2, \ldots, \min(m, n)$, and terminate the algorithm.

T3. Perform the following transformation zero or more times until T_{ij} is zero for all $i > j$: Let T_{kj} be a nonzero element of $\{T_{jj}, T_{(j+1)j}, \ldots, T_{mj}\}$ having the smallest absolute value. Interchange rows k and j of T and of V; interchange columns k and j of U. Then subtract $\lfloor T_{ij}/T_{jj} \rfloor$ times row j from row i, in matrices T and V, and add the same multiple of column i to column j in matrix U, for $j < i \leq m$. ∎

For the stated example, the algorithm yields

$$\begin{pmatrix} 1 & 2 \\ 3 & 4 \end{pmatrix} = \begin{pmatrix} 1 & 0 \\ 3 & 2 \end{pmatrix}\begin{pmatrix} 1 & 2 \\ 0 & -1 \end{pmatrix}, \quad \begin{pmatrix} 4 & 3 \\ 2 & 1 \end{pmatrix} = \begin{pmatrix} 4 & 5 \\ 2 & 3 \end{pmatrix}\begin{pmatrix} 1 & 2 \\ 0 & -1 \end{pmatrix},$$

$$\begin{pmatrix} 1 & 2 \\ 0 & -1 \end{pmatrix} = \begin{pmatrix} 1 & 0 \\ 2 & -2 \end{pmatrix}\begin{pmatrix} 1 & 2 \\ 3 & 4 \end{pmatrix} + \begin{pmatrix} 0 & 0 \\ 1 & 0 \end{pmatrix}\begin{pmatrix} 4 & 3 \\ 2 & 1 \end{pmatrix}.$$

20. It may be helpful to consider exercise 4.6.2–22 with p^m replaced by a small number ϵ.

21. Note that Algorithm R is used only when $m - n \le 1$, and that the coefficients are bounded by (25) with $m = n$. [The stated formula is, in fact, the execution time observed in practice, not merely an upper bound. For more detailed information see G. E. Collins, *Proc. 1968 Summer Inst. on Symbolic Math. Comp.*, Robert G. Tobey, ed. (IBM Federal Systems Center, June 1969), 195–231.

SECTION 4.6.2

1. By the principle of inclusion and exclusion (Section 1.3.3), the number of polynomials without linear factors is $\sum_{k \le n} \binom{p}{k} p^{n-k}(-1)^k = p^{n-p}(p-1)^p$. The probability is therefore $1 - (1 - 1/p)^p$, which is greater than $\frac{1}{2}$.

2. (a) We know that $u(x)$ has a representation as a product of irreducible polynomials, and the leading coefficients of these polynomials must be units, since they divide the leading coefficient of $u(x)$. Therefore we may assume that $u(x)$ has a representation as a product of monic irreducible polynomials $p_1(x)^{e_1} \cdots p_r(x)^{e_r}$, where $p_1(x), \ldots, p_r(x)$ are distinct. This representation is unique, except for the order of the factors, so the conditions on $u(x)$, $v(x)$, $w(x)$ are satisfied if and only if

$$v(x) = p_1(x)^{\lfloor e_1/2 \rfloor} \cdots p_r(x)^{\lfloor e_r/2 \rfloor}, \ w(x) = p_1(x)^{e_1 \bmod 2} \cdots p_r(x)^{e_r \bmod 2}.$$

(b) The generating function for the number of monic polynomials of degree n is $1 + pz + p^2 z^2 + \cdots = 1/(1 - pz)$. The generating function for the number of polynomials of degree n having the form $v(x)^2$, where $v(x)$ is monic, is $1 + pz^2 + p^2 z^4 + \cdots = 1/(1 - pz^2)$. If the generating function for the number of monic squarefree polynomials of degree n is $g(z)$, then by part (a) $1/(1 - pz) = g(z)/(1 - pz^2)$. Hence $g(z) = (1 - pz^2)/(1 - pz) = 1 + pz + (p^2 - p)z^2 + (p^3 - p^2)z^3 + \cdots$. The answer is $p^n - p^{n-1}$ for $n \ge 2$. (Curiously, this proves that $\gcd(u(x), u'(x)) = 1$ with probability $1 - 1/p$; it is the same as the probability that $\gcd(u(x), v(x)) = 1$ if $u(x)$ and $v(x)$ are *independent*, by exercise 4.6.1–5.)

3. Let $u(x) = u_1(x) \cdots u_r(x)$. There is *at most* one such $v(x)$, by the argument of Theorem 4.3.2C. There is *at least* one if, for each j, we can solve the system with $w_j(x) = 1$ and $w_k(x) = 0$ for $k \ne j$. A solution to the latter is $v_1(x)u(x)/u_j(x)$, where $v_1(x)$ and $v_2(x)$ can be found satisfying

$$v_1(x)\big(u(x)/u_j(x)\big) + v_2(x)u_j(x) = 1, \ \deg(v_1) < \deg(u_j),$$

by the extension of Euclid's algorithm (exercise 4.6.1–3).

4. The unique factorization theorem gives us the identity

$$1/(1-z)^{a_1}(1-z^2)^{a_2}(1-z^3)^{a_3}\cdots = 1/(1-pz);$$

after taking logarithms, this can be rewritten

$$\sum_{j\geq 1}\frac{G(z^j)}{j} = \sum_{k,j}\frac{a_k z^{kj}}{j} = \ln\frac{1}{1-pz}.$$

The stated identity then gives us the answer

$$G(z) = \sum_{m\geq 1}\mu(m)\ln\frac{1}{1-pz^m};$$

this formula allows us to calculate the value

$$a_n = \frac{\sum_{d\backslash n}\mu(n/d)p^d}{n}.$$

To prove the stated identity, note that

$$\sum_{n,j\geq 1}\frac{\mu(n)g(z^{nj})}{n^t j^t} = \sum_{m\geq 1}\frac{g(z^m)}{m^t}\sum_{n\backslash m}\mu(n) = g(z).$$

(The identity relating $f(z)$ and $g(z)$ gives us an interesting representation for Pólya's tree enumeration formula, exercise 2.3.4.4–1:

$$A(z) = \sum_{n\geq 1}\frac{\mu(n)}{n}\ln\left(A(z^n)/z^n\right).)$$

5. Since the factors will be distinct with asymptotic probability 1, the desired limit is the limit of the coefficient of z^n in $\sum_k k\cdot G(z/p)^k/k!$, namely

$$\sum_k k\cdot\ln\left(1/(1-z)\right)/k! = \ln\left(1/(1-z)\right)/(1-z).$$

6. For $0\leq s < p$, $x-s$ is a factor of $x^p - x$ (modulo p) by Fermat's theorem. So $x^p - x$ is a multiple of lcm $(x-0, x-1, \ldots, x-(p-1)) = x^p$. (*Note:* Therefore the Stirling numbers $[\begin{smallmatrix}p\\k\end{smallmatrix}]$ are multiples of p except when $k=1$, $k=p$. Equation 1.2.6–41 shows that the same statement is valid for Stirling numbers $\{\begin{smallmatrix}p\\k\end{smallmatrix}\}$ of the other kind.)

7. The factors on the right are relatively prime, and each is a divisor of $u(x)$, so their product divides $u(x)$. On the other hand,

$$u(x)\text{ divides }v(x)^p - v(x) = \prod_{0\leq s < p}(v(x)-s),$$

so it divides the right-hand side by exercise 4.5.2–2.

8. The vector which is output in step N3 is the only output whose kth component is nonzero.

9. For example, start with $x\leftarrow 1$, $y\leftarrow 1$; then repeatedly set $R[x]\leftarrow y$, $x\leftarrow 2x$ mod 101, $y\leftarrow 51y$ mod 101, one hundred and one times.

10. The matrix $Q - I$ below has a null space generated by the two vectors $v^{[1]} = (1, 0, 0, 0, 0, 0, 0, 0)$, $v^{[2]} = (0, 1, 1, 0, 0, 1, 1, 1)$. The factorization is

$$(x^6 + x^5 + x^4 + x + 1)(x^2 + x + 1).$$

$$p = 2$$

$$\begin{pmatrix} 0 & 0 & 0 & 0 & 0 & 0 & 0 & 0 \\ 0 & 1 & 1 & 0 & 0 & 0 & 0 & 0 \\ 0 & 0 & 1 & 0 & 1 & 0 & 0 & 0 \\ 0 & 0 & 0 & 1 & 0 & 0 & 1 & 0 \\ 1 & 0 & 0 & 1 & 0 & 0 & 1 & 0 \\ 1 & 0 & 1 & 1 & 1 & 0 & 0 & 0 \\ 0 & 0 & 1 & 0 & 1 & 1 & 0 & 1 \\ 1 & 1 & 0 & 1 & 1 & 1 & 0 & 1 \end{pmatrix}$$

$$p = 5$$

$$\begin{pmatrix} 0 & 0 & 0 & 0 & 0 & 0 & 0 \\ 0 & 4 & 0 & 0 & 0 & 1 & 0 \\ 0 & 2 & 2 & 0 & 4 & 3 & 4 \\ 0 & 1 & 4 & 4 & 4 & 2 & 1 \\ 2 & 2 & 2 & 3 & 4 & 3 & 2 \\ 0 & 0 & 4 & 0 & 1 & 3 & 2 \\ 3 & 0 & 2 & 1 & 4 & 2 & 1 \end{pmatrix}$$

11. Removing the trivial factor x, the matrix $Q - I$ above has a null space generated by $(1, 0, 0, 0, 0, 0, 0)$ and $(0, 3, 1, 4, 1, 2, 1)$. The factorization is

$$x(x^2 + 3x + 4)(x^5 + 2x^4 + x^3 + 4x^2 + x + 3).$$

12. If $p = 2$, $(x + 1)^4 = x^4 + 1$. If $p = 8k + 1$, $Q - I$ is the zero matrix, so there are four factors. For other values of p we have

$$p = 8k + 3 \qquad\qquad p = 8k + 5 \qquad\qquad p = 8k + 7$$

$$Q - I = \begin{pmatrix} 0 & 0 & 0 & 0 \\ 0 & -1 & 0 & 1 \\ 0 & 0 & -2 & 0 \\ 0 & 1 & 0 & -1 \end{pmatrix} \begin{pmatrix} 0 & 0 & 0 & 0 \\ 0 & -2 & 0 & 0 \\ 0 & 0 & 0 & 0 \\ 0 & 0 & 0 & -2 \end{pmatrix} \begin{pmatrix} 0 & 0 & 0 & 0 \\ 0 & -1 & 0 & -1 \\ 0 & 0 & -2 & 0 \\ 0 & -1 & 0 & -1 \end{pmatrix}$$

Here $Q - I$ has rank 2, so there are $4 - 2 = 2$ factors. (But it is easy to prove that $x^4 + 1$ is irreducible over the integers, since it has no linear factors and the coefficient of x in any factor of degree two must be less than or equal to 2 in absolute value by (28).)

13. $p = 2$: $(x + 1)^4$. $p = 8k + 1$: $(x + (1 + \sqrt{-1})/\sqrt{2})(x + (1 - \sqrt{-1})/\sqrt{2})$ $(x - (1 + \sqrt{-1})/\sqrt{2})(x - (1 - \sqrt{-1})/\sqrt{2})$. $p = 8k + 3$: $(x^2 - \sqrt{-2}x - 1)$ $(x^2 + \sqrt{-2}x - 1)$. $p = 8k + 5$: $(x^2 - \sqrt{-1})(x^2 + \sqrt{-1})$. $p = 8k + 7$: $(x^2 - \sqrt{2}x + 1)(x^2 + \sqrt{2}x + 1)$. The latter factorization also holds over the field of real numbers.

14. Let the polynomials during the calculation be renamed $u_k(x) + sv_k(x)$, for $k = 0, 1, \ldots$, where $v_0 = 0$, $v_1 = -1$. It follows that the sequence of pairs of degrees $(\deg(u_k), \deg(v_k))$ for $k = 0, 1, \ldots$ has the form $(n, -\infty)$, $(n - 1 - a_1, 0)$, $(n - 2 - a_1 - a_2, 1 + a_1)$, $(n - 3 - a_1 - a_2 - a_3, 2 + a_1 + a_2), \ldots$, where the a's are nonnegative integers or $+\infty$. This continues until reaching the first k such that $n - k - 1 - a_1 - \cdots - a_{k+1} \le k + a_1 + \cdots + a_k + 1$, when $\deg(u_{k+1}) \le n/2$; so if $\deg(u_k) > n/2$, the procedure can always be continued further. The desired results can now be derived from an observation of this structure; the case $\deg(u_k) = n/2$ is handled separately. (The method can be pushed a little further by solving quadratic equations in s, as considered in the next exercise.)

15. We may assume that $u \neq 0$ and that p is odd. Berlekamp's method applied to the polynomial $x^2 - u$ tells us that a square root exists if and only if $u^{(p-1)/2} \bmod p = 1$; assume that this condition holds.

Let $p - 1 = 2^t \cdot q$, where q is odd. Zassenhaus's factoring procedure suggests the following square-root extraction algorithm: Set $s \leftarrow 0$. Evaluate

$$\gcd\left((x+s)^q - 1, x^2 - u\right), \; \gcd\left((x+s)^{2q} - 1, x^2 - u\right),$$
$$\gcd\left((x+s)^{4q} - 1, x^2 - u\right), \; \gcd\left((x+s)^{8q} - 1, x^2 - u\right), \ldots,$$

until finding the first case where the gcd is not 1 (modulo p). If the gcd is $x - v$, then $\sqrt{u} = \pm v$. If the gcd is $x^2 - u$, set $s \leftarrow s + 1$ and repeat the calculation.

Notes: If $(x+s)^k \bmod (x^2 - u) = ax + b$, then $(x+s)^{k+1} \bmod (x^2 - u) = (b + as)x + (bs + au)$, and $(x+s)^{2k} \bmod (x^2 - u) = 2abx + (b^2 + a^2 u)$; hence $(x+s)^q$, $(x+s)^{2q}$, ... are easy to evaluate efficiently, and the calculation for fixed s takes $O\left((\log p)^3\right)$ units of time. The square root will be found when $s = 0$ with probability $1/2^{t-1}$; when $s > 0$, the square root will be discovered if and only if the order of $s + \sqrt{u}$ is $2^{t_1} \cdot q_1$ and the order of $s - \sqrt{u}$ is $2^{t_2} \cdot q_2$, where q_1 and q_2 are odd and $t_1 \neq t_2$. Therefore if we assume that the orders of $s + \sqrt{u}$ and $s - \sqrt{u}$ are random and independent, the calculation will be successful with probability $\frac{2}{3} + \frac{1}{3} \cdot 2^{-t} > \frac{2}{3}$, for each fixed $s > 0$.

Another square-root method has been suggested by D. Shanks. When $t > 1$ it requires an auxiliary constant z (depending only on p) such that $z^{2^{t-1}} \equiv -1$ (modulo p). The value $z = n^q \bmod p$ will work for almost one half of all integers n.

S1. Set $y \leftarrow z$, $r \leftarrow t$, $v \leftarrow u^{(q+1)/2} \bmod p$, $w \leftarrow u^q \bmod p$.

S2. If $w = 1$, stop; v is the answer. Otherwise find the smallest k such that $w^{2^k} \bmod p = 1$. If $k = r$, stop (there is no answer); otherwise set $(y, r, v, w) \leftarrow (y^{2^{r-k}}, k, vy^{2^{r-k-1}}, wy^{2^{r-k}})$ and repeat step S2. ∎

The validity of this algorithm follows from the congruences $uw \equiv v^2$, $y^{2^{r-1}} \equiv -1$, $w^{2^{r-1}} \equiv 1$ (modulo p). Step S2 will require about $\frac{1}{4}t^2$ multiplications mod p, on the average.

Still another square root algorithm has been given by D. H. Lehmer, *Studies in Number Theory* (1969), 132–134.

16. (a) Substitute polynomials modulo p for integers in the proof for integers. (b) The proof in exercise 3.2.1.2–16 carries over to any finite field. (c) Since $x = \xi^k$ for some k, $x^{p^n} = x$ in the field. Furthermore, the elements y which satisfy the equation $y^{p^m} = y$ in the field are closed under addition, and closed under multiplication; so if $x^{p^m} = x$, then ξ (being a polynomial in x with integer coefficients) satisfies $\xi^{p^m} = \xi$.

17. If ξ is a primitive root, each nonzero element is some power of ξ. Hence the order must be a divisor of $13^2 - 1 = 2^3 \cdot 3 \cdot 7$, and $\varphi(f)$ elements have order f.

f	$\varphi(f)$	f	$\varphi(f)$	f	$\varphi(f)$	f	$\varphi(f)$
1	1	3	2	7	6	21	12
2	1	6	2	14	6	42	12
4	2	12	4	28	12	84	24
8	4	24	8	56	24	168	48

The probability that two randomly chosen elements have the same order is $(1 + \sum_{f \backslash 168} \varphi(f)^2)/169^2 = 4071/169^2$, about $\frac{1}{7}$. In general, the probability for a field with p^n elements is approximately the product $\prod(1 - 2/(q+1))$, taken over all prime divisors q of $p^n - 1$. (If $m = q_1^{e_1} \cdots q_r^{e_r}$, then

$$\sum_{d \backslash m} \varphi(d)^2 = \prod_{1 \le j \le r} \left(q_j^{2e_j} \left(1 - \frac{2}{q_j + 1} \right) + \frac{2}{q_j + 1} \right).$$

Such summations are readily carried out when m is a power of a prime, and the general case follows from exercise 1.2.4–31.)

18. (a) pp $(p_1(u_n x)) \cdots$ pp $(p_r(u_n x))$, by Gauss's lemma. For example, let

$$u(x) = 6x^3 - 3x^2 + 2x - 1, \quad v(x) = x^3 - 3x^2 + 12x - 36 = (x^2 + 12)(x - 3);$$

then pp $(36x^2 + 12) = 3x^2 + 1$, pp $(6x - 3) = 2x - 1$. (This is a modern version of a fourteenth-century trick used for many years to help solve algebraic equations.)
    (b) Let pp $(w(u_n x)) = \overline{w}_m x^m + \cdots + \overline{w}_0 = w(u_n x)/c$, where c is the content of $w(u_n x)$ as a polynomial in x. Then $w(x) = (c\overline{w}_m/u_n^m)x^m + \cdots + c\overline{w}_0$, hence $c\overline{w}_m = u_n^m$; since $\overline{w}_m$ is a divisor of u_n, c is a multiple of u_n^{m-1}.

19. $|z_0|^n \le |u_{n-1}| |z_0|^{n-1} + \cdots + |u_0| \le \binom{n}{1}\beta|z_0|^{n-1} + \cdots + \binom{n}{n}\beta^n = (|z_0| + \beta)^n - |z_0|^n$, where $\beta = (\sqrt[n]{2} - 1)\alpha$. (*Notes:* $1/(\sqrt[n]{2} - 1)$ is $n/\ln 2 - \frac{1}{2} + O(1/n)$. If we let $\zeta(u)$ be the number β, it is not difficult to show that $(\sqrt[n]{2} - 1) \max (\zeta(u_1), \zeta(u_2)) \le \zeta(u_1 u_2) \le \max (\zeta(u_1), \zeta(u_2))$. If the roots of $u(z)$ are all real and positive, the quantity $\beta_k = |u_{n-k}/\binom{n}{k}|^{1/k}$ is a "generalized mean" of the roots; β_1 is the arithmetic mean, β_n is the geometric mean, and it can be shown that $\beta_1 \ge \beta_2 \ge \cdots \ge \beta_n$.)

20. As in exercise 19, $1 \le |u_{n-1}/z_0| + \cdots + |u_0/z_0^n| < t/(1 - t)$, where $t = \alpha/2|z_0|$.

21. $1 = 1/z_0^2 - 3/z_0^4 - \cdots$ is impossible for $|z_0| \ge 2$, since the absolute value of the quantity on the right would be at most $\frac{1}{4} + \frac{3}{16} + \frac{3}{32} + \frac{8}{64} + \frac{2}{128} + \frac{5}{256} < \frac{3}{4}$. (Perhaps it would be worth while in practice to find a fairly good value of α by starting with the values in exercises 19 and 20 and then using a binary search method to find small values with $1 > |u_{n-1}|/\alpha + \cdots$.)

22. First we prove that if $f(x)$ is any integer polynomial with deg $(f) < $ deg (u), we can find $a_0(x)$, $b_0(x)$, $d_0(x)$ such that deg $(a_0) < $ deg (w), deg $(b_0) < $ deg (v), deg $(d_0) < $ deg (u), and $a_0(x)v(x) + b_0(x)w(x) = f(x) + p^m d_0(x)$. For we can find $a(x)f(x) = w(x)q(x) + a_0(x)$, where deg $(a_0) < $ deg (w), since $w(x)$ is monic; and then we can let $b_0(x) = (b(x)f(x) + v(x)q(x)) \bmod p^m$. Then $a_0(x)v(x) + b_0(x)w(x) = f(x) + p^m d_0(x)$ for some $d_0(x)$. If deg $(b_0) \ge $ deg (v), then deg $(b_0 w) \ge $ deg $(u) > $ deg (f), so $\ell(b_0)$ is a multiple of p^m; but this is impossible, since we have defined $0 \le \ell(b_0) < p^m$. Hence deg $(b_0) < $ deg (v), and deg $(d_0) < $ deg $(vw) = $ deg (u). (For ease in calculation, it suffices to calculate the coefficients of $q(x)$ and $a_0(x)$ with arithmetic modulo p^m.)
    Now let $a_0(x)$, $b_0(x)$ be determined in this way for $f(x) = c(x)$, and let $V(x) = v(x) - p^m b_0(x)$, $W(x) = w(x) - p^m a_0(x)$. Then $V(x)$ and $W(x)$ are monic polynomials, and $V(x)W(x) = u(x) + p^{2m}C(x)$, where deg $(C) < $ deg (u). Also $a(x)V(x) + b(x)W(x) = 1 + p^m f(x)$ for some $f(x)$ with deg $(f) < $ deg (u). If we solve $a_1(x)v(x) + b_1(x)w(x) = f(x) + p^m d_1(x)$ as above, and let $A(x) = a(x) - p^m a_1(x)$, $B(s) =$

$b(x) - p^m b_1(x)$, then $A(x)V(x) + B(x)W(x) = 1 + p^{2m}D(x)$ for some $D(x)$ with deg $(D) <$ deg (u).

(It is not difficult to see, in fact, that $V(x)$, $W(x)$, $A(x)$, $B(x)$ are uniquely determined modulo p^{2m} by the conditions of this exercise; there is only one choice for V and W, and then only one choice for A and B. The result in this exercise is the computational equivalent of a well-known mathematical result called "Hensel's Lemma"; see K. Hensel, *Theorie der Algebraischen Zahlen* (Leipzig: Teubner, 1908), Chapter 4.)

For the polynomial $u(x)$ given by (21), and for $p = 2$, the factorization process would proceed as follows (writing only the coefficients): First exercise 10 gives, mod 2, $v(x) = (1\ 1\ 1)$, $w(x) = (1\ 1\ 1\ 0\ 0\ 1\ 1)$. Euclid's extended algorithm then gives $a(x) = (1\ 0\ 0\ 0\ 0\ 1)$, $b(x) = (1\ 0)$. The factor $v(x) = x^2 + v_1 x + v_2$ must have $|v_1| < 4$, $|v_2| < 4$ by the result of exercise 21. Now mod 2^2 we find that $v(x) = (1\ -1\ -1)$, $w(x) = (1\ 1\ -1\ 0\ 0\ 1\ 1)$, $a(x) = (-1\ 0\ 2\ 0\ 2\ 1)$, $b(x) = (1\ 2)$, using the construction of this exercise; and mod 2^4 we find that $V(x) = (1\ 3\ -1)$, $W(x) = (1\ -3\ -5\ -4\ 4\ -3\ 5)$. So if there is a quadratic factor of $u(x)$ over the integers, it is $x^2 + 3x - 1$; but division shows that the latter polynomial is not a factor, so $u(x)$ is irreducible. (Since we have now proved that this polynomial is irreducible, by using three separate methods, it is unlikely that it has any factors.)

23. The discriminant of $u(x)$ is a nonzero integer, and there are multiple factors mod p iff p divides the discriminant. (In fact, it is not difficult to prove that the smallest prime is $O(n^2 \log N)$, if $n =$ deg (u) and N is a bound on the size of the coefficients of $u(x)$; it is very unlikely that we could not find a "nonramified" prime p very quickly, cf. exercise 2.)

24. Multiply a monic polynomial with rational coefficients by a suitable nonzero integer, to get a primitive polynomial over the integers. Factor this polynomial over the integers, and then convert the factors back to monic. (No factorizations are lost in this way; see exercise 4.6.1–8.)

25. Consideration of the constant term shows there are no factors of degree 1, so if the polynomial is reducible, it must have one factor of degree 2 and one of degree 3. Modulo 2 the factors are $x(x + 1)^2(x^2 + x + 1)$; this is not much help. Modulo 3 the factors are $(x + 2)^2(x^3 + 2x + 2)$. Modulo 5 they are $(x^2 + x + 1)(x^3 + 4x + 2)$. So we see that the answer is $(x^2 + x + 1)(x^3 - x + 2)$.

27. Exercise 4 says that a random polynomial of degree n is irreducible modulo p with rather low probability, about $1/n$. But the Chinese remainder theorem implies that a random monic polynomial of degree n over the integers will be reducible with respect to each of k distinct primes with probability about $(1 - 1/n)^k$, and this approaches zero as $k \to \infty$. Hence almost all polynomials over the integers are irreducible with respect to infinitely many primes; and almost all primitive polynomials over the integers are irreducible. [Another proof has been given by W. S. Brown, *AMM* **70** (1963), 965–969.]

28. First, mod $(y - 0)$, we obtain $q_0(x) = x^2$; this is not a common divisor of the given polynomials, so we must continue. Modulo $(y - 1)$ we obtain $q_1(x) = x + 1$; since this is of degree less than $q_0(x)$, we reject $q_0(x)$. Since $q_1(x)$ is not a common divisor of the polynomials, we calculate modulo $(y - 2)$, obtaining $q_2(x) = x + 2$. The new trial divisor $d(x)$ is the polynomial $x + y$, since this is $\equiv q_j(x)\,(\text{modulo}\,(y - j))$, for $j = 1, 2$; and $d(x)$ works.

SECTION 4.6.3

1. $2^{\lambda(n)}$, the highest power of 2 less than or equal to n.

2. Assume that x is input in register A, and n in location NN; the output is in register X.

01	A1	ENTX	1	1	*A1. Initialize.*
02		STX	Y	1	$Y \leftarrow 1$.
03		STA	Z	1	$Z \leftarrow x$.
04		LDA	NN	1	$N \leftarrow n$.
05		JMP	2F	1	To A2.
06	5H	SRB	1	$L + 1 - K$	
07		STA	N	$L + 1 - K$	$N \leftarrow \lfloor N/2 \rfloor$.
08	A5	LDA	Z	L	*A5. Square Z.*
09		MUL	Z	L	$Z \times Z$
10		STX	Z	L	$\rightarrow Z$.
11	A2	LDA	N	L	*A2. Halve N.*
12	2H	JAE	5B	$L + 1$	To A5 if N is even.
13		SRB	1	K	
14		STA	N	K	$N \leftarrow \lfloor N/2 \rfloor$.
15	A3	LDA	Z	K	*A3. Multiply Y by Z.*
16		MUL	Y	K	$Z \times Y$
17		STX	Y	K	$\rightarrow Y$.
18	A4	LDA	N	K	*A4. N = 0?*
19		JAP	A5	K	If not, continue at step A5. ∎

(It would be better programming practice to change the instruction in line *05* to "JAP", followed by an error indication. The running time is $21L + 17K + 13$, where $L = \lambda(n)$ is one less than the number of bits in the binary representation of n, and $K = \nu(n)$ is the number of one bits in n's representation. The running time could be decreased by $K + 6$ units, by inserting step A4 before step A3 and adding two new instructions to perform A3 when $N = 0$.)

For the serial program, we may assume that n is small enough to fit in an index register; otherwise serial exponentiation is out of the question. The following program leaves the output in register A:

S1	LD1	N	1	$rI1 \leftarrow n$.
	STA	X	1	$X \leftarrow x$.
	JMP	2F	1	
1H	MUL	X	$N - 1$	$rA \times X$
	SLAX	5	$N - 1$	$\rightarrow rA$.
2H	DEC1	1	N	$rI1 \leftarrow rI1 - 1$.
	J1P	1B	N	Multiply again if $rI1 > 0$. ∎

The running time for this program is $14N - 7$; it is faster than the previous program when $n \leq 7$, slower when $n \geq 8$.

3. The sequences of exponents are: (a) 1, 2, 3, 6, 7, 14, 15, 30, 60, 120, 121, 242, 243, 486, 487, 974, 975 [16 multiplications]; (b) 1, 2, 3, 4, 8, 12, 24, 36, 72, 108, 216, 324, 325, 650, 975 [14 multiplications]; (c) 1, 2, 3, 6, 12, 15, 30, 60, 120, 240, 243, 486, 972,

975 [13 multiplications]; (d) 1, 2, 3, 6, 12, 15, 30, 60, 75, 150, 225, 450, 900, 975 [13 multiplications].   (The fewest possible number of multiplications is 12; this is obtainable by combining the factor method with the binary method, since $975 = 15 \cdot (2^6 + 1)$.)

4. $(777777)_8 = 2^{18} - 1$.

5. T1. [Initialize.] Set LINKU[j] $\leftarrow$ 0 for $1 \leq j \leq 2^r$, and set $k \leftarrow 1$, LINKR[0] $\leftarrow$ 1, LINKR[1] $\leftarrow$ 0.

T2. [Change level.] (Now the kth level of the tree has been linked together from left to right, starting at LINKR[0].) If $k = r + 1$, the algorithm terminates. Otherwise set $n \leftarrow$ LINKR[0], $m \leftarrow 0$.

T3. [Prepare for n.] (Now n is a node on level k, and m points to the rightmost node currently on level $k + 1$.) Set $q \leftarrow 0$, $s \leftarrow n$.

T4. [Already in tree?] (Now s is a node in the path from the root to n.) If LINKU[$n + s$] $\neq 0$, go to T6 (the value $n + s$ is already in the tree).

T5. [Insert below n.]   If $q = 0$, set $m' \leftarrow n + s$.   Set LINKR[$n + s$] $\leftarrow q$, LINKU[$n + s$] $\leftarrow n$, $q \leftarrow n + s$.

T6. [Move up.] Set $s \leftarrow$ LINKU[s]. If $s \neq 0$, return to T4.

T7. [Attach group.] If $q \neq 0$, set LINKR[m] $\leftarrow q$, $m \leftarrow m'$.

T8. [Move n.] Set $n \leftarrow$ LINKR[n]. If $n \neq 0$, return to T3.

T9. [End of level.] Set LINKR[m] $\leftarrow 0$, $k \leftarrow k + 1$, and return to T2. ∎

6. Prove by induction that the path to the number $2^{e_0} + 2^{e_1} + \cdots + 2^{e_t}$, if $e_0 > e_1 > \cdots > e_t \geq 0$, is 1, 2, 2^2, ..., 2^{e_0}, $2^{e_0} + 2^{e_1}$, ..., $2^{e_0} + 2^{e_1} + \cdots + 2^{e_t}$; furthermore, the sequences of exponents on each level are in decreasing lexicographic order.

7. The binary and factor methods require one more step to compute x^{2n} than x^n; the power tree method requires at most one more step. Hence (a) $15 \cdot 2^k$; (b) $33 \cdot 2^k$; (c) $23 \cdot 2^k$; $k = 0, 1, 2, 3, \ldots$.

8. The power tree always includes the node $2m$ at one level below m, unless it occurs at the same level or an earlier level; and it always includes the node $2m + 1$ at one level below $2m$, unless it occurs at the same level or an earlier level. (Computational experiments have shown that $2m$ is below m for all $m \leq 2000$, but it appears very difficult to prove this in general.)

10. By using the "FATHER" representation discussed in Section 2.3.3: Make use of a table $f[j]$, $1 \leq j \leq 100$, such that $f[1] = 0$ and $f[j]$ is the number of the node just above j for $j \geq 2$. (The fact that each node of this tree has degree at most two has no effect on the efficiency of this representation; it just makes the tree look prettier as an illustration.)

11. 1, 2, 3, 5, 10, 20, (23 or 40), 43; 1, 2, 4, 8, 9, 17, (26 or 34), 43; 1, 2, 4, 8, 9, 17, 34, (43 or 68), 77; 1, 2, 4, 5, 9, 18, 36, (41 or 72), 77. If either of the latter two paths were in the tree we would have no possibility for $n = 43$, since the tree must contain either 1, 2, 3, 5 or 1, 2, 4, 8, 9.

12. No such infinite tree can exist since $l(n) \neq l^*(n)$ for some n.

13. For Case 1, use a Type-1 chain followed by $2^{A+C} + 2^{B+C} + 2^A + 2^B$; or use the factor method. For Case 2, use a Type-2 chain followed by $2^{A+C+1} + 2^{B+C} + 2^A + 2^B$. For Case 3, use a Type-5 chain followed by $2^A + 2^{A-1}$, or use the factor method. For Case 4, $n = 135 \cdot 2^D$, so we may use the factor method.

14. (a) It is easy to verify that steps $r - 1$ and $r - 2$ are not both small, so let us assume that step $r - 1$ is small and step $r - 2$ is not. If $c = 1$, then $\lambda(a_{r-1}) = \lambda(a_{r-k})$, so $k = 2$; and since $4 \le \nu(a_r) = \nu(a_{r-1}) + \nu(a_{r-k}) - 1 \le \nu(a_{r-1}) + 1$, we have $\nu(a_{r-1}) \ge 3$, making $r - 1$ a star step (lest $a_0, a_1, \ldots, a_{r-3}, a_{r-1}$ include only one small step). Then $a_{r-1} = a_{r-2} + a_{r-q}$ for some q, and if we replace a_{r-2}, a_{r-1}, a_r by $a_{r-2}, 2a_{r-2}, 2a_{r-2} + a_{r-q} = a_r$, we obtain another counterexample chain in which step r is small; but this is impossible. On the other hand, if $c \ge 2$, then $4 \le \nu(a_r) \le \nu(a_{r-1}) + \nu(a_{r-k}) - 2 \le \nu(a_{r-1})$; hence $\nu(a_{r-1}) = 4$, $\nu(a_{r-k}) = 2$, and $c = 2$. This leads readily to an impossible situation by a consideration of the six types in the proof of Theorem B.

(b) If $\lambda(a_{r-k}) < m - 1$, we have $c \ge 3$, so $\nu(a_{r-k}) + \nu(a_{r-1}) \ge 7$ by (22); therefore both $\nu(a_{r-k})$ and $\nu(a_{r-1})$ are ≥ 3. All small steps must be $\le r - k$, and $\lambda(a_{r-k}) = m - k + 1$. If $k \ge 4$, we must have $c = 4$, $k = 4$, $\nu(a_{r-1}) = \nu(a_{r-4}) = 4$; thus $a_{r-1} \ge 2^m + 2^{m-1} + 2^{m-2}$, and a_{r-1} must equal $2^m + 2^{m-1} + 2^{m-2} + 2^{m-3}$; but $a_{r-4} \ge \frac{1}{8}a_{r-1}$ now implies that $a_{r-1} = 8a_{r-4}$. Thus $k = 3$ and $a_{r-1} > 2^m + 2^{m-1}$. Since $a_{r-2} < 2^m$ and $a^{r-3} < 2^{m-1}$, step $r - 1$ must be a doubling; but step $r - 2$ is a nondoubling, since $a_{r-1} \ne 4a_{r-3}$. Furthermore, since $\nu(a_{r-3}) \ge 3, r - 3$ is a star step; and $a_{r-2} = a_{r-3} + a_{r-5}$ would imply that $a_{r-5} = 2^{m-2}$, hence we must have $a_{r-2} = a_{r-3} + a_{r-4}$. As in a similar case treated in the text, the only possibility is now seen to be $a_{r-4} = 2^{m-2} + 2^{m-3}$, $a_{r-3} = 2^{m-2} + 2^{m-3} + 2^{d+1} + 2^d$, $a_{r-1} = 2^m + 2^{m-1} + 2^{d+2} + 2^{d+1}$, and even this possibility is impossible.

16. $l^B(n) = \lambda(n) + \nu(n) - 1$; so if $n = 2^k$, $l^B(n)/\lambda(n) = 1$, but if $n = 2^{k+1} - 1$, $l^B(n)/\lambda(n) = 2$.

17. Let $i_1 < \cdots < i_t$. Delete any intervals I_k which can be removed without affecting the union $I_1 \cup \cdots \cup I_t$. (The interval $(j_k, i_k]$ may be dropped out if either $j_{k+1} \le j_k$ or $j_1 < j_2 < \cdots$ and $j_{k+1} \le i_{k-1}$.) Now combine overlapping intervals $(j_1, i_1], \ldots, (j_d, i_d]$ into an interval $(j', i'] = (j_1, i_d]$ and note that

$$a_{i'} < a_{j'}(1 + \delta)^{i_1 - j_1 + \cdots + i_d - j_d} \le a_{j'}(1 + \delta)^{2(i' - j')},$$

since each point of $(j', i']$ is covered at most twice in $(j_i, i_1] \cup \cdots \cup (j_d, i_d]$.

18. Call $f(m)$ a "nice" function if $\sqrt[m]{f(m)} \to 1$ as $m \to \infty$, that is, $(\log f(m))/m \to 0$. A polynomial in m is nice. The product of nice functions is nice. If $g(m) \to 0$ and c is a positive constant, then $c^{mg(m)}$ is nice; also $\binom{2m}{mg(m)}$ is nice, for by Stirling's approximation this is equivalent to saying that $g(m) \log(1/g(m)) \to 0$.

Now replace each term of the summation by the maximum term which is attained for any s, t, u, v. The total number of terms is nice, and so are $\binom{m+s}{t+v}$, $\binom{t+v}{v} \le 2^{t+v}$, and β^{2v}, because $(t + v)/m \to 0$. Finally, $\binom{(m+s)^2}{t} \le (2m)^{2t}/t! < (4m^2/t)^t e^t$, where $(4e)^t$ is nice; setting t to its maximum value $(1 - \frac{1}{2}\epsilon)m/\lambda(m)$, we have $(m^2/t)^t = (m\lambda(m)/(1 - \frac{1}{2}\epsilon))^t = 2^{m(1-\epsilon/2)} \cdot f(m)$, where $f(m)$ is nice. Hence the entire sum is less than $2^m/\alpha^m$ for large m, if $\alpha = 2^{-\eta}$, $0 < \eta < \frac{1}{2}\epsilon$.

19. (a) $M \cap N$, $M \cup N$, $M \uplus N$, respectively; see Eqs. 4.5.2–6, 4.5.2–7.

(b) $f(z)g(z)$, lcm$(f(z), g(z))$, gcd$(f(z), g(z))$. (For the same reasons as (a), because the monic irreducible polynomials over the complex numbers are precisely the polynomials $z - \zeta$.)

(c) Commutative laws $A \uplus B = B \uplus A$, $A \cup B = B \cup A$, $A \cap B = B \cap A$. Associative laws $A \uplus (B \uplus C) = (A \uplus B) \uplus C$, $A \cup (B \cup C) = (A \cup B) \cup C$, $A \cap (B \cap C) = (A \cap B) \cap C$. Distributive laws $A \cup (B \cap C) = (A \cup B) \cap (A \cup C)$, $A \cap (B \cup C) = (A \cap B) \cup (A \cap C)$, $A \uplus (B \cup C) = (A \uplus B) \cup (A \uplus C)$, $A \uplus (B \cap C) = (A \uplus B) \cap (A \uplus C)$. Idempotent laws $A \cup A = A$, $A \cap A = A$. Absorption laws $A \cup (A \cap B) = A$, $A \cap (A \cup B) = A$, $A \cap (A \uplus B) = A$, $A \cup (A \uplus B) = A \uplus B$. Identity and zero laws $\emptyset \uplus A = A$, $\emptyset \cup A = A$, $\emptyset \cap A = \emptyset$, where $\emptyset$ is the empty multiset. Counting law $A \uplus B = (A \cup B) \uplus (A \cap B)$. Further properties analogous to those of sets come from the partial ordering defined by the rule $A \subseteq B$ iff $A \cap B = A$ (iff $A \cup B = B$).

Notes: Other common applications of multisets are zeros and poles of meromorphic functions, invariants of matrices in canonical form, invariants of finite Abelian groups, etc.; multisets can be useful in combinatorial counting arguments and in the development of measure theory. The terminal strings of a noncircular context-free grammar form a multiset which is a set if and only if the grammar is unambiguous. Although multisets appear frequently in mathematics, they often must be treated rather clumsily because there is currently no standard way to treat sets with repeated elements. Several mathematicians have voiced their belief that the lack of adequate terminology and notation for this common concept has been a definite handicap to the development of mathematics. (A multiset is, of course, formally equivalent to a mapping from a set into the nonnegative integers, but this formal equivalence is of little or no practical value for creative mathematical reasoning.) The author has discussed this matter with many people in an attempt to find a good remedy. Some of the names suggested for the concept were list, bunch, heap, sample, weighted set, collection; but these words either conflict with present terminology, have an improper connotation, or are too much of a mouthful to say and to write conveniently. It does not seem out of place to coin a new word for such an important concept, and "multiset" has been suggested by N. G. de Bruijn. The notation "$A \uplus B$" has been selected by the author to avoid conflict with existing notations and to stress the analogy with set union. It would not be as desirable to use "$A + B$" for this purpose, since algebraists have found that $A + B$ is a good notation for $\{\alpha + \beta \mid \alpha \in A$ and $\beta \in B\}$. If A is a multiset of nonnegative integers, let $G(z) = \sum_{n \in A} z^n$ be a generating function corresponding to A. (Generating functions with nonnegative integer coefficients obviously correspond one-to-one with multisets of nonnegative integers.) If $G(z)$ corresponds to A and $H(z)$ to B, then $G(z) + H(z)$ corresponds to $A \uplus B$ and $G(z)H(z)$ corresponds to $A + B$. If we form "Dirichlet" generating functions $g(z) = \sum_{n \in A} 1/n^z$, $h(z) = \sum_{n \in B} 1/n^z$, the product $g(z)h(z)$ corresponds to the multiset product AB.

20. Type 3: $(b_0, \ldots, b_r) = (A + C - 2, \ldots, C - 2, C - 2, C - 2, \ldots, 0)'$ $(S_0, \ldots, S_r) = (M_{00}, \ldots, M_{r0}) = (\{0\}, \ldots, \{A\}, \{A - 1, A\}, \{A - 1, A, A\}$, $\{A - 1, A - 1, A, A, A\}, \ldots, \{A + C - 3, A + C - 3, A + C - 2, A + C - 2, A + C - 2\})$. Type 5: $(b_0, \ldots, b_r) = (A + C + D, \ldots, C + D, C + D, \ldots, D, D, \ldots, 0)$, $(M_{00}, \ldots, M_{r0}) = (\{0\}, \ldots, \{A\}, \{A - 1, A\}, \ldots, \{A + C - 1, A + C\}$, $\{A + C - 1, A + C - 1, A + C\}, \ldots, \{A + C + D - 1, A + C + D - 1, A + C + D\})$, $(M_{01}, \ldots, M_{r1}) = (\emptyset, \ldots, \emptyset, \emptyset, \ldots, \emptyset, \{A + C - 2\}, \ldots, \{A + C + D - 2\})$, $S_i = M_{i0} \uplus M_{i1}$.

21. For example, let $u = 2^{8q+5}$, $x = (2^{(q+1)u} - 1)/(2^u - 1) = 2^{qu} + \cdots + 2^u + 1$, $y = 2^{(q+1)u} + 1$. Then $xy = (2^{2(q+1)u} - 1)/(2^u - 1)$; if $n = 2^{4(q+1)u} + xy$, we have $l(n) \leq 4(q+1)u + q + 2$ by Theorem F, but $l^*(n) = 4(q+1)u + 2q + 2$ by Theorem H.

22. Underline everything except the $u - 1$ insertions used in the calculation of x.

23. Theorem G (everything underlined).

24. Use the numbers $(B^{a_i} - 1)/(B - 1)$, $0 \leq i \leq r$, underlined when a_i is under-lined; and $c_k B^{i-1}(B^{b_j} - 1)/(B - 1)$ for $0 \leq j < t$, $0 < i \leq b_{j+1} - b_j$, $1 \leq k \leq l^0(B)$, underlined when c_k is underlined, where $c_0, c_1, \ldots$ is a minimum length l^0-chain for B. To prove the second inequality, let $B = 2^m$ and use (3). (The second inequality is rarely, if ever, an improvement on Theorem G.)

25. We may assume that $d_k = 1$. Use the rule $R\,A_{k-1} \cdots A_1$, where $A_j = $ "XR" if $d_j = 1$, $A_j = $ "R" otherwise, and where "R" means take the square root, "X" means multiply by x. For example, if $y = (.1101101)_2$, the rule is R R XR XR R XR XR. (There exist binary square-root extraction algorithms suitable for computer hardware, requiring an execution time comparable to that of division; computers with such hardware could also calculate more general fractional powers with the technique in this exercise.)

26. If we know the pair (F_k, F_{k-1}), then $(F_{k+1}, F_k) = (F_k + F_{k-1}, F_k)$ and $(F_{2k}, F_{2k-1}) = (F_k^2 + 2F_k F_{k-1}, F_k^2 + F_{k-1}^2)$; so a binary method can be used to calculate (F_n, F_{n-1}), using $O(\log n)$ arithmetic operations. Perhaps better is to use the pair of values (F_k, L_k), where $L_k = F_{k-1} + F_{k+1}$ (cf. Section 4.5.4); then $(F_{k+1}, L_{k+1}) = \left(\frac{1}{2}(F_k + L_k), \frac{1}{2}(5F_k + L_k)\right)$, $(F_{2k}, L_{2k}) = (F_k L_k, L_k^2 - 2(-1)^k)$.

27. First form the $2^m - m - 1$ products $x_1^{e_1} \cdots x_m^{e_m}$, where $0 \leq e_j \leq 1$ and $e_1 + \cdots + e_m \geq 2$. Then if $n_j = (d_{j\lambda} \cdots d_{j1} d_{j0})_2$, the sequence begins with $x_1^{d_{1\lambda}} \cdots x_m^{d_{m\lambda}}$ and then we square, and multiply by $x_1^{d_{1i}} \cdots x_m^{d_{mi}}$, for $i = \lambda - 1, \ldots,$ 1, 0. (Straus [*AMM* **71** (1964), 807–808] has shown that $2\lambda(n)$ may be replaced by $(1 + \epsilon)\lambda(n)$ for any $\epsilon > 0$, by generalizing this binary method to a 2^k-ary as in Theorem D. An obvious lower bound on the number of multiplications is

$$l(n_1 + \cdots + n_m).)$$

28. Let us represent a star step i by the symbol S_k if $a_i = a_{i-1} + a_{i-1-k}$; hence S_0 represents a doubling and S_1 stands for a Fibonacci step. Consider the sequence of symbols $S_{k_1} \ldots S_{k_{r-q-1}}$ corresponding to steps $q + 2, \ldots, r$; we will prove that the only sequences possible have the forms S_0^m, $S_0^m S_k S_0^n$, or $S_0^m S_k S_0^n S_1 S_0^p$, where m, n, and p are nonnegative integers.

(a) Let $a_q = x$; then $\lambda(a_{q+i}) = \lambda(x) + i$ for $1 \leq i \leq r - q$, hence $a_{q+i} > 2^{i-1}x$. The latter condition suffices to prove the desired result: After the sequence of opera-tions S_0^m, the addition chain contains the value $a_{q+m+1} = 2^m a_{q+1} \leq 2^{m+1}x$. Sub-sequent operations $S_k S_0^n$ then yield $a_{q+m+n+2} \leq 3 \cdot 2^{m+n}x$; and a subsequent operation S_j for $j \geq 2$ would yield $a_{q+m+n+3} \leq 4 \cdot 2^{m+n}x$, a contradiction. Hence S_1 must follow, and after $S_1 S_0^p$ we have $a_{q+m+n+p+3} \leq 5 \cdot 2^{m+n+p}$. A subsequent operation S_j for $j \geq 1$ would yield $a_{q+m+n+p+4} \leq 8 \cdot 2^{m+n}x$, a contradiction.

(b) It suffices to show that, in the notation of part (a), $\nu(a_r) \leq 8M$, if $\nu(a_i) \leq M$ for $0 \leq i < q$. Since each nondoubling at most doubles the number of one bits, the result is obvious unless steps $q + 2, \ldots, r$ have the form $S_0^m S_k S_0^n S_1 S_0^p$. Let $b_1 = a_q$,

$b_2 = a_{q+1}$, $b_3 = a_{q+m+2}$, $b_4 = a_{q+m+n+3}$. We have $\nu(b_1) \leq 2M$, $\nu(b_2) \leq 3M$, $\nu(b_3) \leq 2\nu(b_2)$, $\nu(b_4) \leq 2\nu(b_3)$; so the result follows unless we have $\nu(b_3) > 4M$, $\nu(b_2) > 2M$. Under the latter assumption, several cases arise:

Case 1, $m = 0$. Then if $b_3 = b_2 + y$, where $y < b_1$, we have $\nu(b_3) \leq 4M$; while $b_3 = b_2 + b_1$ yields $\nu(b_3) \leq 3M$, a contradiction.

Case 2, $m > 0$, $n = 0$. Then $b_3 = 2^m b_2 + y$, $b_4 = b_3 + 2^m b_2 = 2^{m+1} b_2 + y$; hence $\nu(b_4) \leq 2\nu(b_2)$.

Case 3, $m > 0$, $n > 0$. Here $b_3 = 2^m b_2 + y$, for some y, and $b_4 = 3 \cdot 2^{n-1} b_3$. If $y < 2^{m-1} b_2$ then $\lambda(y) \leq \lambda(b_3) - 3$, hence $b_3 < 2^{\lambda(b_3)} + 2^{\lambda(b_3)-2}$ and $\lambda(3b_3) = \lambda(b_3) + 1$, a contradiction. Therefore $b_3 = 3 \cdot 2^{m-1} b_2$, $b_4 = 9 \cdot 2^{m+n-2} b_2$, and $\nu(b_4) \leq 2\nu(b_2)$. (We have strengthened the result of (a), proving that $mn > 0$ in the sequence $S_0^m S_k S_0^n S_1 S_0^p$ implies $k = 1$.)

Reference: *Canad. J. Math.* **21** (1969), 675–683.

30. $n = 31$ is the smallest example; $l(31) = 7$, but 1, 2, 4, 8, 16, 32, 31 is an addition-subtraction chain of length 6. (P. Erdös has stated that Theorem E holds also for addition-subtraction chains.)

SECTION 4.6.4

1. Set $y \leftarrow x^2$, then compute $((\cdots (u_{2n+1}y + u_{2n-1})y + \cdots)y + u_1)x$.

2. One way is to use "Horner's rule" (2), doing *polynomial* multiplication and addition instead of multiplication and addition in the domain of coefficients, with x replaced by $x + c$. This idea can be formalized as follows:

G1. Do step G2, for $k = n$, $n - 1$, $\ldots$, 0 (in this order), and stop.

G2. Set $v_k \leftarrow u_k$, and then set $v_j \leftarrow v_j + cv_{j+1}$ for $j = k$, $k + 1$, $\ldots$, $n - 1$. (When $k = n$, this step simply sets $v_n \leftarrow u_n$.) ∎

The result is $u(x + c) = v_n x^n + v_{n-1} x^{n-1} + \cdots + v_0$, after $n(n + 1)/2$ multiplications and $n(n + 1)/2$ additions.

The method actually used by Horner was rather different. If we write $u(x) = q(x)(x - c) + r$, then $u(x + c) = q(x + c)x + r$, so r is the constant coefficient of $u(x + c)$; and the procedure may be completed by finding the coefficients of $q(x + c)$, where $q(x)$ is a known polynomial of degree $n - 1$. Horner's suggestion may be formulated in the following way:

H1. Start with $v_k \leftarrow u_k$ for $0 \leq k \leq n$; then do step H2 for $k = 0, 1, \ldots, n - 1$ (in this order) and stop.

H2. Set $v_j \leftarrow v_j + cv_{j+1}$ for $j = n - 1$, $\ldots$, $k + 1$, k. ∎

The number of multiplications and additions is the same in each case; in fact, both schemes are computationally equivalent, differing only in the control of the order of the operations.

Note: Horner actually rediscovered a rule previously published by Paolo Ruffini, *Soc. Italiana delle Scienze* **16** (1801), and indeed it may perhaps be a Chinese method from the thirteenth century! See J. L. Coolidge, *Mathematics of Great Amateurs* (Oxford, 1949), Chapter 15.

3. The coefficient of x^k is a polynomial in y which may be evaluated by Horner's rule: $(\cdots(u_{n,0}x + (u_{n-1,1}y + u_{n-1,0}))x + \cdots)x + ((\cdots(u_{0,n}y + u_{0,n-1})y + \cdots)y + u_{0,0})$. (For a "homogeneous" polynomial, such as $u_n x^n + u_{n-1}x^{n-1}y + \cdots + u_1 x y^{n-1} + u_0 y^n$, another scheme is more efficient: first divide x by y, evaluate a polynomial in x/y, then multiply by y^n.)

4. Horner's rule, with $4n$ multiplications and $4n$ additions, is superior to (3), which uses $4n + 2$ multiplications and $4n + 2$ additions.

5. One multiplication to compute x^2; $\lfloor n/2 \rfloor$ multiplications and $\lfloor n/2 \rfloor$ additions to evaluate the first line; $\lceil n/2 \rceil$ multiplications and $\lceil n/2 \rceil - 1$ additions to evaluate the second line; and one addition to add the two lines together. Total: $n + 1$ multiplications and n additions.

6. Also set $c_0 = 0$; $c_j = c_{j-1}x + b_{j-1}$, $1 \le j \le n$. (A similar technique works for $u^{(k)}(x)/k!$; this rule was given by Horner, and in fact it agrees with the first steps of the second procedure in exercise 2.)

7. Use the fact that $u(z) = (z - z_0)(z - \bar{z}_0)q(z) + a_n z + b_n$, where the a's and b's are computed for $z = z_0$ and $q(z) = a_1 z^{n-2} + a_2 z^{n-3} + \cdots + a_{n-1}$; $u'(z_0) = (z_0 - \bar{z}_0)q(z_0) + a_n$. In addition to (3) we therefore let $c_1 = 0$, $d_1 = 0$; $c_j = d_{j-1} + rc_{j-1}$, $d_j = a_{j-1} - sc_{j-1}$, $1 < j \le n$; the final answer is $a_n + (c_n z + d_n)(y + y)i$. This method involves $2n - 3$ extra multiplications, $2n - 2$ extra additions.

A similar method exists to calculate $u(x)$, $u(-x)$, $u'(x)$, $u'(-x)$, generalizing (4).

8. See (16).

9. [*Combinatorial Mathematics* (Buffalo: Math. Ass'n of America, 1963), 26–28.] This formula can be regarded as an application of the principle of inclusion and exclusion (Section 1.3.3), since the sum of the terms for $n - \epsilon_1 - \cdots - \epsilon_n = k$ is the sum of all $x_{1j_1}x_{2j_2}\cdots x_{nj_n}$ for which k values of the j_i do not appear. A direct proof can be given by observing that the coefficient of $x_{1j_1}\cdots x_{nj_n}$ is

$$\sum (-1)^{n-\epsilon_1-\cdots-\epsilon_n}\epsilon_{j_1}\cdots\epsilon_{j_n};$$

if the j's are distinct, this equals unity, but if $j_1, \ldots, j_n \ne k$ then it is zero, since the terms for $\epsilon_k = 0$ cancel the terms for $\epsilon_k = 1$.

To evaluate the sum efficiently, we can start with $\epsilon_1 = 1$, $\epsilon_2 = \cdots = \epsilon_n = 0$, and we can then proceed through all combinations of the ϵ's in such a way that only one ϵ changes from one term to the next. (See "Gray code" in Chapter 7.) The work to compute the first term is $n - 1$ multiplications; the subsequent $2^n - 2$ terms each involve n additions, then $n - 1$ multiplications, then one more addition. Total: $(2^n - 1)(n - 1)$ multiplications, and $(2^n - 2)(n + 1)$ additions. Only $n + 1$ temp storage locations are needed, one for the main partial sum and one for each factor of the current product.

10. $\sum_{1 \le k < n}(k + 1)\binom{n}{k+1} = n(2^{n-1} - 1)$ multiplications and $\sum_{1 \le k < n} k\binom{n}{k+1} = n2^{n-1} - 2^n + 1$ additions. This is approximately half as many arithmetic operations as the method of exercise 9, although it requires a more complicated program to control the sequence. Approximately $\binom{n}{\lceil n/2\rceil} + \binom{n}{\lceil n/2\rceil - 1}$ temporary storage locations must be used, and this grows exponentially large (on the order of $4^n/n^{1.5}$).

The method in this exercise is equivalent to the unusual matrix factorization of the permanent function, given by Jurkat and Ryser in *J. Algebra* **5** (1967), 342–357. It may also be regarded as an application of (12), (13), in an appropriate sense.

13. We obtain the original numbers, multiplied by 2^n. (Cf. Eq. 3.3.4–2, and note that $1/(-1) = (-1)$.) [Reference: *J. Roy. Stat. Soc.*, series B, **22** (1960), 372–375.]

14. (a) Let $m = 2$, $F(t_1, t_2, \ldots, t_n) = F(2^{n-1}t_n + \cdots + 2t_2 + t_1)$, $f(s_1, s_2, \ldots, s_n) = f(2^{n-1}s_1 + 2^{n-2}s_2 + \cdots + s_n)$; note the reversed treatment between t's and s's. Also let $g_k(s_k, \ldots, s_n, t_k)$ be ω raised to the $2^{k-1}t_k(s_n + 2s_{n-1} + \cdots + 2^{n-k}s_k)$ power. It is possible to gain some speed by combining steps k and $k+1$, for $k = 1$, 3, $\ldots$; this expedites several of the computations of sines and cosines. [The fast Fourier transform algorithm is essentially due to C. Runge and H. König in 1924, and it was generalized by J. W. Cooley and J. W. Tukey, *Math. Comp.* **19** (1965), 297–301. Its interesting history has been traced by J. W. Cooley, P. A. W. Lewis, P. D. Welch, *Proc. IEEE* **55** (1967), 1675–1677. Details concerning its use have been discussed by R. C. Singleton, *CACM* **10** (1967), 647–654; M. C. Pease, *JACM* **15** (1968), 252–264; G. D. Berglund, *Math. Comp.* **22** (1968), 275–279; *CACM* **11** (1968), 703–710.]

(b) Let N be the smallest power of 2 which exceeds $2n$, and let $u_{n+1} = \cdots = u_{N-1} = v_{n+1} = \cdots = v_{N-1} = 0$. If $U_s = \sum_{0 \le t < N} u_t \omega^{st}$, $V_s = \sum_{0 \le t < N} v_t \omega^{st}$, $0 \le s < N$, $\omega = e^{2\pi i/N}$, then $\sum_{0 \le s < N} U_s V_s \omega^{-st} = N \sum u_{t_1} v_{t_2}$. The latter sum is taken over all t_1 and t_2 with $0 \le t_1, t_2 < N$, $t_1 + t_2 \equiv t$ (modulo N). The terms vanish unless $t_1 \le n$, $t_2 \le n$, so $t_1 + t_2 < N$; thus the sum is the coefficient of z^t in the product $u(z)v(z)$. If we use the method of (a) to compute the Fourier transforms and the inverse transforms, the number of complex operations is $O(N \log N) + O(N \log N) + O(N) + O(N \log N)$; and $N < 4n$.

Note: The number ω cannot be represented exactly inside a computer, but V. Strassen has shown that it isn't necessary to have too much accuracy to deduce exact results when the coefficients are integers. It is possible to use an *integer* number ω which is of order 2^t, modulo p, to obtain similar results; see part C of Section 4.3.3 for further commentary.

15. By induction on n, we may assume that $\alpha_0 = y_0, \ldots, \alpha_{n-1} = y_{n-1}^{(n-1)}$; and that $v(x) = \beta_0 + \beta_1(x - x_1) + \cdots + \beta_{n-1}(x - x_1) \cdots (x - x_{n-1})$ satisfies the condition $v(x_k) = y_k$ for $1 \le k \le n$, where $\beta_0 = y_1$, $\beta_1 = y_2'$, $\ldots$, $\beta_{n-1} = y_n^{(n-1)}$. We may also write $u^{[n-1]}(x) = \beta_0 + \beta_1(x - x_1) + \cdots + \beta_{n-2}(x - x_1) \cdots (x - x_{n-2}) + \alpha_{n-1}(x - x_1) \cdots (x - x_{n-1})$, since $u^{[n-1]}(x) - v(x)$ is a multiple of $(x - x_1) \cdots (x - x_{n-1})$. Now let $u^{[n]}(x) = u^{[n-1]}(x) + (x - x_0) \cdots (x - x_{n-1})(\beta_{n-1} - \alpha_{n-1})/(x_n - x_0)$; then $u^{[n]}(x_k)$ obviously equals y_k for $0 \le k < n$, and $u^{[n]}(x_n) = u^{[n-1]}(x_n) + (\beta_{n-1} - \alpha_{n-1})(x_n - x_1) \cdots (x_n - x_{n-1}) = u^{[n-1]}(x_n) + y_n - (\beta_0 + \beta_1(x_n - x_1) + \cdots + \beta_{n-2}(x_n - x_1) \cdots (x_n - x_{n-2})) - \alpha_{n-1}((x_n - x_1) \cdots (x_n - x_{n-1})) = y_n$.

16. Carry out the multiplications and additions of (16) as operations on polynomials. (The special case $x_0 = \cdots = x_{n-1} = c$ is considered in exercise 2. We have used this method in step C8 of Algorithm 4.3.3C.)

18. $\alpha_0 = \frac{1}{2}(u_3/u_4 + 1)$, $\beta = u_2/u_4 - \alpha_0(\alpha_0 - 1)$, $\alpha_1 = \alpha_0\beta - u_1/u_4$, $\alpha_2 = \beta - 2\alpha_1$, $\alpha_3 = u_0/u_4 - \alpha_1(\alpha_1 + \alpha_2)$, $\alpha_4 = u_4$.

19. Since α_5 is the leading coefficient, we may assume without loss of generality that $u(x)$ is monic (i.e., $u_5 = 1$). Then α_0 is a root of the cubic equation $40z^3 - 24u_4z^2 + (4u_4^2 + 2u_3)z + (u_2 - u_3u_4) = 0$; this equation always has at least one real root, and it may have three. Once α_0 is determined, we have $\alpha_3 = u_4 - 4\alpha_0$, $\alpha_1 = u_3 - 4\alpha_0\alpha_3 - 6\alpha_0^2$, $\alpha_2 = u_1 - \alpha_0(\alpha_0\alpha_1 + 4\alpha_0^2\alpha_3 + 2\alpha_1\alpha_3 + \alpha_0^3)$, $\alpha_4 = u_0 - \alpha_3(\alpha_0^4 + \alpha_1\alpha_0^2 + \alpha_2)$.

For the given polynomial we are to solve the cubic equation $40z^3 - 120z^2 + 80z = 0$; this leads to three solutions $(\alpha_0, \alpha_1, \alpha_2, \alpha_3, \alpha_4, \alpha_5) = (0, -10, 13, 5, -5, 1)$, $(1, -20, 68, 1, 11, 1)$, $(2, -10, 13, -3, 27, 1)$.

20.

LDA	X	FADD	$=\alpha_1=$
FADD	$=\alpha_3=$	FMUL	TEMP2
STA	TEMP1	FADD	$=\alpha_2=$
FADD	$=\alpha_0-\alpha_3=$	FMUL	TEMP1
STA	TEMP2	FADD	$=\alpha_4=$
FMUL	TEMP2	FMUL	$=\alpha_5=$ ∎
STA	TEMP2		

21. $z = (x + 1)x - 2$, $w = (x + 5)z + 9$, $u(x) = (w + z - 8)w - 8$; or $z = (x + 9)x + 26$, $w = (x - 3)z + 73$, $u(x) = (w + z - 24)w - 12$.

22. $\alpha_6 = 1$, $\alpha_0 = -1$, $\alpha_1 = 1$, $\beta_1 = -2$, $\beta_2 = -2$, $\beta_3 = -2$, $\beta_4 = 1$, $\alpha_3 = -4$, $\alpha_2 = 0$, $\alpha_4 = 4$, $\alpha_5 = -2$. We form $z = (x - 1)x + 1$, $w = z + x$, and $u(x) = ((z - x - 4)w + 4)z - 2$. This takes three multiplications and seven additions; in this special case we see that another addition can be saved if we compute $w = x^2 + 1$, $z = w - x$.

23. (a) We may use induction on n; the result is trivial if $n < 2$. If $f(0) = 0$, then the result is true for the polynomial $f(z)/z$, so it holds for $f(z)$. If $f(iy) = 0$ for some real $y \neq 0$, then $g(\pm iy) = h(\pm iy) = 0$; since the result is true for $f(z)/(z^2 + y^2)$, it holds also for $f(z)$. Therefore we may assume that $f(z)$ has no roots whose real part is zero. Now the net number of times the given path circles the origin is the number of roots of $f(z)$ inside the region, which is at most 1. When R is large, the path $f(Re^{it})$ for $\pi/2 \le t \le 3\pi/2$ will circle the origin clockwise approximately $n/2$ times; so the path $f(it)$ for $-R \le t \le R$ must go counterclockwise around the origin at least $n/2 - 1$ times. For n even, this implies that $f(it)$ crosses the imaginary axis at least $n - 2$ times, and the real axis at least $n - 3$ times; for n odd $f(it)$ crosses the real axis at least $n - 2$ times and the imaginary axis at least $n - 3$ times. These are roots respectively of $g(it) = 0$, $h(it) = 0$.

(b) If not, g or h would have a root of the form $a + bi$ with $a \neq 0$ and $b \neq 0$. But this would imply the existence of at least three other such roots, namely $a - bi$ and $-a \pm bi$, while $g(z)$, $h(z)$ have at most n roots.

24. The roots of u are -7, $-3 \pm i$, $-2 \pm i$, -1; permissible values of c are 2 and 4 (*not* 3, since $c = 3$ makes the sum of the roots equal to zero). Case 1, $c = 2$: $p(x) = (x + 5)(x^2 + 2x + 2)(x^2 + 1)(x - 1) = x^6 + 6x^5 + 6x^4 + 4x^3 - 5x^2 - 2x - 10$; $q(x) = 6x^2 + 4x - 2 = 6(x + 1)(x - \frac{1}{3})$. Let $\alpha_2 = -1$, $\alpha_1 = \frac{1}{3}$; $p_1(x) = x^4 + 6x^3 + 5x^2 - 2x - 10 = (x^2 + 6x + \frac{16}{3})(x^2 - \frac{1}{3}) - \frac{74}{9}$; $\alpha_0 = 6$, $\beta_0 = \frac{16}{3}$, $\beta_1 = -\frac{74}{9}$. Case 2, $c = 4$: A similar analysis gives $\alpha_2 = 9$, $\alpha_1 = -3$, $\alpha_0 = -6$, $\beta_0 = 12$, $\beta_1 = -26$.

25. $\beta_1 = \alpha_2$, $\beta_2 = 2\alpha_1$, $\beta_3 = \alpha_7$, $\beta_4 = \alpha_6$, $\beta_5 = \beta_6 = 0$, $\beta_7 = \alpha_1$, $\beta_8 = 0$, $\beta_9 = 2\alpha_1 - \alpha_8$.

26. (a) $\lambda_1 = \alpha_1 \times \lambda_0$, $\lambda_2 = \alpha_2 + \lambda_1$, $\lambda_3 = \lambda_2 \times \lambda_0$, $\lambda_4 = \alpha_3 + \lambda_3$, $\lambda_5 = \lambda_4 \times \lambda_0$, $\lambda_6 = \alpha_4 + \lambda_5$. (b) $\kappa_1 = 1 + \beta_1 x$, $\kappa_2 = 1 + \beta_2\kappa_1 x$, $\kappa_3 = 1 + \beta_3\kappa_2 x$, $u(x) = \beta_4\kappa_3 = \beta_1\beta_2\beta_3\beta_4 x^3 + \beta_2\beta_3\beta_4 x^2 + \beta_3\beta_4 x + \beta_4$. (c) If any coefficient is zero, the coefficient of x^3 must also be zero in (b), while (a) yields an arbitrary polynomial $\alpha_1 x^3 + \alpha_2 x^2 + \alpha_3 x + \alpha_4$ of degree ≤ 3.

27. Otherwise there would be a nonzero polynomial $f(q_n, \ldots, q_1, q_0)$ with integer coefficients such that $q_n \cdot f(q_n, \ldots, q_1, q_0) = 0$ for all sets $(q_n, \ldots, q_0)$ of real numbers. This cannot happen, since it is easy to prove by induction on n that a nonzero polynomial always takes on some nonzero value. (However, this result is false for *finite* fields in place of the real numbers.)

28. The indeterminate quantities $\alpha_1, \ldots, \alpha_s$ form an algebraic basis for $Q[\alpha_1, \ldots, \alpha_s]$, where Q is the field of rational numbers. Since $s + 1$ is greater than the number of elements in a basis, the polynomials $f_j(\alpha_1, \ldots, \alpha_s)$ are algebraically dependent; this means that there is a nonzero polynomial g with rational coefficients such that $g\big(f_0(\alpha_1, \ldots, \alpha_s), \ldots, f_s(\alpha_1, \ldots, \alpha_s)\big)$ is identically zero.

29. Given $j_0, \ldots, j_t \in \{0, 1, \ldots, n\}$, there are nonzero polynomials with integer coefficients such that $g_j(q_{j_0}, \ldots, q_{j_t}) = 0$ for all $(q_n, \ldots, q_0)$ in R_j, $1 \leq j \leq m$. The product $g_1 g_2 \cdots g_m$ is therefore zero for all $(q_n, \ldots, q_0)$ in $R_1 \cup \cdots \cup R_m$.

30. Starting with the construction in Theorem M, we will prove that $m_2 + (1 - \delta_{0m_1})$ of the β's may effectively be eliminated: If μ_i corresponds to a parameter multiplication, we have $\mu_i = \beta_{2i-1} \times (T_{2i} + \beta_{2i})$; add $c\beta_{2i-1}\beta_{2i}$ to each β_j for which $c\mu_i$ occurs in T_j, and replace β_{2i} by zero. This removes one parameter for each parameter multiplication. If μ_i is the first chain multiplication, then $\mu_i = (\gamma_1 x + \theta_1 + \beta_{2i-1}) \times (\gamma_2 x + \theta_2 + \beta_{2i})$, where $\gamma_1, \gamma_2, \theta_1, \theta_2$ are polynomials in $\beta_1, \ldots, \beta_{2i-2}$ with integer coefficients. Here θ_1 and θ_2 can be "absorbed" into β_{2i-1} and β_{2i}, respectively, so we may assume that $\theta_1 = \theta_2 = 0$. Now add $c\beta_{2i-1}\beta_{2i}$ to each β_j for which $c\mu_i$ occurs in T_j; add $\beta_{2i-1}\gamma_2/\gamma_1$ to β_{2i}; and set β_{2i-1} to zero. The result set is unchanged by this elimination of β_{2i-1}, except for the values of $\alpha_1, \ldots, \alpha_s$ such that γ_1 is zero. (This proof is essentially due to V. Ya. Pan, *Russian Mathematical Surveys* **21** (1966), 105–136.) The latter case can be handled as in the proof of Theorem A, since the polynomials with $\gamma_1 = 0$ can be evaluated by eliminating β_{2i} (as in the first construction, where μ_i corresponds to a parameter multiplication).

31. Otherwise we could add one parameter multiplication as a final step, and violate Theorem C. (The exercise is an improvement over Theorem A, in this special case, since there are only n degrees of freedom in the coefficients of a monic polynomial of degree n.)

32. $\lambda_1 = \lambda_0 \times \lambda_0$, $\lambda_2 = \alpha_1 \times \lambda_1$, $\lambda_3 = \alpha_2 + \lambda_2$, $\lambda_4 = \lambda_3 \times \lambda_1$, $\lambda_5 = \alpha_3 + \lambda_4$. We need at least three multiplications to compute $u_4 x^4$ (see Section 4.6.3), and at least two additions by Theorem A.

33. We must have $n + 1 \leq 2m_1 + m_2 + \delta_{0m_1}$, and $m_1 + m_2 = (n + 1)/2$; so there are no parameter multiplications. Now the first λ_i whose leading coefficient (as a polynomial in x) is not an integer must be obtained by a chain addition; and there must be at least $n + 1$ parameters, so there are at least $n + 1$ parameter additions.

34. Transform the given chain step by step, and also define the "content" c_i of λ_i, as follows: (Intuitively, c_i is the leading coefficient of x in λ_i.) Define $c_0 = 1$. (a) If the step has the form $\lambda_i = \alpha_j + \lambda_k$, replace it by $\lambda_i = \beta_j + \lambda_k$, where $\beta_j = \alpha_j/c_k$; and define $c_i = c_k$. (b) If the step has the form $\lambda_i = \alpha_j - \lambda_k$, replace it by $\lambda_i = \beta_j + \lambda_k$, where $\beta_j = -\alpha_j/c_k$; and define $c_i = -c_k$. (c) If the step has the form

$\lambda_i = \alpha_j \times \lambda_k$, replace it by $\lambda_i = \lambda_k$ (the step will be deleted later); and define $c_i = \alpha_j c_k$. (d) If the step has the form $\lambda_i = \lambda_j \times \lambda_k$, leave it unchanged; and define $c_i = c_j c_k$.

After this process is finished, delete all steps of the form $\lambda_i = \lambda_k$, replacing λ_i by λ_k in each future step which uses λ_i. Then add a final step $\lambda_{r+1} = \beta \times \lambda_r$, where $\beta = c_r$. This is the desired scheme, since it is easy to verify that the new λ_i are just the old ones divided by the factor c_i. The β's are given functions of the α's; division by zero is no problem, because if any $c_k = 0$ we must have $c_r = 0$ (hence the coefficient of x^n is zero), or else λ_k never contributes to the final result.

35. Since there are at least five parameter steps, the result is trivial unless there is at least one parameter multiplication; considering the ways in which three multiplications can form $u_4 x^4$, we see that there must be one parameter multiplication and two chain multiplications. Therefore the four addition-subtractions must each be parameter steps, and exercise 34 applies. We can now assume that only additions are used, and that we have a chain to compute a general *monic* fourth degree polynomial with *two* chain multiplications and four parameter additions. The only possible scheme of this type which calculates a fourth degree polynomial has the form

$$\lambda_1 = \alpha_1 + \lambda_0$$
$$\lambda_2 = \alpha_2 + \lambda_0$$
$$\lambda_3 = \lambda_1 \times \lambda_2$$
$$\lambda_4 = \alpha_3 + \lambda_3$$
$$\lambda_5 = \alpha_4 + \lambda_3$$
$$\lambda_6 = \lambda_4 \times \lambda_5$$
$$\lambda_7 = \alpha_5 + \lambda_6$$

Actually this chain has one addition too many, but any correct scheme can be put into this form if we restrict some of the α's to be functions of the others. Now λ_7 has the form $(x^2 + Ax + B)(x^2 + Ax + C) + D = x^4 + 2Ax^3 + (E + A^2)x^2 + EAx + F$, where $A = \alpha_1 + \alpha_2$, $B = \alpha_1\alpha_2 + \alpha_3$, $C = \alpha_1\alpha_2 + \alpha_4$, $D = \alpha_6$, $E = B + C$, $F = BC + D$; and since this involves only three independent parameters it cannot represent a general monic fourth-degree polynomial.

36. As in the solution to exercise 35, we may assume the chain computes a general monic polynomial of degree six, using only three chain multiplications and six parameter additions. The computation must take one of two general forms

$\lambda_1\ \ = \alpha_1 + \lambda_0$	$\lambda_1\ \ = \alpha_1 + \lambda_0$
$\lambda_2\ \ = \alpha_2 + \lambda_0$	$\lambda_2\ \ = \alpha_2 + \lambda_0$
$\lambda_3\ \ = \lambda_1 \times \lambda_2$	$\lambda_3\ \ = \lambda_1 \times \lambda_2$
$\lambda_4\ \ = \alpha_3 + \lambda_0$	$\lambda_4\ \ = \alpha_3 + \lambda_3$
$\lambda_5\ \ = \alpha_4 + \lambda_3$	$\lambda_5\ \ = \alpha_4 + \lambda_3$
$\lambda_6\ \ = \lambda_4 \times \lambda_5$	$\lambda_6\ \ = \lambda_4 \times \lambda_5$
$\lambda_7\ \ = \alpha_5 + \lambda_6$	$\lambda_7\ \ = \alpha_5 + \lambda_3$
$\lambda_8\ \ = \alpha_6 + \lambda_6$	$\lambda_8\ \ = \alpha_6 + \lambda_6$
$\lambda_9\ \ = \lambda_7 \times \lambda_8$	$\lambda_9\ \ = \lambda_7 \times \lambda_8$
$\lambda_{10} = \alpha_7 + \lambda_9$	$\lambda_{10} = \alpha_7 + \lambda_9$

where, as in exercise 35, an extra addition has been inserted to cover a more general case. Neither of these schemes can calculate a general sixth-degree monic polynomial, since the first case is a polynomial of the form

$$(x^3 + Ax^2 + Bx + C)(x^3 + Ax^2 + Bx + D) + E,$$

and the second case (cf. exercise 35) is a polynomial of the form

$$(x^4 + 2Ax^3 + (E + A^2)x^2 + EAx + F)(x^2 + Ax + G) + H;$$

both of these involve only five independent parameters.

37. Let $u^{[0]}(x) = u_n x^n + u_{n-1} x^{n-1} + \cdots + u_0$, $v^{[0]}(x) = x^n + v_{n-1} x^{n-1} + \cdots + v_0$. For $1 \le j \le n$, divide $u^{[j-1]}(x)$ by the monic polynomial $v^{[j-1]}(x)$, obtaining $u^{[j-1]}(x) = \alpha_j v^{[j-1]}(x) + \beta_j v^{[j]}(x)$. Assume that a monic polynomial $v^{[j]}(x)$ of degree $n - j$ exists satisfying this relation; this will be true for almost all rational functions. Let $u^{[j]}(x) = v^{[j]}(x) - x v^{[j-1]}(x)$. These definitions imply that $\deg(u^{[n]}) < 1$, so we may let $\alpha_{n+1} = u^{[n]}(x)$.

For the given rational function we have

α_j	β_j	$v^{[j]}(x)$	$u^{[j]}(x)$
1	2	$x + 5$	$3x + 19$
3	4	1	5

so $u^{[0]}(x)/v^{[0]}(x) = 1 + 2/(x + 3 + 4/(x + 5))$.

Notes: A general rational function of the stated form has $2n + 1$ "degrees of freedom," in the sense that it can be shown to have $2n + 1$ essentially independent parameters. If we generalize polynomial chains to "arithmetic chains," which allow division operations as well as addition, subtraction, and multiplication, we can obtain the following results with slight modifications to the proofs of Theorems A and M: *An arithmetic chain with q addition-subtraction steps has at most $q + 1$ degrees of freedom. An arithmetic chain with m multiplication-division steps has at most $2m + 1$ degrees of freedom.* Therefore an arithmetic chain which computes almost all rational functions of the stated form must have at least $2n$ addition-subtractions, and n multiplication-divisions; the method in this exercise is "optimal."

38. The theorem is certainly true if $n = 0$. Assume that n is positive, and that a polynomial chain computing $P(x; u_0, \ldots, u_n)$ is given, where each of the parameters α_j has been replaced by a real number. Let $\lambda_i = \lambda_j \times \lambda_k$ be the first chain multiplication step which involves one of $u_0, \ldots, u_n$; such a step must exist because of the rank of A. Without loss of generality, we may assume that λ_j involves u_n; thus, λ_j has the form $h_0 u_0 + \cdots + h_n u_n + f(x)$, where $h_0, \ldots, h_n$ are real, $h_n \ne 0$, and $f(x)$ is a polynomial with real coefficients. (The h's and the coefficients of $f(x)$ are derived from the values assigned to the α's.)

Now change step i to $\lambda_i = \alpha \times \lambda_k$, where α is an arbitrary real number. (We could take $\alpha = 0$; general α is used here merely to show that there is a certain amount of flexibility available in the proof.) Add further steps to calculate

$$\lambda = (\alpha - f(x) - h_0 u_0 - \cdots - h_{n-1} u_{n-1})/h_n;$$

these new steps involve only additions and parameter multiplications (by suitable new parameters). Finally, replace $\lambda_{-n-1} = u_n$ everywhere in the chain by this new element λ. The result is a chain which calculates

$$Q(x; u_0, \ldots, u_{n-1})$$
$$= P\left(x; u_0, \ldots, u_{n-1}, (\alpha - f(x) - h_0 u_0 - \cdots - h_{n-1} u_{n-1})/h_n\right);$$

and this chain has one less chain multiplication. The proof will be complete if we can show that Q satisfies the hypotheses. The quantity $(\alpha - f(x))/h_n$ leads to a possibly increased value of m, and a new vector B'. If the columns of A are $A_0, A_1, \ldots, A_n$ (these vectors being linearly independent over the reals), the new matrix A' corresponding to Q has the column vectors

$$A_0 - (h_0/h_n) A_n, \qquad \ldots, \qquad A_{n-1} - (h_{n-1}/h_n) A_n,$$

plus perhaps a few rows of zeros to account for an increased value of m, and these columns are clearly also linearly independent. By induction, the chain which computes Q has at least $n - 1$ chain multiplications, so the original chain has at least n.

A similar result, which shows further that the possibility of division gives no improvement, was proved by Pan in his survey paper cited in connection with exercise 30. A generalization to the computation of several polynomials in several variables $x_1, \ldots, x_n$ has been given by S. Winograd, *Proc. Nat. Acad. Sciences* **58** (1967), 1840–1842. Winograd's theorem proves in particular that at least mn multiplications are needed to multiply a general $m \times n$ matrix by the vector $(x_1, \ldots, x_n)$.

39. The coefficients can be found in $O(n^4)$ operations if we evaluate the polynomial $\det(xI - A)$ at n points and then take differences to find the coefficients (cf. exercise 16). Another idea is to use "Newton's identities"; we have

$$-a_1 = s_1$$
$$-2a_2 = a_1 s_1 + s_2$$
$$\cdots$$
$$-na_n = a_{n-1} s_1 + \cdots + a_1 s_{n-1} + s_n,$$

where $s_k = \lambda_1^k + \cdots + \lambda_n^k$ and $x^n + a_1 x^{n-1} + \cdots + a_n = (x - \lambda_1) \ldots (x - \lambda_n)$. These formulas follow from the fact that $1 + a_1 x + \cdots + a_n x^n = \exp(-\sum_{k \geq 1} s_k x^k / k)$. as in Eq. 1.2.9–34, combined with the technique for evaluating the exponential of a power series in exercise 4.7–4. Now $s_k = \text{trace } (A^k)$; this is easy to prove for triangular matrices T, and then for any matrix $A = B^{-1} T B$ where B is nonsingular and T is triangular. Therefore Newton's identities allow us to solve for $a_1, a_2, \ldots, a_n$ in $O(n^2)$ operations, once we know trace (A), trace $(A^2), \ldots$, trace (A^n).

A brute-force evaluation of $\text{trace}(A), \ldots, \text{trace}(A^n)$ would require $O(n^{3.81})$ operations, in view of the n matrix multiplications; but, as observed by S. Winograd, this computation can be done in $O(n^{3.31})$ operations: If B and C are given, trace (BC) can be computed in $O(n^2)$ operations, without computing the matrix product BC. Therefore we may form the $2k - 2$ matrices $A, A^2, \ldots, A^k, A^{2k}, A^{3k}, \ldots, A^{(k-1)k}$, where $k = \lceil \sqrt{n} \rceil$, in $O(kn^{2.81})$ operations, and trace (A^{jk+i}) for $0 \leq i, j < k$ in $O(n^3)$ operations; $O(n^{2.5})$ storage locations are required.

Notes: 1. Evaluating $a_1, \ldots, a_n$ and then finding the roots of $x^n + a_1 x^{n-1} + \cdots + a_n = 0$ is *not* the best way to obtain the characteristic roots of a matrix; the techniques suggested here are intended primarily when there is a need to determine the a's themselves. 2. When A is a matrix of integers, the coefficients $a_1, \ldots, a_n$ can be "read off" from the value of $\det(10^M I - A)$ for a suitably large integer M. Although this requires only $O(n^{2.81})$ arithmetic operations, it is obviously impractical since it deals with such large numbers; the operation count isn't the whole story.

40. No; S. Winograd has shown that 13 is the smallest odd n where $\lfloor n/2 \rfloor + 1$ is achievable.

41. $a(c+d) - (a+b)d + i(a(c+d) + (b-a)c)$;
or $ac - bd + i((a+b)(c+d) - ac - bd)$; etc.

SECTION 4.7

1. Find the first nonzero coefficient V_m, as in (4), and divide both $U(z)$ and $V(z)$ by z^m (shifting the coefficients m places to the left). The quotient will be a power series iff $U_0 = \cdots = U_{m-1} = 0$.

2. Yes. When $\alpha = 0$, it is easy to prove by induction that $W_1 = W_2 = \cdots = 0$. When $\alpha = 1$, we find $W_n = V_n$, by the "cute" identity

$$\sum_{1 \le k \le n} \left(\frac{k - (n-k)}{n} \right) V_k V_{n-k} = V_0 V_n.$$

3. If $V(z)$ is of degree n, and m is (say) 2^k, the squaring routines of Section 4.6.3 would form polynomials of respective degrees $n, 2n, 4n, \ldots, 2^k n$; if $O(n^2)$ operations are used to square a polynomial of degree n, then

$O(n^2 + (2n)^2 + \cdots + (2^{k-1}n)^2) = O(m^2 n^2)$ operations are used to form $V(z)^m$.

Equation (9) also uses $O((mn)^2)$ operations, but apparently with a higher constant of proportionality, so it is inferior for this purpose. Essentially the same considerations apply when m is not a power of 2. But Eq. (9) is superior when we want only the first N coefficients, for some fixed N, when m is large; for Eq. (9) will use $O(N^2)$ operations, regardless of the value of m, and this will probably beat squaring a polynomial $\log_2 m$ times modulo z^N.

4. If $W(z) = e^{V(z)}$, then $W'(z) = V'(z)W(z)$; we find $W_0 = 1$, and

$$W_n = \sum_{1 \le k \le n} \frac{k}{n} V_k W_{n-k}, \qquad \text{for} \quad n \ge 1.$$

If $W(z) = \ln(1 + V(z))$, then $W'(z) + W'(z)V(z) = V'(z)$; the rule is $W_0 = 0$, and

$$W_n = V_n + \sum_{1 \le k \le n} \left(\frac{k}{n} - 1 \right) V_k W_{n-k}, \qquad \text{for} \quad n \ge 1.$$

The latter rule is quite similar to (9) when $\alpha = -1$.

5. We get the original series back again. This can be used to test a reversion algorithm.

6. For fixed values of j and k, it is performed $N - j$ times; therefore for fixed k it is performed $(N - k) + \cdots + 1 = \binom{N-k+1}{2}$ times; and hence it is performed $\binom{N}{2} + \binom{N-1}{2} + \cdots + \binom{2}{2} = \binom{N+1}{3}$ times in all. But this quantity should be reduced by $N - 1$, since we forgot to exclude the case $k = j = 1$.

7. $W_n = \binom{2n-2}{n-1}/n$. Cf. Section 2.3.4.4.

8. Input G_1 in step S1, and G_n in step S2. In step S4, the output should be $(U_{n-1}G_1 + 2U_{n-2}G_2 + \cdots + nU_0G_n)/n$. (The running time of the order N^3 algorithm is hereby increased by only order N^2. The value $W_1 = G_1$ might be output in step S1.)

Note: Algorithm T determines $V^{-1}(U(z))$; the algorithm in this exercise determines $G(V^{-1}(z))$, which is somewhat different. Of course, the results of this algorithm and Algorithm T can all be obtained by a sequence of operations of reversion and composition (exercise 11), but it is helpful to have more direct algorithms for each case.

9.

	$n = 1$	$n = 2$	$n = 3$	$n = 4$	$n = 5$
T_{1n}	1	1	2	5	14
T_{2n}		1	2	5	14
T_{3n}			1	3	9
T_{4n}				1	4
T_{5n}					1

10. Form $y^{1/\alpha} = x(1 + a_1x + a_2x^2 + \cdots)^{1/\alpha} = x(1 + c_1x + c_2x^2 + \cdots)$ by means of (9); then revert the latter series. (See the remarks following Eq. 1.2.11.3–11.)

11. Set $(T_0, W_0) \leftarrow (V_0, U_0)$, and set $(T_k, W_k) \leftarrow (V_k, 0)$ for $1 \leq k \leq N$. Then for $n = 1, 2, \ldots, N$, do the following: Set $W_j \leftarrow W_j + U_nT_j$ for $n \leq j \leq N$; and then set $T_j \leftarrow T_{j-1}V_1 + \cdots + T_0V_j$ for $j = n+1, \ldots, N$.

Here $T(z)$ represents $V(z)^n$. A *sequential* power-series algorithm for this problem analogous to Algorithm T, could be constructed, but it would require about $N^2/2$ storage locations. There is also a sequential algorithm which solves this exercise and needs only $O(N)$ storage locations: We may assume that $V_1 = 1$, if U_k is replaced by $U_kV_1^k$ and V_k is replaced by V_k/V_1 for all k. Then we may revert $V(z)$ by Algorithm S, using its output as input to the algorithm of exercise 8 with $G_1 = U_1$, $G_2 = U_2$, etc., thus computing $U((V^{-1})^{-1}(z)) - U_0$.

SECTION A.1

1. Four; each byte would then contain $3^4 = 81$ different values.

2. Five, since five bytes is always adequate but four is not.

3. (0:2); (3:3); (4:4); (5:5).

4. Presumably index register 4 contains a value greater than or equal to 2000, so that after indexing a valid memory address results.

5. "DIV –80,3(0:5)" or simply "DIV –80,3".

6. (a) Sets rA to $\boxed{-\ |\ 5\ |\ 1\ |\ 200\ |\ 15\ }$. (b) Sets rI2 to -200. (c) Sets rX to

$\boxed{+\ |\ 0\ |\ 0\ |\ 5\ |\ 1\ |\ ?\ }$. (d) Undefined, since we can't load such a big value into an

index register. (e) Sets rX to $\boxed{-\ |\ 0\ |\ 0\ |\ 0\ |\ 0\ |\ 0\ }$.

7. Let the magnitude $|rAX|$ before the operation be n, and let the magnitude of V be d. After the operation the magnitude of rA is $\lfloor n/d \rfloor$, and the magnitude of rX is

n mod d. The sign of rX afterwards is the previous sign of rA; the sign of rA afterwards is $+$ if the previous signs of rA and V were the same, and it is $-$ otherwise.

8. rA ← $\boxed{+\ |\ 0\ |\ 617\ |\ 0\ |\ 1}$; rX ← $\boxed{-\ |\ 0\ |\ 0\ |\ 0\ |\ 1\ |\ 2}$.

9. ADD, SUB, DIV, NUM, JOV, JNOV, INCA, DECA, INCX, DECX.

10. CMPA, CMP1, CMP2, CMP3, CMP4, CMP5, CMP6, CMPX. (Also FCMP, floating point.)

11. MOVE, LD1, LD1N, INC1, DEC1, ENT1, ENN1.

12. INC3 0,3.

13. "JOV 1000" makes no difference except time. "JNOV 1001" makes a different setting of rJ in most cases. "JNOV 1000" makes an extraordinary difference, since it may lock the computer in an infinite loop.

14. NOP with anything; ADD, SUB with F $= (0:0)$; HLT (depending on how you interpret the statement of the exercise); any shift with address and index zero; MOVE with F $= 0$; JSJ with address equal to the location of the instruction plus one; any of the INC or DEC instructions with address and index zero; ENTi 0,i for $1 \leq i \leq 6$; SLC or SRC with address a multiple of 10.

15. 70; 80; 120. (record size times 5)

16. STZ 0; ENT1 1; MOVE 0(49); MOVE 0(50) solves both (a) and (b). If the byte size were known to equal 100, only one MOVE instruction would have been necessary, but we are not allowed to make assumptions about the byte size.

17. (a)

```
    STZ   0,2
    DEC2  1
    J2NN  3000
```

(b)

```
          STZ   0
          ENT1  1
          JMP   3004
(3003)    MOVE  0(63)
(3004)    DEC2  63
          J2NN  3003
          INC2  63
          ST2   3008(4:4)
(3008)    MOVE  0
```

(Using assembly language, a slightly faster program which uses "bytesize minus 1" instead of 63 could be written; see the following section.)

18. (If you have correctly followed the instructions, an overflow will occur on the ADD, with minus zero in register A afterward.) Answer: Overflow is set on, comparison is set EQUAL, rA is set to $\boxed{-\ |\ 30\ |\ 30\ |\ 30\ |\ 30\ |\ 30}$, rX is set to $\boxed{-\ |\ 31\ |\ 30\ |\ 30\ |\ 30\ |\ 30}$, rI1 is set to $+3$, and memory locations 0001, 0002 are set to zero.

SECTION A.2

1. ENTX 1000; STX X.

2. The STJ instruction in line 03 resets this address. (It is conventional to denote the address of such instructions by "*", both because it is simple to write, and because it provides a recognizable test of an error condition in a program, where a subroutine has not been entered properly because of some oversight.)

3. Read in 100 words from tape unit zero; exchange the maximum of these with the last one; exchange the maximum of the remaining 99 with the last of those; etc. Eventually the 100 words will become completely sorted into ascending sequence and the result is then written onto tape unit one.

4. Nonzero locations:

3000:	+	0000	00	18	35
3001:	+	2051	00	05	09
3002:	+	2050	00	05	10
3003:	+	0001	00	00	49
3004:	+	0499	01	05	26
3005:	+	3016	00	01	41
3006:	+	0002	00	00	50
3007:	+	0002	00	02	51
3008:	+	0000	00	02	48
3009:	+	0000	02	02	55
3010:	−	0001	03	05	04
3011:	+	3006	00	01	47
3012:	−	0001	03	05	56
3013:	+	0001	00	00	51
3014:	+	3008	00	06	39
3015:	+	3003	00	00	39
3016:	+	1995	00	18	37
3017:	+	2035	00	02	52
3018:	−	0050	00	02	53
3019:	+	0501	00	00	53
3020:	−	0001	05	05	08

3021:	+	0000	00	01	05
3022:	+	0000	04	12	31
3023:	+	0001	00	01	52
3024:	+	0050	00	01	53
3025:	+	3020	00	02	45
3026:	+	0000	04	18	37
3027:	+	0024	04	05	12
3028:	+	3019	00	00	45
3029:	+	0000	00	02	05
0000:	+				2
1995:	+	06	09	19	22	23
1996:	+	00	06	09	25	05
1997:	+	00	08	24	15	04
1998:	+	19	05	04	00	17
1999:	+	19	09	14	05	22
2024:	+				2035
2049:	+				2010
2050:	+				3
2051:	−				499

(the latter two may be interchanged, with corresponding changes to 3001 and 3002)

APPENDIX A

MIX

In many places throughout this book we will have occasion to refer to a computer's "machine language." The machine we use is a mythical computer called "MIX." MIX is very much like nearly every computer now in existence (except that it is, perhaps, nicer). The language of MIX has been designed to be powerful enough to allow brief programs to be written for most algorithms, yet simple enough so that its operations are easily learned.

A.1. DESCRIPTION OF MIX

MIX is the world's first polyunsaturated computer. Like most machines, it has an identifying number—the 1009. This number was found by taking 16 actual computers which are very similar to MIX and on which MIX can be easily simulated, then averaging their numbers with equal weight:

$$\lfloor (360 + 650 + 709 + 7070 + U3 + SS80 + 1107 + 1604 + G20 + B220$$
$$+ S2000 + 920 + 601 + H800 + PDP4 + II)/16 \rfloor = 1009. \qquad (1)$$

This number may also be obtained in a more simple way by taking roman numerals.

MIX has a peculiar property in that it is both binary and decimal at the same time. *The programmer doesn't actually know whether he is programming a machine with base 2 or base 10 arithmetic.* This has been done so that algorithms written in MIX can be used on either type of machine with little change, and so that MIX can be easily simulated on either type of machine. Those programmers accustomed to a binary machine can think of MIX as binary; those accustomed to decimal may regard MIX as decimal. Programmers from another planet might choose to think of MIX as a ternary machine.

Words. The basic unit of information is a *byte*. Each byte contains an *unspecified* amount of information, but it must be capable of holding at least 64 distinct values. That is, we know that any number between 0 and 63, inclusive, can be contained in one byte. Furthermore, each byte contains *at most* 100 distinct values. On a binary computer a byte must therefore be composed of six bits; on a decimal computer we have two digits per byte.

Programs expressed in the MIX language should be written so that no more than sixty-four values are ever assumed for a byte. If we wish to treat the

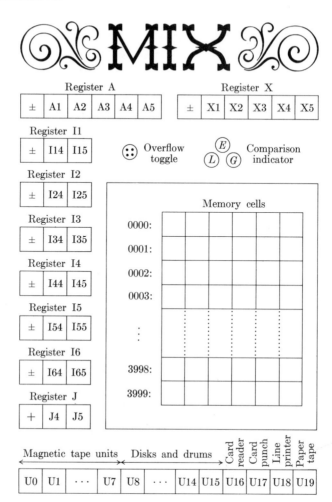

Fig. 1. The MIX computer.

number 80, we should always leave two adjacent bytes for expressing it, even though on a decimal computer one byte is sufficient. *An algorithm in* MIX *should work properly regardless of how big a byte is.* Although it is quite possible to write programs which depend on the byte size, this is an illegal act which will not be tolerated; the only legitimate programs are those which would give correct results with all byte sizes. It is not hard to abide by this ground rule, and we will thereby find that programming a decimal computer isn't so different from programming a binary one after all.

Two adjacent bytes can express the numbers 0 through 4095.

Three adjacent bytes can express the numbers 0 through 262143.

Four adjacent bytes can express the numbers 0 through 16777215.

Five adjacent bytes can express the numbers 0 through 1073741823.

A computer word is five bytes plus a sign. The sign position has only two possible values, + and −.

Registers. There are nine registers in MIX (see Fig. 1):

The A-register (Accumulator) is five bytes plus sign.

The X-register (Extension) is also five bytes plus sign.

The I-registers (Index registers) I1, I2, I3, I4, I5, and I6 each hold two bytes plus sign.

The J-register (Jump address) holds two bytes, and its sign is always +.

We shall use a small letter "r" prefixed to the name, to identify a MIX register. Thus, "rA" means "register A."

The A-register has many uses, especially for arithmetic and operating on data. The X-register is an extension on the "right-hand side" of rA, and it is used in connection with rA to hold ten bytes of a product or dividend, or it can be used to hold information shifted to the right out of rA. The index registers rI1, rI2, rI3, rI4, rI5, and rI6 are used primarily for counting and for referencing variable memory addresses. The J-register always holds the address of the instruction following the preceding "JUMP" instruction, and it is primarily used in connection with subroutines.

Besides these registers, MIX contains

an *overflow toggle* (a single bit which is either "on" or "off"),

a *comparison indicator* (which has three values: less, equal, or greater),

memory (4000 words of storage, each word with five bytes plus sign),

and *input-output devices* (card, tape, etc.).

Partial fields of words. The five bytes and sign of a computer word are numbered as follows:

0	1	2	3	4	5
±	Byte	Byte	Byte	Byte	Byte

.   (2)

Most of the instructions allow the programmer to use only part of a word if he chooses. In this case a "field specification" is given. The allowable fields are those which are adjacent in a computer word, and they are represented by (L : R), where L is the number of the left-hand part and R is the number of the right-hand part of the field. Examples of field specifications are:

(0 : 0), the sign only.

(0 : 2), the sign and the first two bytes.

(0 : 5), the whole word. This is the most common field specification.

(1 : 5), the whole word except for the sign.

(4 : 4), the fourth byte only.

(4 : 5), the two least significant bytes.

The use of these field specifications varies slightly from instruction to instruction, and it will be explained in detail for each instruction where it applies.

Although it is generally not important to the programmer, the field (L:R) is denoted in the machine by the single number $8L + R$, and this number will fit in one byte.

Instruction format. Computer words used for instructions have the following form:

0	1	2	3	4	5
±	A	A	I	F	C

$$(3)$$

The rightmost byte, C, is the operation code telling what operation is to be performed. For example, $C = 8$ is the operation LDA, "load the A register."

The F-byte holds a modification of the operation code. F is usually a field specification $(L:R) = 8L + R$; for example, if $C = 8$ and $F = 11$, the operation is "load the A-register with the $(1:3)$ field." Sometimes F is used for other purposes; on input-output instructions, for example, F is the number of the affected input or output unit.

The left-hand portion of the instruction, ±AA, is the "address." (Note that the sign is part of the address.) The I-field, which comes next to the address, is the "index specification," which may be used to modify the address of an instruction. If $I = 0$, the address ±AA is used without change; otherwise I should contain a number i between 1 and 6, and the contents of index register Ii are added algebraically to ±AA; the result is used as the address of the instruction. This indexing process takes place on *every* instruction. We will use the letter M to indicate the address after any specified indexing has occurred. (If the addition of the index register to the address ±AA yields a result which does not fit in two bytes, the value of M is undefined.)

In most instructions, M will refer to a memory cell. The terms "memory cell" and "memory location" are used almost interchangeably in this book. We assume that there are 4000 memory cells, numbered from 0 to 3999; hence every memory location can be addressed with two bytes. For every instruction in which M is to refer to a memory cell we must have $0 \le M \le 3999$, and in this case we will write CONTENTS(M) to denote the value stored in memory location M.

On certain instructions, the "address" M has another significance, and it may even be negative. Thus one instruction adds M to an index register, and this takes account of the sign of M.

Notation. To discuss instructions in a readable manner, we will use the notation

$$\text{OP}\quad\text{ADDRESS,I(F)}\qquad(4)$$

to denote an instruction like (3). Here OP is a symbolic name which is given to the operation code (the C-part) of the instruction; ADDRESS is the ±AA portion; and I, F represent the I- and F-fields, respectively.

If I is zero, the ",I" is omitted. If F is the *normal* F-specification for this particular operator, the "(F)" need not be written. The normal F-specification for almost all operators is (0:5), representing a whole word. If a different F is standard, it will be mentioned explicitly when we discuss a particular operator.

For example, the instruction to load a number into the accumulator is called LDA and it is operation code number 8. We have

Conventional representation   Actual numeric instruction

Conventional	Actual
LDA 2000,2(0:3)	+ 2000 2 3 8
LDA 2000,2(1:3)	+ 2000 2 11 8
LDA 2000(1:3)	+ 2000 0 11 8
LDA 2000	+ 2000 0 5 8
LDA −2000,4	− 2000 4 5 8

$$(5)$$

To render these in words, the instruction "LDA 2000,2(0:3)" may be read "Load A with the contents of location 2000 indexed by 2, the zero-three field."

To represent the numerical contents of a MIX word, we will always use a box notation like that above. Note that in the word

+	2000	2	3	8

the number +2000 is shown filling two adjacent bytes and sign; the actual contents of byte (1:1) and of byte (2:2) will vary from one MIX computer to another, since byte size is variable. As a further example of this notation for MIX words, the diagram

−	10000	3000

represents a word with two fields, a three-byte-plus-sign field containing -10000 and a two-byte field containing 3000. When a word is split into more than one field, it is said to be "packed."

Rules for each instruction. The remarks following (3) above have defined the quantities M, F, and C for every word used as an instruction. We will now define the actions corresponding to each instruction.

Loading operators

• LDA (load A). C = 8; F = field.
The specified field of CONTENTS(M) replaces the previous contents of register A.

On all operations where a partial field is used as an input, the sign is used if it is a part of the field, otherwise the sign + is understood. The field is shifted over to the right-hand part of the register as it is loaded.

Examples: If F is the normal field specification $(0:5)$, the entire contents of location M is loaded. If F is $(1:5)$, the absolute value of CONTENTS(M) is loaded with a plus sign. If M contains an *instruction* word and if F is $(0:2)$, the "$\pm$AA" field is loaded as

$\pm$	0	0	0	A	A

.

Suppose location 2000 contains the word

$-$	80	3	5	4

; $\qquad\qquad$ (6)

then we get the following results from loading various partial fields:

Instruction	Contents of rA afterwards				
LDA 2000	$-$	80	3	5	4
LDA 2000(1:5)	$+$	80	3	5	4
LDA 2000(3:5)	$+$	0	0	3	5	4
LDA 2000(0:3)	$-$	0	0	80	3
LDA 2000(4:4)	$+$	0	0	0	0	5
LDA 2000(0:0)	$-$	0	0	0	0	0
LDA 2000(1:1)	$+$	0	0	0	0	?

(The last example has a partially unknown effect since byte size is variable.)

• LDX (load X). $C = 15$; $F =$ field.
This is the same as LDA, except that rX is loaded instead of rA.

• LDi (load i). $C = 8 + i$; $F =$ field.
This is the same as LDA, except that rIi is loaded instead of rA. An index register contains only two bytes (not five) plus sign; bytes 1, 2, 3 are always assumed to be zero. The LDi instruction is considered undefined if it would result in setting bytes 1, 2, 3 to anything but zero.

In the description of all instructions, "i" stands for an integer, $1 \le i \le 6$. Thus, LDi stands for six different instructions: LD1, LD2, ..., LD6.

• LDAN (load A negative). $C = 16$; $F =$ field.

• LDXN (load X negative). $C = 23$; $F =$ field.

• LDiN (load i negative). $C = 16 + i$; $F =$ field.
These eight instructions are the same as LDA, LDX, LDi, respectively, except that the *opposite* sign is loaded.

Storing operators.

• STA (store A). $C = 24$; $F =$ field.

The contents of rA replaces the field of CONTENTS(M) specified by F. The other parts of CONTENTS(M) are unchanged.

On a *store* operation the field F has the opposite significance from the *load* operation. The number of bytes in the field is taken from the right-hand side of the register and shifted *left* if necessary to be inserted in the proper field of CONTENTS(M). The sign is not altered unless it is part of the field. The contents of the register is not affected.

Examples: Suppose that location 2000 contains

−	1	2	3	4	5

and register A contains

+	6	7	8	9	0

.

Then:

Instruction	Contents of location 2000 afterwards
STA 2000	+ 6 7 8 9 0
STA 2000(1:5)	− 6 7 8 9 0
STA 2000(5:5)	− 1 2 3 4 0
STA 2000(2:2)	− 1 0 3 4 5
STA 2000(2:3)	− 1 9 0 4 5
STA 2000(0:1)	+ 0 2 3 4 5

- STX (store X). $C = 31$; $F =$ field.
Same as STA except rX is stored rather than rA.

- STi (store i). $C = 24 + i$; $F =$ field.
Same as STA except rIi is stored rather than rA. Bytes 1, 2, 3 of an index register are zero; thus if rI1 contains

±	m	n

,

this behaves as though it were

±	0	0	0	m	n

.

- STJ (store J). $C = 32$; $F =$ field.
Same as STi except rJ is stored, and its sign is always +.

On STJ *the normal field specification for F is* $(0:2)$, *not* $(0:5)$. This is natural, since STJ is almost always done into the address field of an instruction.

- STZ (store zero). $C = 33$; $F =$ field.
Same as STA except plus zero is stored. In other words, the specified field of CONTENTS(M) is cleared to zero.

Arithmetic operators. On the add, subtract, multiply, and divide operations, a field specification is allowed. A field specification of "(0:6)" can be used to indicate a "floating-point" operation (see Section 4.2), but few of the programs we will write for MIX will use this feature; floating-point instructions will be used primarily in the programs written by the compilers discussed in Chapter 12.

The standard field specification is, as usual, (0:5). Other fields are treated as in LDA. We will use the letter V to indicate the specified field of CONTENTS(M); thus, V is the value which would have been loaded into register A if the operation code were LDA.

- ADD. C = 1; F = field.

V is added to rA. If the magnitude of the result is too large for register A, the overflow toggle is set on, and the remainder of the addition appearing in rA is as though a "1" had been carried into another register to the left of A. (Otherwise the setting of the overflow toggle is unchanged.) If the result is zero, the sign of rA is unchanged.

Example: The sequence of instructions below gives the sum of the five bytes of register A.

```
STA   2000
LDA   2000(5:5)
ADD   2000(4:4)
ADD   2000(3:3)
ADD   2000(2:2)
ADD   2000(1:1)
```

This is sometimes called "sideways addition."

- SUB (subtract). C = 2; F = field.

V is subtracted from rA. Overflow may occur as in ADD.

Note that because of the variable definition of byte size, overflow will occur in some MIX computers when it would not occur in others. We have not said that overflow will occur definitely if the value is greater than 1073741823; overflow occurs when the magnitude of the result is greater than the contents of five bytes, depending on the byte size. One can still write programs which work properly and which give the same final answers, regardless of the byte size.

- MUL (multiply). C = 3; F = field.

The 10-byte product of V times (rA) replaces registers A and X. The signs of rA and rX are both set to the algebraic sign of the result (i.e., + if the signs of V and rA were the same, and − if they were different).

- DIV (divide). C = 4; F = field.

The value of rA and rX, treated as a 10-byte number, with the sign of rA, is divided by the value V. If V = 0 or if the quotient is more than five bytes in magnitude (this is equivalent to the condition that $|rA| \geq |V|$), registers A and X are filled with undefined information and the overflow toggle is set on. Otherwise the quotient is placed in rA and the remainder is placed in rX. The sign of rA afterward is the algebraic sign of the quotient; the sign of rX afterward is the previous sign of rA.

Examples of arithmetic instructions: In most cases, arithmetic is done only with MIX words which are single five-byte numbers, not packed with several fields. It is possible to operate arithmetically on packed MIX words, if some caution is used. The following examples should be studied carefully. (The "?" mark designates an unknown value.)

ADD 1000

+	1234	1	150	rA before
+	100	5	50	Cell 1000
+	1334	6	200	rA after

SUB 1000

−	1234	0	0	9	rA before
−	2000	150	0		Cell 1000
+	766	149	?		rA after

MUL 1000(1:1)

−					112	rA before
?	2	?	?	?	?	Cell 1000
−					0	rA after
−					224	rX after

MUL 1000

−	50	0	112	4	rA before
−	2	0	0	0	0	Cell 1000
+	100	0	224		rA after
+	8	0	0	0	0	rX after

DIV 1000

+	0	rA before
+	17	rX before
+	3	Cell 1000
+	5	rA after
+	2	rX after

DIV 1000

−				0	rA before
+	1235	0	3	2	rX before
−	0	0	0	2	0	Cell 1000
+	0	617	?	?	rA after
−	0	0	0	?	2	rX after

(These examples have been prepared with the philosophy that it is better to give a complete, baffling description than an incomplete, straightforward one.)

Address transfer operators. In the following operations, the (possibly indexed) "address" M is used as a signed number, not as the address of a cell in memory.

• ENTA (enter A). $C = 48$; $F = 2$.
The quantity M is loaded into rA. The action is equivalent to "LDA" from a memory word containing the signed value of M. If $M = 0$, the sign of the instruction is loaded.

   Examples: "ENTA 0" sets rA to zeros. "ENTA 0,1" sets rA to the current contents of index register 1.

• ENTX (enter X). $C = 55$; $F = 2$.
• ENTi (enter i). $C = 48 + i$; $F = 2$.
Analogous to ENTA, loading the appropriate register.

• ENNA (enter negative A). $C = 48$; $F = 3$.
• ENNX (enter negative X). $C = 55$; $F = 3$.
• ENNi (enter negative i). $C = 48 + i$; $F = 3$.
Same as ENTA, ENTX, and ENTi, except that the opposite sign is loaded.

   Example: "ENN3 0,3" replaces rI3 by its negative.

• INCA (increase A). $C = 48$; $F = 0$.
The quantity M is added to rA; the action is equivalent to "ADD" from a memory word containing the value of M. Overflow is possible and it is treated just as in ADD.

   Example: "INCA 1" increases the value of rA by one.

• INCX (increase X). $C = 55$; $F = 0$.
The quantity M is added to rX. If overflow occurs, the action is equivalent to ADD, except that rX is used instead of rA. Register A is never affected by this instruction.

• INCi (increase i). $C = 48 + i$; $F = 0$.
Add M to rIi. Overflow must not occur; if the magnitude of the result is more than two bytes, the result of this instruction is undefined.

• DECA (decrease A). $C = 48$; $F = 1$.
• DECX (decrease X). $C = 55$; $F = 1$.
• DECi (decrease i). $C = 48 + i$; $F = 1$.
These eight instructions are the same as INCA, INCX, and INCi, respectively, except that M is subtracted from the register rather than added.

   Note that the operation code C is the same for ENTA, ENNA, INCA, and DECA; the F-field is used to distinguish the various operations in this case.

Comparison operators. The comparison operators all compare the value contained in a register with a value contained in memory. The comparison indicator is then set to LESS, EQUAL, or GREATER according to whether the value of the *register* is less than, equal to, or greater than the value of the *memory cell*. A minus zero is *equal to* a plus zero.

● CMPA (compare A). C = 56; F = field.
The specified field of A is compared with the *same* field of CONTENTS(M). If the field F does not include the sign position, the fields are both thought of as positive; otherwise the sign is taken into account in the comparison. (If F is (0:0) an equal comparison always occurs, since minus zero equals plus zero.)

● CMPX (compare X). C = 63; F = field.
This is analogous to CMPA.

● CMPi (compare i). C = 56 + i; F = field.
Analogous to CMPA. Bytes 1, 2, and 3 of the index register are treated as zero in the comparison.

Jump operators. Ordinarily, instructions are executed in sequential order; i.e., the instruction executed after the one in location P is the instruction found in location P + 1. Several "jump" instructions allow this sequence to be interrupted. Whenever a jump of any kind takes place, the J-register is set to the address of the next instruction (that is, the address of the instruction which would have been next if we hadn't jumped). A "store J" instruction then can be used by the programmer, if desired, to set the address field of another command which will later be used to return to the original place in the program. The J-register is changed whenever a jump actually occurs in a program (except JSJ) and it is never changed except when a jump occurs.

● JMP (jump). C = 39; F = 0.
Unconditional jump: the next instruction is taken from location M.

● JSJ (jump, save J). C = 39; F = 1.
Same as JMP except that the contents of rJ are unchanged.

● JOV (jump on overflow). C = 39; F = 2.
If the overflow toggle is on, it is turned off and a JMP occurs; otherwise nothing happens.

● JNOV (jump on no overflow). C = 39; F = 3.
If the overflow toggle is off, a JMP occurs; otherwise it is turned off.

● JL, JE, JG, JGE, JNE, JLE (jump on less, equal, greater, greater-or-equal, unequal, less-or-equal). C = 39; F = 4, 5, 6, 7, 8, 9, respectively.
Jump if the comparison indicator is set to the condition indicated. For example, JNE will jump if the comparison indicator is LESS or GREATER. The comparison indicator is not changed by these instructions.

● JAN, JAZ, JAP, JANN, JANZ, JANP (jump A negative, zero, positive, nonnegative, nonzero, nonpositive). C = 40; F = 0, 1, 2, 3, 4, 5, respectively.
If the contents of rA satisfy the stated condition, a JMP occurs, otherwise nothing

happens. "Positive" means *greater* than zero (not zero); "nonpositive" means the opposite, i.e., zero or negative.

- JXN, JXZ, JXP, JXNN, JXNZ, JXNP (jump X negative, zero, positive, nonnegative, nonzero, nonpositive). $C = 47$; $F = 0, 1, 2, 3, 4, 5$, respectively.

- JiN, JiZ, JiP, JiNN, JiNZ, JiNP (jump i negative, zero, positive, nonnegative, nonzero, nonpositive). $C = 40 + i$; $F = 0, 1, 2, 3, 4, 5$, respectively.

These are analogous to the corresponding operations for rA.

Miscellaneous operators.

- MOVE. $C = 7$; $F =$ number.

The number of words specified by F is moved, starting from location M to the location specified by the contents of index register 1. The transfer occurs one word at a time, and rI1 is increased by the value of F at the end of the operation. If $F = 0$, nothing happens.

Care must be taken when the groups of locations involved overlap; for example, suppose that $F = 3$ and $M = 1000$. Then if $(rI1) = 999$, we transfer (1000) to (999), (1001) to (1000), and (1002) to (1001). Nothing unusual occurred here; but if (rI1) were 1001 instead, we would move (1000) to (1001), then (1001) to (1002), then (1002) to (1003), so we have moved the *same* word (1000) into three places.

- SLA, SRA, SLAX, SRAX, SLC, SRC (shift left A, shift right A, shift left AX, shift right AX, shift left AX circularly, shift right AX circularly). $C = 6$; $F = 0, 1, 2, 3, 4, 5$, respectively.

These are the "shift" commands. Signs of registers A, X are not affected in any way. M specifies the number of *bytes* to be shifted left or right; M must be nonnegative. SLA and SRA do not affect rX; the other shifts affect both registers as though they were a single 10-byte register. With SLA, SRA, SLAX, and SRAX, zeros are shifted into the register at one side, and bytes disappear at the other side. The instructions SLC and SRC call for a "circulating" shift, in which the bytes that leave one end enter in at the other end. Both rA and rX participate in a circulating shift.

Examples:

		Register A							Register X				
Initial contents	+	1	2	3	4	5		−	6	7	8	9	10
SRAX 1	+	0	1	2	3	4		−	5	6	7	8	9
SLA 2	+	2	3	4	0	0		−	5	6	7	8	9
SRC 4	+	6	7	8	9	2		−	3	4	0	0	5
SRA 2	+	0	0	6	7	8		−	3	4	0	0	5
SLC 501	+	0	6	7	8	3		−	4	0	0	5	0

- NOP (no operation). C = 0.
No operation occurs, and this instruction is bypassed. F and M are ignored.

- HLT (halt). C = 5; F = 2.
The machine stops. When the computer operator restarts it, the net effect is equivalent to NOP.

Input-output operators. MIX has a fair amount of input-output equipment (all of which is optional at extra cost). Each device is given a number as follows:

Unit number	Peripheral device	Record size
t	Tape unit no. t $(0 \le t \le 7)$	100 words
d	Disk or drum unit no. d $(8 \le d \le 15)$	100 words
16	Card reader	16 words
17	Card punch	16 words
18	Printer	24 words
19	Typewriter and paper tape	14 words

Not every MIX installation will have all of this equipment available; we will occasionally make appropriate assumptions about the presence of certain devices. Some devices may not be used both for input and for output. The number of words mentioned in the above table is a fixed record size associated with each unit.

Input or output with magnetic tape, disk, or drum units reads or writes full words (five bytes plus sign). Input or output with units 16 through 19, however, is always done in a *character code* where each byte represents one alphameric character. Thus, five characters per MIX word are transmitted. The character code is given at the top of Table 1, which appears at the close of this section and on the end papers of this book. The code 00 corresponds to "⊔", which denotes a *blank space*. Codes 01–29 are for the letters A through Z with a few Greek letters thrown in; codes 30–39 represent the digits 0, 1, . . . , 9; and further codes 40, 41, . . . represent punctuation marks and other special characters. It is not possible to read in or write out all possible values a byte may have, since certain combinations are undefined. Not all input-output devices are capable of handling all the symbols in the character set; for example, the symbols Φ and Π which appear amid the letters will perhaps not be acceptable to the card reader. When input of character code is being done, the signs of all words are set to "+"; on output, signs are ignored.

The disk and drum units are large external memory devices each containing b^2 100-word records, where b is the byte size. On every IN, OUT, or IOC instruction as defined below, the particular 100-word record referred to by the instruction is specified by the current contents of the two least significant bytes of rX.

- IN (input). C = 36; F = unit.
This instruction initiates the transfer of information from the input unit specified into sequential locations starting with M. The number of locations transferred

is the record size for this unit (see the table above). The machine will wait at this point if a preceding operation for the same unit is not yet complete. The transfer of information which starts with this instruction will not be complete until somewhat later, depending on the speed of the input device, so a program must not refer to the information in memory until then. It is improper to attempt to read any record from magnetic tape which follows the latest record written on that tape.

- OUT (output). C = 37; F = unit.

This instruction starts the transfer of information from memory locations starting at M to the output unit specified. (The machine waits until the unit is ready, if it is not initially ready.) The transfer will not be complete until somewhat later, depending on the speed of the output device, so a program must not alter the information in memory until then.

- IOC (input-output control). C = 35; F = unit.

The machine waits, if necessary, until the specified unit is not busy. Then a control operation is performed, depending on the particular device being used. The following examples are used in various parts of this book:

Magnetic tape: If M = 0, the tape is rewound. If M < 0 the tape is skipped backward $-$M records, or to the beginning of the tape, whichever comes first. If M > 0, the tape is skipped forward; it is improper to skip forward over any records following the one last written on that tape.

For example, the sequence "OUT 1000(3); IOC -1(3); IN 2000(3)" writes out one hundred words onto tape 3, then reads it back in again. Unless the tape reliability is questioned, the last two instructions of that sequence are only a slow way to move words 1000–1099 to locations 2000–2099. The sequence "OUT 1000(3); IOC +1(3)" is improper.

Disk or drum: M should be zero. The effect is to position the device according to rX so that the next IN or OUT operation on this unit will take less time if it uses the same rX setting.

Printer: M should be zero. "IOC 0(18)" skips the printer to the top of the following page.

Paper tape reader: Rewind the tape. (M should be zero.)

- JRED (jump ready). C = 38; F = unit.

A jump occurs if the specified unit is ready, i.e., finished with the preceding operation initiated by IN, OUT, or IOC.

- JBUS (jump busy). C = 34; F = unit.

Same as JRED except the jump occurs under the opposite circumstances, i.e., when the specified unit is *not* ready.

Example: In location 1000, the instruction "JBUS 1000(16)" will be executed repeatedly until unit 16 is ready.

The simple operations above complete MIX's repertoire of input-output instructions. There is no "tape check" indicator, etc., to cover exceptional conditions on the peripheral devices. Any such condition (e.g., paper jam, unit turned off, out of tape, etc.) causes the unit to remain busy, a bell rings, and the skilled computer operator fixes things manually using ordinary maintenance procedures.

Conversion Operators.

• NUM (convert to numeric). $C = 5; F = 0$.
This operation is used to change the character code into numeric code. M is ignored. Registers A, X are assumed to contain a 10-byte number in character code; the NUM instruction sets the magnitude of rA equal to the numerical value of this number (treated as a decimal number). The value of rX and the sign of rA are unchanged. Bytes 00, 10, 20, 30, 40, . . . convert to the digit zero; bytes 01, 11, 21, . . . convert to the digit one; etc. Overflow is possible, and in this case the remainder modulo the word size is retained.

• CHAR (convert to characters). $C = 5; F = 1$.
This operation is used to change numeric code into character code suitable for output to cards or printer. The value in rA is converted into a 10-byte decimal number which is put into register A and X in character code. The signs of rA, rX are unchanged. M is ignored.

Examples:

	Register A						Register X				
Initial contents	−	00	00	31	32	39	+	37	57	47	30	30
NUM 0	−		12977700				+	37	57	47	30	30
INCA 1	−		12977699				+	37	57	47	30	30
CHAR 0	−	30	30	31	32	39	+	37	37	36	39	39

Timing. To give quantitative information as to how "good" MIX programs are, each of MIX's operations is assigned an *execution time* typical for present day computers.

ADD, SUB, all LOAD operations, all STORE operations (including STZ), all shift commands, and all comparison operations take *two units* of time. MOVE requires one unit plus two for each word moved. MUL requires 10 and DIV requires 12 units. Execution time for floating-point operations is unspecified. All remaining operations take one unit of time, plus the time the computer may be idle on the IN, OUT, IOC, or HLT instructions.

Note in particular that ENTA takes one unit of time, while LDA takes two units. The timing rules are easily remembered because of the fact that, except for shifts, MUL, and DIV, the number of units equals the number of references to memory (including the reference to the instruction itself).

The "unit" of time is a relative measure which we will denote simply by u. It may be regarded as, say, 10 microseconds (for a relatively inexpensive computer) or as 1 microsecond (for a relatively high-priced machine).

Example: Execution of the sequence

```
LDA  1000
INCA 1
STA  1000
```

takes a time of 5*u*.

Summary. We have now discussed all of the features of MIX, except for its
"GO button" which is discussed in exercise 1.3.1–26. Although MIX has nearly 150
different operations, they fit into a few simple patterns so they can be easily
remembered. Table 1 summarizes the operations for each C-setting. The
name of each operator is followed by the standard F-field for that operator in
parentheses.

The following exercises give a good review of the material in this section;
most of them are very simple, and the reader should try to do nearly all of them.

EXERCISES

1. [*00*] If MIX were a ternary (base 3) computer, how many "trits" would there be
per byte?

2. [*02*] If a value to be represented within MIX may get as large as 99999999, how
many adjacent bytes should be used to contain this quantity?

3. [*02*] Give the partial field specifications, (L:R), for the (a) address field, (b) index
field, (c) field field, and (d) operation code field of a MIX instruction.

4. [*00*] The last example in (5) is "LDA -2000,4"—how can this be legitimate in
view of the fact that memory addresses should not be negative?

5. [*10*] What is the symbolic notation [as in (4)] corresponding to the word (6)?

▶ **6.** [*10*] Assume that location 3000 contains

+	5	1	200	15

.

What is the result of the following instructions? (State if any of these are undefined
or only partially defined.) (a) LDAN 3000; (b) LD2N 3000(3:4); (c) LDX 3000(1:3);
(d) LD6 3000; (e) LDXN 3000(0:0).

7. [*15*] Give a precise definition of the results of the DIV instruction for all cases in
which overflow does not occur, using the algebraic operations X mod Y and $\lfloor X \rfloor$.

8. [*15*] The last example of the DIV instruction which appears in the text has "rX
before" equal to

+	1235	0	3	2

.

If this were

−	1234	0	3	2

instead, but other parts of that example were unchanged, what would registers A, X
contain after the DIV instruction?

▶ **9.** [*20*] List all the MIX operators which can possibly affect the setting of the overflow toggle. (Do not include floating-point operators.)

10. [*20*] List all the MIX operators which can possibly affect the setting of the comparison indicators.

▶ **11.** [*20*] List all the MIX operators which can possibly affect the setting of rI1.

12. [*10*] Find a single instruction which has the effect of multiplying the current contents of rI3 by two and leaving the result in rI3.

▶ **13.** [*10*] Suppose location 1000 contains the instruction "JOV 1001". This instruction turns off the overflow toggle if it is on (and the next instruction executed will be in location 1001, in any case). If this instruction were changed to "JNOV 1001", would there be any difference? What if it were changed to "JOV 1000" or "JNOV 1000"?

14. [*25*] For each MIX operator, consider whether there is a way to set the $\pm$AA-, I-, and F-portions of the instruction so that the result of the instruction is precisely equivalent to NOP, except that the execution time may be longer. Assume that nothing is known about the contents of any registers or any memory locations. Whenever it is possible to produce a NOP, state how it can be done. *Examples:* INCA is a no-op if the address and index parts are zero. JMP can never be a no-op, since it affects rJ.

15. [*10*] How many *alphameric characters* are there in a typewriter record? in a card-reader or card-punch record? in a printer record?

16. [*20*] Write a program which sets memory cells 0000–0099 all to zero, and which is (a) as short a program as possible; (b) as fast a program as possible. [*Hint:* Consider using the MOVE command.]

17. [*26*] This is the same as the previous exercise, except that locations 0000 through N, inclusive, are to be set to zero, where N is the current contents of rI2. The programs should work for any value $0 \le N \le 2999$; they should start in location 3000.

▶ **18.** [*22*] After the following "number one" program has been executed, what changes to registers, toggles, and memory have taken place? (For example, what is the final setting of rI1? of rX? of the overflow and comparison indicators?)

```
STZ   1
ENNX  1
STX   1(0:1)
SLAX  1
ENNA  1
INCX  1
ENT1  1
SRC   1
ADD   1
DEC1  -1
STZ   1
CMPA  1
MOVE  -1,1(1)
NUM   1
CHAR  1
HLT   1
```

Table 1

Character code:	00	01	02	03	04	05	06	07	08	09	10	11	12	13	14	15	16	17	18	19	20	21	22	23	24
	⊔	A	B	C	D	E	F	G	H	I	Θ	J	K	L	M	N	O	P	Q	R	Φ	Π	S	T	U

00	*1*	01	*2*	02	*2*	03	*10*
No operation		rA ← rA + V		rA ← rA − V		rAX ← rA × V	
NOP(0)		ADD(0:5) FADD(6)		SUB(0:5) FSUB(6)		MUL(0:5) FMUL(6)	
08	*2*	**09**	*2*	**10**	*2*	**11**	*2*
rA ← V		rI1 ← V		rI2 ← V		rI3 ← V	
LDA(0:5)		LD1(0:5)		LD2(0:5)		LD3(0:5)	
16	*2*	**17**	*2*	**18**	*2*	**19**	*2*
rA ← −V		rI1 ← −V		rI2 ← −V		rI3 ← −V	
LDAN(0:5)		LD1N(0:5)		LD2N(0:5)		LD3N(0:5)	
24	*2*	**25**	*2*	**26**	*2*	**27**	*2*
F(M) ← rA		F(M) ← rI1		F(M) ← rI2		F(M) ← rI3	
STA(0:5)		ST1(0:5)		ST2(0:5)		ST3(0:5)	
32	*2*	**33**	*2*	**34**	*1*	**35**	*1 + T*
F(M) ← rJ		F(M) ← 0		Unit F busy?		Control, unit F	
STJ(0:2)		STZ(0:5)		JBUS(0)		IOC(0)	
40	*1*	**41**	*1*	**42**	*1*	**43**	*1*
rA:0, jump		rI1:0, jump		rI2:0, jump		rI3:0, jump	
JA[+]		J1[+]		J2[+]		J3[+]	
48	*1*	**49**	*1*	**50**	*1*	**51**	*1*
rA ← [rA]? ± M		rI1 ← [rI1]? ± M		rI2 ← [rI2]? ± M		rI3 ← [rI3]? ± M	
INCA(0)DECA(1) ENTA(2)ENNA(3)		INC1(0)DEC1(1) ENT1(2)ENN1(3)		INC2(0)DEC2(1) ENT2(2)ENN2(3)		INC3(0)DEC3(1) ENT3(2)ENN3(3)	
56	*2*	**57**	*2*	**58**	*2*	**59**	*2*
rA(F):V → CI		rI1(F):V → CI		rI2(F):V → CI		rI3(F):V → CI	
CMPA(0:5) FCMP(6)		CMP1(0:5)		CMP2(0:5)		CMP3(0:5)	

General form:

C	*t*
Description	
OP(F)	

C = operation code, (5:5) field of instruction
F = op variant, (4:4) field of instruction
M = address of instruction after indexing
V = F(M) = contents of F field of location M
OP = symbolic name for operation
(F) = standard F setting
t = execution time; *T* = interlock time

25	26	27	28	29	30	31	32	33	34	35	36	37	38	39	40	41	42	43	44	45	46	47	48	49	50	51	52	53	54	55
V	W	X	Y	Z	0	1	2	3	4	5	6	7	8	9	.	,	(	)	+	−	*	/	=	$	<	>	@	;	:	'

04	12	05	1	06	2	07	1 + 2F
rA ← rAX/V rX ← remainder DIV(0:5) FDIV(6)		Special NUM(0) CHAR(1) HLT(2)		Shift M bytes SLA(0) SRA(1) SLAX(2) SRAX(3) SLC(4) SRC(5)		Move F words from M to rI1 MOVE(1)	
12	2	13	2	14	2	15	2
rI4 ← V LD4(0:5)		rI5 ← V LD5(0:5)		rI6 ← V LD6(0:5)		rX ← V LDX(0:5)	
20	2	21	2	22	2	23	2
rI4 ← −V LD4N(0:5)		rI5 ← −V LD5N(0:5)		rI6 ← −V LD6N(0:5)		rX ← −V LDXN(0:5)	
28	2	29	2	30	2	31	2
F(M) ← rI4 ST4(0:5)		F(M) ← rI5 ST5(0:5)		F(M) ← rI6 ST6(0:5)		F(M) ← rX STX(0:5)	
36	1 + T	37	1 + T	38	1	39	1
Input, unit F IN(0)		Output, unit F OUT(0)		Unit F ready? JRED(0)		Jumps JMP(0) JSJ(1) JOV(2)JNOV(3) also [*] below	
44	1	45	1	46	1	47	1
rI4:0, jump J4[+]		rI5:0, jump J5[+]		rI6:0, jump J6[+]		rX:0, jump JX[+]	
52	1	53	1	54	1	55	1
rI4 ← [rI4]? ± M INC4(0)DEC4(1) ENT4(2)ENN4(3)		rI5 ← [rI5]? ± M INC5(0)DEC5(1) ENT5(2)ENN5(3)		rI6 ← [rI6]? ± M INC6(0)DEC6(1) ENT6(2)ENN6(3)		rX ← [rX]? ± M INCX(0)DECX(1) ENTX(2)ENNX(3)	
60	2	61	2	62	2	63	2
rI4(F):V → CI CMP4(0:5)		rI5(F):V → CI CMP5(0:5)		rI6(F):V → CI CMP6(0:5)		rX(F):V → CI CMPX(0:5)	

	[*]:		[+]:	
rA = register A	JL(4)	<	N(0)	
rX = register X	JE(5)	=	Z(1)	
rAX = registers AX as one	JG(6)	>	P(2)	
rIi = index reg. i, $1 \le i \le 6$	JGE(7)	≧	NN(3)	
rJ = register J	JNE(8)	≠	NZ(4)	
CI = comparison indicator	JLE(9)	≦	NP(5)	

A.2. THE MIX ASSEMBLY LANGUAGE

A symbolic language is used to make MIX programs considerably easier to read and to write, and to save the programmer from worrying about tedious clerical details which often lead to unnecessary errors. This language, MIXAL ("MIX Assembly Language"), is an extension of the notation used for instructions in the previous section; the main features of this extension are the optional use of alphabetic names to stand for numbers, and a location field for associating names with memory locations.

MIXAL can be readily comprehended if we consider first a simple example. The following code is part of a larger program; it is a subroutine to find the maximum of n elements $X[1], \ldots, X[n]$, according to Algorithm 1.2.10M.

Program M (*Find the maximum*). Register assignments: $rA \equiv m$, $rI1 \equiv n$, $rI2 \equiv j$, $rI3 \equiv k$, $X[i] \equiv \text{cell}(X + i)$.

Assembled instructions	Line no.	LOC	OP	ADDRESS	Times	Remarks
	01	X	EQU	1000		
	02		ORIG	3000		
3000: + 3009 0 2 32	*03*	MAXIMUM	STJ	EXIT	1	Subroutine linkage
3001: + 0 1 2 51	*04*	INIT	ENT3	0,1	1	*M1. Initialize.* $k \leftarrow n$.
3002: + 3005 0 0 39	*05*		JMP	CHANGEM	1	$j \leftarrow n$, $m \leftarrow X[n]$, $k \leftarrow n - 1$.
3003: + 1000 3 5 56	*06*	LOOP	CMPA	X,3	$n - 1$	*M3. Compare.*
3004: + 3007 0 7 39	*07*		JGE	*+3	$n - 1$	
3005: + 0 3 2 50	*08*	CHANGEM	ENT2	0,3	$A + 1$	*M4. Change m.* $j \leftarrow k$.
3006: + 1000 3 5 08	*09*		LDA	X,3	$A + 1$	$m \leftarrow X[k]$.
3007: + 1 0 1 51	*10*		DEC3	1	n	*M5. Decrease k.*
3008: + 3003 0 2 43	*11*		J3P	LOOP	n	*M2. All tested?*
3009: + 3009 0 0 39	*12*	EXIT	JMP	*	1	Return to main program. ∎

This program is an example of several things simultaneously:

a) The columns headed "LOC OP ADDRESS" are of principal interest; they contain a program in the MIXAL symbolic machine language, and we shall explain the details of this program below.

b) The column headed "Assembled instructions" shows the actual numeric machine language which corresponds to the MIXAL program. MIXAL has been designed so that it is a relatively simple matter to translate any MIXAL program into numeric machine language; this process may be carried out by another computer program called an *assembly program*. Thus, a programmer may do all of his "machine language" programming in MIXAL, never bothering to determine the equivalent numeric machine language himself. Virtually all MIX programs in this book are written in MIXAL. Chapter 9 includes a complete description of an assembly program which converts MIXAL programs to machine language in a form that is readily loaded into MIX's memory.

c) The column headed "Line no." is not an essential part of the MIXAL program; it is merely incorporated with the MIXAL programs of this book so that the text can refer to parts of the program.

d) The column headed "Remarks" gives explanatory information about the program, and it is cross-referenced to the steps of Algorithm 1.2.10M. The reader should refer to this algorithm. Note that a little "programmer's license" was used during the transcription of that algorithm into a MIX program; for example, step M2 has been put last. Note also the "register assignments" stated at the beginning of Program M; this shows what components of MIX correspond to the variables in the algorithm.

e) The column headed "Times" will be given for many of the MIX programs in this book; it represents the number of times the instruction on that line will be executed during the course of the program. Thus, line 6 will be performed $n - 1$ times, etc. From this information we can determine the length of time required to perform the subroutine; it is $(5 + 5n + 3A)u$, where A is the quantity which was carefully analyzed in Section 1.2.10.

Now we will discuss the MIXAL part of Program M. Line 1, "X EQU 1000", says that symbol X is to be *equivalent* to the number 1000. The effect of this may be seen on line 6, where the numeric equivalent of the instruction "CMPA X,3" appears as

$$+ \;|\; 1000 \;|\; 3 \;|\; 5 \;|\; 56 \;|,$$

i.e., "CMPA 1000,3".

Line 2 says that the locations for succeeding lines should be chosen sequentially, originating with 3000. Therefore the symbol MAXIMUM which appears in the LOC field of line 3 becomes equivalent to the number 3000, INIT is equivalent to 3001, LOOP is equivalent to 3003, etc.

On lines 3 through 12 the OP field contains the symbolic names of MIX instructions STJ, ENT3, etc. The "OP" in lines 1 and 2, on the other hand, contains "EQU" and "ORIG" which are *not* MIX operators. They are called *pseudo-operators* because they appear only in the MIXAL symbolic program. Pseudo-operators are used to specify the form of a symbolic program; they are not instructions of the program itself. Thus the line "X EQU 1000" only talks *about* the program, it does not signify that any variable is to be set equal to 1000 when Program M is run. Note that no instruction is assembled for lines 1 and 2.

Line 3 is a "store J" instruction which stores the contents of register J into the (0:2) field of location "EXIT", i.e., into the address part of the instruction found on line 12.

As mentioned earlier, Program M is intended to be part of a larger program; elsewhere the sequence

```
ENT1 100
JMP  MAXIMUM
STA  MAX
```

would, for example, jump to Program M with n set to 100. Program M would then find the largest of the elements $X[1], \ldots, X[100]$ and would return to the instruction "STA MAX" with the maximum value in rA and with its position, j, in rI2. (Cf. exercise 3.)

Line 5 jumps the control to line 8. Lines 4, 5, 6 need no further explanation. Line 7 introduces a new notation: an asterisk (read "self") refers to the location of this line; "*+3" ("self plus three") therefore refers to three locations past the current line. Since line 7 is an instruction which corresponds to location 3004, "*+3" appearing there refers to location 3007.

The rest of the symbolic code is self-explanatory; note the appearance of an asterisk again on line 12. (Cf. exercise 2.)

Our next example shows a few more features of the assembly language. The object is to print a table of the first 500 prime numbers, with 10 columns of 50 numbers each. The table should appear as follows:

```
FIRST FIVE HUNDRED PRIMES
      0002 0233 0547 0877 1229 1597 1993 2371 2749 3187
      0003 0239 0557 0881 1231 1601 1997 2377 2753 3191
      0005 0241 0563 0883 1237 1607 1999 2381 2767 3203
       :                                               :
       :                                               :
      0229 0541 0863 1223 1583 1987 2357 2741 3181 3571
```

We shall use the following method.

Algorithm P (*Print table of 500 primes*). This algorithm has two distinct parts: steps P1–P8 prepare an internal table of 500 primes, and steps P9–P11 print the answer in the form shown above. The latter part of the program uses two "buffer" areas, i.e., sections of memory in which a line image is formed; while one buffer is being printed, the other is being filled.

P1. [Start table.] Set PRIME[1] ← 2, N ← 3, J ← 1. (N will run through the odd numbers which are candidates for primes; J keeps track of how many primes have been found so far.)

P2. [N is prime.] Set J ← J + 1, PRIME[J] ← N.

P3. [500 found?] If J = 500, go to step P9.

P4. [Advance N.] Set N ← N + 2.

P5. [K ← 2.] Set K ← 2. (PRIME[K] will run through the possible prime divisors of N.)

P6. [PRIME[K]\N?] Divide N by PRIME[K]; let Q be the quotient and R the remainder. If R = 0, N is not prime, so go to P4.

P7. [PRIME[K] large?] If Q ≤ PRIME[K], go to P2. (In such a case, N must be prime; the proof of this fact is interesting and a little unusual—see exercise 1.3.2–6.)

P8. [Advance K.] Increase K by 1, and go to P6.

P9. [Print title.] Now we are ready to print the table. Advance the printer to the next page. Set BUFFER[0] to the title line and print this line. Set B ← 1, M ← 1.

P10. [Set up line.] Put PRIME[M], PRIME[50 + M], . . . , PRIME[450 + M] in proper format into BUFFER[B].

P11. [Print line.] Print BUFFER[B]; set B ← 1 − B (thereby switching to the other buffer); and increase M by 1. If M ≤ 50, return to P10; otherwise the algorithm terminates. ∎

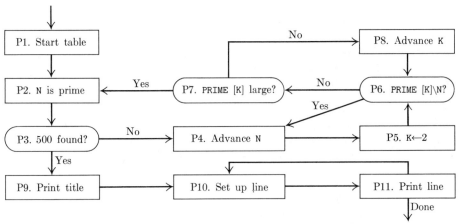

Fig. 2. Algorithm P.

Program P. (*Print table of 500 primes*). This program has deliberately been written in a slightly clumsy fashion in order to illustrate most of the features of MIXAL in a single program. rI1 ≡ J − 500; rI2 ≡ N; rI3 ≡ K; rI4 indicates B; rI5 is M plus multiples of 50.

```
01      * EXAMPLE PROGRAM ... TABLE OF PRIMES
02      *
03      L       EQU   500              Number of primes to find
04      PRINTER EQU   18               Unit number of printer
05      PRIME   EQU   -1               Memory area for table of primes
06      BUF0    EQU   2000             Memory area for BUFFER[0]
07      BUF1    EQU   BUF0+25          Memory area for BUFFER[1]
08              ORIG  3000
09      START   IOC   0(PRINTER)       Skip to new page.
10              LD1   =1-L=            P1. Start table. J ← 1.
11              LD2   =3=                  N ← 3.
12      2H      INC1  1                P2. N is prime. J ← J + 1.
13              ST2   PRIME+L,1            PRIME[J] ← N.
14              J1Z   2F               P3. 500 found?
```

15	4H	INC2	2	P4. *Advance* N.
16		ENT3	2	P5. K ← 2.
17	6H	ENTA	0	P6. PRIME[K]\N?
18		ENTX	0,2	
19		DIV	PRIME,3	
20		JXZ	4B	R = 0?
21		CMPA	PRIME,3	P7. PRIME[K] *large?*
22		INC3	1	P8. *Advance* K.
23		JG	6B	Jump if Q > PRIME[K].
24		JMP	2B	Otherwise N is prime.
25	2H	OUT	TITLE(PRINTER)	P9. *Print title.*
26		ENT4	BUF1+10	Set B ← 1.
27		ENT5	-50	Set M ← 0.
28	2H	INC5	L+1	Advance M.
29	4H	LDA	PRIME,5	P10. *Set up line.* (Right to left)
30		CHAR		
31		STX	0,4(1:4)	
32		DEC4	1	
33		DEC5	50	(rI5 goes down by 50 until
34		J5P	4B	nonpositive)
35		OUT	0,4(PRINTER)	P11. *Print line.*
36		LD4	24,4	Switch buffers.
37		J5N	2B	If rI5 = 0, we are done.
38		HLT		
39		* INITIAL CONTENTS OF TABLES AND BUFFERS		
40		ORIG	PRIME+1	
41		CON	2	First prime is 2.
42		ORIG	BUF0-5	
43	TITLE	ALF	FIRST	Alphabetic information for
44		ALF	FIVE	title line
45		ALF	HUND	
46		ALF	RED P	
47		ALF	RIMES	
48		ORIG	BUF0+24	
49		CON	BUF1+10	Each buffer refers to the other.
50		ORIG	BUF1+24	
51		CON	BUF0+10	
52		END	START	End of routine. ▌

The following points of interest are to be noted about this program:

1. Lines 01, 02, and 39 begin with an asterisk: this signifies a "comment" line which is merely explanatory, having no actual effect on the assembled program.

2. As in Program M, the "EQU" in line 03 sets the equivalent of a symbol; in this case, the equivalent of L is set to 500. (In the program of lines 10–24,

L represents the number of primes to be computed.) Note that in line 05 the symbol PRIME gets a *negative* equivalent; the equivalent of a symbol may be any five-byte-plus-sign number. In line 07 the equivalent of BUF1 is calculated as BUF0+25, namely 2025. MIXAL provides a limited amount of arithmetic on numbers; for another example, see line 13 where the value of PRIME+L (in this case, 499) is calculated by the assembly program.

3. Note that the symbol PRINTER has been used in the F-part on lines 25 and 35. The F-part, which is always enclosed in parentheses, may be a single number or symbol, or two numbers and/or symbols separated by a colon, as in the "1:4" of line 31.

4. MIXAL contains several ways to specify non-instruction words. Line 41 indicates an ordinary constant, "2", using the operation code CON; the result of line 41 is to assemble the word

$$+ \qquad \boxed{\;\; 2}$$.

Line 49 shows a slightly more complicated constant, "BUF1+10", which assembles as the word

$$+ \qquad \boxed{\; 2035}$$.

A constant may be enclosed in equal signs and it then becomes a *literal constant* (see lines 10 and 11). The assembler automatically creates internal names and inserts "CON" lines for literal constants. For example, lines 10 and 11 of Program P would effectively be changed to

```
10            LD1   con1
11            LD2   con2
```

and then at the end of the program, between lines 51 and 52, the lines

```
51a   con1   CON   1-L
51b   con2   CON   3
```

are effectively inserted as part of the assembly procedure for literal constants. Line 51a will assemble into the word

$$- \qquad \boxed{\; 499}$$.

The use of literal constants is a decided convenience, because it means that the programmer does not have to invent a name for the constant and that he does not have to insert that constant at the end of the program; he can keep his mind on the central problems and not worry about such routine matters while writing his programs. Of course, in Program P we did not make an especially good use of literal constants, since lines 10 and 11 would more properly be written "ENT1 1-L; ENT2 3"!

5. A good assembly language should mimic the way a programmer *thinks* about writing programs, so he can express himself fluently. One example of this philosophy is the use of literal constants, as we have just mentioned; another example is the use of "*", which was explained in Program M. A third example is the idea of *local symbols* such as the symbol 2H, which appears in the location field of lines 12, 25, and 28.

Local symbols are special symbols whose equivalents can be *redefined* as many times as desired. A symbol like PRIME has but one significance throughout a program, and if it were to appear in the location field of more than one line an error would be indicated by the assembly program. Local symbols have a different nature; we write, for example, 2H ("2 here") in the location field, and 2F ("2 forward") or 2B ("2 backward") in the address field of a MIXAL line:

> 2B means the closest *previous* location 2H
>
> 2F means the closest *following* location 2H

As examples, the "2F" in line 14 refers to line 25; the "2B" in line 24 refers back to line 12; and the "2B" in line 37 refers to line 28. An address of 2F or 2B never refers to the *same* line; e.g., the three lines

```
2H   EQU   10
2H   MOVE  2F(2B)
2H   EQU   2B-3
```

are virtually equivalent to the single line

```
MOVE *-3(10).
```

The symbols 2F, 2B are never to be used in the location field, and 2H is never to be used in the address field. There are ten local symbols, which can be obtained by replacing "2" in the above examples by any digit from 0 to 9.

The idea of local symbols was introduced by M. E. Conway in 1958, in connection with an assembly program for the UNIVAC 1. Local symbols spare the programmer from the necessity to think of a symbolic name for an address, when all he wants to do is refer to an instruction a few lines away. When making reference to a nearby location in the program there often is no appropriate name with much significance, so programmers have tended to use symbols like X1, X2, X3, etc.; this leads to the danger of using the same symbol twice. That is why the reader will soon find that the use of local symbols comes naturally to him when he writes MIXAL programs, if he is not already familiar with this idea.

6. In lines 30 and 38 the address part is blank. This means the address is to be zero. Similarly, we could have left the address blank in line 17, but the program would have been less readable.

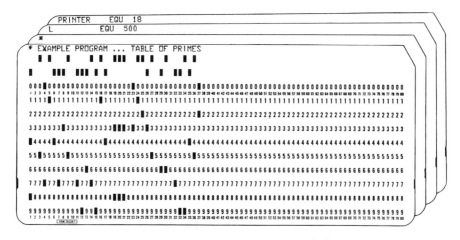

Fig. 3. The first four lines of Program P punched onto cards.

7. Lines 43–47 use the "ALF" operation, which creates a five-byte constant in MIX alphameric character code. For example, line 45 causes the word

+	00	08	24	15	04

to be assembled.

These lines are used as the first 25 characters of the title line. *All locations whose contents are not specified in the* MIXAL *program are ordinarily set to zero* (except the locations which are used by the loading routine, usually 3700–3999); thus there is no need to set the other words of the title line to blanks.

8. Note that arithmetic may be used on ORIG lines, e.g., lines 40, 42, and 48.

9. The last line of a complete MIXAL program always has the operation code END. The address on this line is the location at which the program is to begin once it has been loaded into memory.

10. As a final note about Program P, the reader may observe how the coding has been written so that index registers are counted towards zero, and tested against zero, whenever possible. For example, the quantity J–500, not J, is kept in rI1. Lines 26–34 are particularly noteworthy, although perhaps a bit tricky.

It may be of interest to note a few of the statistics observed when Program P was actually run. The division instruction in line 19 was executed 9538 times; the time to perform lines 10–24 was 182144u.

MIXAL programs can be punched onto cards, as shown in Fig. 3. The following format is used:

Columns 1–10	LOC (location) field,
Columns 12–15	OP field,
Columns 17–80	ADDRESS field and optional remarks,
Columns 11, 16	blank.

However, if column 1 contains an asterisk, the entire card is treated as a comment. The ADDRESS field ends with the first blank column following column 16; any explanatory information may be punched to the right of this first blank column with no effect on the assembled program. (*Exception:* When the OP field is "ALF", the remarks always start in column 22.)

The MIX assembly program (see Section 9.3) accepts card decks prepared in this manner and converts them to machine language programs in loadable form. Under favorable circumstances the reader will have access to a MIX assembly program and MIX simulator, on which various exercises in this book can be worked out.

Now we have seen what can be done in MIXAL. We conclude this section by describing the rules more carefully, and in particular we shall observe what is *not* allowed in MIXAL. The following comparatively few rules define the language.

1. A *symbol* is a string of one to ten letters and/or digits, containing at least one letter. *Examples:* PRIME TEMP 20BY20. The special symbols dH, dF, dB, where d is a single digit, will for the purposes of this definition be replaced by other unique symbols according to the "local symbol" convention described above.

2. A *number* is a string of one to ten digits. *Example:* 00052.

3. Each appearance of a symbol in a MIXAL program is said to be either a "defined symbol" or a "future reference." A *defined symbol* is a symbol which has appeared in the LOC field of a preceding line of this MIXAL program. A *future reference* is a symbol which has not yet been defined in this way.

4. An *atomic expression* is either

a) a number, or
b) a defined symbol (denoting the numerical equivalent of that symbol, see rule 13), or
c) an asterisk (denoting the value of ⊛; see rules 10 and 11).

5. An *expression* is either

a) an atomic expression, or
b) a plus or minus sign followed by an atomic expression, or
c) an expression followed by a binary operation followed by an atomic expression.

The six admissible binary operations are +, −, *, /, //, : ; they are defined on numeric MIX words as follows:

```
C = A+B     LDA AA; ADD BB; STA CC
C = A−B     LDA AA; SUB BB; STA CC
C = A*B     LDA AA; MUL BB; STX CC
C = A/B     LDA AA; SRAX 5; DIV BB; STA CC
C = A//B    LDA AA; ENTX 0; DIV BB; STA CC
C = A:B     LDA AA; MUL =8=; SLAX 5; ADD BB; STA CC.
```

Here AA, BB, CC denote locations containing the respective values of the symbols A, B, C.

Operations within an expression are carried out from left to right. *Examples:*

−1+5	equals 4
−1+5*20/6	equals 4*20/6 equals 80/6 equals 13 (going from left to right)
1//3	equals a MIX word whose value is approximately $(b^5/3)$ where b is the byte size; i.e., a word representing the fraction $\frac{1}{3}$ with decimal point at the left.
1:3	equals 11 (usually used in partial field specification)
*−3	equals ⊛ minus three
***	equals ⊛ times ⊛ !

6. An *A-part* (which is used to describe the address field of a MIX instruction) is either

a) vacuous (denoting the value zero), or
b) an expression, or
c) a future reference (denoting the eventual equivalent of the symbol, see rule 13).

7. An *index part* (which is used to describe the index field of a MIX instruction) is either

a) vacuous (denoting the value zero), or
b) a comma followed by an expression (denoting the value of that expression).

8. An *F-part* (which is used to describe the F-field of a MIX instruction) is either

a) vacuous (denoting the *standard* F-setting, based on the context), or
b) a left parenthesis followed by an expression followed by a right parenthesis (denoting the value of the expression).

9. A *W-value* (which is used to describe a *full-word* MIX constant) is either

a) an expression followed by an F-part [in this case a vacuous F-part denotes (0:5)], or
b) a W-value followed by a comma followed by a W-value of the form (a).

A W-value denotes the value of a numeric MIX word determined as follows: Let the W-value have the form "$E_1(F_1), E_2(F_2), \ldots, E_n(F_n)$" where $n \geq 1$, the E's are expressions, and the F's are fields. The desired result is the final value which would appear in memory location CON if the following hypothetical program were executed: "STZ CON; LDA C_1; STA CON(F_1); ...; LDA C_n; STA CON(F_n)". Here $C_1, \ldots, C_n$ denote locations containing the values of expressions $E_1, \ldots, E_n$. Each F_i must have the form $8L_i + R_i$ where $0 \leq L_i \leq R_i \leq 5$.

Examples:

1	is the word	+ 1
1,−1000(0:2)	is the word	− 1000 1
−1000(0:2),1	is the word	+ 1

10. The assembly process makes use of a value denoted by ⊛ (called the *location counter*) which is initially zero. The value of ⊛ should always be a

nonnegative number which can fit in two bytes. When the location field of a line is not blank, it must contain a symbol which has not been previously defined. The equivalent of that symbol is then defined to be the current value of ⊛.

11. After processing the LOC field as described in rule 10, the assembly process depends on the value of the OP field. There are six possibilities for OP:

a) OP is a symbolic MIX operator (see Table 1 at the end of the previous section). The chart defines the standard C and F values for this operator. In this case the ADDRESS should be an A-part (rule 6), followed by an index part (rule 7), followed by an F part (rule 8). We thereby obtain four values: C, F, A, and I; the effect is to assemble the word determined by the sequence "LDA C; STA WORD; LDA F; STA WORD(4:4); LDA I; STA WORD(3:3); LDA A; STA WORD(0:2)" into the location specified by ⊛, and to advance ⊛ by 1.

b) OP is "EQU". The ADDRESS should be a W-value (see rule 9); if the LOC field is nonblank, the equivalent of the symbol appearing there is set equal to the value specified in ADDRESS. This rule takes precedence over rule 10. The value of ⊛ is unchanged. (As a nontrivial example, consider the line

BYTESIZE   EQU   1(4:4)

which allows the programmer to have a symbol whose value depends on the byte size. This is an acceptable situation so long as the resulting program is meaningful with each possible byte size.)

c) OP is "ORIG". The ADDRESS should be a W-value (see rule 9); the location counter, ⊛, is set to this value. (Note that because of rule 10, a symbol appearing in the LOC field of an ORIG card gets as its equivalent the value of ⊛ before it has changed. *Example:*

TABLE   ORIG *+100

sets the equivalent of TABLE to the *first* of 100 locations.)

d) OP is "CON". The ADDRESS should be a W-value; the effect is to assemble a word, having this value, into the location specified by ⊛, and to advance ⊛ by 1.

e) OP is "ALF". The effect is to assemble the word of character codes formed by columns 17–21 of the card, otherwise behaving like CON.

f) OP is "END". The ADDRESS should be a W-value, which specifies in its (4:5) field the location of the instruction at which the program begins. The END card signals the end of a MIXAL program. Lines are inserted, in arbitrary order, corresponding to all undefined symbols and literal constants (see rules 12 and 13). The LOC field of an END card must be blank.

12. Literal constants: A W-value of 9 characters or less in length may be enclosed between "=" signs and used as a future reference. The effect is as though a new symbol were created and inserted just before the END card (see remark 4 following Program P).

13. Every symbol has one and only one equivalent value; this is a full-word MIX number which is either determined by the symbol's appearance in LOC according to rule 10 or rule 11(b), or else a line, having the name of the symbol in LOC with OP = "CON" and ADDRESS = "0", is effectively inserted before the END card.

Note: The most significant consequence of the above rules is the restriction on future references. A symbol which has not been defined in the LOC field of a previous card may not be used except as the A-part of an instruction. In particular, it may not be used (a) in connection with arithmetic operations; or (b) in the ADDRESS field of EQU, ORIG, or CON. For example,

LDA　2F+1　　　and　　　CON　3F

are both illegal. This restriction has been imposed in order to allow more efficient assembly of programs, and the experience gained in writing this set of books has shown that it is a very mild restriction which rarely makes much difference.

EXERCISES

1. [*00*] The text remarked that "X EQU 1000" does not indicate any instruction which sets the value of a variable. Suppose that you are writing a MIX program in which you wish to set the value contained in a certain memory cell (whose symbolic name is X) equal to 1000. How could you write this in MIXAL?

▶ 2. [*10*] Line 12 of Program M says "JMP *"; since * denotes the location of the line, why doesn't the program go into an infinite loop, endlessly repeating this instruction?

▶ 3. [*23*] What is the effect of the following program, if it is used in conjunction with Program M?

```
START   IN    X+1
        JBUS  *(0)
        ENT1  100
1H      JMP   MAXIMUM
        LDX   X,1
        STA   X,1
        STX   X,2
        DEC1  1
        J1P   1B
        OUT   X+1(1)
        HLT
        END   START
```

▶ 4. [*25*] Assemble Program P by hand; i.e., what are the actual numerical contents of memory, corresponding to that symbolic program?

TABLES OF
NUMERICAL QUANTITIES

Table 1

Quantities which are frequently used in standard subroutines and in analysis of computer programs. (40 decimal places)

$$\sqrt{2} = 1.41421\ 35623\ 73095\ 04880\ 16887\ 24209\ 69807\ 85697-$$
$$\sqrt{3} = 1.73205\ 08075\ 68877\ 29352\ 74463\ 41505\ 87236\ 69428+$$
$$\sqrt{5} = 2.23606\ 79774\ 99789\ 69640\ 91736\ 68731\ 27623\ 54406+$$
$$\sqrt{10} = 3.16227\ 76601\ 68379\ 33199\ 88935\ 44432\ 71853\ 37196-$$
$$\sqrt[3]{2} = 1.25992\ 10498\ 94873\ 16476\ 72106\ 07278\ 22835\ 05703-$$
$$\sqrt[3]{3} = 1.44224\ 95703\ 07408\ 38232\ 16383\ 10780\ 10958\ 83919-$$
$$\sqrt[4]{2} = 1.18920\ 71150\ 02721\ 06671\ 74999\ 70560\ 47591\ 52930-$$
$$\ln 2 = 0.69314\ 71805\ 59945\ 30941\ 72321\ 21458\ 17656\ 80755+$$
$$\ln 3 = 1.09861\ 22886\ 68109\ 69139\ 52452\ 36922\ 52570\ 46475-$$
$$\ln 10 = 2.30258\ 50929\ 94045\ 68401\ 79914\ 54684\ 36420\ 76011+$$
$$1/\ln 2 = 1.44269\ 50408\ 88963\ 40735\ 99246\ 81001\ 89213\ 74266+$$
$$1/\ln 10 = 0.43429\ 44819\ 03251\ 82765\ 11289\ 18916\ 60508\ 22944-$$
$$\pi = 3.14159\ 26535\ 89793\ 23846\ 26433\ 83279\ 50288\ 41972-$$
$$1° = \pi/180 = 0.01745\ 32925\ 19943\ 29576\ 92369\ 07684\ 88612\ 71344+$$
$$1/\pi = 0.31830\ 98861\ 83790\ 67153\ 77675\ 26745\ 02872\ 40689+$$
$$\pi^2 = 9.86960\ 44010\ 89358\ 61883\ 44909\ 99876\ 15113\ 53137-$$
$$\sqrt{\pi} = \Gamma(1/2) = 1.77245\ 38509\ 05516\ 02729\ 81674\ 83341\ 14518\ 27975+$$
$$\Gamma(1/3) = 2.67893\ 85347\ 07747\ 63365\ 56929\ 40974\ 67764\ 41287-$$
$$\Gamma(2/3) = 1.35411\ 79394\ 26400\ 41694\ 52880\ 28154\ 51378\ 55193+$$
$$e = 2.71828\ 18284\ 59045\ 23536\ 02874\ 71352\ 66249\ 77572+$$
$$1/e = 0.36787\ 94411\ 71442\ 32159\ 55237\ 70161\ 46086\ 74458+$$
$$e^2 = 7.38905\ 60989\ 30650\ 22723\ 04274\ 60575\ 00781\ 31803+$$
$$\gamma = 0.57721\ 56649\ 01532\ 86060\ 65120\ 90082\ 40243\ 10422-$$
$$\ln \pi = 1.14472\ 98858\ 49400\ 17414\ 34273\ 51353\ 05871\ 16473-$$
$$\phi = 1.61803\ 39887\ 49894\ 84820\ 45868\ 34365\ 63811\ 77203+$$
$$e^\gamma = 1.78107\ 24179\ 90197\ 98523\ 65041\ 03107\ 17954\ 91696+$$
$$e^{\pi/4} = 2.19328\ 00507\ 38015\ 45655\ 97696\ 59278\ 73822\ 34616+$$
$$\sin 1 = 0.84147\ 09848\ 07896\ 50665\ 25023\ 21630\ 29899\ 96226-$$
$$\cos 1 = 0.54030\ 23058\ 68139\ 71740\ 09366\ 07442\ 97660\ 37323+$$
$$\zeta(3) = 1.20205\ 69031\ 59594\ 28539\ 97381\ 61511\ 44999\ 07650-$$
$$\ln \phi = 0.48121\ 18250\ 59603\ 44749\ 77589\ 13424\ 36842\ 31352-$$
$$1/\ln \phi = 2.07808\ 69212\ 35027\ 53760\ 13226\ 06117\ 79576\ 77422-$$
$$-\ln \ln 2 = 0.36651\ 29205\ 81664\ 32701\ 24391\ 58232\ 66946\ 94543-$$

Table 2

Quantities which are frequently used in standard subroutines and in analysis of computer programs, in *octal* notation. The *name* of each quantity, appearing at the left of the equal sign, is given in decimal notation.

$0.1 =$	$0.06314\ 63146\ 31463\ 14631\ 46314\ 63146\ 31463\ 14631\ 4632$
$0.01 =$	$0.00507\ 53412\ 17270\ 24365\ 60507\ 53412\ 17270\ 24365\ 6051$
$0.001 =$	$0.00040\ 61115\ 64570\ 65176\ 76355\ 44264\ 16254\ 02030\ 4467$
$0.0001 =$	$0.00003\ 21556\ 13530\ 70414\ 54512\ 75170\ 33021\ 15002\ 3522$
$0.00001 =$	$0.00000\ 24761\ 32610\ 70664\ 36041\ 06077\ 17401\ 56063\ 3442$
$0.000001 =$	$0.00000\ 02061\ 57364\ 05536\ 66151\ 55323\ 07746\ 44470\ 2603$
$0.0000001 =$	$0.00000\ 00153\ 27745\ 15274\ 53644\ 12741\ 72312\ 20354\ 0215$
$0.00000001 =$	$0.00000\ 00012\ 57143\ 56106\ 04303\ 47374\ 77341\ 01512\ 6333$
$0.000000001 =$	$0.00000\ 00001\ 04560\ 27640\ 46655\ 12262\ 71426\ 40124\ 2174$
$0.0000000001 =$	$0.00000\ 00000\ 06676\ 33766\ 35367\ 55653\ 37265\ 34642\ 0163$
$\sqrt{2} =$	$1.32404\ 74631\ 77167\ 46220\ 42627\ 66115\ 46725\ 12575\ 1744$
$\sqrt{3} =$	$1.56663\ 65641\ 30231\ 25163\ 54453\ 50265\ 60361\ 34073\ 4222$
$\sqrt{5} =$	$2.17067\ 36334\ 57722\ 47602\ 57471\ 63003\ 00563\ 55620\ 3202$
$\sqrt{10} =$	$3.12305\ 40726\ 64555\ 22444\ 02242\ 57101\ 41466\ 33775\ 2253$
$\sqrt[3]{2} =$	$1.20505\ 05746\ 15345\ 05342\ 10756\ 65334\ 25574\ 22415\ 0303$
$\sqrt[3]{3} =$	$1.34233\ 50444\ 22175\ 73134\ 67363\ 76133\ 05334\ 31147\ 6012$
$\sqrt[4]{2} =$	$1.14067\ 74050\ 61556\ 12455\ 72152\ 64430\ 60271\ 02755\ 7314$
$\ln 2 =$	$0.54271\ 02775\ 75071\ 73632\ 57117\ 07316\ 30007\ 71366\ 5364$
$\ln 3 =$	$1.06237\ 24752\ 55006\ 05227\ 32440\ 63065\ 25012\ 35574\ 5534$
$\ln 10 =$	$2.23273\ 06735\ 52524\ 25405\ 56512\ 66542\ 56026\ 46050\ 5071$
$1/\ln 2 =$	$1.34252\ 16624\ 53405\ 77027\ 35750\ 37766\ 40644\ 35175\ 0435$
$1/\ln 10 =$	$0.33626\ 75425\ 11562\ 41614\ 52325\ 33525\ 27655\ 14756\ 0622$
$\pi =$	$3.11037\ 55242\ 10264\ 30215\ 14230\ 63050\ 56006\ 70163\ 2112$
$1° = \pi/180 =$	$0.01073\ 72152\ 11224\ 72344\ 25603\ 54276\ 63351\ 22056\ 1154$
$1/\pi =$	$0.24276\ 30155\ 62344\ 20251\ 23760\ 47257\ 50765\ 15156\ 7007$
$\pi^2 =$	$11.67517\ 14467\ 62135\ 71322\ 25561\ 15466\ 30021\ 40654\ 3410$
$\sqrt{\pi} = \Gamma(1/2) =$	$1.61337\ 61106\ 64736\ 65247\ 47035\ 40510\ 15273\ 34470\ 1776$
$\Gamma(1/3) =$	$2.53347\ 35234\ 51013\ 61316\ 73106\ 47644\ 54653\ 00106\ 6605$
$\Gamma(2/3) =$	$1.26523\ 57112\ 14154\ 74312\ 54572\ 37655\ 60126\ 23231\ 0245$
$e =$	$2.55760\ 52130\ 50535\ 51246\ 52773\ 42542\ 00471\ 72363\ 6166$
$1/e =$	$0.27426\ 53066\ 13167\ 46761\ 52726\ 75436\ 02440\ 52371\ 0336$
$e^2 =$	$7.30714\ 45615\ 23355\ 33460\ 63507\ 35040\ 32664\ 25356\ 5022$
$\gamma =$	$0.44742\ 14770\ 67666\ 06172\ 23215\ 74376\ 01002\ 51313\ 2552$
$\ln \pi =$	$1.11206\ 40443\ 47503\ 36413\ 65374\ 52661\ 52410\ 37511\ 4606$
$\phi =$	$1.47433\ 57156\ 27751\ 23701\ 27634\ 71401\ 40271\ 66710\ 1501$
$e^\gamma =$	$1.61772\ 13452\ 61152\ 65761\ 22477\ 36553\ 53327\ 17554\ 2126$
$e^{\pi/4} =$	$2.14275\ 31512\ 16162\ 52370\ 35530\ 11342\ 53525\ 44307\ 0217$
$\sin 1 =$	$0.65665\ 24436\ 04414\ 73402\ 03067\ 23644\ 11612\ 07474\ 1451$
$\cos 1 =$	$0.42450\ 50037\ 32406\ 42711\ 07022\ 14666\ 27320\ 70675\ 1232$
$\zeta(3) =$	$1.14735\ 00023\ 60014\ 20470\ 15613\ 42561\ 31715\ 10177\ 0662$
$\ln \phi =$	$0.36630\ 26256\ 61213\ 01145\ 13700\ 41004\ 52264\ 30700\ 4065$
$1/\ln \phi =$	$2.04776\ 60111\ 17144\ 41512\ 11436\ 16575\ 00355\ 43630\ 4065$
$-\ln \ln 2 =$	$0.27351\ 71233\ 67265\ 63650\ 17401\ 56637\ 26334\ 31455\ 5701$

Tables 1 and 2 contain several hitherto unpublished 40-digit values which have been furnished to the author by Dr. John W. Wrench, Jr.

For high-precision values of constants not found in this list, see J. Peters, *Ten Place Logarithms of the Numbers from 1 to 100000*, Appendix to Volume 1 (New York: F. Ungar Publ. Co., 1957); and *Handbook of Mathematical Functions*, ed. by M. Abramowitz and I. A. Stegun (Washington, D.C.: U. S. Govt. Printing Office, 1964), Chapter 1.

Table 3

Values of harmonic numbers, Bernoulli numbers, and Fibonacci numbers for small values of n.

n	H_n	B_n	F_n	n
0	0	1	0	0
1	1	$-1/2$	1	1
2	3/2	1/6	1	2
3	11/6	0	2	3
4	25/12	$-1/30$	3	4
5	137/60	0	5	5
6	49/20	1/42	8	6
7	363/140	0	13	7
8	761/280	$-1/30$	21	8
9	7129/2520	0	34	9
10	7381/2520	5/66	55	10
11	83711/27720	0	89	11
12	86021/27720	$-691/2730$	144	12
13	1145993/360360	0	233	13
14	1171733/360360	7/6	377	14
15	1195757/360360	0	610	15
16	2436559/720720	$-3617/510$	987	16
17	42142223/12252240	0	1597	17
18	14274301/4084080	43867/798	2584	18
19	275295799/77597520	0	4181	19
20	55835135/15519504	$-174611/330$	6765	20
21	18858053/5173168	0	10946	21
22	19093197/5173168	854513/138	17711	22
23	444316699/118982864	0	28657	23
24	1347822955/356948592	$-236364091/2730$	46368	24
25	34052522467/8923714800	0	75025	25

For any x, let $H_x = \sum_{n \geq 1} \left(\dfrac{1}{n} - \dfrac{1}{n+x} \right)$. Then

$H_{1/2} = 2 - 2 \ln 2,$

$H_{1/3} = 3 - \frac{1}{2}\pi/\sqrt{3} - \frac{3}{2} \ln 3,$

$H_{2/3} = \frac{3}{2} + \frac{1}{2}\pi/\sqrt{3} - \frac{3}{2} \ln 3,$

$H_{1/4} = 4 - \frac{1}{2}\pi - 3 \ln 2,$

$H_{3/4} = \frac{4}{3} + \frac{1}{2}\pi - 3 \ln 2,$

$H_{1/5} = 5 - \frac{1}{2}\pi\phi \sqrt{\dfrac{2+\phi}{5}} - \frac{1}{2}(3 - \phi) \ln 5 - (\phi - \frac{1}{2}) \ln (2 + \phi),$

$H_{2/5} = \frac{5}{2} - \frac{1}{2}\pi/\phi\sqrt{2+\phi} - \frac{1}{2}(2 + \phi) \ln 5 + (\phi - \frac{1}{2}) \ln (2 + \phi),$

$H_{3/5} = \frac{5}{3} + \frac{1}{2}\pi/\phi\sqrt{2+\phi} - \frac{1}{2}(2 + \phi) \ln 5 + (\phi - \frac{1}{2}) \ln (2 + \phi),$

$H_{4/5} = \frac{5}{4} + \frac{1}{2}\pi\phi \sqrt{\dfrac{2+\phi}{5}} - \frac{1}{2}(3 - \phi) \ln 5 - (\phi - \frac{1}{2}) \ln (2 + \phi),$

$H_{1/6} = 6 - \frac{1}{2}\pi\sqrt{3} - 2 \ln 2 - \frac{3}{2} \ln 3,$

$H_{5/6} = \frac{6}{5} + \frac{1}{2}\pi\sqrt{3} - 2 \ln 2 - \frac{3}{2} \ln 3,$

and, in general, when $0 < p < q$ (cf. exercise 1.2.9–19),

$$H_{p/q} = \frac{q}{p} - \frac{1}{2}\pi \cot \frac{p}{q} \pi - \ln 2q + 2 \sum_{1 \leq n < q/2} \cos \frac{2\pi np}{q} \ln \sin \frac{n}{q} \pi.$$

INDEX TO NOTATIONS

In the following formulas, letters which are not further qualified have the following significance:

$$j, k \quad \text{integer-valued arithmetic expression}$$
$$m, n \quad \text{nonnegative integer-valued arithmetic expression}$$
$$x, y, z \quad \text{real-valued arithmetic expression}$$
$$f \quad \text{real-valued function}$$

Formal symbolism	Meaning	Section reference
A_n	the nth element of linear array A	
A_{mn}	the element in row m, column n of rectangular array A	
$A[n]$	equivalent to A_n	1.1
$A[m, n]$	equivalent to A_{mn}	1.1
$V \leftarrow E$	give variable V the value of expression E	1.1
$U \leftrightarrow V$	interchange the values of variables U and V	1.1
$(B \Rightarrow E_1; E_2)$	conditional expression: denotes E_1 if B is true, E_2 if B is false	8.1
δ_{jk}	Kronecker delta: $(j = k \Rightarrow 1; \ 0)$	1.2.6
$\sum_{R(k)} f(k)$	sum of all $f(k)$ such that k is an integer and relation $R(k)$ is true	1.2.3
$\prod_{R(k)} f(k)$	product of all $f(k)$ such that k is an integer and relation $R(k)$ is true	1.2.3
$\min_{R(k)} f(k)$	minimum value of all $f(k)$ such that k is an integer and relation $R(k)$ is true	1.2.3
$\max_{R(k)} f(k)$	maximum value of all $f(k)$ such that k is an integer and relation $R(k)$ is true	1.2.3
$j\backslash k$	j divides k: $k \bmod j = 0$	1.2.4

Formal symbolism	Meaning	Section reference
$\gcd(j, k)$	greatest common divisor of j and k: $$(j = k = 0 \Rightarrow 0; \quad \max_{d \backslash j, d \backslash k} d)$$	4.5.2
$\mathrm{lcm}(j, k)$	least common multiple of j and k: $$(j = k = 0 \Rightarrow 0; \quad \min_{\substack{d>0 \\ j \backslash d, k \backslash d}} d)$$	4.5.2
$\det(A)$	determinant of square matrix A	1.2.3
A^T	transpose of rectangular array A: $$A^T[j, k] = A[k, j]$$	1.2.3
x^y	x to the y power, x positive	1.2.2
x^k	x to the kth power: $$\left(k \geq 0 \Rightarrow \prod_{0 \leq j < k} x; \quad 1/x^{-k} \right)$$	1.2.2
$x^{\overline{k}}$	x upper k: $$\left(k \geq 0 \Rightarrow x(x + 1) \cdots (x + k - 1) = \prod_{0 \leq j < k} (x + j); \quad 1/(x + k)^{\overline{-k}} \right)$$	1.2.6
$x^{\underline{k}}$	x lower k: $$\left(k \geq 0 \Rightarrow x(x - 1) \cdots (x - k + 1) = \prod_{0 \leq j < k} (x - j); \right.$$ $$\left. 1/(x - k)^{\underline{-k}} \right) = (-1)^k (-x)^{\overline{k}}$$	1.2.6
$n!$	n factorial: $1 \cdot 2 \cdots \cdot n = n^{\underline{n}}$	1.2.5
$\binom{x}{k}$	binomial coefficient: ($k < 0 \Rightarrow 0$; $x^{\underline{k}}/k!$)	1.2.6
$\binom{n}{n_1, n_2, \ldots, n_m}$	multinomial coefficient, $$n = n_1 + n_2 + \cdots + n_m$$	1.2.6

Formal symbolism	Meaning	Section reference
$\left[\begin{matrix} n \\ m \end{matrix}\right]$	Stirling number of first kind: $$\sum_{0 < k_1 < k_2 < \cdots < k_{n-m} < n} k_1 k_2 \cdots k_{n-m}$$	1.2.6
$\left\{\begin{matrix} n \\ m \end{matrix}\right\}$	Stirling number of second kind: $$\sum_{1 \leq k_1 \leq k_2 \leq \cdots \leq k_{n-m} \leq m} k_1 k_2 \cdots k_{n-m}$$	1.2.6
$/x_1, x_2, \ldots, x_n/$	continued fraction: $1/(x_1 + 1/(x_2 + 1/(\cdots + 1/(x_n) \cdots)))$	4.5.3
$S \backslash T$	set difference: elements of set S not in T	
$(\ldots a_1 a_0 . a_{-1} \ldots)_b$	radix-b positional notation: $\sum_k a_k b^k$	4.1
$\{a \mid R(a)\}$	set of all a for which the relation $R(a)$ is true	
$\{a_1, \ldots, a_n\}$	the set $\{a_k \mid 1 \leq k \leq n\}$	
$\{x\}$	in contexts where a real value, not a set, is required, denotes fractional part: $x \bmod 1$	3.3.3
$[y, z)$	half-open interval: $\{x \mid y \leq x < z\}$	
$\|S\|$	cardinal: number of elements in set S	
$((x))$	sawtooth function	3.3.3
$|x|$	absolute value of x: $(x < 0 \Rightarrow -x; \ x)$	
$\lfloor x \rfloor$	floor of x, greatest integer function: $\max_{k \leq x} k$	1.2.4
$\lceil x \rceil$	ceiling of x, least integer function: $\min_{k \geq x} k$	1.2.4
$x \bmod y$	mod function: $(y = 0 \Rightarrow x; \ x - y\lfloor x/y \rfloor)$	1.2.4
$u(x) \bmod v(x)$	remainder of polynomial $u(x)$ divided by $v(x)$	4.6.1
$x \equiv y \ (\text{modulo } z)$	relation of congruence: $x \bmod z = y \bmod z$	1.2.4
$\log_b x$	logarithm, base b, of x (real positive $b \neq 1$): $x = b^{\log_b x}$	1.2.2
$\ln x$	natural logarithm: $\log_e x$	1.2.2
$\exp x$	exponential of x: e^x	1.2.2

Formal symbolism	Meaning	Section reference
$\langle X_n \rangle$	the infinite sequence $X_0, X_1, X_2, \ldots$ (here n is a letter which is part of the symbol)	1.2.9
$f'(x)$	derivative of f at x	1.2.9
$f''(x)$	second derivative of f at x	1.2.10
$f^{(n)}(x)$	nth derivative: $(n = 0 \Rightarrow f(x)$; $\qquad g'(x) \quad$ where $\quad g(x) = f^{(n-1)}(x))$	1.2.11.2
$H_n^{(x)}$	$1 + 1/2^x + \cdots + 1/n^x = \displaystyle\sum_{1 \le k \le n} 1/k^x$	1.2.7
H_n	harmonic number: $H_n^{(1)}$	1.2.7
F_n	Fibonacci number: $\qquad (n \le 1 \Rightarrow n; F_{n-1} + F_{n-2})$	1.2.8
B_n	Bernoulli number	1.2.11.2
$B(x, y)$	Beta function	1.2.6
$\text{sign}\,(x)$	sign of x: $\qquad (x = 0 \Rightarrow 0; (x > 0 \Rightarrow +1; -1))$	
$\delta(x)$	characteristic function of the integers	3.3.3
$\zeta(x)$	zeta function: $H_\infty^{(x)}$ when $x > 1$	1.2.7
$\Gamma(x)$	gamma function: $\gamma(x, \infty)$; $(x - 1)!$ when x is a positive integer	1.2.5
$\gamma(x, y)$	incomplete gamma function	1.2.11.3
γ	Euler's constant	1.2.7
e	base of natural logarithms: $\displaystyle\sum_{k \ge 0} 1/k!$	1.2.2
∞	infinity: larger than any number	
$\emptyset$	empty set (set with no elements)	
ϕ	golden ratio, $\frac{1}{2}(1 + \sqrt{5})$	1.2.8
$\varphi(n)$	Euler's totient function: $\displaystyle\sum_{\substack{0 \le k < n \\ \gcd(k,n)=1}} 1$	1.2.4
$\mu(n)$	Möbius function	4.5.2
$\Lambda(n)$	von Mangoldt's function	4.5.3
$p(n)$	number of partitions of n	1.2.1

Formal symbolism	Meaning	Section reference
$\Pr(S(n))$	probability that statement $S(n)$ is true, for "random" n	3.5
$O(f(n))$	big-oh of $f(n)$ as $n \to \infty$	1.2.11.1
$O(f(x))$	big-oh of $f(x)$, for small x (or for x in some specified range)	1.2.11.1
(min x_1, ave x_2, max x_3, dev x_4)	a random variable having minimum value x_1, average ("expected") value x_2, maximum value x_3, standard deviation x_4	1.2.10
mean (g)	mean value of probability distribution represented by generating function g: $g'(1)$	1.2.10
var (g)	variance of probability distribution represented by generating function g: $$g''(1) + g'(1) - g'(1)^2$$	1.2.10
deg (u)	degree of polynomial u	4.6
$\ell(u)$	leading coefficient of polynomial u	4.6
cont (u)	content of polynomial u	4.6.1
pp $(u(x))$	primitive part of polynomial u, evaluated at x	4.6.1
▮	end of algorithm, program, or proof	1.1
␣	one blank space	A.1
rA	register A (accumulator) of MIX	A.1
rX	register X (extension) of MIX	A.1
rI1, ..., rI6	(index) registers I1, ..., I6 of MIX	A.1
rJ	(jump) register J of MIX	A.1
(L:R)	partial field of MIX word, $0 \le L \le R \le 5$	A.1
OP ADDRESS,I(F)	notation for MIX instruction	A.1, A.2
u	unit of time in MIX	A.1
*	"self" in MIXAL	A.2
0F, 1F, 2F, ..., 9F	"forward" local symbol in MIXAL	A.2
0B, 1B, 2B, ..., 9B	"backward" local symbol in MIXAL	A.2
0H, 1H, 2H, ..., 9H	"here" local symbol in MIXAL	A.2

INDEX AND GLOSSARY

INDEX AND GLOSSARY

When an index entry refers to a page containing a relevant exercise, see also the *answer* to that exercise for further information; an answer page is not indexed here unless it refers to a topic not included in the statement of the exercise.

Any inaccuracies in this index may be explained by the fact that it has been prepared with the help of a computer. For additional definitions of computer terminology, see the *IFIP-ICC Vocabulary of Information Processing* (Amsterdam: North Holland Publishing Co., 1966).